METHODS in
MICROBIOLOGY

This Series *edited by*

J. R. NORRIS
Borden Microbiological Laboratory,
Shell Research Limited, Sittingbourne, Kent

and

D. W. RIBBONS
Department of Biochemistry,
University of Miami School of Medicine
and Howard Hughes Medical Institute,
Miami, Florida, U.S.A.

METHODS in MICROBIOLOGY

Edited by
C. BOOTH
*Commonwealth Mycological Institute,
Kew, Surrey, England*

Volume 4

 1971

ACADEMIC PRESS
London and New York

ACADEMIC PRESS INC. (LONDON) LTD
Berkeley Square House
Berkeley Square,
London, W1X 6BA

U.S. Edition published by
ACADEMIC PRESS INC.
111 Fifth Avenue,
New York, New York 10003

Library of Congress Catalog Card Number: 68–57745
SBN: 12–521504–5

PRINTED IN GREAT BRITAIN BY
ADLARD AND SON LIMITED
DORKING, SURREY

LIST OF CONTRIBUTORS

G. L. BARRON, *Department of Botany, University of Guelph, Guelph, Ontario, Canada*

F. W. BEECH, *University of Bristol, Research Station, Long Ashton, Bristol, England*

C. BOOTH, *Commonwealth Mycological Institute, Kew, Surrey, England*

HELEN R. BUCKLEY, *Division of Laboratories and Research, New York State Department of Health, Albany, New York, U.S.A.*

M. J. CARLILE, *Department of Biochemistry, Imperial College of Science and Technology, London, England*

J. CROFT, *Department of Genetics, University of Birmingham, Birmingham, England*

T. CROSS, *Postgraduate School of Studies in Biological Sciences, University of Bradford, England*

R. R. DAVENPORT, *University of Bristol, Research Station, Long Ashton, Bristol, England*

R. R. DAVIES, *The Wright–Fleming Institute of Microbiology, St. Mary's Hospital Medical School, London, England*

D. M. DRING, *Royal Botanic Gardens, Kew, Surrey, England.*

L. V. EVANS, *Department of Botany, University of Leeds, Leeds, England*

ISABEL GARCIA ACHA, *Departmento de Microbiologia Facultad de Ciencias and Instituto de Biologia Celular, CSIC, Universidad de Salamanca, Spain*

G. N. GREENHALGH, *Hartley Botanical Laboratories, University of Liverpool, England*

J. L. JINKS, *Department of Genetics, University of Birmingham, Birmingham, England*

E. B. GARETH JONES, *Department of Biological Sciences, Portsmouth College of Technology, Portsmouth, England*

C. M. LEACH, *Department of Botany and Plant Pathology, Oregon State University, Corvallis, Oregon, U.S.A.*

R. L. LUCAS, *Department of Agricultural Science, University of Oxford, Oxford, England*

AGNES H. S. ONIONS, *The Culture Collection, Commonwealth Mycological Institute*

T. F. PREECE, *Agricultural Botany Division, School of Agricultural Sciences, The University, Leeds, England*

v

E. PUNITHALINGHAM, *Commonwealth Mycological Institute, Kew, Surrey, England*

D. H. S. RICHARDSON, *Department of Biology, Laurentian University, Ontario, Canada*

PHYLLIS M. STOCKDALE, *Commonwealth Mycological Institute, Kew, Surrey, England*

MISS G. M. WATERHOUSE, *Commonwealth Mycological Institute, Kew, Surrey, England*

R. WATLING, *Royal Botanic Garden, Edinburgh, Scotland*

JULIO R. VILLANUEVA, *Departmento de Microbiologia Facultad de Ciencias and Instituto de Biologia Celular, CSIC, Universidad de Salamanca, Spain.*

S. T. WILLIAMS, *Hartley Botanical Laboratories, University of Liverpool, England*

ACKNOWLEDGMENTS

For permission to reproduce, in whole or in part, certain figures and diagrams we are grateful to the following—

Evans Electroselenium Ltd; Gallenkamp Ltd; Mullard Ltd; The Radio Corporation of America.

Detailed acknowledgments are given in the legends to figures.

PREFACE

Microbiology involves or impinges upon the study of all micro-organisms although for many years it has become, at least in practice, almost a pseudonym for bacteriology. Yeasts, of course, have been included but the fact that they are fungi has generally been ignored. As a mycologist therefore it gives me considerable pleasure to present a book on Methods in Mycology in a general microbiology Series.

To many, fungi has meant toadstools or things growing on wood whether trees or structural timbers. A specialized few have been involved with their economic importance as pathogens of plants and animals including man, and others in mould deterioration and biodegradation problems. With the onset of the use of fungi for antibiotic production interest has continued to increase and they are now used as a source of protein, of growth promoting substances and enzymes. Their use in bioassay work and for various biochemical syntheses has also expanded. All this has introduced, to a much wider spectrum of workers from other disciplines, an interest in the moulds, their use and cultivation.

The aims and outlines of this Series on Methods in Microbiology has been given in the preface to Volume one and there is no need to reiterate them here. This Volume is intended both as a reference manual and also as an introduction to workers from other fields to the methods used in mycological studies. The work has been planned to cover general methods and media, examination techniques and preservation. Fungi from certain natural groups and the specialized techniques for collecting, isolation and cultivation of these are outlined in specific Chapters. These special Sections also include lichens, slime moulds and the Actinomycetes. These three are not true fungi, the lichens being an association of algae and fungi, but because the methods of investigation are similar to those used for the true fungi they are included here for convenience.

Some of the groups dealt with are based on ecological rather than taxonomic considerations: Thus the special methods required for handling aquatic fungi, soil fungi, dermatophytes and air spora are described in specialized Chapters. The final Section is an introduction to the methods involved in a study of the physical, biochemical and genetic aspects.

Throughout this book where some methods are similar to those used in bacteriology it is the mycological aspect of the method that has been stressed.

My thanks are due to the contributors for their helpful collaboration in producing the Volume and to the staff of Academic Press for their thoroughness and care in preparing the material for press.

CONTENTS

xi

CONTENTS OF PUBLISHED VOLUMES

CHAPTER I

Introduction to General Methods

C. BOOTH

Commonwealth Mycological Institute, Kew, Surrey, England

I. INTRODUCTION

Fungi have a part in the cycle of degeneration of almost all organic matter. They cause spoilage of food-stuffs and occur as human, animal and plant pathogens. Under humid conditions, they can derive sufficient subsistence from inorganic matter or from surface algae to grow over or through many manufactured products. Thus they often cause damage by short-circuiting electronic equipment, by the etching of glass in optical equipment, and damage pictures or decorated surfaces by mould growth. Various chemicals such as standard hypo and the paraffins used as fuel in jet aircraft support fungal growth. Industrial material such as wood pulp is spoilt by the presence of blue-staining fungi and all timber subjected to damp has to be preserved against fungal attack or frequently replaced.

Before studying the effects of fungal attack it is usually necessary to isolate the fungus both for the purpose of identification and in order to determine its growth requirements and the by-products of its metabolism. The methods used for the isolation of fungi and for their cultivation depend largely upon their environment in nature. Methods which are applicable to the cultivation of the common moulds and other fungi imperfecti, which are for the most part conidial states of ascomycetes, are described in this Chapter. Details of methods of isolation and cultivation from more specific groups and from habitats such as soil, water, human and animal tissues, are described in later Chapters under the respective headings.

For the purposes of isolation, fungi can be roughly divided into the parasitic and saprophytic species, although it should be fully understood that this distinction is by no means clear-cut. There is a close relationship between the genetic complement, the health and environment of the host, and the parasitism of many fungi. This may be a question of merely a mechanical barrier in the hosts' protective mechanism preventing access of the fungus. Such protection may be greatly reduced or destroyed following mechanical injury of the host.

In general it is easier to isolate the saprophytic moulds than specific animal or plant pathogens. In either case, isolation is made easier if the fungus is producing either sexual or asexual fructifications so that the isolation can be made from single spores. This makes purification easier and simplifies later handling and identification of the fungus.

II. ISOLATION

A. Isolation from plant tissue

In the case of fungi which cause cankers or necrotic lesions in plants, the problems of isolation are chiefly concerned with separating the disease-causing organism from the many saprophytic species which frequently invade necrotic tissue behind the advancing front of the parasitic mycelium. Similarly fungi invading the vascular tissue of the host plant have to be isolated from the mass of saprophytic species which cover the surface of the host. Other fungi which produce fructifications on or below the surface of the host can be cultured by isolating the spores or part of the fructification such as conidiophores or sometimes from part of the perithecial wall.

Isolation from diseased or infected tissue, therefore, may be considered under two headings: those based on plating out infected host tissue and those based on obtaining spores, mycelium or other fungal tissue from the host or substratum as a starting point for their cultivation.

B. Host tissue transplants

Basically these depend upon the elimination of surface contamination. The general procedure is to remove small pieces of infected tissue from the host. In the case of parasitic species these should be taken from the growing front and not from the necrotic area behind. That is, from the edge of the apparently sound host tissue where it meets the obviously diseased tissue of the lesion. Surface contaminants are removed and the tissue is placed on an agar plate under sterile conditions. With some material where the pathogen is deep-seated, it is possible to remove surface contaminants by slicing off a thin layer of host tissue. With other material such as potatoes it is possible to use the infected tissue as a graft by placing it in a cut made in a healthy tuber. The healthy tissue is often more quickly invaded by the pathogen than by the secondary organisms and it can be isolated from the fresh host tissue after surface sterilization.

The most common means of eliminating surface contamination is by surface sterilization. One of the simplest non-toxic means of reducing surface contamination is by prolonged washing coupled with some form of agitation. It is preferable to use a flask or suitably-sized jar for this washing process so that the material is violently agitated by the inflowing water (Fig. 1). The large outlet is covered with a strainer.

Although this method does not produce surface sterility it does readily remove most surface contaminants and is particularly useful when dealing with species parasitizing the surface tissue and with Phycomycetes and other species which are particularly susceptible to toxic chemicals. Harley and Waid (1955) described their method for the serial washing of roots in their study of root surface fungi. They used distilled water in a series of sterile boiling tubes. Williams (1963) used Perspex boxes fitted with stainless steel sieves of graded sizes for the serial washing of soil. For a controlled method of both washing and surface sterilization one is directed to the somewhat elaborate apparatus used by Slankis (1958). With fungal species

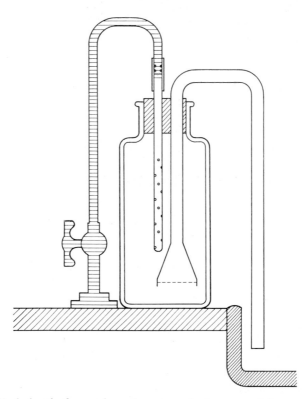

Fig. 1. Wash bottle for surface cleansing of plant material prior to isolation. The outlet is covered with fine gauze.

which are not too sensitive, surface sterilization can be more rapidly and successfully achieved by the use of various chemicals or gases.

The use of fumigants such as propylene oxide or ethylene oxide, frequently used for sterilizing natural media (Table 1), may also be convenient for surface sterilization because of their slow penetration. In this laboratory, small stem sections of woody material 3–4 cm long have been successfully surface-sterilized by placing them in a screw-cap jar into which 2 ml of propylene oxide is poured over a small cotton wool pad. The jar is then closed and left for 30 min. The material is then removed and a small section cut off each end. Further slices are cut off with a razor blade and plated out on suitable nutrient agar. Other chemicals used as surface steriliants include formalin, mercuric chloride, sodium or calcium hypochlorite, hydrogen peroxide, silver nitrate, potassium permanganate, and 70% alcohol. The general method is to wash the material and slice it into suitable pieces about 1 cm long. These are dipped into 75% alcohol to make them readily wettable and immersed in the selected sterilant for a given time. Although the time depends upon the type and density of the material, a period of between 1 and 5 min is usually sufficient. Too long a period of immersion results in excessive penetration of the sterilant and consequent death of the fungus. After immersion the material is thoroughly washed in a rinsing agent and further rinsed in two dishes of sterile water. The ends of the material are cut off with a sterile knife and the centre parts plated out.

TABLE I

Surface sterilizing agent	Concentration (%)	Time (min)	Rinsing agent
Formaldehyde in sol.	51	1–5	70% Ethyl alcohol then sterile water
Hydrogen peroxide	3	1–5	Sterile water
Potassium permanganate	2	1–5	Sterile water
Sodium or calcium hypochlorite[a]	0·35	1–5	Sterile water
Mercuric chloride[b]	0·001	1–5	70% Ethyl alcohol then sterile water
Ethyl alcohol	75	1–5	Sterile water
Silver nitrate	1	1–5	Sterile sodium chloride then sterile water

[a] Normally a 1–10 dilution of commercial solution (Chlorex) is sufficient.
[b] Usually made up from stock solution ($HgCl_2$ 20 g and conc. HCl to make 100 ml) see stock sol. and 995 ml water.

As mentioned, fungi show different susceptibilities to the toxicity of different chemicals used as surface sterilants. Davies (1935) showed that *Ophiobolus graminis* was much more susceptible to mercuric chloride/ethyl alcohol than to the silver nitrate/sodium chloride process. After a period of incubation the fungus will grow out from the ends of the plated host material. As bacteria are frequently present, it is advisable to use a plate in which an antibiotic such as chloramphenicol has been incorporated into the nutrient media; an alternative method is to cut off the hyphal tips and remove them to agar slopes in tubes, this method depends upon the fungal hyphae growing faster than the spread of the bacterial colony.

C. Isolation of spores

Before satisfactory isolations can be made it is often necessary to subject the material to some form of pre-isolation treatment. This may be designed to eliminate surface contamination (although the use of surface sterilants is not always feasible) and secondly to stimulate sporulation. With regard to the latter, the method most frequently employed is to place the material in a damp chamber made simply from a suitably sized plastic box or Petri dish with several layers of blotting paper on the bottom. This is moistened with sterile water and the material placed on it. If necessary the box can be placed in an incubator to stimulate growth. Keyworth (1951) found that the Petri-dish moist chamber often dries out before adequate time had

Fig. 2. The Petri dish moist chamber with strips of paper pulp and xylonite as used in its preparation. (Reproduced by courtesy W. G. Keyworth.)

been allowed for the fungal fructifications to develop. Subsequent rewetting causes variation in humidity and also soaks the plant tissue. This has the general effect of suppressing fungal sporulation. He used a Petri dish as a chamber (Fig. 2), lined with paper pulp which, due to its capacity to absorb large quantities of water, would remain moist for 3–4 weeks. Several discs of filter paper are placed in the bottom of this dish as usual and then strips of pulp are placed round the edge and if necessary diagonally across the dish. The whole is moistened with sterile water before the material is inserted.

If the material is heavily infected by surface contaminants these can largely be removed by vigorous washing in running water as described (II, B). Surface contamination is most profuse under conditions of 100% humidity. Stromatic fructifications, perithecia, pycnidia and sporodochia on wood or herbaceous stems can often be stimulated to complete their development and produce spores by hanging the material in a cool place, and, if the humidity is very low, by periodic spraying with water. This tends to reduce mould growth to a level at which it does not interfere with the isolation of other fungal spores from their fructifications.

1. *Obtaining spores*

Most microfungi form asexual spores readily on the host substrate. These may be borne in sporangia, on aerial conidiophores as in the common moulds, in pustules (acervuli) on the surface, or in enclosed flask-like bodies (pycnidia) which may be superficial or immersed. When examining necrotic lesions on leaves and stems of infected plants or decayed organic material one frequently finds both sexual and asexual fructifications. These are the most preferable sources or starting points for future cultures.

With aerial conidiophores and surface moulds, conidia can easily be transferred into a drop of sterile water on a slide or into a dilution series by the use of a mounted sewing needle. Sporangia of the Mucorales can similarly be picked off on the point of a needle, place in a drop of water and the spores teased out. Slimy-spored fungi, those in which the conidia are coated with a layer of mucus, often form a hard gelatinous crust under dry conditions. This readily becomes mucilaginous if a drop of water is placed on the spore mass, and the spores can then be readily removed on the point of needle. Pycnidia, the globose flasks filled with conidia (pycnidio-spores), occur both superficially and immersed. Under moist conditions they will, when mature, extrude a cirrus of spores or a globule of spores mixed with mucilage around the ostiole. These slimy spores can be picked off with a needle. Alternatively the top of the pycnidium can be sliced off with a razor blade and the horny mass of dry spores picked out of the locule with a needle. When the pycnidia are embedded in wood or in a herbaceous stem, by slicing off a thin surface layer, the contents are exposed. These will appear as a glistening layer if they are in good condition. When the internal walls of the pycnidia are dull and crumbling, they are effete and have discharged their spores. Fungi causing vascular wilts can usually be isolated from the vessels in the plant stem, particularly near the base. In plant material with large vessels, such as banana pseudostems, it is possible after cutting a section to isolate mycelium and even spores directly from the vessels on the point of a needle.

Sections or segments of stems placed in water after surface sterilization or after stripping off the epidermis or periderm will allow spores to diffuse from the vessels into the water. This procedure is rather slow and the concentration of conidia obtained is low. Better results can be obtained by flushing out the conidia. Banfield (1941) described a method of flushing out conidia from the xylem vessels of elm trees. Three to four-foot sections of branches were stripped of bark, the base sharpened to a point and the whole flamed with alcohol. The upper end was covered with wax with the exception of the outer layer of vessels which were left exposed. A metal tube was fitted over this end and sealed on with grafting wax. The branch was hung with the point downwards and the tube filled with sterile water. Gradually the sap was flushed out by gravity and replaced by sterile water. Fifty to five hundred millilitres of sap containing conidia was collected from the basal tips of the logs used.

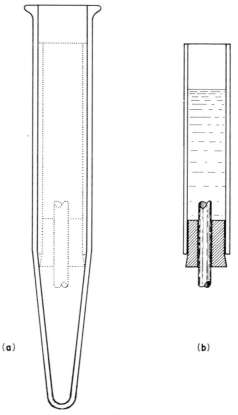

(a)   (b)

FIG. 3. Flushing apparatus for extracting spores from vessels in plant stems or twigs; (a) centrifuge tube with flushing unit; (b) the flushing unit with portion of twig inserted in base through a neoprene stopper and supporting a column of distilled water above. (After Brown and Root, 1964.)

Brown (1963) working on elm diseases described a successful method of flushing conidia from the xylem vessels in small sections of wood by using centrifugal force. His original equipment produced difficulties with regard to sterilization and the following description is taken from an account of the revised equipment described by Brown and Root (1964).

The flushing unit is composed of a glass tube (12×60 mm) into which a twig-holding element is forced (Fig. 3). These twig-holders are constructed from No. 00 neoprene stoppers which are of suitable diameter at the base. The wider upper part is sliced off leaving the stopper about 15 mm long. Holes 1/16, 3/32 or 1/8 in. dia. are drilled through a series of shortened stoppers so as to be available for twig sections of various diameters. In practice the twig sections are first inserted tightly into a suitable stopper and the stoppers are then twisted tightly into the ends of the glass tube. Two millilitres of sterile distilled water is placed in the tube above the twig and the unit is placed in a 16×110 mm centrifuge tube over a 25×12 mm glass sleeve if necessary. Several units can be prepared and the whole centrifuged. Conidia are flushed out of the xylem vessels at speeds of 2500 rpm. Subsequently conidia can be plated out on suitable media.

In the ascomycetes the spores are usually liberated from immature asci when these are placed in water or if necessary the ascus wall can be ruptured with the use of a needle. In species with small tough-walled asci or where the serial isolation of spores is required, liberation of the ascospores without damage can be a problem. Haskins and Spencer (1962) used a preparation of snail enzyme to digest the tough ascus wall of an inoperculate discomycete. This had no harmful effect on the spores, which germinated readily.

D. Selective isolation

All methods of isolating fungi are selective. By opening a Petri dish of nutrient agar on a laboratory bench we are selectively isolating the air spora in the laboratory, in effect, those rapidly germinating fungal spores which can grow on nutrient agar in the presence of other fungal colonies. Therefore, most methods of selective isolation are designed to eliminate or prohibit the growth of the general aerial moulds and common saprophytic soil flora which have the ability to grow in the presence of other moulds and which frequently produce antibiotic or antifungal substances. These prevent the growth of the more slowly germinating spores or specific saprophytic fungi which frequently have need of host or other stimulus before germinating in the presence of other fungi. If one is dealing with spores, then probably the best selective method is by single-spore isolations.

Resting bodies such as ascospores in ascocarps, chlamydospores, oospores or sclerotia, although generally more selective in their requirements for germination, are also more resistant to the effects of heat or toxic chemicals than conidia or mould spores.

Warcup (1951) found soil-steaming effective for the isolation of ascomycetes from soil (Barron, this Volume, p. 405). Later Warcup and Baker (1963) immersed 2·5 g of soil in 60% ethyl alcohol for 6–8 min. The soil and alcohol were then added to water at the rate of 1 : 100 giving an alcohol concentration of less than 1%. This treatment reduced the number of colonies per gramme of soil from 300,000 to 2000–3000 and many of these alcohol-resistant fungi were Ascomycetes.

The use of temperature for selective isolation depends upon the temperature requirements of the fungus. Although throughout fungal species there is a continuous cline, for convenience they may be divided into the categories shown in Table II.

TABLE II

	Temperature (°C)	
	Growth range	Optimum
Psychrophiles	0–40	25 approx.
Mesophiles	10–40	25 approx.
Thermophiles	20–60	+40 approx.

1. *Isolation of psychrophiles*

Although psychrophilic fungi have the ability to grow between 0° and 10°C their optimum temperature for growth is usually much higher. *Fusarium nivale* which will grow between 0°C and 32°C and can cause disease between 0° and 5°C has an optimum growth temperature of about 20°C. Similarly the Basidiomycete which causes snow mould in western Canada can grow between −4° and 26°C. As a method of selective isolation the use of low temperature culture methods may be effective but it is very slow and somewhat uncertain.

2. *Isolation of thermophiles*

The presence of thermophilic fungi is often indicated by the condition of the material under investigation—mouldy hay, hot spots in grain stores or in composted material. The spores are also frequently found in soil although they are not necessarily produced there.

General methods of isolation are by dilution plates using yeast–glucose or yeast–starch agar and incubation at 45°–50°C. The agars mentioned are comparatively resistant to shrivelling at these temperatures, a feature which prohibits the use of potato dextrose or malt agar.

4. *The use of chemicals as selective agents*

The use of prepared cellulose for isolating cellulolytic fungi is discussed on p. 39 and under the Chapter on soil.

The use of other specific substances as baits for selective isolation, such as keratin for dermatophytes, are also most useful; an indication of the bait to use is given by a consideration of the natural habitat if this is known.

Other selective substances may be used because of their depressant or toxic effects. An excellent example of this was given by Parbery (1967) in work with *Cladosporium resinae* which was causing concern due to its growth in aircraft fuel tanks. Efforts to isolate it from soil had generally failed until Christensen *et al.* (1942) used creosote agar. Parbery simplified this by isolating directly from the soil. He placed two decapitated matchsticks soaked in creosote on the surface of a soil sample in each of a series of Petri dishes. With this method Parbery has repeatedly isolated this fungus both in its perfect and imperfect states from soil in Australia, Britain France and Sweden.

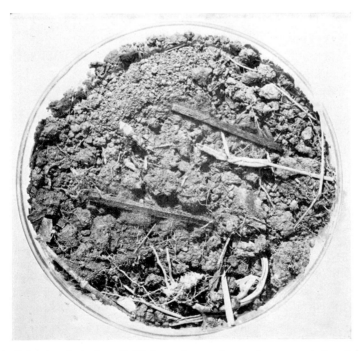

FIG. 4. Selective isolation of *Cladosporium resinae* using creosoted matchsticks. (Reproduced by courtesy D. G. Parbery.)

Antibacterial antibiotics are now widely used in the selective isolation of fungi from soil and contaminated material (Goldberg, 1959). When making selective isolations of fungi, antifungal antibiotics are becoming important tools. The ideal is to select an antifungal antibiotic which has a wide spectrum for fungi in general but to which the fungus under investigation is tolerant. Cycloheximide and nystatin are widely used in medical research for suppression of fungal contaminants in the isolation of bacteria and specific pathogenic fungi. Schneider (1956) reported a selective medium for the isolation of *Graphium*, the conidial state of *Ceratocystis ulmi* which consisted of incorporating 200–300 μg/ml of cycloheximide and 10 μg/ml streptomycin into potato dextrose agar. This medium was found to inhibit *Penicillium*, *Aspergillus* and *Fusarium* species and other common contaminants and yet the growth of the *Graphium* was not inhibited.

Eckert and Tsao (1962) demonstrated the selective nature of nystatin and pimaricin when isolating *Phytophthora* species from plant roots. They recommended the following selective medium for the isolation of *Phytophthora citrophthora*, *P. parasitica* and *P. cryptogea* from infected citrus and alfalfa roots: 100 ppm pimaricin and 50 ppm each of penicillin and polymyxin in standard corn meal agar.

Goldberg (1959) recommended that the following criteria be applied in choosing an antibiotic for use in a selective medium—

  (i) Stability in the medium.
 (ii) Solubility in the medium.
(iii) A highly specific spectrum.
 (iv) Non-toxic to the organism to be cultivated.

There are a large number of antibiotics to meet these requirements. With regard to stability it is preferable that the antibiotic can withstand autoclaving without serious loss of activity. In this respect penicillin does not qualify although it is a most useful antibiotic. Vaartaja (1960) gave a list of 26 antibiotics and fungicides which had different inhibiting effects on ten different test fungi. Later (*in litt.* 1966) he cited Trichlorodinitrobenzene, Endomycin, GS–388 (2(5-nitro-2-furfuryl) mercapto-pyridine-*N*-oxide and an experimental fungicide, HRS 1950, as very selective in their action. The latter was specific to *Pythium* spp., and *Fusarium solani* was tolerant to 500 ppm of GS–388 whereas in most fungi tested, particularly *Pythium* and *Rhizoctonia* spp., growth was inhibited at below 50 ppm.

III. SINGLE-SPORE ISOLATION

After isolation of the fungus it is strongly recommended that the strain should be re-isolated from the isolation plate as a series of single-spore cultures. This ensures not only that one species is present but also only one strain of the species.

Furthermore this gives a clearly defined basis for the starting point of future studies on the isolate and for future pronouncements of its genetic or physiological capabilities.

Single-spore methods used in mycology have largely been adapted from bacteriological techniques, although dilution methods of obtaining single spores as used in bacteriology are considered to be less dependable and far more laborious than the spore selection methods most frequently used in mycology.

A. Spore selection methods

The various techniques employ an isolation plate (Petri dish) and consist basically of three steps: the preparation of a suitably dilute spore suspension in or on the surface of a thin layer of semi-solid media; the selecting or marking from below of suitably dispersed spores and the scanning of the agar surface for contaminants or the proximity of neighbouring spores; the removal of these selected spores to suitable growth media.

1. *Preparation of the dilution plate*

The spore suspension in the isolation can be prepared by the dilution method, by spotting spore suspensions, by streaking, or by the collection of automatically discharged spores.

B. Dilution plate methods

As in bacteriology, the dilution plate methods have stemmed from the work of Lister in 1878.

With the possible exception of some profusely sporulating species such as many *Penicillium* and *Trichoderma* isolates, a 1/10,000 dilution is sufficient to obtain a satisfactory dilution plate. With profusely sporing strains it may be necessary to take the dilution series one step further.

Initially, approximately 1 ml of sterile water, from a tube containing 10 ml, is poured into a tube containing the culture. The culture tube is shaken and the water containing the spores is poured back into the 9 ml of sterile water. After shaking, 1 ml from this tube is poured into 9 ml of sterile water to give a 1/100 dilution. This procedure is repeated with a third tube to give 1/1000 and 1 ml from this is poured into the surface of clear nutrient agar in a Petri dish. This gives approximately a 1/10,000 dilution. It is preferable to slope the plate for about 5 min to allow the spores to settle before pouring off the excess water and incubating the plate.

Hansen and Smith (1932) used direct microscopic observation to assess the spore concentration. Using the low power of a microscope they counted the spores carried by a 2 mm loop. From this assessment they placed 75–100 spores in a test tube containing 100 ml of melted cooled but not set Czapek's media with 0·5% agar. In this method, the low agar concentration makes it possible to cover the bottom of three 9 cm plates, which gives an approximate concentration of 30 spores per plate.

One of the simplest and most versatile methods of obtaining a suitable isolation plate is to prepare the original suspension in a drop of sterile water on a sterile slide placed on the stage of a dissecting microscope.

This allows the suspension to be examined before it is streaked across a plate. Conidiophores can be teased apart with a pair of mounted needles, and clumps of asci can be ruptured to release their spores. When dealing with slimy spores obtained from a sporodochium or pycnidium a dense accumulation of spores can be obtained on the wet tip of a needle. If the tip of the needle is introduced into a drop of water on a slide the spores can be observed as they leave the needle tip and flow into the water. When the suspension is adequate the needle can be withdrawn. Experience of the correct dilution is easily acquired; this is approximately at the point when the spores are clearly distinguishable in the water and are not obscured by overlapping. The spore suspension is then taken up in a loop and streaked across a plate. It is preferable that streaking follows a line previously drawn across the base of the plate.

C. Isolating the spore

The various methods of isolating single spores are adapted to one of the following stages; (i) before germination; (ii) at germination, hyphal primordial stage; (iii) isolation of individual colonies derived from single spores.

Methods involving the examination of droplets of a spore suspension to find drops containing a single spore are considered to be too laborious for general use and unsuitable for isolating large numbers of spores.

1. *Isolation before germination*

These may be summarized as follows: (*a*) capillary tube methods; (*b*) dry needle methods; (*c*) the use of micromanipulators.

(a) *Capillary tube methods.* Micropipettes and capillary tubes, made from glass tubing drawn out into a very fine bore, have long been used in isolating fungal spores, although not to the same extent as in bacteriology. Tubes of 2 mm inside-diameter are drawn out to about 0·5 mm diam. using an ordinary gas jet. They are reheated in a microflame and drawn out to a diameter that is scarcely visible and at the same time broken sharply to leave a flat end. If they are to be used as micropipettes the end can be left straight or bent at right angles about 3–4 mm from the tip.

This method is recommended for the isolation of single zoospores (p. 18).

Hesler (1913) in isolating the ascospores of *Physalospora cydoniae* was probably one of the first to use a capillary tube method. A small glass rod 15–20 cm long with a 3 mm bore was drawn out into a capillary at one end and to the other end a rubber tube was attached. The free end of the rubber was placed in the mouth and the capillary tube was manipulated by hand. By careful handling it was possible to pick out single spores from a microscope slide and transfer them to a drop of sterile water in a Petri dish.

Brown (1924) used capillary tubes to isolate hyphal tips. An isolated hyphal tip was located on the surface of an agar plate and marked. A fine capillary tube was then pushed over the growing hypha and raised to excise the tip and plug the pipette with agar. The hyphal tip was blown out of the pipette onto an agar slope.

Hansen (1926) used fine capillary tubes, slightly larger than the spores to be isolated, in a unique method of obtaining single spores. He made a spore suspension in cooled agar before it set, and filled his lengths of capillary tube from this using capillary attraction. After the agar set the tubes were examined under the microscope and broken up according to the position of the spore. Those containing a single spore were selected, immersed in alcohol to sterilize the outside, washed in sterile water and placed on a suitable nutrient agar. The hyphal primordia grew out from the ends of the tube on to the agar.

(b) *Dry needle method.* The use of a needle to isolate spores before germination relies on the spore adhering to the needle when touched and then being carried to a suitable fresh substrate. Needles modified as spears, narrow scalpels or blades are much more widely used to isolate spores which have just begun to germinate. The spores are transferred, together with a small amount of the nutrient media, to the prepared substrate.

(c) *Mechanical aids.* Simple mechanical aids may be used to isolate both spores in the early stages of germination and ungerminated spores.

One of the simplest and most successful mechanical aids was described by La Rue (1920). This consists of a cutter resembling a microscope objective with the circular blade occupying the position of the first lens. The cutter is screwed into the revolving nosepiece so that, after examination of a suitable spore and the surrounding field by the microscope objective, the cutter can be swung round into the same field and by lowering the cutter a disc of agar bearing the spore is cut out. If the agar is too soft,

the cut disc has to be removed from the cutter by using a needle. If about
3% agar is used the disc is usually cut, left in the plate, and subsequently
removed by a needle. To avoid the difficulty of removing the plug of media
by the use of a needle, Skala (1958) devised a cone-like cutter which
clamped over a short microscope objective. Into the side of the cone a
tube was fitted which allowed sterile compressed air to be introduced to
expel the agar plug with the germinating spore. Possibly the most laborious
part of the procedure in using a La Rue cutter is the constant unscrewing
to sterilize the objective. Keyworth (1959) devised a modified form with

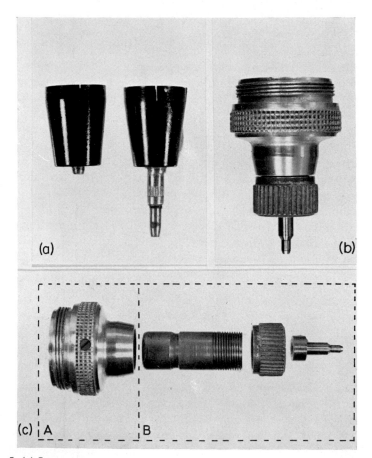

FIG. 5. (a) Lens caps with variable length cutters to match focal length of different lens. (b) The Keyworth isolator. (c) Component parts of the Keyworth;
A, lens type base for microscope nose piece; B, cutting unit, easily removable for
sterilization. (a) courtesy D. W. Fry; (b–c) courtesy W. G. Keyworth, photography
by D. W. Fry).

detachable cutters that could be sterilized independently and clipped in as required.

The cutter (Fig. 5) consists of a stainless steel tip with a collar which is fixed by means of a screw cap to a brass cylinder; this in turn slips into the duralumin body of the apparatus where it is held by a spring-loaded ball engaging a groove. The body then screws into the revolving nosepiece of the microscope.

(d) *Micro-manipulators.* There are available several excellent micromanipulators which allow control of extremely precise movements of the micro-instruments. These microneedles can be used for the isolation of single fungal spores. However, their expense is not warranted if the object is merely to obtain single spore cultures as they are tedious and time consuming to use.

2. *Isolation of germinating spores*

Most fungal spores, if placed on an isolation plate, will germinate overnight to form hyphal primordia after 12–18 h at 25°C. Some will have produced extensive growth by this time and, as these are considered unsatisfactory for isolation, a new plate should be made and kept at a lower temperature overnight or isolated earlier. Certain thermophilic fungi will require a higher temperature, and others, principally aquatic species, require a lower temperature for germination.

Hansen and Smith (1932) after preparation of their isolation plate on weak agar (1%) allowed 12–16 h for spore germination. Under the dissecting microscope the germinating spores were picked out by hooking a very fine steel or glass needle under the germ tube and lifting the spore out of the weak agar. These fine needles, however, could not be sterilized by heat and it was a disadvantage to use chemical means of sterilization.

Khair *et al.* (1966) reported difficulty with this method in removing the spores from the agar on the needle tip. They suggested the following technique. A layer of 4% tap water agar is spread over the centre part of a slide. A piece of sterile dialysis tubing is placed over the agar and secured to the slide at each end by Scotch tape (Sellotape). A dilute suspension of spores is streaked over the surface and the dialysis tubing serves as a base for spore germination. Spores can be readily picked off by an ordinary sewing needle which can be flame-sterilized.

As the aim of single-spore isolation should be, apart from obtaining pure cultures, to damage the germinating spore as little as possible, a modification of the method used by Shear and Wood (1913) has been found to be simple and most effective.

The isolation plate prepared by dilution methods or by streaking is examined from below for germinating spores under a dissecting microscope. Those suitably separated from their neighbours are marked by four dots made by a felt pen on the base of the plate. After a sufficient number have been located the plate is reversed on the stage and the lid removed. The germinating spore is scanned from above either by the higher power of the dissecting microscope or under the low power of a compound microscope to ensure that no other spores or contaminants are present. Then, under the dissecting microscope, and using a needle modified with a flattened spear-like tip (Burrowdale needle), a 2 mm square block of agar is cut around the germinating spore. This is lifted out, with the spore on top, and placed on an agar slope. Experience has shown that this method can be used in the laboratory without the use of a sterile chamber. Even when 40–60 single spores have been taken from one plate, less than 2% contamination has been observed. In fact, the chance of a contaminant falling on the 2 mm square after examination is quite remote.

D. Isolation of colonies derived from a single spore

The general method used here is to make an isolation plate of weak agar with widely dispersed spores. The spores are located from below and those suitably isolated are marked. The plate is incubated and allowed to grow until the colonies are visible. Each colony free from contamination is removed by a small scalpel to suitable nutrient agar. With this technique it is very difficult to locate and mark small-spored fungi.

Schmitthenner and Hilty (1962) described an interesting modification of this method which separated the fungi from bacteria and actinomycetes.

An isolation plate was made with standard nutrient agar, and inoculated by spotting the moist surface with a spore suspension drawn into a capillary pipette. The convex ring of agar round the edge of the plate was removed with a scalpel and the entire agar disc remaining was inverted so that the agar lay perfectly flat on the base of the Petri dish. After incubation the fungi grew through the agar to produce aerial mycelium on the upper surface whereas bacteria and actinomycetes were confined to the interphase between the agar and the glass.

It was also claimed that even when the colonies were grown until the surface colony was mature the origin of the colony could still be determined from the base.

IV. TECHNIQUES FOR OBSERVING GROWTH AND MORPHOGENIC DEVELOPMENT

Although fungi can be grown successfully on various nutrient agar media in Petri dishes or test tubes, these cultures do not readily allow for the critical observation necessary to determine how spores are formed or the nature of the sexual stages initiating sporophore development. Furthermore, to remove perithecial or sporophore initials from the surface of the agar often results in their rupture and disintegration so that it is not possible to determine the relationship of their various parts. Similarly the observation of conidiophore or sporangial development is often obscure on the plate or tube and yet the damage during their removal to a microscope slide does not facilitate later observation. In some cases direct observation of conidium production is possible and the presence or absence of conidium chains can often be observed by direct observation (of the surface of the growing colony) through the low power of a microscope. In most microfungi the disturbance of the material which is necessary to make a microscope slide results in the complete disintegration of the conidium chains, or it leaves small sections of the chains in the mounting media so that the nature of their development, whether acropetalous or basipetalous, cannot be determined. There are many reasons therefore why techniques have been designed which enable direct microscopic observation of the living culture.

One of the simplest methods of making a microscope slide without disturbing the whole colony is to place a sterilized coverslip on the surface of the agar in a Petri dish, setting it near a developing culture so that the hyphae grow over the surface of the coverslip. This method has some advantages because developing mycelium on reaching an area of low nutrition is often stimulated to produce asexual or conidial fructifications. The coverslip can be removed and inverted into a suitable mountant, or examined directly under the microscope. The use of squares of dialysis tubing about the size of a coverslip is in many ways preferable to coverslips. The fungus can be inoculated on to the surface of the membrane when it is lying on the agar. Dialysis tubing allows abundant growth and the mycelium is firmly fixed to the surface of the membrane. Funder and Johannessen (1957) described the use of a molecular membrane filter which was placed on the top of a sterile absorbent pad in the bottom of a Petri dish. The pad was evenly moistened with 2 ml of double-strength sterile yeast water and the fungus streaked across the membrane. Incubation at room temperature for 2–3 days was generally sufficient to produce fructifications. When adequate growth had occurred the membrane was removed from the dish and dried at 50°C for 20–30 min or at room temperature for 3–4 h. Small pieces of the membrane were cut off as required for examination and mounted in a few drops of immersion oil. After a few seconds the membrane became transparent, having the same refractive index as the immersion oil and leaving the fungus clearly visible for examination. Alternatively the colony could be stained before examination.

This method of growth was found suitable for the critical examination of conidial states of fungi imperfecti and members of the Mucorales.

A. Hanging drop cultures

Hanging drop cultures are extremely useful for studying spore germination and to check spores of species such as *Glomerella* for the formation of appressoria. These hanging drop cultures can be made by placing a few spores in water or nutrient broth on a coverslip which is then inverted over the cavity in a cavity slide or over a suitable glass or plastic ring. The slide is placed in a sterile Petri dish with a layer of moist filter paper in the bottom and the lid replaced so that it acts as a damp chamber and the spores can be incubated overnight without drying out.

B. Fixing spores to slides before germination

Thirumalachar and Narasimhan (1953) described the use of mucilage from the stems and flowering stalks of *Tradescantia* sp. to fix spores to slides before germination studies. They mounted the spore material in a droplet of water on a slide and squeezed a tiny drop of viscid mucilage from the cut stem of *Tradescantia*. The mucilage and spore suspension were mixed together, smeared with a needle or scalpel and air-dried. By alternatively moistening and drying the slide, the spores became firmly fixed and subsequent germination was not inhibited. After fixing, the slide could be dipped in a dilute disinfectant to suppress bacteria before being inverted over a water surface in a damp chamber. Alternatively, adhesives such as Haupt's (Dring, this Volume, p. 95) could be used without harm to the spores.

C. Slide cultures

One of the most widely used methods of slide culturing of fungi which would obtain mounts with as little disturbance as possible was described by Riddell (1950). This method is simple and many slide cultures can be obtained from one agar plate (Fig. 6a).

After pouring a plate of suitable agar about 2 mm deep, 6 mm squares are marked out using a sterile scalpel. One square is placed in the centre of a sterile slide and each of the four sides is inoculated with the fungus. A coverslip is placed on top of the square of agar and the slide is put in a damp chamber. This may be made from a large culture dish with filter paper in the bottom on which two glass rods rest to act as supports for the slides. The filter paper is soaked in 20% glycerol in water or merely in water. The dish with the slides is incubated until the mycelium formed from each side of the agar has reached the edge of the cover glass. If the medium

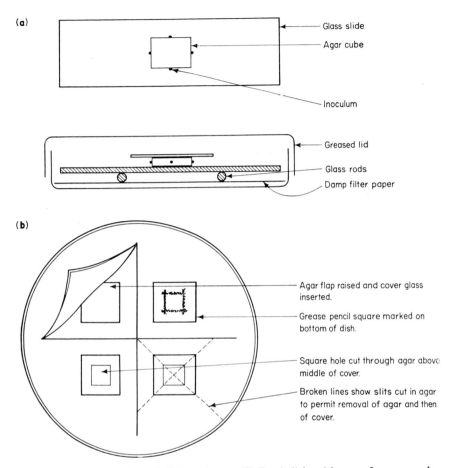

FIG. 6. (*a*) Preparation of slide culture; (*b*) Petri dish with agar for cover glass cultures.

is not too rich, sporulation should have occurred by this time and permanant or semi-permanant mounts can be made of the hyphae and conidiophores, which will be adhering both to the coverslip and the slide round the edge of the agar block.

After culturing, the coverslip is lifted off, a drop of 95% alcohol is placed in the centre of the underside so that it diffuses through the mycelium, and before this has dried, the coverslip is placed in lactophenol or other mountant on a new slide. The square of agar is removed from the culture slide and as with the coverslip a drop of 95% alcohol is placed in the centre to wet the mycelium and a drop of stain or mounting media is placed on the slide and covered by a new coverslip.

Taschdjian (1954) objected to the somewhat elaborate preparations required for this method of slide culture and the danger of contamination. She proposed dipping coverslips in suitable nutrient agar and placing them on the surface of sterile non-nutrient agar in a Petri dish.

Four prepared coverslips are placed in each Petri dish, inoculated with a pinpoint of inoculum in the centre, and incubated. When a suitable stage of growth has been reached, a few drops of 95% alcohol is placed on the coverslip culture and after this has evaporated the coverslips are carefully removed from the plate and inverted into a drop of lactophenol/cotton blue on a clean slide. The chief objection to this method is the presence of agar in the finished mount.

Anthony and Walkes (1962) described a modified slide culture technique that combined to some extent the techniques described by Riddell and by Taschdjian.

They dipped sterile small coverslips into molten nutrient agar and each of these was placed on a slide. The slide was placed on the surface of glycerine–agar in a Petri dish and the four edges of the coated coverslip were inoculated with the fungus under investigation. A large coverslip was placed over the smaller one and the culture incubated for a suitable period. The upper coverslip was removed and treated as in the Riddell method. The agar-coated coverslip was discarded and the culture slide again treated with alcohol and stain. This method was devised because it was considered to be more simple to dip small cover slips in agar than to cut agar squares from a solidified plate; also the solidified non-nutrient agar in the Petri dish was simpler to handle than aqueous glycerine solution.

Knaysi (1957) described a slide culture in which the coverslip, supported on capillary tubes, was raised slightly above the surface of a square of agar. This coverslip was sealed to the surface of the slide by paraffin wax or Vaspar through which two capillary tubes were inserted for gaseous exchange.

This type of culture allows for continuous study of the growing organism although optical conditions are often poor due to the formation of water droplets on the lower side of the cover glass.

These methods of slide culturing all suffer one major setback, which is more important than the somewhat overstressed danger of contamination. This is the short life as a cultural substrate that a small quantity of agar has, even when placed in a chamber with 100% humidity. Dade (1960) described a method which would allow fructifications to develop over the surface of a coverslip which could then be mounted without serious disturbance. In this method the fungus has the maximum use of the agar in the Petri dish during its growth (Fig. 6b). A plate of suitable medium is poured and, after setting, two diametrical slits, at right angles to each other, are cut in the agar with a flamed scalpel. The triangular lips thus formed are lifted one by one, and a sterile coverslip is inserted under each lip (Fig. 6b). The dish is inverted and the position of the coverslips, seen through the base, is marked with a grease pencil by drawing a small ¼ in. square in the middle of each. When the dish is again turned, the blue squares can be seen and the squares of agar above each removed. The medium is then inoculated and the dish incubated. Mycelium and fructifications will eventually cover the exposed centres of the coverslips and when these have proceeded far enough the surrounding agar is removed and the coverslips are mounted in the usual way.

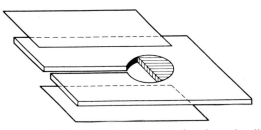

FIG. 7. The Cole and Kendrick thin culture chamber. An "exploded" view showing the drilled and slotted slide, the agar medium, and the two coverslips which seal the upper and lower surfaces. (Reproduced by courtesy G. T. Cole and W. B. Kendrick.)

Cole and Kendrick (1968) described a small microculture chamber which they successfully used for time-lapse photomicrography. The chamber is based on a glass cytology slide $75 \times 50 \times 1$ mm; a hole 18–20 mm dia. is drilled through the slide towards one end and a slot approximately 1 mm wide is cut from the hole to the opposite end of the slide (Fig. 7). A 60×24 mm No. 1 coverslip is sealed to one side of the slide with nail varnish covering the hole and slot. A thin card former is placed across the middle of the shallow circular chamber and the whole sterilized. Then by means of a micropipette sterile molten nutrient agar is injected into the semicircular chamber on the side of the card away from the slot and allowed to solidify. The card is then removed and the vertical surface of the exposed agar is inoculated with the fungus to be investigated. The chamber is then covered by sealing a second 60×20 mm coverslip over the top of the hole and slot (Fig. 7). A sterile Petri dish is lined with filter paper soaked in 50% glycerol and a V-shaped glass rod is laid horizontally on top of the wet paper to support the culture chamber. Cole and Kendrick have produced excellent time-lapse photomicrographic studies on microfungi at high magnification using this culture chamber.

V. STIMULATION OF SPORE GERMINATION

The asexual spores of most fungi germinate readily providing they have a suitable substrate and compatible environment. Other fungal spores, in particular many of those produced as a result of the sexual process, will not germinate even when provided with a suitable substrate, temperature and humidity.

The specific requirements governing the germination of these spores are not well known but in most cases can be related to some extent to their natural environment. Thus many ascospores require a period of dormancy before germination. Others, particularly obligate parasites such as Erysiphales, Meliolas and Rusts, are believed to require the presence of, and presumably some stimulus from, their natural hosts before the spores germinate and begin normal growth.

Some coprophilous species need to pass through the gut of herbivorous animals. The ambrosia fungi may need to be ingested by ambrosia beetles before germination.

A. Removal of substances inhibiting spore germination in relation to the dormancy

It is a common observation that most fungal spores do not germinate in mass. This essential survival factor appears to be achieved by the production of some substance inhibitory to germination which operates under conditions of high spore density such as occur when conidia are produced in pustules or extruded from a fructification. In general these inhibitory substances are removed in nature when the spores are dispersed in moisture droplets or by rain. In the laboratory they are removed when the spore mass is diluted in the preparation of a spread plate or dilution plate. In some cases washing may be required. Dunleavy and Snyder (1962) found that it was necessary to wash the testa of soya bean seeds infected with *Peronospora manshurica* for approximately one week under running tap water before germination of the oospores occurred. However, this excessive washing is probably a means of breaking dormancy rather than the removal of the normal substances inhibiting germination.

Natural dormancy can frequently be broken by the use of some artificial stimuli such as heat, cold, or chemicals. The latter may include acids, alkalis, or enzymes such as those from the gut of animals (Janczewski, 1871) or snails (Voglino, 1895). Natural dormancy in spores should not be confused with the lack of germination in those spores which require a germination trigger. Such a trigger may be the gut of an animal, the temperature of rotting vegetation or a hot-spot in grain. However, for practical purposes, the two types may be considered together because similar treatments may be used for both. The means of breaking dormancy and stimulating germination may be considered under the following:

(i) Heat treatment
(ii) Cold treatment
(iii) Chemical treatment.

Spores which do not respond to any of these methods should be kept as close as possible to their natural environment so that the dormancy period may be allowed to expire.

B. Heat treatment

The necessity of heat treatment before germination is chiefly a characteristic of the ascomycetes, especially the coprophilous species. Warcup and Baker (1963) recorded a number of species in which germination was stimulated by heat treatment. They included members of the Sordariaceae, *Aspergillus* and *Penicillium* species, *Anthracobia* and other Discomycetes. The method used was to steam infected soil samples on plates at temperatures between 50° and 75°C for 30 min. Goddard (1935) found that exposure of ascospores of *Neurospora tetrasperma* to temperatures of 50°–60°C for 5–30 min stimulated germination.

C. Low temperature treatment

The necessity for a period of exposure to low temperature which some spores require before germination is probably related to their ability to overwinter. Blackwell (1935) showed that *Peronospora schleidenii* (P. destructor) required at least a month at 1°–3°C before germination. Sclerotia of *Claviceps purpurea* only germinated after a period of several weeks at 0°–3°C (Kirchoff, 1929). Various workers (Holton, 1943; Meiners and Waldher, 1959) have found that smuts (*Tilletia* species) required prolonged exposure for up to 5 months at 5°–10°C before germination.

The effect of light on spore germination is dealt by Leach (this Volume, Chapter XXIII).

D. Chemical effects

Chemical stimulants to spore germination may act as substitutes for stimulation by heat and light. Sproston and Setlow (1968) use dimethyl sulphoxide in 0.1 M KH_2PO_4 buffer to bring about pigment production and conidium formation in *Stemphylium solani* which normally needs ultraviolet radiation. KH_2PO_4 buffer used in controls had no effect. Ergosterol added to the carriers or solvents increased conidium formation 2–5 times over that of the controls. Kahn (1966) found he could substitute light and induce the formation of conidia of *Sclerotinia fructigena* in the dark by using 50 ppm indolyl-3-acetic acid in ethanol. Adequate conidium formation occurred although Kahn gives no indication of the effects of ethanol alone on sporulation. Sussman *et al.* (1959) found *Neurospora tetrasperma* germinated in response to the presence of aliphatic esters, alcohols and ethers after heat-sensitizing at only 46°C for 30 min. 10 M methanol or ethanol would stimulate satisfactory germination without heat sensitizing. Yu (1954) found *Ascobolus stercorarius*, which normally requires heat sensitization, had an 80% germination in the presence of 0·32% NaOH at 37°C.

Hansen *et al.* (1937) obtained a 3% germination of the ascospores of *Rosellinia necatrix* by suspending them in 2 ml of 5% lactic acid for 15 min and then adding 10 ml of sterile water. Aliquots of 1 ml were poured over Petri dishes containing potato dextrose with 3% agar and incubated at 22°–24°C.

Nutman and Roberts (1962) found high dilutions of fungicides stimulated germination of conidia of *Colletotrichum coffeanum* and the urediniospores of *Hemileia vastatrix*. Organo-mercurials, dithiocarbamates, captan and Cu compounds were used. *Hemileia vastatrix* urediniospores were stimulated to germinate at concentrations of copper between 0·4 and 0·006 ppm of copper sulphate. Conidia of *Colletotrichum coffeanum* were stimulated by between 24 and 0·14 ppm of phenylmercuric acetate which contained 2·5% mercury.

Many of the large number of chemicals which are reported as stimulating spore germination are listed in Sussman and Halvorson (1966), but some of these are of doubtful significance, often merely demonstrating the ability of the spores to germinate in their presence. In this laboratory, the spores of the following species which have been stated to require chemical stimulation for germination have been grown on standard media under standard laboratory conditions: *Aspergillus niger*, *Botrytis cinerea*, *Pyricularia oryzae*, *Sordaria fimicola* and *Gelasinospora calospora*.

E. Stimulation of sporulation

Just as fungal cultures produce inhibitory substances against spore germination there is also evidence of the production of substances that inhibit spore formation. It is obvious that in the artificial environment an agar plate provides for the culture of fungi, there must be present the necessary nutrients and conditions of temperature, light and pH for growth. In addition it should be appreciated that in such an environment autotoxic substances may be formed which in the early stages inhibit spore formation. Park (1961) demonstrated the presence of such autotoxic staling substances in cultures of *Fusarium oxysporum*. These initially prohibit spore formation and at lower concentrations induce chlamydospore development. As the concentration gradually increases their effect on growth becomes more and more marked until even mycelial growth is prohibited.

A simple method of overcoming some of the effects of these autotoxic substances is to remove a small disc of agar from the edge of the colony and place it on a plate of tap-water. Sporulation will frequently occur round the edge of the disc, or as the mycelium grows out onto the nutritionally weak media. Ludwig *et al.* (1962) found that abundant sporulation in *Alternaria solani* could be induced by the following method.

The fungus was grown for two weeks on V8 juice agar in Petri dishes. The aerial mycelium was scraped off and the agar surface washed for 24 h under running water to remove unidentified anti-sporulating factors. The plates were inverted and stacked on a slant so that each plate was partially closed by the one on which it rested. Within two days under laboratory conditions abundant spores were produced, and after washing these off several further crops could be obtained.

Billotte (1963) described a similar method of inducing sporulation in *Alternaria brassicae*. He removed the aerial mycelium and washed the surface of the culture under running water. Abundant sporulation occurred as the opened Petri dishes were allowed to undergo slow desiccation. In fact, spores could be washed off and harvested onto filter papers after 72 h by means of a fine water spray. The filter paper was immediately dried and stored at 4°C.

The effect of light and the presence of light-stimulated sporogenic substances are of major importance in the sporulation of fungi in culture. These effects are dealt with by Leach, this Volume, Chapter XXIII.

F. Effect of the growth media on fungal cultures

In the cultivation of fungi the dependence of the growth form on the nature and concentration of the constituents of the nutrient media should also be appreciated. In general, rich media produce excessive mycelial production and tend to suppress conidium formation, possibly due to the accumulation of staling substances. It is often preferable therefore to use a weak medium such as potato carrot agar to encourage sporulation. For many fungi the surface of the agar is not a suitable substrate for the production of their fructifications and the incorporation of more solid substances, or substances with a different texture, will facilitate their development.

Wheat or other straw cut into sections of 5–8 cm, sterilized by steam or propylene oxide and incorporated in the agar plate by placing it in the Perti dish before pouring the agar will often provide a suitable substrate for the conidiophores of many hyphomycetes such as *Curvularia*, *Helminthosporium* and *Sporodesmium*.

Straw, herbaceous stems such as lupin, or woody twigs, placed in a flask and standing in water, moist sand, or agar, will be found suitable for the production of such fructifications as sporodochia, pycnidia or perithecia.

Strips of filter paper placed on the surface of the agar provide both the cellulose requirements and the substrate for the production of perithecia in *Chaetomium* species. Basidiomycetes usually require moistened sterile sawdust or wood blocks for basidiocarp production.

Many Phycomycetes are extremely susceptible to the toxicity of minute 'quantities of copper and other minerals. Culture media should only be made up with glass-distilled water. The use of seeds, such as hemp, standing in water are also useful cultural substrates (see Waterhouse, this Volume, p. 183).

Fungi which have become adapted to growth on specialized substrates frequently require rich media with a high concentration of sugar for growth. Species in the *Aspergillus glaucus* group sporulate well on malt extract containing 20% concentration of sucrose. *Eremascus albus* requires 40% concentration and *Xeromyces bisporus* only grew in our laboratory on 60% sucrose. Some of the strains of *Aspergillus glaucus*, *A. fumigatus* and *A. versicolor* which cause damage to the lenses of optical instruments in the humid tropics make only very slight growth on normal media but grow well 40% sucrose.

Fungi such as these appear to have some mechanism which protects them from plasmolysis.

Sporendonema sebi has a similar mechanism which allows it to grow and sporulate in the presence of a saturated solution of common salt. When deciding on the cultural medium for a particular fungus the best solution may be suggested by a consideration of the source and the environmental conditions of the original isolate.

Most fungi grow within a comparatively narrow range of cultural conditions which have become accepted as normal.

Within this range, Brown (1925) summarized the effects of culture media on species of *Fusarium*. Due to the unstable nature of many *Fusarium* species media effects are more marked in this genus than in most others. In fact, since Brown's work, these and similar effects of the media on the cultures have been the basis of innumerable scientific papers. Brown summarized his results as follows—

   (i) Increase of phosphate in the form of the neutral salt increases sporulation and diminishes aerial mycelial formation, and vice versa. Acid phosphate produces the opposite effect.

  (ii) The nitrogenous constituent is chiefly responsible for the intensity of the staling reaction shown.

(iii) Glucose and starch have different effects, both quantitatively and qualitatively, on growth-form. Increase in the concentration of glucose tends to produce greater development of aerial mycelium;

a similar increase in the concentration of starch results chiefly in more intense sporulation.

(iv) Colour formation in the medium depends chiefly upon high carbon/ nitrogen ratio of the nutrient. A low concentration of phosphate is also, within limits, favourable to colour formation.

It should be noted that culture pigmentation is directly related to and most strongly effected by the pH of the medium. Most fungi also require a high humidity for growth. Cruickshank (1958) found that *Peronospora tabacina* had an optimum relative humidity of 97–100% and that sporulation dropped to zero as the R.H. dropped to approximately 90% (see Section X).

1. *Pure and mixed cultures*

Although stress has been laid on the necessity of working with pure cultures, when one is dealing with sporulation and in particular when considering the production of perfect states it may be an advantage to use mixed cultures.

In 1909 Heald and Pool found that *Melanospora zamiae* would produce perithecia only when grown with *Nigrospora oryzae*, *Fusarium moniliforme* or less successfully with *Fusarium culmorum*. The association of *Melanospora* species with fusaria is well known. Professor Cettolini found a *Melanospora* associated with a *Fusarium* on wheat in Sardinia and this was described by Saccardo (1895) as *Sphaeroderma damnosum* with the *Fusarium* conidia included as the imperfect state.

Goidanich (1947) returned to Sardinia in 1946 and found the *Melanospora/Fusarium* association still continuing after 50 years in the same locality. Single-spore isolations made by Goidanich and Mezzetti (1947) proved that there was no genetic connection between the *Melanospora* and the *Fusarium* although the presence of the latter certainly stimulated perithecium production in the former.

We have in our laboratory several examples of specific *Melanospora* species which will only produce perithecia in the presence of specific *Fusarium* species. A similar association of fungi has been noted at the C.M.I. between *Acremoniella atra* and a *Cephalosporium* sp.; we also noted *Ceratocystis* sp., which was dependent on the presence of *Gonatobotryum fuscum* before it produced its perithecia.

VI. TRANSFER OF INOCULUM

In general, when subcultures are made from a parent culture on 2% agar, the inoculum is transferred on the point of a sterile needle to the fresh substrate. When standard amounts of inoculum are required, or when it is preferable to transfer mycelium plus agar, it is general practice to use a suitably-sized cork borer to remove a mycelium/agar plug. The plug can be merely cut with the cork borer and removed on the point of a needle, or, if the cork borer is raised with a slight sideways movement, then the disc of agar will be retained in the tip. Alternatively 3% agar can be used, and the plug removed either by a fine needle or by pushing a sterile rod through the centre of the cork-borer.

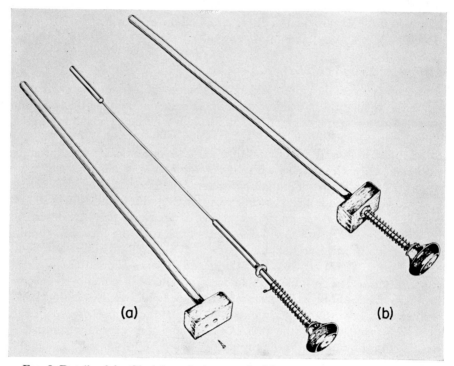

Fig. 8. Details of the Clark inoculating punch; (a) component parts (b) assembled. (After Clark, 1962.)

Clark (1962) described an inoculating punch made in several sizes from various sizes of hypodermic-quality stainless steel tubing with an internal diameter of 0·070 to 0·1730 in. The punch shown in Fig. 8 is about 8 in. long. It is best to use several punches of a given size so that after sterilization by flaming they can be allowed to cool in rotation.

Leach (1964) devised an ingenious punch which allows the mycelium on the surface of the agar to be placed in direct contact with the new substrate to which the inoculum is being transferred (Fig. 9). This avoids the initial period of slow and irregular growth which ensues as the mycelium on the surface of an agar plug is making contact with the new substrate across the chemical barrier of "staled" medium. Where necessary 10–20 plugs (Fig. 10) can be cut in a simple operation, which is time-saving and has a marked advantage in reducing contamination.

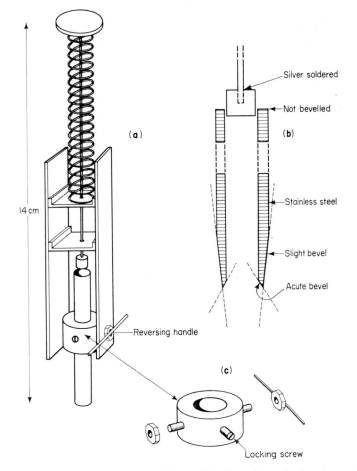

FIG. 9. Plug cutter-extractor-transferer: (*a*) Complete device (not drawn to scale); (*b*) Cross section of cutting tube and plunger; (*c*) Cutting tube holder. (Reproduced by courtesy C. M. Leach.)

The device illustrated above is constructed of brass with the exception of the spring and screw. In later models Leach has also replaced brass cutters with stainless steel as the latter holds a much better cutting edge.

It is suggested that several spare tubes with cutting edges are prepared so that they can be replaced when the edge is dulled. This is not quite so important when using stainless steel.

Garrett (1946) described a multipoint inoculating needle for dealing with the problem of inoculating a large number of plates at several points. The instrument described by Garrett had ten points but this number can be either increased or decreased according to the requirements of the work in progress. The needles were made from ten 15 cm lengths of No. 20

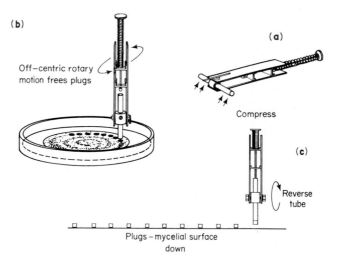

FIG. 10. Procedure for using plug cutter–extractor–transferer. (a) Flaming; (b) cutting and extracting plugs; (c) extruding plugs. (Reproduced by courtesy C. M. Leach.)

gauge steel wire. The inoculating end of each piece of wire was flattened on an anvil to make a spatulate tip. Each wire was bent twice so that it assumes the shape of a spider's leg and the ten pieces are clamped to a wooden holder with the ten "legs" forming a circle 7–8 cm dia.

In use, the number of agar discs required are cut from the inoculum with a sterile cork borer; after cutting, the Petri dish without the lid, is inverted and held above eye level. The agar discs can be removed one by one on each of the spatulate ends of the ten points by slightly twisting the instrument at each stab. This is obviously more laborious than the Leach punch and it is not easy to invert the agar disc onto the new substrate. However, it has the advantage of keeping the discs separate from each other so that no contact is made.

A. Estimation of spore concentration

Fungal spores do not always lend themselves readily to turbidity analysis as it is usually extremely difficult to get spore suspensions free from the mycelium or the sporogenous cells. In general when large quantities of microconidia are required for inoculation work or analysis, liquid culture techniques are used There are however a number of fungi which do not readily sporulate under liquid culture conditions. In these cases the cultures have to be on agar plates, sterilized twigs, or straw, or on filter paper soaked in liquid nutrient (p. 26). In most cultures the spores can be washed off with sterile water, or it may be necessary to scrape them off with a needle or scalpel. In either case, or even in a suspension made from liquid cultures, a considerable amount of mycelium, conidiophores or other non-sporogenous material is usually carried over. To remove this extraneous material it is necessary to filter the suspension through one or more layers of bolting silk or organdie muslin depending on the size and morphology of the conidia. It is preferable to filter the spore suspension even if the spore concentration is to be estimated by a spore count, the method most commonly used. This is carried out by the use of a Neubauer or Petroff–Hausser counting chamber. The Neubauer chamber consists of a special microscope slide with a chamber 0·1 mm deep. The base is marked with a 1 mm square subdivided into 16 squares, each of which is subdivided into 25 smaller squares to give a total of 400 squares in 1 mm.

When using the haemocytometer, a drop of suspension is placed on the engraved grid and a special cover glass is carefully lowered over it so that no air bubbles are trapped between the slide and the cover glass. The cover glass is then slid backwards and forwards until coloured rings (Newton's rings) occur as the two surfaces come into close contact.

The number of spores covering the grid are counted. To calculate the number of spores present per ml. of the suspension the number of spores on the grid is multiplied by a factor of 10,000. If it is necessary to obtain a series of standardized dilutions turbidimetric methods of analysis are used. The spore suspension is diluted until it can be quantitatively measured as optical density with a simple electric colorimeter or nephelometer. Turbidity measurements are usually standardized against a known standard solution such as can be obtained from a Neubauer slide estimation, but additional steps have to be taken to clarify the suspending solution. This can be done first by filtering the spore suspension through bolting silk. The suspension is then centrifuged at 2000 rpm for about 10 min depending on the suspension. The supernanant liquid is poured off and replaced with distilled water. This procedure is repeated until no trace of colour remains in the supernanant, thus avoiding interference with the turbidimetric estimation. The final spore deposit is resuspended in distilled water. Alternatively a suitable filter or monochromator can be placed in some instruments to selectively eliminate colours associated with the growth medium. Uninoculated liquid growth medium can be diluted to the same extent as the culture sample and used in setting the light-transmission value for the instrument. The various electronic methods of cell counting as used in tissue culture laboratories and for blood counts are not generally used in mycology owing to the extremely variable range of spore sizes that occur in fungi.

Quick estimates of spore production in liquid media, which are particularly helpful in determining the suitability of the medium, can be obtained by a rough determination of the spore volume. After filtering through bolting silk the spore suspension is centrifuged in graduated centrifuge tubes. The volume of the sedimented spores at the base of the tubes is a measure of spore production in cubic centimetres.

VII. LIQUID CULTURE TECHNIQUES

For identification and maintenance purposes fungi are normally grown on solidified media as sporulation is usually stimulated by this form of growth. When fungal mycelium is required as opposed to spores, it is preferable to grow fungi in liquid cultures where sporulation is generally suppressed. Quantities of homogeneous mycelium are often required for physiological or bio-assay work, for the extraction of antibiotics, proteins, pigments and enzymes, and for chromatographic, electrophoretic or serological investigations. This is also a useful method of growth for studying nutritional requirements of fungi. In liquid cultures probably the most frequently used medium is Richards but many of the media formulated in Chapter II, 8–43, can be utilized as liquid media if the agar is omitted. For culturing fungi small quantities of the liquid medium are placed in flasks which are then plugged with cotton wool. At this stage they can be sterilized; when cool they are inoculated and incubated at a suitable temperature. If such cultures are grown in a normal incubator a mat of mycelium develops on the surface of the nutrient solution and sporulation takes place on the mat as though it were a solid substrate. To avoid this and to produce homogenous mycelium, liquid cultures are grown as shake cultures which keep the liquid nutrient constantly moving and help aeration. It is preferable to grow liquid cultures on a shaker rather than keeping the medium moving with magnetic stirrers. The latter are not always successful with fungal cultures as mycelial mats tend to form on the surface of the liquid.

Meinecke (1957) described a method of agitating cultures in test tubes by the use of electromagnetically-driven plungers. Each tube contains an iron-cored glass plunger made to rise and fall by an electromagnetic arrangement which permits adjustments. This controls the rate of aeration and the movement of the medium so that surface growth can be prevented.

Simple machines for shaking cultures are available from laboratory suppliers or can be fairly easily constructed as shown by the machine illustrated below. With many cultures grown on a reciprocal shaker, fungal mycelium tends to adhere to and grow on the sides of the flasks or bottles. This can be avoided by the use of shakers with a rotary or swirling action. Kaplan (1956) described a relatively cheap rotary shaking machine which can be built in a laboratory workshop.

Many refinements are available commercially which incorporate the movement in a platform which carries clips to hold various-sized flasks and has a circular orbital motion. This platform is built into an insulated cabinet. For general purposes this is used as a normal incubator but if a

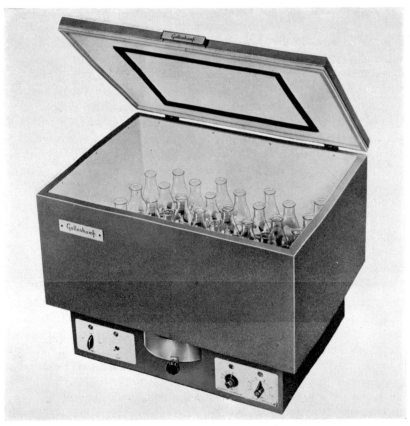

FIG. 11. A versatile orbital incubator by Gallenkamp, London. (Photograph by courtesy of Gallenkamp Ltd.)

sealed lid is also incorporated, atmospheres apart from air can be introduced. An efficient instrument incorporating the above features is Gallenkamp's Orbital Incubator which can be further modified by the incorporation of a bank of fluorescent lamps in the lid (Fig. 11). Facilities are also available for the attachment of a refrigeration system, so that cultures can be grown under illumination at temperatures down to 10°C below ambient.

A. Disintegration

Most fungal mycelium is tough and resilient and requires some kind of grinding or tearing action to bring about disintegration. Ultrasonic disintegration as used for bacterial cells is not very effective. Washed and filtered mycelium may be ground with sand in a mortar. This is laborious and it is usually better initially to homogenize the mycelium using a Waring blender. It requires about 3 min and is carried out with the addition of a phosphate buffer (pH approx. 7). If enzymatic studies are contemplated or if cell wall material is required then acetone is used in place of the buffer. The homogenate can then be further ground with glass beads in a cell disintegrator or forced through a press. Because of the resiliance of fungal mycelium, disintegration is more successful if the homogenate is frozen before grinding. Leis and Ralph (1960) disintegrated frozen cylinders of the homogenate by grinding them against a disc of abrasive paper fixed to a high speed wheel. The cylinders of the homogenate $6 \times 1\frac{1}{2}$–3 in. were deep-frozen for 24 h. They were then immersed in liquid air until quite rigid before being ground at light pressure. The ground material was removed from the wheel as the temperature rose.

VIII. PRELIMINARY SCREENING FOR ANTIBIOTIC ACTIVITY

Screening fungi for the presence of bacterial and fungal antibiotics has been extensively carried out by numerous commercial organizations. However, because of the vast numbers of fungi occurring in nature, and also because the results of assaying for these substances depend largely upon the methods employed, there is no doubt that many still remain to be discovered. Soil fungi are a major source of antifungal and antibacterial substances and in fact the production of some such substance must be a major requirement of any successful soil fungus to enable it to meet the competition of its environment. This statement is borne out by a study of current literature which lists *Penicillium*, *Fusarium*, *Trichoderma*, *Mucor* and *Aspergillus* as the most common soil fungi. Many species in these genera have been shown to produce antibacterial or antifungal substances. The oldest method of surveying fungi for these products is the soil plaque or the enriched soil plaque method. In the latter the soil is moistened with nutrient broth or potato dextrose solution and in the former with water only; it is then spread over the surface of a Petri dish, and incubated for about 7 days at 30°C. Surface examination of the fungal or actinomycete colonies which have developed may show zones of inhibition around the organism.

A modification of this technique, which demonstrates the presence of the antibiotics more clearly, is to flood the original soil plate with cooled nutrient agar and seed the surface with a 1 : 10 dilution of an overnight broth culture of a test bacterium such as *B. cereus* or *E. coli*. The excess is removed, the surface allowed to dry, and the plate incubated. Any antibiotic produced by the organisms in the soil will diffuse into the surface layer and inhibit bacterial growth.

A more satisfactory and also very simple method which produces more easily recognizable results is to first isolate the fungus from the soil plates and streak it across a plate of suitable nutrient agar; Brian and Hemming (1947) used the following—

Glucose	10 g
Peptone	10 g
Lab–Lemco meat extract	3 g
Sodium chloride	5 g
Agar	15 g
Distilled water	1000 ml

After a suitable period of incubation, test fungi or bacteria from broth cultures are streaked at right angles to the established fungal colony and the plate incubated for a further 24–28 h. The degree of antagonism between the new fungus and the test fungi and/or bacteria is represented by the presence or absence, and if present the extent, of the clear zone adjacent to the first fungus.

Various modifications of what is called the cup plate method are also commonly used. A test organism such as *E. coli* or *B. subtilis* is seeded into the cooled medium and this is poured into a Petri dish. For comparative purposes the amount in each dish should be exactly the same. After the agar has set a number of cavities are cut out with a sharp cork borer (10 mm dia). The dics may have to be removed with a sterilized spear-shaped needle. These cavities are filled with equal amounts of culture filtrate or other solution under test. The plate is incubated and inhibition zones around the cups can be measured after 18 or 24 h. These give the measure of antibiotic activity against the test organism.

Jeffreys (1947) described a technique for the demonstration of the production of antifungal substances by fungi. The technique takes about four days to complete and requires a cell made by bending a 7 mm glass rod into a rectangle 1×2 in. This is flame sterilized and placed in a sterile Petri dish, a little agar is placed in the cell so that it flows round the edge, and as it sets the cell is sealed to the base of the dish. The cell is filled to the top with agar and any overflow is removed with a sterile scalpel. The upper surface of the cell is inoculated with a streak of the fungus under test and incubated at 25°C for 2–3 days or until good growth has taken place. The cell is prised off the bottom of the dish with a scalpel, picked up with forceps, turned over and transferred to a second Petri dish. The newly exposed surface is inoculated with a standard test fungus. After a further 16 to 18 hours' incubation the upper surface is examined under the low power of a microscope to determine the degree of germination or the degree of inhibition of germination by the proximity of the fungus A.

Culture conditions such as temperature, pH, and nutritional constituents of the media have often a very marked effect on the production of both antifungal and antibacterial substances. Hanus et al. (1967) demonstrated that even the agar may have a marked effect on the activity of some antibiotics. Under their experimental conditions this interference effect could be reduced by purifying the agar with water extractions. The optimum requirements for each organism can only be found by test (Brian et al., 1953). Further simple methods in common use for assaying antibacterial and antifungal activity include the agar diffusion or cup method and its modifications (Abraham et al., 1941; Vincent and Vincent, 1944) and the turbidimetric method (Foster and Wilkes, 1943).

The next step after determining the presence of an antibiotic is to estimate its activity. This is usually carried out by various serial dilution methods. One of the simplest ways of testing the effectiveness of a fungal extract against bacteria is to use a series of test bacteria such as E. coli, B. subtilis, B. mycoides, B. cereus and S. aureus. A solution containing the antibiotic is obtained from the culture filtrate either directly or after extraction and concentrated by ethanol or chloroform. This solution is added to five Petri dishes in the following proportion: 1·0 ml, 0·3 ml, 0·1 ml, 0·03 ml and 0·01 ml. 10 ml of agar cooled to 45°C is added to the plate and the two are mixed by gently rocking the plate before the agar sets. After setting, each plate is marked from below into five segments and each segment is streaked with three 1 cm streaks of are of the five test organisms. The three streaks on each sector are made without recharging the needle which is flamed and recharged between plates.

The plates are incubated at 28°C for 18–20 h, after which the bacterial growth on each plate is compared with the growth on a control plate. The end point is the highest dilution at which growth is completely or almost completely inhibited.

IX. ENZYMES AND DEGRADATION POWERS OF FUNGI

A. Cellulolytic enzymes

The ability of many fungi to bring about the rapid disintegration of cellulose has serious economic effects on products manufactured from cotton, including military equipment. For practical purposes the loss of the tensile strength of the cotton fibres is the important feature. Thus many tests for the cellulolytic activity of fungi are designed to test such loss of strength. A simple method for assessing the relative cellulolytic activity of a number of fungi is to place a strip of standardized white cotton duck in a 1 in. test tube containing Schultz liquid medium or something similar. The medium is inoculated with the fungus under investigation and the culture incubated. After a given period, the cotton duck is removed and tested for loss of strength against a unmoulded control sample using a tensile-testing machine.

A more rapid method was described by Hazra *et al.* (1958) based on the lysis of cellulose pulp incorporated into nutrient agar. Single-colony cultures of various fungi were made at different points on the surface of the media and the plates incubated for 48 h at room temperature. The plates were then flooded with chloro-iodide of zinc when the uncoloured zone gave a measure of the cellulolytic power of the fungi.

A modification of this technique was described by Eggins and Pugh (1962) primarily for the study of cellulolytic fungi in soil, the medium being poured over soil crumbs on a plate. This medium contains a very finely powdered cellulose as the major carbon source so that the medium is white and opaque when set. Growth of cellulolytic fungi is apparent by the clearing of the medium as the fungus grows.

The medium is made up as follows—

Ammonium sulphate	0·5 g
L-asparagine	0·5 g
Potassium dihydrogen phosphate	1·0 g
Potassium chloride	0·5 g
Crystalline magnesium sulphate	0·2 g
Calcium chloride	0·1 g
Difco yeast extract	0·5 g
Agar	20 g
Ball-milled cellulose	10 g
Water to 1 litre	

The cellulose is prepared as a 4% suspension in water of Whatman's standard-grade cellulose powder for chromatography (derived from cotton), ball-milled for 72 h. This suspension of ball-milled cellulose is added to the other constituents of the medium after they have been steamed, and before autoclaving for 20 min at 10 lb pressure. The final pH of the medium is 6·2.

Savory *et al.* (1967) considered the rather poorly defined clearance zone in this method to be a serious disadvantage, particularly when testing fungi with dark-coloured mycelia. They solved the problem by growing the fungus under test on a cellulose/mineral agar plate for about 14 days at 25°C. A new plate of the same medium was then poured and a series of spaced holes were cut in the agar using the Leach (Fig. 9) stainless steel cutter. The same cutter was used to cut plugs from the culture, and these plugs, with the surface mycelium uppermost, were inserted into the holes in the new plate. The physical check to growth of the fungus caused by the cutting enabled the enzymes from it to diffuse into the surrounding test medium faster than any obscuring fungal growth, and a well-defined clearance growth appeared if any cellulase was present in the original culture. With rapidly growing fungi, growth over the new medium can be prevented by incorporating 0·005 mole/litre of sodium azide in the test medium. The azide acts on the respiratory enzymes of the fungus from the plug, but it does not affect the diffusion of the cellulase into the surrounding medium.

In testing for cellulase the influence of the pH of the medium can be important, and optimum pH for each culture has to be determined by test. Also of importance is the composition of the medium and temperature of growth. Simpson and Marsh (1964) demonstrated that fungi such as *Aspergillus niger*, believed to have little or no ability to decompose cellulose, have in fact a very marked effect if incubation takes place in the presence of suitable amounts of glucose or other soluble carbon source.

Keratinaceous materials such as wool, hair or feathers are not subject to such rapid decomposure as cotton. White *et al.* (1950) studied the effect of fungi on the degradation of woollen fabrics.

Strips of 18 oz olive-drab, wool serge approximately 6 × 1 in. were prepared and each one placed in a 200 × 25 mm test tube containing 25 ml mineral salts (Richards formula A, Greathouse *et al.*, 1942) so that the lower half of each strip was submerged. The tubes were plugged and autoclaved for 20 min at 15 lb pressure. Inoculation was by spore suspension and the tubes were incubated at 85°F for 10–14 days. Following incubation the strips were removed and washed in 1/1000 mercuric chloride, rinsed in water, and tested for breaking pressure. Wool was found to be less subject to fungus attack than cotton. Species attacking wool are limited, most are apparently dermatophytes such as *Microsporum gypseum*, or species related to dermatophytes.

B. Pectolytic enzymes

Enzymes concerned with fungal pathogenicity have also attracted attention. For instance, there is considerable evidence that extracellular pectic enzymes play a part in many plant diseases. The production of such extracellular enzymes as pectinesterase in liquid culture may depend on the cultural conditions, in particular the pH. Isolates to be assayed for this enzyme are best grown in 1% pectic-dox liquid media and after incubation for 6–7 days the culture filtrate can be assayed for the presence of pectinesterase or polygalacturonase by the cup-plate method of Dingle *et al.* (1953).

For the assay of polygalacturonase (Mann, 1962) a square plastic culture dish is filled with a 2·5 mm layer of the following medium—0·7% ammonium oxalate, 0·01% salicylanilide, 2·0% Difco agar and 1·0% purified sodium polypectate in 0·2 M acetate buffer at pH 5·0 (when obtained from the suppliers, both pectin and sodium polypectate may be contaminated by hexoses which can be removed by washing in acidified alcohol). Cups 9 mm dia are cut from the plate with a cork borer and 1 ml of the filtrate is placed in each cup. Boiled filtrate can be placed in the cups used as controls. The plate is incubated for 16 h and sprayed with 5 N hydrochloric acid. A clear zone with a white halo will be present around the cups containing filtrate with active polygalacturonase.

For pectinesterase assay the cup plate medium contains 1·0% purified pectin, 0·01% salicylanilide, methyl red and 2·0% Difco agar in distilled water. The pH of both substrate medium and filtrate is adjusted to 6·0 with sodium hydroxide solution. Activity is indicated by a red zone around the cup caused by production of acid.

For methods demonstrating the presence of laccase, peroxidase and tyrosinase see Lyr (1958). The presence of a wide range of enzymes can be demonstrated by electrophoresis using acrylamide gels.

Of the growth enzymes produced by fungi gibberellic acid produced by certain strains of *Gibberella fujikuroi* (*Fusarium moniliforme*) is the best known. This enzyme is also present in the culture filtrate from liquid cultures of this species. It is a highly active substance and can be effective in concentrations as low as 1 ppm. Drops of culture filtrate placed in the axil of a leaf of a growing seedling will demonstrate its presence by pronounced elongation of the adjacent stem and leaf as compared to the controls. Alternatively the filtrate can be made into a paste with lanolin or other inactive substance (Barton, 1956; Kato, 1955; Lockhart, 1956;

Phinney, 1956). For an excellent account of the cell wall degradation by enzymes see Wood (1967).

C. Proteolytic enzymes

Proteolytic organisms in soil and plant residues are chiefly concerned with the decomposition of vegetable protein. Many examinations of these organisms have been carried out using a medium based on animal protein such as casein, gelatin or albumen.

Grossbard and Hall (1962) recommended the use of plant protein rather than animal protein. The plant protein was prepared as follows—

Juice expressed from fresh minced lucerne was heated to 55°C and centrifuged, 2 N sulphuric acid was added to precipitate the cytoplasmic protein, which was then washed with hot water, alcohol and acetone. On drying this yielded a white powder of 14% nitrogen content. The dried powder was ground in demineralized water (25 g/500 ml) in a ball mill for 5 h to give a uniform suspension. This was steam-sterilized on three consecutive days and simultaneously shaken to prevent coagulation.

The protein–agar medium was prepared by adding the plant protein suspension to mineral salts agar containing 0·01% oxoid yeast extract to give a final concentration of 8% protein suspension. The pH was adjusted to 7 by the addition of sodium carbonate solution before the addition of the protein suspension.

These plates were used after a dilution series and incubated for 6 days at 25°C. Proteolytic organisms showed clear, sometimes transparent, zones underneath and around the colony. The relative proteolytic activity was expressed in terms of area of lysis in mm².

X. ENVIRONMENTAL CONTROL

The effect of temperature and humidity on spore germination and growth of fungal cultures has already been discussed. Temperature control is simple, as controlled temperature chambers are provided by the standard incubators which are now readily available, giving a controlled temperature both above and below the ambient and covering a range greater than any possible fungal growth.

The effect of light and light control is discussed by Leach, this Volume, Chapter XXIII.

A. Relative humidity

The relative humidity is extremely important in relation to fungal growth, and most species require an R.H. of 95–100% for sporulation (Cruikshank, 1958). In laboratory experiments it is generally preferable to use small humidity chambers with chemical solutions rather than the physical injection of humidified air as is used in large-scale work. Different humidities can be achieved by different concentrations of various chemicals in water. Initially, many experimental humidity chambers relied for their relative humidity on a series of aqueous solutions of sulphuric acid based on tables by Hastings (1909). The most serious disadvantage in the use of sulphuric acid is the possible presence of SO_3 in the overlying atmosphere. The removal of this by bubbling the air through water prevents initial R.H. levels being maintained. Also in studies in R.H. it is desirable to eliminate the mechanical effects of air currents. Because of these disadvantages, salt solutions are often used in place of sulphuric acid.

TABLE III

Relative humidity values at 25°C (%)	NaCl (g %)	$CaCl_2$ (g %)	Glycerine (g %)	H_2SO_4 (g %)
80	32	22·25	51·0	26·79
85	24	19·03	44·0	22·88
90	16	14·95	32·5	17·91
95	8	9·33	12·5	11·02
100	0	0	0	0

However, Carson (1931) and Hopp (1936) pointed out that solutions of salts, particularly concentrated solutions, can be very sensitive to temperature change, and glycerol or sucrose solutions are now widely used particularly when studying small changes of R.H. as they are comparatively insensitive to temperature and do not give off any volatile substances (Cruikshank, 1958).

Studies on the effect of relative humidity on spore release or conidiophore formation often necessitate rapid changes in the humidity of the cultural environment. This is brought about either by transfer of the culture between two vessels, such as Petri dishes, or by the replacement of air in the vessel. With the former, cultures in Petri dishes are transferred to a new atmosphere by placing the inverted culture into a new lid containing a different humidifying liquid. The disadvantage in this method is that the culture is exposed to the air of the laboratory during the transfer, and, as demonstrated by Jarvis (1960) with *Botrytis cinerea*, many imperfecti can respond within a few seconds of exposure to a new environment. When changing humidity by bubbling air through humidifying liquids, one either introduces mechanical effects due to air currents or the exchange is so slow as to be inadequate for the study fo atmospheric response. To

C. BOOTH

Fig. 12. The Jarvis chamber for hydroscopic studies, with details of the component parts. (Reproduced by courtesy W. R. Jarvis.)

overcome these effects Jarvis designed an apparatus in which cultures could be observed in atmospheres of different relative humidity without exposure to the air of the laboratory.

The apparatus is constructed in Perspex and is in the form of a cylinder 6 in. dia × 3¼ in. high divided into eight equal sectors (Fig. 12) by vertical walls. These are continuous through a wheel-like structure which carries the cultures, and through the shallow lid. Once the lid is in position the wheel is free to rotate about the central axis. Between the eight spokes of the wheel shallow Perspex pans are mounted in which cultures can be grown or leaf discs can be floated in mannitol solution (Cruikshank, 1958). Above each pan a circular hole is cut in the lid and a ¾ in. cover glass is sealed over the hole so that the culture can be observed under a dissecting microscope. The whole wheel can be turned over so that the culture pans face downwards, and released spores can be collected on coverslips coated with glycerine jelly and supported on the tops of the recessed and removable Perspex columns. All sliding surfaces are ground and lubricated with petroleum jelly or silicone grease. Humidifying liquids may be glycerol or someother solutions. Using different concentrations in the various sectors, fungal cultures can be transferred rapidly to a new atmosphere without exposure to the general laboratory atmosphere and with the concomitant transfer of a minimal volume of air.

ACKNOWLEDGMENTS

I wish to express my thanks to the Authors and Editors of *Mycologia*, the *Report of the Forestry Laboratory, Madison* and the *Transactions of the British Mycological Society* for permission to reprint the respective figures.

REFERENCES

Abraham, E. P., Chain, E., Fletcher, C. M., Florey, H. W., Gardner, A. D., Heatley, M. G., and Jennings, M. A. (1941). *Lancet*, **241**, 177.
Anthony, E. H., and Walkes, A. C. (1962). *Can. J. Microbiol.*, **8** (6), 929–930.
Banfield, W. M. (1941). *J. agric. Res.*, **62**, 637–681.
Barton, L. V. (1956). *Contr. Boyce Thompson Inst. Pl. Res.*, **18**, 311–317.
Billotte, J. M. (1963). *C.r. hebd. Séanc. Acad. Agric. Fr.*, **49**, 1056–1061.
Blackwell, E. (1935). *Nature, Lond.*, **135**, 546.
Brian, P. W., and Hemming, H. G. (1947). *J. gen. Microbiol.*, **1**, 158–167.
Brian, P. W., Hemming, H. G., Moffatt, J. S., and Unwin, C. H. (1953). *Trans. Br. mycol. Soc.*, **36**, 243–247.
Brown, W. (1924). *Ann. Bot.*, **38**, 402–444.
Brown, W. (1925). *Ann. Bot.*, **39**, 154.
Brown, M. F. (1963). *Phytopathology*, **53**, 347.
Brown, M. F., and Root, R. A. (1964). *Pl. Dis. Reptr*, **48**, 654–655.
Carson, F. T. (1931). *Paper Trade J.*, **93**, 71–74.
Christensen, C. M., Kaufert, F. H., Schmitz, H., and Allison, J. L. (1942). *Am. J. Bot.*, **29**, 552–558.

Clark, J. W. (1962). *Rep. Forest Prod. Lab., Madison,* 2262, 2 (4 fig.).
Cole, G. T., and Kendrick, W. B. (1968). *Mycologia,* 60, 340–344.
Cruikshank, I. A. M. (1958). *Aust. J. Biol. Sci.,* 11, 162–170.
Dade, H. A. (1960). *Herb. I.M.I. Handbook,* 64–65.
Davies, F. R. (1935). *Can. J. Res.,* 13, Sec C, 168–173.
Dingle, J., Reid, W. W., and Solomons, G. L. (1953). *J. Sci. Fd. Agric.,* 4, 149–155.
Dunleavy, J., and Snyder, G. (1962). Abs. in *Phytopathology,* 52, 8, 1962.
Eckert, J. W., and Tsao, H. P. (1962). *Phytopathology,* 52, 771–777
Eggins, H. O. W., and Pugh, G. J. F. (1962). *Nature, Lond.,* 193, 94–95.
Foster, J. W., and Wilkes, B. L. (1943). *J. Bact.,* 46, 377.
Funder, S., and Johannessen, S. (1957). *J. gen. Microbiol.,* 17, 117–119.
Garrett, S. D. (1946). *Trans. Br. mycol. Soc.,* 29, 171–172.
Goddard, D. R. (1935). *J. gen. Physiol.,* 19, 45–60.
Goidanich, G. (1947). *Italia agric.,* 1, 1–7.
Goidanich, G., and Mezzetti, A. (1947). *Annali. Sper. agr.,* 1, 123–129.
Goidanich, G., and Mezzetti, A. (1947). *Annali. Sper. agr.,* 1, 123–129; 2, 489–514.
Goldberg, H. S. (1959). "Antibiotics". D. van Nostrand, New York.
Greathouse, G. A., Klemme, D., and Barker, H. D. (1942). *J. Ind. Eng. Chem. Anal. Ed.,* 14, 614–620.
Grossbard, Erna, and Hall, D. M. (1962). *Nature, Lond.,* 196, 1119–1120.
Hansen, H. N. (1926). *Science N.Y.,* 64, 384.
Hansen, H. N., and Smith, R. E. (1932). *Phytopathology,* 22, 11.
Hansen, H. N., Thomas, H. E., and Thomas, H. Earl. (1937). *Hilgardia,* 10, 561–564.
Hanus, F. J., Sands, J. G., and Bennett, E. O. (1967). *Appl. Microbiol.,* 15, 31–34.
Harley, J. L., and Waid, J. S. (1955). *Trans. Br. mycol. Soc.,* 38, 104–118.
Haskins, R. H., and Spencer, J. F. T. (1962). *Can. J. Microbiol.,* 8, 279–281.
Hastings, M. M. (1909). Circ. U.S. Dep. Agric. Anim. Indust. No. 149.
Hazra, A. K., Bose, S. K., and Guha, B. C. (1958). *Sci. Cult.,* 24, 39–40.
Heald, F. D., and Pool, V. W. (1909). *Rep. Neb. agric. Exp. Sta.,* 22, 130–132.
Hesler, L. R. (1913). *Phytopathology,* 3, 290–295.
Holton, C. S. (1943). *Phytopathology,* 33, 732–735.
Hopp, H. (1936). *Bot. Gaz.,* 98, 25–44.
Janczewski, de, E. G. (1871). *Bot. Ztg.,* 29, 257–262.
Jarvis, W. R. (1960). *Trans. Br. mycol. Soc.,* 43, 525–528.
Jeffreys, E. G. (1947). *Trans. Br. mycol. Soc.,* 31, 246–248.
Kahn, M. (1966). *Nature, Lond.,* 212, 640.
Kaplan, L. (1956). *Mycologia,* 48, 609–611.
Kato, Y. (1955). *Bot. Gaz.,* 117, 16–24.
Keyworth, W. G. (1951). *Trans. Br. mycol. Soc.,* 34, 291–292.
Keyworth, W. G. (1959). *Trans. Br. mycol. Soc.,* 42, 53–54.
Khair, J., Fleischmann, G., and Dinoor, A. (1966). *Photopathology,* 56, 346.
Kirchoff, H. (1929). *Zentbl. Bakt. ParasitKde,* Abt. II, 77, 310–369.
Knaysi, G. (1957). *J. Bact.* 73, 431–435.
La Rue, C. D. (1920). *Bot. Gaz,* 20, 319–320.
Leach, C. M. (1964). *Mycologia,* 56, 926–928.
Leis, E., and Ralph, B. J. (1960). *Aust. J. Sci.,* 22, 348–349.
Lockhart, J. A. (1956). *Pl. Physiol.,* 31, (Suppl. 12).
Lyr. H. (1958). *Arch. Mikrobiol.,* 28, (3) 310–324.

Ludwig, R. A., Richardson, L. T., and Unwin, C. H. (1962). *Can. Pl. Dis. Surv.*, **42**, 149–150.

Mann, B. (1962). *Trans. Br. mycol. Soc.*, **45** (2), 169–178.

Meinecke, G. (1957). *Zentbl. Bakt. ParasitKde, Abt* **2**, 184–193.

Meiners, J. P., and Waldher, J. T. (1959). *Phytopathology*, **49**, 724–728.

Nutman, F. J., and Roberts, F. M. (1962). *Trans. Br. mycol. Soc.*, **45**, 449–456.

Parbery, D. G. (1967). *Trans. Br. mycol. Soc.*, **50**, 682–685.

Park, D. (1961). *Trans. Br. mycol. Soc.*, **44**, 367–390.

Phinney, B. O. (1956). *Proc. natn. Acad. Sci. U.S.A.*, **42**, 185–189.

Riddell, R. W. (1950). *Mycologia*, **42**, 265–270.

Saccardo, in Saccardo, P. A., and Berlese, A. N. (1895). *Riv. Patol. veg. Padova*, **4**, 56–65.

Savory, J. G., Mathur B, Maitland, C. C., and Selby, K. (1967). *Chem. Ind.*, 153–154.

Schmitthenner, A. F., and Hilty, J. W. (1962). *Phytopathology*, **52**, 582–583.

Schneider, I. R. (1956). *Pl. Dis. Reptr*, **40**, 9.

Shear, C. L., and Wood, A. K. (1913). *U.S. Dept. Agric. Bus. Pl. Ind. Bul.*, **252**, 1–110.

Simpson, M. E., and Marsh, P. B. (1964). *Tech. Bull. U.S. Dept. Agric.* 1303.

Skala, J. (1958). *Česká. Mykol.*, **12**, (3), 189–190.

Slankis, V. (1958). *Can. J. Bot.*, **36**, 837–842.

Sproston, T., and Setlow, R. B. (1968). *Mycologia*, **60**, 104–114.

Sussman, A. S. and Halvorson, H. O. (1966). "Spores: Their Dormancy and Germination". Harper and Row, New York.

Sussman, A. S., Lowry, R. J., and Tyrrell, E. (1959). *Mycologia*, **51**, 237–247.

Taschdjian, C. L. (1954). *Mycologia*, **46**, 681–683.

Thirumalachar, M. J., and Narasimhan, M. J. (1953). *Mycologia*, **45**, 461–466.

Vaartaja, O. (1960). *Phytopathology*, **50**, 820–823.

Vincent, J. G., and Vincent, H. W. (1944). *Proc. Soc. exp. Biol. Med.*, **55**, 162.

Voglino, P. (1895). *Nuovo G. bot. ital.*, **27**, 181–185.

Warcup, J. H. (1951). *Trans. Br. mycol. Soc.* **34**, 515.

Warcup, J. H., and Baker, K. F. (1963). *Nature, Lond.*, **197**, 1317–1318.

White, W. L., Mandels, G. R., and Siu, R. G. H. (1950). *Mycologia*, **42**, 199–223.

Williams, S. T. (1963). Proc. Colloqu. Soil Organism, Oosterbeck, The Netherlands, pp. 158–159.

Wood, R. K. S. (1967). "Physiological Plant Pathology" Blackwell Scientific Publications, Oxford, England.

Yu, C. C. C. (1954). *Am. J. Bot.*, **41**, 21–30.

CHAPTER II

Fungal Culture Media

C. BOOTH

Commonwealth Mycological Institute, Kew, Surrey, England

I. GENERAL CULTURAL CONCEPTS

A. Agar

Agar is now made in many countries, some of which are self-supporting in production or nearly so. The type of seaweed from which agar is produced is different in each country. The main sources are given in Table I though other species are used to a lesser extent to augment supplies. Conditions of harvesting vary from country to country since the best yields

TABLE I

Main sources of agar

Country	Source
Australia	*Gracilaria confervoides*
British Isles	*Gigartina stellata* Batt., *Chondrus cripus* (L.) Stackh.
India	*Gracilaria lichenoides*
Japan	*Gelidium corneum, G. amansii*
New Zealand	*Pterocladia lucida, P. capillacea*
S. Africa	*Gracilaria confervoides, Gelidium cartilagineum*
U.S.A.	*Gelidium cartilagineum, Gracilaria confervoides, Chondrus* sp.
U.S.S.R.	*Ahnfeltia plicata, Phyllophora rubens*

of agar come from weed bearing the sexual organs. Hence in the British Isles there is only one short harvest period (September), but in the U.S.A. (California) it extends from May to September and in New Zealand goes on all the year round with a peak in May–June.

Agars of different origin differ considerably in chemical composition and also to a greater or lesser extent in gelling capacity, melting point, hardness (% needed for a certain set) and viscosity. The differences depend on the type of seaweed used—and most countries use mixtures—the proportions of the different species, the time of harvesting (condition of the weed), and the weather conditions of each individual year on which the growth of the weed will depend. Differences also arise in processing which entails cleaning, weather bleaching, pounding, boiling, blending, acidification during boiling, addition of previous boilings, chemical bleaching, straining, setting, alternate freezing and thawing, and drying to give the product which will be used as a basis for the different types of finished agar.

It is evident, therefore, that agars—even those from one country—will differ materially and this fact should be taken into account when crucial experiments, such as growth rates under different conditions or the effect of nutrients on sporulation, germination, etc., are planned. Tests at the Commonwealth Mycological Institute have shown that growth rates of a single *Pythium* isolate on two different (plain water) agars varied from 0 to 10 mm in 96 h and varied also according to whether tap or distilled water was used to make up the agar. Emge (1963) found that germination of *Puccinia striiformis* on 1–5% water agar varied from 5–20% and on specially purified agar from 40–70%. Washing the granulated agar in three washes of distilled water plus one wash in redistilled water yielded consistently higher germination values and an acid rinse of the germination plate prior to the final distilled water rinse increased germination still further. In a series of tests, spores on washed agar in acid-washed plates showed 61% germination; those on washed agar in non-acid-washed plates had a 13% germination; whereas those on non-washed agar showed only 6% germination. Klemmer and Lenney (1965) showed that it was necessary to add certain lipids to Difco cornmeal agar to obtain sex organs of *Pythium* spp., but this addition is not necessary (though it enchances production) if cornmeal is made up with Japanese agar. Some "specially purified" agars have adverse effects on the growth and sporulation of some oomycetes but it is not certain yet whether this effect is due to substances removed during purification or to the presence of an inhibitor.

Hanus and Bennett (1964) showed that Difco Bacto agar contained at least two factors which inhibited or reduced the activity of fatty amines and this had serious implications when using the standard agar plate for screening the new antimicrobial agents.

It is clear that any account which aims to record the critical behaviour of a fungus on agar should specify the type of agar used. Whether the production of "standard" agar is at all feasible remains to be seen but it is certainly a goal to be aimed at.

Agar clarification

(i) Dissolve 25·0 g agar in one litre of distilled water. Cool the medium to approximately 45°C and stir into it one stiffly beaten egg white. Place in steamer until egg white is completely coagulated (approximately ½ hour). Carefully pour the medium away from the egg white so as not to break the coagulum and filter the medium through a Büchner funnel containing a layer of macerated filter paper (or filter pulp) about $\frac{3}{16}$ in. thick. Before using, thoroughly wash the filter and paper with hot distilled water (the filtering is best done in a steamer). Tube and autoclave.

(ii) Dissolve 1% w/v New Zealand Agar in distilled water. Filter twice through well-rinsed glass wool in a Büchner funnel to remove larger particles. (Addition of 1% w/v sodium azide is optional.)

To the filtrate, add 1 or 2% w/v of a mixture containing equal parts of bentonite and "Hyflo Super-Cel" (Johns-Manville Co.) and shake vigorously. (Using 2% speeds clarification but the volume of agar recovered is smaller.) Store at 56°C for several days, inverting the bottles daily. When all the cloudy flocculum has been completely carried down, decant the agar. Using a heated funnel, filter through Whatman No. 5 paper into a bottle standing in hot water. (Return the first 25·0 ml, or so, for refiltering.)

B. Formulation and sterilization of mycological media

The selection of a satisfactory medium for stimulating growth and sporulation of a particular fungus can only be found by test. A few general principles that may guide ones choice are set out as follows.

(i) Most fungi grow well on media having a pH of 6–6·5. Any rich carbohydrate source will support fungal growth and the following are the most commonly used media. PDA, potato dextrose agar; CMA, corn-meal agar; CzA, Czapek Dox agar; OA, oatmeal agar.

(ii) A medium too rich in nutrients tends to produce too much mycelium at the expense of fructifications. It is often better to grow fungi on nutritionally weak media such as potato carrot agar when the purpose of the cultures is to study the fructifications. Garrett (1954) found with *Armillaria mellea* that rhizomorph initials were produced most abundantly when an inoculum disc of minimal nutrient content is laid on a fresh agar substrate. Plain water agar was used to provide such a disc.

The inclusion of sucrose rather than dextrose in potato sucrose agar for

Fusarium cultures means that the carbohydrate is not so readily available to the fungus and sporulation is thereby stimulated.

(iii) Synthetic or semi-synthetic media are advisable for assay work, enzymatic or other biochemical studies.

(iv) Many fungi need a solid substrate on which to produce their fructifications and the inclusion of pieces of sterilized twigs, straw or lupin stems will often supply this need.

(v) Light is also important and the simulatory effect of near UV should be appreciated. Petri dishes incorporating straw or twigs in the agar and exposed to an alternate 12 h on/off sequence of near UV light has proved to be the best standard method of producing fructifications in mass sporulating cultures at the Commonwealth Mycological Institute.

(vi) Antibiotics should be included in media when bacterial contamination needs to be suppressed. Penicillium and Streptomycin have to be added to the cooled agar just before it sets. Chloramphenicol may be added to the agar before it is autoclaved.

1. *Sterilization*

(a) *Dry heat.* Laboratory glassware may be sterilized in a hot air oven. To ensure sterility after removal, it should be well wrapped in paper or placed inside a suitable container. Space should be allowed between the various packages. Heating for 1–2 h at 160°C is sufficient to allow penetration and ensure sterility.

(b) *Moist heat.* The standard autoclave whether on a steam line or with an independent heat supply is standard equipment for sterilization. A domestic pressure cooker is also a suitable alternative.

TABLE II

Autoclave pressures and approximate temperatures

Pressure, psi	Temperature, °C
5	107
7	110
10	115
15	121
20	126

Twenty minutes at 15 psi is sufficient to sterilize most equipment. Batches of soil or grain may require up to 1h. It is often necessary to autoclave soil on 3 or 4 successive days.

(c) *Surface sterilization.* Klarman and Craig (1960) demonstrated that 10 ml of propylene oxide in an open dish placed inside an airtight bell jar containing contaminated poured plates will sterilize the agar in the plates even when the lids are left on the plates. See also Booth, this Volume p. 4.

C. Control of infestation of cultures

Cultures kept in a state of active metabolism rather than under induced dormancy, as when maintained as lyophils or in liquid nitrogen, are susceptible to infestation by mites. In fact, this is one of the most constant sources of trouble in a mycological laboratory. Once having gained access to the laboratory, mites make their way into the Petri dishes or through the cotton-wool plugs into the culture tubes. Here they feed on the growing fungi and in particular on the spores and fructifications. Apart from destroying the fructifications they introduce fungal and bacterial contaminants into the cultures. The fact that they can wander quite rapidly from culture to culture means that, following a mite infestation, a whole collection of fungal cultures may be contaminated and an immense amount of work is required to obtain pure cultures again.

Mites occur in nature on decaying plant material, on grain, flour, cheese and in dust and soil particles. Those infecting fungal cultures most commonly belong to such genera as *Tyroglyphus* and *Tarsonemus*. They are most easily controlled by high standards of hygiene in laboratories which have a screening procedure for incoming cultures and which do not have to handle organic material. In laboratories which have to handle fresh plant material, soil, batches of grain or manufactured food stuffs, precautions must be taken against mites, particularly in humid climates. These precautions may consist of one of the following—

(i) Storage of the cultures in an atmosphere of an acaricidal chemical.
(ii) Addition of some acaricidal chemical to the tube which is not also fungicidal.
(iii) Cold storage.
(iv) Chemical or physical barrier between the growing culture and the surrounding atmosphere.

1. *Storage of fungi in the presence of an acaricidal chemical*

Camphor and later paradichlorobenzene (PDB) have long been used to keep beetles away from dried herbarium material and these substances have also been used to keep mites away from cultures.

Crude tractor vapourizing oil was for a long time found to be an effective deterrent against mites at the Commonwealth Mycological laboratories. However, the methods of purification now used for petroleum products appears to have robbed tractor fuel of this property. Jewson and Tattersfield (1922) found pyridine to be an extremely effective acaricide, however, its obnoxious smell and probable mammalian toxicity prohibits its general use in the laboratory. Jewson and Tattersfield recommended that test-tube cultures should be placed overnight under a bell jar containing about 20 ml of pyridine in a flat dish. The pyridine would penetrate the plugs, and one treatment was usually sufficient, although a second treatment given after about 14 days will ensure the death of any mites which have hatched from eggs that survived the first treatment.

Crowell (1941) recommended a similar method using 7 g dichloricide crystals placed in a watch glass under a stoppered bell jar sealed with Vaseline to a sheet of glass.

2. *The addition of acaricidal chemicals to test tube cultures*

Smith (1967) after tests with seven chemicals commonly used in mite control found that Cypro (active ingredients 1·18% pyrethrin and 11·87% piperonyl butoxide), Kelthane (active ingredient 1·8% of di[*p*-chlorphenyl] trichloromethyl carbinol) and paradichlorobenzene (PDB) were extremely effective acaracides. Crypto and Kelthane, used as two drops of the concentrate on the inside end of the cotton-wool plug, were effective acaricides and yet had very low fungistatic or fungicidal effects. Crystals of about 0·05 g of PDB inserted into culture tubes also effectively killed mites, but they had a marked fungistatic effect.

3. *Control of mites by low temperature*

Temperatures of 2°–5°C suitable for the maintenance of fungal cultures in the refrigerator have little acaricidal effect although they do tend to slow down the mites and prevent their movement from one culture to another. Smith (1967) found that no eggs hatched at 5°C during a period of 4 weeks' observation.

4. *Chemical and physical barriers against mite infestation*

An early method of protecting cultures against mites was to stand them on trays coated with oil or Vaseline. Barnes (1933) recommended keeping culture plates on a tripod standing in water. These methods are effective against crawling mites but they do not protect the cultures against mites carried by insects, in dust, or by the laboratory workers.

A simple method for long-term maintenance of fungal cultures is to cover the culture growing on an agar slope in a test tube or bottle with sterile mineral oil such as liquid paraffin. This has a marked fungistatic effect on the growth of the fungus and also prevents infestation by mites. Although fungal cultures have been shown to survive for many years in this condition (Agnes Onions, this Volume, p. 113), it is not suitable for cultures under investigation.

The treatment of the cotton-wool plugs with mercuric chloride is also successful in killing mites moving in or out of the tubes. This chemical is very poisonous to fungi and to the laboratory personnel and should be used with great care.

Snyder and Hansen (1946) described the sealing of tubes with cigarette papers to provide a barrier against mites which would allow air to diffuse through for the growth of the culture. This is a most effective method which has no effect on the fungal cultures and after many years of use the writer discounts the opinion that it is a laborious method (Smith, 1967). The material required consists of a copper sulphate gelatine as adhesive and standard cigarette papers.

The adhesive consists of 20% gelatine in water to which is added 2% copper sulphate. About 25 ml of this is poured into a Petri dish and allowed to solidify. The cigarette papers are unfolded and cut in half. If thought necessary, they may be sterilized either by dry heat or by propylene oxide. After the culture has been made, the cotton-wool plug is flamed and pushed down into the tube. The rim of the tube is again flamed and whilst still hot is pressed into the gelatine copper sulphate mixture, removed, and immediately pressed on to the centre of half a cigarette paper. The paper adheres to the end of the tube and it can be pressed down if any imperfections are observed. The tubes are placed with the top over the edge of the bench, and when a series is complete and the gelatine adhesive has set the surplus cigarette paper can be flamed off to leave a very neat unobtrusive seal.

5. *Removal of bacterial contamination*

The inhibition of bacterial contamination and growth has in the past been a major problem in the culturing of fungi. This problem has now been greatly simplified by the use of antibiotics, especially such antibiotics as Chloramphenicol which can be incorporated into the media before sterilization. Nevertheless one should not have to rely entirely on antibiotics for bacteria-free cultures and some of the earlier control measures are still applicable.

Many water moulds and certain other fungi can be grown at temperatures below those normally required for bacterial growth and therefore rapidly outgrow any bacterial contamination. Fungal mycelium penetrates agar gels more rapidly than bacteria and if the gel in an agar plate is inverted (Booth, this Volume, p. 18) then fungal mycelium penetrates the gel and sporulates on the upper surface. Fungi in general have a greater tolerance of acid condition than bacteria and the use of Rose Bengal to acidify the media to prevent bacterial growth has long been used. Similarly antiseptics are also much less effective against fungi than bacteria. Blank and Tiffney (1936) described the use of UV to inhibit bacterial growth in the presence of fungi. In fact little bacterial growth is observed when fungi are grown under UV (black-light), see Leach, this Volume, Chapter XXIII Details of concentrations of the most commonly used bacterial antibiotics which can be incorporated in most media are given by Buckley, this Volume, p. 461.

Dextrose-Phytone with Aureomycin and Rose bengal (Cooke, 1954)

Dextrose	10 g
Phytone (or peptone)	5 g
Potassium dihydrogen phosphate (KH_2PO_4)	1 g
Magnesium sulphate ($MgSO_4.7H_2O$)	0·5 g
Rose bengal	0·035 g
Agar	20 g
Water	1 litre
Aureomycin (chlortetracycline)	35 $\mu g/ml$

Autoclave at 15 psi for 20 min. To prepare antibiotic: dissolve 1 g aureomycin in 150 ml distilled water and store in refrigerator. Pipette 0·05 ml of this solution into 10 ml tubes of cool medium before each plating.

Littman's medium (Littman, 1947)

Ox bile, dehydrated	15 g
Peptone	10 g
Dextrose	10 g
Agar	12 g
Water	1 litre

Dissolve by heat. Adjust to pH 7·0 and add 10 ml of 0·1% crystal violet. Sterilize at 10 psi for 10 min. Before pouring plates, cool to 50°C and add 300–500 μg streptomycin per 100 ml medium.

Martin's medium modified (Snyder et al., 1959)

Peptone	15 g
Magnesium sulphate ($MgSO_4.7H_2O$)	0·5 g
Potassium hydrogen phosphate (K_2HPO_4)	1 g
Agar	25 g
Water	1 litre
Rose bengal	1 : 30,000

Autoclave in bottles in 100 ml lots, and at time of pouring add 10 drops of 10% sodium taurocholate and streptomycin to give 300 ppm to each bottle.

II. FORMULAE FOR MEDIA

1. *Actinomycete enzyme assay media*, see Williams and Cross, this Volume, p. 295.
2. *Actinomycete media—general*, see Williams and Cross, this Volume, p. 295.

3. *Alphacel medium for stimulation of sporulation in Ascomycetes and Fungi Imperfecti (Sloan et al., 1961)*

†Alphacel	20 g
Magnesium sulphate (MgSO₄.7H₂O)	1 g
Potassium dihydrogen phosphate (KH₂PO₄)	1·5 g
Sodium nitrate (NaNO₃)	1 g
‡Coconut milk	50 ml
Agar	12 g
Distilled water	1 litre

Adjust pH to 5·6; autoclave at 20 psi for 20 min.
† Non-nutritive cellulose
‡ Filter coconut milk through several layers of muslin, autoclave at 15 psi for 15 min and store at 6°C until required.

4. *Alphacel medium (modified)*

Same as above but with the addition of—

Tomato paste	10 g
Oatmeal	10 g

also for stimulation of sporulation.

Weitzman and Silva-Hutner modified this medium, excluding the alphacel, for dermatophytes, see Stockdale, this Volume, p. 429.

5. Aphanomyces euteiches, *synthetic medium for study of sulphur nutrition (Haglund and King, 1962)*

D-Glucose	5·0 g
L-Asparagine	0·75 g
Monobasic potassium phosphate (KH₂PO₄)	2·0 g
Magnesium chloride (MgCl₂)	0·05 g
Manganese chloride (MnCl₂)	0·005 g
Zinc chloride (ZnCl₂)	0·005 g
Ferric chloride (FeCl₃)	0·005 g
Distilled water	1 litre

Adjust pH to 5·5. *A. euteiches* requires a reduced form of sulphur for growth and the addition of 5–20 ppm of the amino-acids L-methionine or L-cystine is necessary for growth.

6. *Apple leaf decoction agar for perithecial production in* Venturia inaequalis *(Keitt and Langford, 1941)*

Steam 25 g air-dried apple leaves (collected during autumn or winter) in about 500 ml distilled water for 30 min. Make up to 1 litre and dissolve 5 g malt extract and 17 g agar. Autoclave at 15 psi for 20 min.

7. *Asthana and Hawker medium A for perithecial formation of* Sordaria destruens
(*Asthana and Hawker, 1936*)

Glucose	5·0 g
Potassium nitrate (KNO₃)	3·5 g
Potassium dihydrogen phosphate (KH₂PO₄)	1·75 g
Magnesium sulphate (MgSO₄)	0·75 g
Agar	15 g
Distilled water	1 litre

8. *Badcock's medium for growth of Basidiomycetes* (*Badcock, 1941*)

The medium consists of dry beech or spruce sawdust mixed with 5% by weight of the accelerator and then moistened to at least 17% moisture content based on the oven-dry weight.

The accelerator—

Maize meal	50 g
Bone meal (containing 3·75% organic nitrogen)	30 g
Potato starch	17 g
Sucrose	2 g
Wood ash (from combustion of Scots pine sap wood)	1 g

The following modification of Badcock's medium is used at the **Forest Products Research Laboratories, U.K.**—

Sawdust	20 g (10–20 mesh 0·6–1·2 mm particle size)
Maize meal	0·6 g
Bone meal	0·4 g
Water	40–60 ml to give 200–300% moisture content

This amount fills one boiling tube. Hardwood fungi are grown on beech sawdust and softwood fungi on either Norway or Siberian spruce sawdust. In fact any sawdust of an easily decaying wood may be substituted.

9. *Barnes' agar* (*Gwynne-Vaughan and Barnes, 1927*)

Potassium phosphate (K₃PO₄)	1 g
Ammonium nitrate (NH₄NO₃)	1 g
Potassium nitrate (KNO₃)	1 g
Glucose	1 g
Agar	25 g
Distilled water	1 litre

Melt the agar in half of the water in a water bath, then add the other constituents dissolved in the remainder of the water. Autoclave at 15 psi for 15 min.

10. *Barnett maltose casamino medium for production of perithecia of* Ceratocystis fagacearum (*Barnett, 1953*)

Maltose	5 g
Difco Casamino Acids (tech, grade)	1·0 g
Potassium dihydrogen phosphate (KH₂PO₄)	1·0 g
Magnesium sulphate (MgSO₄.7H₂O)	0·5 g
Zn ⎫	0·2 mg ⎤
Fe ⎪ (as sulphates)	0·2 mg ⎪
Mn ⎭ —or use microelement solution	0·1 mg ⎦
Biotin (0·5 g yeast extract may be substituted)	5 μg
Agar	20 g
Distilled water	1 litre

Adjust pH to 6·0.

11. *Basidiomycetes, liquid medium for* (*Robbins and Hervey, 1959*)

8 ml of wood or tomato extract is added to each 100 ml of basal medium. *Basal medium per litre of distilled water*

Potassium dihydrogen phosphate (KH₂PO₄)	1·5 g
Magnesium sulphate (MgSO₄.7H₂O)	0·5 g
Dextrose	20 g
Casein hydrolysate	2 g
Adenine sulphate	8·09 mg
Cytosine	2·22 mg
Guanine HCl	4·11 mg
Hypoxanthine	2·72 mg
Thymine	0·25 mg
Uracil	2·24 mg
Xanthine	0·06 mg
Choline Cl	5·58 mg
Orotic acid	0·62 mg
Thiamine HCl	0·337 mg
Riboflavin	0·376 mg
Pyridoxine	0·205 mg
Nicotinic acid	0·123 mg
Calcium pantothenate	0·476 mg
para-aminobenzoic acid	0·137 mg
m-inisitol	216·19 mg
Folic acid	1 mg
Biotin	0·01 mg
Vitamin B₁₂	0·01 mg
B	0·005 mg
Cu	0·02 mg
Fe	0·1 mg
Ga	0·01 mg
Mn	0·01 mg
Mo	0·01 mg
Zn	0·09 mg

Neutralize with calcium carbonate (CaCO₃).

Wood extract
Autoclave 500 g dry beech wood (ground in a mill) with 10 : 1 distilled water. Reduce the filtrate on a hot plate to 1 : 1 and filter through celite.

Tomato extract
Filter canned tomato juice through muslin and celite.

12. *Bean juice agar, for conidial formation of* Colletotrichum lindemuthianum (*Romanowski and Kuć, 1962*)

Bean Juice from canned green beans	215 ml
Agar	10·0 g
Water	285 ml

13. *Bean meal agar for* Phytophthora cinnamomi (*Royle and Hickman, 1964*)

Ground dwarf bean seed (Variety Canadian Wonder)	30 g
Agar	20 g
Water	1 litre

Boil ground seed for 5 min and filter, add liquified agar. After 8 days growth discs are cut out of the plate and immersed in Wills non-sterile soil extract.

14. *Beef agar for* Pilobolus (*Swartz, 1934*)

Boiling beef	210 g
Agar	15 g
Water (distilled)	1 litre

Boil beef in water until thoroughly cooked. Strain the broth through several thicknesses of muslin; resore to original volume. Add agar and dissolve. Autoclave at 15 psi for 15 min.

15. *Beef extract agar*

Beef extract	5 g
Peptone (BDH Bact.)	10 g
Common salt	5 g
Agar	20 g
Water	1 litre

Dissolve beef extract in 300 ml water; mix peptone and salt to a paste with 200 ml water at 60°C. Mix the two liquids and steam for 45 min. Add the remainder of the water and agar and dissolve. Adjust pH to 8; autoclave to 10 psi for 20 min.

16. *Beer wort agar*

Beer wort	1 litre
Agar	30 g

Melt agar in beer wort by boiling in water bath for 15 min. Adjust pH to 5·0–5·5, autoclave at 10 psi for 10 min.

Wort agar (according to Biourge, see Thom, 1930)
Select a pale, unhopped wort (from the brewery), autoclave for 15 min at 115°–120°C, filter in the boiling condition, distribute in tubes or flasks and sterilize for 15 min at 120°C. The density at 4·8°–5·6°C should be 12°–14° Balling.

To prepare wort–gelatine agar, dissolve 1·5% agar in the wort by autoclaving at 120°C for ½ hour, then add an equal quantity of wort containing 10% of gelatine, and sterilize the mixture at 110°C for 15–20 min.

17. *Bianchi medium*, see Richardson, this Volume, p. 267.

18. *Bilai medium, modification by Joffe (Joffe, 1963)*

Potassium dihydrogen phosphate (KH_2PO_4)	1·0 g
Potassium nitrate (KNO_3)	1·0 g
Magnesium sulphate ($MgSO_4$)	0·5 g
Potassium chloride (KCl)	0·5 g
Starch powder	0·2 g
Glucose	0·2 g
Sucrose	0·2 g
Water	1 litre

Strips of cellulose lens paper are added. According to Joffe, this medium always induces conidial formation in *Fusaria*.

19. *Botrytis separation agar (Netzer and Dishon, 1967)*

Potassium chloride (KCl)	1 g
Potassium dihydrogen phosphate (KH_2PO_4)	1·5 g
Sodium nitrate ($NaNO_3$)	3 g
Magnesium sulphate ($MgSO_4$)	0·5 g
Casein hydrolysate	5 g
Yeast extract	3 g
Glycerol	5 g
L-Sorbose	2·5 g
Agar	20 g
Tapwater	1 litre

A selective medium to distinguish between *Botrytis allii* and *B. cinerea* on onion. The former species is severely restricted by sorbose.

20. *Bread crumb agar (Berliner, 1961)*

Commercial bread crumbs (without preservatives)	100 g
Bacto-agar	18 g
Tap water	1 litre

Used to permit the luminescence of various fleshy fungi.

21. *Brown's agar, for* Sclerotium rolfsii—basidia *(Brown, 1926)*

Glucose	2 g
Asparagine	2 g
Neutral potassium phosphate (K_2HPO_4)	1·25 g
Magnesium sulphate ($MgSO_4.7H_2O$)	0·75 g
Agar	20 g
Water (distilled)	1 litre

22. *Burkholder's trace element solution*, see under Lukens and Sisler (No. 94).

23. *Cantino PYG agar for Blastocladiella* (*Horenstein and Cantino, 1961*)

Peptone	1·25 g
Yeast extract	1·25 g
Glucose	3·0 g
Agar	20 g
Water (distilled)	1 litre

If grown as liquid cultures it is necessary to add bromocresol purple indicator so that intermittent neutralization with NaOH can be carried out to maintain a pH of 6.8.

24a. *Carboxy-methyl-cellulose agar* (*Jefferys et al., 1953*)

Ammonium tartrate ((NH_4)-$C_4H_4O_6.4H_2O$)	2·0 g
Potassium dihydrogen phosphate (KH_2PO_4)	1·0 g
Magnesium sulphate ($MgSO_4.7H_2O$)	0·5 g
Potassium chloride (KCl)	0·5 g
Ferrous sulphate ($FeSO_4$)	0·01 g
Sodium carboxy-methyl-cellulose	10·0 g
Distilled water	1 litre

24b. *Carboxy-methyl-cellulose agar* (*Gams, W., 1960*)

Czapek agar (No. 45) plus trace elements together with $CaCl_2$(50·0 mg); $MnSO_4$ (5·0 mg).

Sodium carboxy-methyl-cellulose	10·0 g
Difco yeast extract	0·5 g
(See also No. 35)	

25. *Carlile's semi-defined medium* for Physarum polycephalum, see Carlile, this Volume, p. 237.

26. *Carrot agar* (*Smith, 1960*)

Whole carrot	300 g
Agar	30 g
Distilled water	1500 ml

Lightly cook carrot until tender in 500 ml distilled water, macerate, and autoclave at 15 psi for 30 min. Add 1000 ml distilled water and 30 g agar and dissolve. Autoclave at 15 psi for 15 min.

27. *Cellulose yeast extract agar*

Filter paper	12 g
Difco Yeast Extract	4 g
Agar	10 g
Tap water	1 litre

Tear paper into small pieces and macerate in some of the water until the fibres are separated. Add this to the remainder of the water, and dissolve the yeast extract and agar. Autoclave at 20 psi for 20 min.

28. *Powdered cellulose agar* (Eggins and Pugh 1962), see Booth, this Volume, p. 39.

29. *Cerelose ammonium nitrate medium* (*Scheffer and Walker, 1953*)

Cerelose	50 g
Ammonium nitrate (NH_4NO_3)	10 g
Potassium dihydrogen phosphate (KH_2PO_4)	5 g
Magnesium sulphate ($MgSO_4.7H_2O$)	2·5 g
Ferric chloride ($FeCl_3.6H_2O$)	0·02 g
Water	1 litre

Liquid media for growth of *Fusarium oxysporum* f. sp. *lycopersici* without shaking.

30. *Charcoal water* (*Wills, 1954*)

Charcoal	20 g
Tap water	1 litre

For inducing sporulation of *Phytophthora parasitica*, after growth on potato dextrose broth.

31. *Cherry agar* (*CBS formula*)

†Cherry extract	300 ml
Agar	20 g
Water	700 ml

Dissolve agar in water, add cherry extract. Distribute into pre-sterilized bottles or tubes and sterilize at 102°C for 5 min. Overheating should be avoided, as the acid reaction will prevent the setting of the medium. pH 3·8–4·6.

†*Cherry extract*
Stone cherries (red variety). To each 200 g pulp, add 1 litre water. Bring to the boil and simmer gently for 2 h. Strain through cloth, and sterilize at 110°C for 1 h. Store in stock bottles or flasks.

32. *Chick-pea sucrose agar for* Phytophthora infestans (*Keay, 1953*)

Chick-pea (*Cicer arietinum*)	250 g
Sucrose	20 g
Agar	15 g
Distilled water	1 litre

Wash seed in tap water for 1 h then soak in distilled water overnight. Drain off water and mash seeds with pestle and mortar. Add 1 litre of water and steam for 1 h, and strain through surgical gauze. Sucrose and agar are dissolved separately and combined, making up to volume. Medium is tubed and sterilized for 15 min at 10 psi. Dried garden peas (*Pisum sativum*) are a satisfactory alternative.

33. *Chytrid Medium*, see Gareth Jones, this Volume, p. 335.

34. *Claussen's* (*1912*) *Agar for Ascomycetes, also known as Johansen's medium for* Pyronema confluens (*McLean and Cook, 1941*)

Potassium dihydrogen phosphate (KH_2PO_4)	0·05 g
Ammonium nitrate (NH_4NO_3)	0·05 mg
Magnesium sulphate ($MgSO_4.7H_2O$)	0·02 g
Ferrous phosphate ($Fe_3(PO_4)_2$)	0·001 g
Agar	3·0 g
Distilled water	100·0 ml
Inulin	2·0 g

Fill the lower half of a Petri dish with the above medium. Place the dish inside another dish of somewhat greater diameter but of the same height, and fill the space surrounding the inner Petri dish with the same medium minus the inulin. Inoculate the inulin agar and cover the whole. Fructifications will occur in a few days on the inulin-free portion.

35. *CMC medium for stimulation of macroconidial formation of* Gibberella zeae (*Capellini and Peterson, 1965*)

Carboxymethylcellulose (CMC7MP-Hercules Powder Co.)	15·0 g
Ammonium nitrate (NH_4NO_3)	1·0 g
Potassium dihydrogen phosphate (KH_2PO_4)	1·0 g
Magnesium sulphate ($MgSO_4.7H_2O$)	0·5 g
Yeast extract	1·0 g
Distilled water	1 litre

The formation of abundant conidia occurs in shake culture after 4 days at 24°C.

36. *Colloidal chitin medium*, see Williams and Cross, this volume, p. 295.

37. *Conn's Agar*

Potassium nitrate (KNO_3)	2 g
Magnesium sulphate ($MgSO_4.7H_2O$)	1·2 g
Potassium dihydrogen phosphate (KH_2PO_4)	2·7 g
Maltose	7·2 g
Potato starch	10 g
Agar	15 g
Water	1 litre

38. *Coon's medium (for* Fusarium)

Saccharose	7·2 g
Dextrose	3·6 g
Magnesium sulphate ($MgSO_4$)	1·23 g
Potassium dihydrogen phosphate (KH_2PO_4)	2·72 g
Potassium nitrate (KNO_3)	2·02 g
Agar	12·0 g
Water	1 litre

Add to this malachite green to make 1 : 40,000 solution or gentian violet to make 1 : 26,000 solution.

39. *Corbaz's half-strength nutrient agar* for thermophilic actinomycetes, see Williams and Cross, this Volume, p. 295.

40. *Corn (maize) meal agar*

Cornmeal	30 g
Agar	20 g
Water	1 litre

Place the cornmeal in the water (if meal is not available, break up 30–35 g of grain and pass through mill). Heat in water bath until boiling, stirring occasionally, for 1 h. Filter the decoction through muslin, add agar, and boil until dissolved. Autoclave at 15 psi for 20 min.

41. Corn (maize) steep agar

Corn steep liquor	90 ml
Agar	25 g
Water	1 litre

Adjust pH to 6·5.

Soak maize in water overnight. Boil in water bath for 1 h. Strain, add agar and boil until melted. Cool to 55°C and add switched whites of 4–5 eggs. Autoclave at 5 psi for 1 h. Filter through damp filter paper (stand filtering apparatus in a steamer to keep the medium molten). Autoclave at 15 psi for 20 min.

42. Cornmeal (maize) peptone yeast-extract agar (Benjamin, 1958)

Cornmeal	20 g
Dextrose	10 g
Peptone	10 g
Yeast extract (Difco)	4 g
Agar	20 g
Water	1 litre

Boil cornmeal in 700 ml of water for 10 min without water bath. Strain through cloth. Dissolve peptone in 300 ml of water at 60°C. Add this, together with dextrose, yeast extract and agar to the filtrate and cook over a water bath until agar is dissolved. Autoclave at 15 psi for 20 min.

43. Crabill's medium, said to be favourable for Phyllosticta pyrina (Coniothyrium pyrinum)

Ammonium nitrate (NH_4NO_3)	10·0 g
Potassium hydrogen phosphate (K_2HPO_4)	5·0 g
Magnesium sulphate ($MgSO_4$)	2·5 g
Sucrose	50·0 g
Agar	10·0 g
Water	1 litre

44. Craveri and Pagani's medium for thermophilic actinomycetes, see Williams and Cross, this Volume, p. 295.

45. Czapek agar with nitrate replaced by urea for increased perithecial production in Aspergillus nidulans (Acha and Villanueva, 1961)

Urea, or ammonium oxalate	3 g
Potassium hydrogen phosphate (K_2HPO_4)	1 g
Magnesium sulphate ($MgSO_4.7H_2O$)	0·5 g
Potassium chloride (KCl)	0·5 g
Ferric sulphate ($Fe_2(SO_4)_3$)	0·05 g
Sucrose	30 g
Agar	20 g
Distilled water	1 litre

46. *Czapek–Dox agar—Method 1*

Sodium nitrate (NaNO₃)	2 g
Potassium hydrogen phosphate (K₂HPO₄)	1 g
Magnesium sulphate (MgSO₄.7H₂O)	0·5 g
Potassium chloride (KCl)	0·5 g
Ferrous sulphate (FeSO₄)	0·01 g
Sucrose	30 g
Agar	20 g
Distilled water	1 litre

If glass distilled water is used, to each litre of the above add 1 ml of each of the following—

$$\left.\begin{array}{l}\text{1g Zinc sulphate}\\ \text{0·5 g Copper sulphate}\end{array}\right\} \text{dissolve each in 100 ml distilled water}$$

Boil all chemicals in water bath for 15 min, add sucrose and agar when cool, melt agar; autoclave at 25 psi for 20 min. Do not filter.

Webster's formula: as above, with addition of 5 g dried yeast.

47. *Czapek–Dox agar—Method 2*

Stock soln. A	50 ml
Stock soln. B	50 ml
Sucrose	30 g
Agar	20 g
Distilled water	900 ml

Dissolve sucrose in soln. A diluted to about 500 ml; add soln. C, make up to 1 litre, add agar and melt. Autoclave at 15 psi for 20 min.

Stock solutions

Solution A

Sodium nitrate (NaNO₃)	40 g
Potassium chloride (KCl)	10 g
Magnesium sulphate (MgSO₄)	10 g
Ferrous sulphate (FeSO₄)	0·2 g
Disolve in 1 litre distilled water.	

Solution B

Dissolve 20 g potassium hydrogen phosphate (K₂HPO₄) in 1 litre distilled water.

48. *Czapek–Dox broth* modified by Weary and Graham (1966) for Dermatophytes, see Stockdale, this Volume, p. 429.

49. *Czapek-malt agar (for* Penicillium)

Stock Czapek soln. A	50 ml
Stock Czapek soln. B	50 ml
Sucrose	30 g
Malt extract	40 g
Agar	20 g
Distilled water	900 ml

Dissolve malt extract and agar in water. Add solutions A and B and sucrose and heat in water bath until dissolved. Autoclave at 15 psi for 20 min.

50. *Daniel and Rusch medium for* Physarum polycephalum (*Daniel and Rusch, 1961*)

Tryptone (Difco)	10 g
Yeast Extract (Difco)	1·5 g
Glucose, anhydrous	10 g
Potassium dihydrogen phosphate (KH_2PO_4)	2 g
Calcium chloride ($CaCl_2.2H_2O$)	0·6 g
Magnesium sulphate ($MgSO_4.7H_2O$)	0·6 g
Ferrous chloride ($FeCl_2.4H_2O$)	0·06 g
Manganese chloride ($MnCl_2.4H_2O$)	0·084 g
Zinc sulphate ($ZnSO_4.7H_2O$)	0·034 g
Citric Acid.H_2O	0·48 g
Hydrochloric acid (HCl, conc.)	0·06 ml
Distilled water	to 1 litre
Calcium carbonate ($CaCO_3$)	3 g
†Chick embryo extract	15 ml

† Difco ampoule containing 2 ml of a lyophilized 50% extract reconstituted with 8·3 ml distilled H_2O.

51. *Dextrose peptone yeast (modified) DPYA, for evaluation of antimicrobial agents in soil (Papavizas and Davey, 1959)*

Dextrose	5·0 g
Peptone	1·0 g
Ammonium nitrate (NH_4NO_3)	1·0 g
Potassium hydrogen phosphate (K_2HPO_4)	1·0 g
Magnesium sulphate ($MgSO_4.7H_2O$)	0·5 g
Ferric chloride ($FeCl_3.6H_2O$)	trace
Yeast extract	2·0 g
Oxgall	5·0 g
Sodium propionate	1·0 g
Agar	20·0 g
Water	1 litre
†Chlortetracycline	20 mg
†Streptomycin	30 mg

Sterilized at 11 psi for 15 min.
† Fresh solutions to be added after the media has been sterilized and cooled to approximately 45°C.

52. *Dick's agar* for the Saprolegniaceae, see Gareth Jones, this Volume, p. 335.

53. *Dox agar*

Potassium nitrate (KNO_3)	1 g
Magnesium sulphate ($MgSO_4.7H_2O$)	0·25 g
Potassium hydrogen phosphate (K_2HPO_4)	0·5 g
Ferrous chloride ($FeCl_2$)	0·025 g
Sucrose	7·5 g
Agar	10 g
Water	1 litre

Add chemicals to water and boil in water bath for 15 min. Cool, add agar and sucrose and melt agar. Correct pH and autoclave at 25 psi for 20 min.

54. *Dung agars* (*Langeron, 1952*)

About 500 g of partially dried horse, cow or rabbit dung is soaked in cold water for 3 days. The supernatant is decanted and diluted to a straw colour before adding 2% agar and autoclaving.

Horse dung

Horse dung or other Ungulates for culturing and conservation of filamentous fungi.

Use fresh dung, uncontaminated with soil or other manure. Introduce pieces 7–8 cm high with spatula into large test tubes without touching the upper surface. Add water to cover the dung and plug tube. Sterilize twice at 15 psi for 30 min with an interval of 24 hr. (See also Lange's medium.)

Rabbit dung agar

Bottle whole pellets with plain tap water agar (15 g agar in 1 litre water)—about 6 pellets to each 15 ml agar. Autoclave at 20 psi for 20 min.

Sheep dung agar for Pilobolus (*Swartz, 1934*)

Sheep dung	300 g
Agar	15 g
Water (dist.)	1 litre

Boil the dung in the water until dung is broken down. Filter and restore liquid to original volume. Add agar and dissolve. Autoclave at 15 psi for 15 min.

55. *Egg-yolk potato medium for the yeast-like phase of* Histoplasma capsulatum (*Titsworth and Grunberg, 1950*)

Potato base

Potato (peeled and finely ground)	200 g
Glycerol	60 ml
Citric Acid	0·2 g
Bacto-haemoglobin	5 g
Water	1 litre

Autoclave at 15 psi for 30 min. Filter through a double layer of muslin. Flask in 500 ml quantities and autoclave at 15 psi for 30 min.

Egg-yolk filtrate

Clean eggs, soak in 80% alcohol for 2 h. Remove eggs and flame. Separate whites from yolks into a sterile container. A proportion of one whole egg to eleven yolks is used. Filter mixture through double layer of sterile muslin.

Mix 500 ml of the sterile egg-yolk filtrate with 500 ml of sterile potato base. Add 10 ml of sterile 10% congo red. Dispense the mixture into tubes or dishes and inspissate in free-flowing steam for 1h.

56. *Elliott's agar* (*Elliott, 1917*)

Potassium dihydrogen phosphate (KH_2PO_4)	1·36 g
Sodium carbonate (Na_2CO_3)	1·06 g
Magnesium sulphate ($MgSO_4$)	0·50 g
Dextrose	5·0 g
Asparagin	1·0 g
Agar	15·0 g
Distilled water	1 litre

57. Entomophthora muscae *medium* (*Srinivasan* et al., *1964*)

Wheat grain extract	30 g
Peptone	20 g
Yeast extract	10 g
Glycerine	10 g
Agar	20 g
Tap water	1 litre

Boil wheat grain in 500 ml of water for 1 h. Decant extract and add other ingredients, except agar. Melt agar in the other 500 ml water, mix the two liquids, filter and adjust pH to 7 with sodium hydroxide soln. before sterilization.

58. Erysiphe graminis: *culturing on partially isolated epidermal tissue*
(*Dueck* et al., *1965*)

Inoculate the inside epidermis of split barley coleoptiles and incubate in darkness on 0·01 M Ca (NO₃)₂. After 36–48 h the inoculum will have produced immature haustoria. Clamp the coleoptile halves between two plastic wafers, one of which has a slot.

Remove part of the outer epidermis and mesophyll through this slot and place the newly exposed side of the epidermis in contact with water or nutrient solution and further development can be observed through a hole in the plastic on the colonized side. The colonies should grow for 2–5 days and will often sporulate. Growth is prolonged by yeast extract or sucrose and it shows powdery mildews can be supported by epidermal tissue alone.

59. *Fell and von Uden Hemiascomycete media*, see Gareth Jones, this Volume, p. 335.

60. *Filter paper yeast agar*

Filter paper is dispersed in water for perithecia formation of *Sordaria*. (Ingold and Dring 1957; see also *Cellulose Yeast Extract Agar* (*No. 27*)

Filter paper	12 g
Yeast extract (Difco)	4 g
Agar	24 g
Tap water	1 litre

Filter paper is dispersed in the water by the use of a blender.

61. *Fish broth medium—Method 1*

Modification of fomula given by Höye for cultivation of *Sporendonema sebi*.

†Fish broth	500 ml
‡Flour	400 g
Sodium chloride (NaCl)	50 g

† The fish broth is merely water in which fish has been boiled.

‡ Packeted flour as sold in shops usually contains fungicides. It may be necessary therefore to grind up whole wheat grain.

62. *Fish broth medium—Method 2*

Fresh cod fish	1 kg
Peptone	5 g
Sodium chloride (NaCl)	200 g
Wheat flour	400–500 g
Water	1 litre

Macerate the fish in the water and allow to stand for 4 h. Strain through muslin. Add peptone and sodium chloride and autoclave at 15 psi for 20 min. Add the flour and, if necessary, strain through muslin. Autoclave at 7 psi for 35 min. The medium will set during autoclaving.

63. *Flentje's soil extract agar (for* Corticium praticola *basidia)*

Soil	1000·0 g
Water	1 litre
Extract	1 litre
Sucrose	1·0 g
Potassium dihydrogen phosphate (KH$_2$PO$_4$)	0·2 g
Dried yeast	0·1 g
Agar	25·0 g

† Agitate frequently for a day or two. Filter through glass wool. Make up to 1 litre.

64. Fomes annosus *selective isolation medium (Kuhlman and Hendrix, 1962)*

Bacto-peptone	5·0 g
Magnesium sulphate (MgSO$_4$)	0·25 g
Potassium hydrogen phosphate (KHPO$_4$)	0·5 g
Pentachloronitrobenzene (PCNB)	190 ppm
Agar	20 g
Water	1 litre
Streptomycin	100 ppm
Lactic Acid (50%)	2 ml
Ethyl alcohol (95%)	20 ml

† Add after sterilization when cooled to 41°–45°C.

65. *Freezing agar (Kuehn, Orr and Ghosh, 1961)*

Diced potatoes	200 g
Dextrose	8·0 g
Yeast extract	0·5–1·0 g
Activated charcoal	0·5 g
Agar	20·0 g

Adjust to pH 7.

Add diced potatoes to 500 ml tap water and autoclave at 15 psi for 15 min. Mash potatoes and filter through 4 layers of cheesecloth, bring the volume to 1 litre with distilled water and add the dextrose, yeast extract and activated charcoal.

66. *Fries medium modified (Ryan, 1950)*

Ammonium tartrate [(NH$_4$)$_2$C$_4$H$_4$O$_6$.4H$_2$O]	5 g
Ammonium nitrate (NH$_4$NO$_3$)	1 g
Potassium dihydrogen phosphate (KH$_2$PO$_4$)	1 g
Magnesium sulphate (MgSO$_4$.7H$_2$O)	0·5 g
Sodium chloride (NaCl)	0·1 g
Calcium chloride (CaCl$_2$)	0·1 g
Sucrose	10 g
d-Biotin	4 mg
Distilled water to make	1 litre

The following amounts of trace elements per litre should also be added: B, 10 mg; Cu, 100 mg; Fe, 200 mg; Mn, 20 mg; Mo, 20 mg; Zn, 200 mg.

67. *Fuller's modification of Vishniac's medium,* see Gareth Jones, this Volume, p. 335.

68. *Garlic agar (for* Sclerotium—*basidia)*

Garlic	300 g
Agar	20 g
Distilled water	1 litre

Peel and cut up the garlic and boil in the water for 1 h. Filter and restore to original volume with more distilled water. Add agar and sterilize at 15 psi for 20 min.

69. *Georg and Camp's basal medium* for vitamin testing of dermatophytes, see Stockdale, this Volume, p. 429.

70. *Glucose alanine agar for production of fruiting bodies of* Coprinus lagopus *(Madelin,* 1956)

Glucose	10 g
dl α-Alanine	1 g
Thiamin HCl	500 μg
Potassium hydrogen phosphate (K_2HPO_4)	2 g
Magnesium sulphate ($MgSO_4.7H_2O$)	0·2 g
Agar	20 g
Distilled water	1 litre

71. *Glucose asparagine agar (Krainsky's Medium) for stimulating production perithecia in* Penicillium luteum—*but originally for Actinomycetes*

Glucose	10 g
Asparagine	0·5 g
Potassium hydrogen phosphate (K_2HPO_4)	0·5 g
Agar	15 g
Water	to 1 litre

Dissolve glucose, asparagine and potassium phosphate in the water and add agar. Boil until dissolved. Autoclave at 7·5 psi for 30 min.

72. *Glucose nitrate medium (Hendrix, 1965)*

Glucose	5·4 g
Sodium nitrate ($NaNO_3$)	1·5 g
Potassium dihydrogen phosphate (KH_2PO_4)	1·0 g
Magnesium sulphate ($MgSO_4.7H_2O$)	0·5 g
Agar	17 g
Thiamine hydrochloride	(2 ml of a 1000 ppm stock solution)
Distilled water	1 litre

Adjust pH to 6·0. Autoclave at 15 psi for 10 min. This media was used by Hendrix to study the effect of sterols on the growth and reproduction of *Pythium* and *Phytophthora* species. Add 0·5 mg sterol in ether solution to surface of agar plate. Sterols such as ergosterol, phytosterol or cholesterol can be applied in this way and the plates stored for several hours before inoculation.

73. *Glucose–peptone agar* (*Goos and Tschersch, 1962*)

Peptone	2 g
Glucose	10 g
Magnesium sulphate ($MgSO_4.7H_2O$)	0·5 g
Potassium dihydrogen phosphate (KH_2PO_4)	0·5 g
Agar	15 g
Water	1 litre

Autoclave at 15 psi for 15 min.

74. *Glucose yeast extract agar for* Harposporium (*Aschner and Kohn, 1958*

Peptone	5 g
Glucose	20 g
"Difco" yeast	2 g
Agar	20 g
Water	1 litre

75. *Glycerol-arginine medium*, see Williams and Cross, this Volume, p. 295.

76. *Glycerol asparagine agar* (*for Actinomycetes*)

Glycerol	10 g
Asparagine	1 g
Dipotassium hydrogen phosphate (K_2HPO_4)	1 g
Agar	20 g
Water	1 litre

Dissolve glycerol, asparagine and potassium phosphate in the water and add agar. Boil until dissolved. Adjust pH to 7. Autoclave at 7 psi for 30 min.

77. *Grain*

Wheat, barley, rice (use unpolished rice in preference to polished).

Pack the bottom of boiling tubes with a 2–3 cm layer of absorbent cotton wool. Above this place a layer of grain about 2 cm deep. Add enough water to moisten the cotton wool and the base of the grain. In the case of rice more water is necessary —enough to cover 75% of the grain. Sterilize twice at 15 psi for 20 min with an interval of 24 h (see Langeron, 1952).

Alternatively, sprinkle sterilized grains on to plates of tap water agar (15 g agar, 1 litre water).

78. *Hansen's medium for yeasts*

Peptone	1 g
Maltose	5·9 g
Potassium dihydrogen phosphate (KH_2PO_4)	0·3 g
Magnesium sulphate ($MgSO_4.7H_2O$)	0·2 g
Water	1 litre

See also Beech and Davenport, this Volume, p. 153.

79. *Hay extract agar*

Hay	200 g
Glucose	5 g
Agar	20 g
Water (distilled)	1 litre

Weigh out the hay and boil for 30 min in 500 ml of the water. Decant and add the glucose to the liquid. Make up to 1 litre, then add the agar and heat until dissolved. Sterilize at 15 psi for 20 min.

80. *Hay infusion agar* (*Raper and Thom, 1949*)

Distilled water	1 litre
Decomposing hay	50·0 g

Autoclave for 30 min at 15 psi. Filter.

Infusion filtrate	1 litre
Potassium hydrogen phosphate (K₂HPO₄)	2·0 g
Agar	20·0 g

Adjust to pH 6·2.

81. *Haricot bean agar* (*for* Phytophthora)

Dried haricot beans	30 g
Agar	10 g
Water	500 ml

Split beans and break into pieces and pass through mill (avoid grinding to powder). Add to water in a beaker; immerse in water bath, bring to boiling point, stirring mixture occasionally. Add agar and boil for 5 min. Strain through muslin, squeezing out as much as possible. Correct pH and autoclave at 20 psi for 20 min. Various kinds of beans can be used in making this medium; 10, 15, 20 or 30 g of agar are used depending on the required firmness of the medium.

82. *Hayduk's solution* (*Thom, 1930*)

Dipotassium phosphate (K₂HPO₄)	1·0 g
Magnesium sulphate (MgSO₄.7H₂O)	0·32 g
Asparagine	0·80 g
Sucrose	80·0 g
Water	1 litre

83. *Honey peptone medium* (*Backus and Stauffer, 1955*)

Honey	60 g
Difco Bacto-peptone	10 g
Agar	20 g
Water	1 litre

The pH is favourable to fungal growth and inhibitory to most bacteria. Agar is dispersed in water before the other ingredients are added to prevent caramelization of the sugar.

84. *Horne and Mitter's medium for* Fusarium (*1927*)

Glucose	2·0 g
Potato starch	10·0 g
Asparagine	2·0 g
Potassium tribasic phosphate (K₃PO₄)	1·25 g
Magnesium sulphate (MgSO₄.7H₂O)	0·75 g
Agar	15·0 g
Water	1 litre

85. *Kauffman's agar*

Maltose	5·0 g
Magnesium sulphate (MgSO$_4$.7H$_2$O)	0·10 g
Calcium nitrate (Ca(NO$_3$)$_2$.4H$_2$O)	0·50 g
Potassium dihydrogen phosphate (KH$_2$PO$_4$)	0·25 g
Agar	15·0 g
Water	1 litre

86. *Kerr and Sussman's medium*, See Carlile, this Volume, p. 237.

87. *Kirk's medium* for marine ascomycetes, see Gareth Jones, this Volume, p. 335.

88. *Knop's solution for* Chaetomium

Calcium nitrate (CaNO$_3$)	0·5 g
Potassium nitrate (K$_2$NO$_3$)	0·125 g
Magnesium sulphate (MgSO$_4$.7H$_2$O)	0·125 g
Potassium phosphate (K$_2$HPO$_4$)	0·125 g
Ferrous chloride (FeCl$_2$)	0·005 g
Water (distilled)	1 litre

Filter paper saturated with Knop's solution (with agar) is used for maintaining *Chaetomium* spp.

89. *Knox-Davis*: *the use of peanut agar plus longwave ultraviolet light for production of pcynidia in* Macrophomina phaseoli (*Knox-Davis, 1965*)

A small quantity of peanut meal—prepared by milling shelled peanuts—is added to a 2% water agar and autoclaved at 15 psi for 15 min.

Inoculated plates are subjected to longwave ultraviolet light (Leach, this Volume, Chapter XXIII). This is a most effective method.

90. *Kuehner's basal medium (liquid) for studying the effects of added vitamins and micrometabolic substances on growth and ethyl acetate production by* Hansenula anomala (*Kuehner, 1951*)

Dextrose A.R.	20 g
Asparagine A.R.	2 g
Potassium dihydrogen phosphate (KH$_2$PO$_4$)A.R.	1·5 g
Magnesium sulphate (MgSO$_4$.7H$_2$O) A.R.	0·5 g
Calcium chloride (CaCl$_2$(Anhyd.)) A.R.	0·33 g
Ammonium sulphate (NH$_4$)$_2$SO$_4$ A.R.	2 g
Potassium iodide (KI) A.R.	0·1 mg
Double distilled water	1 litre

A.R. indicates specially purified products.

91. *Lactose casein hydrolysate medium* (*Malca and Ullstrup, 1962*)

Lactose	37·5 g
Casein hydrolysate	3·0 g
Potassium dihydrogen phosphate (KH$_2$PO$_4$)	1·0 g
Magnesium sulphate (MgSO$_4$.7H$_2$O)	0·5 g
Microelements (p. 78)	2 ml
Agar	10 g
De-ionized water	1 litre

Adjust pH to 6·0. A good sporulation medium for *Helminthosporium turcicum* and *H. carbonum* and probably for other *Helminthosporium* species.

92. *Lange's* (*modified Kauffman's medium* (*Lange, 1952*)

Maltose	5·0 g
Magnesium sulphate (MgSO$_4$.7H$_2$O)	0·5 g
Calcium nitrate (CaNO$_3$)	0·5 g
Potassium hydrogen phosphate (K$_2$HPO$_4$)	0·25 g
Peptone	0·1 g
Distilled water	900·0 ml
Horsedung decoction	100·0 ml

Dung decoction is prepared by boiling 1 fresh "horseapple" in 150 ml water for 1–2 min; 100·0 ml of the filtrate is added to the medium.

93. *Leonian's agar* (*Bonar's modification*)

Potassium dihydrogen phosphate (KH$_2$PO$_4$)	1·2 g
Magnesium sulphate (MgSO$_4$.7H$_2$O)	0·6 g
Peptone	0·6 g
Maltose (or glucose)	6·0 g
Malt extract	6·0 g
Agar	20·0 g
Distilled water	1 litre

94. *Lukens and Sisler synthetic medium* (*Lukens and Sisler, 1958*)

Glucose	20 g
Ammonium sulphate ((NH$_4$)$_2$SO$_4$)	3·0 g
Magnesium sulphate (MgSO$_4$.7H$_2$O)	0·25 g
Monobasic potassium phosphate (KH$_2$PO$_4$)	3·0 g
Glycine	1·0 g
Thiamine HCl	20 μg
Niacin	20 μg
Biotin	1 μg
I-inositol	200 μg
Pyridoxine	10 μg
Folic acid	10 μg
Boron	0·01 ppm ⎤
Mn	0·01 ppm ⎟
Zn	0·07 ppm ⎬ †
Cu	0·01 ppm ⎟
Mo	0·01 ppm ⎟
Fe	0·05 ppm ⎦
Distilled water	to make 1 litre

† Burkholder's (1943) trace element supplement.

Adjust to pH 6·0 with KOH. Lukens (1960) used this liquid medium in conjunction with filter paper to stimulate formation of conidia of uniform size and age in *Helminthosporium vagans* and *Alternaria solani.*

95. *Lutz Medium (modified) general medium for agarics and boletes*
(Singer, 1962)

Vitrums malt extract†	10·0 g
Ammonium nitrate (NH_4NO_3)	1·0 g
Ammonium phosphate ((NH_4)$_2$HPO$_4$)	1·0 g
Agar	25·0 g
Magnesium sulphate ($MgSO_4.7H_2O$)	0·1 g
Ferric sulphate ($Fe_2(SO_4)_3$)	0·1 g
Manganese sulphate ($MnSO_4$)	0·025 g

† Obtained from Apoteksvarucentral, Stockholm, Sweden.

96. *Machlis' medium* for Trichomycetes, see Gareth Jones, this Volume, p. 335.

97. *Malachite green–captan medium, a modification of Czapek-Dox used for selective isolation of* Fusarium *from soil*

Sodium nitrate ($NaNO_3$)	2·0 g
Dipotassium hydrogen phosphate (K_2HPO_4)	1·0 g
Magnesium sulphate ($MgSO_4.7H_2O$)	0·5 g
Potassium chloride (KCl)	0·5 g
Ferrous sulphate ($FeSO_4$)	0·01 g
Sucrose	30 g
Distilled water	1 litre
Malachite green	50 mg
Captan	100 mg
Dicrysticin (mixture of Streptomycin sulphate Procain penicillin G and Sodium penicillin G)	0·75 mg

Add fresh solutions of malachite green, captan and dicrysticin after autoclaving and before pouring the plates.

98. *Malt extract agar*

Malt extract	20 g
Agar	20 g
Water	1 litre

Heat the malt extract in water until dissolved; add agar and dissolve. Normal pH is 3 to 4, and this should be adjusted to 6·5 with NaOH. Autoclave at 15 psi for 20 min. May be modified for Basidiomycetes by using 5% malt extract and adding 0·5% malic acid.

99. *Malt extract agar*

With 20% sucrose for organisms requiring high osmotic pressure for sporulation. With 40% sucrose for hygrophobes.

Malt extract	20 g
Sucrose	200 g (or 400 g)
Agar	20 g
Water	1 litre

Add malt extract and agar to the water and heat in a water bath until dissolved. Add sucrose and dissolve. Autoclave at 15 psi for 20 min.

100. *Malt carrot agar (for Yeasts)*

Carrot	250 g
Malt extract	5 g
Agar	25 g
Water	1 litre

Boil carrots in water until quite soft; filter through muslin or rub through sieve. Dissolve agar, then add malt extract. Autoclave at 15 psi for 20 min.

101. *Malt salt agar (for organisms requiring high osmotic pressure*

Malt extract	100 g
Sodium chloride (NaCl)	100 g
Agar	20 g
Water	1 litre

Dissolve the ingredients in the water and autoclave at 15 psi for 20 min.

102. *Malt/yeast extract agar* (Lilly and Barnett, 1951), see Richardson, this Volume, p. 267.

103. *Maltose tartrate medium for* Melanconium fuligineum (*Timnick, Barnett and Lilly, 1952*)

Maltose	20 g
Ammonium tartrate ($(NH_4)_2C_4H_4O_6.4H_2O$)	2·8 g
Potassium dihydrogen phosphate (KH_2PO_4)	1 g
Magnesium sulphate ($MgSO_4.7H_2O$)	0·5 g
Thiamine	50 μg
Zinc	0·2 mg
Iron	0·2 mg
Manganese	0·1 mg
Agar	20 g
Distilled water	1 litre

104. *Mannite agar*

Mannite	15·0 g
Potassium dihydrogen phosphate (KH_2PO_4)	0·2 g
Magnesium sulphate ($MgSO_4.7H_2O$)	0·2 g
Sodium chloride (NaCl)	0·2 g
Calcium sulphate ($CaSO_4$)	0·1 g
Agar	15·0 g
Distilled water	1 litre

105. *Menzies and Dade—selective indicator medium* for *Streptomyces* (see Williams and Cross, this Volume, p. 295).

106. *Meyers and Richards—yeast extract seawater broth*, see Gareth Jones, this Volume, p. 335.
107. *Meyers and Simms medium for marine ascomycetes*, see Gareth Jones, this Volume, p. 335.

108. *Microelements (used by Lilly and Barnett, 1951)*

$Fe(NO_3)_3 . 9H_2O$	723·5 mg
$ZnSO_4 . 7H_2O$	439·8 mg
$MnSO_4 . 4H_2O$	203·0 mg

Dissolve one at a time in a litre of water. Add sulphuric acid to yield a clear solution.

109. *Miller's medium for zoospore isolation*, see Gareth Jones, this Volume, p. 335.

110. *Mineral salts agar FA No. 5 (Berk et al., 1957)*

Potassium dihydrogen phosphate (KH_2PO_4)	0·7 g
Dipotassium hydrogen phosphate (K_2HPO_4)	0·7 g
Magnesium sulphate ($MgSO_4 . 7H_2O$)	0·7 g
Ammonium nitrate (NH_4NO_3)	1·0 g
Sodium chloride (NaCl)	0·005 g
Ferrous sulphate ($FeSO_4 . 7H_2O$)	0·002 g
Zinc sulphate ($ZnSO_4 . 7H_2O$)	0·002 g
Manganese sulphate ($MnSO_4 . 7H_2O$)	0·001 g
Agar (Difco)	15 g
Water (distilled)	1 litre

pH, 6·4. Autoclave for 20 min at 10 psi

111. *Molybdenum medium for identification of* Candida albicans *(MacLaren, 1961)*

Proteose-peptone (Difco)	10 g
Sucrose	40 g
Agar	15 g
Distilled water	1 litre

Adjust pH to 7·6, autoclave at 10 psi for 15 min. Cool to 50°–55°C, add 15 ml of a 12·5% aqueous soln. of Merck phosphomolybdic acid. (Final concentration 1·9 mg/ml.)

112. Mucor-*synthetic medium (Hesseltine, 1954)*

Dextrose	40 g
Asparagine	2 g
Potassium dihydrogen phosphate (KH_2PO_4)	0·5 g
Magnesium sulphate ($MgSO_4$)	0·25 g
Thiamine chloride	0·005 g
Agar	15 g
Distilled water	1 litre

113. *Myxomycete media*, see Carlile, this Volume, p. 237.

114. *Nash and Snyder PCNB medium* (*Nash and Snyder, 1962*)

Difco peptone	15 g
Potassium dihydrogen phosphate (KH_2PO_4)	1·0 g
Magnesium sulphate ($MgSO_4.7H_2O$)	0·5 g
Streptomycin	300 ppm
Agar	20 g
†Pentachloronitrobenzene (PCNB)	1 g
Water	1 litre

For the isolation of *Fusarium* from soil.

† 75% wettable powder, Terraclor, obtained from Olin Mathieson Chem. Corp., Baltimore 3, Maryland, U.S.A.

115. Neurospora "*minimal*" *medium* (*Beadle and Tatum, 1945*)

Ammonium tartrate (($CH(OH).COO.NH_4)_2$)	5·0 g
Ammonium nitrate (NH_4NO_3)	1·0 g
Potassium dihydrogen phosphate (KH_2PO_4)	1·0 g
Magnesium sulphate ($MgSO_4.7H_2O$)	0·5 g
Sodium chloride (NaCl)	0·1 g
Calcium chloride ($CaCl_2$)	0·1 g
Sucrose	15·0 g
Biotin	5×10^{-6} g
Bo	0·01 mg
Cu	0·1 mg
Fe	0·2 mg
Mn	0·02 mg
Mo	0·02 mg
Zn	2·0 mg
Distilled water	1 litre

pH 5·6

Extensive aerial growth of *Neurospora* can be prevented by changing concentration of sucrose to 1 g and adding 0·8 g of *l*-sorbose. (Tatum *et al.*, 1949).

116. Neurospora "*complete*" *medium* (*Beadle and Tatum, 1945*)

Glucose	5·0 g
Sucrose	5·0 g
Hydrolized casein	5·0 ml
Difco yeast extract	2·5 g
Spray-dried malt syrup	5·0 g
Vitamin sol	10 ml
Agar	15 g
Water	1 litre

Vitamin sol

Thiamin	100 mg/litre
Roboflavin	50 mg/litre
Pyridoxin	50 mg/litre
Pantothenic acid	200 mg/litre
p-Aminobenzoic acid	50 mg/litre
Nicotinamide	200 mg/litre

(*continued*)

Choline	200 mg/litre
Inositol	400 mg/litre
Alkali hydrolised yeast nucleic acid	500 mg/litre
Folic acid	4 μg pure substance

The casein hydrolysate is prepared by HCl hydrolysis and made up to the equivalent of 50 mg casein per litre.

117. *Nette's modified medium* for Actinomycetes, see Williams and Cross, this Volume, p. 295.

118. *Niger seed-creatinine medium* for the isolation of Cryptococcus neoformans see Buckley, this Volume, p. 461.

119. *Nutrient agar*

Oxoid nutrient agar granules	14 g
Agar	10 g
Water	1 litre

The following weak nutrient agar is recommended for thermophilic Actinomycetes.

"Lab-Lemco" Beef Extract	0·5 g
Yeast extract (Oxoid L 20)	1·0 g
Peptone (Oxoid L 37)	2·5 g
Sodium chloride (NaCl)	2·5 g
Agar	20 g
Water	1 litre

120. *Nutrient agar*

Beef extract	3 g
Peptone	10 g
Agar	15 g
Water	1 litre
Carbohydrate (if desired)	10 gm

121. *Oak wilt agar for perithecial production (Barnett, 1953)*

Maltose	5·0 g
Difco Casamino acids (tech. grade)	1·0 g
Potassium dihydrogen phosphate (KH_2PO_4)	1·0 g
Magnesium sulphate ($MgSO_4.7H_2O$)	0·5 g
Zinc sulphate ($ZnSO_4$)	0·2 mg ⎤ see micro-
Ferrous sulphate ($FeSO_4$)	0·2 mg ⎬ element
Manganese sulphate ($MnSO_4.7H_2O$)	0·1 mg ⎦ solution
Biotin (0·5 g yeast extract may be substituted for some isolates)	5 μg
Agar	20 g
Distilled water	1 litre

Adjust pH to 6·0 before autoclaving at 15 psi for 15 min.

122. *Oat agar*, see Carlile, this Volume, p. 237.

123. *Oatmeal agar*

Powdered oatmeal	30 g
Agar	20 g
Water	1 litre

Add oatmeal to water and gradually heat to boiling in a water bath and boil for 1 h. Strain through muslin and make up liquor to 1 litre with more water. Add agar and dissolve. Autoclave at 15 psi for 20 min.

124. *Onion-asparagine agar* (*Mundkur, 1934*)

Onion	100 g
Asparagine	0·25 g
Protease peptone (Bacto)	0·5 g
Agar (Bacto)	15 g
Distilled water	1 litre

Peel and cut up onions and boil in water bath in 500 ml of the water. Strain and add agar and remainder of ingredients dissolved in the other 500 ml water. Autoclave at 10 psi for 15 min.

125. *Orange fluid medium for* Penicilliopsis

Glucose	50 g
Ammonium citrate ((NH4)2COOH)	1·9 g
Potassium dihydrogen phosphate (KH2PO4)	1 g
Potassium chloride (KCl)	0·5 g
Magnesium sulphate (MgSO4.7H2O)	0·5 g
Ferrous sulphate (FeSO4.7H2O)	0·01 g
Extract of 1 orange to	1 litre

pH will be approx. 6·7 and should be adjusted to 4·5 by the addition of HCl, before making up to final volume.

Orange extract: For 1 litre medium, macerate one good sized orange in 250 ml water. Make up to 500 ml and boil for ten minutes with constant stirring. Strain through muslin. Stir up pulp with another 250 ml water and strain again. (The extract can be partially cleared by filtration using a filter pump, if necessary.)

126. *Pailey, Stafamak, Olson and Johnson's medium for* Aspergillus "aureus"

Glycerol	7·5 g
Cane sugar	7·5 g
Peptone	5 g
Magnesium sulphate (MgSO4.7H2O)	0·05 g
Potassium dihydrogen phosphate (KH2PO4)	0·06 g
Sodium chloride (NaCl)	4 g
Agar	20 g
Water	1 litre

Add glycerol, cane sugar, magnesium sulphate, potassium bisulphate to 200 ml water. Mix peptone and salt to a paste with 200 ml water, at 60°C and add to the first mixture. Dissolve agar in 600 ml water, then mix all together. pH does not need adjustment. Autoclave at 15 psi for 20 min.

127. Papulospora *medium* (*Hotson, 1942*)

Starch A.R.	30 g
Malt (Bacto)	10 g
Agar (Bacto)	15 g
Peptone (Bacto)	5 g
Dextrose A.R.	10 g
Distilled water	1 litre

128. *Park's medium* (*liquid*) *for* Fusarium (*Park, 1964*)

Glucose	0·7 g
Magnesium sulphate ($MgSO_4.7H_2O$)	0·5 g
Potassium dihydrogen phosphate (KH_2PO_4)	0·2 g
Ammonium nitrate NH_4NO_3	0·1 g
Distilled water	1 litre

This may be solidified by adding 10 g Oxoid No. 3 agar.

129. *Pasteur's solution* (*modified*)

Potassium dihydrogen phosphate (KH_2PO_4)	1 g
Calcium phosphate ($CaHPO_4$)	0·1 g
Magnesium sulphate ($MgSO_4.7H_2O$)	0·1 g
Ammonium tartrate ($CH(OH).COO.NH_4)_2$	5 g
Glucose	75 g
Distilled water	1 litre

130. *Pea agar* (*Gwynne-Vaughan and Barnes, 1927*)

Dried peas	400 (100 g)
Agar	25·0 g
Water	1 litre

400 dried peas are boiled for an hour and the liquid made up to one litre.

131. *Pea seed—for the production of zoospores of* Phytophthora infestans (*Thurston, 1957*)

Soak dried yellow peas overnight. Place 1 in. layer of soaked peas in 250 ml flasks. Add enough water to cover the peas.
See also Chick pea medium.

132. *Petri solution for the production of sporangia of* Phytophthora

Calcium nitrate ($Ca(NO_3)_2$)	0·4 g
Magnesium sulphate ($MgSO_4.7H_2O$)	0·15 g
Potassium dihydrogen phosphate (KH_2PO_3)	0·15 g
Potassium chloride (KCl)	0·06 g
Water	1 litre

The solution should be kept in a refrigerator and used unsterile to obtain sporangia from infected host material. See Holliday and Mowat, 1963.

133. *Perrott's medium* for Monoblepharis, see Gareth Jones, this Volume, p. 335.

134. *Phytone dextrose agar for* Stemphylium bolicki
(Sobers and Seymour, 1963)

Phytone	15 g
Dextrose	15 g
Yeast extract	1 g
Agar	17 g
Water	1 litre

135. Physarum (*Myxomycete*) *medium*, see Carlile, this Volume, p. 237.

136. *Piefer, Humphrey and Acree's medium (for wood-destroying fungi)*

Glucose	40·0 g
Potassium phosphate (K_2HPO_4)	4·0 g
Asparagine	4·0 g
Ammonium dibasic phosphate ((NH_4)$_2HPO_4$)	2·0 g
Magnesium sulphate ($MgSO_4.7H_2O$)	2·0 g
Calcium carbonate ($CaCO_3$)	0·25 g
Calcium chloride ($CaCl_2$)	0·1 g
Agar	15·0 g
Distilled water	1 litre

137. *Pontecorvo minimal medium (Pontecorvo, 1953)*

Sodium nitrate ($NaNO_3$)	6·0 g
Potassium chloride (KCl)	0·52 g
Magnesium sulphate ($MgSO_4.7H_2O$)	0·52 g
Ferrous sulphate ($FeSO_4.7H_2O$)	10·0 mg
Potassium dihydrogen phosphate (KH_2PO_4)	3·8 g
Zinc sulphate ($ZnSO_4.7H_2O$)	1·0 g
Glucose	10·0 g
Water	1 litre

Adjust pH to 6·5 with NaOH before sterilization at psi for 10 min. For perithecial production in *Aspergillus nidulans* reduce the sodium nitrate to 1·0 g and increase the glucose to 20·0 g.

138. *Potato agar*

Potatoes (peeled)	250 g
Agar	25 g
Water	1 litre

Gently boil the chopped potato in the water for 30 min. Allow to cool and settle, decant the fluid and make up to 1 litre. Add agar, heat in a water bath until dissolved, and autoclave at 15 psi for 15 min.

139. *Potato carrot agar—a weak medium suitable for conservation*
(Langeron, 1952)

†Potato (washed, peeled and grated)	20 g
Carrot (washed, peeled and grated)	20 g
Agar	20 g
Water	1 litre

† Avoid new potatoes.

Boil potato and carrot in the water for 1 h, then pass through a fine sieve and add the agar to the liquid. Boil until agar is all dissolved. Autoclave at 15 psi for 20 min.

140. *Potato dextrose agar*

†Potato (scrubbed and diced)	200 g
Dextrose	15 g
Agar	20 g
Water	1 litre

† Avoid new potatoes.

Rinse potato under running water, then add to water. Boil in double saucepan for 1 h, then pass through a fine sieve, squeezing through as much pulp as possible. Add agar and boil until dissolved; remove from heat, add dextrose and stir until dissolved. Autoclave at 15 psi for 20 min.

N.B. Lacy and Bridgmon (1962) described the use of dehydrated potato for the preparation of potato dextrose agar. Using the formula of 22 g dehydrated potato mixed with 178 ml distilled water added, with 20 g dextrose and 17 g agar, to 1 litre distilled water and autoclaved at 15 psi for 20 min, they found with nine test organisms that growth was as good as that on media made from fresh potatoes and better than that on commercial PDA.

At this time W. L. Gordon in Winnipeg also used a similar potato source for the preparation of potato sucrose agar for *Fusarium* culture. He had similar favourable results and obtained more consistency in colony appearance than with media made from batches of potatoes which inevitably varied throughout the year.

141. *Potato dextrose agar modified for* Venturia inaequalis *conidia* (*Boone and Keitt, 1956*)

Potatoes	40·0 g
Dextrose	5·0 g
Agar	17·0 g
Water	1 litre

Apple leaf decoction may be added to this for perithecia production.

142. *Potato-malt agar with cellulose for the Chaetomiaceae* (*Ames, 1961*)

Potatoes	60 g
Fleischman's Dry Malt Syrup	10 g
Difco Bacto-Agar	30 g
Distilled water	2 litres

The sliced potatoes are cooked until soft in 250 ml of the distilled water, the clear liquid decanted and added to the other constituents. Autoclave at 120°C for 20–30 min. Strips of sterilized white blotting paper are placed on the surface of the cool medium in Petri dishes or tubes.

143. *Potato sucrose agar* (*for Fusarium*)

Potato extract	500 ml
Sucrose	20 g
Agar	20 g
Distilled water	500 ml

Add potato extract, sucrose and agar to water and heat to dissolve agar. Autoclave at 15 psi for 20 min. Adjust pH to 6·5 if necessary with calcium carbonate.

(*continued*)

Potato extract

Potatoes (peeled and diced)	1800 g
Water	4500 ml

Suspend potato in muslin in the water and boil for 10 min. Discard the potato, and autoclave the liquor in large glass containers at 15 psi for 20 min. Store in refrigerator.

144. *Prune agar*

Prunes	30 g
Sucrose	40 g
Agar	30 g
Water	1 litre

Boil prunes in water, then pass through sieve to include as much pulp as possible. Add agar, boil to dissolve. Remove from heat, add sucrose and stir until dissolved. Correct pH; autoclave at 20 psi for 20 min.

145. *Prune lactose yeast agar or pigmentation in* Verticillium

Prune extract (at pH 5·8–6)	5 g
Lactose	5 g
Difco yeast extract	1 g
Agar	30 g
Water	1 litre

146. *Raper's medium for* Achlya ambisexualis (*Raper, 1940*)

Soluble starch	3 g
Peptone (Difco)	1 g
Hot water extract of	10 g lentils
Agar (Difco-Bacto)	20 g
Water (containing salts, see below)	1 litre

To 1000 cc of glass-distilled water add—

Potassium dihydrogen phosphate (KH_2PO_4)	0·0045 g
Magnesium sulphate ($MgSO_4.7H_2O$)	0·003 g
Calcium chloride ($CaCl_2$)	0·001 g
Ferric chloride ($FeCl_3$)	0·00016 g
Zinc sulphate ($ZnSO_4.7H_2O$)	0·00003 g

147. *Raulin's solution* (*Raulin, 1870*)

Cane sugar	70 g
Tartaric acid	4·0 g
Ammonium nitrate (NH_4NO_3)	4·0 g
Potassium carbonate (K_2CO_3)	0·6 g
Ammonium hydrogen phosphate ($(NH_4)_2HPO_4$)	0·6 g
Magnesium carbonate ($MgCO_3$)	0·4 g
Ammonium sulphate ($(NH_4)_2SO_4$)	0·25 g
Zinc sulphate ($ZnSO_4.7H_2O$)	0·07 g
Ferrous sulphate ($FeSO_4.7H_2O$)	0·07 g
Potassium silicate	0·07 g
Distilled water	1500 ml

148. *Raulin-Thom solution*

Same as Raulin's solution except that ammonium nitrate is substituted by ammonium tartrate.

149. *Raulin-Dierckx neutral solution*

(1) Dissolve 0·04 g magnesium carbonate (MgCO₃) with 0·71 g of tartaric acid in 100 ml of distilled water.

(2) In 800–900 ml distilled water dissolve—

Saccharose	46·6 g
Ammonium nitrate (NH₄NO₃)	2·66 g
Ammonium hydrogen phosphate ((NH₄)₂HPO₄)	0·4 g
Potassium carbonate (K₂CO₃)	0·4 g
Ammonium sulphate ((NH₄)₂SO₄)	0·16 g
Zinc sulphate (ZnSO₄.7H₂O)	0·04 g
Ferrous sulphate (FeSO₄.7H₂O)	0·04 g

(3) Add to solution (2) 66·7 ml of solution (1) and make up to 1 litre.

150. Rhizoctonia *agar (Diehl, 1916)*

Saccharose	10·0 g
Potassium hydrogen phosphate (K₄HPO₄)	1·0 g
Agar	20·0 g
Water	1 litre

151. *Richards' agar (McLean and Cook, 1941)*

Potassium nitrate (KNO₃)	10·0 g
Potassium dihydrogen phosphate (KH₂PO₄)	5·0 g
Magnesium sulphate (MgSO₄.7H₂O)	0·25 g
Ferric chloride (FeCl₃)	0·02 g
Potato starch	10 g
Sucrose	50·0 g
Agar	20 g
Water	1 litre

152. *Sabouraud's glucose (or maltose) "proof" medium*, see Stockdale, this Volume, p. 429

153. *Sabouraud's "conservation" medium*, see Stockdale, this Volume, p. 429.

154. *Sach's Agar used with corn stalks or leaves to induce perithecia of* Helminthosporium *(Luttrell, 1958)*

Calcium nitrate Ca (NO₃)₂	1·0 g
Dipotassium hydrogen phosphate (K₂HPO₄)	0·25 g
Magnesium sulphate (MgSO₄)	0·25 g
Ferric chloride (FeCl₃)	trace
Calcium carbonate (CaCO₃)	4·0 g
Agar	20·0 g
Water	1 litre

155. *Sartory's defined medium* for dermatophytes, see Stockdale, this Volume, p. 429.

156. *Sawdust, beech or spruce, etc. (for Basidiomycetes)*

See also Badcock's medium (No. 8).

Sawdust	100·0 g
Maize meal	10·0 g

Mix thoroughly and moisten with water.

157. *Schopfer's medium (liquid) for* Eremothecium ashbyi *(Yaw, 1952)*

Glucose A.R.	10 g
Magnesium sulphate (MgSO$_4$.7H$_2$O)	0·5 g
Potassium dihydrogen phosphate (KH$_2$PO$_4$)	1·5 g
Asparagine (twice recrystallized)	1 g
Biotin	2·5 μg
B$_1$	400 μg
Inositol	40 mg
Distilled water	1 litre

158. *Shaker's medium (liquid)*

Yeast extract	10 g
Peptone (commercial bacteriological)	20 g
Glucose	20 g
Distilled water	1 litre

This can be solidified by adding 20 g agar. No pH adjustment is necessary.

159. *Soil*, see also Barron, this Volume, p. 405.

160. *Media for the primary isolation of fungi from soil, water, etc.*
(Kaufman et al., 1963)

Medium 1

Glucose	5 g
Yeast extract	2 g
Sodium nitrate (NaNO$_3$)	1 g
Magnesium sulphate (MgSO$_4$.7H$_2$O)	0·5 g
Potassium phosphate (monobasic) (KH$_2$PO$_4$)	1 g
Streptomycin sulphate	50 mg
Chloromycetin (chloramphenicol)	50 mg
Oxgall	1 g
Sodium propionate	1 g
Agar	20 g
Distilled water	1 litre

Autoclave at 11 psi for 15 min.

Medium 2

Glucose	10 g
Sodium nitrate (NaNO$_3$)	1 g
Potassium phosphate (K$_2$HPO$_4$) (dibasic)	1 g
Rose bengal	65 mg
Agar	20 g
Soil extract (see below)	1 litre

Autoclave at 15 psi for 15 min.

Medium 3

Glucose	10 g
Peptone	5 g
Potassium phosphate (KH$_2$PO$_4$) (monobasic)	1 g
Magnesium sulphate (MgSO$_4$.7H$_2$O)	0·5 g
Rose bengal	33 mg
Streptomycin	30 mg
Agar	15 g
Distilled water	1 litre

Autoclave at 11 psi for 15 min.

Medium 4

Glucose	1 g
Potassium phosphate (KH$_2$PO$_4$) (dibasic)	0·5 g
Streptomycin sulphate	50 mg
Chloromycetin (chloramphenicol)	50 mg
Agar	20 g
Soil extract (see below)	100 ml
Tap water	900 ml

Autoclave at 11 psi for 15 min.

Soil extract: Steam 1000 g soil with 1200 ml water for 1 h, then filter through a doubled layer of Whatman No. 1 filter paper.

161. *Soil extract, Smith and Dawson's Medium*

Glucose	10·0 g
Sodium nitrate (NaNO$_3$)	1·0 g
Potassium hydrogen phosphate (K$_2$HPO$_4$)	1·0 g
Agar	15·0 g
Rose bengal	0·067 g
	(1 : 15,000)
Soil extract	1 litre

To prepare soil extract, autoclave 500 g loam in 1000·0 ml water for 1 h, filter through paper and make up to 1000·0 ml

162. *Soil extracts (Wills, 1954)*

Soil extracts prepared by boiling 300 g of air dry soil for 1 h in 1 litre water. The solids were then removed by vacuum filtration.

163. *Soil extract agar*, see Williams and Cross, this Volume, p. 295.

164. *Soil extract agar acidified with lactic acid for isolation of Heterosporium from soil (Atkinson, 1952)*

Soil	1000 g
Potassium hydrogen phosphate (K$_2$HPO$_4$)	0·2 g
Agar	15 g
Water	1 litre

Add soil to water and autoclave at 15 psi for 30 min. While still hot, filter under suction through No. 3 Whatman filter paper. Make up to 1 litre and add agar and K$_2$HPO$_4$ (buffer). Autoclave at 15 psi for 20 min. Acidify with lactic acid just before using.

(Flentje's formula for promotion of formation of basidia in *Corticium praticola* includes 1 g sucrose and 0·1 g dried yeast.)

165. *Soil as a storage medium*

Half fill McCartney bottles with fine sieved loam and autoclave at 20 psi for 30 min. Autoclave again after a week in order to destroy any heat resisting organisms which may have survived the first sterilization. Inoculate the soil and moisten at the same time with 2–3 ml of a suspension of spores and/or tissue in sterile distilled water. Incubate the cultures for 10–14 days ,then store at 3°–6°C. Retrieve by sprinkling a few soil grains onto an agar plate.

166. *Sporulation medium (NRRL, Peoria)*

Glycerol	7·5 g
Brer rabbit molasses (Orange label)	7·5 g
Curbay B.G. (U.S.I.)	2·5 g
Peptone	5·0 g
Magnesium sulphate (MgSO₄.7H₂O)	0·05 g
Potassium dihydrogen phosphate (KH₂PO₄)	0·06 g
Sodium chloride (NaCl)	4·0 g
Agar	25 g
Water	1 litre

167. *Starch agar*

Soluble starch	40 g
Marmite (yeast extract)	5 g
Agar	20 g
Water	1 litre

Place all the constituents in water, and heat in water bath until dissolved. Bottle and sterilize. (pH is 6·5–7 and requires no adjustment).

168. *Starch-casein medium*, see Williams and Cross, this Volume, p. 295.

169. *Sucrose proline agar (SPA) for* Drechslera, *perithecial* Helminthosporium *(Shoemaker, 1962)*

Sucrose	6·0 g
Proline	2·7 g
Dipotassium hydrogen phosphate (K₂HPO₄)	1·3 g
Potassium dihydrogen phosphate (KH₂PO₄)	1·0 g
Potassium chloride (KCl)	0·5 g
Magnesium sulphate (MgSO₄)	0·5 g
Ferrous sulphate (FeSO₄)	10 mg
Zinc sulphate (ZnSO₄)	2 mg
Manganese chloride (MgCl₂)	1·6 mg
Agar	20 g
Water	1 litre

170. *Tapwater agar*

Agar	15 g
Tapwater	1 litre

Dissolve agar in water for ½ h. Sterilize at 15 psi for 20 min. Sterile wheat straw or rice grains may be added to this. Many sensitive fungi spore well on this medium.

171. *Tauroglycocholate Medium* (Martin and Scott, 1952), see Stockdale, this Volume, p. 429.

172. *Trace salts solution* (*Shirling and Gottlieb, 1966*)

$FeSO_4.7H_2O$	0·1 g
$MnCl_2.4H_2O$	0·1 g
$ZnSO_4.7H_2O$	0·1 g
Distilled water to	100 ml

Usually used as 1 ml/litre.

173. *Trebouxia medium*, see Richardson, this Volume, p. 267.

174. *Tubeuf's medium for* Merulius lacrymans

Ammonium nitrate (NH_4NO_3)	10·0 g
Potassium tribasic phosphate (K_3PO_4)	5·0 g
Magnesium sulphate ($MgSO_4.7H_2O$)	1·0 g
Lactic acid	2·0 g
Distilled water	1 litre

175. *Ullscheck's agar*

Sucrose	50·0 g
Potassium nitrate (KNO_3)	10·0 g
Dipotassium hydrogen phosphate (K_2HPO_4)	8·0 g
Magnesium sulphate ($MgSO_4$)	5·0 g
Agar	30·0 g
Distilled water	1 litre

Sterilization is by heating to 100°C for 30 min on three successive days. pH is 5·4.

176. *Vegetable plugs* (*Potato, carrot, turnip, etc*)

Punch out cylinders with apple corer or cork borer, and cut into two pieces diagonally. Put a piece of saturated cotton wool at bottom of each tube and insert plug on top. Sterilize at 20 psi for 10 min.

177. *Vishniac's medium*, see Gareth Jones, this Volume, p. 335.

187. *Waksman's egg albumen agar for Actinomycetes*

Dextrose	10 g
Potassium hydrogen phosphate (K_2HPO_4)	0·5 g
Magnesium sulphate ($MgSO_4$)	0·2 g
Ferric sulphate ($Fe_2(SO_4)_3$)	trace
Egg albumen (dried)	0·15 g
Agar	15 g
Water	1 litre

Albumen is dissolved in N/10 sodium hydroxide until neutral to phenolphthalein, then added to the warm mixture.

179. *Waksman's special medium for counting soil fungi*

Glucose	10·0 g
Peptone	5·0 g
Potassium dihydrogen phosphate (KH₄PO₄)	1·0 g
Magnesium sulphate (MgSO₄.7H₂O)	0·5 g
Agar	25·0 g
Water	1 litre

Adjust to pH 4·0 by addition of N H_2SO_4 or H_3PO_4.

180. *Wieringa's modified medium* for *Streptomyces*, see Williams and Cross, this Volume, p. 295.

181. *Yeasts, defined, differential, selective and standard media*, see Beech and Davenport, this Volume, p. 153.

182. *Yeast extract agar* (*for* Chaetomium *and some other Ascomycetes*)

Yeast extract	4·0 g
Malt extract	10·0 g
Dextrose	4·0 g
Agar	15·0 g
Distilled water	1 litre

183. *Yeast extract agar* for thermophilic actinomycetes, see Williams and Cross, this volume, p. 295.

184. *Yeast phosphate soluble starch agar,* "*YpSS agar*" (*Emerson, 1941*)

Yeast extract (Difco)	4 g
Soluble starch	15 g
Dipotassium hydrogen phosphate (K₂HPO₄)	1 g
Magnesium sulphate (MgSO₄.7H₂O)	0·5 g
Agar	20 g
Water	1 litre

185. *Zehner and Gorham's medium* (1960), see Richardson, this Volume, p. 267.

ACKNOWLEDGMENTS

My thanks are due to Miss G. M. Waterhouse for her notes on the sources of agar-agar, and to Mrs. G. B. Butterfill for help with the compilation and standardization of the various formulae.

REFERENCES

Acha, I. G., and Villanueva, J. R. (1961). *Nature, Lond.*, **189**, No. 4761, 328.
Ames, L. M. (1961). "A Monograph of the Chaetomiaceae". The United States Army Research and Development Series No. 2.
Aschner, M., and Kohn, S. (1958). *J. gen. Microbiol.*, **19**, 183.
Asthana, R. P., and Hawker, L. E. (1936). *Ann. Bot.*, **50**, 325–343.
Atkinson, R. G. (1952). *Mycologia*, **44**, 816.
Backus, M. P., and Stauffer, J. F. (1955). *Mycologia*, **47**, 429–463.
Badcock, E. C. (1941). *Trans. Br. mycol. Soc.*, **25**, 200–205.

Barnes, B. (1933). *Trans. Br. mycol. Soc.*, **18**, 172–173.
Barnett, H. L. (1953). *Mycologia*, **45**, 450–457.
Beadle, G. W., and Tatum, E. L. (1945). *Am. J. Bot.*, **32**, 678–686.
Benjamin, R. K. (1958). *Aliso*, **4**, 150.
Berk, S., Ebert, H., and Teitell, L. (1957). *Ind. Engng Chem.*, **49**, 1117.
Berliner, M. D. (1961). *Mycologia*, **53**, 84–90.
Blank, I. H., and Tiffney, W. N. (1936). *Mycologia*, **28**, 324–329.
Boone, D. M., and Keitt, G. W. (1956). *Am. J. Bot.*, **43**, 227.
Brown, W. (1926). *Ann. Bot.*, **40**, 224.
Burkholder, P. R. (1943). *Am. J. Bot.*, **30**, 206–211.
Capellini, R. A., and Peterson, J. L. (1965). *Mycologia*, **57**, 962–966.
Claussen, P. (1912). *Z. Bot.*, **4**, 1.
Cooke, W. B. (1954). *Antibiotics Chemother.*, **4**, 657–662.
Crabill, C. H. (1912). *Science*, **36**, 155–157.
Crowell, I. H. (1941). *Mycologia*, **33**, 137.
Daniel, J. W., and Rusch, H. P. (1961). *J. gen. Microbiol.*, **25**, 47.
Diehl, W. W. (1916). *Phytopath.*, **6**, 336–340.
Dueck, J., Bushnell, W. R., and Rowell, J. B. (1965). (Abstracts) *Am. J. Bot.*, **52**, 613–656.
Elliott, J. A. (1917). *Am. J. Bot.*, **4**, 439–476.
Emerson, R. (1941). *Lloydia*, **4**, 77.
Emge, E. G. (1963). (Abstract) *Phytopathology*, **53**, 745–746.
Gams, W. (1960). *Sydowia*, Ser. II, **14**, 300.
Garrett, S. D. (1953). *Ann. Bot.*, **17**, 63–79.
Goos, R. D., and Tschersch, M. (1962). *Mycologia*, **54**, 353–367.
Gwynne-Vaughan, H. C. I., and Barnes, B. (1927). "The Structure and Development of the Fungi". Cambridge University Press, England.
Haglund, W. A., and King, T. H. (1962). *Phytopathology*, **52**, 315–317.
Hanes, F. J., and Bennett, E. O. (1964). *Antonie van Leeuwenhoek*, **30**, 412–416.
Hendrix, J. W. (1965). *Phytopathology*, **55**, 790–797.
Hesseltine, C. W. (1954). *Mycologia*, **46**, 362.
Holliday, P., and Mowat, W. P. (1963). *Phytopath. Pap.*, No. 5.
Horenstein, E. A., and Cantino, E. C. (1961). *Trans. Br. mycol. Soc.*, **44**, 185–198.
Horne, A. S., and Mitter, J. H. (1927). *Ann. Bot.*, **41**, 519–547.
Hotson, H. H. (1942). *Mycologia*, **34**, 392.
Ingold, C. T., and Dring, V. G. (1957). *Ann. Bot. N.S.*, **21**, 465–477.
Jefferys, E. G., Brian, P. W., Hemming, H. G., and Lowe, D. (1953). *J. gen. Microbiol.*, **9**, 314–341.
Jewson, S. T., and Tattersfield, F. (1922). *Ann. appl. Biol.*, **9**, 213–240.
Joffe, A. Z. (1963). *Mycologia*, **55**, 271–282.
Kaufman, D. D., Williams, L. E., and Sumner, C. B. (1963). *Can. J. Microbiol.* **9**, 6, 743–744.
Keay, M. A. (1953). *Pl. Path.*, **2**, 103.
Keitt, G. W., and Langford, M. H. (1941). *Am. J. Bot.*, **28**, 805–820.
Klarman, W. L., and Craig, J. (1960). *Phytopathology*, **50**, 868.
Klemmer, H. W., and Lenney, J. F. (1965). *Phytopathology*, **55**, 320–323.
Knox–Davis, P. S. (1965). *S. Afr. J. agric. Sci.*, **8**, (1), 205.
Kuehn, H. H., Orr, G. F., and Ghosh, G. R. (1961). *Mycopath. Mycol. appl.*, **14**, 215–229.
Kuehner, C. C. (1951). *Mycologia*, **43**, 390.

Kuhlman, E. G., and Hendrix, F. F., Jr. (1962). *Phytopathology*, **52**, 1310–1312.
Lacy, M. L., and Bridgmon, G. H. (1962). *Phytopathology*, **52**, 173.
Lange, M. (1952). *Dansk bot. Ark.* **14**, 6, 20.
Langeron, M. (and Vanbreuseghem, R.) (1952). "Précis de Mycologie", Ed. 2. Masson et Cie, Paris.
Lilly, V. G., and Barnett, H. L. (1951). "Physiology of the Fungi". McGraw-Hill, New York.
Littman, M. L. (1947). *Science*, **106**, 109.
Lukens, R. J. (1960). *Phytopathology*, **50**, 867–868.
Lukens, R. J., and Sisler, H. D. (1958). *Phytopathology*, **48**, 235–244.
Luttrell, E. S. (1958). *Phytopathology*, **48**, 281–287.
MacLaren, J. A. (1961). *Mycologia*, **52**, 149.
Madelin, M. F. (1956). *Ann. Bot.*, **20**, 307–330.
Malca, I., and Ullstrup, A. J. (1962). *Bull. Torrey bot. Club*, **89**, 240–249.
Marshall, S. M., Newton, L., and Orr, A. P. (1949). "A Study of Certain British Seaweeds and their Utilisation in the preparation of Agar". Her Majesty's Stationery Office, London.
McLean, R. C., and Cook, W. R. Ivimey. (1941). "Plant Science Formulae". Macmillan & Co., London.
Mundkur, B. B. (1934). *Indian J. agric. Sci.*, **4**, 779.
Nash, S. M., and Snyder, W. C. (1962). *Phytopathology*, **52**, 567–572.
Netzer, D., and Dishon, I. (1967). *Phytopathology*, **57**, 795–796.
Newton, L. (1951). "Seaweed Utilisation". Sampson Low, London.
Papavizas, G. C., and Davey, C. B. (1959). *Soil. Sci.*, **88**, 112–117.
Park, D. (1964). *Trans. Br. mycol. Soc.*, **47**, 541.
Pontecorvo, G. (1953). *Adv. Genet.*, **4**, 141–238.
Raper, J. R. (1940). *Mycologia*, **32**, 714.
Raper, K. B., and Thom, C. (1949). "Manual of the Penicillia". The Williams & Wilkins Co., Baltimore, U.S.A.
Raulin, J. (1870). "Études Chimiques sur la Végétation". Thèse Fac. sc. Paris.
Robbins, W. J., and Hervey, A. (1959). *Mycologia*, **50**, 745.
Romanowski, R. D., and Kuć, J. (1962). *Phytopathology*, **52**, 1259–1263.
Royle, D. J., and Hickman, C. J. (1964). *Can. J. Bot.*, **42**, 311–318.
Ryan, F. J. (1950). *Meth. med. Res.*, **3**, 51–75.
Scheffer, R. P., and Walker, J. C. (1953). *Phytopathology*, **43**, 116–125.
Shirling, E. B., and Gottlieb, D. (1966). *Int. J. Syst. Bacteriol.*, **16**, 313–340.
Shoemaker, R. A. (1962). *Can. J. Bot.*, **40**, 809–836.
Singer, R. (1962). "The Agaricales". Hafner, New York.
Sloan, B. J., Routien, J. B., and Miller, V. P. (1961). *Mycologia*, **52**, 47–63.
Smith, G. (1960). "An Introduction to Industrial Mycology", 5th Ed. Edward Arnold Ltd., London.
Smith, R. S. (1967). *Mycologia*, **59**, 600.
Snyder, W. C., and Hansen, H. M. (1946). *Mycologia*, **38**, 455–462.
Snyder, W. C., Nash, S. M., and Trujillo, E. E. (1959). *Phytopathology*, **49**, 310.
Sobers, E. K., and Seymour, C. P. (1963). *Phytopathology*, **53**, 1443–1446.
Srinivasan, M. C., Narasimhan, M. J., and Thirumalachar, M. J. (1964). *Mycologia*, **56**, 683.
Swartz, D. (1934). *Mycologia*, **26**, 193.
Tatum, E. L., Barratt, R. W., and Cutter, V. M. (1949). *Science*, **109**, 509–517.
Thom, C. (1930). "The Penicillia". Baillière, Tindall and Cox, London.

Thurston, H. D. (1957). *Phytopathology*, **47**, 186.
Timnick, M. B., Barnett, H. L., and Lilly, V. G. (1952). *Mycologia*, **44**, 141–149.
Titsworth, E. H., and Grunberg, E. (1950). *Mycologia*, **42**, 298.
Wills, W. H. (1954), *J. Elisha Mitchell Scient. Soc.*, **70**, 235–243.
Yaw, K. E. (1952). *Mycologia*, **44**, 308.

CHAPTER III

Techniques for Microscopic Preparation

D. M. Dring

Royal Botanic Gardens, Kew, Surrey, England

In fungi, as with other organisms, for the study of detail and for accurate measurements of spores and other structures, microscopic preparations must be made. The nature of the fungus itself and its host, if any, determine the method to be employed in making the preparation. Thus, internal plant-parasites are often examined by techniques appropriate to the host; external parasites may be embedded in a film of cellulose acetate and stripped off the epidermis of the host; saprophytic moulds are often grown on the microscope slide on which they are to be examined; and so on. It is impossible in the space of a short introduction to give a complete account of all the techniques which have been found suitable for the examination of fungi. However, it is felt that the following notes will serve for most routine and many specialized investigations.

I. DIRECT MOUNTING

A. General techniques

For rapid and routine examination of almost all types of fungi the material which is to be examined, usually spores and spore-bearing structures, is teased out on the slide in a drop of mounting fluid and a cover-glass placed over the preparation, which is then ready for examination.

The excision of a fragment of fungal material and its subsequent teasing out are conveniently accomplished by using a pair of mounted steel needles, not nichrome inoculating needles, one of which has a small knife-edge in addition to a finely pointed tip. Teasing out should result in a flat, even preparation in which the original spatial relationships of the various parts are not completely obliterated. It is often carried out under a dissecting microscope.

Placing of the object directly in the mountant, particularly if this be lactophenol, may result in the inclusion of too many air bubbles, and it is often advisable to put the fragment on a slide and moisten it with a drop of alcohol, most of which should be allowed to evaporate before being replaced by a rather larger drop of mountant. Many detached spores are usually floated off by this process, which is also an advantage since in too great numbers they obscure other parts of the preparation. Alternatively air may be expelled by gentle heating. The medium should not be allowed to boil if the slide is needed for critical, comparative work as boiling tends to change the size, reaction to dyes, etc., of the material significantly. This is particularly important if iodine or cotton blue are being used in histochemical studies. Air bubbles may also be removed by placing the slide in a vacuum chamber. In this case it is advisable to have a little excess mountant around the edges of the cover-glass.

Note, however, that in some cases a few air bubbles may serve a useful purpose. They are indispensable for detecting very fine ornament of spore walls. Thus, *Trichoderma viride* may be distinguished from its close relatives by the very fine warts on its conidia. These warts are barely discernable in liquid mounts but a few spores accidentally enclosed in an air bubble on the slide show the rough surface unmistakably.

It is sometimes desirable to apply gentle pressure to the cover-glass. When carefully done this has the advantage of spreading the elements to be examined and placing them in the same optical plane. It also frequently disperses air bubbles. However, it may have the effect of bending the cover-glass so that when, after a few hours, its shape is regained, air bubbles are sucked into the preparation. Sizes and shapes of some of the elements may, of course, be changed by pressure, and for a comparative series of slides it is well to adopt a roughly standard procedure for squashing. Techniques for macerating host tissues are dealt with in the section on whole mounts.

B. Mounting media and stains

1. *Lactophenol*

The question of the mounting medium is one of some complexity and it must be admitted at the outset that the ideal liquid medium for mounting fungi has not yet been evolved. Without doubt the most popular medium is Amann's lactophenol. Though originally devised for other cryptogams, this medium, particularly with the addition of cotton blue as a stain, has become the standard mountant for fungi. Countless observations of fungi mounted in this medium have been published and thus a degree of uniformity has been attained.

The recipe for Amann's lactophenol is as follows—

Phenol (pure crystals)	20 g
Lactic acid (sp.gr. 1·21)	20 g
Glycerol	40 g
Water	20 ml

Warm the phenol in the water to dissolve it, then add lactic acid and glycerol. A little stain may be added if desired, e.g. 0·05 g cotton blue or trypan blue per 100 ml, a saturated aqueous solution of picric acid instead of the water. Details of other stains which can be used in conjunction with lactophenol are given by Maneval (1936).

Lactophenol is best stored in brown glass bottles particularly if used only occasionally.

Lactophenol shrinks unfixed protoplasts and the spores of some moulds disastrously, though not their hyphal walls, the former difficulty can be overcome by the prior use of a drop of alcoholic fixative and the latter can sometimes be cured by the application of a little heat. Old herbarium material in which the fungal structures have shrunk can often be reflated by heating in lactophenol on the slide.

The main disadvantage of lactophenol, however, is that its refractive index, about 1·45, is very close to that of fungal hyphae, which are therefore rendered more difficult to observe and measure accurately. Measurements of the same object by the same observer, taken in water and lactophenol, may differ by almost 1 μm. The addition of stain hardly helps, since as all the dyes which are compatible with lactophenol tend to stain the cytoplasm rather than the cell wall, stained objects are apt to appear even smaller than when unstained.

Against this very grave disadvantage must be set the fact that lactophenol has come to be universally used, if not understood, by mycologists and this is the mountant in which most of them will have carried out the bulk of their microscopic observations. In addition to this it must be admitted that very satisfactory, durable slides can easily be made with this medium. The idea that lactophenol mounts are necessarily "semi-permanent" is largely fallacious.

2. *Sealing lactophenol mounts*

In order to make a good, virtually permanent preparation, care should be taken to use precisely the right quantity of mountant so that when the cover-glass is put in place the liquid just flows to the edge of the cover, not out on to the uncovered slide, or sealing will be impossible. The cover-glass used should be as small as possible.

The edge of the cover-glass is now sealed to the slide with cosmetic nail-varnish, a thin layer of colourless varnish being run round the edge of the cover, using a turn-table if the cover-glass is round. The first coat is allowed to dry and a second placed on top. Subsequent coats should be of coloured varnish, which contains more body and is easier to see, and should be added until the sharp edge of the cover-glass is no longer visible as a projection in the coat of varnish. Slides carefully sealed in this way will keep for many years, the only sign of deterioration being that lactophenol dissolves some of the chemical components of the fungus and these tend to be precipitated as crystals or oil droplets outside the hyphae.

Proprietary sealing compounds of a similar composition to nail-varnish are now available in most countries under a series of trade names. These have, for the most part, the merit of retaining their plasticity for a very long period, thus avoiding cracks in the seal due to shrinkage.

Dade (1968) renders these preparations more robust by reinforcing the varnish with an epoxy resin. Only one suitable kind is known, Araldite AZ.107, which when applied over the first two coats of varnish sets to a seal that can only be broken with steel instruments.

3. *Other mountants and stains*

In view of the disadvantages of lactophenol, for comparison, it is always advisable to mount part of the material in other fluids. The two most useful are ordinary water and Melzer's iodine. In order to prevent the water from causing living protoplasts to swell by osmosis, and to eliminate air bubbles, it is useful to precede its application by a drop of 70% alcohol or to use 5% saline instead of distilled water. Water mounts evaporate quickly and are unsuitable for sealing. Various stains may be incorporated, e.g. 0·5% phloxine. Shoemaker (1964) recommends 1% aqueous azure A for staining annelations on conidiophores. The width of these structures is near the limit of optical resolution and they are therefore difficult to demonstrate otherwise. These mounts may be made permanent by replacing the aqueous solution with a drop of saturated azure A in glycerol and sealing with nail-varnish.

The recipe for Melzer's reagent is as follows—

Chloral hydrate	100 g
Potassium iodide	5 g
Iodine	1·5 g
Distilled water	100 ml

This fluid has excellent optical properties and clears and stains the fungus tissue as well. It is used as the basis of the amyloid reaction in basidiomycete histochemistry. Features of the cell wall and other parts of the fungus which have been overlooked in lactophenol preparations often become immediately obvious in Melzer's iodine. Though an iodine mount will last for several days, it is unsuitable for permanent preparations. The iodine may, however, be replaced by lactophenol and the preparation sealed.

Fleming and Smith (1944) recommended picro-nigrosin stain for moulds. A saturated aqueous solution of picric acid containing 2% of nigrosin is used in place of water in making lactophenol. After staining on a slide the picro-nigrosin is removed and replaced by colourless lactophenol. The fungus is stained grey, which does not dissolve out as do most other stains when the coloured lactophenol is replaced by colourless.

A method for staining spore-walls is given by de Silva (1965). The stain is the supernatant obtained by centrifuging at 2000 rpm for 15 min, equal volumes of 10% aqueous tannic acid, which serves as a mordant, and 1% aqueous basic fuchsin. The speed and time of centrifugation are critical. The material is placed in a drop of the stain on a slide and excess stain washed away with 50% glycerol, in which the object is mounted and sealed. Good differentiation between spore-walls and cytoplasm is given by this staining technique. When rapid examination of delicate, hyaline spores and mycelium of species belonging to genera such as *Fusarium, Cylindrocladium*, and *Gliocladium* is required, an excellent stain to use is 0·1% erythrosin in 10% ammonia. This produces a clear mount showing septation and wall-structure immediately.

Cresyl blue is frequently used for spore-walls of basidiomycetes. Being metachromatic, it differentially stains the various layers of the spore wall. Loquin (1952) gives directions for use of a variable monochromatic light source, in the form of a series of interference filters, which permits the effects of this stain to be better observed in thin-walled spores.

Mucilaginous deposits on the surface of spores and hyphae, though not easily stainable, can be demonstrated by so-called negative staining. Thin preparations are mounted and examined in Indian ink or nigrosin, when the mucilage shows up as colourless areas against the dark background.

Lowry (1963) gives an aceto-iron haematoxylin technique for staining chromosomes of agarics—

(i) Fix pieces of lamellae in Newcomer's fluid (see V.D, 1).
(ii) Hydrolyse and mordant in aqueous N HCl containing 2% each of aluminium alum, chrome alum and iodic acid, gradually increasing the temperature to 60°C, which is maintained for 10–15 min.
(iii) Wash in three changes of distilled water; ½ h each.

(iv) Stain in Whittmann's aceto-iron haematoxylin; 2 h (4% haematoxylin+1% iron alum in 45% acetic acid).

(v) Tease out piece of tissue in a drop of stain on a slide, apply cover-glass, squash out basidia and blot off excess stain.

(vi) Heat almost to boiling point to intensify stain.

(vii) Seal with nail-varnish.

Singh (1969) gives a method for staining the nuclei of moulds. The schedule is as follows—

(i) Stick conidia or hyphal fragments to slide with Haupt's adhesive (see V.C).

(ii) Fix in Newcomer's fluid (see V.D, 1).

(iii) Hydrolyse in 0·17M HCl; 20 min.

(iv) Wash.

(v) Mordant in 2% $LiCO_3$; 4 h.

(vi) Place a drop of stain on the slide, add cover-glass and warm.

The stain is made by boiling 0·5 g powdered carmine in 40% aqueous propionic acid to which 2–3 drops of saturated ferric chloride solution are added. Permanent preparations may be made by dissolving excess stain with 1 : 3 propionic acid in absolute alcohol, and proceeding through absolute alcohol to balsam in the usual way.

Glycerine jelly has often been used for making semi-permanent mounts, though it now finds little favour and is included here for the sake of completeness. The recipe is as follows—

Gelatin	1 g
Glycerol	7 g
Water	6 ml

The jelly is applied to the slide after having been melted in a water-bath, a cover-glass placed in position and the preparation allowed to harden. Slides made in this way will keep for several years without the necessity of sealing.

II. TECHNIQUES FOR FUNGI ON LEAF SURFACES

The following methods are particularly suitable for superficial parasites such as Erysiphaceae and Microthyriaceae, and for saprophytes on leaf surfaces. Although pieces of fungi growing superficially on leaves or other surfaces may be scraped off with a needle and mounted as described above, it is often preferable to remove them in more or less intact sheets. This is easily done, either by embedding them in, or sticking them to, transparent films.

The simplest method is to press a short length of adhesive, transparent tape such as Sellotape, Scotch Tape, etc., over the leaf surface on which the fungus is growing. When the tape is removed, the fungus adheres to it and is thus stripped off the leaf. The tape is then placed on a drop of lactophenol on a slide, another drop of lactophenol is placed on top of the tape and a cover-glass added. Other mountants may render the tape opaque.

A more refined modification of this method is given by Bretz and Berry (1964). The adhesive is dissolved from 1 m of tape with xylene or other suitable solvent, a drop of the solution is spread on a cover-glass and the xylene allowed to evaporate. The treated cover can then be pressed directly on the fungus, then placed over a drop of lactophenol on a slide and examined.

A solution of cellulose acetate in acetone, such as "Necol", may be dropped on to the fungus on the leaf surface, allowed to dry for about 15–30 min, and then carefully peeled off with the aid of a scalpel, as a thin, colourless film in which the fungus is embedded. It may be mounted in pure glycerol, or placed on a slide and a little acetone dropped on it. The result of the latter is to dissolve the acetate film, which tends to run towards the periphery of the preparation, where it can be blotted off. After washing with several drops of acetone and similarly blotting, the specimen may be directly mounted either in an aqueous mountant or a balsam.

"Necol" may be made up as follows:

Acetone	4 parts
Diacetone alcohol (containing 1% benzyl abietate and 1% triacetin)	1 part

Add cellulose acetate until of suitable consistency.

III. MOUNTING FUNGI FROM AGAR MEDIA

Slide cultures provide an admirable method for the examination of moulds. In these techniques the fungus is induced to grow on a small quantity of agar medium on a microscope slide and the agar is then discarded leaving the sporing structures and aerial mycelium more or less firmly adhering to the glass. This is particularly useful for moulds such as *Penicillium* and *Aspergillus*, whose conidiophores and chains of conidia are very fragile and do not lend themselves to teasing. Chapter I gives the details of slide culture methods.

Mounting media and stains used for this type of preparation are the same as those for direct mounts.

Fungi growing beneath the surface of agar media may be mounted by squashing. However, this tends to result in great distortion, and in particular, curling of the hyphae. It is far preferable for a block of agar taken from the culture to be melted on the slide and the liquid agar replaced by mounting fluid. A refinement of this technique is described by Feder and Hutchins (1964). A small block of agar containing the fungi to be examined is placed right side up on the slide. If required, a little stain may be dropped on the agar, allowed to penetrate for a few minutes, the excess washed away, and the agar blotted. A few drops of Fisher's "Permount" are placed on the block and then covered by a cover-glass tilted so that one side is in contact with the slide. The agar is then gently melted without boiling the Permount and the cover-glass allowed to fall under its own weight. The result is that a thin film of agar containing the fungi is surrounded by a ring of Permount which hardens in 24 h. Details of *Phytophthora* sporangia and nematode-trapping rings of *Dactylella* are examples of delicate structures preserved at least semi-permanently by this technique. Presumably Permount could be replaced by any resin immiscible with water and with a boiling point slightly above the melting point of agar.

IV. WHOLE MOUNTS

It is frequently desirable to prepare whole mounts of plant tissues invaded by fungi so that the morphological relationships between the fungus and its host can be clearly seen. This method is particularly applicable to leaves and fine roots, anthers, and thalli of host crytopgams. Larger portions of host tissue must be sectioned.

Material of the host is first trimmed to a convenient size for mounting either on an ordinary or a cavity slide. It is then fixed in FAA, FPA, or preferably weak chrome–acetic (see V, D). Objects from marine habitats must be fixed with reagents made up in normal saline. Fixation often needs to be carried out *in vacuo* to ensure removal of air, and to increase speed of penetration by the fixative.

The material must now be decolorized. This may usually be done with any chlorine bleach. However, material which contains tannins may need to be decolorized by boiling, or if necessary autoclaving, in 10% potassium hydroxide. This treatment should be followed by immersion in 10% hydrochloric acid. The longer and more brutal the treatments in alkali and acid the greater the degree of maceration of the tissues which will be obtained. Partial maceration is often desirable with larger objects and may be made more complete by substituting 5% chromic for the hydrochloric acid. Decolorized material must be washed thoroughly in running water for several hours. It is cleared in chloral hydrate solution (64 g in 40 ml water) and then stained and mounted in lactophenol–cotton blue. Better staining may be achieved by boiling the tissue for about 2 min in a mixture of 95% alcohol, 2 parts and lactophenol–cotton blue, 1 part (Shipton and Brown, 1962).

An alternative method is to decolorize in pyridine, which is compatible with lactophenol and therefore need not be washed out before mounting.

Isaac (1960) gives a whole mount technique suitable for thick sections, leaves, small roots and stems. Material is fixed in 95 : 5 dioxan–propionic acid (reagent grade) in a closed vessel at 60°C until the leaves are colourless. The material is washed for three hours and mounted in haem–gum staining mountant (Isaac 1958), which has the following constitution—

Formic acid (85–90%)	10 ml
Glycerol	20 ml
Gum Arabic	20 g
Chloral hydrate	30 g
Haematoxylin	0·5 g
Iron alum	1·5 g
Chrome alum	0·5 g
Bismark brown	0·15 g
Distilled water	50 ml

The initial reddish-purple of the stained material gradually changes to a dense blue-black as the formic acid evaporates. Mounts made in this medum are at least semi-permanent.

Hering and Nicholson (1964) used the following technique for staining fungi in whole mounts of fragments of leaf-litter. Pieces up to 1 cm square were bleached at 60°C for 4–6 h in 0·3 g sodium chlorite in 40 ml of 10% acetic acid, then dehydrated by immersion for 20 min each in 50%, 70%, 95% and absolute alcohol under reduced pressure. The material was cleared in methyl salicylate and stained for 15 sec in 20% safranin in methyl salicylate. It was then washed in salicylate and mounted in Canada balsam. Of the common stains tried, only safranin and crystal violet were satisfactorily soluble in methyl salicylate.

V. MICROTOME SECTIONS

In mycology, microtomed sections are generally used for observations of parasitic fungi in their host tissues or for histological and cytological examination of the fruiting bodies of the larger fungi. It should be mentioned in passing that hand-sections made with a cut-throat razor or sometimes a stiff razor blade held in a special holder are not to be despised, in fact with host materials of very uneven texture this may prove to be the only way to obtain a satisfactory section. This approach is warmly to be recommended to anyone wishing to study fungi in palm tissues for instance.

A. Freezing microtome

This is the most usual method and is suitable for a wide range of material providing that it is not too hard and serial sections are not required.

Material which is hard or brittle, particularly dried herbarium material,

can be softened and reflated to more or less its original dimensions in a
variety of ways. The most usual is to soak it in 10% potassium hydroxide.
Some workers prefer to use dilute ammonia, or hypochlorite solutions such
as Parazone. An alternative is to heat the material in water until it is soft:
it should not be boiled. Placing it in a beaker under a slowly running hot
water tap for a few hours is a convenient, if wasteful, method which also
often removes preservatives which may interfere with subsequent staining.
Black precipitates of mercurial preservatives, often encountered in herbarium material, and rendering microscopic examination difficult, may be
removed by the iodine–sodium thiosulphate sequence. The process is to
wash in the following mixture—

Potassium iodide	3 g
Iodine	2 g
Alcohol (70%)	100 ml

until no more of the iodine is decolorized, then transfer to—

Sodium thiosulphate	0·75 g
Alcohol (96%)	10 ml
Distilled water	90 ml
Thymol	1 crystal

to remove the iodine.

Material to be sectioned should be soaked in gum arabic solution. The
strength of the solution is not critical provided that it is not too thick and the
commercial gums used for sticking paper are of about the right consistency.
In practice, the length of time for which the material is soaked in gum aften
depends on the urgency with which the section is needed. Fresh material
will sometimes be immediately ready for cutting, but better results are
obtained by first fixing and then placing it at least overnight in gum. Watch
glasses, etc., must be kept covered to prevent evaporation. The process of
impregnation can be speeded up by putting the specimen and gum in a
closed tube in a 30°C incubator.

The most common form of freezing microtome consists of a simple bench
microtome with movable blade and fixed stage. Freezing is accomplished
by a jet of carbon dioxide from a cylinder, which plays on the underside of
the stage. To cut the sections a thin layer of gum is placed on the stage and
frozen with a blast of CO_2. The material, suitably trimmed, is then mounted
and orientated on this base and covered with gum, the whole being frozen
with CO_2. Several layers of gum may be required to build up a complete
case for the material. The ideal temperature of the frozen block varies
according to the material, but in general too cold a block will result in
cracking of the material, too warm a block in torn sections.

The height of the stage is adjusted so that the top of the material is level
with the blade, the microtome set for the desired thickness of section, the

blade lubricated with a little water and the material cut. Most operators prefer a fairly quick blade movement. Better sections often result when a number are removed in quick succession.

Sections are lifted from the blade with a fine brush either into a watch-glass, or, if the sections are large, fragile and valuable, directly to a drop of aqueous mountant on a slide.

The fully cryostatic microtome, in which the object is first frozen to the stage in a jet of CO_2 and then cut on a rocking type microtome enclosed in a refrigerating unit, has not been found suitable for either host-plant or fungal tissues, because the resulting sections break up disastrously in every mounting fluid tried. Even if this fault should be overcome, it is difficult to see what advantages the machine would have over those less elaborate, except, perhaps, in preparation of parasitized animal tissue.

B. Sledge microtome

An instrument which deserves to be much more fully employed by mycologists is the sledge microtome. This instrument is, of course, almost a necessity for preparing microtome sections of hard wood infected by fungi, but it is equally applicable to most other material. Much fresh or herbarium material needs merely to be fixed, soaked-up if necessary, and mounted directly into the chuck-stage of the machine, lubricated with a little alcohol, and cut. Leaves, etc., need to be supported between blocks of dry *Sambucus* pith. This is, of course, a much simpler procedure than embedding in gum, and the much greater rigidity of the sledge microtome, and its fixed blade, give infinitely better sections than those obtained with the bench microtome.

The sledge may also be adapted as a freezing microtome. The most convenient way to do this is by use of the Pelcool freezing stage, which, working somewhat after the manner of a reversed thermocouple, requires only a supply of electricity and reasonably cool running water, thus obviating the inconvenient need for CO_2, and maintaining the material at a constant, pre-selected temperature. There is no doubt that a sledge microtome with a detachable freezing stage of the Pelcool type is the apparatus of choice for mycological section-cutting of all kinds because of its great simplicity and versatility, and the high quality of the sections produced.

The electrically cooled stage can also be used with all other types of microtome, including the bench type described above. A special adaptor is required to attach it to each type of microtome.

C. Staining and mounting of freezing and sledge microtome sections

Most of the methods of mounting and staining given in the section on direct mounts are suitable for use with the freezing microtome and sledge microtome sections. In addition, if they are first fixed to slides, they may be subjected to the more complicated staining and mounting techniques applicable to wax-embedded sections. Freezing microtome sections may be fixed to slides by the use of Haupt's adhesive. This is more tenacious and has less tendency to absorb dyes than have the older but more frequently employed adhesives, egg albumen and saliva.

Haupt's adhesive is prepared by dissolving 1 g pure granulated gelatine in 100 ml distilled water at 30°C. When dissolved, add two crystals of phenol and 15 ml glycerol. Smear the solution evenly on a slide and allow to evaporate. Then, immediately flood the slide with 4% aqueous formalin. Float the section on the solution and arrange in place, heat the slide over a small flame or hot plate at 40°C until the formalin has evaporated. The gelatin is thus coagulated and the section should now be firmly fixed to the slide.

D. Preparation of paraffin sections

Fruit-bodies of fungi, and diseased plant material can be embedded in wax and cut by the more conventional methods. Difficulties have often arisen because of the different properties of the fungus hyphae and the host plant tissue, but there is no reason why these cannot be overcome by proper attention to fixation and dehydration procedures.

1. *Fixation*

When going to the trouble of cutting paraffin sections it is well to give careful attention to fixation of the material. Though formalin–acetic–alcohol (FAA) is frequently used—

Ethyl alcohol (50 or 70%)	90 ml
Acetic acid (glacial)	5 ml
Formalin	5 ml

better results may usually be obtained with fungi by substituting proprionic for acetic acid in the same proportions (FPA). 18 h is usually considered to be the minimum time for fixation in this fluid. If the tertiary butyl alcohol dehydration series is to be employed the material may be transferred directly to the 70% stage of that method. With other dehydration methods the FAA or FPA should be washed out in two changes of 70% ethyl alcohol.

Gelatinous fungus fruit bodies should, to prevent undue shrinkage, be fixed in weak chrome–acetic acid—

Chromic acid (10% aqueous)	2·5 ml
Acetic acid (10% aqueous)	5 ml
Distilled water	to 100 ml

A modified chrome–acetic formula is recommended by Lu (1962) for fixing chromosomes, nucleoli, and centrioles—

Chromic acid (10% aqueous)	20 ml
Acetic acid (glacial)	40 ml
Normal butyl alcohol	60 ml

However, most acetic alcohol mixtures cause fungus nuclei to lose their staining ability after prolonged immersion, chromosomes are fixed atrociously and mitochondria completely dissolved. Therefore the fixative should be washed away after about 6 h if it is desired to stain these structures.

Newcomer's fluid overcomes this fault, probably because of its rapid penetration—

Dioxane	10 ml
Acetone	10 ml
Petroleum ether	10 ml
Proprionic acid	30 ml
Isopropyl alcohol	60 ml

Fixation of small pieces of tissue is accomplished in one hour, after which the material may safely be stored in the fluid at 4°C.

2. *Paraffin embedding*

It is recommended that those wishing to use this technique and not already familiar with the basic principles of it consult a more general work, such as Johansen (1940), Peacock (1966), or Purvis, Collier and Walls (1964), in order to obtain the background information necessary to carry out properly the various steps given in the outline below.

The most satisfactory dehydration series for fungi is the tertiary butyl series, which causes a minimum of distortion of fungal tissues. Material fixed in FAA or FPA is transferred directly to the 70% stage of the schedule; that fixed in other mixtures is introduced at an appropriate point of alcoholic concentration. Picric acid, if any, should be washed out in alcohol and mercuric deposits removed with iodine as described for freezing microtome preparations.

Material from water should be taken up by two-hour stages through 10%, 25%, and 50% aqueous ethyl alcohol and then through the following series of ethyl–tertiary butyl mixtures as perfected by Johansen (1940)—

	70%	85%	95%	100%
Distilled water	30	15		
Ethyl alcohol (95%)	50	50	45	
Tertiary butyl alcohol	20	35	55	75
Ethyl alcohol (100%)				25
Time of immersion	overnight	1–2 h	1–2 h	1–2 h

From the 100% stage material is transferred to two changes of tertiary butyl alcohol, in one of which it should remain overnight.

Material is now ready for infiltration with wax and should be transferred to a mixture of equal parts paraffin oil and tertiary butyl alcohol for 1–2 h and then to a mixture of about $\frac{1}{3}$ melted paraffin wax to $\frac{2}{3}$ tertiary butyl alcohol in an embedding oven. Allow the alcohol to evaporate slowly, at the same time increasing the concentration of wax by adding wax chippings a few at a time. After a minimum of 6 h (usually the material is left overnight at this stage) the mixture is replaced by two changes of clean melted wax for 6 h each, until there is no odour of butyl alcohol. Individual workers have their preferences as to the type of wax used; some add a small quantity of beeswax to paraffin wax, others use specially prepared wax containing 0·5% petroleum ceresin. The melting point of the wax should be as low as climatic conditions allow: too hard a finished block will be difficult to cut.

In the block-making process the material is poured into suitable containers such as folded paper trays, arranged conveniently for sectioning, using a heated needle, and the wax cooled as rapidly as possible to prevent crystalization. If this occurs and is visible as white opacities in the block, it should be melted and repoured.

When hard, the block should be trimmed square and in such a way that the material will face up to the microtome blade at the correct angle, and mounted on the microtome stage with melted paraffin. The block is then cut, successive sections adhering edge-to-edge so as to form a ribbon. In centrally heated laboratories it is often difficult to prevent the ribbon, which collects a static electrical charge from the knife, from attaching itself permanently to the microtome or other nearby object. A large bowl of steaming water placed by the microtome helps to keep exasperation to a minimum.

Ribbons are placed in sequence on sheets of paper, and covered if necessary, to keep them free from dust until mounting. Convenient lengths of ribbon are fixed to the slide with Haupt's adhesive by the method described for freezing microtome sections. When the slide is flooded with formalin and heated, the ribbon expands and the sections become flattened. Too strong heating melts the wax, disintegrating the sections, whereas at too low a temperature the sections remain crumpled.

After the slides have dried, the wax is removed by soaking them in xylene for about 10 min. Remove slides carefully from the jar and transfer to a jar of equal parts xylene and absolute ethyl alcohol for 5 min. This is followed by immersion for 5 min each in absolute, 95%, 85%, 70% and 50% alcohols, after which staining schedules may be commenced.

E. Staining of paraffin sections

1. *General and cytological*

For sections of fungal tissue alone the choice of dyes is very wide and only a few schedules can be given here. A very useful and simple stain which should perhaps be tried first is Newton's crystal violet.

(i) Stain sections in fresh, filtered 1% crystal violet, 10 min.
(ii) Wash quickly in two changes of distilled water.
(iii) Post-mordant in 1% iodine in 1% potassium iodide in 80% ethyl alcohol, 30 sec.
(iv) Pass through 95% ethyl alcohol, rapidly.
(v) Dehydrate in 100% ethyl alcohol, rapidly.
(vi) Clear and differentiate under the microscope in clove oil, 2 parts; absolute alcohol, 1 part; xylene, 1 part.
(vii) Wash in xylene and mount in Canada balsam.

If other mountants, such as euparal or its successor sandeural, are used, euparal or sandeural essence should be used instead of xylene, which may cause collapse of tissue. Alternatively, chloroform may be used instead of xylene, here and in other schedules.

Another frequently used stain, for sections of 15 μ thick and under, is Heidenhein's haematoxylin. Individual workers have their own preferences with regard to the exact details but a suggested schedule is as follows—

(i) Mordant at 30°C in freshly prepared 2% aqueous iron alum, 10–20 min.
(ii) Wash in running water, 5 min.
(iii) Rinse in distilled water (to remove traces of metals).
(iv) Overstain at 30°C in Heidenhein's haematoxylin, at least 20 min.
(v) Wash in distilled water.
(vi) Differentiate in 2% iron alum observing the intensity of the stain under an old microscope from time to time.
(vii) Wash in running water, 1 h.
(viii) Dehydrate, clear and mount.

Recipe for Heidenhein's haematoxylin—

Haemotoxylin	0·5 g
Absolute ethyl alcohol	10 ml
After haemotoxylin has dissolved add	
Distilled water	90 ml

Leave to "ripen" for a few days, during which the haematoxylin oxidizes to haematin. The solution will remain usable while it is the colour of red wine or until a scum forms.

Harris's haematoxylin resembles the better known Delafield's in its effects but is generally more suitable for fungi—

(i) Bring slides to water.
(ii) Stain in Harris's haematoxylin, 20 min.
(iii) Rinse out excess stain in distilled water.
(iv) Destain in slightly acidified water, about 5 sec.

(v) Wash in tap water (this should be made alkaline if necessary) to blue the stain.

(vi) Dehydrate, clear and mount.

Recipe for Harris's haematoxylin—

Haematoxylin crystals	5 g
Aluminium ammonium sulphate	3 g
Ethyl alcohol (50%)	1 litre

2. *Fungi in plant tissue*

Although the above methods are all useful for fungi in plant tissue it is often convenient to have the host tissue stained a different colour from the hyphae. This may be accomplished by one of the following techniques.

Stoughton's (1930) technique relies on staining in thionin, 0·1 g in 5% aqueous phenol, 100 ml, and counterstaining in a saturated solution of orange G in absolute alcohol at the appropriate points in the ordinary dehydration schedule.

Dring's (1955) modification of the periodic acid–Schiff (PAS) method not only differentiates the fungus from the host tissue but stains the cell walls of the fungus rather than the protoplast.

(i) From water, oxidize in 1% periodic acid, 2–3 min (if 3 min is exceeded the host tissue will stain).

(ii) Wash in running water, 10 min.

(iii) Stain in Schiff's reagent (by which time the reaction is complete but colour has not developed).

(iv) Transfer to two changes of the following mixture—

Anhydrous potassium metabisulphite (10% aqueous)	5 ml
Normal hydrochloric acid	5 ml
Distilled water	90 ml

for a total of 10 min. (Keep stocks of this solution stoppered, discard when smell of SO_2 begins to wane.)

(v) Wash in running water (when colour develops), 10 min.

(vi) Dehydrate to 85% alcohol.

(vii) Counterstain in light green SF in 95% alcohol.

(viii) Dehydrate in absolute alcohol.

(ix) Clear and mount.

Schiff's reagent (leuco-basic fuchsin) is prepared by pouring 100 ml boiling distilled water over 0·5 g basic fuchsin or pararosanilin to dissolve it, cooling to 50°C, filtering, and adding 10·0 ml normal hydrochloric acid and 0·5 g anhydrous potassium metabisulphite. Leave overnight to decolourize. The reagent should be stored in the dark, preferably in a refrigerator.

Pianese IIIb (Vaughan, 1914) has often been used for staining fungi in host tissue. A suggested schedule is as follows—

(i) Take slides to 30% alcohol.
(ii) Stain in Pianese IIIb, 15–45 min.
(iii) Wash in 50% alcohol and dehydrate to 85%.
(iv) Differentiate in alcohol (95%), 100 ml+hydrochloric acid (conc.), 1 ml.
(v) Wash in 95% alcohol.
(vi) Dehydrate, clear and mount.

Pianese IIIb is compounded as follows—

Martus yellow	0·01 g
Malachite green	0·5 g
Acid fuchsin	0·1 g
Alcohol (95%)	50 ml
Distilled water	150 ml

3. *Fungi in wood*

Some of the methods outlined above (e.g. the PAS technique) are useful for staining fungi in woody tissues. An additional method is given by Gram and Jørgensen (1953), employing 0·1% fast green+0·3% safranin in 60% alcohol. The sections are dehydrated and mounted in the usual way.

REFERENCES

Bretz, T. W., and Berry, F. H. (1964). *Plant. Dis. Reptr.*, **48**, 514.
Dade, H. A. (1968). *Vict. Nat.*, **85**, 41–47.
de Silva, E. M. (1965). *Stain Technol.*, **40**, 253–257.
Dring, D. M. (1955). *New Phytol.*, **54**, 277–279.
Feder, W. A., and Hutchins, P. C. (1964). *Phytopathology*, **54**, 863–864.
Fleming, A., and Smith, G. (1944). *Trans. Br. mycol. Soc.*, **28**, 13–19.
Gram, K., and Jørgensen, E. (1953). *Friesia* **4**, 262–266.
Hering, T. F., and Nicholson, C. B. (1964). *Nature, Lond.*, **201**, 942–943.
Isaac, P. K. (1958). *Stain Technol.*, **33**, 261–264.
Isaac, P. K. (1960). *Phytopathology*, **50**, 474–475.
Johansen, D. A. (1940). "Plant Microtechnique". McGraw-Hill, New York.
Loquin, M. (1952). *Bull. Soc. mycol. Fr.*, **68**, 170–171.
Lowry, R. J. (1963). *Stain Technol.*, **38**, 199–200.
Lu, B. C. (1962). *Can. J. Bot.*, **40**, 843–847.
Maneval, W. E. (1936). *Stain Technol.*, **11**, 9–11.
Peacock, H. A. (1966). "Elementary Microtechnique", 3rd Ed. Arnold, London.
Purvis, M. J., Collier, D. C. and Walls, D. (1964). "Laboratory Techniques in Botany". Butterworths, London.
Shipton, W. A., and Brown, J. F. (1962). *Phytopathology*, **52**, 1313.
Shoemaker, R. A. (1964). *Stain Technol.*, **39**, 120–121.
Singh, P. (1969). *Mycopath. Mycol. Appl.*, **87**, 142–144.
Stoughton, R. H. (1930). *Ann. appl. Biol.*, **17**, 162–164.
Vaughan, R. E. (1914). *Ann. Mo. bot. Gard.*, **1**, 241–243.

CHAPTER IV

Preservation of Fungi

AGNES H. S. ONIONS

*The Culture Collection, Commonwealth Mycological Institute,
Kew, Surrey, England*

As soon as any serious work with fungi is undertaken it becomes desirable to retain some form of reference material, both for use during the work and later as a permanent record. This material should be maintained both as living cultures and dried reference material.

If the fungi have been used in the production of an antibiotic or for other biochemical research it would be desirable or perhaps essential to retain it in the living form and in such a condition that its physiological activity remains constant. If the work is of a taxonomic nature it might be sufficient or only possible (many fungi have not yet been grown in artificial culture) to retain a record of its structure by means of slides and dried specimens. However, the tendency today is as far as possible, to retain both living and preserved material.

I. MAINTENANCE OF LIVING FUNGI

With the increasing importance of microfungi to industry (in biochemical and antibiotic production, microbial assay and as spoilage organisms), human, animal and plant pathologists, geneticists, taxonomists and teachers, there has developed a need for culture collections. There are several large public service collections which serve as repositories for cultures and as sources of their distribution.

The best known of these are the Centraalbureau voor Schimmelcultures (C.B.S.), founded in 1906, the American Type Culture Collection (A.T.C.C.), founded 1925, and the collection of the Commonwealth Mycological Institute (C.M.I.), founded 1947. Several other countries are developing their own national collections, and there are large collections belonging to industrial concerns as well as specialized government departments.

However, any worker with living material must at least temporarily maintain his own cultures during the course of his studies or until they are ready for deposit in one of these major collections. This depositing of important strains is most desirable as in the past many organisms which have been the subject of intensive investigation have been discarded at the end of the work or on the death of the worker, and much valuable material has been lost.

A. Aims and hazards

The aim of the curator of a culture collection is to maintain the organisms alive and healthy in a condition as near as possible, both physically and physiologically, to their condition at the time of deposit. In some cases it is necessary to work on the cultures before incorporation to find the best methods for maintaining them.

The running of the collection and the methods of maintenance employed are designed to minimize the following hazards to which cultures are exposed—

(i) By repeated transfer selection can occur, either of a mutant strain or of a purely vegetative non-sporulating form. The transfer should therefore be done as far as possible by an expert with an eye for the wild strain. However, the fewer transfers made the less the risk.

(ii) Some strains sometimes tend to become attenuated under the artificial conditions of culture. Others deteriorate to wet slimy disintegrated mycelium or spores. Simmons (1963) suggests that this may be due to virus infections and there is considerable evidence to support this. He thinks it is possible that there may be internal fungal parasites and we have seen internal bacterial infections, which are very difficult to eradicate.

(iii) The maintenance processes to which the fungus is subjected are selective and only adaptive strains survive. These may have somewhat atypical characteristics.

(iv) Cultures are subject to contamination, infection with mites and adverse conditions of temperature, light, humidity, etc. These latter may arise through breakdown of apparatus, or by incomplete understanding of the organism.

(v) Adequate documentation of the strains must be made. In a culture collection of long standing the strains may well survive several generations of mycologists, so to assist in maintaining them in their original condition a clear description of the cultural characteristics supported by dried cultures should be made at the time of receipt.

B. History and development of modern methods

The early maintenance of fungus cultures was empirical and a form of micro-gardening and in part of most collections it is still largely so. Westerdijk (1947) and van Beverwijk (1959) describe the history of the early culture collections and the methods used. The cultures were grown and kept healthy by transfer from one substrate to another at fairly frequent intervals. This is still the normal practice for ordinary culture growth in the laboratory. Much work was done searching for the most suitable media on which to grow the fungi. The disadvantage of this method is the frequency with which transfers to new substrata have to be made, due to the staling of the media and ageing of the cultures, with the consequent hazards of selection, variation and infection. There are, however, strains available derived from the Kral Collection established around 1900 and some of these are still healthy.

Attempts were made to improve on this by varying the method of growth and substrate, then by increasing the period between transfers, and by storing the grown fungi in refrigerators or under oil (Sherf, 1943). However, the cultures had ultimately to be regrown, so methods of induced dormancy were tried. Spores were suspended in a suitable medium (St. John-Brookes and Rhodes, 1936), dried and stored under vacuum. This was followed by the technique known as lyophilization or drying from the frozen state (Raper and Alexander, 1945). This latter method has proved most satisfactory for many spore-bearing fungi and tends at present to be the most popular means of preservation. The material is dried when newly isolated, or at least while the cultures are still young and healthy, and the spores remain dormant until time for revival. They are sealed in ampoules so that there is no risk of contamination. Survivals for very long periods have been recorded, although a fall off in viability is expected. Some fungi — mycelial forms, some species with extra large or delicate spores, and many Phycomycetes do not survive the process.

An alternative method now being used is to store the fungi at ultra-low temperatures under or above liquid nitrogen. At these low temperatures metabolism is assumed to be more or less at a standstill (Meryman, 1956), and if the fungus survives it should remain in its original condition indefinitely. This is provided it survives the shocks of freezing and thawing, which could be selective. Most reports indicate that, provided care is taken in freezing and thawing, many strains that do not survive lyophilization do well under liquid nitrogen.

All the above methods of maintenance are still used. The simple early ones of micro-gardening are convenient for ordinary laboratory work and available to the smallest establishments. Indeed some fungi only survive under these cultural conditions. The more complex methods are preferable because they tend to prevent degeneration or change, but they are suitable for use in the larger more specialist laboratories.

C. Methods of culture maintenance

1. *Micro-gardening*

(a) *Substratum.* The usual method of culture maintenance is on a sterile nutrient agar jelly contained in a suitable vessel. Fungi are often grown in nutrient solution for biochemical or similar work, but a surface growth on jelly is more satisfactory for maintenance work, because it can be more readily observed.

The number of possible nutrient decoctions is enormous and almost every mycologist has some medium he favours or which he considers most suitable for a particular fungus (for media see Booth, this Volume, Chapter 2).

Although pure media of biochemical salts are desirable to give a standard form of growth, e.g., Czapek's Agar for *Penicillium* species (see Thom, 1930; Raper and Thom, 1949; G Smith, 1969), it must be remembered that growth in pure culture on jelly is for most fungi an unnatural condition and to obtain the fullest development the conditions encountered in nature should be simulated as far as possible. It is not possible in artificial pure culture to imitate the normal succession of fungi and the effect of the presence of other organisms and their metabolites, but the use of natural media or vegetable decoctions are helpful (Dade, 1960; Snyder and Hansen, 1947). Animal and human pathogens grow better on media containing some protein or digested protein as is supplied by the peptone in the famous Sabouraud's media.

A few parasitic fungi appear to be obligate parasites and will only grow in tissue cultures or in the presence of their host. If the host is a microfungus as in the case of *Piptocephalis* it is possible to grow the one fungus on the other. Others such as *Dictyostelium* require bacteria, and these can be grown in mixed culture (Dade, 1960). If care is not taken one can be left with the host and no pathogen. However, these fungi are the exceptions and a good general purpose agar is Potato Dextrose agar, which is rich and contains most necessary nutrients. Malt agar, Sabouraud's medium and Czapek's Biochemical medium are other well known and useful media. Other substrata can be used for micro-gardening, such as soil, various parts of plant tissue, cereal grains, twigs, wheat straw, pieces of potato and various fruits. These may be used by themselves or set in agar jelly.

The object in using these various substrata is to obtain a satisfactory growth that is most suitable for storage. It has been observed at the C.M.I. (Dade, 1960), that it is frequently best to have restricted growth or production of very sparse mycelium with a relatively high proportion of spores, as is obtained on a starvation medium such as the Potato Carrot agar of Langeron and Vanbreuseghem (1952), which subsequently results in very healthy cultures on transfer to a richer medium. Westerdijk (1947), considered a change of diet from a weak to strong medium desirable for continued healthy growth and it is still the practice at the C.B.S. to make two subcultures of each strain at the time of culture maintenance, one on a normal medium and one on a weak medium, and at the next subculture, changing them round, putting the one from normal medium on to a weak medium and the one from the weak on to a normal medium. Considering the age and quality of this collection this practice must be of great value.

Sensitivity to pH and osmotic pressure must be considered. Most fungi grow at a pH 5–7. There are a few fungi termed osmophiles that thrive on media with a high osmotic pressure which can be produced by increasing the sugar or salt concentration of the media.

Besides a suitable substratum other factors effect the growth of cultures for storage.

(b) *Light intensity*. Light intensity is very important to fungi (see Leach, this Volume, Chapter XXIII). Most strains respond well to normal daylight with its daily variations, but some will produce their typical sporing apparatus better in the dark, e.g., *Aspergillus paradoxus*. Others gain an extra boost from direct sunlight, e.g., *Pyronema domesticum* produces good fruiting bodies when grown in front of a south-facing window. The use of "black light" to induce fungi to produce spores when they are reluctant to do so (as in dematiaceous cultures, e.g., *Epicoccum* and *Helminthosporium* after the first few transfers) is becoming of increasing importance, and with the introduction of plastic bottles is now available for normal culture maintenance (see Leach, this Volume, Chapter XXIII.

(c) *Temperature*. The majority of the fungi grow well at a room temperature of about 20–22°C although many incubators are maintained at 25°C;

some fungi are more sensitive to temperature and the necessary conditions for these thermophiles (Cooney and Emerson, 1964) must be provided. Chytrids and other water moulds are happier at temperatures below 20°C and should be grown in cool conditions at about 15°C. They tend to be very specialized in their growth requirements.

(d) *Humidity*. Little consideration is normally given to this physical character when growing for a culture collection. The culture tubes and bottles tend to produce their own local climate and most fungi thrive at a fairly high humidity.

(e) *Standard growth conditions*. When subculturing a large collection standard media and conditions of growth are employed as a matter of convenience.

The procedure at the C.M.I. consists of the inoculation on to one of a relatively small selection of media, followed by growth at room temperature in glass cabinets to allow the penetration of light. The cabinets away from windows have extra illumination and some have simple heating elements to raise the temperature to about 25°C. The demands of particular genera, species and strains are provided for separately. When new strains are incorporated into the collection their needs and peculiarities are investigated and recorded for future reference, so that specialized media, suitable temperature and appropriate illumination can be used at later subculturing. The Saprolegniaales, for example, have to be given special attention (for methods see Dick, 1965), and are kept separately.

(f) *Methods of transfer*

(i) *Mass spore transfer*. The quantity of fungus transferred has been a matter of discussion over the years. Fennell (1960), is firmly convinced that mass spore transfer, wherever possible, remains the most reliable technique. This is usually done by means of a sterile nichrome wire held in a suitable handle.

(ii) *Mycelium transfer*. The transfer of mycelium should only be used where there are no spores available, and this should be made from the growing edge of the colony. Pythiaceae are best subcultured by removal of the basal felt.

(iii) *Single spore cultures*. At the other extreme the use of single spore cultures has received much favour, especially in the culture of *Fusarium* species (Gordon, 1952). By this technique, especially when six cultures are made, the wild type sporing strain is maintained by selection if necessary when dealing with unstable strains. This is of particular use where variant strains are likely, due to their more rapid growth, to outgrow the sporing forms.

(g) *Culture vessels*. For retention in a culture collection it is usual to keep the stock cultures in standard bacteriological test tubes or McCartney bottles, but it is a matter of personal preference. There is no reason why the cultures should not be grown in flasks or any other receptacle which can be sterilized, protects from contamination and yet allows the free passage of air.

(i) *Bacteriological test tubes*. These are well made and of good quality glass, so that the cultures can be examined quite easily through the glass. Many fungi seem to thrive in them. They will not stand up on their own and various methods of stacking them have been devised. Compartmented stainless steel baskets are used at the C.B.S. and slanted wooden racks with pin separators are used at the C.M.I. In the past much use was made of compartmented wooden boxes. The tubes with straight sides instead of lips are easier to pack. Cotton wool plugs are usually used to close the tubes. These allow satisfactory passage of air and when well made will form quite a good physical barrier to mites. To prevent mites entering the culture the cotton plugs can be safely flamed and pushed down the tube, which can then be sealed with a cigarette paper above the plug (Snyder and Hansen, 1946), or they can be treated with a drop of a coloured solution of Mercury salts (Fennell, 1960). Many workers find the cotton wool plugs and tubes easy to handle, but making plugs can be time consuming, although excellent machines for their manufacture are now available.

(ii) *McCartney bottles*. The 1-oz McCartney bottle came into considerable use in medical pathological laboratories during the war 1939–1945. There are various advantages in their use. They will stand on their own, making packing easy. Media can be prepared months in advance and, with the lids screwed up tightly, can be stored for long periods. They will withstand repeated sterilization. Reference numbers can be written on the lid so that location of strains is easy. They are, however, made of coarse glass, which makes it difficult to see culture details. Some cultures do not seem to thrive so well, which may be a matter of light penetration. Mites can easily penetrate the gap between the bottle and the lid, which must be left loose to allow for ventilation. Similarly there is a risk of contamination by dust and airborne spores. Some workers prefer to use the smaller or $\frac{1}{2}$-oz version of these bottles as an economy of storage space (Carmichael, 1962).

(iii) *Plastic bottles*. Disposable plastic bottles are now available. They are relatively cheap and strong and unlikely to break if dropped. They allow better penetration of light than glass and can be used to expose cultures to stimulation by "black light". They are received sterile, but have to be filled with sterile agar, using sterile precautions, instead of sterilizing as a whole after filling. Most of them can only be used once. They present no more barrier to mites and infection than metal caps and glass bottles.

(iv) *Racks*. Although expensive racks with individual compartments are not a necessity, they are highly desirable for test tubes and would prove useful for storing bottles. Those in stainless steel will withstand repeated washing and sterilization, and maintain an orderly arrangement which tremendously facilitates retrieval. The possible range of shape, size and overall design of racks is enormous and a matter of personal requirements and preference.

(v) *Labels*. Tubes can be labelled with the strain number by the use of glass ink or grease pencil. However, stick-on labels can carry more information and are of a more permanent nature. In a large collection the use of printed labels with the name and number of the culture has great advantages, both in clarity and time saving.

(h) *Storage*. Once a good culture has been obtained and is growing well in a suitable container, it must be stored. The method of storage employed depends largely on the facilities available, but cultures must be stored with regard to their individual requirements.

(i) *At room temperature*. The simplest method of storage is to keep the cultures at room temperature. These are best protected from draught, dirt, dust or aerial contamination by placing in a suitable box or cupboard, although shelves with plastic curtains or similar protection may serve. Simple wooden cupboards with good basal ventilation have been found satisfactory at the C.M.I. (Dade, 1960). Wood is a poor conductor so that the cultures are unlikely to be subjected to rapid changes of temperature. When new cupboards are installed great care must be taken, as new wood tends to be damp and may easily carry mites or at least produce an atmosphere in which they thrive. All new cupboards or shelves, etc., should be well dried out and treated with an acaricide and if necessary insect and mite repellants placed in them for several weeks before use. Cultures stored at room temperature require constant care and vigilance in their upkeep. They tend to dry out rapidly, depending on the local climate, and most cultures require to be transferred to fresh medium every three to six months. Some cultures, e.g., the water moulds, human and animal pathogens, require to be transferred more frequently. Both in the tropics, and in low humidity climates, they require even more frequent attention (Fennell, 1960). Therefore this method of storage is most suited to use in temperate climates and has been used satisfactorily for many years at the C.B.S. and other collections. The risks of variation and accident through frequent transfer are great, and constant skilled supervision is required. However, it costs nothing in extra apparatus, and is so far one of the most reliable general methods of upkeep for many sensitive strains.

Refinements of this method of storage consist of keeping the strains in a cool basement or a room with controlled temperature and humidity. In a room with constant temperature metal shelves would be quite satisfactory. If the room faces north and receives no sunlight the conditions are probably easier to control. A temperature of 16°–17°C and relative humidity of about 60% are considered satisfactory (C.B.S.). If the relative humidity is too high growth on the outside of containers can occur, and cultures become cross infected. Too low a relative humidity results in rapid drying out of the cultures.

(ii) *Refrigeration or cool storage.* An easy method of storing cultures is at 5°–8°C in a refrigerator. A domestic-type refrigerator is quite satisfactory for this, although somewhat larger models are usually used. G. Smith at the London School of Hygiene and Tropical Medicine kept the collection of strains for the department of biochemistry by this means for many years with great success. Under these conditions, the rate of growth, the over-growth of aerial mycelium and drying out are reduced. Mites are immobilized, although on transfer to room temperature they can become active again. Most moulds and common fungi, e.g., Penicillia, Aspergilli, Mucorales, survive well at 5°–8°C, but some are sensitive to the cold, e.g., *Piptocephalis*. It is therefore necessary to keep duplicate cultures by some other means for at least one or two transfers before relying on cool storage. The interval between transfers can be longer than at room temperature and a period of 6–8 months is usual.

The methods of racking the cultures in the refrigerator are various and the refrigerator may well have to be adapted for this. Flat trays in which the cultures can be examined comparatively at a glance (Raper and Thom, 1949, p. 80) are very pleasant to use, but compartmented boxes, or racks of various types, are quite satisfactory. Trouble was encountered in the early days of the collection at the C.M.I. (Dade, 1960), when cultures grown in bottles were tightly packed in a refrigerator. As the refrigerator was in constant use, the doors were frequently opened and condensation occurred on the outside of the cool bottles. The humidity inside the refrigerator became high and a fungus grew on the surface of the bottles, finally penetrating between the lids and bottles, and infecting the cultures. This could possibly have been prevented by packing the cultures less tightly, abandoning the use of bottles for plugged tubes or opening the refrigerator less frequently, but the method was discontinued except for a restricted number of strains. The presence of an ice box may also help to keep the internal humidity down. Most workers have found the method satisfactory, and strains of *Fusarium*, grown in test tubes, are now routinely stored in this way at the C.M.I. Refrigeration storage is illustrated in Fig. 1.

(iii) *Deep freeze.* There have been several reports (Hamilton and Weaver, 1943; Meyer, 1955; Carmichael, 1956; Kramer and Mix, 1957; Webster *et al.*, 1958; Carmichael, 1962), of cultures being successfully preserved by

FIG. 1. Refrigeration storage. A refrigerator showing cultures grown on agar in tubes and cultures grown on soil in bottles. (Photograph by D. W. Fry.)

placing in a deep freeze at temperatures of about $-10°$ to $-20°C$. An ordinary domestic deep freeze appears satisfactory. The method has been particularly applied to storage of human and animal pathogens (Carmichael, 1956). These fungi are notoriously short lived, and deep freeze storage appears to have prolonged their life considerably. Carmichael does not recommend thawing and refreezing but Kramer considered it possible to remove cultures, take a subculture and return the parent to the deep freeze. One of the main objections to deep freeze storing of any living material is the risk of breakdown and subsequent loss of specimens, but in the case of fungi, thawing and refreezing does not necessarily result in death of the strains. However, an immediate reculturing of all the cultures in a freezer should be regarded as essential following breakdown.

(iv) *Mineral Oil.* Healthy cultures ready for storage can be covered with mineral oil, and will survive for long periods, growing at a very reduced rate. Of 2000 strains maintained under oil for 10 years at the C.M.I. only forty-five were lost. Other workers have had as good, or better and longer, revivals (Hesseltine *et al.*, 1960).

Oil storage is illustrated in Fig. 2. This method of preservation is cheap and easy and requires no special apparatus or skills. It was first extensively used by Buell and Weston (1947).

The oil used should be of good quality; British Pharmacopoeia medicinal paraffin oil of specific gravity 0·865–0·890 is quite satisfactory. At the C.M.I. this is sterilized in McCartney bottles for 15 min at 15 lb/in.² However, Fennell (1960) insists that the oil be autoclaved at 15 lb/in.² for 2 h and dried in the oven at 170°C for 1–2 h. Simmons (1963) also stresses the need for high quality oil, initial sterility and dryness.

The cultures are grown on agar until good growth and sporulation are obtained and then covered, using sterile technique, with mineral oil to a depth of about 1 cm above the top of the agar slant. If a short slant of agar is used less oil is required. The depth of 1 cm is fairly critical (Fennell, 1960) as the oxygen transmission by layers of mineral oil in excess of 1 cm becomes less favourable. If less oil is used, strands of mycelium may be exposed which allow the cultures to dry out (Dade, 1960; Fennell, 1960). The method depends on reduced rate of metabolism and prevention of drying, not on a completely arrested metabolism. In fact some strains grow quite considerably under oil. If the McCartney bottles are used the rubber liners should be removed from the metal caps as the oil tends to dissolve the rubber and this can be toxic to the cultures. The oil itself should be added as individual doses from separate containers as blow up of spores from the culture being treated can contaminate the receptacle from which the oil is being poured (Dade, 1960, discusses the procedure in great detail). If the method is used as a main method of preservation, it is desirable to

FIG. 2. Oil storage. Trays of bottles in a wooden cupboard, with a bottle of oil, a culture grown on a short slope ready for covering with oil and a culture covered with oil. (Photograph by D. W. Fry.)

keep a fairly regular check on the viability of the strains and this should be done every two to three years. A culture is taken from the oldest oiled culture in the collection and, if it grows, put back into the collection. This is done until about four generations are preserved, but there should always be a culture as near to the original as possible and one not more than 2–3 years old. If the parent culture is dead or infected then the later generations should be viable. Retrieval is by draining off as much oil as possible and streaking the inoculum onto agar in plates or tubes. The first subculture is often slow in starting as the growth seems to be retarded by the presence of the oil and these cultures are often sticky looking, so that a second generation is often necessary to give suitable cultures. If the oil appears to be blocking growth, growth on a slant in an upright position may allow the oil to drain away, leaving the fungus free to grow, or efforts may be made to wash off the excess oil in sterile water before streaking.

There is some risk to personnel of contamination and infection due to spatter when sterilizing the inoculating needles and great care should be exercised especially when working with human pathogens (Fennell, 1960).

As with all methods of preservation some strains seem rather sensitive to oil storage However, it is a method of particularly wide application and strains of mycelial forms, *Phytophthora* and related genera do well under oil. Some fungi, such as *Saprolegniaceae* and other water moulds, will last 12–30 months (Reischer, 1949) under oil, but not as long as the majority of common moulds.

The oil cultures can be stored at room temperature, in a refrigerator or cold room. If low temperatures are employed, it should be borne in mind that fungi which are normally sensitive to cold will still be sensitive to cold.

The method is cheap and easy to apply. Mites will not penetrate the cultures; the fungi can go for long periods without regrowing, thus cutting down work, and the risks of selection, mutation and variation. However, oil is messy to handle, liable to contamination and the tubes and bottles must be stored in the upright position. Growth although retarded still continues and variation of both biochemical and morphological characters sometimes occurs in strains preserved for long periods.

In some collections one tube of each strain is put under oil at the time of accession and then kept as an emergency duplicate requiring little supervision in case of accident. It also appears an ideal method of storage for a busy laboratory with limited funds and facilities and a relatively small collection. Many workers have reported experience with the oiled culture maintenance technique (Sherf, 1943; Norris, 1944; Wernham, 1946; Buell and Weston, 1947; Edwards *et al.*, 1947; Wernham and Miller, 1948; Stebbins and Robbins, 1949; Schulze, 1951; Weiss and Oteifa, 1953; Schneider, 1957; Braverman and Crosier, 1966; Little and Gordon, 1967).

(v) *Soil, etc.* Preservation of fungi in soil can be of two kinds. Inoculation of soil with a spore suspension, and growth on this soil and final preservation of the culture (Bakerspigel, 1953, 1954), or the placing of dry fungal spores in dry soil or similar substrate and subsequent storage of this dry material. This latter method is described under Preservation of Original Material and is fully discussed by Fennell (1960).

The method of soil storage in use at the C.M.I. consists of inoculating about 5 g of garden loam (20% moisture and particular attention to sterility; autoclaved at least twice) with 1 ml of a water spore suspension and subsequent growth for about 10 days at room temperature before storage, which is preferably in a refrigerator. This has proved useful for the preservation of *Fusarium* sp. (Cormack, 1951; Gordon, 1952) and is used by Booth at the C.M.I. for the same purpose. The period of survival is greater than on agar and strains tend to remain typical, but the strain has undergone growth from the original and as no definite growth characters are evident changes or infections are not observed until recultivation. As many subcultures can be made from each specimen these cultures give a useful uniform base for mass inoculations.

2. *Maintenance by suspended metabolism*

(a) *Drying*. If cultures are left to dry in the laboratory most of them die, but it is surprising how long a small minority of cultures remain viable. Fennell (1960) cites several records of this type of longevity including the viability of *Aspergillus oryzae* spores after 30 years (McCrea, 1923, 1931).

Spore suspensions prepared in a suitable protective medium were dried under vacuum using desiccants and sealed off while still under vacuum (St. John-Brookes and Rhodes, 1936; Rhodes, 1950; Barmenkov, 1959) with some viable results.

Goldie-Smith (1956) had surprisingly good results with the Blastocladiaceae by slow drying of liquid cultures on filter paper strips and records survivals of strains of *Allomyces arbuscula* for 14 and 17 years.

Ainsworth (1962) reports on the specimens of *Schizophyllum* which were dried and sealed under vacuum by Buller in 1909 and 1910 and opened by Bisby (1945), who obtained spore discharge and growth after 35 years. Ainsworth himself opened further tubes of Buller's and obtained a spore discharge and growth after 50 years.

Dry spores or mycelium can be dispersed in dry sand, soil, silica gel, etc. The tubes are usually evacuated and sealed before storage. Spore-forming fungi survive well under these conditions and such specimens can be used repeatedly and produce a very uniform supply of inoculum for biochemical studies and industrial processes (Fennell, 1960).

Perhaps one of the widest applications of this type of drying is the preservation of strains on silica gel at the Fungal Genetics Stock Centre and described by Perkins (1962) with later modifications by Ogata (1962) and Barratt *et al.* (1965).

C. F. Roberts of the Department of Genetics at the University of Leicester has stored cultures by this method for several years and has retained viability and genetical characteristics. He three-quarters fills screw cap containers with gel (silica-gel purified, without indicator, 6–22 mesh) and sterilizes them for a minimum of 90 min at 180°C and stores them in a dry atmosphere. The silica gel containers are stood in an ice bath for at least 30 min as considerable heat is evolved when the gel is wetted and viability is lost if the temperature is allowed to rise. A heavy spore suspension is prepared from a good sporing culture by tipping cooled 5% sterile skim milk on to the slant and using a long sterile wire to scrape off the conidia. The cool conidial suspension is tipped on to the cold gel and the whole returned to the ice-bath and kept there for at least 15 min. The gel should not be saturated, but about three-quarters wetted. The gels are kept at room temperature until the crystals readily separate when shaken (about a week). The sample is checked for viability and the cap screwed down tightly. The tubes are stored over indicator silica gel in a desiccator in a cold room, although good revivals have also been obtained from samples stored at room temperature.

(b) *Lyophilization or freeze-drying.* Lyophilization or freeze-drying as a method for the preservation of micro-organisms consists of drying cultures or a spore suspension from the frozen state under reduced pressure. This may be done in several ways and various forms of apparatus have been devised to this end. When cells are dried under these conditions they remain dormant for long periods, but on reconstitution and return to normal media usually grow well.

The process of lyophilization was first applied to microfungi on a large scale by Raper and Alexander in 1942, who successfully processed cultures at the Northern Regional Research Laboratory at Peoria reporting on this work in 1945. Since this time a whole series of papers has come from this department, in which the continued and increasingly long survivals of fungus spores by this and other methods of preservation are recorded and compared (Fennell *et al.*, 1950; Hesseltine *et al.*, 1960; Mehrotra and Hesseltine, 1958; Ellis and Roberson, 1968). It is still considered at the department of the N.R.R.L. that lyophilization is the most satisfactory method of long-term preservation for the majority of sporing fungi.

Most workers seem to have similar results, although some fungi do not survive the process.

The majority of moulds, Penicillia, Aspergilli, and Mucorales survive well, whereas Pythiaceae, Entomophthorales and mycelial forms seldom survive the initial treatment. Large spored forms tend to fragment, for example the Helicosporae. If the initial treatment is survived the strains are likely to remain viable for periods of 10 to 20 years (Ellis and Roberson, 1968).

The spores or cells are suspended in various protective media. Serum was used at first (Raper and Alexander, 1945), followed by other media including sugars (Blackwood, 1955), but skim milk is now considered satisfactory and has the advantage of being more readily available than serum. It is possible to dry pieces of cultures (Last *et al.*, 1969), or mini-cultures without a protective medium (Bazzigher, 1962) but revivals at C.M.I. have been unpredictable. The suspension is distributed in small quantities into ampoules using sterile technique and these are connected to a vacuum system usually incorporating a desiccant, such as phosphorous pentoxide, silica gel or a freezing trap, and lowered into a freezing mixture of dry ice and ethyl acetate, for example. The vacuum pump is turned on and the ampoules evacuated till drying is complete, after which they can be sealed off. The details of the methods used vary from one laboratory to another. At the Commonwealth Mycological Institute a centrifugal two stage freeze dryer is used. In this the cooling is by evaporation and no freezing mixture is required. After drying the ampoules can be filled with dry nitrogen instead of sealing under vacuum.

Best revivals are obtained when using a good healthy strain that is sporing well. It is the spores which best withstand the treatment and survive. If the spore suspension is weak revival tends to be poor. In some cases a poorly sporing and fluffy strain may improve after lyophilization, presumably due to the mycelial element being killed (Simmons, 1963, and personal observation). Care must be taken with the sterility of the suspending media, which may carry heat resistant spores. A small protective plug is incorporated in the ampoules before evacuation to prevent infection at the time of opening due to the inrush of air, and for this reason the ampoules are sometimes filled with sterile nitrogen before sealing, or small plugged tubes placed inside larger tubes which are evacuated and sealed.

The methods of revival vary from one laboratory to another. The dry pellet may be transferred to a suitable liquid and allowed to dissolve before it is streaked out on agar. At the Commonwealth Mycological Institute a volume of sterile water equal to the original volume of the spore suspension is placed in the ampoule at room temperature, and the ampoule then left for about 20–30 min for the water to be absorbed slowly before streaking out. This delay seems to produce more satisfactory cultures. The degree of viability is assessed visually or by spore counts.

After a test to see that the fungus has survived the initial shock very little attention is required. Checking of viability can be at long intervals, either when cultures are required or by routine sampling. It is usual to make many replicates so that material from a constant source can be supplied over a considerable period. It makes possible the "Seed stock" system used at the American Type Culture Collection (Clark and Loegering, 1967) for conserving living reference micro-organisms over long periods. Clark and Loegering describe this method as follows:

> "In this system ampoules of freeze-dried or frozen culture material from original stocks are set aside as seed stock, and stored under optimum temperature conditions. This material is never distributed. Periodically an ampoule of seed stock is opened, and the culture grown from it refrozen or re-lyophilized in quantity for distribution. It is also used to prepare new seed stock if the viability of original material declines or the original seed stock becomes nearly depleted. In this manner original material can be conserved over a long period."

The lyophilized cultures are stored in small ampoules and once a filing system has been devised storage is easy and compact. The ampoules can be stuck to filing cards, placed in envelopes or sealed in plastic bags and then stored in drawers, cupboards or filing cabinets, etc. (see Fig. 3), at room temperature (C.M.I.), in a cold room or refrigerator at about 7°C (A.T.C.C. and N.R.R.L.) or low temperature basement at about 15°C (C.B.S.). So far most cultures have been stored in the dark.

As the ampoules are sealed there is no risk of contamination or infection with mites. The ampoules' small size makes them ideal for postage.

Lyophilization cuts down the number of transfers and aims to maintain the fungus unchanged from the original culture. However, some people (Dade, 1960) feel that the violent treatment is almost certain to be selective. Others believe the process not to be selective, normal and variant spores surviving equally well. Hesseltine supports this latter belief and Mehrotra (1967) on his instigation conducted a series of experiments by which certain strains of fungi were lyophilized, revived and re-lyophilized for many generations and showed no change in the biochemical and physiological character at the end of the process.

(c) *Nitrogen storage or storage at ultra-low temperatures.* The effect of low temperatures on fungi was considered many years ago by Buller (1912) and Lipman (1937), but it was the breakthrough by Polge, Smith and Parks (1949), in which they found it possible to cool semen protected by glycerol to ultra-low temperatures and again revive it, that made the preservation of biological materials by this means a possibility. This led to much research into the methods of freezing, especially of spermatozoa and blood (see reviews by Harris, 1954; Smith, 1961; Meryman, 1966).

Meryman (1956) cited the temperature of − 130°C as that below which no

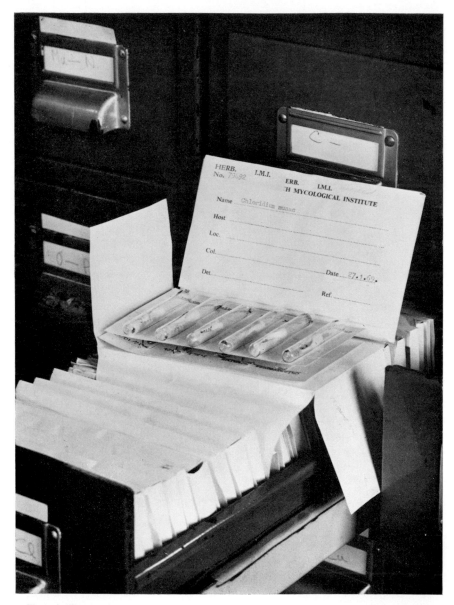

FIG. 3. Freeze-dried cultures in ampoules on cards in envelopes stored alphabetically according to genera in standard filing drawers. (Photograph by D. W. Fry.)

biochemical activity should take place, arguing that the temperature at which water mobility is too low to permit crystallization could logically be expected to be a cut-off point for biological changes as well. Meryman (1966) warned that this is not the complete situation and other factors such as radiation could affect the activity during long-term storage at low temperatures. However, assuming that metabolism is more or less at a standstill at $-130°C$ and below, fungi, which will survive cooling and subsequent thawing, should when stored at the temperature of liquid nitrogen survive more or less indefinitely. This method of storage after preliminary trials (Hwang, 1960, 1961) has been under test at the A.T.C.C. since 1961 and in routine use since 1965 (Hwang, 1966, 1968) with very satisfactory results.

Hwang protects her material with 10% glycerol and cools it slowly at a rate of about 1°C per min to a temperature of $-35°C$; thereafter the rate is uncontrolled and faster. The procedure and apparatus are fully described (Hwang, 1966). She is able to preserve mycelial forms, Saprolegniaceae, Pythiaceae, Entomophthoraceae and other fungi which would not survive the vigorous process of freeze-drying as well as the more resistant moulds.

The cultures are suspended as a spore suspension, as finely broken-up particles of mycelium, or as a piece of fungus mycelium in the protective medium. This is distributed into ampoules, which must be resistant to cold shock, and the ampoules drawn out and sealed. A check should be made for cracks or faulty sealing and this can be done by including a dye in the medium of the precooling bath.

Some strains will survive freezing and thawing without any protective medium, but revivals are better when one is used. Wellman and Walden (1964) used mini-agar-cultures alone. The protective medium is usually glycerol and this was used in the original work of Polge et al. (1949). Dimethyl sulphoxide has also been used as a protective medium (Lovelock and Bishop, 1959). Davis (1965) used it when preserving Myxomycetes. Hwang (1968) and Hwang and Howells (1968) compared the use of dimethyl sulphoxide with glycerol for the protection of some strains which had been found sensitive to freezing and obtained considerably increased viabilities. As dimethyl sulphoxide is less pleasant to handle it would appear most satisfactory to use glycerol as a suspending medium as a routine practice and only use dimethyl sulphoxide when revivals are poor. There are other protective substances available and protective substances and their mechanism are discussed by Nash (1966).

Hwang (1966) recommends precooling to 7°C before freezing is begun. The method of freezing can be by plunging the ampoules straight into the liquid, by suspending them over the liquid for a short period and then lowering into the liquid, or by controlled cooling; this latter is favoured by Hwang

(1966). The rates, effects, methods, and theory of freezing are discussed by various authors in Meryman (1966).

At the Commonwealth Mycological Institute the ampoules containing the fungus suspended in 10% glycerol are arranged on metal canes and lowered into the gas phase above the liquid nitrogen and suspended there for a period of about 20 minutes before lowering into the liquid. For general routine work reasonable revivals are obtained by this means. This method is illustrated in Fig. 4.

There are many interacting factors concerned in the choice of method of freezing and it will depend largely on the organisms to be frozen, the degree of revival required and the apparatus available. The more resistant fungi appear to survive quite well when plunged straight into the liquid nitrogen, others need more careful handling.

Techniques of thawing. Mazur (1956) working with *Aspergillus flavus* and Meryman (1966) favour rapid thawing and this is practised by Hwang (1966, 1968). The ampoules are removed from the nitrogen and immediately agitated in a water bath at 37°–40°C. They can be temporarily stored in a solid carbon-dioxide and ethanol freezing mixture if it is inconvenient to make the straight transfer. Goos *et al.* (1967) made a study of the effect of warming rates on the recovery of fungus spores frozen and stored at liquid nitrogen temperatures. Recoveries approximated those of the pre-frozen controls from both slow and rapidly frozen cultures when thawed rapidly. If thawing was slow, then there appeared to be a greater sensitivity to the method of cooling and slow cooling seemed preferable. The results suggested that in most fungi the method of freezing is not significant if warming is rapid. Loegering and Harmon (1962) reported induced dormancy in uredospores, which could be broken by heating to 40°–45°C, either during thawing or after thawing at a lower temperature.

Once the fungi are in the nitrogen they need no further attention till they are required. However, it is necessary immediately or a few days after freezing to thaw out at least one ampoule and grow cultures from it to check that the fungus has survived the freezing process.

Precautions. Cracked or faulty ampoules are dangerous, as the liquid nitrogen may penetrate and fill the ampoules with the result that at the time of thawing the ampoules may explode due to the sudden expansion of the nitrogen into gas. The storage nitrogen may also become infected. If the thawing is done in water and there is an explosion the water will tend to hold the fragments of glass, but it is recommended that gloves and an eye or face shield be worn when thawing out ampoules. Tuite (1968) sealed his cultures in polyester film instead of ampoules before freezing. This film does not shatter in the same way as glass and incidentally much more material can be stowed by this means in the same space.

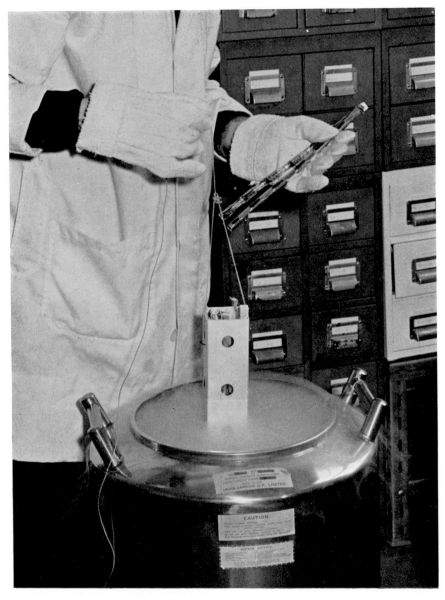

FIG. 4. Nitrogen storage. Spore suspensions in glycerol in glass ampoules held on metal canes and a metal storage canister being removed from a 36-litre liquid nitrogen refrigerator. (Photograph by D. W. Fry.)

The nitrogen itself should be handled carefully because of the risk of cold burns. The danger from this is not so great as might be expected as a gas layer tends to form between the liquid and the skin, but if it should become caught in a shoe, unpleasant burns can be received. Great care should be taken not to handle any metal parts in direct contact with the liquid nitrogen.

As the gas is odourless and colourless there is a small risk of asphyxiation when handling it in confined spaces, as build up of nitrogen in the atmosphere tends to go unnoticed. The nitrogen containers should therefore be housed in a well-ventilated room.

The advantage of this method of storage is that at least in theory no subculturing is required and living material of a type which will not normally grow in culture and would not be preserved in a culture collection can be retained in a viable state, including mycelial forms of Basidiomycetes, obligate parasites, and other fungi at present difficult to grow in culture, as for example the rusts in the Collection of Plant Rust Fungi at the A.T.C.C. The preservation of rusts by this means has been investigated by Loegering (1965), Bugbee and Kernkamp (1965), Leath *et al.* (1966), and Loegering *et al.* (1966). From the biochemical standpoint mycelium can be stored so that storage can be without change of generation and therefore without mutation. This could be of particular importance when storing highly specialized strains used in industrial processes. The ampoules are sealed and therefore not open to contamination or invasion by mites. The drawbacks to the process are probably the expense of the apparatus and the necessity for a reliable source of nitrogen.

The apparatus used in maintenance of microfungi in liquid nitrogen consists of liquid nitrogen refrigerators and containers, ampoules, canes and boxes. One of the problems is retrieval of specimens.

Liquid nitrogen refrigerators. The refrigerators used will depend on the number of fungi to be stored and the money available.

For this work it is usual to use metal containers or dewars. If ordinary glass vacuum flasks are used there is some risk of cracking the glass by cold shock when filling with liquid nitrogen, but some glass containers are available which are designed to be used at ultra-low temperatures, and these are somewhat cheaper than metal ones.

The metal containers are available for biological purposes in the U.K. from Union Carbide Ltd. (a subsidiary of Linde & Co., U.S.A.), the British Oxygen Company Ltd., and from Spembly Technical Products Ltd. The usual refrigerators used for this work are—

   (i) The 10 litre, which will hold about 250 ampoules on canes in canisters, and is suited to storage on a small scale and experimental work. It is relatively small and can be carried about, weighing about 14 lb empty and 27 lb full.

(ii) The 35 litre, which will hold about 900 ampoules in canisters, is rather heavy, and best supplied on a trolley.

(iii) Larger containers are often used. At the C.M.I. a 185 litre storage container is used as the "bank" for processed fungi and it is hoped to store in it about 12,000 ampoules or 1000 strains with 12 replicates, but this is probably very optimistic. At the A.T.C.C. containers of 250 litres and 650 litres are used, which are said to store about 14,000 and 40,000 ampoules respectively (these figures are those given by the manufacturers and as in practice it is difficult to pack the containers to capacity, perhaps an estimate of half this number should be considered). These are circular, made of stainless steel on the outside and supplied on wheeled bases for mobility, as they are very heavy and bulky. Rectangular chests may well become popular in the future as they are so much more convenient.

The larger the refrigerator the more economical it becomes, both in storage space and nitrogen consumption. The loss of nitrogen depends to a great extent on the diameter of the top opening. If a small diameter lid is used less nitrogen is required. However, with a circular container and a small lid the specimens become less accessible, so the choice is a balance between consumption and accessibility. Thus the 36 litre container which has a wide neck will hold a large number of accessible specimens, but will boil off as much nitrogen as a 650 litre container. A 36 litre refrigerator is a useful wide-necked "pot" in which to conduct experimental work before transferring to a larger storage refrigerator.

Liquid nitrogen containers. Although the refrigerators can be filled directly, it is convenient to have some stock or reserve of liquid nitrogen, so that some storage containers would be useful. The most usual of these are 25 and 35 litre gas containers. They have narrow necks, can be supplied on wheels and are relatively strong. A 160 litre liquid nitrogen container, which can be moved about on a trolley and is self pressurizing is used at the C.M.I. to overcome the various local problems. The room in which the nitrogen apparatus is housed is inaccessible to a delivery lorry and there is no lift. The container is large enough to act as a storage tank and therefore to receive nitrogen deliveries, in economical quantities, but it can still be wheeled from the delivery point to where it is required. The vessel is self-pressurizing and the refrigerators are filled by means of an insulated pipe through the window of the laboratory. Larger storage tanks are available.

Ampoules. The specimens are held in ampoules, which are resistant to cold shock. At the A.T.C.C. 1·2 ml Wheaton "Goldband" prescored borosilicate glass ampoules have been found satisfactory. Johnson and Jorgensen's 1 ml, 34/11/1504 artificial insemination (A.I.) ampoules of white neutral glass, specially developed for storage of bull semen in liquid nitrogen, are used at the C.M.I. The manufacturers warn that they must not be overfilled (less than 0·5 ml being recommended), that they must be drawn sealed, care being taken to get a perfect seal, and that they do not touch on the canes.

Canes. The ampoules are usually clipped on metal (aluminium) canes one above the other, six to a cane. The canes can be made locally or they are available from various suppliers.

Boxes. The canes are packed in metal boxes or canisters (aluminium), which hold about twenty canes. These are perforated to allow the free running of the liquid nitrogen. Square boxes can be arranged more compactly than round ones. Some form of handle simplifies retrieval.

If at all possible it is easiest and cheapest to obtain the ampoules, canes and boxes used locally by the A.I. centres as they will be buying a standard product in bulk.

Retrieval. One of the problems of nitrogen storage is finding the specimen required when it is packed tightly with many others under liquid nitrogen. It is not possible to hunt around in the liquid. Water vapour tends to form in the neck of the vessel and obscure vision. If the canisters are removed they rapidly become covered with snow, obliterating any markings. In any case they can not be withdrawn for more than a minute without the temperature rising dangerously high.

Some workers map the contents of the refrigerator, but although it may be fairly easy to reach specimens centrally placed, those at the edges take some locating and may be difficult to remove and replace.

At the C.M.I. the refrigerator is segmented. Strings can be attached to the canisters, and these hang out of the neck. These strings are used when lifting the canisters from the refrigerator and have labels attached which indicate the contents. Coloured canes and discs attached to the boxes are easily seen.

Other apparatus. Other apparatus required includes a water bath to hold at 37°C, asbestos gloves, face shields, insulated tongs and other standard laboratory equipment.

Slow coolers. These are rather expensive pieces of apparatus. At the A.T.C.C. two different makes are used, the Canaleo Slow Freeze and the Linde BF-3-2. In the U.K. controlled slow freezers are produced by Mathburn Research Ltd. and Spembley Technical Producs Ltd. If the department has a skilled workshop it might be possible to produce a home-made freezer. If a controlled freezer is used a temperature recorder is required to record the actual rate of freezing.

II. MAINTENANCE OF HERBARIUM MATERIAL
OR PRESERVATION OF DEAD MATERIAL

It is usual when describing a new fungus to deposit it in one of the main herbaria as a dead specimen on which the description is based; this is designated "the Type" of the species. Thus it is possible to refer to the actual specimens on which many old and famous species have been based. The date for the first accepted names of fungi has been laid down in the International Code of Botanical Nomenclature (Lanjouw, ed., current edition 1966). For Hyphomycetes and most other fungi this begins with Fries (1821) and many of his species are still available in herbaria all over Europe. Similarly much of the original material of Corda and Persoon is still in existence.

The deposit of fungal material in herbaria is not of course restricted to "Type" material. When any work with interesting fungi or strains of fungi is started it becomes desirable to retain a permanent record of the organism. Preserved specimens first kept as part of a small private collection can be transferred later to one of the larger herbaria.

Most specimens prepared for permanent preservation consist of dried material.

The preparation of this dried material has until recently followed the normal botanical herbarium techniques and as the substrate of leaves, twigs, fruits, etc., have been dried any microfungi growing on them have been preserved at the same time. Some specimens have from time to time been stored by placing in various preservative fluids, but this material is messy, tends to deteriorate with time, and spores, etc., often float off. Recently somewhat more sophisticated methods of drying and preservation have been devised. When storing fungi in a herbarium specimens supported by full documentation should be kept systematically according to their kind or some definite scheme.

The practice in many of the larger and older herbaria, if the material can be reduced to a reasonable size, is to follow the traditional means of storage for plant material and to place it in standard-sized packets or envelopes. These are fixed with glue to standard-sized herbarium sheets of paper ($16\frac{1}{2} \times 10\frac{1}{2}$ in. is a convenient size). Specimens of one species are collected together in a white paper "species folder" which is in its turn included with other "species folders" in a "genus folder" of stout brown paper (Anon., 1968; Ellis, 1960). The size and shape of the packets and folders varies among Herbaria. Colour variables are sometimes used to indicate anything special, for example red borders to the folders are used to indicate the presence of "Types" both at Kew and the British Museum. Once a system has been adopted it is almost essential to adhere to it, any changes in system causing enormous trouble. The Herbarium folders are usually arranged in suitable shelved cupboards. Wood cupboards used to be favoured, but metal with a dust-proof seal are more frequently used. Smaller numbers of specimens may be kept in packets in "shoe boxes" or any convenient boxes, as was the practice with many old-time mycologists, or on shelves, but they should be protected from dust. These latter methods of storage become cumbersome as the collection begins to grow. If a consistent form of material is involved as with dried cultures, standard packets and slides may be stored in normal filing cabinets. Field specimens should be trimmed to a reasonable size, small enough to store and yet large enough to show the original growth form. If it will grow in culture a dried specimen of the original isolation and a series of dried subcultures on various media is desirable. Slides, slide cultures, sections and any other relevant material should also be included. In practice the material available to a herbarium may consist of any one of these.

A. Preparation of herbarium material

1. *Dried specimens*

Except in extremely humid climates the specimen can be left to dry slowly in the air. Infected leaves may be pressed in the conventional manner between sheets of absorbent paper (special botanical drying paper, blotting paper or newspaper) in a plant press and allowed to dry. The press may be of slatted wood or metal mesh and is kept flat by weights or straps. Frequent changes of paper are required at first, and gentle heat may be helpful. Particular care must be taken in hot wet climates to prevent overgrowth with moulds (see Deighton, 1960). If there is no intention of isolating the fungi from the material at a later date, the specimens of twigs, etc., may be placed in an incubator or oven at about 50°C and dried for 2–3 days. This treatment usually prevents growth of moulds and kills insects (Sutton, personal communication). However, it usually kills the specimen as well.

2. *Dried cultures*

Cultures for drying at the Commonwealth Mycological Institute* (Anon. 1968) are grown on agar until they are in good condition. The cultures are then killed by placing a piece of filter paper soaked in formalin solution in the culture tube or dish for two days or placing the cultures over formalin in a desiccator. A smooth surface such as hardboard, ground glass or plastic laminate is flooded with 1% tapwater agar. The culture to be dried is removed from the tube or plate and placed, fungus colony uppermost, on the still semi-liquid tapwater agar (Fig. 5). Excess agar is cut away from slope cultures or other cultures on a deep or uneven layer of agar, before

* A modification of Pollack's (1967) method (see below) is now in use at the C.M.I.

Fig. 5. Preparation of dried agar cultures. Placing agar culture on tapwater agar on a drying board. (Photograph by D. W. Fry.)

placing the superficial layers with the fungus colony on the tapwater agar. The whole is protected from draughts and dust with a loose cardboard cover. The specimen is left to dry at room temperature and humidity and this usually takes about 2–3 days in Great Britain. At the C.M.I. a special cupboard is used to dry the cultures (Fig. 6), with a small through draught vented outside the building to reduce the risk of air contamination in the laboratory. Perhaps it would be ideal to dry at a controlled temperature and humidity. When dry the specimen is easily pealed off the board and is ready for storage. If buckled it can be softened and straightened by placing in a damp chamber for a few hours. The specimen is trimmed and stuck down in a box with glue or, if the back of the culture is to be examined, it is stuck at the edge to a ring of cardboard (Fig. 7) which snaps into a space in the cardboard box or similar container and is so arranged that there is plenty of free space above the colony so that any felt and fruiting structures are undamaged when closing the lid. These cultures when packaged in standard boxes can be stored in filing cabinets, and thus give a very compact form of dried reference material particularly suitable for use in a culture collection.

Various modifications of this process have been suggested. Pollack (1967) at the A.T.C.C. has considerably simplified the method. She dried her cultures in the lid of the Petri dish used to grow it. It is necessary when doing this to use plastic Petri dishes as the cultures tend to stick to glass. She also uses a tapwater agar containing 2·5% glycerol as her base and this gives a smoother, stronger and more pliable product.

Laundon (1968) uses PVA instead of tapwater agar and a Perspex acrylic tile for drying purposes, and dries in a refrigerator. Flemming and Smith (1944) recommended placing a piece of cellophane on the surface of agar and growing the fungus on this and subsequently peeling off the cellophane and drying it. This tended to buckle. Carmichael (1963) overcame this by securing the cellophane film to a specially designed Perspex drying block. This has a slight flange which allows enough give to prevent trouble from shrinkage. The blocks are designed to stack, so that quite a large number of specimens can be prepared in a relatively small space. Although this produces satisfactory specimens for storage, the cultures have to be specially grown. It is possible, however, to place agar cultures on damp stretched cellophane and dry them in this way.

3. *Lyophilization*

Freeze-drying of larger fungi is becoming quite popular and can be applied to microfungi and cultures of microfungi.

This is done by freezing in a special freezing chamber and then drying from the frozen state by evacuation. Specimens preserved in this way are

Fig. 6. Preparation of dried agar cultures. Drying cultures in a drying cupboard. This gives more controlled drying and prevents dispersal of the spores into the laboratory. (Photograph by D. W. Fry.)

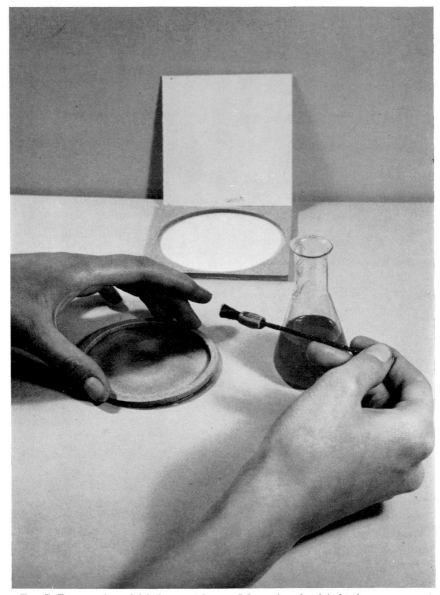

FIG. 7. Preparation of dried agar cultures. Mounting the dried culture on a cardboard ring prior to fixing in protective cardboard box. (Photograph by D. W. Fry.)

particularly fine and look very like the original, showing little loss of shape and colour (Haskins, 1960; Davies, 1962). The apparatus can be obtained as a complete commercial product or produced at less cost from parts at least some of which may already be available in the laboratory. Taylor (1968) describes such an apparatus for drying fleshy-fungi. Bebbington and Burrell (1968) coated their dried or freeze-dried specimens with clear polyurethane, which forms a hard, durable and transparent coating and does not obscure fine detail. The coated specimens can be handled freely without damage. This was done by dipping the specimens into polyurethane diluted 2 : 1 with white spirit and drying in an oven at 50°–60°C, care being taken to ensure that the specimen was completely covered.

4. *Slides*

The methods of preparing mycological slides are fully described in the chapter on technique (see Dring, this Volume, p. 96). However they are stained or mounted they must be adequately sealed and packeted for storage.

5. *Liquid methods of preservation*

The preservation of microfungi by liquid methods is not usually very successful and therefore not popular especially as most fungi are adequately preserved by drying. In the Hyphomycetes the conidia tend to wash off.

However, the methods in normal use in botanical laboratories are usually applicable. There are three principal means for this type of preservation and they and their application are fully discussed in the manuals on botanical technique such as Purvis *et al.* (1966). These consist of preservation in (*a*) alcohol solution, which can range from 70–90%; (*b*) formalin solution ranging from 2–5%, or a solution containing a combination of these with the possible addition of glacial acetic acid; (*c*) various solutions containing copper in order to maintain the green colour of plant parts. Other solutions can be used to preserve the colour in fruits and fungi. Formulae for some of these are given in Ainsworth (1961), Anon. (1968) and Purvis *et al.* (1966).

When making these preparations care should be taken to ensure that all air bubbles have been removed from the specimens before enclosing in the museum jars. The jars must either be kept topped up with solution or adequately sealed. The jars themselves must have at least one flat side to avoid a distorted image. Until recently jars were made of glass but jars made of Perspex are quite satisfactory, and Dade (1960) found Perspex micro-jars very useful.

The preservation of animal tissues containing fungal mycelium or symptoms of mycoses is a specialized procedure and particularly suited to liquid preservation. This is done according to the techniques employed in the morbid anatomy departments of the medical and veterinary laboratories.

B. The herbarium

The room or building in which the collection of dried or preserved material is housed is traditionally known as a Herbarium. A cool dry room is preferable for this purpose.

1. *Hazards and their control*

The greatest danger is from insects, mites and moulding, and of course fire and water. In most of the larger herbaria there are extensive fire precautions and care is taken to store material above any possible flood level.

(a) *Moulding*. Moulding is easily prevented by storing in a dry room. This is no great problem in temperate climates where most normal rooms are quite satisfactory. Under moist tropical conditions it may not be so easy and air conditioning must be considered, with particular control of humidity. If this is not possible great attention must be given to ventilation and airing of the cupboards as well as periodic drying of the specimens (see Deighton, 1960).

(b) *Insects and mites*. The attack of the specimens by insects is one of the greatest hazards to a dried collection. The herbarium beetle and other insects can survive in surprisingly dry conditions. If a specimen is attacked it can be completely cleared of sporing structures and in particular the conidial apparatus of Hyphomycetes erased.

Many collectors make lavish use of naphthalene, paradichlorbenzene and other insecticides and repellents, but they must be periodically renewed. Too great a concentration of these can be unpleasant for workers in the room.

Regular fumigation of the specimens with a poisonous gas is in the long run probably the most efficient method of control. The specimens are exposed to methyl bromide in a fumigation chamber for 24–48 h at the C.M.I. (Ellis, 1960). Other poisonous gases could be used.

All new material should be treated before incorporation in the herbarium and any folders or specimens exposed to outside contamination should be gassed before returning to the cupboards. An effort should be made to gas all the specimens and folders from time to time. This is easily done in a large collection by regular systematic gassing of a few folders each week.

New, damp, fresh material received in the department is best unwrapped and examined quite separately and then dried and gassed before bringing into the herbarium.

If wooden cupboards are used, they should be well dried and treated with insecticides before use as new wood tends to be damp and may carry insects.

C. Arrangement of specimens

The arrangement of the specimens is a matter of personal preference, but it is necessary to have some definite system and to adhere to it strictly. At the C.M.I. the arrangement of the folders is according to the systematic groups and then alphabetically according to genera and species. At the Royal Botanic Gardens, Kew, the fungi are mostly arranged according to Saccardo's Sylloge and then alphabetically. Some herbaria keep the collections of individual workers separately.

Herbarium cupboards in use at the C.M.I. are shown in Fig. 8.

III. DOCUMENTATION

The actual means of keeping records of the specimens stored in an herbarium or culture collection is again a matter of personal preference, needs and resources available. The information can be stored in books, looseleaf books, files, cards, visible indexes or punch cards or by any other means of data processing, including computers, but whatever the means of documentation certain categories of information will be required sooner or later. It is easiest to record this information at the time of receipt of the specimens and to have it well cross referenced.

A. Accession numbers

When dealing with several strains of the same species or with any material it is useful to be able to designate it accurately. Thus the giving of an individual number to each specimen at the time of receipt, coupled in the record book with certain descriptive information such as the name at the time of receipt, the name of the sender with his number for it, the place of isolation and the host has great advantages. This accession number (Ellis, 1960) refers to the one specimen, however it may be named, renamed and moved from one species or genus and hence folder to another. If, as at the C.M.I. there is a culture collection as well as a herbarium it is convenient to have only one system of numbering and specimens and subcultures are retained under the same number in both collections.

B. Information recorded

As much information as possible should be recorded about each specimen and this should include—

(i) The name of the person who collected the organism and his number for it.

(ii) The date of collection.

(iii) Where it was collected.

IV   7

Fɪɢ. 8. Herbarium cupboards with the list of genera contained in them shown on the outside and a herbarium folder laid open to display the packets containing the specimens attached to the herbarium sheets. (Photograph by D. W. Fry.)

(iv) The substrate or host from which it was collected or isolated.

(v) Any hands through which it has passed, including the numbers under which it has been held.

(vi) The name of the person who identified it, if this is different from the collector.

(vii) The date on which it was received.

(viii) Any additional information available, such as any papers in which it is described or mentioned (if it is a "Type" of a species or has been identified by the author of the species this should be particularly noted), peculiarities or activities such as the production of interesting biochemical compounds or enzymes; or its use in any industrial process or genetical studies.

In a herbarium this full information can go on the packet or card containing the organism. In a culture collection this is not possible and it must be kept on a separate sheet, card or "history sheet". Cultures also require another record sheet for each strain, on which all cultural treatments of the fungus while in the collection are recorded and notes on any special requirements for maintenance are listed.

C. Indexes

The fungi are best indexed in at least three ways: (*a*) by the name assigned to the specimen or culture; (*b*) by the host or substratum; (*c*) by taxonomic groups.

1. *Index of generic and specific names*

This can be a straight index of names of genera, with species arranged alphabetically within the genera. Under each name the specimens deposited are listed by their accession numbers. Additional information can be included here such as the name of the host or substrate and place of origin. Provision must be made for fungi not fully identified. As in an active herbarium there are frequent transfers of specimens from one species or genus folder to another, there must be provision for full cross reference and record of such movement, whether it is the result of taxonomic revision or change of opinion. This is done at the C.M.I. by the inclusion of a section "Misdet. and Syn." at the end of each genus in the fungus book, and in this such transfers and their destination are listed.

In addition to the name index an index of all specific epithets in which the history of a fungus name is recorded with references to the relevant literature is kept at the C.M.I. A simple index to the literature is useful.

When a culture is first received in a culture collection information concerning its cultural characters should be recorded. This should include a full description and photographs of the colour, texture, type of growth and the degree and type of sporing. The inclusion of photographs, drawings, and measurements of spores and spore-producing apparatus is desirable and can be supported by a slide collection and dried cultures.

This is most important when dealing with living material as a check on variation and contamination.

If time and opportunity allow, similar records could be kept of dried herbarium material. If, however, this has been adequately prepared it should remain unchanged, so there is not such an urgent need to record its condition, as it can be examined by specialists when convenient.

2. *Host index*

An alphabetical index of the host or substratum often proves very useful. This is easy to keep if it is done as part of the system of incorporating the fungus strains into the collection.

3. *Index to taxonomic groups*

It is convenient to be able to find what material is available according to taxonomic groups. At the C.M.I. this is done by the arrangement of the specimens in the herbarium, but other documentation can be used. If the putting away is done by non-specialists and for the benefit of visitors it is helpful to have a key to the species according to their taxonomic groups and a map as to where to locate them.

D. Computers

It seems probable that with the upsurge in the use of computers and other data-processing machines the arrangement, organization and documentation of the stored material will rapidly assume a new look. At present by the use of cards, loose-leaf books, etc., much data is available and with the expenditure of considerable labour catalogues of the species held, host lists, and lists of species from one area can be and are produced. However, assuming that the information is fully processed at the time of accession these lists could be produced at the "press of a button". This topic was one of the subjects discussed at the International Conference on Culture Collections in Tokyo, in 1968.

There are at present several moves to produce regional and international lists of cultures available. It seems highly probable that this data will be stored in some form of computer, and will perforce be centred near an available computer rather than in conjunction with a culture collection or herbarium. Some of the first compilations of this type, though not produced by computer, are the "Directories of Collections and Lists of species maintained in various Commonwealth Countries" produced by the permanent Committee of the British Commonwealth Collections of Micro-organisms in London; the first appeared in 1960. The future availability of information as to the location of different strains of microfungi on an international scale will no doubt prove of great help to all interested in fungal culture. Such a scheme could be later applied to dried herbarium material.

However well the specimens of microfungi dead or alive may be listed and documented the problems of keeping them in good condition and overcoming the hazards of biological attack, deterioration and variation will still remain.

REFERENCES

Ainsworth, G. C. (1961). *In* "Ainsworth and Bisby's Dictionary of the fungi", p. 245. Commonwealth Mycological Institute, Kew.

Ainsworth, G. C. (1962). *Nature, Lond.*, **195**, 4846, 1120–1121.

Anon. (1968). "Plant Pathologists Pocketbook", pp. 169–170, 247–248. Commonwealth Mycological Institute, Kew.

Bakerspigel, A., (1953). *Mycologia*, **45**, 596–604.

Bakerspigel, A. (1954). *Mycologia*, **46**, 680–681.

Barmenkov, A. S. (1959). *Mikrobiologiya*, **28**, 444–446.

Barratt, R. W., Johnson, G. B., and Ogata, W. N. (1965). *Genetics, Princeton*, **52**, 233–246.

Bazzigher, G. (1962). *Phytopath, Z.*, **45**, 53–56.

Bebbington, R. M., and Burrell, M. M. (1968). *Bull. Br. mycol. Soc.*, **2**, 75.

Bisby, G. R. (1945). *Nature, Lond.*, **155**, 732–733.

Blackwood, A. C. (1955). *In* "Symposium on the maintenance of cultures of micro-organisms". *Bact. Rev.*, **19**, 280–283.

Braverman, S. W., and Crosier, W. F. (1966). *Pl. Dis. Reptr.*, **50**, 321–323.

Bugbee, W. M., and Kernkamp, M. F. (1965). *Phytopathology*, **55**, 1052.

Buell, C. B., and Weston, W. H. (1947). *Am. J. Bot.*, **34**, 555–561.

Buller, A. H. R. (1912). *Trans. Br. mycol. Soc.*, **4**, 106–112.

Carmichael, J. W. (1956). *Mycologia*, **48**, 378–381.

Carmichael, J. W. (1962). *Mycologia*, **54**, 432–436.

Carmichael, J. W. (1963). *Mycologia*, **55**, 283–288.

Clark, W. A., and Loegering, W. Q. (1967). *A. Rev. Phytopath.*, **5**, 319–342.

Cooney, D. G., and Emerson, R. (1964). "Thermophilic Fungi". Freeman & Co., San Francisco and London.

Cormack, C. W. (1951). *Can. J. Bot.*, **29**, 32–45.

Dade, H. A. (1960). *In* "Herb", I.M.I. Handbook, pp. 40–69. Commonwealth Mycological Institute, Kew.

Davies, D. A. L. (1962). *Trans. Br. mycol. Soc.*, **45**, 424–428.

Davies, E. E. (1965). *Mycologia*, **57**, 986–989.

Deighton, C. T. (1960). *In* "Herb". I.M.I. Handbook, pp. 78–83. Commonwealth Mycological Institute, Kew.

Dick, M. W. (1965). *Mycologia*, **57**, 828–830.

Edwards, G. A., Buell, C., and Weston, W. H. (1947). *Am. J. Bot.*, **34**, 551–555.

Ellis, J. J., and Roberson, Jane A. (1968). *Mycologia*, **60**, 399–405.

Ellis, M. B. (1960). *In* "Herb". I.M.I. Handbook, pp. 24–36. Commonwealth Mycological Institute, Kew.
Fennell, D. I. (1960). *Bot. Rev.*, **26**, 79–141.
Fennell, D. I., Raper, K. B., and Flickinger, M. H. (1950). *Mycologia*, **42**, 135–147.
Fleming, A., and Smith, G. (1944). *Trans. Br. mycol. Soc.*, **27**, 13–19.
Greene, H. C., and Fred, E. B. (1934). *Ind. Eng. Chem.*, **26**, 1297–1298.
Goldie-Smith, E. K. (1956). *J. Elisha Mitchell Scient. Soc.*, **72**, 158–166.
Goos, R. D., Davis, E. E., and Butterfield, W. (1967). *Mycologia*, **59**, 58–66.
Gordon, W. L. (1952). *Can. J. Bot.*, **30**, 209–251.
Hamilton, J. M., and Weaver, L. O. (1943). *Phytopathology*, **33**, 612–613.
Harris, R. J. C., ed. (1954). "Biological Applications of Freezing and Drying". Academic Press, New York.
Haskins, R. H. (1960). *Mycologia*, **52**, 161–164.
Hesseltine, C. W., Bradle, B. J., and Benjamin, C. R. (1960). *Mycologia*, **52**, 762–774.
Hwang, S. (1960). *Mycologia*, **52**, 527–529.
Hwang, S. W. (1961). *Mycologia*, **52**, 527–529.
Hwang, S. W. (1966). *Appl. Microbiol.*, **14**, 784–788.
Hwang, S. (1968). *Mycologia*, **60**, 613–621.
Hwang, S., and Howells, Ann (1968). *Mycologia*, **60**, 622–626.
Kramer, C. L., and Mix, A. J. (1957). *Trans. Kans. Acad. Sci.*, **60**, 58–64.
Langeron, M., and Vanbreuseghem, R. (1952). *In* "Précis de Mycologie", p. 408. Masson et Cie, Paris.
Lanjouw, J., ed. (1966). "International Code of Botanical Nomenclature", Utrecht, Netherlands.
Laundon, G. F. (1968). *Trans. Br. mycol. Soc.*, **51**, 603–604.
Last, F. T., Price, D., Dye, D. W., and Hay, E. M. (1969). *Trans. Br. mycol. Soc.*, **53**, 328–330.
Leath, K. J., Romig, R. W., and Rowell, J. B. (1966). *Phytopathology*, **56**, 570.
Lipman, C. B. (1937). *Bull. Torrey bot. Club*, **64**, 537–546.
Little, G. N., and Gordon, M. A. (1967). *Mycologia*, **59**, 733–736.
Loegering, W. Q., and Harmon, D. L. (1962). *Pl. Dis. Reptr.*, **46**, 299–302.
Loegering, W. Q., Mekinney, H. W., Harmon, D. L., and Clark, W. A. (1961). *Pl. Dis. Reptr.*, **45**, 384–385.
Loegering, W. Q., (1965). *Phytopathology*, **55**, 247.
Loegering, W. Q., Harmon, D. L., and Clark, W. A. (1966). *Pl. Dis. Reptr.*, **50**, 502–506.
Lovelock, J. E., and Bishop, M. W. H. (1959). *Nature, Lond.*, **183**, 1349–1395.
Mazur, P. (1956). *J. gen. Physiol.*, **39**, 869–888.
McCrea, A. (1923). *Science*, **58**, 426.
McCrea, A. (1931). *Mich. Acad. Sci. Proc.*, **13**, 165–167.
Mehrotra, B. S. (1967). *In* "Studies on survival and possible genetic change in industrially useful microorganisms subjected to lyophilization". Final Technical Report. PL-480. Project FG-In-122. University of Allahabad.
Mehrotra, B. S., and Hesseltine, C. W. (1958). *Appl. Microbiol.*, **6**, 179–183.
Meryman, H. T. (1956). *Science*, **124**, 515–521.
Meryman, H. T. (1966). *In* "Cryobiology", pp. 2–114. Academic Press, London and New York.
Meyer, E. (1955). *Mycologia*, **47**, 664–668.
Nash, T. (1966). *In* "Cryobiology", pp. 179–213. Academic Press, London and New York.

Norris, D. (1944). *J. Aust. Inst. agric. Sci.*, **10**, 77.

Ogata, W. N. (1962). *Neurospora News Letter*, **1**, 13.

Parks, A. S. (1957). *Proc. R. Soc.*, *B.*, **147**, 423–557.

Perkins, D. D. (1962). *Can. J. Microbiol.*, **8**, 591–594.

Polge, C., Smith, A. V., and Parks, A. S. (1949). *Nature, Lond.*, **164**, 666.

Pollock, Flora, G. (1967). *Mycologia*, **59**, 541–544.

Purvis, M. J., Collier, D. C., and Walls, D. (1966). "Laboratory techniques in botany", 2nd. ed., pp. 209–215. Butterworths, London.

Raper, K. B., and Alexander, D. F. (1945). *Mycologia*, **37**, 499–525.

Raper, K. B., and Thom, C. (1949). "A Manual of the Penicillia", pp. 64–65. Bailiere, Tindall and Cox, London.

Rhodes, M. (1950). *Trans. Br. mycol. Soc.*, **33**, 35–39.

Reischer, H. S. (1949). *Mycologia*, **41**, 177–179.

St. John-Brookes, R., and Rhodes, M. (1936). Rep. Proc. II In. Cong. Microbiol., p. 43.

Schneider, C. L. (1957). *Phytopathology*, **47**, 453–454.

Schulze, K. L. (1951). *Brauweissenschaft*, **1951**, 161–165.

Sherf, A. F. (1943). *Phytopathology*, **33**, 330–332.

Simmons, E. G. (1963). *In* "Culture Collections. Perspectives and Problems", pp. 100–110. Toronto Press, Toronto.

Smith, G. (1969). *In* "An Introduction to Industrial Mycology", pp. 233–234. Edward Arnold Ltd., London.

Smith, A. U. (1961). "Biological Effects of Freezing and Supercooling". E. Arnold, London.

Snyder, W. C., and Hansen, H. N. (1946). *Mycologia*, **38**, 455–462.

Snyder, W. C., and Hansen, H. N. (1947). *Phytopathology*, **37**, 420–421.

Stebbins, M. E., and Robbins, W. J. (1949). *Mycologia*, **41**, 632–636.

Taylor, L. D. (1968). *Trans. Br. mycol. Soc.*, **51**, 600–603.

Thom, C. (1930). *In* "The Penicillia", pp. 42–43. Bailiere, Tindall and Cox, London.

Tuite, J. (1968). *Mycologia*, **60**, 591–594.

Van Beverwijk, A. L. (1959). *Antonie van Leeuwenhoek*, **25**, 1–20.

Webster, R. E., Drechsler, C., and Jorgensen, H. (1958). *Pl. Dis. Reptr.*, **42**, 233–234.

Weiss, F. A., and Oteifa, B. A. (1953). *Phytopathology*, **43**, 407.

Wellman, A. M., and Walden, D. B. (1964). *Can. J. Microbiol.*, **10**, 585–593.

Wernham, C. C. (1946). *Mycologia*, **38**, 691–692.

Wernham, C. C., and Miller, J. J. (1948). *Phytopathology*, **38**, 932–934.

Westerdijk, J. (1947). *Antonie van Leeuwenhoek*, **12**, 222–231.

CHAPTER V

Isolation, Purification and Maintenance of Yeasts

F. W. Beech and R. R. Davenport

University of Bristol, Research Station, Long Ashton, Bristol, England

I. INTRODUCTION

The objective of any isolation programme should be defined clearly before any attempt is made to examine the habitat. Whether it is to isolate all the micro-organisms present therein, the yeasts alone, or only those yeasts possessing a particular characteristic, determines the technique to be chosen. This technique should be such that the biochemical, nutritional and physical properties of the isolates are not changed during the process of isolation. Further, the pure cultures produced should be grown and stored without changing their essential properties.

The techniques for isolating and storing yeasts are, in general, similar to those described for bacteria but references have been chosen that illustrate the special requirements of yeasts. It has been assumed that the reader will be familiar with the methods normally employed for taking sterile samples.

II. SAMPLING FROM THE HABITAT

A. Techniques

It is doubtful whether any sampling method yields an exact measure of the total yeast flora of a habitat; a possible exception is the membrane filtration of liquids or dilute suspensions. Often, all that can be measured is the number of yeasts removed from a sample. This can be satisfactory if the results on replicate samples are statistically significant (Gibbs and Stuttard, 1967). Enrichment techniques, often necessary when a yeast forms only a minute proportion of the total flora, make enumeration impossible, since growth of the major components is precluded.

It is important that any sampling programme should be preceded by a survey of the habitat, when any direct observations are compared with results from several sampling methods. (Direct microscopic observation of the organism in infected tissue is probably the only certain way of determining whether a particular yeast has any pathological significance—Buckley, 1967.) Replicates of dilutions should be plated on a wide range of media and incubated at several temperatures. A rational programme can then be developed. The dilutions prepared during the survey should be examined for yeast inhibitors, whether derived from the raw material (Lüthi, 1958) or applied prior to sampling, e.g., preservatives or fungicides (von Schelhorn, 1958; Sudario, 1958; Minárik and Rágala, 1966). At all times the sample should be chilled as soon as it is removed from the habitat or the numbers of yeasts will change (Phaff et al., 1966). In the extreme case of polar soils, di Menna (1960) kept the samples deep frozen in transit and all laboratory manipulations were carried out at 4°C.

1. Quantitative

(a) *Solids*. In the following methods, known weights of sample are examined for their total or internal yeast flora.

Total flora. Mechanical maceration, using a Waring blender, or any similar machine equipped with autoclavable jars, offers the best method of liquifying samples of reasonable texture (Clark *et al.*, 1954). Yeast growth during comminution can be prevented by adding an equal weight of chilled sterile water to the sample in a chilled sterile blendor jar (Bowen and Beech, 1964). The jar and contents should be re-chilled if a slurry is not formed in 3 min. The first ten-fold dilution is similarly blended; further dilutions can be prepared normally.

Hand-pulping of sliced tissues was used by Marshall and Walkley (1951) but it is difficult to obtain a uniform suspension by this method. Very small objects, e.g., *Drosophila* spp. can be squashed directly on to an agar medium and spread with a moistened wire loop: the number and types of yeast present need to be limited if this is to be successful.

Menzies (1960) proposed a novel gradient elution method for soils and, presumably, any fine particulate matter. The magnetically-agitated sample is diluted logarithmically with flowing sterile water, so that plates can be prepared at any calculated dilution.

Internal flora. Normally the sample is sterilized externally, and the surface layers removed aseptically to expose the internal organs (Davenport, 1967). Van Uden and Carmo Sousa (1957, 1962a) examined the intestinal yeast flora of swine and bovines by alcohol-swabbing the required parts of the intestines of freshly opened carcasses and removing sections or internal contents. Dissected samples can be macerated or squashed and diluted with water before plating, as described above. It is not often appreciated that fruits can have an internal yeast flora (Beech, 1957).

External flora. The number of yeasts on a surface is expressed in terms of unit area: replicate samples are made with a template or a device with a known sampling area (Reuter, 1963). Areas of round objects, e.g., apples, can be calculated from their diameters in two planes (Marshall and Walkley, 1951). Only the Keratotome ensures that all the organisms on a known area of skin are removed (Castroviejo, 1959; Blank et al., 1961); all other methods remove but a portion. Gibbs and Stuttard (1967) considered that repeated self-washing with sterile water was suitable for human skin; Clark's method (1965a and b) can be used for other surfaces. Swabbing brings in the extra problem of removing organisms from the swabs (Walters, 1967), even using alginate wool with Calgon as a solvent. Flat surfaces can be sampled very conveniently with the malt agar sausage of ten Cate (1965), which is particularly suitable for control of food plant hygiene (Greig, 1966; Beech, 1967). Irregular surfaces can be examined with malleable pads (Holt, 1966), velvet pads (Gentles, 1956) and agar surfaces on gauze supports (Foster, 1960). Similarly the Sellotape technique of Endo (1966) is very convenient, particularly with the tape impressator of Woodworth and Newgard (1963), which applies a known area of tape with a standardized pressure. Two modifications have been used extensively in this laboratory (Davenport, 1967) to examine vegetative parts of apple trees. A portion of tape can be placed, impressed side downwards, on a thin block of agar and examined by direct microscopy. Alternatively, known areas of tape can be punched out, pressed at random on the surface under examination and covered with a plug of agar medium. Incubation in a moistened Petri dish allows viable cells to develop so that they can be identified more readily. Permanent preparations can be made by treatment with lactophenol and sealing with nail varnish.

Shaking samples with water has been used extensively. Crosse (1959) shook numerous sub-samples for 2 min. every 30 min. for 4 hours but, as shown in Table I, this can lead to anomalous results.

Di Menna (1957) shook soils with water for 15 min. without any increase in numbers.

(b) *Liquids.* Measured volumes of liquid are examined either directly or after reduction of solids to dilute slurries. Standard serial dilutions can be

TABLE I

Yeast counts per g of macerated and shaken apple leaves

	Yeasts and moulds	Bacteria
Macerated leaves	238	231
Leaves shaken for 5 min	413	121
4 h	5630	< 2000

TABLE II

The effect of centrifuge speed on the separation of yeasts and bacteria in apple juice

| | Yeasts in supernatant liquid | |
Speed (rpm)	Numbers/ml	% of original count
0	806,000	100
500	610,000	76
1000	380,000	47
2200	25,000	3

Speed applied for 4 min.
After Millis (1951).

prepared (Report, 1956) and either spread on agar media or spotted using the technique of Miles and Misra (1938) or Reed and Reed (1948): the spotting method is suitable mainly for restricted microfloras. Comparisons of the two methods have been given by Pikulska (1953) and Zellner *et al.* (1963). Minute numbers of yeasts in liquid samples can be concentrated to some extent by centrifugation, usually 15 min. at 2000 rpm. Suggestions that yeasts can be separated from bacteria by differential centrifugation have rarely proved successful under practical conditions (Table II).

The membrane filter is the obvious method of choice for examining liquid samples: there is an extensive bibliography on this subject (e.g., Probst, 1955; Haas, 1956; Sykes and Hooper, 1959; Millipore, 1965; Beech, 1967) and the technique has been developed extensively for a wide range of products—as seen in the technical literature of the manufacturing companies (Millipore (UK) Ltd, Oxoid Ltd, Sartorius Membranfilter GmbH). The method is particularly valuable when the liquid carrier contains a preservative or traces of a fungicide such as Captan. Two membranes have been proposed in such cases (Lagodsky, 1960), the upper for subsequent growth of the organisms deposited thereon and the lower for checking that all the inhibitor has been removed by the washing procedure. Otherwise, as found by Hislop (1967) for washings of Captan-sprayed leaves, normal dilutions down to 10^{-3} showed the usual tenfold reduction of the flora, whereas the plate of the 10^{-4} dilution was covered with a pure culture of a yeast absent from all plates of lesser dilutions.

(c) *Gases*. The assessment of the number and type of organism in known volumes of air is also the subject of a voluminous literature (Goetz, 1955; Wolf *et al.*, 1959; Gregory, 1961; Zhukova, 1962). The Perkins Rotobar or Rotorod sampler (Davies, this Volume p. 367) (Asai, 1960; Carter, 1961) has proved valuable in this laboratory as a portable device for examining the aerial flora of apple orchards and cider factories. Membrane filtration of media, through which known volumes of air have been bubbled, is becoming more widely used. The organisms can be removed from the membrane by mechanical agitation (Miller, 1963) or by ultrasonic vibration for 2 min. in a liquid, or by placing the membrane on an agar medium or on a pad soaked in nutrient medium as usual. Electrostatic precipitators, the Manning slit sampler (di Menna, 1955) and direct collection on plates of agar media (Hasegawa, 1959; Adams, 1963) have also been used for surveying airborne yeasts. No one method is completely satisfactory.

2. *Qualitative*

Techniques to encourage the growth of yeasts forming only a minute proportion of the flora are usually designed for isolating fermenting yeasts, since their presence can be detected by the formation of gas bubbles in a liquid medium. Mrak and McClung (1940) and Domercq (1956) allowed pulped fruit to ferment either alone or with the addition of sterile juice or nutrient media (Hansen, 1881). Van Zyl and du Plessis (1961) placed whole grapes in sterile grape juice to encourage the growth of yeasts on the surface. Hesseltine *et al.* (1952) used a similar technique with aureomycin in the medium to demonstrate the presence of yeasts in soil. Similarly, investigations on yeasts in the nectar of flowers have usually included a pre-incubation of the florets in a moist chamber (Zinkernagel, 1929; Niethammer, 1942) or in liquid media (Lockhead and Heron, 1929). Phaff *et al.* (1966) have pointed out the dangers of attempting to define the major yeasts in the flora with any experiments of this type.

Ellison and Doran (1961) have used an interesting method for isolating non-flocculent yeasts from a mass of flocculent brewers' yeast. The yeast is incubated in an inclined aspirator: non-flocculent strains are isolated from any haze forming in the supernatant beer. They claimed to be able to detect this type of yeast when diluted to a ratio of 1 cell in 16×10^6 culture yeast cells.

B. Culture media

The media used for isolating yeasts can be considered under several headings—

Standard   Differential   Selective

The first should allow the growth of virtually every yeast. Differential media are needed if the dilution to be plated is excessively contaminated with moulds or bacteria. Selective media allow the growth of a very restricted number of species or allow their presence to be deomonstrated on the plate, either by colour formation or by physical changes in the agar.

1. Standard media

The primary isolation medium should contain the same sugars as are in the substrate (Kamiński, 1958) or glucose alone (Scheda and Yarrow, 1966); a mixture of amino-acids and simple peptides, possibly supplemented with ammonium salts (but not relying on NH_4^+ alone—Thorne, 1945, 1950; Sims and Folkes, 1964); all the major B Group vitamins and a balance of inorganic salts, particularly potassium, magnesium, phosphate and sulphate (Joslyn, 1951; Morris, 1958). Yeasts vary in their nutritional requirements, even closely related species within a genus (Takahashi, 1954; Brady, 1965), and a natural medium is more likely to supply these until the special requirements of the purified cultures are known. Thus, wort of a standard gravity from the brewery (10° Balling, sp. gr. 1·040, Lund, 1956) with the addition of 2% agar is used widely in the brewing industry. Wickerham (1951) dissolved 20 g powdered malt extract in 400 ml hot distilled water containing 12 g agar and sterilized at 120°C for 15 min. A clear liquid extract or solid agar can be obtained by removal of heat-labile or chill haze proteins prior to the final sterilization (Walters, 1943). 200 g spray-dried malt extract is dissolved in 1 litre tap water, the pH adjusted to 5·4 and the mixture autoclaved at 120°C for 15 min. The extract is then boiled gently under reflux for 1 hour, cooled and chilled in a refrigerator, preferably overnight. The hazy liquid can be clarified with kieselguhr and a Seitz K5 or Carlson-Ford No. 5 filter pad. A fibre glass and/or coarse grade membrane can be substituted for the filter pad and obviates the necessity for discarding the first part of the filtrate from the latter. The clarified extract is adjusted to pH 5·4 and specific gravity 1·060 for liquid media and 1·040 (plus 4% agar) for solid. Both are then dispensed and steam-sterilized for 30 min on three successive days.

Glucose yeast extract agar (Capriotti, 1955) was used for many years but has been replaced very largely by media based on glucose and peptone (di Menna, 1960), which di Menna (1957) found superior to soil agar for isolating yeasts from soil. Sometimes glucose-peptone is supplemented with yeast extract (van Uden and Carmo Sousa, 1957) beef extract and grape juice (Adams, 1964).

The choice of a standard medium should not be haphazard and, where possible, the medium should be related to the composition of the habitat. The medium should remain unchanged for a reasonable period of storage and an agar made from it should set firmly. A detailed account of the development of one such medium for investigating the yeast flora of apple orchards, fruit, juice and fermented cider will perhaps illustrate these points. Marshall and Walkley (1951) diluted apple juice to 1–2% sugar content and added 0·2% ammonium sulphate and 2% agar. The agar was often sloppy after sterilizing, due to the low pH of the juice, virtually unchanged by dilution. It must have been deficient in vitamins and amino-acids since neither these authors nor Clark *et al.* (1954) isolated *Kloeckera apiculata*. Williams *et al.* (1956), who also used the same medium, isolated *Kloeckera* spp. only 3 times out of 222 isolates. Millis (1951) used natural strength, depectinized, cider apple juice, fortified with 1% (Difco) yeast extract, adjusted to pH 4·5, solidified with 3% agar and sterilized at 115°C for 15 min. The liquid form of this medium soon became cloudy because of the precipitation of tannin components from the apple juice (Millis, 1956). Carr (1956) and Beech (1957) overcame these problems by using depectinized juice of the low tannin variety Bramley's Seedling, later Cox's Orange Pippin, both of which have much higher nitrogen contents (20–30 mg N/100 ml of juice compared with an average of 5 mg N/100 for most cider apple juices). For use with yeasts the juice was fortified with 0·001% thiamin, pH adjusted to 4·8, solidified with 3% Ionagar and sterilized at 110°C for 15 min, to improve the setting power of the agar. The liquid medium remains bright for at least 2 months when kept at 5°C.

This type of medium must be available all the year round, whereas the raw material, apple juice, is available for only part of the year. The juice can be stored after depectinizing by filling into winchesters, covering the surface with ½ in. layer of medicinal grade liquid paraffin, sealing and pasteurizing the submerged containers at 65°C for 30 min: the jars are stored at 0°C until required. Alternatively the pectin-free juice can be concentrated under vacuum to sp. gr. 1·330 and the concentrate stored in sealed bottles at 5°C. More commonly, the enzymed and clarified juice is deep frozen in plastic containers and thawed out when required. For laboratories without hydraulic presses, the juice can be extracted by keeping sliced fruit stirred at 40°C in the presence of the enzyme Rohament P (Rohm and Hass GmbH) and a commercial depectinizing enzyme. After 2 hours the juice can be squeezed out through muslin.

2. Differential media

It is almost standard practice to reduce the pH of the medium to restrict the growth of all but acid-tolerant bacteria. Most yeasts will grow down to pH 2·5 except for *Schizosaccharomyces pombe* whose lower growth limit occurs at 5·45 (Battley and Bartlett, 1966a). Difficulties occur in preserving the set of the agar at very low pH levels so that the level is adjusted to 4·0 by the addition of a calculated amount of HCl between autoclaving and pouring the plates. Alternatively, adjustment to 4·8 when making up the medium and the use of 3% high strength agar removes the need for the second addition (Carr, 1956; Beech, 1957).

Mould growth can be restricted on the plates, without restricting the growth of yeasts, by the addition of 0·25% sodium proprionate (Lund, 1956) or 0·025% of the calcium salt (Bowen, 1962; Bowen and Beech, 1967). The lower the pH, the more undissociated acid in the medium and the greater the fungistatic action of the propionate: the pH of Lund's medium was 5·5, Bowen's 4·8. Diphenyl (Hertz and Levine, 1942; Beech and Carr, 1955) is not as effective in restricting the development of fungal hyphae. Adams (1960) followed Miller and Webb (1954) by incorporating 0·003% rose bengal in the medium to inhibit bacteria and to limit but not prevent growth of moulds. Similarly, Burman (1965) used 0·7% rose bengal and 100 mg/ml kanamycin (see also Martin, 1950).

Further inhibitors are necessary when acid-tolerant bacteria are present in large numbers or are likely to inhibit yeast growth (Suzuki, 1956; Gilliland and Lacey, 1966; Motoc and Dimitriu, 1966). Beech and Carr (1955, 1960) tested the effects of a large number of compounds on yeast and bacteria. They recommended the addition of 2 p.p.m. actinomycin and 50 p.p.m. aureomycin for inhibition of acid-tolerant bacteria. This mixture has proved very satisfactory for the last 10 years in our laboratory. Other combinations of antibiotics can be used when the bacterial flora is more restricted. Van Uden and Carmo Sousa (1957, 1962) incorporated 60 units/ml penicillin and 100 units/ml streptomycin in their medium. Buckley *et al.* (1969) used glucose peptone agar containing 20 units/ml penicillin and 40 units/ml streptomycin to isolate yeasts suspected as causative agents in human and animal mycoses (*Candida albicans*, *C. tropicalis*, *C. parapsilosis* and *Torulopsis glabrata*). Tubes were inoculated in duplicate, with and without the addition of cycloheximide. This antibiotic can be of assistance in isolating many strains of these yeasts, but some are sensitive (Negroni and Daglio, 1962), hence the need for duplicate tubes. Ross and Morris (1965) followed Fell *et al.* (1960) and used a mixture of chloramphenicol, streptomycin and chlortetracyline. Richards and Elliott (1966) suggested that the routine use of 40 p.p.m. streptomycin in any medium for isolating yeasts should be reconsidered since some species can be inhibited.

Motoc and Dimitriu (1966) showed that this antibiotic inhibited and growth of *Saccharomyces cerevisiae* var. *ellipsoideus* but not its alcohol production.

Differential inhibitors can also be used to restrict the yeast flora to certain species (Green, 1955). Van der Walt and van Kerken (1961) incorporated aureomycin, chloromycetin and actidione in their medium to inhibit bacteria and strong fermenting yeasts, allowing the isolation of *Brettanomyces* spp. A high concentration of vitamins, particularly biotin and thiamin, was necessary to encourage rapid growth of this group of yeasts (van der Walt and van Kirken, 1960). Beech and Carr (1955) found that *S. cerevisiae* var. *ellipsoideus* was generally more resistant to inhibitors than its parent species. In addition to *Brettanomyces bruxellensis*, they found that *S. fragilis*, *Hanseniaspora valbyensis*, *Lipomyces starkeyi*, *K. apiculata*, *Trigonopsis variabilis* and *Rhodotorula glutinis* resisted 500 p.p.m. actidione (see also Negroni and Daglio, 1962). It is more difficult to isolate *Saccharomyces* spp. only from a mixture with yeasts of other genera. Sugama (1966) proposed the addition of 2·5% ethyl acetate to a defined medium, adjusted to pH 4·0 with acetic acid and held in a hermetically sealed Petri dish: the first colonies to appear were said to be *S. cerevisiae* (see also Silhánkova, 1963).

3. *Selective media*

The simplest method of demonstrating the presence of a particular yeast is to induce it to form a specific colour on solid media. Thus, the addition of 0·5% ferric ammonium citrate to MYPG agar (Wickerham, 1951), apple juice/yeast extract agar (Beech, 1957) or to van der Walt's medium (1952), causes colonies of *Candida pulcherrima* to assume shades of maroon varying from pale to intense with a metallic iridescence. Colonies of yeasts once described by Wickerham (1955) as *Dekkeromyces*—but not validly published—also produce this pigment, but have a much more restricted distribution in nature. The medium should not be deficient in biotin as certain yeasts can also produce pulcherrimin under these conditions (Cutts and Rainbow, 1950; Chamberlain *et al.*, 1952; van der Walt, 1952). *Sporobolomyces* and *Rhodotorula* spp. forming pink colonies can be distinguished from *C. pulcherrima* by growing all pigmented isolates in synthetic media (Difco yeast nitrogen base plus 2% glucose, Wickerham, 1951). Colony pigment soluble in organic solvents is carotenoid, while pulcherrimin is soluble in ethanolic potash.

There are many selective media based on single carbon (Skinner and Bouthilet, 1947; Green and Stone, 1952), nitrogen or vitamin sources (Green and Sullivan, 1959), with all other nutrient requirements being satisfied. The basal yeast nitrogen or yeast carbon bases are normally used for this purpose (Table III).

TABLE III

Composition of chemically defined media for growing yeasts

	Yeast Nitrogen base grams	Yeast Carbon Base grams
Carbon source		
D-glucose	none[a]	10
Nitrogen source		
(NH₄)₂SO₄	5·0	none[b]
Salts		
KH₂PO₄	1·0	1·0
MgSO₄.7H₂O	0·5	0·5
NaCl	0·1	0·1
CaCl₂.2H₂O	0·1	0·1
Amino-acids	milligrams	milligrams
L-histidine.HCl.H₂O	10	1·0
DL-methionine	20	2·0
DL-tryptophan	20	2·0[c]
Compounds supplying		
trace elements	micrograms	micrograms
H₃BO₃	500	500
CuSO₄.5H₂O	40	40
KI	100	100
FeCl₃.6H₂O	200	200
MnSO₄.1H₂O	400	400
Na₂MoO₄.2H₂O	200	200
ZnSO₄.7H₂O	400	400
Vitamins		
Biotin	2	2
Calcium pantothenate	400	400
Folic acid	2	2
Inositol	2000	2000
Niacin	400	400
Para-aminobenzoic acid	200	200
Pyridoxine.HCl	400	400
Riboflavin	200	200
Thiamine.HCl	400	400

Amounts are given per litre of distilled water.

[a] The desired carbon source must be added.

[b] The desired nitrogen source must be added.

[c] The nitrogen contained in these three amino-acids is insufficient to support visible growth.

Thus, *S. lactis* can be isolated preferentially by using lactose as sole source of carbon and an incubation temperature of 45°C. Walters and Thiselton (1953) advocated the use of 1(+)-lysine as sole source of nitrogen for detecting yeast contaminants in cultures of *S. cerevisiae* and *S. carlsbergensis*: the method has been widely adopted by many workers. Van der Walt (1962) proposed ethylamine as an alternative to lysine since, with one exception, the ability to utilize ethylamine by 162 yeast strains belonging to 32 species coincided with an ability to utilize lysine.

Attempts to use an isolation medium based on nitrate as sole source of nitrogen have not proved successful in this laboratory. The colonies formed were very similar in appearance. Some that were subsequently identified as *C. pulcherrima* gave negative results with Wickerham's nitrate test (1951). Others proved to be *Aureobasidium pullulans* which normally has a very distinctive colonial appearance. Nitrite has been proposed (Wickerham, 1957) for the selective isolation of *Debaryomyces* spp., although some *Brettanomyces* spp. will also use this nitrogen source (van der Walt, 1963).

Debaryomyces spp. and soy yeasts will tolerate very high concentrations of salt—18–20% (Phaff *et al.*, 1966; Onishi, 1963), whereas the tolerance of other yeasts tested by Battley and Bartlett (1966b) ranged from 3 to 11% NaCl. This property of being able to resist high osmotic pressures has long been used for the selective isolation of halo- and osmophilic yeasts. At high salt concentrations other yeast properties change. Thus, the pH range for growth of soy yeasts, normally 3·0–8·0 in salt-free media, becomes 4·0–5·0. Even salt concentrations as low as 2 to 3% NaCl stimulate the growth of *D. hansenii*, which also requires either thiamin or biotin (Rose, 1963), urea or ammonia nitrogen: it grows faster on fructose than glucose (Merdinger and Shair, 1962). Ross and Morris (1962) found that marine yeasts could be maintained on MYPG agar in which 15% of the water had been replaced by "aged" sea water (Zobell, 1946), and the pH adjusted to 5·0. They also found that *Debaryomyces* spp. were more halo-tolerant than species of most other genera: yeasts of marine origin were more halo-tolerant than those of terrestrial origin. There are, however, no distinctive metabolic differences ascribable to environmental influence (Ahearn *et al.*, 1962). A complex organic nitrogen source could stimulate growth and even increase the maximum salt tolerance, e.g., peptone gave a greater response than fish extract > asparagine > $(NH_4)_2SO_4$ > urea. These factors are important when planning media for the selective isolation of halophilic yeasts.

Ingram (1959a) divided yeasts resistant to high sugar concentrations into semi-osmophiles that ferment sucrose and osmophiles that tolerate high sugar concentrations. Scarr and Rose (1966) defined the latter group more closely as yeasts than can grow at concentrations over 65° Brix (dissolved solids % w/v at 25°C). *S. rouxii* and *S. mellis* grow on traces of invert sugar present in impure concentrated sucrose solutions producing organic acids that decrease the pH, so hydrolysing sucrose for further growth (Scarr, 1951). Although these yeasts lack invertase, Scarr and Rose (1966) have now found a group of *Torulopsis* spp. that can ferment concentrated sucrose because they possess this enzyme. These yeasts were found by picking colonies off osmophilic agar, transferring to filter-sterilized 20% w/v sucrose—yeast water in McCartney bottles containing Durham tubes. After 3 weeks at 27°C the centrifuged contents of any positive bottles were transferred to 65° Brix sucrose solution containing 0·5% Difco peptone, 0·1% $MgSO_4$ and 0·2% KH_2PO_4. Cultures obtained after 3 weeks at 27°C were streaked and stored on osmophilic agar. The latter medium, a modification of de Whalley and Scarr's medium (1947), consists of Difco wort agar dissolved in 45° Brix syrup containing 35 parts sucrose and 10 parts glucose (Scarr, 1959). Colonies are clear and develop within 4–5 days at 27°C. It can be remelted 3 times without serious colour formation, unlike Ingram's medium (1959b) which cannot be remelted once it has set. The latter consists of 50% glucose, 1% citric acid, 1% Bacto tryptone, 2% agar; the components are autoclaved separately and mixed immediately afterwards.

Such media are still not entirely satisfactory for very low concentrations of osmophilic yeasts, i.e., 0–100/100 g sugar product. Devillers (1957) membrane filtered the dissolved product and incubated the membranes on a medium containing 5 g yeast extract, 2 g peptone, 2 g soluble starch, 2 g glycerol, 1 g ammonium chloride, 20 g glucose, 400 g raw sugar, 25 g agar and 500 ml water. The final sugar concentration should be between 45 to 50° Brix for good colony formation: it was suggested that colonial appearance should be supplemented with microscopical observations. Scarr (1959) proposed black Oxoid membranes incubated on Sabourand broth adjusted to pH 4·4 with lactic acid; presumably her samples did not contain any non-osmophilic yeasts. Halden *et al.* (1960) differentiated between the two types of yeasts by using membranes with 0·45 μ and 0·80 μ pore sizes. These quantitative methods for osmophilic yeasts should be supplemented with counts on normal media for counting sugar intolerant yeasts accompanying the osmophiles (Ingram, 1959b).

C. Inoculation and cultural conditions

1. *Inoculation of media*

Normally dilutions of the sample are spread on solid media since colonial differentiation is more certain than if pour plates are used. Samples are collected in screw-capped bottles so that the contents can be shaken vigorously. Serial dilutions are prepared as described for the Presumptive Coliform Count (Report, 1956), using a fresh delivery pipette for each dilution

(Ingram and Eddy, 1953). 0·2 ml of the appropriate dilution is pipetted on to the surface of the agar and spread uniformly with a bent piece of thin glass rod (Drigalski spatula). The plates are incubated the right way up for the first 24 hours, to allow the liquid to be absorbed into the agar. Thereafter the plates are inverted to prevent drops of condensed moisture falling on to the agar and causing the colonies to spread. The period of incubation is dependent upon the incubation temperature, but is normally continued until the yeast colonies are well defined.

Samples for membrane filtration are sucked or forced through a sterile membrane, followed by sterile water to wash organisms off the sides of the funnel and to remove any inhibitors originally present in the liquid. Membranes are normally white and printed with a grid; those for osmophilic yeasts are black to show up the almost transparent colonies. The pore size of the membrane is usually 1·2 μ for routine work but for very small yeasts 0·45 μ may be necessary, with consequent reduction in flow rate. The funnel is removed and the membrane placed, with sterilized forceps, on to a sterile pad soaked in a rich nutrient medium; too dilute a medium will give low yeast counts. The pads are held in glass or metal dishes and are sealed before incubating. The technical literature of the manufacturing companies should be consulted for details of specialized techniques and membranes.

2. *Incubation conditions*

Yeasts are said to be psychrophilic, mesophilic or thermophilic according to their optimum growth temperatures. Ingraham (1958) defines a psychrophile as a micro-organism that grows well at 0°C; they always grow more rapidly at higher temperatures than they do at 0°C. The upper limit of growth for a few is 15°C or less, but 20°C, 30°C or 45°C is not unknown, for bacteria at least (Elliott and Michener, 1965). Sinclair and Stokes (1965) isolated 4 obligate psychrophilic yeasts that grew in the range 0–20°C, had an optimum of 15°C and died after exposure to 30°–40°C. Mackenzie and Auret (1963) found that *Rhodotorula nitens* was viable at 26°C but did not grow; growth was just observed at 24°C, the growth range was 4°–20°C with an optimum of 14°C. Hagen and Rose's pyschrophilic cryptococcus grew well at 3°C (1961). Di Menna (1966a) defined an obligate psychrophile as one that grew poorly or not at all at 20°C. Her primary isolation plates were incubated at 4°C for 4–5 weeks before counting and sub-culturing. All isolates were sub-cultured at 4°C until their maximum growth temperature was known. When this was 15°, sub-cultures were grown at 5°, when it was 20°, at 15°, and when the maximum was greater than 20° at 20° or room temperature. Times of incubation decreased in proportion to the increase in temperature. These are very sound criteria to follow. Even with samples taken in an English orchard, some of the isolates were incapable of growth at 25°C (Table IV).

The upper limit for growth is usually taken as 37°C for yeasts found in the intestines of birds and animals. Weathered bird droppings found in an apple orchard contained yeasts that grew vigorously at this temperature.

TABLE IV

Effect of medium and incubation temperature on viable yeast count (per ml)

| | Incubation Temperature | | | |
| | 15°C | | 25°C | |
Sample	Medium A	Medium B	Medium A	Medium B
1	30	Too many	1	54
2	296	Too many	161	Too many
3	50	0	0	0
4	9	50	0	3
5	30	0	18	0
6	36	0	0	0

Loginova *et al.* (1966) studied yeasts growing at 40°–45°C; high concentrations of oleic acid were needed for growth at elevated temperatures, but this could be overcome by induced adaptation (Sherman, 1959; Kates and Baxter, 1962). The majority of yeasts—so-called mesophiles—grow well between 20°–25°C; some die at 30° (Richards, 1934; Phaff *et al.*, 1966). Normally they are incubated at 25°C for 5 days, but observations should be continued until at least the fifteenth day as the development of *Brettanomyces* spp. is very slow, unless extra biotin and thiamin have been added. In the initial isolation programme, therefore, plates or membranes should be incubated in triplicate at 15°, 25° and 37°C respectively. Any yeasts growing readily at 37° should be handled with care, since two of them, *Cryptococcus neoformans* (Kreger-van Rij, 1961) and *Candida albicans*, produce pathological symptoms in humans.

Normally plates are incubated in darkness but Gurinovich *et al.* (1966) considered that 24 hours of exposure to light encouraged optimum carotenoid production when interposed between two periods of 3 days of darkness. Openoorth (1957) found that light encouraged sporulation in *S. cerevisiae* but other observations are rare.

D. Bibliography of sampling methods for different habitats

(a) *Vegetable sources*

Apples. Marshall and Walkley (1951), Williams *et al.* (1956), Bowen and Beech (1964), Davenport (1967).

Cacao. Quesnel (1960), Griffiths (1961), Berger (1964).

Coffee bean. Agate and Bhat (1966).

Grain. Nichols and Leaver (1966).

Grapes. Mrak and McClung (1940).

Olives. Cancho (1957), Santa Maria (1962).

Phyllosphere. di Menna (1959), Ruinen (1961, 1963), Last and Deighton (1965).

Rhizosphere. Bab'eva and Savel'eva (1963).

Silage. Endo (1957), Shchelokova and Mubarakova (1965).

(b) *Animal sources*

Birds. Kawakita and van Uden (1965).

Bovines. van Uden and Carmo Sousa (1957), Clarke and di Menna (1961)

Earthworms. Parle (1963).

Fish. Potter and Baker (1961), Ross and Morris (1965).

Humans. Connell and Skinner (1953), Connell *et al.* (1954), Huxley and Hurd (1956), MacKenzie (1961), Gibbs and Stuttard (1967).

Insects. Shifrine and Phaff (1956), Baker and Kreger van Rij (1964), Phaff *et al.* (1964).

Pigs. van Uden and Carmo Sousa (1962a and b).

Shrimp. Phaff *et al.* (1952), Spencer *et al.* (1964).

(c) *Mineral sources*

Breakdown of herbicides. Baldwin *et al.* (1966).

Petroleum. Foster (1962), Tsuru (1963), Cameron and Terry (1964), Iizuka *et al.* (1965), Miller and Johnson (1966).

Soils. Chesters (1949), Capriotti (1955), di Menna (1957, 1960, 1962 and 1966b), Phaff *et al.* (1960), Adams (1960), Dommerques *et al.* (1965), Sinclair and Stokes (1965).

(d) *Air, water and effluents*

Air. Goetz (1953), di Menna (1955), Wolf *et al.* (1959), Gregory (1961), Zhukova (1962), Adams (1963 and 1964).

Corn steep liquor. Kuznetsov (1957).

Effluents. Cooke and Matsura (1963).

Fresh water. van Uden and Ahearn (1963).

Sea water. Zobell (1946), van Uden and Castelo-Branco (1961), Ahearn and Roth (1962), van Uden and Zobell (1962), Fell and van Uden (1963), Kriss *et al.* (1967).

Tidal waters. Kaplovsky (1957), Wilkinson (1959), Eden and Melbourne (1959), Fell *et al.* (1960).

(e) *Foods and beverages*

Beer. Walters (1943), Wiles (1949).

Butter. Sivadjian *et al.* (1956), Masek *et al.* (1956), Belova (1960), Schwarz and Ciblis (1965), Muys and Willemse (1965), Skorodumova (1965), Cerna and Krisova (1966), Ritter and Eschmann (1967).

Canned foods. Herson and Hulland (1964).

Cheese. Olson and Bonner (1957), Bonner *et al.* (1957), Stadhouders (1958, 1959), Proks *et al.* (1959)

Chocolate and cocoa. Powell and Harris (1964), Kleinert (1966).

Cider, Clark *et al.* (1954), Beech (1957, 1958), Bowen (1962), Bowen and Beech (1967).

Compressed yeast. Windisch *et al.* (1958).

Confectionery. Mansvelt (1964).

Fruit juices. Marshall and Walkley (1951), Recca and Mrak (1952), Lüthi (1959).

General. Mossel *et al.* (1962).

Grape wine. Domercq (1956), van Zyl and du Plessis (1961), Mosiashvili (1956), Rigone de Pritz (1958), Melas-Joannidis *et al.* (1958), van der Walt and van Kerken (1960, 1961), Teal *et al.* (1961), Minárik (1964), Zhuravleva and Timuk (1965).

Honey. Lochhead and Heron (1929).

Ice cream. Nutting *et al.* (1959).

Lunch meats. Wickerham (1957).

Palm wine. Bassir (1962).

Pickles. Etchels *et al.* (1952), Dakin and Day (1958).

Poultry. Walker and Ayres (1959), Mountney *et al.* (1965).

Rice wine (Saké). Inoue *et al.* (1962), Takeda and Tsukahara (1965), Sugama (1966).

Soft drinks. Witter *et al.* (1959), Mossel and Scholts (1964), Chevalier (1967).

Sour-dough. Schulz and Stephan (1958), Spicer and Fouda (1958).

Sugar products. Mrak and Phaff (1948), Scarr (1951), von Schelhorn (1951), Ingram (1959b), El-Tabey Shehata (1960), Scarr and Rose (1966).

(f) *Industrial plant*

Wiles (1959), Smith (1956), Lewis and Johar (1958), Vidal-Leira (1966), Thomas *et al.* (1966), Beech (1967), Rice (1967).

An important source of references on the occurrence of yeasts in nature in the U.S.A. was written by Miller *et al.* (1961). Lund (1954, 1956) has published information on similar habitats in Denmark.

III. ISOLATION AND PURIFICATION

A. From culture media

1. *Isolation*

(a) *Standard method*. The backs of the plates of incubated media should be marked off into 12 to 16 numbered squares using Indian ink containing a little Teepol. Plastic Petri dishes etched with squares can be purchased; similarly membrane filters are available printed with a grid. The plate is then examined with a plate microscope (× 10 magnification) by transmitted and reflected light (reflected light for membranes) and the colonies examined systematically in every square. The characteristics of each colony type should be described exactly, preferably using a standard set of terms (Salle, 1961). Features examined include colour, surface texture, degree of glossiness, colonial cross-section, plan and margin. Anyone unfamiliar with the appearance of yeast colonies should refer to the excellent photographs of de Becze (1959b) and Etchells *et al*. (1953). Some yeast colonies, particularly those capable of forming pseudomycelia, produce ciliate-edged colonies that resemble the juvenile stage of *Aureobasidium pullulans* (Cooke, 1962). In cases of doubt it is better to make a tentative isolation of colonies that are intermediate between yeasts and moulds and to abandon any that develop aerial spores on prolonged incubation.

It is advisable to have a duplicated sheet on which can be recorded colonial appearance, the number of colonies having this appearance, the microscopic appearance of a wet mount (examined at × 1000 magnification) and the code number of an isolate made from one of them. (Photographs of yeast cells are given by de Becze (1959b) and Phaff *et al*. (1966)). Space should also be available to record the total count, dilution, size and identity of the sample.

The loopful of culture picked off from a representative colony should be streaked on to a plate of the same medium for purification. It is by no means certain that all yeasts having the same colonial appearance belong to the same species, but the number of colonies may be so great that only one representative of each colony can be isolated and purified for further study. With smaller numbers more isolates can be chosen. Sometimes isolates from colonies with slightly different appearances are found to belong to the same species, but this is quite fortuitous. It must be emphasized that the first examination of the primary isolation plates is most important, since some of the yeasts may exhibit characteristics that do not appear on subculturing. Thus the ability to form ascospores can soon be lost. Photomicrographs of any sporulating isolates should be taken so that there is a permanent record of the number and shape of the ascospores. The formation of a mirror image of ballistospores on the inner side of the Petri dish lid indicates the presence of *Sporobolomyces* and *Bullera* spp. Absence of ballistospores would lead to the first group of yeasts being placed in the genus *Rhodotorula*. The isolation plates should not be discarded but should be left inverted at the original incubation temperature to allow any slow growing yeasts to develop and to ensure that the cultures obtained are viable. When the search is for a yeast with a particular characteristic (e.g., osmophilism) growing on selective media, such a detailed search may not be necessary. But an assessment cannot be made of the importance of the selected strain without a quantitative estimate of the accompanying yeast flora.

(b) *Protection of saprophytic associations.* It might be assumed that the accompanying microflora of moulds and bacteria is unimportant, but this would be a fatal assumption if the product being examined was produced by a mixture of micro-organisms. Thus sour-dough consists of two species of lactobacilli besides the yeast flora (Spicer and Fouda, 1958); the inoculum of Shao-Shing wine consists of a solid culture of saccharifying moulds, yeasts and bacteria (Liu *et al.*, 1959); kvass consists of yeasts and lactic acid bacteria (Fedorov and Zhupikova, 1964); kumiss, fermented mare's milk, of *Lactobacillus bulgaricus*, *Streptococcus thermophilus*, *Strep. lactis* and *Saccharomyces lactis* (Makhanta, 1960 and 1961; Manta, 1964); boza, apparently still made by methods used in ancient Mesopotamia and Egypt, requires a mixed culture of yeasts and lactic acid bacteria (Pamir, 1961), while the tea fungus culture of the East Indies and Russia requires a yeast and an acetic acid bacterium (Sukiasyan, 1957; Abadie, 1965). It would be necessary to isolate all organisms present on isolation plates of samples from such habitats. Beech and Carr (1955, 1960) have detailed suitable media for the selective isolation of lactic and acetic acid bacteria. With the products described above, Koch's postulates (Wilson and Miles, 1948) can only be served by inoculating the sterile substrate with the correct proportions of all the components of the microflora.

(c) *Possible future methods.* The techniques described have hardly changed since the days of Hansen in the 1880's, one reason being that only the minimum number of organisms can be isolated, because of the labour of identifying the purified isolates. Techniques such as those outlined by Beech *et al.* (1968) reduce this labour a great deal but larger numbers of isolates can be taken only when identification methods are automated. If such identification methods were available, and obviously they would be based on biochemical rather than on morphological characteristics, the initial isolates could be prepared in at least two ways—

(i) by physically removing all colonies growing on a range of media at several temperatures, or
(ii) by pressing a random sampler (possibly based on the multipoint inoculator of Beech *et al.* (1955) on each of multiple plates of these same media and then inoculating tubes of liquid media.

The number of plates and isolates per plate would need to be determined statistically in order to ensure that no yeast forming, say, 1% of the total microflora was missed. Both methods would still require purification of the cultures before they could be identified and this stage would also need to be automated, before the process from plate to computer-identification of each isolate could be completed.

Alternatively isolates could be identified directly by the use of serological (Tsuchiya *et al.*, 1965; Buckley, 1967) or immunofluorescent (Report, 1967) techniques, which in some cases would obviate the need to isolate the organism. The numerous tests for identifying yeasts would no longer be necessary. Needless to say a considerable amount of both research and development work will be needed before these two methods can be brought into general use.

2. *Purification*

It is essential that the media on which the isolates are streaked should not differ appreciably from the original. Otherwise there will be changes in colonial form (apart from the differences due to spreading and streaking), that will make identity checking more difficult. Even more important, acquired characteristics can be lost quite readily in the wrong media (Scarr, 1951). For example, wine yeasts soon lose any specially high tolerance to alcohol if grown on alcohol-free media. With this proviso, the yeasts on the purification media are compared with the microscopic and macroscopic descriptions of the original. A colony of the selected type is streaked twice more to ensure purity and a colony from the last plate grown in the same medium free of agar ready for storage of the culture.

B. From existing pure cultures

All stocks of pure cultures should be examined at intervals for evidence of purity. Working stocks of industrial cultures also need replenishment at intervals, usually from the mass culture itself, rather than from the laboratory stock (Thorne, 1962; Stevens, 1966).

With such cultures it is essential to test for the presence of "wild" yeasts or respiration-deficient strains. Usually spread plates are prepared on a suitable medium and replicated with the velveteen technique (Lederberg and Lederberg, 1952) on to Walter and Thiselton's lysine agar (1953) and on to the media of Czarnecki and van Engel (1959), Nagai (1963, 1965) or Kleyn and Vacano (1963). The presence of yeasts other than *S. cerevisiae* and *S. carlsbergensis* is indicated by their growth on the lysine medium. Respiration-deficient yeasts on the other media are either brightly coloured or fail to reduce tellurite or TTC. Knowing the positions of these aberrant colonies on the original plate enables them to be avoided when collecting a mass of normal colonies on a wire loop. This mass of colonies is then checked for the presence of bacteria by streaking duplicate wort agar plates containing actidione and 8-hydroxy-quinoline (Beech and Carr, 1960). One plate is incubated aerobically to show up acetic acid bacteria and the other anaerobically (nitrogen and carbon dioxide) for lactic acid bacteria. Further tests may be needed for cultures contaminated by specialized bacteria (e.g., *Flavobacterium proteus*, Strandskov and Bockelman, 1956).

The mass of colonies, now proved free of contaminants is suspended in nutrient media as a "mother culture" and kept for the following tests (Stevens, 1966). Fifty single cell cultures are prepared from the mother culture using standard micro-manipulator techniques or by the older method of Lindner (1893), as described by Jørgensen and Hansen (1948). Giant colonies are prepared from these single cells (Stevens, 1967), and isolates made of each different colony type. If the giant colonies of the first 50 cells are all similar, further single cell isolates need to be made from the "mother culture" until a number of colony types has been collected. Any colony forming sectors or pronounced papillae should be avoided as this indicates a pronounced ability to produce mutants. The cultures prepared from the selected giant colonies can now be subjected to pilot scale tests applicable to the industry they are meant to serve, e.g., brewing, wine making, baking, etc. As an example, the following tests could be used for brewing yeasts: flocculation (Burns, 1941; Gilliland, 1951; Hough, 1957; Stevens, 1966), the ability to react with finings and fermentation efficiency (Bishop and Whitley, 1943; Cook, 1963; Stevens, 1966). The final test is a taste panel's judgment on the flavour of the beer produced under both pilot- and large-scale conditions.

IV. STORAGE

The simplest method of storing yeast cultures is on slants of wort or malt extract agar (Lodder and Kreger-van Rij, 1952). Atkin *et al.* (1949) stored a number of yeasts on malt extract slants at 5°C for 97 to 302 days. The viability of the stock cultures was not reported upon, but the authors stated that the vitamin requirements of the yeasts remained virtually unchanged. Daily transfer for 13 to 32 times on slants of the same medium caused a reduction in the extent of growth when tested on one or more of their vitamin deficiency media (see also Reusser, 1963). Dehydration of the storage medium can be prevented by storing the slants under sterile medicinal grade liquid paraffin (Henry, 1947). Mrak and Phaff (1948) reported that they had used this method for 9 years to store 600 cultures representing all genera except *Trigonopsis* and *Pityrosporum*, cultures being held at 5°C and transferred at yearly intervals. It was satisfactory except for *Brettanomyces* spp, which needed storage on chalk agar (e.g., MYPG agar plus 0·5% re-precipitated chalk; also Custers' medium, 1940). Experience at this laboratory has shown that *Kloeckera* spp. need the same treatment because of their production of acid. Cultures with special characteristics, e.g., osmophiles (Scarr, 1951), alcohol tolerance etc. should be stored on media containing the appropriate additions. There is now a growing tendency to abandon malt-based agar for stock cultures and to use glucose–peptone agar instead. This follows the work of Scheda and Yarrow (1966) who insisted that glucose should be the only sugar present in the storage medium (see also Kosikov and Bocharov, 1961).

Kirsop (1954) maintained yeasts in the National Collection of Yeast Cultures in MYPG liquid at 0°C and sub-cultured every 4 months, on the grounds that this inhibited spore formation and, presumably, reduced any danger of mutation. Storage on both liquid and solid media have been used for some years in this laboratory with the modification that the storage tubes have screw caps sealed with self-shrinking Viskrings, to reduce chances of mould infection (Beech, 1957).

In spite of the advantages of these methods, more and more yeast collections are changing over to storing cultures in the freeze-dried state, mainly in order to reduce the labour of repeated sub-culturing, necessary with a large collection. Wickerham and Flickinger (1946), following the technique of Wickerham and Andreasen (1942), who freeze-dried cultures in sterile horse serum, reported that 98% of 1000 cultures preserved by this method were alive after 2 years' storage. Their technique for testing viability, by inoculating the whole contents of the ampoule into a tube of nutrient broth, has been criticized on the grounds that only a single organism need survive. Kirsop (1955) determined the viability of cultures before and after freeze-drying by plating and serial dilutions. Although there was a considerable loss of viability as a result of freeze-drying there was little further loss after 2 years' storage. Further experience with freeze-drying or lyophilization at the NCYC has been reported by Brady (1960). Both Atkin *et al.* (1949) and Kirsop (1955) noted changes in the vitamin requirements of the lyophilized cultures but neither Kirsop nor Wickerham (1951) found any changes in their basic physiological characteristics, such as their fermentation patterns, etc. Kirsop also re-examined the method for determining vitamin requirements and concluded that yeasts could not be clearly divided into those requiring and not requiring a particular vitamin, so that this test had only limited taxonomic value. In a final paper Haynes *et al.* (1955) gave very detailed descriptions of methods for maintaining cultures of industrially important organisms, including yeasts, using sterile-filtered bovine serum as the suspending medium. The lyophilized cultures were stored at 5°C and transferred every 2 years. They reported rare failures with *Cryptococcus*, *Schizosaccharomyces*, and *Saccharomycodes*. In this laboratory, Carr's modification (1956) of Lord Stamp's suspending medium (1947) has also proved satisfactory with a large collection of yeast cultures. This medium consists of 1% gelatin, 1% Difco yeast extract, 0·5% glucose, 0·25% ascorbic acid, adjusted to pH 5·5 and sterilized by Tyndallization. Scott (1958) reported that 0·1 M lysine added to his suspending medium increased survival from 30 to 100%. Brady (1960) and Pedersen (1965) considered that malt extract was the best medium for reconstituting freeze-dried cultures. But Scheda and Yarrow (1966) found that the culture should be revived in a medium with glucose as the sole sugar, not by virtue of improved viability but because the culture does not thereby acquire the ability to ferment or utilize sugars that the parent culture could not. Further aspects of the lyophilization technique have been discussed by Muggleton (1963), Martin (1964), Greaves and Davies (1965), Lichtensztejn and Blechert (1965). There has been a number of reports on the alternative method of preservation in liquid nitrogen (Tsuji, 1966). This latter method needs further investigation into its applicability for a wider range of yeast genera.

While lyophilization offers a convenient method for storing large numbers of cultures, it is by no means the perfect method for storing yeasts with completely unchanged characteristics.

REFERENCES

Abadie, M. (1965). *Revue Mycol.*, **30**, 27–41.
Adams, A. M. (1960). *Rep. hort. Exp. Stn Prod. Lab. Vineland, for* 1959–60, 79–82.
Adams, A. M. (1963). *Rep. hort. Exp. Stn Prod. Lab. Vineland, for* 1963, 68–71.
Adams, A. M. (1964). *Can. J. Microbiol.*, **10**, 641–646.
Agate, A. D., and Bhat, J. V. (1966). *Appl. Microbiol.*, **14**, 256–260.
Ahearn, D. G., and Roth Jr, F. J. (1962). *Devs ind. Microbiol.*, **3**, 163–173.
Ahearn, D. G., Roth, F. J., and Meyers, S. P. (1962). *Can. J. Microbiol.*, **8**, 121–132.
Asai, G. N. (1960). *Phytopathology*, **50**, 535–541.
Atkin, L., Gray, P. P., Moses, W., and Feinstein, M. (1949). *Proc. Eur. Brew. Conv., Lucerne*, pp. 96–112. Elsevier, Holland.
Atkin, L., Moses, W., and Gray, P. P. (1949). *J. Bact.*, **57**, 575–578.
Bab'eva, I. P., and Savel'eva, N. D. (1963). *Mikrobiologiya*, **32**, 86–93.
Baker, J. M., and Kreger-van Rij, N. J. W. (1964). *Antonie van Leeuwenhoek*, **30**, 433–441.

Baldwin, B. C., Bray, M. F., and Geoghegan, M. J. (1966). *Biochem. J.*, **101**, 15P.
Bassir, O. (1962). *W. Afr. J. biol. Chem.*, **6**, 20–25.
Battley, E. H., and Bartlett, E. J. (1966a). *Antonie van Leeuwenhoek*, **32**, 245–255.
Battley, E. H., and Bartlett, E. J. (1966b). *Antonie van Leeuwenhoek*, **32**, 256–260.
Beech, F. W. (1957). Ph.D. Thesis, University of Bristol.
Beech, F. W. (1958). *J. appl. Bact.*, **21**, 257–266.
Beech, F. W. (1967). *Rep. agric. hort. Res. Stn Univ. Bristol, for* 1966, 239–245.
Beech, F. W., and Carr, J. G. (1955). *J. gen. Microbiol.*, **12**, 85–94.
Beech, F. W., and Carr, J. G. (1960). *J. Sci. Fd Agric.*, **11**, 35–40.
Beech, F. W., Carr, J. G., and Codner, R. C. (1955). *J. gen. Microbiol.*, **13**, 408–410
Beech, F. W., Davenport, R. R., Goswell, R. W., and Burnett, J. K. (1968). *In* "Identification methods for microbiologists. Part B. No. 2", pp. 151–175. Academic Press, London.
Belova, T. S. (1960). *Nauchnye Zapiski Stalinskii Institut Sovetskoi Torgovli*, No. 10, 351–355 (*CA.*57 : 14245d).
Berger, F. (1964). *Pharmazie*, **19**, 348–361.
Bishop, L. R., and Whitley, W. A. (1943). *J. Inst. Brew.*, **49**, 223.
Blank, H., Rosenberg, E. W. and Sarkany, I. (1961). *J. invest. Derm.*, **36**, 303–304.
Bonner, M. D., Harmon, L. G., and Smith, C. K. (1957). *J. Dairy Sci.*, **40**, 1360–1364.
Bowen, J. F. (1962). Ph.D. Thesis, University of Bristol.
Bowen, J. F., and Beech, F. W. (1964). *J. appl. Bact.*, **27**, 333–341.
Bowen, J. F., and Beech, F. W. (1967). *J. appl. Bact.*, **30**, 475–483.
Brady, B. L. (1960). *In* "Recent research in freezing and drying" (Ed. A. S. Parkes and A. V. Smith), pp. 243–247. Blackwell, Oxford.
Brady, B. L. (1965). *Antonie van Leeuwenhoek*, **31**, 95–102.
Buckley, H. (1967). Personal communication.
Buckley, H., Campbell, C. K. and Thompson, J. C. (1969). *In* "Isolation methods for microbiologists. Part A. No. 3" (Ed. E. Shapton and G. W. Gould). Academic Press, London.
Burman, N. P. (1965). *Proc. Soc. Wat. Treat. Exam.*, **14**, 27–33.
Burns, J. A. (1941). *J. Inst. Brew*, **47**, 10.
Cameron, J. L., and Terry, B. M. (1964). *Mater. Protect.*, **3**, 60–67.
Cancho, F. G. (1957). *Grasas aceit.*, **8**, 258–266.
Capriotti, A. (1955). *Antonie van Leeuwenhoek*, **21**, 145–156.
Carr, J. G. (1956). Ph.D. Thesis. University of Bristol.
Carter, M. V. (1961). *Rep. Rothamsted exp. Stn* 1960, 125.
Castroviejo, R. (1959). *Am. J. Ophthal.*, **47**, 226–230.
Cerna, E., and Krisova, M. (1966). *Prûm. Potravin*, **17**, 417–420. (*CA.* 66 : 1667).
Chamberlain, N., Cutts, N. S., and Rainbow, C. (1952). *J. gen. Microbiol.*, **7**, 54–60.
Chesters, C. G. C. (1949). *Trans. Br. mycol. Soc.*, **32**, 197–216.
Chevalier, P. (1967). *Écho Brass.*, **23**, 203, 204, 206.
Clark, D. S. (1965a). *Can. J. Microbiol.*, **11**, 407–413.
Clark, D. S. (1965b). *Can. J. Microbiol.*, **11**, 1021.
Clark, D. S., Wallace, R. H., and David, J. J. (1954). *Can. J. Microbiol*, **1**, 145–149.
Clarke, R. T. J., and di Menna, M. E. (1961). *J. gen. Microbiol.*, **25**, 113–117.
Connell, G. H., and Skinner, C. E. (1953). *J. Bact.*, **66**, 627–633.
Connell, G. H., Skinner, C. E., and Hurd, R. C. (1954). *Mycologia*, **46**, 12–15.
Cook, A. H. (1963). *Proc. Eur. Brew. Conv.*, *Brussels*, p. 477. Elsevier, Holland.
Cooke, W. B. (1962). *Mycopath. Mycol. appl.*, **17**, 1–43.

176   F. W. BEECH AND R. R. DAVENPORT

Cooke, W. B., and Matsura, G. S. (1963). *Protoplasma*, **57**, 163–187.
Crosse, J. E. (1959). *Ann. appl. Biol.*, **47**, 306–317.
Custers, T. J. (1940). Ph.D. Thesis. University of Delft.
Cutts, N. S., and Rainbow, C. (1950). *Nature, Lond.*, **166**, 1117.
Czarnecki, H. T., and van Engel, E. L. (1959). *Brewers' Dig.*, **34**, 52–56.
Dakin, J. C., and Day, P. M. (1958). *J. appl. Bact.*, **21**, 94–96.
Davenport, R. R. (1967). *Rep. agric. hort. Res. Stn Univ. Bristol, for* 1966, 246–248.
de Becze, G. I. (1959a). *Wallerstein Labs Commun.*, **22**, 103–117.
de Becze, G. I. (1959b). *Wallerstein Labs Commun.*, **22**, 199–217.
Devillers, P. (1957). *Industries alimentaires et agricoles*, **74**, 269–271.
de Whalley, H. C. S., and Scarr, M. P. (1947). *Chemy. Ind.*, 351.
di Menna, M. E. (1955). *Trans. Br. mycol. Soc.*, **38**, 119–129.
di Menna, M. E. (1957). *J. gen. Microbiol.*, **17**, 678–688.
di Menna, M. E. (1959). *N.Z. Jl agric. Res.*, **2**, 394–405.
di Menna, M. E. (1960). *J. gen. Microbiol.*, **23**, 295–300.
di Menna, M. E. (1962). *J. gen. Microbiol.*, **27**, 249–257.
di Menna, M. E. (1966a). *Antonie van Leeuwenhoek*, **32**, 29–38.
di Menna, M. E. (1966b). *N.Z. Jl agric. Res.*, **9**, 567–589.
Domercq, S. (1956). Ph.D. Thesis. University of Bordeaux.
Dommergues, Y., Mutaftschiev, S., Dusansoy, M., and Lalloz, P. (1965). *Annls Inst. Pasteur, Paris (Suppl.)*, No. 3, 112–120.
Eden, G. E., and Melbourne, K. V. (1959). *Chemy Ind.*, 220–222.
Elliott, R. P., and Michener, H. D. (1965). *In* "Factors affecting the growth of psychrophilic microorganisms in foods". Tech. Bull. No. 1320, U.S. Dept. Agric.
Ellison, J., and Doran, A. H. (1961). *Proc. Eur. Brew. Conv., Vienna*, pp. 224–234. Elsevier, Holland.
El-Tabey Shehata, A. M. (1960). *Appl. Microbiol.*, **8**, 73–75.
Endo, A. (1957). *Nihon Nôgei Kagaku Kaishi.* **31**, 845–849 (*CA.* 53 : 20614f).
Endo, R. M. (1966). *Mycologia*, **58**, 655–659.
Etchells, J. L., Bell, T. A., and Jones, I. D. (1953). *Farlowia*, **4**, 265–304.
Etchells, J. L., Costilow, R. N., and Bell, T. A. (1952). *Farlowia*, **4**, 249–264.
Fedorov, A. F., and Zhupikova, T. G. (1964). *Fermentnaya i Spiritovaya Promyshlennost*, **30**, 17–18 (*CA.* 62 : 4574a).
Fell, J. W., and van Uden, N. (1963). *In* "Symposium on marine microbiology" (Ed. C. H. Oppenheimer), pp. 329–340. C. C. Thomas, Springfield, Illinois.
Fell, J. W., Ahearn, D. G., Meyers, S. P., and Roth, F. J. (1960). *Limnol. Oceanogr.*, **5**, 366.
Foster, J. W. (1962). *Antonie van Leeuwenhoek*, **28**, 241–274.
Foster, W. D. (1960). *Lancet, i*, 670–673.
Gentles, J. C. (1956). *J. clin. Path.*, **9**, 374–377.
Gibbs, B. M., and Stuttard, L. W. (1967). *J. appl. Bact.*, **30**, 66–77.
Gilliland, R. B. (1951). *Proc. Eur. Brew. Conv., Brighton*, pp. 35–58. Elsevier, Holland.
Gilliland, R. B., and Lacey, J. P. (1966). *J. Inst. Brew.*, **72**, 291–303.
Goetz, A. (1955). *Am. ind. Hyg. Ass. Q.*, **16**, 113–120.
Greaves, R. I. N., and Davies, J. D. (1965). *Ann. N.Y. Acad. Sci.*, **125**, 548–558.
Green, S. R. (1955). *Wallerstein Labs Commun.*, **18**, 239–251.
Green, S. R., and Stone, I. (1952). *Wallterstein Labs Commun.*, **15**, 347–361.
Green, S. R., and Sullivan, P. J. (1959). *Wallerstein Labs Commun.*, **22**, 285–295.

Gregory, P. H. (1961). "The microbiology of the atmosphere". Leonard Hill, London,

Greig, J. R. (1966). *Publ. Hlth Insp.*, **73**, 170–173.

Griffiths, L. A. (1961). *Rep. Cacao Res.*, 42–46.

Gurinovich, E. S., Lukashik, A. N., and Koroleva, I. F. (1966). *Issled. Fiziol. Biokhim. Rast. Inst. Eksp. Bot. Akad. Nauk. Beloruss, SSR.*, 165–170. (*CA.* 66 : 27679).

Haas, G. J. (1956). *Wallerstein Labs Commun.*, **19**, 7–22.

Hagen, P. O., and Rose, A. H. (1961). *Can. J. Microbiol.*, **7**, 287–294.

Halden, H. E., Leethem, D. D., and Eis, F. G. (1960). *J. Am. Soc. Sug. Beet. Technol.*, **11**, 137–142.

Hansen, E. Chr. (1881). *Meddr Carlsberg Lab.*, **1**, 293.

Hasegawa, T. (1959). *J. gen. appl. Microbiol., Tokyo*, **5**, 30–34.

Haynes, W. C., Wickerham, L. J., and Hesseltine, C. W. (1955). *Appl. Microbiol.* **3**, 361–368.

Hesseltine, C. W., Hauk, M., Ten Hagen, M., and Bohonos, N. (1952). *J. Bact.*, **64**, 55–61.

Henry, B. S. (1947). *J. Bact.*, **54**, 264.

Herson, A. C., and Hulland, E. D. (1964). "Canned foods: an introduction to their microbiology". Chemical Publishing Co., New York.

Hertz, M. R., and Levine, M. (1942). *Fd Res.*, **7**, 430–441.

Hislop, E. (1967). Personal communication.

Holt, R. J. (1966). *J. appl. Bact.*, **29**, 625–630.

Hough, J. S. (1957). *J. Inst. Brew.*, **63**, 483–487.

Huxley, M. J., and Hurd, R. C. (1956). *J. Bact.*, **71**, 492–493.

Iizuka, H., Sulkane, M., and Nakajima, Y. (1965). *J. gen. appl. Microbiol., Tokyo*, **11**, 153–159.

Inoue, T., Takaoka, Y., and Hata, S. (1962). *Hakko Kogaku Zasshi*, **40**, 505–511. (*CA.* 62 : 16917d).

Ingraham, J. L. (1958). *J. Bact.*, **76**, 75–80.

Ingram, M. (1958). *In* "The chemistry and biology of yeasts". (Ed. A. H. Cook) pp. 603–633. Academic Press, London.

Ingram, M. (1959a). *Revue Ferment. Ind. aliment.*, **14**, 23–33.

Ingram, M. (1959b). *J. appl. Bact.*, **22**, 234–247.

Ingram, M., and Eddy, B. P. (1953). *Lab. Pract.*, **2**, 11–13.

Jørgensen, A., and Hansen, A. (1948). *In* "Microorganisms and fermentation", 15th Ed., pp. 150–151. Griffin and Co., London.

Joslyn, M. A. (1951). *Mycopath. Mycol. appl.*, **5**. 260–276.

Kamiński, J. (1958). *Acta microbiol. pol.*, **7**, 209–219.

Kaplovsky, A. J. (1957). *Sewage ind. Wastes*, **29**, 1042–1053.

Kates, M., and Baxter, R. M. (1962). *Can. J. Biochem. Physiol.*, **40**, 1213–1227.

Kawakita, S., and van Uden, N. (1965). *J. gen. Microbiol.*, **39**, 125–129.

Kirsop, B. (1954). *J. Inst. Brew.*, **60**, 210–213.

Kirsop, B. (1955). *J. Inst. Brew.*, **61**, 466–471.

Kleinert, J. (1966). *Gordian*, **66**, 3–7.

Kleyn, J. G., and Vacano, L. N. (1963). *Am. Brew.*, **96**, 26–34.

Kosikov, K. V., and Bocharov, S. N. (1961). *Trudy Inst. Genet.*, No. 28, 217–227.

Kreger-van Rij, N. J. W. (1961). *Antonie van Leeuwenhoek*, **27**, 59–64.

Kriss, A. E., Mishustina, I. E., Mitskevitch, I. N., and Zemlsova, E. V. (1967). "Microbial population of oceans and seas". Edward Arnold Ltd, London.

Kuznetsov, V. D. (1957). *Mikrobiologiya*, **26**, 367–373.
Lagodsky, H. (1960). *C. r. Séanc. Soc. Biol.*, **154**, 1435.
Last, F. T., and Deighton, F. C. (1965). *Trans. Br. mycol. Soc.*, **48**, 83–99.
Lederberg, J., and Lederberg, E. M. (1952). *J. Bact.*, **63**, 399–406.
Lewis, Y. S., and Johar, D. S. (1958). *Fd Sci.*, **7**, 285.
Lichtensztejn, A., and Blechert, A. (1965). *Pr. Instw. Lab. badaw. Przem. roln. spozyw.*, **15**, 43–49. (*CA.* 64 : 20519h).
Lindner, P. (1893). *Wschr. Brau.*, **10**, 1354.
Liu, P.-W., Chen, S. H., Ch'iu, T.-H., and Chen, C.-C. (1959). *Chemistry, Taipei*, 181–188. (*CA.* 53 : 22722i).
Lochhead, A. G., and Heron, D. A. (1929). *Bull. Dep. Agric. Dom. Can.*, No. 116.
Lodder, J., and Kreger-van Rij, N. J. W. (1952). "The yeasts." North Holland Publishing Co., Amsterdam.
Loginova, L. G., Posmogova, I. N., and Seregina, L. M. (1966). *Mikrobiologiya*, **35**, 1024–27.
Lund, A. (1954). Ph.D. Thesis. University of Copenhagen.
Lund, A. (1956). *Wallerstein Labs Commun.*, **19**, 221–236.
Lüthi, H. (1958). *Symp. Fruit Juice Concentrates, Bristol.*, pp. 391–401. Juris-Verlag, Zurich.
Lüthi, H. (1959). *In* "Advances in Food Research" (Ed. C. O. Chichester, E. M. Mrak, and G. F. Stewart), pp. 221–284. Academic Press, New York.
Mackenzie, D. W. R. (1961). *Sabouraudia*, **1**, 8.
Mackenzie, D. W. R., and Auret, B. J. (1963). *J. gen. Microbiol.*, **31**, 171–177.
Makhanta, K. Ch. (1960). *Nauch. Dokl. Vyssh. Shk.*, No. 4, 190–194. (*CA.* 55 : 11694a).
Mkhanta, K. Ch. (1961). *Nauch. Dokl. Vyssh. Shk.*, No. 1, 177–181. (*CA.* 55 : 27685d).
Mansvelt, J. W. (1964). *Confect. Pros.*, **30**, 33–35, 37, 39.
Manta, K. C. (1964). *Indian J. Dairy Sci.*, **17**, 51–55.
Marshall, C. R., and Walkley, V. T. (1951). *Fd Res.*, **16**, 448–458.
Martin, J. P. (1950). *Soil Sci.*, **69**, 215–232.
Martin, S. M. (1964). *A. Rev. Microbiol.*, **18**, 1–16.
Masek, J., Moxa, V., and Vedlich, M. (1956). *Int. Dairy Congr. 14th Rome*, **2**(1), 237–243.
Melas-Joannidis, Z., Carni-Catsadimas, I., Verona, O., and Picci, G. (1958). *Annali Microbiol.*, **8**, 118–137.
Menzies, J. D. (1960). *Can. J. Microbiol.*, **6**, 583–589.
Merdinger, E., and Shair, S. (1962). *Can. J. Microbiol.*, **8**, 213–220.
Miles, A. A., and Misra, S. S. (1938). *J. Hyg., Camb.*, **38**, 732–749.
Miller, E. J. (1963). *J. appl. Bact.*, **26**, 211–215.
Miller, E. J., and Webb, N. S. (1954). *Soil Sci.*, **77**, 197–204.
Miller, M. W., Phaff, H. J., and Snyder, H. E. (1961). *Mycopath. Mycol. appl.*, **16**, 1–18.
Miller, T. L., and Johnson, M. J. (1966). *Biotechnol. Bioengng*, **8**, 549–565.
Millipore (1965). "Bibliography of references concerning application of millipore filters". Millipore (UK) Ltd, Middlesex.
Millis, N. F. (1951). Ph.D. Thesis. University of Bristol.
Millis, N. F. (1956). *J. gen Microbiol.*, **15**, 521–528.
Minárik, E. (1964). *Mitt. höh. Bundeslehr- u. VersAnst. Wein- Obst- u. Gartenb.*, Ser. A. **14**, 306–315.

Minárik, E., and Rágala, P. (1966). *Mitt. höh. Bundeslehr-u. Vers Anst. Wein-Obst-u. Gartenb.*, **16**, 107–114.
Morris, E. O. (1958). *In* "The chemistry and biology of yeasts" (Ed. A. H. Cook), pp. 251–321. Academic Press, New York.
Mosiashvili, G. I. (1956). *Microbiologiya*, **25**, 484–488.
Mossel, D. A. A., and Scholts, H. H. (1964). *Annls. Inst. Pasteur Lille*, **15**, 11–30.
Mossel, D. A. A., Visser, M., and Mengerink, W. H. J. (1962). *Lab. Pract.*, **11**, 109–112
Motoc, D., and Dimitriu, C. (1964). *Lucr. Inst. Cerc. aliment.*, **7**, 393–417 (*CA.* 65 : 7967f).
Motoc, D., and Dimitriu, C. (1966). *Ind. aliment. Buc.*, **17**, 5–12. (*CA* 65 : 7670h).
Mountney, G. J., Blackwood, U. B., Kinsley, R. N., and O'Malley, J. E. (1965). *Poult. Sci.*, **43**, 778–780.
Mrak, E. M., and McClung, L. S. (1940). *J. Bact.*, **40**, 395–406.
Mrak, E. M., and Phaff, H. J. (1948). *A. Rev. Microbiol.*, **2**, 1–70.
Muggleton, P. W. (1963). *In* "Progress in industrial microbiology" (Ed. D. J. D. Hockenhull), **4**, pp. 189–214. Gordon and Breach, Inc. New York.
Muys, G. T., and Willemse, R. (1965). *Antonie van Leeuwenhoek*, **31**, 103–112.
Nagai, S. (1963). *J. Bact.*, **86**, 299–302.
Nagai, S. (1965). *J. Bact.*, **90**, 220–222.
Negroni, P., and Daglio, C. A. N. (1962). *An. Soc. cient. argent.*, **173**, 69–72.
Nichols, A. A., and Leaver, C. W. (1966). *J. appl. Bact.*, **29**, 566–581.
Niethammer, A. (1942). *Arch. Mikrobiol.*, **13**, 45.
Nutting, L. A., Lomot, P. C., and Barber, F. W. (1959). *Appl. Microbiol.*, **7**, 196.
Olson, H. C., and Bonner, M. D. (1957). *Tech. Bull. Okla. agric. Expt. Stn*, No. T71, 1–24.
Onishi, H. (1957). *Bull. agric. chem. Soc. Japan*, **21**, 137–142 (*CA.* 51 : 15062a).
Onishi, H. (1963). *Adv. Fd Res.*, **12**, 53–94.
Openoorth, W. F. F. (1957). *Proc. Eur. Brew. Conv., Copenhagen*, 222–240.
Pamir, M. H. (1961). *Zir. Fak. Yayinl. Ankara Univ.*, **176**, 1–60.
Parle, J. N. (1963). *J. gen. Microbiol.*, **31**, 1–11.
Pedersen, T. A. (1965). *Antonie van Leeuwenhoek.* **31**, 232–240.
Peterson, W. J., Bell, T. A., Etchells, J. L., and Swart, W. W. G. (1954). *J. Bact.*, **67**, 708–713.
Phaff, H. J., Miller, M. W., and Cooke, W. B. (1960). *Antonie van Leeuwenhoek*, **26**, 182–188.
Phaff, H. J., Miller, M. W., and Mrak, E. M. (1966). "The life of yeasts". University Press, Harvard.
Phaff, H. J., Miller, M. W., and Spencer, J. F. T. (1964). *Antonie van Leeuwenhoek*, **30**, 132–140.
Phaff, H. J., Mrak, E. M., and Williams, O. B. (1952). *Mycologia*, **44**, 431–451.
Pikulska, A. (1953). *Acta microbiol. poll.*, **11**, 34–43.
Potter, L. F., and Baker, G. E. (1961). *Can. J. microbiol*, **7**, 595–605.
Powell, B. D., and Harris, T. L. (1964). *In* "Kirk-Othmer encyclopaedia of chemical technology" 2nd ed., **5**, 363–402.
Probst, E. (1955). *Revue Brass.*, **66**, 35–37.
Proks, J., Doležálek, J., and Pech, Z. (1959). *Int. Dairy Congr. Proc. 15th, London*, **2**, 729–735.
Quesnel, V. C. (1960). *Chemy Ind.*, 101–102.

Recca, J., and Mrak, E. M. (1952). *Fd Technol.*, *Champaign*, 6, 450–454.
Reed, R. W., and Reed, G. B. (1948). *Can. J. Res. Sect. E Med.* 26, 317–326.
Report (1956). *No.* 71 *on Public Health and Medical Subjects*, p. 27. HMSO, London.
Report (1967). *J. Inst. Brew.*, 73, 324.
Reusser, F. (1963). *Adv. appl. Microbiol.*, 5, 189–215.
Reuter, H. (1963). *Fleischwirtschaft*, 15, 483.
Rice, F. G. R. (1967). *J. appl. Bact.*, 30, 101–105.
Richards, M. (1967). *J. Inst. Brew.*, 73, 162–166.
Richards, M., and Elliott, F. R. (1966). *Nature, Lond.*, 209, 536.
Richards, O. W. (1934). *Cold Spring Harb. Symp. quant. Biol.*, 2, 157.
Rigone de Pritz, M. J. A. (1958). *Univ. nacl Cuyo (Mendoza), Fac. cienc. agrar.*, *Bol. ext.*, No. 18, 1–62 (*CA.* 54 : 2654g).
Ritter, P., and Eschmann, K. H. (1967). *Alimenta*, 6, 39–40.
Rose, A. H. (1963). *J. gen. Microbiol.*, 31, 151–160.
Ross, S. S., and Morris, E. O. (1962). *J. Sci. Fd Agric.*, 13, 467–475.
Ross, S. S., and Morris, E. O. (1965). *J. appl. Bact.*, 28, 224–234.
Ruinen, J. (1961). *Pl. Soil*, 15, 81–109.
Ruinen, J. (1963). *Antonie van Leeuwenhoek*, 29, 425–438.
Salle, A. J. (1961). *In* "Fundamental principles of bacteriology", 5th ed., p. 226. MacGraw-Hill, London.
Santa Maria, J. (1962). *J. gen. Microbiol.*, 28, 375–378.
Scarr, M. P. (1951). *J. gen. Microbiol.*, 5, 704–713.
Scarr, M. P. (1959). *J. Sci. Fd Agric.*, 10, 678–681.
Scarr, M. P., and Rose, D. (1966). *J. gen. Microbiol.*, 45, 9–16.
Scheda, R., and Yarrow, D. (1966). *Arch. Mikrobiol.*, 55, 209–225.
Schulz, A., and Stephan, H. (1958). *Brot Gebäck*, 12, 22–27.
Schwarz, G., and Ciblis, E. (1965). *Kieler. milchw ForschBer.*, 17, 137–173 (*CA.* 66 : 27796).
Scott, W. J. (1958). Australian patent 212,235. Jan. 17.
Shchelokova, S. S., and Mubarakova, K. Yu. (1965). *Uzbek. biol. Zh.*, 9, 16–20. (*CA.* 63 : 18942g).
Sherman, F. (1959). *J. cell. comp. Physiol.*, 54, 29–35.
Shifrine, M., and Phaff, H. J. (1956). *Mycologia*, 48, 41–55.
Silhánková, L. (1963). *Folia Mikrobiol.*, 8, 102–108.
Sims, A. P., and Folkes, B. F. (1964). *Proc. R. Soc. B.*, 159, 479–502.
Sinclair, N. A., and Stokes, J. L. (1965). *Can. J. Microbiol.*, 11, 259–269.
Sivadjian, R. A., Varma, K., Laminaryana, H., and Iya, K. K. (1956). *Int. Dairy Congr.* 14th Rome, 2(1), 427–434.
Skinner, C. E., and Bouthilet, R. (1947). *J. Bact.*, 53, 37–43.
Skorodumova, A. M. (1965). *Mikrobiologiya*, 34, 912–917.
Smith, P. D. (1956). *J.A.S.T. Jl.*, 19, 58–66 (*CA.* 53 : 20855c).
Spencer, J. F. T., Phaff, H. J., and Gardner, N. R. (1964). *J. Bact.*, 88, 758–762.
Spicer, G., and Fouda, M. A. (1958). *Brot Gebäck*, 12, 27–30.
Stadhouders, J. (1958). *Missel's zuivelbereid. en-Hand.*, 64, 567.
Stadhouders, J. (1959). *Missels zuivelbereid, en-Hand.*, 65, 93–94.
Stamp, Lord (1947). *J. gen. Microbiol.*, 1, 251–265.
Stevens, T. J. (1966). *J. Inst. Brew.*, 72, 369–373.
Strandskov, F. B., and Bockelman, J. B. (1956). *J. agric. Fd Chem.*, 4, 945–947.
Sudario, E. (1958). *Riv. Vitic. Enol.*, 11, 61–69 (*CA.* 53 : 5582f).
Sugama, S. (1966). *J. Soc. Brew. Japan*, 61, 164. (*Yeast News Letter*, 15 (2), (18–19).

Sukiasyan, A. O. (1957). *Trudy erevan. zoovet. Inst.*, No. 21, 195–202 (*CA*. 53 : 1475h).
Suzuki, M. (1956). *Nippon Saikingaku Zasshi*, **11**, 823–834 (*CA*. 51 : 16689d).
Sykes, G., and Hooper, M. C. (1959). *J. Pharm. Pharmac.*, **11**, *Suppl.* 235T–239T.
Takahashi, M. (1954). *Nihon Nôgei Kagaku Kaishi*, **28**, 395–398 (*CA*. 52 : 18640i *et seq.*).
Takeda, M., and Tsukahara, T. (1965). *Hakko Kyokaishi*, **23**, 352–360 (*CA*. 63 : 168327a).
Teal, B. I., Varela, V. A., Bravo, F., and Llanguno, C. (1961). *Revta Agroquim. Tecnol. Alimentos*, **1**, 11–17. (*CA*. 56 : 14740a).
ten Cate, L. (1965). *J. appl. Bact.*, **28**, 221–223.
Thomas, S. B., King, K. P., and Davies, A. (1966). *J. appl. Bact.*, **29**, 423–429.
Thorne, R. S. W. (1945). *J. Inst. Brew.*, **51**, 114–126.
Thorne, R. S. W. (1950). *Wallerstein Labs Commun.*, **13**, 319–340.
Thorne, R. S. W. (1962). *In* "Colloque sur les levures", pp. 85–93. École de Brasserie, Nancy.
Tsuchiya, T., Fukazawa, Y., and Kawakita, S. (1965). *Mycopath. Mycol. appl.*, **26**, 1–15.
Tsuji, K. (1966). *Appl. Microbiol.*, **14**, 456–461.
Tsuru, N. (1963). *Biol. Sci. Tokyo*, **15**, 146–150 (*CA*. 63 : 12933f).
van der Walt, J. P. (1952). Ph.D. Thesis. University of Delft.
van der Walt, J. P. (1962). *Antonie van Leeuwenhoek*, **28**, 91–96.
van der Walt, J. P. (1963). *Antonie van Leeuwenhoek*, **29**, 52–56.
van der Walt, J. P., and van Kerken, A. E. (1960). *Antonie van Leeuwenhoek*, **26**, 292–296.
van der Walt, J. P., and van Kerken, A. E. (1961). *Antonie van Leeuwenhoek*, **27**, 81–90.
van Uden, N., and Ahearn, D. C. (1963). *Antonie van Leeuwenhoek*, **29**, 308–312.
van Uden, N., and Carmo Sousa, L. do (1957). *J. gen. Microbiol.*, **16**, 385–395.
van Uden, N., and Carmo Sousa, L. do (1962a). *J. gen. Microbiol.*, **27**, 35–40.
van Uden, N., and Carmo Sousa, L. do (1962b). *Antonie van Leeuwenhoek*, **28**, 73–77.
van Uden, N., and Castelo-Branco, R. (1961). *J. gen. Microbiol.*, **26**, 141–148.
van Uden, N., and Zobell, C. E. (1962). *Antonie van Leeuwenhoek*, **28**, 275–283.
van Zyl, J. A., and du Plessis, L. de W. (1961). *S. Afr. J. agric. Sci.*, **4**, 393–404.
Vidal-Leiria, M. (1966). *Antonie van Leeuwenhoek*, **32**, 447–449.
von Schelhorn, M. (1951). *Adv. Fd Res.*, **3**, 429–482.
von Schelhorn, M. (1958). *Z. Lebensmittelunters. u. -Forsch.*, **107**, 212–215.
Walker, H. W., and Ayres, J. C. (1959). *Appl. Microbiol.*, **7**, 251–255.
Walters, A. H. (1967). *J. appl. Bact.*, **30**, 56–65.
Walters, L. S. (1943). *J. Inst. Brew.*, **49**, 245–256.
Walters, L. S., and Thiselton, M. R. (1953). *J. Inst. Brew.*, **59**, 401–404.
Wickerham, L. J. (1951). "The taxonomy of yeasts". Tech. Bull. No. 1029. U.S. Dept. Agric., Washington.
Wickerham, L. J. (1955). *Nature, Lond.*, **176**, 22.
Wickerham, L. J. (1957). *J. Bact.*, **74**, 832–833.
Wickerham, L. J., and Andreasen, A. A. (1942). *Wallerstein Labs Commun.*, **5**, 165–169.
Wickerham, L. J., and Flickinger, M. H. (1946). *Brewers' Dig.*, **21**, 55–59.
Wiles, A. E. (1949). *J. Inst. Brew.*, **55**, 165–172.
Wilkinson, L. (1959). *N.Z. Jl Sci.*, **2**, 196–207.

Williams, A. J., Wallace, R. H., and Clark, D. S. (1956). *Can. J. Microbiol.*, **2**, 645–648.

Wilson, G. S., and Miles, A. A. (1948). *In* "Topley and Wilson's Principles of Bacteriology and Immunity", 3rd ed., p. 1002. Edward Arnold, London.

Windisch, S., and Herbst, A. M. (1958). *Branntwein-wirtschaft*, **80**, 101–108.

Witter, L. D., Berry, J. M., and Folinazzo, J. F. (1959). *Fd Res.*, **23**, 133–142.

Wolf, H. W., Skaly, P., Hall, L. B., Harris, M. M., Decker, H. M., Buchanan, L. M., and Dahlgren, C. M. (1959). "Air sampling. Monograph of apparatus available in 1959". Publ. Hlth Monogr. No. 60, U.S. Dept. Agric., Washington.

Woodworth, H. H., and Newgard, P. M. (1963). *Res. Ind., Stanford*, **15**, 8.

Zellner, S. R., Gustin, D. F., Buck, J. D., and Meyers, S. P. (1963). *Antonie van Leeuwenhoek*, **29**, 203–210.

Zhukova, A. I. (1962). *Mikrobiologiya*, **31**, 745–757.

Zhuravleva, V. P., and Timuk, O. E. (1965). *Izv. Akad. Nauk. turkmen. SSR. Ser. Biol. Nauk.* No. 1, 36–40. (*CA*. 63 : 3581d).

Zinkernagel, H. (1929). *Zentbl. Bakt. ParasitKde Abt. II.* **78**, 191.

Zobell, C. E. (1946). "Marine Microbiology", p. 58. Chronica Botanica Co., Waltham, Mass.

CHAPTER VI

Phycomycetes

MISS G. M. WATERHOUSE

Commonwealth Mycological Institute, Kew, Surrey, England

I. CHYTRIDIOMYCETES AND OOMYCETES

A. Collection and isolation

1. *Aquatic saprophytes*

Aquatic saprophytes, i.e. those producing zoospores and living in water and/or soil, see Gareth Jones (this volume, p. 335).

2. *Non-obligate pathogens—isolation from plant parts*

(a) *Roots*. External signs are softening of the tissues, sloughing-off of the cortex, or complete rotting away.

The root system should be well washed in running water, overnight if much bacterial contamination is present, and the whole system (if small) or representative parts (if large) submerged in shallow water (1 pond: 2 glass distilled autoclaved at 15 lb psi at 121°C). Surface sterilization is not recommended unless contamination is very heavy. After 24 h and on successive days the roots are searched under a stereoscopic dissecting microscope for *Phytophthora-* and *Pythium*-like hyphae and sporangia, which grow out readily into the water, and (internally) for sex organs and chytrid reproductive organs in the tissues. Pieces of root 2–3 mm long bearing such organs are cut out and *either* put into fresh water and "baited" with 1 in. pieces of grass blade (boiled for 10 min; Emerson, 1958), which is best for *Pythium*, small seeds (cold sterilized with propylene oxide; Hansen and Snyder, 1947), or seedlings, *or*, if they appear to contain only one species, plated 2 in. apart on dry plain water agar (see I, B, 1). If more than one species is present "baiting" should be repeated and the "bait" removed to a fresh dish of water after early infection until it appears to bear a single strain when it can be plated. First isolation is achieved more readily if zoospores are trapped on "bait" than if an attempt is made to plate zoospores. Mixtures of species can be separated readily if they produce, or can be induced to produce (e.g. by growth at different temperatures) sporangia and zoospores at different times (p. 187).

(b) *Aerial parts.* After thorough washing, thin slices are cut at what appears to be the advancing edge of the mycelium (usually marked by a line of discoloration) and floated in shallow water. When hyphae grow out the slices are treated in the same way as for root pieces (see above). For typical aerial species on leaves or fruits placing in a damp chamber for 24 h will encourage sporulation and enable sporangia to be picked or washed off and used for initial culturing.

Such parasitic species can readily be recovered from soil by burying or partially burying fruits, leaf pieces, etc., *in situ* in the field or in soil samples brought into the laboratory or by growing seedlings in the soil (avocado fruit, Zentmyer *et al.*, 1960; pineapple leaf and crown, Anderson, 1951, Klemmer and Nakano, 1962; sisal leaves, Wienk and Peregrine, 1965; seedlings, Barton, 1958, Chee and Newhook, 1965; irrigation water, McIntosh, 1966). The soil samples may be used damp, or better still, mixed with water and treated as an aquatic habitat. Seeds left in contact with soil samples for 48 h and then plated often yield Pythiaceae. These fungi may also be recovered from soil by packing it into a small cavity in an apple which is then sealed. Later the fungus can be recovered from the flesh some distance away (Campbell, 1949).

The time of year (temperature) and season of growth (maturity of host and pathogen) may be reasons for failure to isolate from hosts.

3. *Obligate parasites*

(a) *On lower plants.* Chytrids parasitic in or on algae or on other fungi are collected by means of the methods given by Gareth Jones (this Volume, p. 335) or by means of a plankton sampler.

(b) *On higher plants.* Collection, isolation and preservation of infected plant parts follows the usual procedure (see Booth, this Volume, p. 1, and Punithalingam, this Volume, p. 193; CMI Handbook).

Synchytrium causes distortion of aerial vegetative parts and sometimes underground parts forming brown or orange galls or pustules (sori), the latter being like rust caeomata or sori (Karling, 1964). They contain one or few large thick-walled or many small thinner-walled spherical spores (resting spores and prosporangia), the latter resembling smooth aecidiospores.

Physoderma causes brown streaks or patches on aerial parts, and galls on roots, particularly of aquatic and marsh plants. Under low power of the microscope the brown or golden resting sporangia can be seen through the transparent epidermis lying in masses in the cells of the tissues, not in sori. In shape they are flattened spheroids with the dehiscence lid making one convex side bulge more than the other. (Rather similar is *Protomyces*, but the dark or orange brown spherical resting spores scattered in the tissues are very thick-walled.)

Of the Peronosporales, *Albugo* is very easily recognized by its white or creamish blisters or pustules of zoosporangia, usually on the lower leaf surface, indicated by paler patches on the upper, and also on the stems. Infection may be accompanied by distortion. Sometimes the dark brown oospores, usually with thick ornamented walls, are present in and around the pustules, but may be missed as they are deeper in the tissues or even in separate pustules. On Cruciferae *Albugo* is frequently accompanied by *Peronospora* and then it is difficult to determine to which member the associated oospores belong.

Other Peronosporales (*Pseudoperonospora, Plasmopara, Bremia, Bremiella, Basidiophora*, and aerial species of *Phytophthora*) also cause a pallor of the upper leaf surface at first, usually in vague patches but sometimes as angular leaf spots in leaves with marked veins, e.g., *Vitis, Urtica*, later spreading and involving larger areas or even the whole leaf, perhaps with distortion. In some plants the patches may become brown or black (*Solanum, Ranunculus*). The sporangiophores are usually confined to the lower surface but may spread to the upper and to the stem in bad infections. A few species affect flower parts. The sporangiophores form a mass of white, off-white, grey, or greyish mauve "down" or a white "frost" (*Bremia*) which can easily be seen with a hand lens, though in dry weather very little may be visible until the material has been kept for 24 h in a damp chamber.

CAUTION: white or whitish fructifications in tiny close groups may be those of *Ramularia* and greyish ones those of *Botrytis*. Powdery mildews usually look "mealy" rather than downy and also occur more commonly on the upper leaf surface and as a uniform infection rather than in spots or patches.

Again, oospores may be buried in the tissues and away from the seat of infection, e.g., those of *Peronospora viciae* line the tissues bordering the pith, and should always be searched for. For permanent slides to show oospores sections are best, although whole pieces of thin leaves cleared in lactophenol or potash show very clearly the dark brown sculptured walls.

Normally obligate parasites cannot be brought into culture on agar media but recently tissue cultures have been used to isolate and grow them (see following section).

B. Purification and culturing

1. *Agar cultures*

Purification of those fungi able to grow saprophytically may be achieved fairly readily by plating the primary material on plain water agar. Japanese agar is the most satisfactory; specially purified agars inhibit the growth of these fungi. The agar should be made up with non-toxic distilled water or tap water if it is non-toxic. If the plates have been poured a day or two beforehand, the surface will be dry and this will keep down bacteria. *Pythium* and *Phytophthora* will usually outgrow bacteria on such a medium and as the hyphae grow fairly widely apart the tips of those clear of contaminants can easily be cut out and replated. Where bacterial contamination is heavy, e.g., in the tropics or in isolations from soil, it may be necessary to use an antibiotic in the medium, e.g., pimaricin (Eckert and Tsao, 1960), or to reduce the temperature (10°–15°C).

For maintenance, oat agar for *Phytophthora* and cornmeal agar for *Pythium*, both with added wheat germ oil (500 mg dry substance/litre; Klemmer and Lenney, 1965) have proved very satisfactory over the years. For prolonged storage, slopes covered with mineral oil (see Agnes Onions, this Volume, p. 113) will survive for many years; even slow growing species (*Phytophthora infestans*) will cover a slope rapidly under oil and remain viable for at least 15 years.

For those species which do not normally develop sporangia on solid media, production may be induced in *Phytophthora* by one of the following methods—

    (i) Submerging small cubes (2–5 mm³) or discs (5 mm dia.) cut from a young culture in a shallow layer of Petri solution for 3 days (or less if sporangia appear sooner) and then transferring to glass distilled water or pond : distilled (1:2).

   (ii) Infecting small cold-sterilized seeds by contact with a culture for 24 h and then treating as in (i),

  (iii) Using either type of material in soil leachate (Mehrlich, 1935) or non-sterilized pond water (Goode, 1956).

It should be emphasized that each species has different requirements: what works well for one species (or even isolate) may not work for another.

Fresh isolates usually produce sporangia more readily than those that have been long in culture and it may be necessary to put the latter back into a host. Temperature should be near the optimum for growth. For example, Vujičić and Colhoun (1966) found that for *P. erythroseptica* 3–5 day mats from pea broth cultures at 22°C gave the best sporulation when placed in shallow Petri solution at 18°–22°C with some daylight and optimum pH of 7–7·5. Optimum zoospore emission occurred when these cultures were transferred to 8°–13°C for 1½ h then back to room temperature. There was no beneficial effect from extra aeration. *P. parasitica* on the other hand needed 3–4 days in the dark before being transferred to the light and the temperature to be reduced from 31° to 25°C. *P. infestans* did best on rye extract agar (Caten and Jinks, 1968) at 20°C (no specific lighting) and produced maximum zoospores from ten-day-old cultures in sterile distilled water refrigerated at 8°C for 3 h. *P. megasperma* var. *sojae* (Ho and Hickman, 1967) produced sporangia when five-day-old cultures were immersed in non-sterile stream water at 25°C for 14 h followed by two changes of water at hourly intervals.

For obtaining large quantities of sporangia or zoospores for inoculation the above methods can be adapted. Klotz and De Wolfe (1960), however, grew their *Phytophthora* isolates on autoclaved alfalfa (lucerne) sticks for 4–5 days at 26°C and induced soprangia by washing in a stream of aerated water at 20°–25°C for 18–24 h. Zoospore formation was induced by lowering the water temperature to 16°–18°C.

For species that produce deciduous sporangia, e.g., *P. infestans, P. palmivora, P. cactorum, P. nicotianae* etc., it is easier to use sporangia for inoculation, a suspension being obtained by washing them off slopes or inoculated host parts, particularly fruits.

For *Pythium* spp. small agar cubes or discs with mycelium or 1 in. pieces of boiled grass blade infected by contact with a culture for 12 h (fast-growing spp.) or 24 h (slow-growing) are placed in pond: distilled water renewed daily.

If sex organs are not produced in monocultures with wheat germ oil in the oat or cornmeal medium or on hemp seed agar, they may develop when two species or two strains of one species are grown together. Inocula are plated about 1 in. apart and left in the dark at room temperature. If the two are compatible, within 10 days sex organs in large numbers will be seen along the meeting line if the reverse of the plate is examined with strong illumination from below.

The following pairs of species of *Phytophthora* are known to be compatible: *cinnamoni–cryptogea, cinnamoni–palmivora, cambivora–nicotianae* (and its var. *parasitica*), and many others (Savage *et al.* 1968). And the following have compatible (±) strains: *capsici, cinnamomi, cryptogea, drechsleri, infestans, palmivora* ("cacao" and "rubber"), *nicotianae* and its var. *parasitica*. For further information on compatibility see Savage *et al.* (1968).

Species of *Pythium* which have compatible strains are: *intermedium, sylvaticum, heterothallicum,* and *splendens*.

A technique in which glass-fibre tape was used to study the formation and germination of *Phytophthora* oospores in soil was described by Legge (1952). The position of the tape can easily be marked in the soil; it does not rot and so can be recovered complete. Barton (1958) extended its use for the observation of the behaviour of *Pythium* hyphae in soil. Pieces 75 mm wide and of any convenient length are soaked in 70% alcohol for 1 h and autoclaved at 15 lb for 1 h. They are placed on agar plates inoculated with the fungus. After hyphae have grown onto the tape, the latter is placed in soil, enough pieces being used so that representative samples can be taken out and examined over the required interval of time. Barton used steaming Rose Bengal for 1 min to stain hyphae.

2. *Host and tissue cultures*

(a) *Chytrids*. The algal hosts of parasitic chytrids may be grown in liquid culture, on agar, or in a soil–water mixture. There are various recipes for culture liquids (Pringsheim, 1949; Brunel *et al.*, 1950); most contain—

$Ca(NO_3)_2$	40–60 mg
K_2HPO_4	5–10 mg
$MgSO_4$	2·5–5 mg
Na_2SiO_3	20–25 mg
Ferric citrate + citric acid	1 + 1 mg per litre

Soil water is useful for gross uni-algal cultures to start from and is made by autoclaving 1 kg of garden loam in 1 litre water for 1 h at 15 lb and using 30 ml in 300 ml distilled water plus 1 ml 5% KNO_3. It is fairly easy to initiate infection from chytrid-infected algae but so far such cultures have been used mostly for host range tests (Cook, 1963; Barr and Hickman, 1967). Maintenance of infected algae in culture for longer than a few days seems to present problems.

Dual cultures of chytrids obligate on filamentous fungi are achieved fairly readily (Emerson, 1958; Slifkin, 1962) and can be maintained for six weeks or more. Yeast–starch–peptone agar is a good general purpose medium for these cultures. For example, Saprolegniaceae with internal chytrid parasites are grown on sterilized hemp (*Cannabis sativa*) seeds (Slifkin, 1962). When bacteria-free a young culture actively discharging zoospores is left for 4 h in water with a fresh hemp seed bearing a culture of the uninfected host. This freshly infected seed is then blotted on to a sterile plain agar plate. The infected host quickly grows out and is subcultured after 24 h on to cornmeal agar. The parasite is still viable in such cultures after 2 months.

Chytrids obligate on higher plants may be maintained either in seedlings or in tissue cultures (see following section). *Olpidium brassicae*, for example, may be cultured in sterile seedlings of lettuce (*Lactuca sativa*), *Phaseolus aureus*, or *Brassica* spp. grown in sterile sand (Kassanis and Macfarlane, 1964). When the seedlings are two weeks old the medium is flooded with a zoospore suspension. Intermittent flooding spreads infection. Good zoospore suspensions are obtained from plants 1–2 weeks after inoculation by washing them and placing in diluted (1 : 20) Hoagland's solution (Hoagland and Snyder, 1933).

Plasmodiophora brassicae, *Spongospora subterranea* and *Polymyxa graminis* have all been grown in seedling culture or tissue culture.

(b) *Filamentous fungi*. The following medium serves as a general medium for tissue cultures but it may require modifications to secure the optimum growth of individual hosts—

Na_2SO_4	800 mg
Na_2HPO_4	33 mg
KNO_3	80 mg
KCl	65 mg
$MgSO_4.10H_2O$	180 mg
$Ca(NO_3)_2$	400 mg
KI	1·5 mg
$ZnSO_4$	0·3 mg
H_3BO_3	1·5 mg
$MnCl_2$	0·2 mg
Glycine	3·0 mg
Thiamine	0·1 mg
Nicotinic acid	0·5 mg
Pyridoxine	0·8 mg
Ca pantothenate	2·5 mg
Na_2EDTA 0·0143 g + $FeSO_4.7H_2O$ 0·01069 g	25 mg
2,4-Dichlorophenoxyacetic acid	6·0 mg
α-Naphthylacetic acid	0·1 mg
Sucrose	20 mg
Difco yeast extract	1·0 mg
Coconut milk	130 ml

Fungi which have been grown in tissue cultures are: *Phytophthora infestans* on potato, *Pseudoperonospora humuli* on hop, *Peronospora parasitica* on *Brassica*.

II. ENTOMOPHTHORALES: ENTOMOPHTHORACEAE

A. Collection

Dead insects are searched for in their normal environment. Infected ones usually die clinging to the host or substratum or are fixed to it by the rhizoids of the fungus. They are best collected as soon as possible after death as full fructification usually occurs then, but sluggish living insects near infected ones may give earlier stages; such insects may be kept and killed at the right stage of development of the fungus. They are collected with a portion of the substratum, particularly if they are attached by rhizoids since these are diagnostic.

The insects are best stored dry in paper packets or small boxes with a suitable insect deterrent. If not quite dry they readily develop other moulds If later sectioning is intended preservation in 96% alcohol is recommended (Gustafsson, 1965). Clearing and mounting in lactophenol cotton blue follows the usual procedure.

B. Isolation and culture

It is only comparatively recently that parasitic members of this family have been cultured successfully. Even now only about twenty species have been isolated on artificial media. There is no doubt that these fungi have specific physiological requirements and when these peculiarities have been elucidated many more species will be brought into culture and the growth of those already cultured, often very slow, will be enhanced. Srinivasan et al. (1964) summarize the history of attempts to grow them in culture.

For isolation, Müller-Kögler's egg yolk medium (Müller-Kögler, 1959; Gustafsson, 1965) is recommended. The yolk is transferred from surface sterilized eggs to a sterile plate; 5 ml aliquots are conveyed by means of a sterile, wide-mouthed, glass syringe to small plates (4·5 cm dia.) and autoclaved at 80°–90°C for 40–50 min, after which the plates are sealed with tape to prevent drying. A small piece of agar (e.g. Sabouraud) is fixed inside the lid and the infected insect (if small) or a portion (if large) is slightly embedded in it. The lid is replaced over the egg yolk for 12 h when the agar plus insect is removed. When the deposited spores have germinated, colonies are cut out and transferred to Sabouraud maltose or dextrose agar with peptone in place of meat for cleaning up.

Gustaffson and others have found that Sabouraud gives the best growth but other media, e.g., peptone glucose agar, with or without yeast extract, may give better sporulation. Srinivasan et al. (1964) obtained the most rapid growth of Entomophthora and Conidiobolus on a wheat grain extract medium of the following composition: wheat grain extract (made from steeping 30 g grain in 500 ml water) 3%, peptone 2%, yeast extract 1%, glycerin 1%, agar 2%, pH 6·8–7.

Culture tubes are best stored at a slope so that discharged spores fall back on the medium to germinate and enlarge the colony. Different species have different tempertaure (Hall and Bell, 1961) and light (Ege, 1965) requirements. Conidial production in three species was higher under continuous light than with various periods of darkness, and in one species was enhanced by red light.

L-Asparagine and peptone were the best nitrogen sources, glucose, mannose and trehalose good carbon sources. Ability to sporulate may be lost after some time in ordinary culture, therefore special methods of conservation are recommended (see Agnes Onions, this Volume, p. 113). If sporulation is poor, addition of fresh, filtered, autoclaved coconut milk to the medium (50 ml/litre; Sloan *et al.*, 1960) may increase it. The period of maximum sporulation after inoculation varies from species to species and ranged from 1–7 days (Gustafsson, 1965). Germination of conidia was found to be dependent on moisture and temperature. Germination of azygospores, previously dried, occurred under high humidity.

III. ZOOPAGACEAE

A. Collection

1. *Soil inhabiting*

Samples of soil and decomposing plant debris or any substratum likely to harbour eelworms and protozoa, e.g., dung, compost, leaf mould, are collected in corked tubes or closed tins to keep moist (Drechsler, 1950, and many other papers since; Duddington, 1957).

2. *Aquatic*

Water samples rich in protozoa, particularly rotifers, and eelworms are collected as described by Gareth Jones (this Volume, p. 335). Habitats with much decomposing organic matter in the water give the best collections.

B. Isolation

A small amount (a pinch) of the material or isolated plant pieces are scattered in a Petri dish and cooled maize meal agar or rabbit dung agar is poured over it (Duddington, 1957). The plates are kept moist and should be retained at room temperature (temperate areas) for at least 3 months. Uninfected eelworms, amoebae or hosts from stock cultures added to the plates from time to time or even 2 or 3 pinches of friable leaf mould placed firmly on the surface increase the numbers of predaceous fungi. Conidia are picked off with a needle and used to start pure cultures on maize agar. (For purification see Chapter 1.)

REFERENCES

Anderson, E. J. (1951). *Phytopathology* **41**, 187–189.
Barr, D. J. S., and Hickman, C. J. (1967). *Can. J. Bot.*, **45**, 423–430.
Barton, R. (1958). *Trans. Br. mycol. soc.*, **41**, 207–222.

Brunel, J., Prescott, G. W., and Tiffany, L. H. (1950). "The Culturing of Algae—a Symposium". Charles F. Kettering Foundation, Yellow Springs, Ohio.
Campbell, W. A. (1949). *Pl. Dis. Reptr*, **33**, 134–135.
Caten, C. E., and Jinks, J. L. (1968). *Can. J. Bot.*, **46**, 329–348.
Chee, K.-H., and Newhook, F. J. (1965). *N.Z. J. agric. Res.*, **8**, 88–95.
C.M.I. Handbook (1960). Commonwealth Mycological Institute, Kew.
Cook, P. W. (1963). *Am. J. Bot.*, **50**, 580–588.
Drechsler, C. (1941). *Biol. Rev.*, **16**, 265–290.
Duddington, C. L. (1957). "The Friendly Fungi". Faber and Faber, London.
Eckert, J. W., and Tsao, P. H. (1960). *Pl. Dis. Reptr*, **44**, 660–661.
Ege, O. (1965). *Arch. Mikrobiol.*, **52**, 20–48.
Emerson, R. (1958). *Mycologia*, **50**, 589–621.
Goode, P. M. (1956). *Trans. Br. mycol. Soc.*, **39**, 367–377.
Gustafsson, M. (1965). *Lantbr. Högsk. Annlr*, **31**, 103, 212, 405–457.
Hall, I. M., and Bell, J. V. (1961). *J. Insect Pathol.*, **3**, 289–296.
Hansen, H. N., and Snyder, W. C. (1947). *Phytopathology*, **37**, 369–371.
Ho, H. H., and Hickman, C. J. (1967). *Can. J. Bot.*, **45**, 1963–1981.
Hoagland, D. R., and Snyder, W. C. (1933). *Proc. Am. Soc. hort. Sci.*, **30**, 288.
Karling, J. S. (1964). "Synchytrium". Academic Press, New York and London.
Kassanis, B., and Macfarlane, I. (1964). *J. gen. Microbiol.*, **36**, 79–93.
Klemmer, H. W., and Lenney, J. F. (1965). *Phytopathology*, **55**, 320–323.
Klemmer, H. W., and Nakano, R. Y. (1962). *Phytopathology*, **52**, 955–956.
Klotz, L. J., and DeWolfe, T. A. (1960). *Pl. Dis. Reptr*, **44**, 572–573.
Legge, B. J. (1952). *Nature, Lond.*, **169**, 759.
McIntosh, D. L. (1966). *Can. J. Bot.*, **44**, 1591–1596.
Mehrlich, F. P. (1935). *Phytopathology*, **25**, 432–435.
Müller-Kögler, E. (1959). *Entomophaga*, **4**, 261–274.
Pringsheim, E. G. (1946). "Pure Cultures of Algae: Their Preparation and Maintenance". Cambridge University Press, London.
Savage, E. J., Clayton, C. W., Hunter, J. H., Berneman, J. A., Laviola, C., and Tallegly, M. E. (1968). *Phytopathology*, **58**, 1004–1021.
Slifkin, M. K. (1962). *Mycologia*, **54**, 105–106.
Sloan, B. J., Routien, J. B., and Miller, V. P. (1960). *Mycologia*, **52**, 47–63.
Srinivasan, M. C., Narasimhan, M. J., and Thirumalachar, M. J. (1964). *Mycologia*, **56**, 683–691.
Vujičić, R., and Colhoun, J. (1966). *Trans. Br. mycol. Soc.*, **49**, 245–254.
Wienk, J. F., and Peregrine, W. T. H. (1965). *Rep. Tanganyika Sisal Grow. Ass.*, 1964–1965, Part 2, 51–52.
Zentmeyer, G. A., Gilpatrick, J. D., and Thorn, W. A. (1960). *Phytopathology*, **50**, 87.

CHAPTER VII

Basidiomycetes : Heterobasidiomycetidae

E. Punithalingam

Commonwealth Mycological Institute, Kew, Surrey, England

I. INTRODUCTION

The heterobasidiomycetidae constitute a specialized and distinctive group which manifests a very high degree of evolutionary development in many directions. It includes the rust and smut fungi (i.e., Uredinales and Ustilaginales) which are of great economic importance because of the damage they cause to many crops each year in both temperate and tropical regions. Besides these, the heterobasidiomycetidae are represented by fungi parasitic on scale insects (e.g., *Septobasidium* and *Uredinella*) and saprophytic on dead wood (e.g., *Auricularia* and *Dacrymyces*). So far only the Ustilaginales, and the root parasite *Helicobasidium*, have been extensively studied on artificial culture media and the literature on this is vast (Buddin and Wakefield, 1927; Fischer, 1951a). Although the Uredinales have been under investigation for many years (Eriksson, 1894; 1902; Plowright, 1889; McAlpine, 1906; Stakman, 1914; Arthur, 1921) this group of fungi has never been cultured except on living tissues of the host until the appearance of a report by Hotson and Cutter (1951) and the recent demonstration by Williams *et al.* (1966) of the growth of *Puccinia graminis* on synthetic culture media. The direct culturing of *P. graminis* is an important step towards solving the most puzzling question concerning obligate parasitism in the Uredinales. If, as has been demonstrated with *P. graminis* (Williams *et al.*, 1966, 1967; Bushnell, 1968), the problem of culturing of all Uredinales is essentially nutritional, then inducing rust fungi to grow on synthetic media may prove to be relatively simple.

II. COLLECTION AND PRESERVATION

Most members of the heterobasidiomycetidae make good herbarium material but collections should be made with care, in adequate amount and include the various spore stages. After thoroughly drying, a part of each collection should be preserved for future reference. When making collections, a strong knife, secateurs, scalpel, forceps, parchment bags, brown paper bags, a hand lens and a field notebook are invaluable equipment for the collector. Each collection should be given a separate serial number and particular care be taken to ensure that the collections contain both the mature and immature elements of the fungus. Since the heterobasidiomycetidae are represented by fungi which have several different spore stages in their life cycle it is desirable that collecting should be continued over a period of time until each species has been represented by the various spore stages. Each collection should be accompanied with details of the locality, date of collection, the host plant or other substrate, the name of the collector and the person making the identification. It is essential that the host plant is identified correctly, and therefore each collection should include specimen material of the uninfected host to allow for later confirmation or revision of the identification. It is important that only well-dried specimens free of insect pests (Deighton, 1960) should be preserved in mycological packets (Arthur, 1929) as reference collections. Often, an identification requires slide preparations and it is a useful practice to include such preparations with the specimen. Slide preparations of rust (Uredinales) spores are made according to the methods described by Arthur (1929). The mounting medium has considerable taxonomic significance in smuts (Ustilaginales), particularly in the genus *Tilletia* where Shear's mounting medium described by Chupp (1940) and modified by Graham (1960), is widely used (Duran and Fischer, 1961). The formula for Shear's mounting fluid as modified by Graham is—

Potassium acetate (2%) in pH 8 McIlvaine's buffer (0·2 M)	300 ml
Glycerine	120 ml
Ethyl alcohol (95%)	180 ml

Mount spores on a glass slide in 2 drops of the buffered medium, fix with a cover slip and heat slide over an alcohol flame. The mount is now ready for examination. Only fresh and mature specimens are suitable for cultural studies both on synthetic media and in the greenhouse.

III. CULTURE TECHNIQUES

In general the heterobasidiomycetidae thrive well on the living host plant; the smuts (Ustilaginales) and all species of *Helicobasidium* can readily be grown on many standard culture media in the laboratory. The most interesting feature of this group is that the methods essential in culturing are for the most part simple. Since the methods used in culturing rusts (Uredinales) and smuts in the greenhouse and on synthetic media do not follow the same procedure it is appropriate to deal with them separately.

Problems in the culturing of rusts

Establishment of rust cultures in the greenhouse has helped in furthering the knowledge regarding life histories, heteroecium, and physiologic specialization. The various culture methods employed at present are practically the same as those used by early investigators (Carleton, 1903; Kern, 1906; Melhus, 1912; Rosen, 1918; Klebahn, 1923; Mains, 1924).

As the source of inoculum plays an important part in culture work, under no circumstances should material collected from distant localities be covered with wet cloth, cotton wool, or placed in polythene bags as this will result in the growth of secondary moulds over the rust pustules. The rusted straw or leaves should be dried overnight and placed in Manila envelopes with the collection data and this can be mailed without additional wrappings. Manners (1951) recommends packing a few infected leaves in a tin box if the specimen has to be mailed. Another method advocated by Oliveira (1939) is to use living plants grown in test tubes on agar inoculated with rust as a means of transporting specimens. Collections, when not required for immediate culture work, should be stored in a refrigerator at a temperature of 5°–10°C and a relative humidity of 50% (Peltier 1925; Gassner and Straib, 1931; Newton and Johnson, 1932; Stakman *et al.*, 1944; Schuster, 1956) until required.

One of the problems encountered in the culturing of a rust is that spores of certain species remain dormant for a period of time before they will germinate (Reed and Crabill, 1915; Melhus and Durrell, 1919; Mains, 1924; Fukushi, 1925). When such species are to be used in culture work, it is essential that the spores are brought out of their dormancy into a germinable condition. The dormancy period can be terminated either by placing the rusted spore bearing part of plants in coarse cheese cloth bags or in pots containing finely sifted sand. Finally the pots, covered with wire netting, are left outside during winter to be subjected to the usual weather changes (Fischer, 1898; Klebahn, 1923). Dormancy can also be shortened and the spores brought to a germinable condition by alternate wetting and drying or prolonged soaking in tap water (Klebahn, 1914; Mains, 1916; Maneval, 1922). It is usual to find that teliospores produced in the greenhouse fail to germinate even after they have been overwintered out of doors. Such teliospores are brought to a germinable condition by the technique devised by Johnson (1931). According to this method, wheat plants approaching maturity are maintained at 60°F during the formation of teliospores. Immediately after this, the culms bearing the teliospores are cut into several pieces and frozen for a period of 2 weeks in blocks of ice and maintained at −5°C. The blocks of ice are then thawed and the culms bearing teliospores are fixed firmly to a wooden frame. They are then submerged in cold tap water or placed under cold running water for one week. Finally they are alternatively dried and wetted (dry 2 days, wet 2 days) until germination commences.

Before spores are used as inoculum, they have to be tested for their germinability and there are several methods of ascertaining this (Mains, 1916; Melhus and Durrell, 1919; Theil and Weiss, 1920; Hursh, 1922; Weber, 1922; Manners, 1950). When agar plates are used for studying germination tests, the leaf bearing the sori is placed over solidified agar and a spore print obtained. This enables one to assess separately the germinability of spores from any particular sorus (Dodge, 1923).

1. *Establishing rust cultures in the greenhouse*

Establishing rust cultures in the greenhouse involves the inoculation of healthy plants; the method of inoculation primarily depends on the nature of the investigation and the amount of inoculum available. It is desirable first to establish the rust collected from the field on the same host in the greenhouse before attempting further investigation.

(a) *Mass inoculation.* Different methods of inoculation are employed by different investigators. When there is plenty of inoculum available, dusting young greenhouse-grown seedlings with spores is an effective method of inoculation. Large batches of plants can be inoculated by placing the inoculum in a glass tube with a bulb, then blowing it upon the plants (Durrell and Parker, 1920) or by spraying a spore suspension in 0·1% agar with an atomizer (Gassner and Straib, 1931). In most rust laboratories the "cyclone" technique developed by Tervet and Cassell (1951) is used both for collecting spores from the field and for inoculating plants in the greenhouse (Hughes and Macer, 1964). The procedure in the cyclone technique is as follows:

about 0·5 g of talc is first collected in a cyclone 8 cm long and 2 cm in diameter. Urediospores from the field sample are sucked into the cyclone and mixed with the talc. A further amount of talc is added (about 0·5–1·0 g) and the spore/talc material is thoroughly mixed. By reversing the flow of air the spore talc mixture is distributed upon the plants in an incubator. With minor adjustments the rate of dusting can be controlled and this method has an additional advantage in that even a small quantity of inoculum can be evenly dispersed on several plants. In general, grasses and cereals are inoculated after the appearance of the seedling leaf or when the seedling has produced about 4–6 leaves (Newton and Johnson, 1932; Brown, 1937; Bean et al., 1954). Before the application of the inoculum, each leaf is gently rubbed between the moistened fingers to remove the waxy bloom which prevents moisture adhering to the leaf. The inoculum is then applied to the lower surface with a sterile scalpel or a flattened needle with its metal tip wrapped in a thin layer of cotton wool. After inoculation the plants are sprayed with water and then placed in the incubation chamber for 48 h. There are several types of incubation chambers, and the various types often used in culture work have been discussed in detail by Arthur (1929), and Newton and Johnson (1932).

Heavier infections can often be obtained with cereals by an inoculation technique devised by Zehner and Humphrey (1929). According to this method, a spore suspension in distilled water is injected into the host plant tissue above the uppermost node by means of a hypodermic syringe. During this process the syringe should be held parallel to the plant so that the needle passes through two or three leaf sheaths and the culms or heads. This method gives heavier infection than when the inoculum is deposited outside the tissue (Newton and Brown, 1934). When quantitative leaf rust infection studies of cereals are required, the settling tower technique designed by Eyal et al. (1968) will be of considerable use.

When the fungus is transmitted by seed, inoculation can be effected either by mixing surface-disinfected seeds with spores and sowing these seeds in pots as demonstrated by Prasada and Chothia (1950) with safflower rust (*Puccinia carthami*) or by dipping seeds in fresh undiluted egg albumen and then heavily coating them with teliospores prior to sowing in steam sterilized soil (Calvert and Thomas, 1954). On the other hand if it is hypocotyl infection that is required it can be produced, as in the case of *Carthamus tinctorius*, by suspending teliospores in water and pouring the suspension on soil in flats containing emerging seedlings (Zimmer, 1962).

Woody plants are inoculated under the bark, beneath the cortical tissue, by first cutting with a thin narrow scalpel blade and then introducing the spores within (Meinecke, 1920). A more standard way of inoculating trees is by the cork-borer inoculation method (Wright, 1933; Clapper, 1944).

The inoculum is inserted under a disc of bark tissue, cut out by a cork borer or similar instrument to replace a similar disc of healthy tissue removed from the tree to be inoculated. With white pine (*Pinus sorbus*) uniform stem infection can be obtained several weeks after inoculation by various methods of grafting tissues containing the rust fungus (*Coronartium ribicola*) from one pine tree to another (Ahlgren, 1961; Boyer, 1964; Patton, 1962; Van Arsdel, 1962). This inoculation technique has not only considerable value in establishing rust cultures but can also serve as a useful technique for screening individual trees for resistance to rust.

A relatively simple method of inoculating white pine consists of wrapping *Ribes* leaves covered with a mat of telia around the young shoots. The pine tree with *Ribes* leaves around the shoots is then left at 16°C in a fog chamber for a period of 72 h (Van Arsdel, 1968). Pine trees inoculated by this technique develop blister rust cankers within three years.

Obtaining a single spore culture is relatively simple once the rust fungus has been established on its host plant by mass inoculation. Since cultures originating in the field often contain more than one physiologic form, single spore culturing is essential in order to ensure a pure strain of the fungus.

(b) *Single spore cultures.* There are a number of ways of establishing single spore cultures in the greenhouse. Methods often recommended are the agar method (Pieschel, 1931), Newton and Johnson's method (1932) and a method devised by Oliveira (1939). The procedure followed by Newton and Johnson (1932) is as follows: spores are dusted sparsely onto a sterile glass slide which is then placed under the low power objective of a microscope. A needle with a very fine point and fitted to a holder is sterilized by passing through an alcohol flame. The needle is then dipped into Vaseline and passed through one or two layers of thin paper. Single spores are picked up on the point of the needle and deposited in a small droplet of water on a seedling leaf grown under spore-proof conditions. The needle is sterilized before each transfer and the inoculated plants are kept under spore-proof covers throughout the investigation.

According to the technique devised by Oliveira (1939) a capillary tube, first dipped in sterile water, is used to pick up single spores. Single spores are then transferred to a seedling leaf by blowing gently through the mouthpiece of a pipette (see Chapter 1). The pipettes are sterilized between inoculations. The seedlings are kept at room temperature until the drops of water have evaporated and the plants are then incubated for 48 h. When the flecks of pustules appear, all except one fleck are covered with a layer of Vaseline on both sides of the leaf. Each leaf is then numbered and covered with a test tube kept in position by a support. Three days after the appearance of the pustules spores are taken from each pustule and transferred to new seedlings grown under absolutely spore-proof conditions. Cultures thus derived from single spores serve as pure cultures. It is standard practice to use such cultures in the investigation of rust races.

2. Cultures on detached leaves

Apart from culturing on the living host plant many species of rust fungi can successfully be grown on detached leaves in Petri dishes. Experiments conducted with species of *Puccinia* and *Uromyces* show that with due care, cultures on detached leaves can be maintained in good condition for 5–6 weeks (Waters, 1928). For culturing rusts on detached leaves Waters employed a technique in which leaves were thoroughly washed with water, placed in a sterile Petri dish, inoculated with a spore suspension, and incubated in high humid atmosphere for 48 h. After incubation the petioles were cut back to facilitate intake of nutrient. Finally they were floated on 5–7% unsterilized solution of commercial cane sugar (sucrose). Of the ten species cultured by this technique nine produced uredia and telia. Recently using a modified technique *Puccinia sorghi* (maize rust) has been cultured on detached leaves and maintained through its full life cycle by Hooker and Yarwood (1966). This technique consists of cutting 1–2 cm long pieces from the second leaf of seedlings in the third leaf stage and inoculating them. After incubation the leaf sections are floated on 5% non-sterilized sucrose solution containing 20 ppm non-sterilized N^6-benzyladenine or a sterile water solution of kinetin (6-furfuryl aminopurine).

Rust cultures on detached leaves have the advantage over entire plants in that greater control over the environment and purity of the pathogen can be maintained with ease. This method appears to be the answer to all greenhouse problems and where maximum economy has to be applied. The response, however, of a detached leaf to infection is not a true indication of the symptom expression of a well established plant.

3. Culturing rusts on synthetic media

For almost half a century repeated attempts by several workers to culture rust fungi on synthetic media proved futile and it was assumed from the time of De Bary that the vegetative stages of these fungi would not continue independent growth apart from their respective host cells. Nevertheless Ray (1901) and Gretschushikoff (1936) reported the growth of rust fungi under artificial conditions, though these reports have not subsequently been confirmed.

The growth of systemically infected plant tissue with rust fungi on synthetic media, using tissue culture technique, was the first step in the culturing of rusts (Hotson and Cutter, 1951; Cutter, 1959; Walkinshaw *et al.*, 1965; Koenigs, 1968). This led to the isolation in axenic culture (i.e., in the complete absence of host cells) of *Gymnosporangium juniperi-virginianae* (Hotson and Cutter, 1951) and subsequently several more strains were isolated by Cutter (1951; 1959). Cutter (1961) using a technique similar to that used for isolating *G. juniperi-virginianae* was able to obtain five strains of *Uromyces ari-triphylli* in axenic culture. The principle in the isolation of fungi causing systemic infection is the establishment of rust-infected tissue on synthetic media (free of saprophytes) and this is followed by the isolation of the fungus when it grows out into the medium. The details are as follows: first the rust infected tissue (e.g., with the telial *Gymnosporangium* gall) is immersed in undiluted clorox (sterilizing agent) for 5–20 min and then rinsed in sterile distilled water. Following this treatment the tissue is sliced aseptically and the cubes approximately 5 mm square are then dipped in 1% ascorbic acid and planted in Pyrex tubes containing 10 ml of Gautheret's agar medium (1942) prepared according to the formulation by Cutter (1959).

Gautheret's solution as modified by Cutter (1959)

Green-coconut milk,	150 ml
Trace element (Burkholder and Nickell's) solution	1 ml
Complete vitamin solution	10 ml
Sucrose	30 g
Agar (Difco Noble)	10 g
Indoleacetic acid, 0·01 solution (filter sterilized)	10 ml

(Naphthalene acetic may be substituted; see Hotson, 1953).
Half strength Knop's solution to 1 litre. The medium is mixed and autoclaved at 15 lb pressure for 20 min. Approximate pH 5·3.

Burkholder and Nickell's trace element solution, mg/litre

H_3BO_3	570
$MnCl_2\ 4H_2O$	360
$ZnCl_2$	625
$CuCl_2$	268
$NA_2MoO_4.2H_2O$	252
$Fe_2(C_4H_4O_6)_3$	1825

Half strength Knop's solution, mg/litre

$Ca(NO_3)_2.4H_2O$	720
KNO_3	125
$MgSO_4$	125
KH_2PO_4	125
$FeCl_3$	1

Vitamin solution, mg/litre

Thiamin	100
Riboflavin	50
Pyridoxine,	50
Calcium pantathenate	200
Para-amino-benzoic acid	50
Choline	200
Inositol	200
Yeast nucleic acid	500
Folic acid	5
Biotin	1
Nicotinamide	200

As the proper orientation of the tissue segment produces enhanced callus tissue it is recommended that the tissue segment be inverted and thrust partially below the surface of the solidified culture medium. Inoculated culture tubes are maintained at 20°C with 50% relative humidity and under constant warm white fluorescent illumination (350 or 890 foot candle) until the development of callus, which, in normal circumstances takes about 6 weeks. When the callus develops the tubes are opened and the entire culture removed to sterile Petri dishes and portions of the callus excised and returned aseptically to fresh media. Periodically callus cultures should be examined for the presence of rust mycelium. When the rust mycelium grows out from the callus tissue into the surrounding nutrient substrate it is then transferred to fresh nutrient medium. After the newly established mycelial isolate has reached a reasonable colony size, transfers can be made onto a wide variety of media. Finally the identity of the isolate should be confirmed by reinfecting their host in tissue culture and by greenhouse inoculation or field inoculation as demonstrated by Cutter (1959, 1961).

So far, only a limited number of rust fungi causing systemic infection have been isolated in axenic culture by the elaborate tissue culture technique described above. The direct rust-culturing technique developed by Williams *et al.* (1966, 1967) and confirmed by Bushnell (1968) is the most promising method available at present. According to the method described by Williams *et al.*, *Puccinia graminis* can be made to grow directly on well-defined synthetic media and the procedure is as follows: wheat seedlings are first sprayed with water, dusted with urediospores and then incubated for 20–24 h at 26°–28°C. At the early flecking stage of disease development (i.e., 5–6 days after incubation) the leaves are harvested and surface sterilized by immersing for 4 min in a solution of calsol (3% available chlorine) containing 0·0001% Tween 80. The leaves are then washed in sterile distilled water and placed on nutrient solution (Turel and Ledingham, 1957) solidified with 2% Difco bacto agar. The leaf cultures are incubated in a desiccator over calcium chloride at 17°–18°C under white fluorescent light (400 foot candle). As the sori open the uncontaminated spores are collected and used to seed the culture medium containing Czapek's minerals, sucrose, 0·1% Difco yeast extract and 1% Difco bacto-agar which has been previously autoclaved at 121°C for 15 min (pH 6·4). A slightly modified nutrient medium which incorporates 0·1% Evans peptone has been found more suitable for the growth *in vitro* of *P. graminis* (Bushnell, 1968). Rust colonies grow slowly and when single colonies develop they are transferred to fresh media. Single colonies established in this manner take approximately 2–4 weeks for the production of spores and *P. graminis* produces urediospores and teliospores on culture media. Only the Australian isolates of four races of *Puccinia graminis* have been grown on synthetic media and attempts to culture non-Australian isolates have so far met with no success (Bushnell, 1968). As this direct rust culture technique is a very recent method, it may be some time before it can be adapted to culture other rust fungi.

IV. MAINTENANCE AND PRESERVATION OF RUST FUNGI

The maintenance and preservation of a large number of cereal rust cultures in viable condition at rust laboratories require the passage of the rust fungi through the appropriate host every 4 months and the storage of spores in glass vials under refrigeration during intervals. This routine procedure involves considerable labour, risk from contamination, and consequently the loss of some isolates. The whole procedure has now been made simple by the lyophilization technique widely used for bacteria (Flosdorf, 1949) and successfully applied to a number of fungi (Raper and Alexander, 1945; Atkins *et al.*, 1949). Because of its success lyophilization has now become the basis for the establishment of permanent culture collections in many laboratories. The basic principle of lyophilization is the freezing of spores in some type of suspending medium such as blood serum, skimmed milk or a solution of gelatin or sucrose and subsequent sublimation of the water at reduced pressure. With urediospores of *Puccinia graminis*, better results can be obtained by omitting the suspending medium during lyophilization (Sharp and Smith, 1952). The two common methods developed by these workers to preserve rust spores are as follows. According to one method urediospores are frozen at −45° to −50°C and then dried under vacuum ranging from 20–150 μm: during the first drying period, which lasts 2–3 h, the temperature is raised to −10°C and finally to room temperature during the second drying period lasting one hour. In the second method urediospores are dried under vacuum at room temperature and subsequently stored under refrigeration. By further modifying these methods Sharp and Smith (1957) have been able to preserve urediospores of *P. graminis* and *P. coronata* for a period of $5\frac{1}{2}$ years in a viable state.

Stewart (1956), on the other hand, reports that the urediospores of *P. graminis* mixed with recrystallized hemin and dried for a period of 30 min under vacuum equivalent to 3 in of mercury and then sealed at room temperature gives good results. This technique in every detail is only a modification of the non-freeze vacuum drying technique described previously by Sharp and Smith (1952).

Another reliable vacuum drying technique is now used for the storage of all rust cultures at plant breeding stations, and spores processed by the application of this technique remain viable and without any change in pathogenicity for 4 years (Flor, 1967; Hughes and Macer, 1964). Such a technique used for storing rust spores by the British Plant Breeding Institute was described by Hughes and Macer as follows. Urediospores collected from greenhouse grown seedlings are placed in 0·5 ml glass ampoules and vacuum dried in an Edwards centrifugal freeze drier at a pressure of $0·05 \pm 0·01$ mm Hg without any suspending medium, using phosphorus pentoxide as the desiccating agent. The dried spores are sealed under vacuum and stored *in vacuo* at 1°C (see Onions, this Volume p. 113).

Although the preservation of rust spores by lyophilization and vacuum drying are now applied exclusively in many laboratories, the percentage germination of spores after a five-year period is usually low. The relentless search for a long term preservation technique has yielded a simple and more promising method whereby frozen spores are stored in liquid nitrogen (Loegering *et al.*, 1961). The promising results shown by this method have led to the American Type Culture Collection adopting it for the preservation of rust cultures (Loegering, 1965). The first step in liquid nitrogen storage is to seal 1 mg of air dry urediospores free of any additives (Davis *et al.*, 1966) in 6 in. tubes made from 7 mm borosilicate glass tubing with a cross-fire oxygen torch. Sealed tubes are attached to a wire cane by surgical tape, immersed in liquid nitrogen ($-196°C$) for 20 min and then transferred to a liquid nitrogen refrigerator ($-160°$ to $-196°C$). The sealed tubes can also be directly placed in a liquid nitrogen refrigerator without first immersing in liquid nitrogen (Loegering and Harmon, 1962). When it is desired to use the spores a sealed tube is removed from the refrigerator and plunged immediately into water kept at 40°–50°C for 2–5 min.

As this method is only in its experimental stage, germination tests have to be carried out on agar (Loegering, 1941) and by inoculating seedlings grown in the greenhouse (Loegering *et al.*, 1961) at one or two-year intervals. The length of time rust spores can be kept stored in liquid nitrogen is still under investigation, but present results are extremely promising.

V. PRINCIPLES AND PROBLEMS IN THE
CULTURING OF SMUTS

It is well known that a highly germinable inoculum is prerequisite to culture work and knowledge of the factors that influence spore germination is required in the establishment of cultures of smut species. One of the primary factors that influence germination is spore maturity. Smut spores should be allowed to reach maturity on the living host plant prior to harvesting. Premature dispersal should be prevented by covering the sori with parchment bags. Spores from dusty sori should be collected free of host tissue by passing through a series of sieves from 20–60 mesh. If the spores are produced in submerged sori, the infected parts of the plant should be macerated in water and strained through a cheese cloth (Fischer, 1940; Fischer and Holton, 1943; Kreitlow, 1945). Dry, sieved spores of long-lived smuts can be stored in loosely stoppered bottles at 5°–10°C.

Many smut species require an after-ripening period before the spores can be made to germinate. But species of *Entyloma* and *Dossansia* produce spores which germinate *in situ* at maturity thus causing fresh infection of leaves. There is, however, no one standard treatment for reducing the after-ripening period. Kreitlow (1943b) reduced the after-ripening period required for the germination of spores of *Ustilago striiformis* by growing the plants at 32°C and storing the smutted leaves in a moist chamber at 35°C: but Leach *et al.* (1946) found this method unreliable. Davis (1924) found exposure to chloroform vapour for 1 min or to a 10% citric acid solution for 5 min reduced considerably the after-ripening period. Holton (1943) induced bunt spores to germinate by soaking in tap water for three months or more at 4°C. A similar form of treatment is effective in reducing the after-ripening period in other smut species (Noble, 1923; Fischer and Holton, 1943). The duration of this treatment is entirely dependent on maturity and source of spores, as in the case of *Tilletia caries* which can germinate in 16 days at 3°C (Gassner and Niemann, 1954a). In contrast, others find an after-ripening period unnecessary for the germination of *U. striiformis* (Fischer, 1940) and *U. maydis* (Schmitt, 1940). In view of the fact that no two workers conducted their investigation with material obtained from the same source and considering that *U. striiformis* is a composite species, it is not surprising that the results are variable.

Environmental factors like light and temperature also influence germination and the requirements depend on the smut species (Walker and Welman, 1926; Landen, 1939; Ling, 1940; Hulea, 1947; Zscheile, 1965). For an extensive review on dormancy and spore germination, see Sussman (1966).

In all culture work it is routine to assess spore germinability before inoculation and there are several methods for carrying out germination tests.

One simple method consists of dispersing spores in drops of distilled water on a microscope slide containing four paraffin-wax wells and incubating in a Petri dish moist chamber at optimum conditions for 24 h (Kreitlow, 1943a). Another way of testing germination is to spread a dilute spore suspension on the surface of 2% solidified water agar and incubate the spores at the optimum temperature known for the species (Holton, 1943; Holton and Siang, 1953; Baylis, 1955). For the germination of spores of certain smut species special media are required. Soil extract agar (made by adding 500 ml of boiling water to 75 g of garden soil placed within a funnel on filter paper and mixing the filtrate with 25 g of previously prepared agar and making the final volume up to 1 litre) is most suitable for the germination of bunt spores (Kienholz and Heald, 1930; Meiners and Waldher, 1959). Lacy (1967) recommends the use of another medium, having the composition in g/litre—

Malt extract	10
Peptone	5
KH_2PO_4	1·5
KCl	0·5
$MgSO_4$	0·5
$FeSO_4$	0·01
Agar	15

for the germination of *Urocystis colchici* spores. Further useful details on spore germination are scattered throughout the literature (Walker and Welman, 1926; Noble, 1934; Kaiser, 1936; Lobik and Dahlstrem, 1936; Ling, 1940; Fischer and Hirschhorn, 1945; Thirumalachar and Dickson, 1953), and an extensive list of references has been compiled by Fischer (1951a).

A. Establishment of cultures on the host

The culturing of smuts on the living plant in the greenhouse is essential for crop improvement work: identifying races, eliminating susceptible strains of plants and selecting plants with desirable characters including smut resistance. To achieve these objectives healthy plants have to be infected with smuts and cultured in the greenhouse. There are several methods of producing infection depending primarily on the natural site of infection. On account of the economic interest, only techniques for producing infection with cereal smuts have been worked out. These methods can be adapted for other smuts when the site of infection is known.

In general smut infection can be brought about by inoculating plants in the seedling stage, flowering stage, or through the shoots with a hypodermic needle.

1. *Seedling infection*

Dusting formaldehyde-treated seeds with spores is standard procedure in smut inoculation (Rodenhiser and Holton, 1937; Luttrell *et al.*, 1964; Kendrick, 1961). Wheat seeds are inoculated with bunt by shaking 100 g of clean grains with 1 g of bunt spores (Heald, 1921; Heald and Boyle, 1923) and sowing then in soil at a depth of $1\frac{3}{4}$ in and 50% saturated with water with a reaction of pH 5·5–7·5 (Rodenhiser and Taylor, 1940). Many workers prefer to dust dehulled barley and oats for obtaining high percentage smut infection (Sampson, 1929; Western, 1936; Brandwein, 1938; Sampson and Western, 1938; Schafer *et al.*, 1962a, b). On the other hand, removal of pales in barley lowers the percentage germination (Tisdale, 1923). The sand-paper treatment (Aamodt and Johnston, 1935) and dehulling with sulphuric acid also cause injury (Briggs, 1927; Johnston, 1934; Woodward and Tingley, 1941). In view of this drawback wet methods of inoculating grains in the husk are often employed.

According to one method, seeds to be inoculated are added to a standard spore suspension (usually 1 g of spores in 1 litre of water) and shaken for $\frac{1}{2}$ min and allowed to soak for 15 min. Later the suspension is decanted and the vial inverted over blotting paper. Subsequently the seeds are packed in tightly covered tin boxes lined with moist blotting paper and incubated for 24 h at 20°C. Following incubation, the seeds are transferred to wide open packets and allowed to dry for 3 days. Finally the seeds are sown in dry soil at about 15°C (Leukel, 1936; Tapke and Bever, 1942).

A highly satisfactory method of inoculating wheat and barley seeds consists essentially of agitating seeds while they are immersed in a water suspension of smut spores for a period of 10–25 sec with a high-speed homogenizer. Following agitation the seeds are poured into a sieve and drained free of the suspension. The inoculated seeds are packed immediately in envelopes and dried for 2 days at room temperature (Popp and Cherewick, 1953). Although this technique is an improvement on the method described by Tapke and Bever (1942) for physiologic race studies with large batches of seed samples, it can be adapted for inoculating small samples (Cherewick, 1965). According to Cherewick's technique, partially dehulled seeds are inoculated with a suspension of compatible sporidia of a smut race. After the inoculated seeds have been kept in a damp atmosphere at 22°C for 24 h, they are sown in moist soil. Any watering is done only after commencement of germination.

For producing high-percentage bunt infection a technique described by Meiners (1959) is much used (Hoffmann *et al.*, 1962; Singh and Trione, 1967). At first bunt spores are surface sterilized in a 5% solution of bleaching agent (5·25% sodium hypochlorite) for 1 min, washed in sterile distilled water and plated on 2·5% soil extract agar. The plates are incubated at

5°C and exposed to both daylight and artificial light. During inoculation 2 ml of distilled water and 100 seeds are placed in each plate containing spores at various stages of germination and stirred with a glass rod. When the seeds are well covered with the germinating bunt spores, they are placed on a 2 cm layer of moist vermiculite in a plastic dish and covered with an additional layer of moist vermiculite. Finally the dish is covered with a plastic lid and kept in a greenhouse at 10°–15°C until the seedlings reach the stage when they begin to press the cover. Generally it takes 1–2 weeks for the seedlings to reach this stage. The seedlings are then transplanted in a potting mixture ($\frac{1}{2}$ soil, $\frac{1}{4}$ sand and $\frac{1}{4}$ peat moss) and retained in a greenhouse until they are well established.

Fairly consistent results can be obtained by inoculating grains under the hull by the partial vacuum method. This method is quicker and more effective than the dusting method (Haarring, 1930; Leukel, 1936; Western, 1937; Leukel et al., 1938; Holton, 1964). The technique as described by the American Phytopathological Society (1944) is as follows. First a spore suspension is made by shaking thoroughly 2 g of 60 mesh spore material in 1 litre of distilled water. Next 500 ml of clean seeds are immersed in this spore suspension in a high pressure desiccator and evacuated for 10 min at 5 in of mercury. On releasing the vacuum the seeds are drained and dried for 24 h. Following this, the seeds are kept at 20°C and 80–90% humidity for 20 h. Finally they are dried with a fan and stored until required for sowing. The partial vacuum method has been successfully used for inoculating cereals and forage grasses with *Urocystis* (Fischer and Holton, 1943).

Among other inoculation techniques developed for producing bunt infection, contaminating the surface of the soil in which wheat seeds have been sown (at a depth of 0·5–1 in.) with bunt spores (Röder, 1953), or applying spores to the seeds in open drill (Baylis, 1955), are effective in producing high infection. Several other methods of producing smut infection in the seedling stage are often employed and the choice depends purely on the scope of the investigation (Fischer, 1940, 1951b, 1953; Fischer and Holton, 1957). Infection results obtained by various workers with cereal smuts by different methods have been varied. In view of this, the American Phytopathological Society has compiled details primarily for infecting wheat, oats and barley in the greenhouse.

2. *Flower infection*

Infection of the florets is brought about by depositing spores either on the ovary or stigma during anthesis (Tapke, 1935; Hanna, 1937). There are several ways of carrying out this operation. Some workers prefer to spray the florets with a spore suspension, and this is very effective with wheat and barley (Grevel, 1930; Nahmmacher, 1932; Roemer *et al.*, 1937).

It is quicker to inoculate ears of cereals by the partial vacuum method designed by Moore (1936), and with certain modifications several ears can be inoculated (Oort, 1939, 1940; Vanderwalle, 1945). Another method of inoculating the florets is to inject a spore suspension (made with 1% glucose solution) by means of a hypodermic needle one or two days after the exsertion of the inflorescence (Poehlman, 1945). But Shands and Schaller (1946), after testing several inoculation methods used for infecting barley with *Ustilago nuda*, found injecting dry smut spores into florets to be most effective.

3. *Shoot infection*

Inoculation of active meristems with suitably paired haploid or with diploid lines is a way of determining pathogenicity and sexual compatibility among biotypes. This consists of forcing the inoculum (i.e., cultures grown in a solution of 1% malt) into the growing point of the stem by hypodermic injection (Tisdale and Johnston, 1926; Stakman and Christensen, 1927; Hanna, 1929) or directly into the plumule (Stevens *et al.*, 1946). One-week-old maize seedlings when inoculated and maintained at 27°–32°C develop galls in 3–4 weeks (Schmitt, 1940). It is necessary to suspend the inoculum (sporidia) in a medium having low surface tension to obtain good infection results (Davies, 1935; Wilkinson and Kent, 1945). Leach *et al.* (1946) found that grass seedlings (*Poa*) inoculated through the sheaths at the base of the stem with *Ustilago striiformis* gave good infection results. The data presented by various workers undoubtedly indicate the usefulness of the hypodermic injection method. Other methods of shoot inoculation include spraying seedlings, when their coleoptiles have reached 2 mm, with a suspension of sporidia (Gassner and Niemann, 1954b), or dipping wounded seedlings (2–5 mm long) in a suspension of germinating spores and fused sporidia (Niemann, 1955). Inoculated seedlings are incubated at 5°C for 2 weeks and then established in pots (Gassner and Niemann, 1954b). For additional information on inoculation through the coleoptile, see Kiesling (1962) and Hodges and Britton (1969).

Many investigators however, find the partial vacuum method of inoculating active maize meristems most suitable for obtaining uniform results (Wilkinson and Kent, 1945; Rowell and DeVay, 1952, 1953). Essentially this technique consists of germinating disinfected seeds aseptically. When the coleoptile reaches 1 cm in length, 2 mm of it is cut off to open the tissues surrounding the plumule and growing point. The seedlings are then immersed in the inoculum in test tubes and subjected to maximum partial vacuum with a water aspirator for 5 min. Finally the inoculum is drained off and the seedlings planted in soil or vermiculite. Although this is an efficient method for inoculating corn seedlings it can be slightly modified for producing high percentage smut infection in barley and wheat (Kavanagh, 1961).

B. Cultures on synthetic media

Smut fungi readily grow on many standard synthetic culture media. For routine investigation and maintenance potato glucose agar, potato sucrose agar, unstrained heavy oatmeal agar, modified Czapek's agar and meat extract agar are most suitable (Sartoris, 1924; Kienholz and Heald, 1930; Sampson and Western, 1938; Schmitt, 1940; Holton and Heald, 1941; Govindu and Fischer, 1957; Kendrick, 1957; Trione and Metzger, 1962). Generally cultures are derived from chlamydospores or sporidia but isolation from mycelium within the infected host tissue is also possible (Leach *et al.*, 1946; Trione, 1964; Singh and Trione, 1967).

1. *Isolation from mycelium within the host*

Establishing smut cultures from mycelium has not been widely applied, nevertheless *Tilletia caries*, *T. contraversa*, and *Ustilago striiformis* have been obtained in pure culture from infected tissue.

Cultures of *Tilletia* species are obtained by surface sterilizing infected kernels or infected stem sections (approximately 1 mm) with 0·5% sodium hypochlorite and plating them on a special medium formulated by Trione—

KH_2PO_4	613 mg
$MgSO_4.7H_2O$	246 mg
K_2HPO_4	114 mg
$CaCl_2$	55·5 mg
Chelated Fe (sodium ferric diethylene-triaminopentaacetate)	20 mg
$ZnSO_4.7H_2O$	3·52 mg
$CuSO_4.5H_2O$	0·38 mg
$MnSO_4.H_2O$	0·031 mg
$Na_2MoO_4.2H_2O$	0·025 mg
Thiamin HCl	5 mg
L-Asparagine	3 g
Sucrose	20 g
Distilled water	1 litre
Acidity adjusted to pH	6·0

Next, the plates are incubated at 18°C in continuous light from a 25 W incandescent lamp. *T. contraversa* cultures obtained in this manner are capable of producing spores similar to those produced on the host plant (Trione, 1964).

Cultures of *U. striiformis* are obtained by surface sterilizing smut infected leaves during early stages of infection, sectioning them aseptically and pouring melted agar over the leaf segments. When colonies develop, they are transferred to potato dextrose agar on which they grow well and subsequently produce spores.

Although sporulating cultures can be derived from mycelium present within the infected host, it is not a satisfactory method for investigating the saprophytic behaviour of smuts because of the existence of physiologic races, heterothallism and segregation of gametophytic characters. For a complete study of even one physiologic race, it is essential to grow the dicaryophyte as well as its component haplonts.

2. Isolation from spores and sporidia

Smut cultures are derived either from a single spore or by mass isolation. Cultures of *Ustilago striiformis* are established by aseptically sectioning surface sterilized unopened pustules and pouring melted agar over the segments. After incubation under optimum conditions, single germinating spores are picked out and transferred aseptically to fresh potato dextrose agar for further growth (Leach *et al.*, 1946). *U. maydis* cultures are obtained by dusting sieved, washed and dried spores on agar media containing sucrose and casamino acids (Caltrider and Gottlieb, 1966), or adding a spore suspension in sterile distilled water plus 0·01% Tween 20 (polyethylene sorbitan monolaurate), allowing the spores to germinate and aseptically transferring the germinating spores to fresh medium. Alternatively, cultures can be derived by germinating spores on a modified Czapek's medium—

$MgSO_4$	0·30 g
KH_2PO_4	1·25 g
KCl	0·50 g
$FeSO_4$	0·01 g
Asparagine	1·00 g
$NaNO_3$	0·50 g
Sucrose	30·00 g
Agar	15·00 g
Distilled water	1 litre

and transferring the sporidia with a flat-tip needle to fresh medium or Difco cornmeal agar (Schmitt, 1940). Since fewer mutants are produced on these media, they are useful for maintaining stock cultures of *U. maydis*. Often the modified Czapek's medium is preferred. Many other *Ustilago* species can be brought into culture by similar isolation techniques and maintained on media suitable for the species under investigation. Meat extract agar (1%) is suitable for maintaining stock cultures of oat smuts, while Knop solution agar is favourable for sporidial fusions (Sampson and Western, 1938). Spore-producing binucleate sporidial cultures of wheat bunt fungi can be established by germinating bunt spores on water agar, allowing primary fused sporidia to appear and transferring them to potato–sucrose agar with a micromanipulator. If the inoculated plates are incubated at 5°C and subjected to a 12 h photoperiod from a 100 foot candle incandescent lamp, colonies producing spores (very similar to those seen on the host) can be expected (Trione, 1964; Trione and Metzger, 1962).

Sporidial cultures of foliar smuts (e.g., *Entyloma* species) are established by attaching portions of fresh infected leaves to Petri dish lids, allowing the sporidia to fall as discharged on the surface of clear solidified agar, and subculturing the desired colonies on potato–dextrose agar.

Cultures derived from a single spore or by mass isolation are suitable for general studies, but for genetical analysis it is necessary to obtain monosporidial isolates from a single chlamydospore. An elaborate procedure for obtaining monosporidial isolates in *U. maydis* (*U. zeae*) and *Sphacelotheca reiliana* is described by Hanna (1928). But it is relatively simple to pick single sporidia from the promycelium of one germinating spore with a micromanipulator and to establish monosporidial cultures.

Growth and appearance of smut species in culture depend on the medium on which they are grown (Fleroff, 1923). It is advisable wherever possible to consult published plate illustrations of cultural types (e.g., for *Ustilago maydis*, see Stakman *et al.*, 1929; Christensen 1931; for *U. avenae* and *U. hordei*, see Dickinson, 1931; Western, 1936). If testing compatibility of monosporidial lines in *Sphacelotheca sorghi* by sporidial fusion, a slightly alkaline medium (3% malt, 2% agar) can be used (Tyler, 1938). A liquid medium containing 1% malt extract is suitable for testing sporidial fusion in *Ustilago maydis* (Bowman, 1946). The well-known Bauch test (Bauch, 1927, 1932) is often applied as a test for the compatibility of monosporidial lines, but success of this to some extent depends on the choice of the medium. Fischer (1940) obtained satisfactory results using a non-nutrient medium with *U. striiformis*. Lade (1967) found V-8 juice agar (pH 6·3) extremely useful as a differential medium for determining mating types in *U. hordei*.

VI. MAINTENANCE AND PRESERVATION OF SMUT FUNGI

Smut cultures are maintained in an active state by transferring to fresh media every 4 weeks. The merits of various media have already been discussed (see V B). Spore collections obtained from the field are capable of remaining viable for many years and the classic examples are *Tilletia foetida* (25 years), *T. caries* (18 years) *Ustilago hordei* (23 years), *U. avenae* (13 years), *Sphacelotheca sorghi* (13 years) and *Entyloma dahliae* (10 years) (Fischer, 1936). Noble (1934) germinated *Urocystis tritici* spores after storing at low humidity for 10 years. In view of the encouraging results obtained from longevity tests it is not surprising that many workers did not attempt to devise special methods of preservation. Nevertheless, attempts to preserve smut spores by lyophilization have proved successful (Vanderwalle, 1953; Kondo, 1961; Hughes and Macer, 1964). Their success has, however, no practical application since smut spores collected in the field and stored at low temperature are as good as spores preserved either by lyophilization or vacuum-drying. If spores are harvested and stored under controlled conditions, there appears to be no need for any elaborate methods of preservation.

VII. CULTURING "HYMENOMYCETOUS HETEROBASIDIAE"

Early taxonomists included the Exobasidiales and the Tulasnellaceae (Corticiaceae) under "Hétérobasidiés" (see Donk, 1966). Since then the situation has been reviewed and these groups of fungi formerly placed in the "Hétérobasidiés" are now accommodated in the Homobasidiomycetidae (Ainsworth, 1961). If one leaves the controversy to the taxonomists and accepts the present view regarding the status of these groups, then there appears to be no need for discussing them in this chapter. But for completeness a brief survey is, however, made of the isolation techniques. The most abundantly cultured "hymenomycetous heterobasidiomycetidae" are the *Rhizoctonia* state of *Corticium* and species of *Exobasidium*. *Rhizoctonia* species are often associated with the damping-off of seedlings, and cultures are obtained by plating portions of surface-sterilized damped-off seedlings on nutrient agar. Isolation from the hypocotyl region can be effected by plating portions of affected tissues after rinsing several times with sterile distilled water. Pure cultures can be established by hyphal tip transfers and kept in an active state by renewing every 4 weeks (Storey, 1941). When spores occur in nature, cultures can be established from single spores. For general culture work, potato–dextrose, malt, and potato–marmite–dextrose agar are useful (Wellman, 1932; Storey, 1941; Houston, 1945; Flentje, 1956). But for producing fructifications in culture Flentje recommends special soil extract agar (prepared by autoclaving 1 kg air-dried soil with 1 litre of water for 30 min at 1 atm, allowing it to stand overnight and adding to the filtered supernatant liquid 1·0 g dextrose, 0·1 g yeast extract, 0·2 g KH_2PO_4, 20 g agar, distilled water to make up the volume to one litre and finally adjusting the pH to 7 before sterilizing the medium).

Wolf and Wolf (1952) obtained cultures of *Exobasidium camelliae* var. *gracilis* by surface sterilizing affected leaves with 1 : 1000 $HgCl_2$ and plating pieces of it onto malt agar: isolations were also made from basidiospores. Graafland (1953) however had little success with these methods but succeeded in isolating four species of *Exobasidium* by two other methods. In the first, hypertrophic parts were incubated at high humidity in a Petri dish. At the end of 24 h germinating spores were transferred with a needle onto nutrient agar slants. In the second method, portions of hypertrophic parts were fixed to the inside of Petri dish lids with Vaseline so that the discharged sporidia fell onto the surface of solidified agar. When colonies developed they were transferrred to fresh agar slants. This method was used successfully for establishing single spore cultures by Sundström (1960). There is, however, no difficulty in maintaining cultures as species of *Exobasidium* grow well on many ordinary media.

VIII. SIGNIFICANCE OF CULTURE STUDIES

It is now abundantly clear from extensive studies made by several workers that cultures have much significance in epidemiological studies, testing purity and identity of host varieties and studying disease resistance. Hence all investigations other than those dealing with the determination of races in mixtures (i.e., field collections) should be by pure cultures. This is the only way of obtaining reliable information about the pathogenic performance of any particular species. Since physiologic specialization occurs in the Uredinales and Ustilaginales, adequate information should be secured before any attempt is made to identify a physiologic race or biotype differing in pathogenicity and other physiologic characters. Physiologic race identification is essentially based upon the reactions of a group of host varieties to different strains of a fungus. In designating races a uniform system should be adopted and wherever possible the leading publication should be consulted (e.g., Hanna, 1937, for loose smut of wheat; Churchward, 1938, for bunt in wheat; Reed, 1940, for oat smuts; Thomas, 1958, for safflower rust; Oliveira and Rodrigues, 1959, for coffee rust; Waterhouse, 1952, for physiologic race determinations in cereal rusts; Stakman *et al.*, 1962, for identification of physiologic races of *Puccinia graminis* var. *tritici*; Fletcher, 1963, for mint rust. There can be little doubt that detailed knowledge of the behaviour of races in culture could help to understand and control plant disease.

REFERENCES

Aamodt, O. S., and Johnston, W. H. (1935). *Can. J. Res.*, **12**, 590–613.
Ahlgren, C. E. (1961). *J. For.*, **59**, 208–209.
Ainsworth, G. C. (1961). "Ainsworth and Bisby's Dictionary of the Fungi", 5th Ed. Commonwealth Mycological Institute, Kew, Surrey, England.
American Phytopathological Society (1944). *Phytopathology*, **34**, 401–404.
Arthur, J. C. (1921). *Mycologia* (1899–1917), **13**, 230–262.
Arthur, J. C. (1929). "The Plant Rusts (Uredinales)". Wiley, New York.
Atkins, L., Moses, W., and Gray, P. H. (1949). *J. Bact.*, **57**, 575–578.

Bauch, R. (1927). *Biol. Zbl.*, **47**, 370–383.
Bauch, R. (1932). *Phytopath. Z.*, **5**, 315–321.
Baylis, R. J. (1955). *Pl. Dis. Reptr*, **39**, 159–160.
Bean, J., Brian, P. W., and Brooks, F. T. (1954). *Ann. Bot.*, **18**, 129–142.
Bowman, D. H. (1946). *J. agric. Res.*, **72**, 233–243.
Boyer, M. G. (1964). *Can. J. Bot.*, **42**, 335–337.
Brandwein, P. F. (1938). *Bull. Torrey bot. Club*, **65**, 477–483.
Briggs, F. N. (1927). *Phytopathology*, **17**, 747–748.
Brown, M. R. (1937). *Ann. appl. Biol.*, **24**, 504–526.
Buddin, W., and Wakefield, E. M. (1927). *Trans. Br. mycol. Soc.*, **12**, 116–139.
Bushnell, W. R. (1968). *Phytopathology*, **58**, 526–527.
Caltrider, P. G., and Gottlieb, D. (1966). *Phytopathology*, **56**, 479–484.
Calvert, O. H., and Thomas, C. A. (1954). *Phytopathology*, **44**, 609.
Carleton, M. A. (1903). *Jour. appl. Microsc. Lab. Meth.*, **6**, 2109–2114.
Cherewick, W. J. (1965). *Phytopathology*, **55**, 1368–1369.
Christensen, J. J. (1931). *Phytopath, Z.*, **4**, 129–188.
Chupp, C. (1940). *Mycologia*, **32**, 269–270.
Churchward, J. G. (1938). *J. Proc. R. Soc. N.S.W.*, **71**, 362–384.
Clapper, R. B. (1944). *Phytopathology*, **34**, 761–762.
Cutter, V. M. (1951). *Trans. N. Y. Acad. Sci.*, **14**, 103–108.
Cutter, V. M. (1959). *Mycologia*, **51**, 248–295.
Cutter, V. M. (1961). *Mycologia*, **52**, 726–742.
Davies, G. N. (1935). *Iowa St. Coll. J. Sci.*, **9**, 505–507.
Davis, W. H. (1924). *Phytopathology*, **14**, 251–267.
Davis, E. E., Hodges, F. A., and Goos, R. D. (1966). *Phytopathology*, **56**, 1432–1433.
Deighton, F. C. (1960). *In* "Herb. IMI Handbook", p. 81. Commonwealth Mycological Institute, Kew, Surrey, England.
Dickinson, S. (1931). *Proc. roy. Soc. B.*, **108**, 395–423.
Donk, M. A. (1966). *Persoonia*, **4**, 145–335.
Dodge, B. O. (1923). *J. agric. Res.*, **25**, 209–242.
Duran, R., and Fischer, G. W. (1961). "The Genus *Tilletia*". Washington State University, U.S.A.
Durrell, L. W., and Parker, J. H. (1920). *Res. Bull. Iowa agric. Exp. Stn*, **62**, 27–56.
Eriksson, J. (1894). *Ber. dt. bot. Ges.*, **12**, 292–331.
Eriksson, J. (1902). *Zentbl. Bakt. ParasitKde*, Abt. II, **9**, 590–607.
Eyal, Z., Clifford, B. C., and Caldwell, R. M. (1968). *Phytopathology*, **58**, 530–531.
Fischer, E. (1898). *Beitr. KryptogFlora Schweiz*, **1**, 1–120.
Fischer, G. W. (1936). *Phytopathology*, **26**, 1118–1127.
Fischer, G. W. (1940). *Phytopathology*, **30**, 93–118.
Fischer, G. W. (1951a). "The Smut Fungi". Ronald Press Company, New York.
Fischer, G. W. (1951b). *Phytopathology*, **41**, 839–853.
Fischer, G. W. (1953). *Phytopathology*, **43**, 547–550.
Fischer, G. W., and Hirschhorn, E. (1945). *Mycologia*, **37**, 236–266.
Fischer, G. W., and Holton, C. S. (1943). *Phytopathology*, **33**, 910–921.
Fischer, G. W., and Holton, C. S. (1957). "Biology and Control of the Smut Fungi". Ronald Press Company, New York.
Flentje, N. J. (1956). *Trans. Br. mycol. Soc.*, **39**, 343–356.
Fleroff, B. K. (1923). *Trans. Myc. Phytopath. Sec. Russian Bot. Soc.*, I, *Trans. Moscow Branch*, pp. 23–36.
Fletcher, J. T. (1963). *Trans. Br. mycol. Soc.*, **46**, 345–354.

Flor, H. H. (1967). *Phytopathology*, **57**, 320–321.
Flosdorf, E. W. (1949). "Freeze Drying". Reinhold, New York.
Fukushi, T. (1925). *J. Coll. Agric. Hokkaido imp. Univ.*, **15**, 269–307.
Gassner, G., and Niemann, E. (1954a). *Phytopath. Z.*, **21**, 367–394.
Gassner, G., and Niemann, E. (1954b). *Phytopath. Z.*, **22**, 109–124.
Gassner, G., and Straib, W. (1931). *Züchter*, **3**, 240–243.
Gautheret, R. J. (1942). "Manual Technique de la Culture des Tissues Vegetaux". Masson et Cie, Paris.
Govindu, H. C., and Fischer, G. W. (1957). *Phytopathology*, **47**, 522.
Graafland, W. (1953). *Acta bot. neerl.*, **I**, 516–522.
Graham, S. O. (1960). *Mycologia*, **51**, 477–491.
Gretschushnikoff, A. I. (1936). *C.r. Acad. Sci. U.R.S.S.*, **11** (N.S. 2) No. **8**, 335–340.
Grevel, F. K. (1930). *Phytopath. Z.*, **2**, 209–234.
Haarring, F. (1930). *Bot. Arch.*, **29**, 444–473.
Hanna, W. F. (1928). *Phytopathology*, **18**, 1017–1021.
Hanna, W. F. (1929). *Phytopathology*, **19**, 415–442.
Hanna, W. F. (1937). *Can. J. Res.* Sect. C., **15**, 141–153.
Heald, F. D. (1921). *Phytopathology*, **11**, 269–278.
Heald, F. D., and Boyle, L. W. (1923). *Phytopathology*, **13**, 334–337.
Hodges, C. F., and Britton, M. F. (1969). *Phytopathology*, **59**, 301–304.
Hoffmann, J. A., Kendrick, E. L., and Meiners, J. P. (1962). *Phytopathology*, **52**, 1153–1157.
Holton, C. S. (1943). *Phytopathology*, **33**, 732–735.
Holton, C. S. (1964). *Phytopathology*, **54**, 660–662.
Holton, C. S., and Heald, F. D. (1941). "Bunt or Stinking Smut of Wheat". Burgess Publishing Co., Minneapolis, Minnesota.
Holton, C. S., and Siang, W. N. (1953). *Phytopathology*, **43**, 219–220.
Hooker, A. L., and Yarwood, C. E. (1966). *Phytopathology*, **56**, 536–539.
Hotson, H. H. (1953). *Phytopathology*, **43**, 360–363.
Hotson, C. S., and Cutter, V. M. (1951). *Proc. natn. Acad. Sci. U.S.A.*, **37**, 400–403.
Houston, B. R. (1945). *Phytopathology*, **35**, 371–393.
Hughes, H. P., and Macer, R. C. F. (1964). *Trans. Br. mycol. Soc.*, **47**, 477–484.
Hursh, C. R. (1922). *Phytopathology*, **12**, 353–361.
Hulea, A. (1947). *Pub. Inst. Cerc. agron. Român.* No. 99.
Johnson, T. (1931). *Bull. Dep. Agric. Dom. Can.* No. 140, N.S.
Johnston, W. H. (1934). *Can. J. Res.*, **11**, 458–473.
Kaiser, W. (1936). *Angew. Bot.*, **18**, 81–131.
Kavanagh, T. (1961). *Phytopathology*, **51**, 175–177.
Kendrick, E. L. (1957). *Phytopathology*, **47**, 674–676.
Kendrick, E. L. (1961). *Phytopathology*, **51**, 537–540.
Kern, F. D. (1906). *Proc. Indiana Acad. Sci.* (1905), 127–131.
Kienholz, J. R., and Heald, F. D. (1930). *Phytopathology*, **20**, 495–512.
Kiesling, R. L. (1962). *Phytopathology*, **52**, 16–17.
Klebahn, H. (1914). *Z. PflKrankh.*, **24**, 1–32.
Klebahn, H. (1923). *Handb. biol. ArbMeth.*, Abt. **11**, Teil 515–688.
Koenigs, J. W. (1968). *Phytopathology*, **58**, 46–48.
Kondo, W. T. (1961). *Phytopathology*, **51**, 407.
Kreitlow, K. W. (1943a). *Phytopathology*, **33**, 707–712.
Kreitlow, K. W. (1943b). *Phytopathology*, **33**, 1055–1063.
Kreitlow, K. W. (1945). *Phytopathology*, **35**, 152–158.

Lacy, M. L. (1967). *Phytopathology*, **57**, 818.

Lade, D. H. (1967). *Phytopathology*, **57**, 818.

Landen, E. W. (1939). Abstract in *Rev. appl. Mycol.*, **19**, 84.

Leach, J. G., Lowther, C. V., and Ryan, M. A. (1946). *Phytopathology*, **36**, 57–72.

Leukel, R. W. (1936). *Phytopathology*, **26**, 630–642.

Leukel, R. W., Stanton, T. R., and Stevens, H. (1938). *J. Am. Soc. Agron.*, **30**, 878–882.

Ling, Lee (1940). *Phytopathology*, **30**, 579–591.

Lobik, V. I., and Dahlstrem, A. F. (1936). *Summ. sci. Res. WK Inst. Pl. Prot. Leningr.* (1935), **177**–178.

Loegering, W. Q. (1941). *Phytopathology*, **31**, 952–953.

Loegering, W. Q. (1965). *Phytopathology*, **55**, 247.

Loegering, W. Q., and Harmon, D. L. (1962). *Pl. Dis. Reptr*, **46**, 299–302.

Loegering, W. Q., McKinney, H. H., Harmon, D. L., and Clark, W. A. (1961). *Pl. Dis. Reptr*, **45**, 384–385.

Luttrell, E. S., Craigmiles, J. P., and Harris, H. B. (1964). *Phytopathology*, **54**, 612.

Mains, E. B. (1916). *Rep. Mich. Acad. Sci.* (1915), **17**, 136–140.

Mains, E. B. (1924). *Proc. Indiana Acad. Sci.* (1923), 241–257.

Maneval, W. E. (1922). *Phytopathology*, **12**, 471–488.

Manners, J. G. (1950). *Ann. appl. Biol.*, **37**, 187–214.

Manners, J. G. (1951). *Indian Phytopath.*, **4**, 21–24.

McAlpine, D. (1906). "The Rusts of Australia". Govt. Press, Melbourne.

Meinecke, E. P. (1920). *Phytopathology*, **10**, 279–297.

Meiners, J. P. (1959). *Phytopathology*, **49**, 4–8.

Meiners, J. P., and Waldher, J. T. (1959). *Phytopathology*, **49**, 724–728.

Melhus, I. E. (1912). *Phytopathology*, **2**, 197–203.

Melhus, I. E., and Durrell, L. W. (1919). *Res. Bull. Iowa agric. Exp. Stn*, 49, 113–144.

Moore, M. B. (1936). *Phytopathology*, **26**, 397–400.

Nahmmacher, J. (1932). *Phytopath. Z.*, **4**, 597–630.

Newton, M., and Brown, A. M. (1934). *Can. J. Res.*, **11**, 564–581.

Newton, M., and Johnson, J. (1932). *Bull. Dep. Agric. Dom. Can.*, No. 160. N.S.

Niemann, E. (1955). *Z. PflBau PflSchutz*, **6**, 217–225.

Noble, R. J. (1923). *Phytopathology*, **13**, 127–139.

Noble, R. J. (1934). *J. Proc. roy. Soc. N.S.W.*, **68**, 403–410.

Oliveira, Br. D' (1939). *Ann. appl. Biol.*, **26**, 56–82.

Oliveira, Br. D', and Rodrigues, C. J., Jr. (1959). *Garcia de Orto*, **7**, 279–292.

Oort, A. J. P. (1939). *Phytopathology*, **29**, 717–728.

Oort, A. J. P. (1940). *Meded. Inst. Phytopath. Wageningen*, No. 92.

Patton, R. F. (1962). *Phytopathology*, **52**, 1149–1153.

Peltier, G. L. (1925). *Res. Bull. Neb. agric. Exp. Stn*, No. 34.

Pieschel, E. (1931). *Phytopath. Z.*, **3**, 89–100.

Plowright, G. B. (1889). "A Monograph of the British Uredineae and Ustilagineae". Kegan Paul, Trench & Co., London.

Poehlman, J. M. (1945). *Phytopathology*, **35**, 640–644.

Popp, W., and Cherewick, W. J. (1953). *Phytopathology*, **43**, 697–699.

Prasada, R., and Chothia, H. P. (1950). *Phytopathology*, **40**, 363–367.

Raper, K. B., and Alexander, D. F. (1945). *Mycologia*, **37**, 499–525.

Ray, J. (1901). *C.r. hebd. Séanc. Acad. Sci., Paris*, **133**, 307–309.

Reed, G. M. (1940). *Am. J. Bot.*, **27**, 135–143.

Reed, H. S., and Crabill, C. H. (1915). *Tech. Bull. Va agric. Exp. Stn*, 9.

Rodenhiser, H. A., and Holton, C. S. (1937). *J. agric. Res.*, **55**, 483–496.

Rodenhiser, H. A., and Taylor, J. W. (1940). *Phytopathology*, **30**, 400–408.

Röder, K. (1953). *NachrBl. dt. Pflschutzdienst.*, *Stuttg.*, **5**, 140–141.

Roemer, T., Fuchs, W. H., and Isenbeck, K. (1937). *Kuhn-Arch.*, **45**, 1–427.

Rosen, H. R. (1918). *Phytopathology*, **8**, 581–583.

Rowell, J. B., and De Vay, J. E. (1952). *Phytopathology*, **42**, 17.

Rowell, J. B., and De Vay, J. E. (1953). *Phytopathology*, **43**, 654–658.

Sampson, K. (1929). *Ann. appl. Biol.*, **16**, 65–85.

Sampson, K., and Western, J. H. (1938). *Ann. appl. Biol.*, **25**, 490–505.

Sartoris, G. B. (1924). *Am. J. Bot.*, **11**, 617–647.

Schmitt, C. G. (1940). *Phytopathology*, **30**, 381–390.

Schafer, J. F., Dickson, J. G., and Shands, H. L. (1962a). *Phytopathology*, **52**, 1157–1161.

Schafer, J. F., Dickson, J. G., and Shands, H. L. (1962b). *Phytopathology*, **52**, 1161–1163.

Schuster, M. L. (1956). *Phytopathology*, **46**, 591–595.

Shands, H. L., and Schaller, C. W. (1946). *Phytopathology*, **36**, 534–548.

Sharp, E. L., and Smith, F. G. (1952). *Phytopathology*, **42**, 263–264.

Sharp, E. L., and Smith, F. G. (1957). *Phytopathology*, **47**, 423–429.

Singh, J., and Trione, E. J. (1967). *Phytopathology*, **57**, 1009.

Stakman, E. C. (1914). *Bull. Minn. agric. Exp. Stn*, No. 138.

Stakman, E. C., and Christensen, J. J. (1927). *Phytopathology*, **17**, 827–834.

Stakman, E. C., Christensen, J. J., Eide, C. J., and Peturson, B. (1929). *Tech. Bull. Minn. agric. Exp. Stn*, No. 65.

Stakman, E. C., Levine, M. N., and Loegering, W. Q. (1944). *U.S. Dep. Agric. B.E.P.Q.*, E. 617.

Stakman, E. C., Stewart, D. M., and Loegering, W. Q. (1962). *U.S.D.. Agric. Res Serv.*, E 617.

Stevens, K., Melhus, I. E., Semeniuk, G., and Tiffany, L. (1946). *Phytopathology*, **36**, 411.

Stewart, D. M. (1956). *Phytopathology*, **46**, 234–235.

Storey, I. F. (1941). *Ann. appl. Biol.*, **28**, 219–223.

Sundström, K. R. (1960). *Phytopath. Z.*, **40**, 213–217.

Sussman, A. S. (1966). *In* "The Fungi" (G. C. Ainsworth and A. S. Sussman, Eds.) Vol. 2, pp. 733–764. Academic Press, New York and London.

Tapke, V. F. (1935). *J. agric. Res.*, **51**, 491–508.

Tapke, V. F., and Bever, W. M. (1942). *Phytopathology*, **32**, 1015–1021.

Tervet, I. W., and Cassell, R. C. (1951). *Phytopathology*, **41**, 286–290.

Theil, A. F., and Weiss, F. (1920). *Phytopathology*, **10**, 448–452.

Thirumalachar, M. J., and Dickson, J. G. (1953). *Phytopathology*, **43**, 527–535.

Thomas, C. A. (1958). *Pl. Dis. Reptr*, **42**, 1089–1090.

Tisdale, W. H. (1923). *Phytopathology*, **13**, 551–554.

Tisdale, W. H., and Johnston, C. O. (1926). *J. agric. Res.*, **32**, 649–668.

Trione, E. J. (1964). *Phytopathology*, **54**, 592–596.

Trione, E. J., and Metzger, R. J. (1962). *Phytopathology*, **52**, 366.

Turel, F. L. M., and Ledingham, G. A. (1957). *Can. J. Microbiol.*, **3**, 813–819.

Tyler, L. J. (1938). *Tech. Bull. Minn. agric. Exp. Stn*, No. 133.

Van Arsdel, E. P. (1962). *Pl. Dis. Reptr*, **46**, 306–309.

Van Arsdel, E. P. (1968). *Phytopathology*, **58**, 512–514.

Vanderwalle, R. (1945). *Parasitica*, **1**, 58–63.
Vanderwalle, R. (1953). *Parasitica*, **9**, 139–144.
Walker, J. C., and Welman, F. L. (1926). *J. agric. Res.*, **32**, 133–146.
Walkinshaw, C. H., Jewell, F. F., and Walker, N. M. (1965). *Pl. Dis. Reptr*, **49**, 616–618.
Waterhouse, W. L. (1952). *Proc. Linn. Soc. N.S.W.*, **77**, 209–258.
Waters, C. W. (1928). *Phytopathology*, **18**, 157–213.
Weber, G. F. (1922). *Phytopathology*, **12**, 89–97.
Wellman, F. L. (1932). *J. agric. Res.*, **45**, 461–469.
Western, J. H. (1936). *Ann. appl. Biol.*, **23**, 245–263.
Western, J. H. (1937). *Phytopathology*, **27**, 547–553.
Wilkinson, R. E., and Kent, G. C. (1945). *Iowa St. Coll. J. Sci.*, **19**, 401–413.
Williams, P. G., Scott, K. J., and Kuhl, J. L. (1966). *Phytopathology*, **56**, 1418–1419.
Williams, P. G., Scott, K. J., Kuhl, J. L., and Maclean, D. J. (1967). *Phytopathology*, **57**, 326–327.
Woodward, R. W., and Tingley, D. C. (1941). *J. Am. Soc. Agron.*, **33**, 632–642.
Wolf, F. T., and Wolf, A. F. (1952). *Phytopathology*, **42**, 147–149.
Wright, E. (1933). *Phytopathology*, **23**, 487–488.
Zehner, M. G., and Humphrey, H. B. (1929). *J. agric. Res.*, **38**, 623–627.
Zimmer, D. E. (1962). *Phytopathology*, **52**, 1177–1180.
Zscheile, F. P. Jr. (1965). *Phytopathology*, **55**, 1286–1292.

CHAPTER VIII

Basidiomycetes: Homobasidiomycetidae

ROY WATLING

Royal Botanic Garden, Edinburgh, Scotland

I. INTRODUCTION

The growth of the Basidiomycetes under artificial conditions has always been the poor sister to the culture of the Ascomycetes and related micro-fungi. Although some members of the Homobasidiomycetidae have assisted in both genetical and physiological studies the species exploited have been limited and are perhaps rather specialized ones at that, e.g., *Coprinus cinereus* and *Schizophyllum commune.* The growth and fructification of members of this group in the laboratory is fraught with seemingly inexplicable problems, even mystics, which are also reflected in the now ancient methods of culti-vation of edible agarics in different parts of the world today. Brefeld (1877 1888; 1889), however, and many early workers (e.g., Constantin, 1891; Constantin and Matriechot, 1889; Ward, 1897; Biffin, 1898) had remarkable success in germinating spores and producing fruit-bodies, but instead of progressing during the lasty seventy-five years we have stood still if not retrogressed. Because the study of the higher fungi in culture can assist in their systematics, in understanding their biology and the composition of wild populations, as well as the economics of timber production, etc., it is very desirable that we find procedures which give reproducible results.

II. COLLECTING TECHNIQUES

As with other fungi it is always necessary to bear in mind when collecting material for culture that the specimens should be suitable not only for laboratory work but also for later preservation as herbarium voucher material In this way a record is kept of the starting collection for future reference. Unlike microfungi, where several stages of development are present on the same substrate at any one time, it is desirable with the macromycetes to collect a selection of fruit-bodies covering various ages from buttons or conks to the mature fruit-body. Toadstools and mushrooms (agarics) should be dug up complete, with the aid of a strong knife or fern trowel and placed in tins, or the more delicate specimens in tubes or grease-proof paper twists; for larger specimens, such as polypores, brown paper bags are admirable. Lignicoles should be cut off the wood carefully along with a small portion of that substrate; this will ensure that the basal structures are retained intact and also allows the substrate to be re-examined and re-determined later if found necessary. As in the case of terrestrial specimens they should be handled as little as possible. The substrate must always be noted and if the fungus was growing on or under a tree the tree-species should be recorded. Note any characters which may change on the journey back to the laboratory, such as smell, fresh colours and stickiness. Coprophilous fungi can be conveniently collected simply by taking some of the substrate, air drying carefully and then incubating in a damp chamber when time permits; the fruit-bodies will develop normally. Short notes on the fresh specimen accompanied by a sketch, coloured if at all possible, are really necessary in order to make full use of the material; if in doubt as to the identity of a specimen expert opinion should always be sought. For cultural purposes a fresh fruit-body in active growth is always preferable.

III. PROBLEMS AND PRINCIPLES OF ISOLATION

A. Isolation from fruit-body tissue

1. *Isolation from hymenial and cap tissue*

Just as the most successful media for growth of fungi depends on the species, even on the individual used, so the area from which the tissue should be selected for culture may differ one species to another. The purpose of the fruit-body is to produce basidiospores and, although much of the fruit body of a Basidiomycete may die or become dormant, that area which is most active for the longest period is often the actual spore-producing tissue (see Hiromoto, 1961). The basidia and related structures are produced on the hymenophoral trama and this in the larger fungi is frequently the best place to seek the tissue plug which will be utilized as an inoculum. The fruit-body should be cut, or preferably broken, under as sterile conditions as possible and a piece of tissue about 3 mm square taken with a sterile scalpel from the central area of the gill, tube wall or fold where it joins the main flesh of the fungus. With the woodier specimens it may be necessary to cut away the external skin with a sterile scalpel first and with a second sterile scalpel cut out a small piece of tissue. With soft, watery fungi, although the procedure is the same, bacteria are invariably present; because of the very nature of the slow growing bracket fungus, bacteria and other fungi (see IV, B) may grow within the fruit-body and great care must be taken in culturing from them.

Agarics are admirable for giving bacteria-free cultures but with less amenable fruit-bodies such as resupinates badly contaminated cultures are constantly obtained. In such cases either other methods must be resorted to or the fruit-body can be first washed free of unwanted debris, etc., by using a water softener (Warcup and Talbot, 1966) and then teased out in a drop of sterile water to which has been added a little antibiotic (see IV, B) and shaken; the fragments can then be plated out.

Sometimes the hyphae, particularly if the tissue has been taken from a mycorrhizal fungus, do not grow out from the block if placed on solid agar, but when the tissue is floated on the edge of a small piece of sterilized paper on a nutrient solution hyphal filaments can be observed (Norkrans, 1949).

2. *Isolation from stem tissue and velar fragments*

Some fungi have little or no cap tissue and extraction of the smallest piece of tissue would mean also taking tissue in contact with the environment; in such cases stem tissue can be very suitable material. The stem (or stipe) of the fruit-body when present acts both as a water transporter and a support for the usually more delicate apparatus producing the basidiospores. Thus there are areas of the stem which may be quite dead but others which are in active growth.

With a sterile scalpel the fresh fruit-body should be cut or broken to expose a clean inner surface, a small block of tissue should be then taken from the stem apex, from just within the cortex or if hollow away from the air spaces, or from the stem base where active connections exist with the substrate.

With more delicate fungi the stem can be either plated out directly on to agar or can be washed in vials of water treating them as if they were rhizomorphs (see below). The hyphae of the stem or stem fragments on plating out will proliferate; contamination appears to decrease or even disappear upwards from the basal section.

Many fungi possess a veil which protects the developing spore-producing layer. Fragments of this veil from an actively growing fungus can be treated as delicate stems and plated out directly on to agar or if more leathery treated as segments of a rhizomorph. The inner surface of the ring of a toadstool which has yet to expand is very suitable as are the filamentous or subglobose cells of certain Coprini.

B. Isolation from vegetative phase

1. *Isolation from soil*

Dilution plates and hyphal isolation techniques (Warcup, 1957) have resulted in the culture of several previously little known Homobasidiomycetidae (Warcup and Talbot, 1962) and such techniques have been adequately described above by Barron (see Chapter XIV). Sclerotia can be obtained either by actually digging amongst the bases of fructifications or for smaller ones by sieving, or by utilizing nematological techniques (Warcup, 1959). Active mycelium can be obtained from these resting-bodies by slicing or transferring the softer medulla to agar, e.g., *Lentinus* (Gallymore, 1949) or by surface sterilizing and burying the sclerotium in sterile sand flushed with nutrient solution or embedding it in nutrient agar.

Rhizomorphs or mycelial fragments can be directly picked from the soil and if dry, shaken free of soil debris, or if moist, washed in sterile water (Warcup, 1959). However, it has been found advantageous to treat the rhizoids of *Agrocybe* spp. as if they were mycorrhizal short roots and wash as outlined by Harley and Waid (1955). The rhizomorph is first washed several times, then cut into segments and finally plated out.

Vegetative mycelium can be found attached to the base of the fruit-body and in an actively fruiting population this is frequently a very reliable place from which cultures of otherwise temperamental fungi can be obtained. Washing several times with water (see above) has been found adequate although surface sterilization with 0·1% hydrochloric acid for 10 sec has been recommended by Levisohn (1955).

2. *Isolation from wood samples*

Many basidiomycetes can be directly isolated from wood specimens provided the samples are taken from the edge of the decay where the hyphae are active and in profusion. The wood sample should be either washed with

1/1000 w/v mercuric chloride solution, flamed lightly (Cartwright and Findlay, 1958) or washed with chloros. A clean surface should be exposed with a sterilized scalpel and a small piece of wood cut out and plated on nutrient agar. The mycelium may take some time to develop and the sample should be left for at least 4 weeks. For more sophisticated techniques of cutting wood and taking samples see Cartwright and Findlay (1958).

3. Isolation from mycorrhiza and root associations

Downie (1959) and later workers surface sterilized orchid roots with 0·1% mercuric chloride for 2 min, washed them in sterile water, cut the roots aseptically into two or three segments and plated them out for the growth of thanatephoroid fungi. With subtle manipulation however coils of the endophyte can actually be taken out of orchid roots squashed in sterile water and these can then be plated out. For ectotrophic mycorrhizal fungi, a group which includes a large range of basidiomycetous species, the short roots are washed several times in distilled water (Harley and Waid, 1955) cut into portions and plated out.

Isolation from roots can be also accomplished simply by dissecting the roots in sterile water into two parts (Waid, 1957), an outer cortex and an inner stele. These are either plated out direct or incorporated into nutrient agar after further fragmentation; dispersal in the latter case is carried out by shaking and rotating the plate before the agar solidifies.

C. Use of propagules

1. Basidiospores

All the techniques described above result in the preparation of dicaryotic colonies, but often it is necessary to commence with the monocaryon. Unless one carries out lengthy screening experiments with various chemicals inducing dedicaryotization or mechanical disruption of the hyphae the isolation of monocaryotic components is best achieved by using basidiospores; these are, however, traditionally erratic in germination. Although many aspects have been reviewed (Madelin, 1966) the plain fact is we still need to know a lot more about their physiology. Only with great difficulty have some spores been induced to germinate (Gottlieb, 1950; Kauffman, 1934; Ferguson, 1902; Duggar, 1901; etc.) and still spores of some species have failed to co-operate. It is impossible to indicate all the possible conditions which have been and must be tried to induce germination, but they include cold and heat treatment, proximity to hyphae of the same or different taxon (Lösel, 1964), proximity to other micro-organisms (Fries, 1941), specific metabolites (Petersen, 1960; Lösel, 1967), animal and plant extracts (see V) or refined chemical substrates (Robbins, 1950). However, in many cases these treatments (Fries, 1966) are needed only when the spores are initially allowed to dry out. Completely imbibed spores can be obtained by cutting a macroscopically clean piece of hymenial tissue from a fructification and placing it on a block of agar or vaseline on the underside of a Petri-dish lid, mounting it so that the fruiting surface is over the agar. In the case of *Exobasidium* and similar plant parasites the actual tissue containing the parasite can replace the tissue segments. The fruiting surface is then allowed to discharge its spores; this will take anything from one quarter to a forty-eight hour period. Longer periods are very unsatisfactory for they increase the possibility of contamination. When a deposit of spores is found on the agar, the lid is replaced by a second sterilized one and the dish incubated; during the spore-shedding it is often useful to rotate the lid periodically in order to sow the spores over a wider area of agar surface. Germination may ensue immediately or may take a further period.

This technique is cumbersome in the field and thick glass test tubes have been used at the United States Department of Agriculture laboratory by Miller (personal communication) and subsequently in Edinburgh and have been found to be much more satisfactory. In this technique the first 30 mm of an agar slope is cut off and rotated and pushed down so as to be immediately above the top of the remaining area of the slope. The fungus tissue is placed on this small block of agar and the basidia allowed to shed their spores; on completion of spore-shedding the block of agar and tissue is removed intact and discarded.

For more critical work more sophisticated procedures are called for; thus spore-tetrads can be isolated directly from the basidia (Papazain, 1950b) or after discharge (Moore, 1966)—such techniques are common in genetical research. In some circumstances, e.g., when using herbarium material, specimens received during routine identification-service, it may be necessary to use spores which have dried out; some species have spores whose germination is unaffected by excess free water and in this case a spore suspension can be diluted and added to the agar in the normal way. However, a higher percentage germination is obtained by incubating the dry spores in a Petri dish overnight in a humid atmosphere and in some species this is the only way in which germination may be induced (Watling, 1963). Producing a spore suspension or plating out thereafter gives acceptable germination; the induced dormancy patterns seem to follow those described for seeds (Barton and Crocker, 1948). Thus spores from a spore-print or the crushed parts of the hymenial tissue of a dried specimen can be effectively used by employing this last technique. The spore-print can be obtained from the fructification by lying it hymenial surface down on a glass-slide.

Spores of most coprophiles and many lignicolous fungi germinate immediately when plated on agar; some do not, possibly due to the presence of self inhibitors produced by a high concentration of spores. In this latter case the spore-print is cut out, agar and all, and mixed in a microblender; the debris is then plated out thus distributing the spores over a wider area. If the spores are easily wettable it is simply necessary to wash the spores off the agar with sterile water.

2. *Asexual spores*

Homobasidiomycetidae produce asexual stages far more frequently than many microbiologists, indeed mycologists, appreciate; these can be used to advantage. Asexually produced spores can be of a number of different forms but what is important is that they can be either dicaryotic or monocaryotic. One of the asexual processes found in the Basidiomycetes is chlamydospore production and this may be either on or in the fruit-body, or on the vegetative hyphae (e.g., Kligman, 1942). Frequently the chlamydospores are large enough to be picked off by the dry-needle technique (see Booth, Chapter I) or dissected out in a drop of sterile water. Other asexual spores are known, some formerly being called oidia, and in the future there is little doubt we will recognize more. Bayliss as early as 1908 used oidia for starting cultures of wood rotting fungi (cf. Brodie, 1931). Many of the asexual stages appear already in the schemes of classification of the Fungi Imperfecti, e.g., *Oedocephalum* in *Heterobasidion*, *Oidium* in *Botryobasidium*. The techniques described by Booth (Chapter I) are suitable here.

Under this same heading one can treat the easily removable veil cells of certain agarics, e.g., *Cystoderma*, *Phaeolepiota*, and *Coprinus*, which in nature may even act as propagules. In this instance the powdery surface from the epithelium can be rubbed across the agar and allowed to develop further.

IV. CULTIVATION AND MAINTENANCE

A. Procedure after isolation

Once the hyphae have begun to grow out from the inoculum tissue, wood, root, etc., they should be cut off and placed on a clean plate. In the case of spores it is best to pick out several germinating spores (check with fruit-body) and allow them to intermingle on fresh agar; this will ensure that an anastomosing colony with genetic constitution suitable for a balanced secondary mycelium (Buller, 1922; Kniep, 1928) is formed. Because in the majority of fungi the apical cells contain several nuclei, only the tip of a single hyphae is required (four or five cells) in order to give a balanced thallus. The colonies so produced should be grown on general or specialized media and the colony examined periodically for sectors and contaminants (see below). "Clean" areas, 5 mm^2, can then be cut out and maintained as a pure stock.

Downie (1959) isolated *Thanatephorus* spp. on P.D.–Marmite agar, Warcup (1959) used Dox + yeast agar for sclerotia, etc., and Cartwright and Findlay (1958) 1·5–2·5% prune agar and malt extract variously adjusted with 0·5% malic acid (acidic) or $Na_2H PO_4$ (alkaline) for wood rotting fungi; Nobles (1948) has used malt agar, Lange (1952) dung extract agar, etc. In Edinburgh it has been found sufficient to isolate the fungus on a potato extract agar, e.g., P.D.A., P./C.A. or a malt extract agar (1·5–3%). In all cases the agars have been made up on the spot from raw material in preference to "made-up" media and only when this substrate has failed have more specialized media been considered (see V). Incubation conditions should fall into line with the ecology of the fungus but room temperature, 25°C or 27°C, in the dark with light at weekly intervals is favourable to the majority of Homobasidiomycetidae.

B. Restrictions and purification

Clamp-connections characterize the Basidiomycetes, so the textbooks inform the student, but there are many members of the Homobasidiomycetidae which lack them. When clamp-connections are absent only experience can teach the investigator how to recognize the mycelium belonging to a basidiomycete; obviously mould contaminants can be recognized by virtue of their characteristic fructifications. Nevertheless several fungi, e.g., *Chaetomium globosum* and *Gymnoascus* spp., appear to reach maturity very slowly and it may be several subcultures after the primary isolation before they fruit. *Fusarium sporotrichoides* is frequently isolated from the tissue of *Hygrophorus* spp.; *Trichoderma viride* growing as a parasite is also frequently isolated along with basidiomycete hyphae. What is more surprising is the publication (Griffith and Barnett, 1967) recently of results describing the hyphae of different species of Basidiomycete parasitizing one another. It is therefore doubly necessary to be sure the right fungus has been isolated, particularly as perennial fruit-bodies even though still growing may be permeated at a later stage by its own mycelium, as in *Ganoderma applanatum*, or by the mycelium of an alien fungus. The phenomenon is not confined to woody fungi for in the field several records are now available of intimacy between hyphae of unrelated basidiomycetes, e.g. *Suillus bovinus* and *Gomphidius roseus* (Watling, 1964), *Rhizopogon parasitica* and *Brauniellula nancyae* (Smith, 1966). This may be the case in the laboratory also (Kemp, personal communication) where *Coprinus pellucidus* has been isolated from the stipe tissue of other unrelated species of *Coprinus*. Although at first extremely perturbing, the fact still remains and has now been reproduced several times; it is however understandable when one considers that several coprophilous fungi when grown in culture together do not separate out; the advancing front of the colony is composed of the hyphae of the various species.

For control of bacterial contamination several suggestions have been put forward over the years ranging either from the addition to the agar of inorganic salts, to the use of antibiotics, or simply to a change in culture technique. Wood-decaying fungi are frequently able to decompose (and utilize) phenolic compounds some of which are toxic to both bacteria and other fungi. It was with this in mind that Russell (1956) suggested a media which contained 0·006% of σ-phenyl phenol and 0·004% α-naphthol for growing certain Basidiomycetes.

In Edinburgh we have tried to do without antibiotics in case any slow change is induced, for unlike the ascomycetous fungi we are ignorant of many aspects of the physiology of these higher fungi. However, sometimes antibiotics must be used and a measure of success has been obtained using Rose Bengal (0·035 g/litre) and potassium tellurite (0·1 g/litre). Streptomycin (30 μg) (Martin, 1950) and similar broad spectrum antibiotics have been also used.

Warcup and Talbot (1962) have found that inoculation of the basidiomycete on to sterile, wet, wheat straw will allow rapid separation of bacteria and fungi because of preferential colonization. A further way is to allow the mycelium to grow round a glass tube on to the medium (Cartwright and Findlay, 1958) or to cut the colony out and place the mycelium face down on a fresh agar surface. The mycelium in the first case will grow over the tube and if held horizontally 2–3 mm from an agar slope will spread on to that agar; in the second case the hyphae can be re-isolated on the upper surface of the agar block because the mycelium has grown through more rapidly than the bacteria have been able to colonize the block.

In general Homobasidiomycetidae produce a rather fluffy to silky colony frequently aggregating in areas to form strands or knots of mycelium. This fluffy colony is particularly true of wood-rotting fungi; several coprophiles, however, produce a submerged, greasy-moist colony with very little aerial growth. Commencing from spores of certain species one frequently can isolate two distinct types of colony, a fast-growing colony and a small, slower-growing colony. These appear to be inherent properties of the fungus, but there are some isolates which although commencing healthy and active become lazy with hyphal degeneration. This appears in gross characters similar to a virus disease of cultivated mushrooms described by Holling, Gandy and Last (1963). Such cultures the author has always discarded but techniques which may be applicable have been now developed to free mushrooms of virus by heat treatment (Holling, 1962).

C. Long-term maintenance

Subculturing is best carried out approximately every six months on to wooden blocks, into sterile dung, etc., the substrate chosen depending on the ecology of the original isolate (see V). This is because when the cultures are maintained on agar there is a fall off in mycelial characters and ability to fruit due to of lack of certain growth factors. Carbon/nitrogen relationships of the media on which Homobasidiomycetidae are maintained, as in natural environments, are often critical (see Cooke, 1968). Mycorrhizal fungi are generally more difficult to maintain on semi-synthetic media (Lilly, 1966) and vitamins such as thiamine or one of its constituent moieties are often required (Melin, 1953); Robbins (1950) has surveyed the growth requirements of Basidiomycetes. Other techniques, e.g., use of oil, parallel in every way those adopted for Ascomycetes and Fungi Imperfecti.

V. INDUCING FRUCTIFICATION

Many more macromycetes of all ecological types are in culture on synthetic or natural media now than even ten years ago and there is little doubt this number will increase. Fructification of this same number, however, has been very poor and in almost all cases has been confined to dung- or wood-inhabiting species; media have always been resorted to which contain extracts of such material as yeast, malt, dung, etc. No single standard method can be expected to induce fructification in all the basidiomycetous fungi equally for they behave very differently one to another even in their natural habitats.

It has always been a strong personal conviction that it is necessary in mating experiments to go from fructification to fructification and not simply to rely on changes in the vegetative phase (see VI, B). One criticism which could be justly lodged by the geneticists and/or physiologists is that the production of fruit-bodies is difficult and often extremely erratic except in the few species already selected, although these species are not as few as one would be led to believe from Emerson and Davis (1966). It is now up to the mycologist to improve his techniques in order to bring about consistent results.

Flammulina velutipes, *Schizophyllum commune* and species of *Agrocybe*, *Coprinus*, *Panus*, *Psathyrella*, *Psilocybe*, etc., produce normal fruit-bodies in culture; others, however, remain sterile or produce abnormal or abortive fructifications. Many fungi, including mycorrhizal species, produce fruit-body initials but these come to a halt in their development for little explored reasons and a common feature of many fungi is that the original isolate fruits but subsequent subcultures do not.

When using the stipe of *Coprinus* spp., fruit-bodies often appear directly on the stipe tissue, whilst other *Coprinus* spp. fruit directly from the pad of hyphae produced from a spore or spore group. In the case of certain strains of *Coprinus pellucidus* fructification is only obtained when bacteria are also present in the culture; a three-spored strain of *Coprinus narcoticus* has up to now only fruited from the centre of a bacterial colony (cf. Urayama 1961).

Generally for the production of fructifications a supply of rich, moist, well-aerated media with high relative humidity and exposure to light of moderate intensity are all required (Badcock, 1941; 1943) these facts must always be considered in the light of the ecology of the fungus—preferably the original fruit-body from which the isolate was made. With this in mind a culture system has been designed in Edinburgh to induce the fructification of members of the *Bolbitiaceae* (Watling ,1963); the media was prepared to simulate something more resembling the soil system than the familiar plaque of agar in a Petri dish. Thus the shredded, moist paper pulp loosely interwoven and mixed with nutrient agar incorporated into the Edinburgh system gave air cavities, pockets of liquid and plugs of solid medium. The thin layers of nutrient agar are quickly colonized and have successfully produced the necessary food reserves required for fructification. This media is a sophistication for terrestrial fungi of a medium originally suggested for lignicolous fungi by Badcock (1941; 1943) and which consisted of sawdust impregnated with an accelerator consisting primarily of bone and maize meals. Etter (1929) has used a mixture of corn meal, corn starch and powdered wood kept moist by the addition of a solution of malt extract (2·5%). Vermiculite kept within porous, hollow, soft tiles and moistened with nutrient solutions have also successfully been used for *Schizophyllum* (Papazain, 1950a). A constant, fairly high relative humidity is necessary for fructification and an effective although simple procedure is to cut a disc of agar from a colony in a Petri dish and fill it with either water or a nutrient solution, e.g., dung or malt extract; a useful technique to bear in mind when inducing spore germination.

Flentje (1957) and Warcup and Talbot (1962) have found it necessary to use casing soil, either natural or sterilized, to induce fructification and this has been found a requisite in some of the Edinburgh experiments. However, this technique is normal practice in cultivating mushrooms (p. 15). Soil containing high quantities of organic material when sterilized, however, often produces unsuitable chemicals, toxic compounds often inhibiting fruiting (Melin, 1948; Dawson *et al.*, 1965). Toxic material, particularly volatile substances, are known to accumulate in culture. Some may discourage fructification and, even inhibit growth, while others are stimulatory (McTeague *et al.*, 1959; Lösel, 1964); these compounds may be removed by forced aeration (cf. Plunkett, 1956 and see Cartwright and Findlay, 1958; see VI).

The techniques of Badcock, Etter, and others, do not explore temperature ranges, etc.; it is well known that some fungi require high temperatures for growth whilst others need only room temperatures. Similarly some fungi require heat treatment for fructification, e.g., *Coprinus delicatulus* (Apinis, 1965), and others cold treatment, e.g., *Flammulina velutipes* (Kinugawa and Furukawa, 1965) and *Flammula* spp. (= *Pholiota*) (Deneyer, 1960). It cannot be over-emphasized therefore that ecological studies should always run parallel with culture studies.

Lohwag (1952) has indicated all those species which have been described in the literature as having been induced to fruit, but since that date several other species have been taken to sporulation. Up until Lohwag's time many decoctions had been relied on; celery, beet, alfalfa, prune, carrot and dried fruit solidified with agar (or gelatine in the earliest work) (Lyman, 1907; Baylis, 1908; Long and Harsch, 1918; Hein, 1930; Lutz, 1925a, b). Sterile or unsterile dung (Schenck, 1919), blocks of sterile wood (Johnson, 1920; Price, 1913; Hopp, 1938; Falck, 1909; Brooks, 1911; Glaser and Sosna, 1956; Macrae, 1955; etc.) have also been used; see also VI, C for cultivation of *Lentinellus edodes*. Malt agar and Hagem's medium have been very useful and served mycologists well throughout the period of "trial and error" to the present day (cf. Koch, 1958; Bayliss, 1908; Aschan, 1954; Miller, 1967; Modess, 1941).

Several fungi, particularly *Coprinus*, will fruit on plates or in flasks containing semi-synthetic media provided some natural extract, decided upon by consideration of the ecology of the taxa, has been added. In the coprophilous fungi Hesseltine *et al.* (1953) consider a growth substance required by many fungi is present in dung and named it coprobin. However, as far as many species of *Conocybe*, *Coprinus* and *Bolbitius* are concerned the dung can be replaced by an extract of soil; Lange (1952) successfully fruited thirteen members of the Setulosi section of *Coprinus* and incorporated the information obtained in the taxonomic treatment of this group (see VI, B). He found, and this has subsequently been confirmed in Edinburgh for other species of *Coprinus*, for *Conocybe*, *Psathyrella* and *Panaeolus* spp., that the best medium was of rather loose constitution with a gel concentration of no more than 1·5%. Lange used a modification of an original recipe by C. H. Kauffman which was based on a fairly well defined medium with the addition of dung extract and peptone. This and media made from a mixture of potato/carrot (+dung,) corn meal (+dung), malt extract (+dung) have all been utilized with success.

Whole groups of the higher fungi have failed to fruit in culture but on considering physiological and ecological aspects of a single taxon or groups of taxa there is little doubt many of the mysteries about Basidiomycetes will gradually disappear. Some of the initial work is already available in the literature as physiological exercises (see below), for some such procedures were necessary preliminaries to studies with a more genetic or systematic bias (e.g., Deneyer, 1960).

Ectotrophic mycorrhizal fungi are even more coy to fruit than the most diffcult of the Coprini and it is more by chance that those which have fructified in culture have done so; the experimental techniques at the moment do not give reproducible results in all laboratories. Some researchers have recorded producing the short-roots and subsequent fruiting of the suspected mycorrhizal fungus by exposing plants grown under sterile conditions to the spores or mycelium of the fungus associate under observation. This technique has been successfully used for the gastromycete *Scleroderma aurantium* (Thapar *et al.*, 1967) a member of a group of fungi very difficult to grow, let alone fruit and the hymeromycetes. A similar technique is used in ignorance in the semi-artificial cultivation of truffles by peasants in Southern France (Singer, 1961). Modess (1941), Pantidou (1961b; 1962; 1964) and McLaughlin (1964) have all used the semi-synthetic technique of culture and successfully raised fruit-bodies from those primordial initials which are frequently found in culture but which, unless special care is taken, fail to develop further. Shock waves can induce or inhibit primordial development; atmosphere content and flow are also important as is quantity and quality of light, and its periodicity. Some of the boletes which have been taken to completion in culture are border-line cases in a family characteristically mycorrhizal and cover fungi which are distinctly lignicolous *Boletus sulphureus* and *Boletus lignicola* (Pantidou 1961b, 1962); we have failed to fruit as yet a good *Suillus* or a true *Boletus* (i.e., *Tubiporus*), although primordia have been seen (Modess, 1941; Pantidou, personal communication). In this family there is a grading of dependence from tree-relationship to less specialization and it is pleasing that the successful cultivation of those at the lower end has now been performed, something many were sceptical about even 20 years ago.

VI. SPECIFIC PROBLEMS

A. Physiological and genetical studies

Much has been written on the physiology of the higher fungi in culture and now we have an amazing amount of information. Much of this is confined to the vegetative stage (Fries, 1949; Norkrans, 1949; Modess, 1941) and less, although still a significant amount comes from studies on the effects of light, growth substances, atmosphere content, humidity, etc., on the production of fructifications (Plunkett, 1956; Borriss, 1934; Billie-Hansen, 1953; Lu, 1965; etc.). Several media have been specially designed for studying various aspects of growth, e.g., Norkrans–Melin media (Norkrans,

1949), and both special and sophisticated techniques have been developed, e.g., running medium, for specific projects.

Few higher fungi have been used to study the genetics of the fungi as a whole and it is rather unfortunate that those which have been chosen are not always typical of the group, e.g., *Schizophyllum*, or that there has been some confusion in the identity of the fungus used, e.g., *Coprinus cinereus* (as *C. lagopus*, Day, 1959). Much work unfortunately has been based primarily on the production of clamp-connections in culture and rarely has the fungus been taken to the ultimate conclusion, i.e., basidium production. This has many drawbacks for it can be demonstrated in several taxa that production of clamp-connections is dependent on the medium used and in some cases they may be totally absent, although present in the original isolation and in the original fruit-body. Many active field mycologists know that in a single fruiting population clamp-connections may be consistently present, absent or variable from one area of the fruit-body to the other and from vegetative to spore-producing tissue; further critical study of the use of the clamp-connection in taxonomy has frequently been advocated (Singer, 1962; Smith, 1963; Hesler and Smith, 1963; Watling, 1967; etc.). However, further cultural studies of the way in which clamp-connections vary will assist in taxonomic studies (see below). Day (1959), working with *Coprinus*, produced a semi-synthetic medium on which to carry out his mating experiments but fructification was always induced on sterile dung (also see end of references).

B. Taxonomic implications

Brefeld (1877–89) realized the importance of mycelial characters in the classification of the higher fungi. By careful and controlled growth the use of cultures in identification of the higher fungi has been successfully employed by those studying the wood-rotting members of the Aphyllophorales. From the time of Falck in Möller (1902–9) and Lyman (1907), many researchers have used this technique (Fritz, 1923; Campbell, 1938; Refshauge and Proctor, 1936; McKeen, 1952; Maxwell, 1954; Nobles, 1948, 1965; etc). Keys have been prepared not only to genera but to species on phenoloxidase activity, clamp-connection characters and colony texture. Many lignicolous fungi possess laccase and/or tyrosinase or neither as first noticed by Bavendamm (1928), and by incorporating tannic acid into the media following Davidson *et al.* (1938) it has been possible to demonstrate this character visually.

In the above work a rigorous procedure is adopted. From an original inoculum the fungus is grown for one week, then inocula from the colony so produced are placed at the edge of a Petri dish containing about 30 ml tannic acid agar; the plates are incubated for six weeks. This technique, but using 2% malt agar in place of the tannic acid agar, has been found useful for saprophytic fungi also, allowing certain genera (and in some cases species) to be recognized, e.g., *Coprinus, Conocybe, Psilocybe* and *Panaeolus*. The technique, although only widely used in distinguishing commercially important fungi, has very great potential in understanding better the restrictions (or limits) of species in the rather chaotic Agaricales (e.g. Pantidou, 1961a; Pantidou and Groves, 1966). Sundström (1964) has used parallel techniques in studying the taxonomy of members of the Exobasidiales. The utilization of culture characters if they can be found to be constant will be indeed welcome, for any additional evidence to assist the identification of a fungus in a group of organisms which possess so few characters is most valuable.

The accounts above have really dealt solely with hyphal characters although mention has been made of fruiting surfaces when they appeared. Lange (1952) as reported previously, however, has fruited several species of the *Coprinus* section *Setulosi*, confirming the reliability of characters such as spore shape, size, etc., cystidial distribution and morphology. Warcup and Talbot (1966) have successfully fruited cultures of fungi whose perfect stages were unknown until their work; this reflects the great potential which the mycologist can look forward to seeing realized in the future.

C. Mushroom growing

Singer (1961) has covered the necessary procedures of isolation and growth of mycelium, and the production of fruit-bodies of the species of *Agaricus*, *Volvariella* and *Lentinellus* grown for human consumption. His work supersedes earlier commercial and government bulletins which frequently only deal with *Agaricus hortensis* s. lato., e.g., Duggar (1905); Atkins (1956); Jackson (1951).

Treschow (1944) has considered certain physiological aspects of mushroom culture adequately, as has Mader (1943); further information may be sought in articles appearing in *Mushroom Growers' Journal, Mushroom Science*, etc.

REFERENCES

Aschan, K. (1954). *Physiologia Pl.*, **7**, 571–591.
Ainsworth, G. C., and Sussman, A. S. (1965). "The Fungi", Vol. I. Academic Press, London and New York.
Ainsworth, G. C., and Sussman, A. S. (1966). "The Fungi", Vol. II. Academic Press, London and New York.
Apinis, A. E. (1965). *Trans. Br. mycol. Soc.*, **48**, 653–656.
Atkins, F. C. (1956). "Mushroom Growing Today" Faber & Faber, London.
Badcock, E. C. (1941). *Trans. Br. mycol. Soc.*, **25**, 200–205.
Badcock, E. C. (1943). *Trans. Br. mycol. Soc.*, **26**, 127–132.

Barton, L. V., and Crocker, W. (1948). "Twenty Years of Seed Research". Faber & Faber, London.

Bavendamm, W. (1928). *Z. Pfl-Krank. Pfl-Schutz.*, **38**, 257–276.

Bayliss, J. S. (1908). *J. Econ. Biol.*, **3**, 1–24.

Biffin, R. H. (1898). *J. Linn. Soc.*, **34**, 147–162.

Billie-Hansen, E. (1953). *Bot. Tidsskr.*, **50**, 81–85.

Borriss, H. (1934). *Planta*, **22**, 28–69.

Brefeld, O. (1877). "Bot. Untersuchungen uber schimnelpilze". A. Felix, Leipzig.

Brefeld, O. (1888). "Unter aus dem Gesammtgebiete der Mykologie". A. Felix, Leipzig.

Brefeld, O. (1889). "Unter aus dem Ges. der Mykol". A. Felix, Leipzig.

Brodie, H. J. (1931). *Ann. Bot.*, **45**, 315–344.

Brooks, F. T. (1911). *J. agric. Sci., Camb.*, **4**, 133–144.

Buller, A. H. R., (1922). "Researches in the Fungi". Longmans, London.

Campbell, W. A. (1938). *Bull. Torrey bot. Club*, **65**, 31–78.

Cartwright, K. St. G., and Findlay, W. P. K. (1958). "Decay of Timber and Its Prevention". H.M.S.O., London.

Constantin, M. J. (1891). *Revue gen. Bot.*, **3**, 497–511.

Constantin, M. J., and Matriechot, L. (1899). *C.r. Acad. Sci. Paris*, **109**, 752–770.

Cooke, W. B. (1968). *Mycopath. Mycol. appl.*, **34**, 305–316.

Davidson, R. W., Campbell, W. A., and Blaisdell, D. J. (1938). *J. Agric. Res.*, **57**, 683–695.

Davis, R. H. (1966). In "The Fungi" (G. C. Ainsworth and A. S. Sussmann, Eds.), Vol. II, pp, 567–588. Academic Press, London and New York.

Dawson, J. R., Johnson, R. A., Adams, P., and Last F. T. (1965). *Ann. appl. Biol.*, **56**, 243–484.

Day, P. R. (1959). *Heredity*, **13**, 81–88.

Deneyer, W. B. G. (1960). *Can. J. Bot.*, **38**, 909–920.

Downie, D. G. (1959). *Trans. Proc. bot. Soc. Edinb.*, **38**, 16–29.

Duggar, B. M. (1901). *Bot. Gaz.*, **31**, 38–66.

Duggar, B. M. (1905). *Bull. U.S. Dept. Agric.*, **85**, 1–60.

Emerson, S. (1966). *In* "The Fungi" (G. C. Ainsworth and A. S. Sussmann, Eds.), Vol. II, pp. 513–566. Academic Press, London and New York.

Etter, B. E. (1929). *Mycologia*, **21**, 197–203.

Falck, R. (1909; 1912; 1921). *In* "Möller", *Hausschwam forschungen*, **3**, 6–9.

Ferguson, M. C. (1902). *Bull. U.S. Dept. Agric.*, **16**, 1–43.

Flentje, N. T. (1957). *Trans. Br. mycol. Soc.*, **40**, 322–336.

Fries, N. (1941). *Arch. Microbiol.*, **12**, 266–284.

Fries, N. (1949). *Svensk Bot. Tid.*, **43**, 316–342.

Fries, N. (1966). See Madelin below.

Fritz, C. (1923). *Trans. roy. Soc. Can.*, Ser. 5., **17**, 191–288.

Galleymore, B. (1949). *Trans. Br. mycol. Soc.*, **32**, 315–317.

Glaser, T., and Sosna, Z. (1956). *Acta Soc. Bot. Poloniae*, **25**, 385–303.

Gottlieb, D. (1950). *Bot. Rev.*, **16**, 229–257.

Griffith, N. T., and Barnett, H. L. (1967). *Mycologia*, **59**, 149–154.

Harley, J. L., and Waid, J. S. (1955). *Trans. Br. mycol. Soc.*, **38**, 104–118.

Hein, I. (1930). *Am. J. Bot.*, **17**, 882–915.

Hesler, L. R., and Smith, A. H. (1963). "North American Species of Hygrophorus". University of Tennessee Press, Knoxville.

Hesseltine, C. W., Whitehill, A. R., Pidacks, C., Ten Hagen, M., Bohonos, N.,

Hutchings, B. L., and Williams, J. H. (1953). *Mycologia*, **45**, 7–19.
Hiromoto, K. (1961). *Bot. Mag. Tokyo*, **74**, 154–159.
Hollings, M. (1962). *Nature, Lond.*, **196**, 962–965.
Hollings, M., Gandy, D. G., and Last, F. T. (1963). *Endeavour*, **22**, 112–117.
Hopp, H. (1938). *Phytopathology*, **28**, 356–358.
Jackson, R. L. O. (1951). "Mushroom Growing—a Practical Manual". English, Universities Press, London.
Johnson, M. E. M. (1920). *Trans. Br. mycol. Soc.*, **6**, 348–352.
Kauffman, F. H. O. (1934). *Bot. Gaz.*, **96**, 282–297.
Kligman, A. M. (1942). *Am. J. Bot.*, **29**, 304–307.
Kinugawa, K., and Furukawa, H. (1965). *Bot. Mag. Tokyo*, **78**, 240–244.
Kniep, H. (1928). "Die Sexualitat der niederen Pflanzen". Fischer, Jena.
Koch, W. (1958). *Arch. Mikrobiol.*, **30**, 407–420.
Lange, M. (1952). *Dansk. Bot. Ark.*, **14** (6), 1–164.
Levisohn, I. (1955). *Nature Lond.*, **176**, 519.
Lilly, V. G. (1966). In "The Fungi" (G. C. Ainsworth and S. C. Sussman, Eds.), Vol. II. Academic Press, London and New York.
Lohwag, K. (1952). *Sydowia*, **6**, 323–335.
Long, W. H., and Harsch, R. M. (1918). *J. Agric. Res.*, **12**, 33–82.
Lösel, D. M. (1964). *Ann. Bot.* (*N.S.*) **28**, 465–478.
Lösel, D. M. (1967). *Ann. Bot.* (*N.S.*) **31**, 417–425.
Lu, D. C. (1965). *Am. J. Bot.*, **52**, 432–437.
Lutz, C. (1925a). *C.r. Acad. Sci. Paris*, **180**, 532–534.
Lutz, C. (1925b). *Bull. Soc. mycol. Fr.*, **41**, 310–312.
Lyman, G. R. (1907). *Proc. Boston Soc. Nat. Hist.*, **33**, 125–210.
Macrae, R. (1955). *Mycologia*, **47**, 812–820.
McKeen, C. G. (1952). *Can. J. Bot.*, **30**, 764–787.
McLaughlin, D. J. (1964). *Mycologia*, **56**, 136–138.
McTeague, D. M., Hutchinson, S. A., and Reid, R. I. (1959). *Nature, Lond.*, **183**, 1736.
Madelin, M. F. (1966). "The Fungus Spore", Colston Papers No. 18. Butterworths, London.
Mader, E. O. (1943). *Phytopathology*, **43**, 1134–1145.
Martin, J. P. (1950). *Soil. Sci.*, **69**, 215–232.
Maxwell, M. B. (1954). *Can. J. Bot.*, **32**, 259–280.
Melin, E. (1948). *Trans. Br. mycol. Soc.*, **30**, 92–99.
Melin, E. (1953). *A. Rev. Pl. Physiol.*, **4**, 325–346.
Miller, O. K. (1967). *Can. J. Bot.*, **45**, 1939–1943.
Ministry of Agriculture, Food & Fisheries (1960). Mushroom Growing Bulletin No. 34. H.M.S.O., London.
Modess, O. (1941). *Symb. bot. upsal.*, **5**, 1–147.
Moore, D. (1966). *Nature, Lond.*, **209**, 1157–1158.
Nobles, M. K. (1948). *Can. J. Res. C.*, **26**, 281–431.
Nobles, M. K. (1965). *Can. J. Bot.*, **43**, 1097–1139.
Norkrans, B. (1949). *Svensk. bot. Tidskr.*, **43**, 485–490.
Pantidou, M. E. (1961a). *Can. J. Bot.*, **39**, 1149–1162.
Pantidou, M. E. (1961b). *Can. J. Bot.*, **39**, 1163–1167.
Pantidou, M. E. (1962). *Can. J. Bot.*, **40**, 1313–1319.
Pantidou, M. E. (1964). *Can. J. Bot.*, **42**, 1147–1157.
Pantidou, M. E., and Groves, J. W. (1966). *Can. J. Bot.*, **44**, 1371–1392.
Papazain, H. P. (1950a). *Bot. Gaz.*, **112**, 138.
Papazain, H. P. (1950b). *Bot. Gaz.*, **112**, 139.

Petersen, R. (1960). *Mycologia*, **52**, 513.

Plunkett, B. E. (1956). *Ann. Bot.*, **20**, 563–586.

Price, S. R. (1913). *New Phytol.*, **12**, 269–281.

Refshauge, L. D., and Proctor, E. M. (1936). *Proc. Roy. Soc. Victoria*, **48** (N.S.), 105–123.

Robbins, W. J. (1950). *Mycologia*, **42**, 470–476.

Russell, P. (1956). *Nature Lond.*, **177**, 1038–1039.

Schenck, E. (1919). *Beih. bot. Zbl.*, **36**, 335–413.

Singer, R. (1961). "Mushrooms and Truffles—Botany, Cultivation and Utilization". Leonard Hill, London

Singer, R. (1962). "The Agaricales in Modern Taxonomy", 2nd Ed. J. Cramer, Weinheim.

Smith, A. H. (1963). *Mycologia*, **55**, 691–697.

Smith, A. H. (1966). *Mem. New York Bot. Gdn.*, **14**, No. 2.

Sundström, K. R. (1964). *Symb. bot. upsal.*, **18** (3), 1–89.

Steinberg, R. A. (1939). *Bot. Rev.*, **5**, 327–350.

Steinberg, R. A. (1950). *Bot. Rev.*, **16**, 208–228.

Thaper, H. S., Balwait Singh, and Bakshi, B. K. (1967). *Indian Forester*, **93**, 756–760.

Treschow, C. (1944). *Dansk. Bot. Ark.* **11** (6), 1–180.

Urayama, T. (1961). *Bot. Mag. Tokyo*, **74**, 56–59.

Waid, J. S. (1957). *Tran. Br. mycol. Soc.*, **40**, 391–406.

Warcup, J. H. (1957). *Trans. Br. mycol. Soc.*, **40**, 237–264.

Warcup, J. H. (1959). *Trans. Br. mycol. Soc.*, **42**, 45–52.

Warcup, J. H., and Talbot, P. H. B. (1962). *Trans. Br. mycol. Soc.*, **45**, 495–518.

Warcup, J. H., and Talbot, P. H. B. (1966). *Trans. Br. mycol. Soc.*, **49**, 427–435.

Ward, M. (1897). *Phil. Trans. Roy. Soc.*, **189B**, 123–130.

Watling, R. (1963). *Nature, Lond.*, **197**, 717–718.

Watling, R. (1964). *Trans. Proc. bot. Soc. Edinb.*, **39**, 475–488.

Watling, R. (1967). *Notes R. bot. Gdn. Edinb.*, **29** (1).

Appendix

Since this article went to Press a most useful bibliography on nutritional regulation of basidiocarp formation has appeared:—Volz, P. A., and Beneke, E. S., (1969). *Mycopath. & Myc. appl.* **37**, 225–253.

CHAPTER IX

Myxomycetes and other Slime Moulds

M. J. CARLILE

*Department of Biochemistry, Imperial College of Science and Technology,
London, England*

I. INTRODUCTION

The term "slime mould" was first applied to the Myxomycetes (i.e.,
"slime fungi"), a group of micro-organisms having as a characteristic
phase in their life-cycle the *plasmodium*, an irregular mass of protoplasm
containing many nuclei but not sub-divided into cells. Their life-cycle also
includes an amoeboid phase, the amoebae being capable of developing
flagella under appropriate conditions. Some distinguished students of the
Myxomycetes have regarded them as protozoa and others as fungi.

Subsequently, several other groups of micro-organisms, correctly or
mistakenly believed to produce plasmodia, were also referred to as slime
moulds, the best known of these being the Plasmodiophorales, Acrasiales
and Labyrinthulales. The Plasmodiophorales are obligate parasites, producing a plasmodium within host cells, and are commonly regarded as fungi.

The Acrasiales have an amoeboid phase, the amoebae later undergoing aggregation to form a structure capable of migration and resembling a minute slug. This was at first incorrectly interpreted as a plasmodium, and subsequently, when found to consist of numerous uninucleate amoebae which retained their identity, renamed the *pseudoplasmodium*. The Labyrinthulales are unicellular organisms which secrete slimy tracks along which they glide, the resulting pattern of cells and tracks having at one time been misguidedly designated the "net plasmodium". A useful account of all four groups of organisms can be found in the mycological textbook by Alexopoulos (1962), the Myxomycetes and Plasmodiophorales being dealt with as lower fungi, and the Acrasiales and Labyrinthulales, very tactfully, as organisms of uncertain affinity.

The term slime mould is an unfortunate one, as it suggests close taxonomic relationships between probably very distantly related organisms, and has in fact led to confusion, even among microbiologists, between the strikingly different Myxomycetes and Acrasiales. The zoological term Mycetozoa (i.e., fungus animals) has also become confusing as it has been used by some authors as synonymous with Myxomycetes and by some to include all slime moulds.

The present article will be devoted largely to the Myxomycetes, which have recently been the subject of much research but relatively few reviews. The Acrasiales, also the subject of much work, will be considered briefly, as many excellent reviews, often with critical discussion of methods, have been devoted to them. Finally, the recently discovered Protostelida, not mentioned above, and the Labyrinthulales will be discussed, necessarily briefly, as these groups have so far attracted little attention. The Plasmodiophorales will not be considered, as these obligate parasites have so far been immune to microbiological investigation.

The discussion of the various groups of slime moulds in a single article is not taxonomically justifiable, but is reasonable on the basis of comparable methodology. Most slime moulds which have been studied in detail have proved amenable to two-membered culture with common bacteria or yeasts. but few species have yet been grown in pure culture. Thus, for the microbiologist, these different groups can be handled by similar techniques and present similar problems.

II. THE MYXOMYCETES*

Most work on Myxomycetes has been carried out by individuals, often mycologists by training, who have become interested in the intriguing

---

* A book devoted exclusively to the Myxomycetes, with a chapter on methods, has been published [Gray, W. D., and Alexopoulos, C. J. (1968). "Biology of the Myxomycetes". Ronald Press, New York].

features of these organisms, and have become specialists devoting themselves to the group. One species, however, *Physarum polycephalum*, has
become more widely known. The plasmodium of this species has been used
for studies on the rheological properties of protoplasm, especially protoplasmic streaming (reviews by Kamiya, 1959 and Jahn, 1964), and, since
its pure culture by Daniel and Rusch, for studying the synthesis of DNA,
RNA and protein during the different stages of the nuclear cycle (e.g.,
Mittermeyer *et al.*, 1966; Cummins and Rusch, 1968).

A. Life cycle

There is still a great deal to be learned about the details of Myxomycete
life-cycles, but the main features of that of *Ph. polycephalum*, the most
thoroughly studied species, are well known. Its life-cycle (fig. 1) will now
be described and some of the ways in which other species differ will be noted.
More detailed accounts of the Myxomycete life-cycle are provided by
Alexopoulos (1962, 1963, 1966).

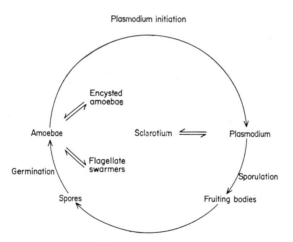

FIG. 1. Life cycle of a Myxomycete.

The plasmodium (fig. 2) of *Ph. polycephalum* is bright yellow and may
cover several square centimeters and contain many thousands of nuclei
which in *Ph. polycephalum* and some other species have been shown to
divide synchronously. A system of prominent channels ("veins") is present
in which rapid protoplasmic streaming occurs at rates of up to one milli-
meter per second, with reversals in the direction of flow taking place at
approximately one minute intervals. If nutritional and environmental
conditions are favourable, plasmodia tend to stay put, but will migrate

rapidly if conditions are unfavourable or if nutrients become exhausted. The plasmodia of a large group of Myxomycetes, the Physarales, resemble that of *Ph. polycephalum* in their major features. Such plasmodia are known as phaneroplasmodia (i.e., visible plasmodia) as they are readily visible to the naked eye. Other types of plasmodia are known, however, such as the delicate and transparent aphanoplasmodia (i.e., invisible plasmodia) and the tiny protoplasmodia. Alexopoulos (1960, 1966) describes the various types of plasmodia in detail.

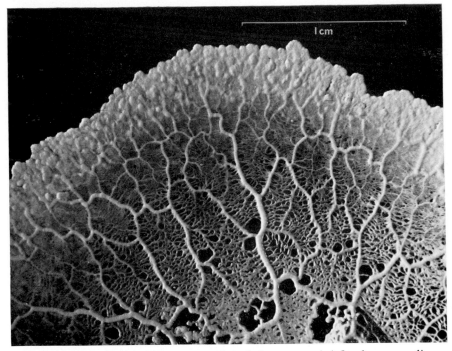

Fig. 2. Plasmodium of *Physarum polycephalum* on semi-defined agar medium. Note "veins".

As indicated above, lack of food and unfavourable nutritional conditions tend to provoke migration of the plasmodium of *Ph. polycephalum*. If, however, migration does not lead to improved conditions, the plasmodium gives rise to a resting phase, the sclerotium. The sclerotium is resistant to such adverse conditions as cold and desiccation and can survive for a year or more. It is built up of small spherical walled subunits, known as spherules (fig. 3). The production of sclerotia and their reconstitution into a new plasmodium under favourable conditions, has been described in detail by Jump (1954).

Provided other conditions are favourable, however, food exhaustion leads not to sclerotium formation but to sporulation (fig. 4a, b), in which numerous commonly uninucleate spores are produced within sporangia. In *Ph. polycephalum* several sporangia are borne on a single stalk—hence the specific name. The form, structure and arrangement of the sporangia (fruiting bodies) vary greatly among the Myxomycetes and are the main criteria used in their classification. The dry, powdery spores are readily dispersed through the air and can survive for long periods.

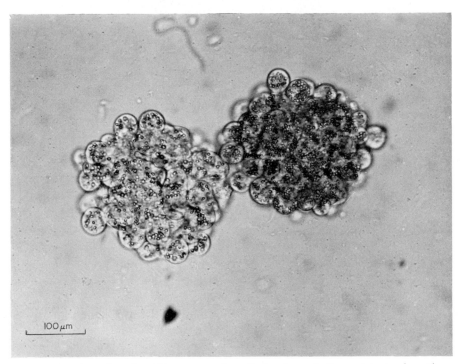

FIG. 3. Microsclerotia of *Physarum polycephalum* formed after nutrient exhaustion in shaken liquid culture, semidefined medium. Note constituent spherules.

Under favourable conditions spores germinate to give rise to small uninucleate amoebae, which do not differ markedly from other small amoebae but are sometimes termed myxamoebae. These amoebae have prominent contractile vacuoles, feed on bacteria or yeasts and multiply by binary fission. Exhaustion of food leads to encystment (microcyst formation). Encysted amoebae can survive for many months and germinate again when conditions are favourable. When immersed in water the amoebae develop anterior flagella, the flagellate cells being known as swarmers.

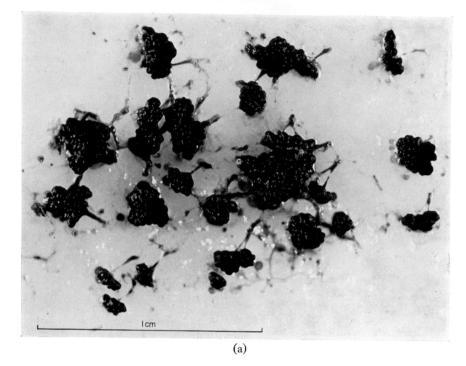

(a)

(b)

FIG. 4. Fruiting bodies of *Physarum polycephalum* formed after nutrient exhaustion and exposure to light on semi-defined agar medium. (a) Seen from above (b) Side view.

Usually two flagella are present, but only one is conspicuous. Conditions influencing myxamoeba–swarmer interconversion require further study but in general it seems that free water favours the development of flagella and relatively dry conditions the amoeboid state. In some species, the amoeboid condition is more commonly seen whereas in others the flagellate condition predominates.

The initiation of the plasmodial state requires further study. *Ph. polycephalum* is heterothallic and plasmodia will not arise in clones of amoebae, fusion of amoebae of differing mating types being essential (Dee, 1966b). Some other Myxomycetes have been shown to be heterothallic but in others plasmodia will arise in amoeba clones, indicating either a homothallic condition or the initiation of plasmodia without the occurrence of mating.

The amoeboid phase of heterothallic species is haploid and the plasmodial phase, arising from the fusion of amoebae of differing mating types is thought to be diploid, with meiosis restoring the haploid condition during spore formation. In species in which plasmodia can arise in amoeba clones it is likely that the amoeboid and plasmodial phases do not differ in ploidy and that nuclear fusion and meiosis do not occur (Kerr, 1968).

B. Habitat

Myxomycetes obtain nutrients mainly by attacking and digesting other micro-organisms, such as bacteria, yeasts, and fungi. In the active state they require moisture, are aerobic and do not usually tolerate near-freezing conditions, although in the resting state cold and dry conditions are tolerated. Hence they may be found in the active condition in almost any damp situation where there is sufficient decaying vegetation to support the microflora on which they feed, and in the resting state almost anywhere where there is vegetation. Truly aquatic species are, however, unknown; it is presumably their essentially aerobic metabolism that confines them to a terrestrial environment.

Myxomycetes are especially prominent in damp woodland in the autumn, particularly on or under the bark of decaying logs and on fallen leaves. Many other habitats have a characteristic Myxomycete flora, however, such as mountain turf exposed by melting snow, and dead herbaceous plants. The habitats of individual species are listed in taxonomic monographs, such as those of Lister (1925) and Martin (1949).

C. Collection

Myxomycetes may be recognized in their natural habitat and collected as plasmodia, fruiting bodies or occasionally as sclerotia. To avoid injury they are taken along with a portion of the substratum to which they are attached, a sharp penknife or similar tool being used to cut away specimens on wood. Plasmodia are very readily injured by crushing (but not cutting) or by desiccation and thus may be killed while being transported to the laboratory; fruiting bodies are also fragile and may arrive in the laboratory too battered to be identified. Hence transport from field to laboratory is the main problem in collecting Myxomycetes. Collection in plastic bags is not usually satisfactory. Perhaps the best method is the use of a metal box (Mr. Bruce Ing, personal communication) with a hinged lid and a layer of cork in the bottom, on which the specimens are fixed with bead-headed pins. Another useful procedure is that of Howard (1931) who took Petri dishes containing 1·5% plain agar into the field and placed plasmodia directly on the agar as they were collected.

A valuable alternative to collecting Myxomycetes in the field is to collect appropriate substrates, and maintain them in the laboratory under humid conditions until Myxomycetes appear. Such "moist chamber cultures" have been described by Gilbert and Martin (1933), Alexopoulos (1964) and more romantically, as "slime-mould gardens" by Nauss (1947). Suitable material (e.g., bark, dead wood, leaves) is soaked overnight in distilled water, and then placed on wet filter paper in Petri dishes or other suitable covered dishes, and examined daily, preferably with a dissecting microscope. Alexopoulos (1964) has observed Myxomycetes within a few days by this method, although more often one to two weeks or even longer, may be required for their development. The method, applied to slivers of bark from living trees, has yielded many species having minute plasmodia or fruiting bodies and hence readily overlooked in field collecting. Recently, several common Myxomycetes have been obtained from banana peel by this method (Davis and Butterfield, 1967).

D. Identification

The identification of Myxomycetes is based almost entirely on the morphology of the fruiting bodies. Fortunately, plasmodia brought from nature into the laboratory, or developing on natural substrata in the laboratory, will usually sporulate satisfactorily, thus permitting their identification. The classical monograph on the Myxomycetes is that of Lister (1925), now unobtainable except in long-established libraries. The monograph of North American species by Martin (1949) has the advantage of dichotomous keys, and owing to the cosmopolitan distribution of many Myxomycetes, its usefulness is not restricted to N. America. Alexopoulos (1963) has discussed the taxonomic literature and provides a list of taxa, with authorities, published subsequently to Martin's monograph or omitted by him. See also Martin and Alexopoulos, 1969.

Fruiting bodies dried slowly over a radiator or in the sun retain their form and colour. Such fruiting bodies may be sent to appropriate authorities by post for identification, wrapped in tissue paper and enclosed in a small box. The preparation of permanent microscope slide and herbarium specimens is described by Lister (1925).

E. Maintenance in crude culture on artificial media

The pure culture of Myxomycetes requires a knowledge of their nutritional needs which is so far available for very few species. Continued maintenance in moist chambers and "slime-mould gardens" on the other hand is dependent on a supply of appropriate substrates, such as decayed wood or the fruiting bodies of suitable Basidiomycetes. A convenient compromise is crude culture on artificial media. In such crude cultures no attempt is made to free the plasmodial inoculum from bacteria but the artificial medium provided is sterile. If the medium is not so rich as to lead to excessive bacterial growth and is otherwise suitable for the Myxomycete, vigorous growth may occur, the slime mould probably deriving its nutrients partly from the medium and partly from the bacteria present. Crude cultures have been extensively employed for studies on Myxomycete life-cycles and for some types of physiological work. They are also a useful preliminary to establishing pure cultures or two-membered cultures. Alexopoulos (1963) lists, with appropriate literature citations, 20 members of the Physarales and 8 other species which have been grown in crude culture on artificial media from "spore to spore", that is through the entire life cycle.*

One of the most satisfactory media for the crude culture of many Myxomycetes is oat agar, introduced for this purpose by Howard (1931), who autoclaved a mixture of 3% rolled oats and 1·5% agar. Some workers prefer to sterilize oats and agar separately and mix immediately before pouring into Petri dishes, or merely to sprinkle a few sterile oats on plain agar. Commercial preparations of rolled oats and porridge oats differ in their suitability: the present author uses "Scott's Porage Oats". Camp (1937) cultured plasmodia on a filter paper bridge dipping into water, and daily sprinkled a few sterile oats on the filter paper. The exact way in which oats are used will depend on the species and problem being investigated.

Crude cultures are commonly started from plasmodia, which may be freed from troublesome moulds by migration on plain agar. Plasmodia may also be obtained by mass-sowings of spores from fruit bodies on corn-meal agar; we find 4% Difco Bacto Corn Meal Agar satisfactory. Spores of some species are wetted with difficulty and Elliott (1949) has advocated the use

* A more recent list (Martin and Alexopoulos, 1969) indicates that 38 members of the Physarales and 19 other species have been grown from "spore to spore".

of sodium taurocholate as a wetting agent to facilitate wetting and germination. Sclerotia should be soaked overnight before placing on agar media. Sub-culturing of plasmodia can be carried out at appropriate intervals. Alternatively, sclerotia, commonly formed after nutrient exhaustion, may be stored for periods of about a year.

Many Myxomycetes will sporulate in crude culture, although the literature available is often contradictory about the conditions required. Light appears to be essential for the sporulation of many species, but not for others (Gray, 1938). Nutrient exhaustion also appears to be a necessary condition for sporulation; hence it is probably advisable to use rather weak media in which nutrient exhaustion can occur without excessive accumulation of toxic metabolic products.

F. Growth in two-membered culture and in pure culture

Many Myxomycetes can readily be grown in two-membered culture with a single, known bacterial or yeast species. A few Myxomycetes have been maintained for long periods in pure (axenic) culture, but none has been grown throughout its life cycle under such conditions. The plasmodia of *Ph. polycephalum*, for example, grow excellently in pure culture, but two-membered culture is required for the amoeboid phase. Methods of handling the various phases of the Myxomycete life cycle will now be described.

1. *Plasmodia*

(a) *Purification.* Cohen (1939), who was the first to achieve undoubtedly pure cultures of Myxomycetes, describes two procedures for freeing plasmodia from contaminants, the migration and the enrichment methods. A plasmodium on a non-nutrient agar will migrate, leaving behind contaminating micro-organisms. The further it migrates—provided it does not cross its own track, clearly visible from deposited slime—the more thorough will be the decontamination. Hence, repeated migration across Petri dishes of plain agar will often eliminate contaminants. Sometimes, however, the starving plasmodium may die or become a sclerotium before adequate migration has taken place. If, with a particular species, this is a problem, the migration method can be supplemented by the enrichment method. Cohen (1939) used Petri dishes of non-nutrient agar streaked with heavy suspensions of washed yeast. Plasmodia will migrate along the streak of yeast consuming it as they go (fig. 5). A vigorous plasmodium, contaminated only with the yeast species employed, is thus obtained and the yeast subsequently eliminated by the migration method. Essentially, this procedure utilizes two-membered culture as a preliminary to pure culture. Many Myxomycetes favour acid conditions, so the acidification of the plain agar may facilitate elimination of bacteria. We find that a single migration across plain agar at pH 5·0 is generally sufficient to free plasmodia of *Ph. polycephalum* from *Escherichia coli*. The utilization of antibiotics for purifying plasmodia (for example, by incorporation in the migration medium) has been reported by Sobels and Cohen (1953) and Hok (1954), penicillin and streptomycin being those most readily tolerated by the Myxomycetes employed. Lazo (1960) claimed, however, that *Ph. polycephalum* underwent permanent changes as a result of prolonged growth on media containing streptomycin, so it may be best to avoid their use unless other methods fail.

The elimination of contaminants should be confirmed by inoculating plasmodia into tubes of liquid media. The use of at least two test media is desirable—one based on the medium to be used for routine pure culture of the Myxomycete, and one suitable for vigorous growth of the most likely contaminant (commonly the species used in the enrichment procedure). Testing for purity is described by a number of authors, for example, Cohen (1939), Sobels and Cohen (1953), Hok (1954), Daniel and Rusch (1961) and Scholes (1962).

(b) *Nutritional requirements of plasmodia of* Ph. polycephalum. The growth of a Myxomycete plasmodium (*Ph. polycephalum*) in pure culture on a soluble medium was first described by Daniel and Rusch (1961). The medium employed contained glucose, an enzymic protein hydrolysate (tryptone), mineral salts, yeast extract and chick embryo extract. Subsequently, it was shown that the chick embryo extract could be replaced by haematin (Daniel *et al.*, 1962) and yeast extract by biotin and thiamin (Daniel *et al.*, 1963). The replacement of tryptone by a mixture of amino-acids was also reported (Daniel *et al.*, 1963) thus achieving a completely defined medium. This final observation we have been unable to confirm; media in which enzymic protein hydrolysates were replaced by acid-hydrolysed proteins or by amino-acid mixtures were unsatisfactory. This discrepancy between our results and those of Daniel *et al.* (1963) is difficult to explain; possibly Daniel and his co-workers may have been particularly fortunate in the strain they employed.

Daniel and Baldwin (1964) describe a semi-defined medium and three completely defined, "synthetic", media. In our experience semi-defined media normally gave excellent growth with high growth rates and final yields. We found, however, that with shake cultures, particularly when small inocula were used, occasional batches of media were unsatisfactory, no growth occurring, or more frequently, normal growth being preceded by a lag phase of several days. These effects were found to be due to the partial or complete destruction of the inoculum, possibly through trace metal toxicity, and the medium was modified in a semi-empirical manner in order to avoid these effects. The semi-defined medium given below closely resembles that described by Daniel and Baldwin (1964), but has yielded uniformly excellent growth both in our laboratory and elsewhere. The most important modifications are probably the inclusion of a chelating agent, disodium ethylene diamine tetracetic acid (Na_2 EDTA) and the omission of manganese. The replacement of yeast extract by biotin and thiamin leaves peptone as the only undefined constituent. Ross (1966) has also devised a medium which is essentially a modification of that of Daniel and Baldwin, and which contains tryptone as the only undefined constituent.

Semi-defined medium for Physarum polycephalum

Glucose	10 g
Peptone, bacteriological (Oxoid)	10 g
Citric acid.H_2O	3·54 g
KH_2PO_4	2·0 g
$CaCl_2.6H_2O$	0·9 g
$MgSO_4.7H_2O$	0·6 g
$Na_2.EDTA$	0·224 g
$FeCl_2.4H_2O$	0·06 g
$ZnSO_4.7H_2O$	0·034 g
Thiamin hydrochloride	0·0424 g
Biotin	0·005 g
Haemin (e.g., Sigma Equine Hemin)	0·005 g
Distilled water	1 litre

Dissolve the various components, other than haemin, in distilled water, adjust the pH to 4·6 with 10% NaOH, and sterilize. Prepare haematin solution by dissolving haemin in 1% NaOH to give an 0·05% solution (e.g., 50 mg. haemin in 100 ml 1% NaOH solution). The sterilized haematin solution can be stored at 5°C for a week and is added aseptically to sterile medium (1 ml haematin solution/100 ml medium) at room temperature immediately prior to inoculation. Autoclaving for 20 minutes at 10 lb/sq. inch pressure is suitable for sterilizing both the haematin and other components (haematin is heat-stable under alkaline conditions).

We have been unable to obtain satisfactory growth on the three defined media advocated by Daniel and Baldwin (1964). We find that peptone can be replaced by a variety of other enzymic protein hydrolysates, for example, tryptone (Difco) or casitone (Difco), but not by acid-hydrolysed proteins such as casein hydrolysate or by amino-acid mixtures. Such a situation formerly existed with *Lactobacillus casei*, but it was found that by adjustments in the levels of the amino-acids employed the apparent requirement for an enzymic hydrolysate of casein was eliminated (Guirard and Snell, 1962). Possibly, a similar laborious adjustment of amino-acid levels could eliminate the apparent requirement for enzymic protein hydrolysate and result in the development of a completely defined medium for *Ph. polycephalum*.

(c) *Culture techniques for* Ph. polycephalum. The plasmodia of *Ph. poly-cephalum* may be grown either in surface culture or in shake cultures. A useful account of appropriate methods is provided by Daniel and Baldwin (1964).

Agar media for surface culture are prepared by including agar (e.g., 2% Oxoid Ionagar No. 2 or Difco Bacto-Agar) in the semi-defined medium described above. Haematin is added to small batches (250 ml or less) of agar at 40°C immediately prior to pouring Petri dishes. We normally incubate cultures at 24°C, at which temperature a plasmodial inoculum 0·5 × 0·5 cm cut from a previous agar culture will cover the Petri dish in about 4 days

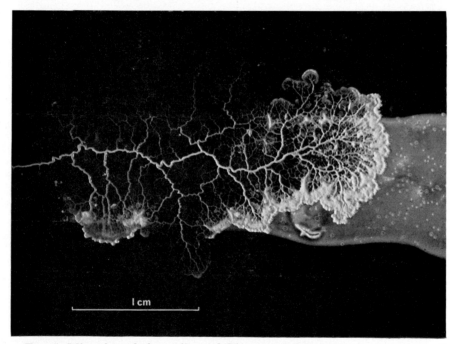

| 1 cm |

FIG. 5. Migration of plasmodium of *Physarum polycephalum* along a streak of washed yeast (*Saccharomyces cerevisiae*) on plain agar.

(fig. 2) and continue to remain active (i.e., to show streaming) for about another week, after which sclerotium formation occurs. Alternatively, agar media may be inoculated with microplasmodia from liquid cultures. Faster growth may be obtained by means of higher temperatures up to about 28°C and slower growth by lower temperatures down to about 12°C. Daniel and Baldwin (1964) describe surface culture on filter paper supported on glass beads in Petri dishes, liquid media being pipetted beneath the filter paper. This method is invaluable when the presence of agar is undesirable,

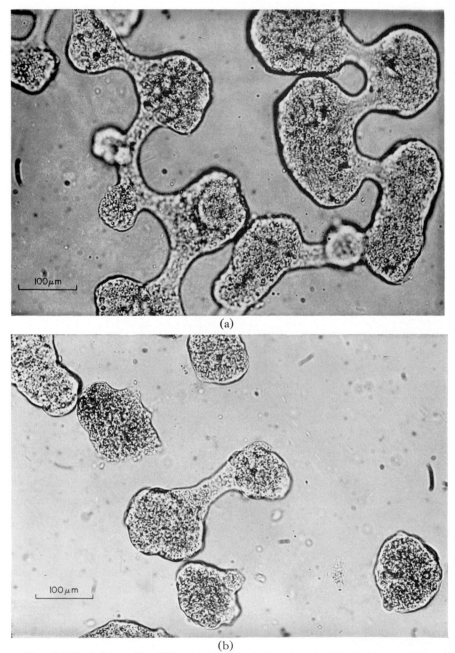

(a)

(b)

FIG. 6. Microplasmodia of *Physarum polycephalum* from a shaken culture on semi-defined liquid medium. (a) Young culture (b) Older culture.

and is employed for establishing plasmodia with nuclear synchrony (see below).

Growth in shake cultures is in the form of microplasmodia (Fig. 6), of which several hundred thousand per ml are present when maximum growth is attained. We employ 50 ml of semi-defined medium in 500 ml Erlenmeyer flasks and use a rotary shaker with a radius of gyration of 4·5 cm and a speed of 200 rpm. Equipment with these characteristics happened to be available; Daniel and Baldwin (1964) have successfully employed both rotary and reciprocating shakers with other operating characteristics and different culture volumes. Shake cultures may be inoculated with plasmodia from either agar or liquid cultures. If material from agar cultures is used it is permitted to migrate on to the surface of the medium before shaking is commenced. Such inocula result in a considerable lag period before vigorous growth occurs. Once shake cultures are established, the most suitable method of inoculation is to pipette medium with suspended microplasmodia from a vigorously growing culture. We find that at 24°C inoculation of 50 ml of medium with 0·5 ml inoculum will yield maximum growth in about 4 days and 2·5 ml in about 3 days. After a lag of about 12 hours, growth is approximately exponential with a doubling time of less than 12 hours until growth approaches maximum, after which dry weight falls as microsclerotium formation occurs (fig. 3). Maximum plasmodium dry weights obtained with 50 ml of medium containing 500 mg glucose and 500 mg peptone are 400–500 mg.

Microplasmodia may also be cultured in small fermenters (Brewer et al., 1964). We use 3·5 litres of medium in 5 litre Labro Ferm (New Brunswick Scientific Company) glass fermenter vessels (20 cm diameter × 44·5 cm height) with baffles. The medium is sparged with 2–3·5 litres/min of air and stirred with two sets of impeller blades. Foaming of the medium is prevented by the inclusion of 0·01% silicone RD antifoam (Midland Silicones Ltd.) and the aseptic addition of further antifoam during the fermentation if necessary. Yields similar to those in shaken cultures were obtained with shaft speeds and impeller diameters (e.g., 275 rmp × 7·8 cm, 350 × 6·4 cm, 500 rpm × 5·0 cm) which provided impeller tip speeds of 5000–10,000 cm/min. Poor growth was obtained outside this range. The main problem in culturing microplasmodia in fermenters appears to be determining agitation rates which will give adequate aeration without causing physical damage. The growth of microplasmodia in fermenters of various sizes containing 2–100 litres of medium has been reported by Brewer et al. (1964).

Microplasmodia pipetted on to agar media or on to other suitable surfaces rapidly fuse to give a large plasmodium in which mitosis is synchronized throughout the plasmodium. The details of the production of such plasmodia, which have been extensively used for the study of metabolism through the various phases of the nuclear cycle, are provided by Guttes and Guttes (1964).

The preservation of plasmodia of *Ph. polycephalum* by liquid nitrogen refrigeration has recently been reported (Boder and Johnson, 1967). Plasmodia grown in semidefined medium are harvested by centrifugation (800 *g*) and resuspended in fresh medium to which 5% dimethylsulphoxide has been added. The preparation is sealed in ampoules, slowly frozen and then stored in liquid nitrogen. We find Boder and Johnson's procedure effective for storing sclerotia but not plasmodia; possibly their cultures were old enough for some sclerotia to be present.

(d) *Measurement of growth of* Ph. polycephalum *plasmodia*. Growth measurements on *Ph. polycephalum* are most readily carried out with liquid media. The procedure preferred will depend on the problem being investigated. A rough indication of growth may be obtained by centrifugation in graduated tubes and noting the *packed volume*. If injury to the plasmodia is to be avoided, rather gentle treatment (e.g., 250 *g*) must be used; if subsequent viability is unimportant, centrifuging can be more vigorous. *Dry weight* is a more accurate measurement of growth. The plasmodia are centrifuged, the culture medium decanted, the plasmodia resuspended in distilled water, centrifuged again and dried overnight at 110°C. Daniel and Baldwin (1964) make use of *protein content* as an estimate of growth; an advantage of this procedure is that it may be carried out on small samples taken from cultures to be used for other purposes. The same authors also estimate growth by spectrophotometric determination of *extracted pigment*. *Total counts* of microplasmodia/ml may be made with a haemocytometer; we find that the relatively large size and irregular shape of microplasmodia render the operation somewhat tedious and inaccurate. *Viable counts* give figures that are similar and of higher internal consistency. One ml of medium with suspended microplasmodia is added to 9 ml sterile water, shaken, and the operation repeated until an appropriate dilution—usually 10,000 × or 100,000 × — is obtained. One-ml samples are then placed in Petri dishes, agar medium at 45°C is added and the medium swirled around to give good mixing. Counts are best carried out after about 4–5 days' incubation; earlier counts are likely to be in error through small plasmodia being overlooked, and later counts in error through fusion of plasmodia. Serial dilution is also of value as a method for obtaining cultures from single micro-plasmodia.

We have generally found good agreement between estimations based on packed volume, dry weight, total count and viable count. However, as the time of maximum growth approaches, the microplasmodia may fragment, leading to a large increase in plasmodial number unaccompanied by a proportionate increase in plasmodial mass.

(e) *Plasmodia of other species*. The first convincing report of the growth of Myxomycete plasmodia in pure or two-membered culture was that of Cohen (1939) who grew several species with the yeast *Saccharomyces ellipsoideus* and in pure culture, with autoclaved baker's yeast streaked on plain agar. Cohen (1941) considered in more detail the relationship between plasmodia, bacteria and substrate in two-membered culture, and both Sobels (1950) and Hok (1954) published detailed accounts of two-membered and pure culture of various Myxomycetes by methods similar to those of Cohen. A useful review is that of Sobels and Cohen (1953).

The literature of attempts to grow plasmodia in pure culture is often contradictory. For example, Lazo (1961a) reported pure culture of *Fuligo septica* on oat agar, Cohen (1939) found oat agar unsatisfactory for the pure culture of Myxomycetes including *F. septica* but grew this species on autoclaved yeast, and Scholes (1962) was unable to grow it satisfactorily in pure culture at all. Presumably with further intensive studies on single species contradictions of this sort will be resolved.

Two-membered culture remains the most practicable procedure for many Myxomycetes. Kerr and Sussman (1958) for example, have carried out a range of studies of *Didymium nigripes* in two-membered culture with *Aerobacter aerogenes*. An interesting recent development is the culture of *Physarum didermoides* and *Fuligo cinerea* by Lazo (1961b) in an apparently symbiotic relationship with the alga *Chlorella*. Most plasmodia tested engulfed and digested the various pure-cultured algae to which they were introduced, but three species of *Chlorella* survived ingestion by *Ph. didermoides* and *F. cinerea* and rendered these plasmodia bright green. Such plasmodia, if illuminated, were able to thrive under conditions unsuitable for plasmodia lacking the alga.

Pure culture on killed micro-organisms also remains a useful procedure. Considine and Mallette (1965) used autoclaved *E. coli* for the culture of *Physarum gyrosum* for experiments on antibiotic production by plasmodia. Methods of killing that avoid the use of high temperatures may sometimes be advisable; Hok (1954), for example, advocated the use of yeast killed with nitrogen-mustard.

The media and methods advocated by Daniel and Baldwin (1964) for *Ph. polycephalum* will doubtless ultimately be adapted to the culture of other plasmodia. This has already been done for *Physarum flavicomum* and *Physarella oblonga* by Ross (1964), appropriate media for Petri dish culture on agar and shake cultures having been devised.

2. *Sclerotia*

Shake cultures of *Ph. polycephalum* form microsclerotia (fig. 3), consisting of clusters of spherules, within 1–2 days of maximum growth being attained. Such microsclerotia remain viable for a considerable period, and give rise, after a few days lag phase, to microplasmodia when inoculated into shaken liquid media. We normally store microsclerotia with suspending media in screw-cap bottles at 5°C, under which conditions they remain viable for about a year. The review of Daniel and Baldwin (1964) contains a useful account of work on the formation and storage of sclerotia. Ross (1964) mentions microsclerotium formation in shaken cultures of *Physarum flavicomum* and *Physarella oblonga*.

The formation of sclerotia in agar culture after nutrient exhaustion or exposure to various conditions, such as slow desiccation, has been discussed by many authors (e.g., Alexopoulos, 1963). Such sclerotia often retain viability when stored, although they are less convenient to handle than microsclerotia formed in liquid culture.

3. *Sporulation*

Ph. polycephalum can be induced to sporulate in pure culture. We find that if cultures on semi-defined agar medium (page 248) are exposed to daylight instead of being kept in an incubator, sporulation will occur a few days after the plasmodium has completely covered the Petri dish. A procedure has also been described (Guttes *et al.*, 1961; Daniel and Rusch, 1962a,b; Daniel and Baldwin, 1964) for obtaining synchronous sporulation at a specified time on agar-free media. Microplasmodia from a shaken culture on semi-defined medium are harvested by gentle centrifugation (e.g., 250 g for 2 min), resuspended in distilled water and pipetted onto filter paper supported on glass beads. In a few hours the microplasmodia fuse into a single plasmodium and a salts solution containing 0·1% (w/v) nicotinic acid and 0·1% (w/v) nicotinamide is pipetted beneath the filter paper (Daniel and Baldwin, 1964). The plasmodium is then incubated in darkness for 4–5 days. At this stage, two hours' exposure to light (e.g., from 40 watt fluorescent tubes) will induce the production of sporangia 12–16 hours later. Care must be taken to avoid excessive heating.

A few other Myxomycetes have been induced to sporulate in pure or two-membered culture, for example, *Didymium nigripes* in the presence of *Aerobacter aerogenes* (Kerr and Sussman, 1958) but relatively little information is yet available on appropriate methods. It is probable that relatively poor media which lead to the onset of starvation without excessive accumulation of toxic metabolites are desirable, and for many species exposure to light is essential.

4. *Spore Germination*

There is an extensive literature on Myxomycete spore germination, based mostly on studies with spores from fruiting bodies collected from nature. Smart (1937) records the results of germination experiments with a wide range of species on a variety of media, and showed that species differed greatly both in the times taken to germinate (15 minutes to 18 days) and the proportion germinating (0 to 100%). Marked differences in behaviour also occur within species. Collins (1961) found that germination of spores of *Didymium iridis* taken from different fruiting bodies on agar media varied from 0–100%. It is not surprising therefore that the reports of different authors on conditions optimal for germination conflict. Usually, however, some germination occurs and hence cultures of amoebae may be established. Alexopoulos (1963) provides a useful review of literature on spore germination.

5. *Amoebae and microcysts*

(a) *Elimination of contaminants.* The problem of the elimination of contaminants from amoebae is often avoided by starting amoeboid phase cultures with spores produced by plasmodia in pure culture or two-membered culture, as plasmodia are much more readily freed from contaminants than amoebae. In most other reports of the two-membered culture of amoebae, the elimination of contaminants is not discussed, and tests to establish the absence of other organisms are not mentioned; presumably what are effectively two-membered cultures were readily established, and were adequate for the author's purposes. The elimination of contaminants from amoebae is, however, worthy of consideration, as it is not always possible to obtain sporulation in pure culture.

Two-membered cultures of members of the Acrasiales are established by permitting the amoebae to eat their way along streaks of non-nutrient agar of the bacterial species to be used for two-membered culture (Raper, 1951); preliminary experiments suggests that this method can be used to establish two-membered cultures with some Myxomycete amoebae. Ross (1964) found that amoebae of *Badhamia obovata* (*Ba. curtisii*) migrated rapidly away from an inoculum drop on agar, and when individuals were transferred to fresh media with a micro-knife commonly proved to be uncontaminated. The method was not successful when tried with the smaller and less active amoebae of other species. Schuster (1964) briefly mentions that *D. nigripes* can be freed from contaminants by means of an antibiotic mix and culture procedures earlier used to established pure cultures of *Naegleria gruberi* (Schuster, 1961). Kerr (1963) found that spores of *D. nigripes* could only occasionally be freed from contaminants by antibiotic treatment, and resorted to the elimination of contaminants from plasmodia prior to sporulation. Clearly more work on the topic is needed. A useful approach might be to establish as a preliminary to pure culture two-membered culture with a species readily eliminated by antibiotics or other means.

(b) *Two-membered culture.* Bacteria, especially *E. coli* and *A. aerogenes*, are normally employed for the two-membered culture of Myxomycete amoebae. Some species with relatively large amoebae will readily ingest yeasts, and preliminary experiments suggest that, with such species, yeasts as well as bacteria are useful for two-membered cultures.

Methods for the routine culture of the amoebae of *Ph. polycephalum*, with *Pseudomonas fluorescens* and later with *E. coli* were developed by Dee (1962, 1966a, b). Spores, or amoebae (active or encysted) are suspended in distilled water and a suspension of *E. coli* is also prepared. About 0·1 ml of each suspension is spread with a bent glass rod on to the surface of liver infusion agar (Oxoid liver infusion, 0·05%, Agar 2·0%). Germination of spores (about 5%) or encysted amoebae (about 100%) takes place and the amoebae feed on the bacteria and multiply. If an appropriately dilute inoculum is employed, each spore or amoeba gives rise to a separate colony, visible to the naked eye as a neat circular transparent plaque in the opalescent film of bacteria. Encystment of amoebae takes place when nutrient exhaustion occurs, i.e., when almost all the bacteria have been consumed. Provided that clones of amoebae are employed, the amoeboid phase may be propagated indefinitely without plasmodium formation, as *Ph. polycephalum* is heterothallic. The procedures advocated for the reliable production of plaques from spores and from amoebae, and for plasmodium initiation differ; Dee's publications should be consulted for details. If, however, all that is required is efficient two-membered culture of amoebae, a variety of procedures based on those of Dee are effective.

Kerr and Sussman (1958) describe procedures for the two-membered culture of *D. nigripes* with *A. aerogenes*. A suspension of spores or amoebae together with a few drops of a culture of *A. aerogenes* are spread on a medium of the following composition—

Medium

Bacto-peptone	2·0 g
Glucose	2·0 g
Yeast extract	0·2 g
K_2HPO_4	0·2 g
KH_2PO_4	0·3 g
$MgSO_4.7H_2O$	0·2 g
Agar	20 g
Distilled water	1 litre

pH 6·0–6·3

Spore germination and amoeba viability were 100%, and an appropriately dilute inoculum resulted in plaques, permitting cloning. In *D. nigripes* plasmodium formation readily occurs, even in clones of amoebae, but it was found that it could be prevented by the inclusion of 2% (w/v) glucose or 0·2% (w/v) brucine (a compound very toxic to man) in the medium.

(c) *Pure culture.* As yet there are few reports of the pure culture of Myxomycete amoebae. Kerr (1963) has reported the growth of those of *D. nigripes* in a liquid medium containing peptone, yeast extract, glucose, and formalin-killed *A. aerogenes*, and Schuster (1965) the growth of *D. nigripes* and *Physarum cinereum* on *A. aerogenes* killed by one minute at 100°C. The unusually large amoebae of *Ba. obovata* are exceptional in having been cultured (Ross, 1964) on a semi-defined medium based on that of Daniel and Rusch (1961).

(d) *Amoeba-flagellate transformation.* Studies on the transformation of Myxomycete amoebae into the flagellate conditions have been carried out on *D. nigripes* (Kerr, 1960, 1965b; Schuster, 1965). The transformation can be brought about (Kerr, 1960) by removing amoebae from an agar surface, washing and resuspending in distilled water or a salts solution. Immersion in water appears effective in other species also, but further studies are needed. A very detailed study on methods for obtaining closely controlled synchronous conversion of amoebae into the flagellate condition has been carried out *Na. gruberi* (Fulton and Dingle, 1967) and the procedures developed may well be applicable to Myxomycete amoebae.

(e) *Preservation of amoebae, microcysts and spores.* Encysted amoebae (microcysts) of Myxomycetes survive for long periods. If amoebae are grown on agar slopes in screw-capped (McCartney) bottles, the caps can be screwed down when sufficient time for encystment has elapsed, to prevent desiccation of the agar. Dee (1966a) advocated the preservation of encysted amoebae of *Ph. polycephalum* on agar slopes in test-tubes by covering the slope with autoclaved liquid paraffin (B.P.). Kerr (1955a) describes a lyophilization (freeze-drying) procedure suitable for laboratories without specialized equipment and mentions that spores and amoebae of Myxomycetes have been stored successfully for over 5 years by the method. Davis (1965) of the American Type Culture Collection, Rockville, Maryland, reports the successful preservation of encysted amoebae of *D. iridis* and *Physarum pusillum* by both controlled freezing (1°C per minute in the range +25°C to −35°C) in 10% glycerol solution followed by liquid nitrogen refrigeration, and also by means of lyophilization in skimmed milk.

G. Maintenance of isolates in a genetically uniform state

The Myxomycete literature is full of contradictory assertions by different workers studying the same species. It is likely that the source of these contradictions is usually the variability, both genetic and environmental, shown by Myxomycetes. Environmental variability may be controlled by achieving more precisely defined conditions—replacing crude culture by two-membered or preferably pure culture, and replacing particulate substrates such as oat flakes by soluble, and as far as possible, defined media. Pure culture, although leading to a more closely controlled environment, can, however, lead to a genetically worsened situation. This is shown by the cytological work of Ross (1966), who found that the amoebae of *Ba. obovata* (*Ba. curtisii*) in two-membered culture with bacteria gave a chromosome count of approximately 80, whereas those in pure culture had chromosome counts with a modal value of 80 but ranging from about 20 to about 300. Probably these aberrations resulted from a not entirely adequate medium; Ross obtained uniform chromosome counts from pure-cultured plasmodia of *Ph. flavicomum* and one of the two strains of *Ph. polycephalum* he studied. These results are, however, important in indicating that the step from satisfactory two-membered culture to pure culture on an inadequate medium may be cytologically, and hence, genetically, disastrous. So although the ultimate objective for controlled environmental conditions is pure culture on a defined medium, for routine work it is essential that the conditions used do not lead to obvious abnormalities.

In order to achieve genetical uniformity the use of cloned material (i.e., material originating from a single nucleus) is essential. Since the only unquestionably uninucleate phase in the Myxomycete life-cycle is the amoeba (as spores, and the spherules from microsclerotia, are not invariably uninucleate) this involves the use of cultures originating from single amoebae. Such cultures may be established from encysted amoebae transferred by micromanipulation, or from plaques produced by single spores, as in the procedures of Dee (1962, 1966a, b) and Kerr and Sussman (1958) described above. Storage of cultures as microcysts (see above) avoids the risks of genetic change involved in frequent sub-culture.

Genetically uniform plasmodia may be produced in heterothallic species (e.g., *Ph. polycephalum*) by bringing together amoeba clones of appropriate mating type (Dee, 1966b; Poulter and Dee, 1968). In homothallic or apogamic species it is sufficient to provide conditions suitable for plasmodium development in a culture of cloned amoebae (Kerr and Sussman, 1958).

Changes in the behaviour of plasmodia after long periods in pure culture, including a decline in the ability to sporulate, have been described by Daniel and Baldwin (1964). We have frequently observed, after prolonged pure culture of *Ph. polycephalum* on agar media, a rapid deterioration, leading to very slow growth, abnormal morphology and copious slime production, and disintegration and death of part of the plasmodium. Such cultures may ultimately prove impossible to maintain, or, alternatively, from part of the culture vigorous growth occurs and subculture leads to the recovery of a normal growth rate and morphology. Deterioration of the above kind rarely occurs in shaken liquid cultures; presumably on agar media deleterious characteristics spread throughout the culture, whereas in liquid media natural selection eliminates any microplasmodia with abnormally slow growth rates. It seems, therefore, that prolonged subculture on agar media is hazardous. Hence, liquid culture (if practicable) and storage by liquid nitrogen refrigeration (Boder and Johnson, 1967) are to be recommended.

III. THE ACRASIALES (CELLULAR SLIME MOULDS)

The Acrasiales have received a great deal of attention from biologists interested in morphogenesis. A brief consideration of the life cycle of the most intensively studied species, *Dictyostelium discoideum*, reveals the features that have attracted this attention.

The amoebae of *Di. discoideum* feed on bacteria, grow, and divide about every three hours if conditions are favourable. During this phase the amoebae are not very different from other small amoebae. Nutrient exhaustion leads to a remarkable aggregation process, in which streams of amoebae converge on one or more centres. This aggregation process, shown to be controlled by a hormone ("acrasin") and contact guidance, has been the subject of much research and is discussed in detail by Shaffer (1962). The aggregated amoebae become organized into a pseudoplasmodium which after migration differentiates into a stalk bearing a sorus containing numerous spores. During these morphogenetic events there is no further food intake and little cell division. It is thus possible to study differentiation in the absence of growth, and much work has been carried out on the pseudoplasmodium and its conversion into the fruiting body.

Recently, the Acrasiales have been the subject of a book (Bonner, 1967) which deals with all major aspects of their biology, devotes most of a chapter to practical information on laboratory methods, and provides a complete bibliography of the Acrasiales up to and including 1965. Other recent reviews, which include information on methods, are those of Gregg (1966), Raper (1963) Shaffer (1962, 1964), Sussman (1966) and Wright (1964); an earlier review by Raper (1951) on the isolation, cultivation and conservation of slime moulds remains valuable. In view of the quantity and quality of this reviewing activity, the remainder of the present section will be little more than a guide to the literature, arranged under headings similar to those in the Myxomycete section.

A. Habitat

Quantitative sampling methods have been devised and the distribution of the Acrasiales in nature studied by Cavender and Raper (1965a, b, c).

Dictyostelium and *Polysphondylium* were found to be particularly abundant in the humus layer and in moist decaying leaves in deciduous forests, whereas *Acrasis* occurred in dry leaf litter. The Acrasiales are also found in cultivated soils, compost, dung and rotting wood—in fact "in almost any situation where vegetable matter is undergoing aerobic decomposition" (Raper, 1951), thus providing bacteria for the nourishment of the amoeboid phase.

B. Collection

Cavender and Raper (1965a) advocate the collection of materials (e.g., soils, vegetable matter) for investigation in plastic vials with wide mouths and snap-on covers. If necessary, samples may be stored a little above freezing point (e.g., 4°C) for a few weeks, but drying out of the sample must be avoided.

C. Isolation

Members of the Acrasiales may be isolated from soil by spreading pregrown *E. coli* (or other suitable species) and a $25 \times$ dilution of soil on hay infusion agar (Cavender and Raper, 1965a). The procedure can be used for quantitative studies and with materials other than soil. Alternatively, particles of soil or plant fragments may be placed on the surface of hay infusion agar or other dilute media (Raper, 1951) resulting in the development of slime moulds on or close to the inoculum. The various methods result in the production of fruiting bodies within a few days, permitting identification and if necessary, sub-culture and elimination of contaminants (see below).

D. Identification

No monograph on the cellular slime moulds has been published since that of E. W. Olive in 1902. There are, however, relatively few genera and species, and all members of the Acrasiales known in 1965 are described in the book by J. T. Bonner, with references to the original taxonomic work (Bonner, 1967). A few species have been described subsequently (Raper and Fennell, 1967; Nelson *et al.*, 1967).

E. Two-membered culture

1. *Elimination of contaminants*

The fruiting bodies that develop on the original isolation plates may be used as a source of spores for establishing two-membered cultures. A suitable bacterium is streaked on an appropriate agar medium and a fruiting body placed at one end of it. The spores germinate, and the amoebae produced migrate along the streak consuming bacteria as they go. When the amoebae reach the end of the streak they are transferred to a fresh plate and the process repeated. A few repetitions of the process lead to the elimination of contaminants and the establishment of a two-membered culture with the chosen bacterial species. The method, and various modifications of it, are described in detail by Raper (1951). Often the elimination of contaminants can be facilitated by the careful transfer of spores from solitary fruiting bodies, as the sori of such fruiting bodies, especially of long-stalked species, are commonly bacteria-free.

2. *Routine culture on agar media*

Members of the Acrasiales will readily complete their entire life cycle from spore to spore in two-membered culture with bacteria on agar media. Suitable media and methods are described in detail by Raper (1951) and more recent developments reviewed by Bonner (1967). Bacteria may be grown separately from the slime mould, transferred to a non-nutrient agar and inoculated with slime mould spores (Singh, 1946). Alternatively, bacteria and slime-mould spores may be inoculated together on to a nutrient medium that permits simultaneous growth of both species, such as hay-infusion agar or lactose-peptone (both constituents at 0·1% w/v) agar (Raper, 1951). The most extensively employed bacterial species are *E. coli* and *A. aerogenes*, but many others are satisfactory. It is desirable that the bacterial associate should be maintained separately from the slime mould to provide the bacterial inoculum each time the slime mould is sub-cultured; reliance solely on bacteria carried over with the slime mould inoculum leads through natural selection to the development of inedible strains.

3. *Growth in shaken liquid cultures*

Several members of the Acrasiales have been grown with *E. coli* or *A. aerogenes* in shaken liquid cultures, using either a nutrient medium (Sussman, 1961) or pre-grown bacteria suspended in buffer (e.g., Hohl and Raper, 1963a). The topic is discussed by Bonner (1967).

F. Pure culture

The growth of *Di. discoideum* on heat-killed *E. coli* streaked on lactose-peptone agar is briefly described by Raper (1951). A number of species have been grown in shaken liquid culture with heat-killed *E. coli* suspended in a buffer, but only *Po. pallidum* grew as well on dead as on living bacteria (Hohl and Raper, 1963a). Subsequently, the amoebae of *Po. pallidum* were grown on a soluble medium in shaken liquid culture (Hohl and Raper, 1963b, c): such amoebae will aggregate and produce fruiting bodies if transferred to plain agar. Growth of amoebae of *Po. pallidum* on a defined medium in static but not shaken liquid culture has been achieved by Goldstone *et al.* (1966).

G. Maintenance of isolates in a genetically uniform state

It is generally held that sexuality does not occur in the Acrasiales, and hence that a major source of variability present in the Myxomycetes is lacking. It would appear that many isolates in culture have been established by inoculations from a single sorus. However, the amoebae which aggregate to form a fruiting body may themselves be of varied origin and hence the spores in a sorus may differ genetically, as was established by Filosa (1962). Clones of amoebae may readily be obtained by diluting a spore suspension adequately, spreading on agar along with bacteria, and subculturing amoebae from the resulting plaques (Filosa, 1962).

The amoebae of many members of the Acrasiales do not encyst, and cannot be preserved, but as spores are readily produced, this is not usually a problem. Raper (1951) advocates sub-culture of isolates every 3 or 4 months, with storage at 3–6°C as soon as growth and development is complete (5–6 days). Preservation of cultures for longer periods may be achieved by covering them with liquid paraffin or by lyophilization (Raper, 1951; Kerr, 1965a).

IV. THE PROTOSTELIDA

The Protostelida are a group of amoeboid organisms producing spores singly on slender stalks. They may be obtained by placing particles of soil, humus or dead plant parts on weakly nutrient agar media (e.g., lactose-yeast extract or hay infusion agar) and can be grown in two-membered culture on such media with *E. coli* and *A. aerogenes* and other bacteria, yeasts and fungi. The topic, including methods, is reviewed by Olive (1967).

V. THE LABYRINTHULALES

A comprehensive review and bibliography of *Labyrinthula*, the most important genus of the Labyrinthulales, has been published by Pokorny (1967). The two-membered culture of *Labyrinthula* spp. on bacteria and yeasts is considered by Aschner (1958, 1961) and Aschner and Kogan (1959). The pure culture of marine *Labyrinthula* spp. on liquid and agar serum–sea water media after elimination of contaminants by means of penicillin and streptomycin is described by Watson and Ordal (1957) and Watson and Raper (1957). A partially defined liquid medium was devised by Vishniac and Watson (1953) and a completely defined medium by Vishniac (1955a); one species was shown to have a steroid requirement (Vishniac 1955b).

I wish to thank Professor E. B. Chain, F.R.S., for helpful discussion and Miss Susan Ford for photography.

REFERENCES

Alexopoulos, C. J. (1960). *Mycologia*, **52**, 1–20.
Alexopoulos, C. J. (1962). "Introductory Mycology", 2nd ed. Wiley, London.
Alexopoulos, C. J. (1963). *Bot. Rev.*, **29**, 1–78.
Alexopoulos, C. J. (1964). *SWest Nat.*, **9**, 155–159.
Alexopoulos, C. J. (1966). *In* "The Fungi—an Advanced Treatise" (Ed. G. C. Ainsworth and A. S. Sussman), Vol 2, pp. 211–234. Academic Press, New York.
Aschner, M. (1958). *Bull. Res. Coun. Israel.* **6D**, 174–179.
Aschner, M. (1961). *Bull. Res. Coun. Israel*, **10D**, 126–129.
Ashner, M., and Kogan, S. (1959). *Bull. Res. Coun. Israel*, **8D**, 15–24.
Boder, G. B., and Johnson, R. W. (1967). *J. Boct.*, **94**, 1257.
Bonner, J. T. (1967) "The Cellular Slime Moulds", 2nd ed. Princeton University Press, Princeton, N.J.
Brewer, E. N., Kuraishi, S., Garver, J. C., and Strong, F. M. (1964). *Appl. Microbiol.*, **12**, 161–164.
Camp, W. G. (1937). *Bull. Torrey bot. Club*, **64**, 307–335.
Cavender, J. C., and Raper, K. B. (1965a). *Am. J. Bot.*, **52**, 294–296.
Cavender, J. C., and Raper, K. B. (1965b). *Am. J. Bot.*, **52**, 297–302.
Cavender, J. C., and Raper, K. B. (1965c). *Am. J. Bot.*, **52**, 302–308.
Cohen, A. L. (1939). *Bot. Gaz.*, **101**, 243–275.
Cohen, A. L. (1941). *Bot. Gaz.*, **103**, 205–224.
Collins, O. R. (1961). *Am. J. Bot.*, **48**, 674–683.
Considine, J. M., and Mallette, M. F. (1965). *Appl. Microbiol.*, **13**, 464–468.
Cummins, J. E., and Rusch, H. P. (1968). *Endeavour*, **27**, 124–129.
Daniel, J. W., Babcock, K. L., Sievert, A. H., and Rusch, H. P. (1963). *J. Bact.*, **86**, 324–331.
Daniel, J. W., and Baldwin, H. H. (1964). *In* "Methods in Cell Physiology" (Ed. D. M. Prescott), Vol. 1, pp. 9–41. Academic Press, New York.
Daniel, J. W., Kelley, J., and Rusch, H. P. (1962). *J. Bact.*, **84**, 1104–1110.
Daniel, J. W., and Rusch, H. P. (1961). *J. gen. Microbiol.*, **25**, 47–59.
Daniel, J. W., and Rusch, H. P. (1962a). *J. Bact.*, **83**, 234–240.
Daniel, J. W., and Rusch, H. P. (1962b). *J. Bact.*, **83**, 1224–1250.
Davis, E. E. (1965). *Mycologia*, **57**, 986–988.
Davis, E. E., and Butterfield, W. (1967). *Mycologia*, **59**, 935–937.
Dee, J. (1962). *Genet. Res.*, **3**, 11–23.
Dee, J. (1966a). *Genet. Res.*, **8**, 101–110.
Dee, J. (1966b). *J. Protozool.*, **13**, 610–616.
Elliott, E. W. (1949). *Mycologia*, **41**, 141–170.
Filosa, M. F. (1962). *Am. Nat.*, **96**, 79–91.
Fulton, C., and Dingle, A. D. (1967). *Devl Biol.*, **15**, 165–191.
Gilbert, H. C., and Martin, G. W. (1933). *Stud. nat. Hist. Iowa Univ.*, **15**, 3–8.
Goldstone, E. M., Banerjee, S. D., Allen, J. R., Lee, J. J., Hutner, S. H., Bacchi, C. J., and Melville, J. F. (1966). *J. Protozool.*, **13**, 171–174.
Gray, W. D. (1938), *Am. J. Bot.*, **25**, 511–522.

Gregg, J. H. (1966). *In* "The Fungi—an Advanced Treatise" (Ed. G. C. Ainsworth and A. S. Sussman), Vol II, pp. 235–281. Academic Press, New York.

Guirard, B. M., and Snell, E. E. (1962). *In* "The Bacteria" (Ed. I. C. Gunsalus and R. Y. Stanier), Vol. IV, pp. 33–93. Academic Press, New York.

Guttes, E., and Guttes, S. (1964). *In* "Methods in Cell Physiology" (Ed. D. M. Prescott), Vol. I, pp. 43–54. Academic Press, New York.

Guttes, E., Guttes, S., and Rusch, H. P. (1961). *Devl. Biol.*, **3**, 588–614.

Hohl, H. R., and Raper, K. B. (1963a). *J. Bact.*, **85**, 191–198.

Hohl, H. R., and Raper, K. B. (1963b). *J. Bact.*, **85**, 199–206.

Hohl, H. R., and Raper, K. B. (1963c). *J. Bact.*, **86**, 1314–1320.

Hok, K. A. (1954). *Am. J. Bot.*, **41**, 792–799.

Howard, F. L. (1931). *Am. J. Bot.*, **18**, 624–628.

Jahn, T. L. (1964). *Biorheology*. **2**, 133–152.

Jump, J. A. (1954). *Am. J. Bot.*, **41**, 561–567.

Kamiya, N. (1959). *Protoplasmatologia*, **8** (3a), 1–199.

Kerr, N. S. (1960). *J. Protozool.*, **7**, 103–108.

Kerr, N. S. (1963). *J. gen. Microbiol.*, **32**, 409–416.

Kerr, N. S. (1965a). *BioScience*, 469.

Kerr, N. S. (1965b). *J. Protozool.*, **12**, 276–278.

Kerr, N. S., and Sussman, M. (1958). *J. gen. Microbiol.*, **19**, 173–177.

Kerr, S. (1968). *J. gen. Microbiol.*, **53**, 9–15.

Lazo, W. R. (1960). *Mycologia*, **52**, 817–819.

Lazo, W. R. (1961a). *J. Protozool.*, **8**, 97.

Lazo, W. R. (1961b). *Am. Midl. Nat.*, **65**, 381–383.

Lister, A. L. (1925). "A Monograph of the Mycetozoa", 3rd ed. Br. Mus. Nat. Hist., London.

Martin, G. W. (1949). *N.Am. Flora*, **1** (1), 1–190.

Mittermayer, C., Braun, R., Chayka, T. G., and Rusch, H. P. (1966). *Nature, Lond.*, **210**, 1133–1137.

Nauss, R. N., (1947). *N.Y. bot. Gdn.*, **48**, 101–109.

Nelson, N., Olive, L. S., and Stoianovitch, C. (1967). *Am. J. Bot.*, **54**, 354–358.

Olive, L. S. (1967). *Mycologia*, **59**, 1–29.

Pokorny, K. L. (1967). *J. Protozool.*, **14**, 697–708.

Poulter, R. T. M., and Dee, J. (1968). *Genet. Res.*, **12**, 71–79.

Raper, K. B. (1951). *Q. Rev. Biol.*, **26**, 169–190.

Raper, K. B. (1963). *Harvey Lect.*, **57**, 111–141.

Raper, K. B., and Fennell, D. I. (1967). *Am. J. Bot.*, **54**, 515–528.

Ross, I. K. (1957). *Am. J. Bot.*, **44**, 843–850.

Ross, I. K. (1964). *Bull. Torrey bot. Club.*, **91**, 23–31.

Ross, I. K. (1966). *Am. J. Bot.*, **53**, 712–718.

Schuster, F. L. (1961). *J. Protozool.*, **8**, (suppl) 19.

Schuster, F. L. (1964). *Rep. Argonne natn. Lab. biol. med. biophys. Div.*, **6971**, 70–74.

Schuster, F. L. (1965). *Expl Cell Res.*, **39**, 329–345.

Scholes, P. M. (1962). *J. gen. Microbiol.*, **29**, 137–148.

Shaffer, B. M. (1962, 1964). *Adv. Morphogenesis*, **2**, 109–182; **3**, 301–322.

Singh, B. N. (1946). *Nature, Lond.*, **157**, 133–134.

Smart, R. F. (1937). *Am. J. Bot.*, **24**, 145–159.

Sobels, J. C. (1950). *Antonie van Leeuwenhoek*, **16**, 123–243.

Sobels, J. C., and Cohen, A. L. (1953). *Ann. N.Y. Acad. Sci.*, **56**, 944–948.

Sussman, M. (1961). *J. gen. Microbiol.*, **25**, 375–378.
Sussman, M. (1966). *In* "Methods in Cell Physiology" (Ed. D. M. Prescott) Vol. II, pp . 397–410. Academic Press, NewYork.
Vishniac, H. S. (1955a). *J. gen. Microbiol.*, **12**, 455–463.
Vishniac, H. S. (1955b). *J. gen. Microbiol.*, **12**, 464–472.
Vishniac, H. S., and Watson, S. W. (1953). *J. gen. Microbiol.*, **8**, 248–255.
Watson, S. W., and Ordal, E. J. (1957). *J. Bact.*, **73**, 589–590.
Watson, S. W., and Raper, K. B. (1957). *J. gen. Microbiol.*, **17**, 368–377.
Wright, B. E. (1964). *In* "Biochemistry and Physiology of Protozoa" (Ed. S. H. Hutner) Vol. III, pp. 341–381. Academic Press, New York.

CHAPTER X

Lichens

D. H. S. RICHARDSON

Department of Biology, Laurentian University, Ontario, Canada

I. INTRODUCTION

Lichens are one of the outstanding examples of symbiosis. In each lichen two micro-organisms, an alga and a fungus become closely associated and function as a single unit in nature.

Within the last decade lichens have been shown to be plants of increasing scientific and economic importance. They may absorb radioactive fallout

(Svenson and Kurt, 1965; Hanson and Palmer, 1965), indicate the presence of air pollution (Gilbert, 1965; Pearson and Skye, 1965; Skye, 1968) and produce antibiotics (Brightman, 1960; Korzybski *et al.*, 1967). Some species grow so slowly that relatively accurate dating of rock surfaces can be achieved for up to a thousand years (Beschel, 1961; Follmann, 1965). Recently, with the advent of radioactive tracer techniques, lichens have proved to be excellent material for studying certain general problems of symbiosis because the complete plant as well as both isolated symbionts may be used in experiments (see figs. 1 and 2).

Most of the mycological techniques for studying free-living fungi can be employed to examine thallus structure, ascocarp development, ascus and ascospore morphology. It is also possible to isolate the two components of a large number of lichens into pure culture (Ahmadjian, 1961, 1967a), and a number of specialized techniques have been developed to do this.

II. ISOLATION OF THE FUNGAL SYMBIONT

The great majority of lichen fungi belong to the class Ascomycetes and differ culturally from many free-living Ascomycetes in that they are (a) extremely slow growing, (b) nearly always require biotin and thiamine, (c) grow better at 18° to 21°C than at 25°C, (d) and hardly ever form conidia or other reproductive structures in culture. These fungi may be isolated either from lichen ascospores or from hyphae within the complete thallus.

A. Isolation from ascospores

This is carried out in essentially the same way as for free-living fungi. An ascocarp is stuck on to the lid of a Petri dish with a little moist cotton wool and petroleum jelly. The ejected spores are allowed to germinate on the agar and are then picked off with a fine needle. Bailey and Garrett (1968) found that *Lecanora conizaeoides* showed a tendency to discharge spores more readily at low temperature and it may take as much as 96 h for some lichen ascospores to germinate, e.g. *Xanthoria aureola* (Richardson and Smith 1968b). The germinated ascospores are placed on agar slopes in McArtney bottles and visible colonies usually develop after about two months at 18°C. The most widely employed medium for lichen fungi is malt/yeast extract agar of the following composition—

Malt-yeast extract agar (Lilly and Barnett, 1951)

Malt extract	20 g
Yeast extract	2 g
Agar	15 g
Distilled water	1 litre

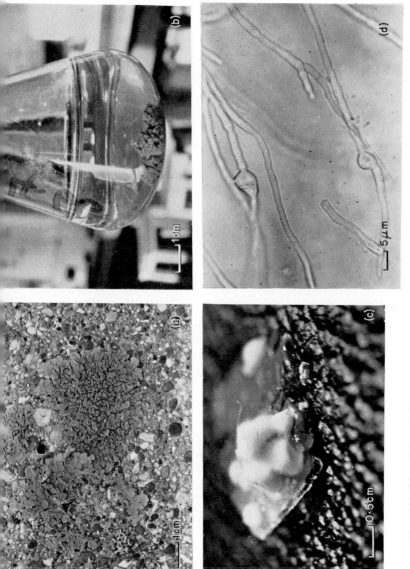

Fig. 1(a). Two foliose lichens growing on a concrete post in front of the Hatherly Laboratories, University of Exeter. On the left is *Xanthoria parietina*, on the right *Xanthoria aureola*. (b) The fungus isolated from *Xanthoria aureola* growing in liquid (Bianchi) media, shake culture at 18°C. (c) The same fungus growing on agar medium. The colony was photographed against a dark background. (d) A photomicrograph of the hyphae of the same fungus growing on an agar medium (Bianchi, 1964). This shows the typical intercalary swellings which have been thought to be conidia (Tomaselli *et al.*, 1963).

However, Bertsch and Butin (1967) report that the germination of ascospores in *Endocarpon pusillum* is strongly inhibited by malt extract and it might be better to use the more defined media given below.

B. Isolation from the thallus

In some cases it is very difficult to obtain germination and subsequent growth of lichen ascospores; for instance Scott (1957) found that vitamins present in agar caused the germinated ascospores of *Peltigera polydactyla* to burst. In such cases one method available is to grind up a fragment of the lichen thallus and pick up short lengths of hyphae with a needle or micromanipulator.

Two drawbacks are associated with this technique: (a) it is possible to isolate a free-living fungus growing epiphytically on the lichen or a para-symbiont (an extra fungus growing within the lichen thallus), and (b) the fungus isolated is probably heterokaryotic which may not be desirable in nutritional and genetic studies. Another technique which has the same disadvantages was used by Bertsch and Butin (1967) who initially germina-ted the ascospores of *Endocarpon pusillum* which grew rapidly (germ tubes up to 50 μm per day) for a few days but then stopped. They therefore isolated the fungus from the complete lichen. Mature thalli were cultured in little pots of sterile soil and hyphae grew out from the edge. Approximately 0·5 mm portions of the hyphae were removed to malt agar (3% malt extract, 1·8% agar) containing 0·3% supracillin (from the firm of Grunenthal) and this prevented bacterial growth. Within three weeks the isolates had produced small white colonies about 2 mm diameter. They assume this to be the symbiotic fungus as it contained the typical vesicles (10 μm diameter subtended by a stalk 2·5–5·3 μm long) seen in hyphae from the germin-ating ascospores and mature thalli.

C. Continued growth of the lichen fungi

As stated earlier, lichen fungi are extremely slow growing both on solid media (Ahmadjian, 1961) and in liquid media. For instance Quispel (1943) obtained only 35·1 mg dry weight of *Xanthoria parietina* fungus in 25 ml of enriched Czapek Dox medium after three months' growth. There are a number of reports of fast-growing lichen fungi. It is doubtful whether these rapid growing isolates from *Buellia stillingiana* (Hale, 1957), *Baeomyces rufus* (Tilden Smith, 1957) and *Peltigera aphthosa* (Bednar, 1963) are the true lichen fungus.

The isolated fungi of *Acarospora smaragdula*, *Acarospora fuscata* and *Cladonia cristatella* are of interest as they produce antibiotic substances effective against *Staphylococcus aureus* and *Bacillus subtilis* but not against *Escherichia coli* (Ahmadjian and Reynolds, 1961).

During investigations on the physiology of *Xanthoria aureola* (Richardson and Smith, 1968b) it was necessary to grow quantities of the lichen fungus in liquid medium. In cultural studies, a defined medium (Bianchi, 1964) was used rather than the semi-natural malt-yeast extract agar. It proved possible to get much better yields of fungus than those obtained by Quispel in Bianchi medium supplemented with biotin and thiamine only.

Bianchi Medium

NH_4 tartrate	5·0 g
NH_4NO_3	1·0 g
KH_2PO_4	1·0 g
$MgSO_4 . 7H_2O$	0·5 g
NaCl	0·1 g
$CaCl_2 . 2H_2O$	0·1 g
Sucrose	10 g
Trace element solution	1 ml
Biotin	10 μg
Thiamine	0·5 mg
Distilled water	1 litre

For rapid growth, shake culture was essential and at 18°C, in 50 ml conical flasks, about 100 mg dry weight could be obtained in eight weeks. The fungus grew well in media in which the carbohydrate present was ribitol, mannitol, glucose or sucrose.

III. ISOLATION OF THE ALGAL SYMBIONT

Ahmadjian (1967b) reports that there are eight genera of blue-green algae, seventeen genera of green algae and one genus of yellow-green alga symbiotic in lichens. In the British Isles only 8–9% of the lichen flora contains blue-green algae. It is vital in many instances to culture lichen algae as then cell division and reproductive states can be examined which help in identification at the specific level. The size and shape of the cells may change in pure culture, e.g. *Myrmecia sp.* In addition, the character of the wall may vary in culture; for example *Nostoc sp.* from *Peltigera polydactyla* develops a thick gelatinous sheath, *Coccomyxa* produces quantities of slime, and *Trebouxia* is reported to form a mucilaginous coat which is not present on algae within the thallus (Drew, 1966; Ahmadjian, 1959).

The physiology of these symbiotic algae also seems to differ from free living forms. Zacharias (1900) observed that symbiotic strains of *Nostoc* lacked the cyanophycin granules and Peat (1968) finds that they do not have the α granules which are believed to be polyglucosidic in nature. *Trentepholia* when in symbiosis often lacks the haematochrome which occurs in fat globules around the chloroplast and is thought to be a form of protein reserve (Fritsch, 1961). The carbohydrate metabolism of symbiotic algae appears to change in culture; for example the production of glucose by strains of *Nostoc* from *Peltigera polydactyla* and sugar alcohols by symbiotic green algae is greatly reduced after a period in pure culture (Drew and Smith, 1967b; Richardson *et al.*, 1968).

It is thus essential in both taxonomic and physiological studies to obtain the symbiotic algae in pure culture and also in quantity directly from the lichen thallus. This has been possible since Drew (1966) discovered that he could obtain *Nostoc* cells free from significant amounts of fungal filaments, by centrifuging a thallus macerate at various speeds. Similar techniques have been used to obtain green algae from lichen thalli. Such suspensions of lichen algae can also provide a convenient starting point for isolating single cells in the preparation of pure cultures.

A. Quantitative isolation of algae direct from lichen thalli

1. *Blue-green algae*

The lichen is first washed free of adhering dirt and then ground up without added abrasives in a pestle and mortar with distilled water. The thallus macerate is centrifuged for three minutes at about 375 g and the green algal zone obtained in the precipitate is scraped off and resuspended. This is then centrifuged at 125 g for 30 seconds and the supernatant again decanted. This procedure is repeated using increasing centrifuging times up to 90 seconds. Under the optimum centrifuging regime, preparations of the alga are obtained almost free of fungal fragments. Microscopic examination of the supernatants followed by adjustment of the centrifuging regime enables very pure preparations of directly isolated algae to be made. Drew (1966) was able to obtain about 30 mg dry weight of *Nostoc* alga from 5 g dry weight of *Peltigera polydactyla*. This was sufficient for studies using radio-active tracers but only about 5% of the algae originally present in the thallus are recovered in the final algal suspension. This is due to incomplete maceration and loss in the centrifuging process. A quantitative technique for the isolation of algae from lyophilised cephalodia and thallus of *Peltigera aphthosa* has recently been developed using density gradient centrifugation, (Millbank and Kenshaw, 1969).

2. *Green algae*

A similar technique to that described above has proved successful for the direct isolation of green algae such as *Trebouxia*, *Coccomyxa* and *Myrmecia*. A thallus macerate is centrifuged for ten seconds at approximately 60 g. This precipitates the heavier thallus fragments and the supernatant is recentrifuged at 375 g for five minutes. The cell fragments are then discarded and the algal cells and fungal fragments resuspended in distilled water and centrifuged for five minutes. Each particular lichen species or different lichen alga requires a slightly different centrifuging regime which can be worked out in conjunction with microscopic examination of the supernatant and precipitate at each stage of the isolation. Some lichen algae, e.g. *Coccomyxa* are easy to isolate in quantity in a very pure suspension because the algae are very small and can be easily separated from fungal fragments.

B. Pure culture of lichen algae

1. *Blue-green algae*

These algae can be isolated from an aqueous suspension of macerated thallus by removing single cells from the suspension with a micro-pipette and placing them in media used for the culture of free-living forms. However, studies on free-living blue-green algae have shown that it is particularly difficult to get pure cultures from single cells or single filaments. One technique which may be employed instead, is to surface sterilize the lichen with a solution of hypochlorite (Scott, 1960) to help kill the algal epiphytes. The thallus is then ground up and a suspension of cells produced as described in the previous section. This suspension is washed with sterile distilled water and inoculated into nitrogen free media. Drew and Smith (1967b) used such a technique for the isolation of *Nostoc* from *Peltigera polydactyla* but the resulting algal cultures were not bacteria-free and had to be sub-cultured every six weeks to prevent substantial contamination. The growth media they used was No. 32 of Zehnder and Gorham (1960) modified by the omission of nitrogen sources. This is prepared as follows: six 500 ml stock solutions are made, $NaNO_3$, 4·25 g; K_2HPO_4, 6·0 g; $MgSO_4.7H_2O$, 6·25 g; $CaCl_2.6H_2O$, 2·5 g; Na_2CO_3, 1·0 g; $Na_2SiO_3.9H_2O$, 5·0 g. Ten ml. of each solution is added to 940 ml distilled water and the medium autoclaved.

2. *Green algae*

Four main kinds of technique have been used to isolate green algae from lichen.

Firstly, Ahmadjian (1967b) has isolated a large number of algae with the following technique. He picked single algal cells from a suspension of ground-up thallus with a micro-pipette. The algal cell was then ejected into a drop of sterile water on a sterile slide. The micro-pipette is then used to retrieve the algal cell from the drop of water. It is transferred to a second drop of sterile water and the process repeated four or five times. Before the last two transfers, steam is passed over the micro-pipette to help kill contaminating micro-organisms and then the algal cell is placed on sloped agar medium and left for several weeks at about 18°C with an illumination of not more than 250 ft candles.

A second technique may be used if a micromanipulator is available as cells can get lost or damaged by the previous method. A suspension of directly isolated algae is prepared using the method described above in Section A2. The suspension is washed with sterile distilled water, resuspended in the same medium, and placed in a sterile tube. After several such washings a drop of this suspension is then placed on agar medium in a Petri dish and spread out to about one inch diameter. The dish is left for a short while to allow the liquid to be absorbed by the agar. Using a micromanipulator and binocular microscope single algal cells with a small piece of adhering hypha (indicating that they are algae closely associated with the lichen fungus and not epiphytes) are picked up with a fine glass loop. These are transferred to the centres of 2 mm cores that have been cut in the agar by an agar cutter in a different part of the plate to that on which the drop of suspension had been placed.

Each core with a single algal cell in its centre is then placed in a McArtney bottle containing modified Trebouxia medium (Starr, 1964) made up as follows: six 400 ml stock solutions are prepared each containing one of the salts in the amount shown below. $NaNO_3$, 10·0 g; $CaCl_2$, 1·0 g; $MgSO_4 . 7H_2O$, 3·0 g; K_2HPO_4, 3·0 g; KH_2PO_4, 7·0 g; $NaCl$, 1·0 g. Ten ml of each stock solution is added to 940 ml water. To this is added 1 drop 1% $FeCl_3$ and 2 ml of micro-element solution (Trelease and Trelease, 1935). To every 860 ml of this medium 140 ml coconut milk, 20 g of glucose, 10 g proteose peptone and 15 g of agar are added. The medium is autoclaved. If bacteria or yeasts are transferred with the algal cell on to this rich medium they soon become evident. After about 6 weeks at 18°C with an illumination of about 150 ft. candles visible colonies of the alga develop.

A third technique involves incubating in the light a small piece of lichen thallus in a Petri dish containing sterile distilled water. If the lichen thallus is examined after several days lichen algae may be seen growing out of the cut edge. These can be carefully removed to a suitable medium. One problem here is that it is quite possible to isolate a faster growing epiphytic alga rather than the true algal symbiont.

Finally, there are a few genera of lichens in which the ascocarp hymenium contains not only asci but also symbiotic algae, e.g., *Endocarpon* and *Staurothele*. As the ascospores are ejected these algae adhere to them. Bertsch and Butin (1967) found that only a few of the ejected ascospores from *Endocarpon pusillum* carried no alga. If the ascospores from this lichen are caught on a rich medium similar to that mentioned above, the algae outgrow the fungus and can thus be isolated.

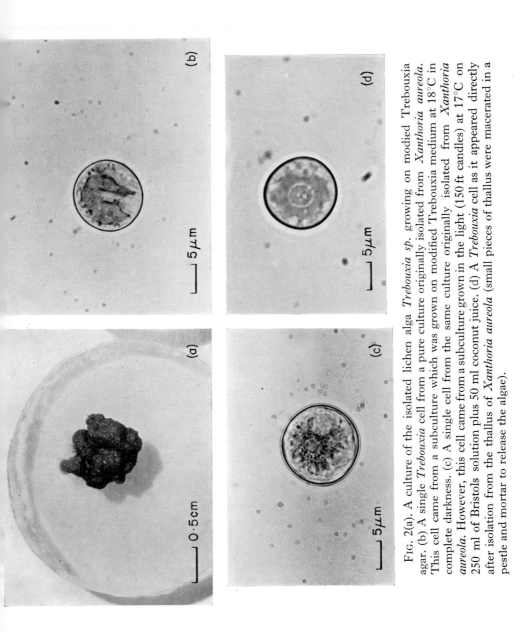

FIG. 2(a). A culture of the isolated lichen alga *Trebouxia sp.* growing on modied Trebouxia agar. (b) A single *Trebouxia* cell from a pure culture originally isolated from *Xanthoria aureola.* This cell came from a subculture which was grown on modified Trebouxia medium at 18°C in complete darkness. (c) A single cell from the same culture originally isolated from *Xanthoria aureola.* However, this cell came from a subculture grown in the light (150 ft candles) at 17°C on 250 ml of Bristols solution plus 50 ml coconut juice. (d) A *Trebouxia* cell as it appeared directly after isolation from the thallus of *Xanthoria aureola* (small pieces of thallus were macerated in a pestle and mortar to release the algae).

C. Continued growth of pure cultures of lichen algae

1. *Blue-green algae*

It is usually desirable to have rapid growth of the isolated alga. Drew and Smith (1967b) found that continuous shaking of cultures in conical flasks (100 ml), plugged with cotton wool, at 20°C with an illumination of 425 ft candles provided reasonable growth rates of *Nostoc* from *Peltigera polydactyla*.

Kratz and Myers (1955) found that blue-green algal cultures were limited in growth by the lack of sufficient carbon dioxide in plugged flasks and showed that most rapid growth was obtained by bubbling 0.5% CO_2 through the culture medium. Under such conditions the following medium was most satisfactory—

Medium

$MgSO_4.7H_2O$	0·25 g
K_2HPO_4	1·0 g
$Ca(NO_3)_2.4H_2O$	0·025 g
KNO_3	1·0 g
Na citrate.$2H_2O$	0·165 g
$FeSO_4.6H_2O$	0·004 g
Micro-elements	1 ml
Distilled water	1 litre

Ahmad and Winter (1968) using the same medium found that 10^{-5} to 10^{-9} M indole-3-acetic acid greatly stimulated the growth of blue-green algae.

2. *Green algae*

The commonest green alga of temperate lichens is *Trebouxia*. This alga, which is seldom found in the free-living state (Degelius, 1964; Ahmadjian, 1967a) is remarkable because in culture it is so slow growing. From the earliest studies (Beijerinck, 1890) it has been shown to require added carbohydrates in the growth medium for good growth. Most strains also grow better on organic nitrogen sources (1% Casamino-acids) than on nitrate, but only if glucose is also present in the medium (Ahmadjian, 1966; Fox, 1967) and these algae can grow quite satisfactorily in complete darkness on media containing both sugar and peptone. Di Benedeto and Furnari (1962) found that 50–100 g of IAA per litre stimulated the growth of *Trebouxia* but giberellic acid (5–15 mg/litre) had no effect. An early report (Quispel, 1943) that ascorbic acid (1 mg/litre) in a hydrogen atmosphere with 5% carbon dioxide enabled *Trebouxia* cells to grow on inorganic media at a rate comparable to growth on media with added sugar, does not seem to have been subsequently exploited or confirmed. For optimum growth light intensities of 200 to 250 ft candles are required; higher light intensities usually result in the loss of pigmentation by the cells especially when the medium does not contain added carbohydrates.

IV. RESYNTHESIS OF THE LICHEN

During the last century there were a number of reports of successful resyntheses of lichens in culture from the constituent alga and fungus; for example Rees (1871) with *Collema sp.*, Stahl (1877) with *Endocarpon sp.*, Bonnier (1889) with *Xanthoria sp.*, and Moller (1887) with *Calicium sp.* There have been many subsequent attempts to repeat their results, but only one, that of *Endocarpon pusillum*, has been verified (Bertsch and Butin, 1967). As a result doubts have been raised by many authors (e.g., Quispel, 1943; Ahmadjian, 1962b) as to the purity of the cultures made by the early workers and whether apparent synthesis could have resulted from airborne contamination. Some of the published illustrations by these workers of the resynthesized structures do not resemble very closely the complete lichen growing naturally, nor do they show mature reproductive structures.

More recently Thomas (1939) claimed complete synthesis in a sterile culture, on elder pith, of the lichen *Cladonia pyxidata*. Upright structures (podetia) were formed bearing cups (scyphae) that in nature develop apothecia. However, he was unable to repeat this result with a further 800 cultures.

Within the last decade Ahmadjian (1962a, 1963, 1966) has attempted to resynthesize the crustaceous lichen *Acarospora fuscata* on agar media and rock chips. He found that on minimal agar media and rock chips the fungus was only able to survive when the alga was present. A pseudoparenchyma was formed round the alga by the fungus but a characteristic thallus was not developed in that the upper cortex was absent, nor were reproductive structures formed. Herisset (1946) tried to resynthesize crustaceous lichens belonging to the Graphidiales which contain the alga *Trentepohlia*. While he too managed to get an association between the symbionts and saw haustoria, he did not observe the production of apothecia in any cultures.

Ahmadjian (1966b) had more success with the resynthesis of the fruiticose lichen *Cladonia cristatella*; podetia were formed in response to the joint effect of drying and poor nutrient conditions. These bore pycnidia and apothecia but did not contain asci. Scott (1956, 1960) used a different approach trying to maintain the symbiosis in disks cut from mature thalli of *Peltigera praetextata*. The disks were kept in illuminated culture tubes (150 ft candles at 20°C, 12 hours light out of 24) and he discovered that changes in nutrient supply, light and moisture upset the symbiotic balance. Anderson and Ahmadjian (1962), using Scott's technique, were able to get the development of podetia from disks of *Cladonia coniocraea* on which the initial stages of apothecia were found. Thus there is evidence that humidity, water relations and temperature of the lichen environment are critical if the established symbiosis is to be maintained. In this connection it is noteworthy that Pearson and Skye (1965) showed that thalli of *Xanthoria parietina* kept constantly moist at 15°C for several days showed abnormal photosynthetic patterns whereas those samples which had been initially moistened, kept illuminated but not covered (so that they dried slowly) for three days at this temperature and then placed in another illuminated "thallotron" at 27°C for a further three days before being returned to the lower temperature and remoistened, retained the normal photosynthetic pattern for six months in this cycle.

A very important step in the resynthesis of lichen thalli has been due to the work of Bertsch and Butin (1967) who have repeated the reported successful synthesis of the lichen thallus and reproductive structures in *Endocarpon pusillum* by Stahl (1877). On water agar, hyphae from germinating ascospores of this lichen wrapped round the algal cells but no thallus was formed. However, on small pots of fine sterile soil pH 6·9 covered with a Petri dish and illuminated continuously at 400 ft candles at 15°C∓1, thallus formation took place. The pots were inoculated with ascospore cultures transferred at a very early stage from the agar to the soil with as little agar as possible. Alternatively the ascospores could be shot directly on to the soil. After only about 14 days thallus initials could be seen with the naked eye as green spots on the soil. The formation after about 4 weeks of thallus scales from these is promoted if conditions are not too wet. If they are, lumps of algae develop which are only partially penetrated by the lichen fungus. Bertsch and Butin noted that once the balance of the symbiosis is upset it is not possible to restore it by drying the pots. As months pass bigger and bigger scales are formed and large thalli result from the confluence of several initials. It seems that at least two initials are required for successful thallus formation. Perithecia were developed after about six months and ascospores formed which were similar in size and shape to those from thalli collected in the natural habitat. Ascospores from the cultured thalli were germinated successfully and thus verified that this lichen can be cultured from spore to spore in the laboratory.

Many of the problems associated with resynthesizing lichens are now known. The main difficulties are the slow growth of the cultures and the delicate balance of conditions required to maintain both partners in a healthy condition. This delicate balance is not only one of physical factors but evidence is accumulating to show that chemical interactions are involved as well. Studies on the physiology of the complete thallus dealt with in the next section should help to elucidate this aspect.

V. TECHNIQUES FOR STUDYING THE COMPLETE THALLUS

A. Collection

The way in which lichen material must be collected and handled depends upon the subsequent experiments to be carried out in the laboratory. If the lichen is to be used for studies of metabolic processes such as respiration, assimilation or uptake of dissolved substances, it should be transported in a moist condition in a polythene bag. Under these conditions the rates of such processes do not seem to change much within a few days after collection, although it is always desirable that lichens should not be kept too long in the dark at room temperature. Ideally they should be kept at low temperature in the light. If transit will take longer than a few days, the material should be air dried at about 20°C and treated on arrival at the laboratory as described below. However, lichens from aquatic or very moist habitats quickly show abnormal metabolic patterns if dried and so should be sent in a moist condition.

When collecting, it is often difficult to remove foliose lichens from the substrate without damaging the lower cortex. If water is thrown over the lichens and substrate, a few minutes before collection, the material can be removed with almost no damage. It is also important to collect samples from a number of thalli, especially when examining seasonal variation of metabolites. Finally the growth rate of lichens is particularly slow, on average 0·1 to 1·3 cm/year. In the interests of conservation as little material as possible should be collected if the lichen in question is at all rare.

B. Killing material

If quantitative studies on the levels of compounds such as storage products are to be made, the material should be killed rapidly, if possible at the site of collection. The importance of this was realized in experiments on the polyol and reducing sugar content of lichens. For example in *Peltigera polydactyla* the mannitol content fell by 40% during starvation on distilled water in the dark at 25°C after 48 h (Drew, 1966). In *Xanthoria aureola* the total polyol content fell by a similar amount in this time and the pentitols which had constituted almost 50% of the total, were no longer detectable (Richardson and Smith 1967). Solberg (personal communication) found only minute quantities of reducing sugars in *Stereocaulon sp.* and this was possibly due to collecting large amounts of material and allowing it to air dry slowly. For such experiments 80% ethanol, preferably hot, is one of the most effective killing and extracting agents.

C. Seasonal variation

Both seasonal variation in metabolite concentration and physiological activity have been found in *Peltigera polydactyla*. Lestang Laisné (1966) and Smith and Ozin (unpublished), found the highest mannitol content in autumn and the lowest in winter in *Lichina pygmaea* and *Peltigera polydactyla*. In *Xanthoria aureola* there was a winter low value but the highest records for polyol content were in early spring. Stålfelt (1939a, b) showed that the rate of photosynthesis per unit dry weight of eleven non-crustaceous species was much greater in winter (December to January) than in summer (May to June) at any given temperature. Differences in respiration rate were usually much smaller, so that carbon assimilation was also higher in winter. The optimum temperature for maximum net assimilation in the light was lower in winter than in summer by an average of 4°C; thus the month of the year when an experiment is carried out should be noted as it may later explain unusual results. Some substances are highly toxic to lichens and experiments should be done in laboratories free from organic solvent vapours, ammonia and coal gas.

D. Treatment of material sent by mail

Material arriving in a dry condition should be washed clean with distilled water. This is quickly absorbed so that the lichen becomes soft instead of being brittle. It can then be quickly surface dried and sampled as appropriate. It is evident from the work of Reid (1960a, b) that lichens should never be used for respirometry or assimilatory studies immediately after soaking with water. Barrett and Smith (unpublished) have shown, for example, that it may take nine hours for the respiration rate of *Peltigera polydactyla* to return to normal levels after resoaking and the abnormally high respiration during this period is accompanied by high consumption of carbohydrate reserves. It seems best after sampling to place material sent by mail in Petri dishes on moist filter paper at about 5°C with an illumination of about 250 ft. c. for 24 h (Smith and Jackson Hill, unpublished). Under these conditions the assimilatory and other processes are resumed but the carbohydrate reserves are only slowly depleted (Drew, 1966; Richardson, 1967a).

If material is sent by mail in a moist condition, it only requires washing clean with distilled water and sampling before use. When excess material is available it is best stored in the light at 5°C; under these conditions bacterial contamination does not seem to be a problem and the disadvantages of trying to surface sterilize the thalli far outweigh any advantages.

E. Sampling material

1. *Large foliose lichens*

Harley and Smith (1956) developed a method for sampling *Peltigera polydactyla* by punching disks out of thalli. They obtained 7 mm disks though in some lichens with narrower lobes 5 mm disks have been used (Richardson *et al.*, 1968) for example *Lobaria amplissima* and *Roccella fuciformis*. No detailed studies have been done on the effect of disk size on the rate of physiological processes such as absorption. This "disk" sampling has the advantage that samples have equal surface area and cut edge which is important in uptake experiments and Warburg respirometry. In uptake experiments in 20–25 ml of solution, not using radioactive tracers, Smith (1960) used 30 to 50, 7 mm disks per sample. In short term experiments using tracers, ten 7 mm disks were employed to study the movement of carbohydrate from the alga to the fungus in *Peltigera polydactyla*.

2. *Small foliose lichens*

For thalli with small or irregular lobes, sampling by fresh weight rather than by surface area has been proved best (see next Section). It is often preferable to use only the marginal lobes of such thalli which are less dirt encrusted and seem to be physiologically more active. In fresh weight sampling, the washed thallus lobes are first placed in distilled water until they are fully saturated. The time taken for this to happen varies with the lichen but is 1 to 90 min (Reid 1960). The lobes are then surface dried between filter paper and quickly weighed into samples of 100–400 mg, which is an appropriate sample size for most experiments.

3. *Crustose lichens*

These are especially difficult. Thicker thalli such as *Lecanora conizaeoides* may be scraped off the substrate, the thallus crumbs washed, dried and sampled by fresh weight. In other cases it is necessary to make small chips of rock with the lichen growing thereon or to cut the rock into pieces with geological saws. In such cases samples containing equal amounts of lichen are practically impossible to prepare. Some of the techniques used for sampling various lichens are given by Richardson *et al.* (1968).

F. Sampling errors

Experiments with lichens are usually limited by the supply of material and the fact that preparing samples is laborious. Smith (1960c) found that expressing the results on a surface area basis (30×7 mm disks) when dealing with *Peltigera polydactyla* gave satisfactory replication within experiments. This is a useful parameter especially for expressing the properties of the algal layer which is of a more or less constant thickness. The medulla on the other hand varies widely so that the dry weight of the discs with similar amounts of alga can vary considerably. The difficulties of expressing results on a dry weight basis are (a) much material is needed to devise a satisfactory method for evaluating the initial dry weight of a sample and (b) indirect evidence suggests the absence of a simple relationship between dry weight and the uptake of sugars and other substances by lichens. It was found that 30, 7 mm disk samples of *Peltigera polydactyla* had a coefficient of variation of about 5% in absorption experiments. This is a value close to that calculated by Stålfelt (1939a, b).

In *Xanthoria aureola*, 100 mg fresh weight samples of thallus were allowed to photosynthesize on solutions of [^{14}C] sodium bicarbonate. The ^{14}C fixed during photosynthesis was expressed in terms of either fresh weight, surface area or extracted dry weight. Extracted dry weight is the dry weight of a sample after the ethanol soluble substances have been removed with hot 80% ethanol. Fresh weight proved to be the parameter on which the smallest coefficient of variation (about 10%) was shown.

In cases where the dry weight of lichen material is to be determined, Smith showed that a temperature of 75°C for 24 h enabled drying to a constant weight without the charring which occurred at 100°–105°C, especially if the lichen samples were initially not completely air dried.

G. Warburg respirometry

This technique has been used to measure respiration rates in lichens (Smith, 1960a, b; Feige, 1967). Smith used samples of twenty-five, 7 mm disks floated on 2 ml of medium in a Warburg flask. The medium was buffered with M/100 phthalate at pH 5·5. Concentrations of 10 mM sugars or M/40 sodium fluoride were used to stimulate or inhibit the output of carbon dioxide. The temperature used was 25°C which was perhaps rather high for lichen material. A long equilibration period of at least an hour was necessary and suitable reading intervals are about 15 min. The rate of respiration appeared steady, from one half hour after the addition of respiratory substrate, for up to two hours. The addition of such substances as glucose and asparagine considerably stimulated oxygen uptake by disks of this lichen. More recently, Smith and Barrett (unpublished) have used a lower temperature (20°C) and have found that the basal rate of respiration in *Peltigera polydactyla* stays steady showing only about a 10% decline over nine hours. In studies on the effect of resoaking on the respiration rate of *Peltigera polydactyla* disks, Smith and Barrett placed 15 air dried disks in a Warburg flask without liquid and then added 1·5 ml water from the side arm. Two minutes after adding the water, the taps were closed and readings commenced.

H. Incubation on liquids

In laboratory investigations it is very convenient to carry out experiments on photosynthesis and uptake by lichen samples in liquid media. In such experiments the samples have to be contained in vessels that are transparent, have a small liquid/air interface and which can be sealed from the outer atmosphere. This is especially necessary when volatile radioactive compounds such as $NaH^{14}CO_3$ or acetate are being used. The vessels which fulfil these requirements best are 2×1 in. specimen tubes sealed with a 1 in. coverslip kept in place by a ring of Vaseline on the lip of the specimen tube. Aluminium foil is placed under the tube (to reflect light upwards), which is then clipped into an illuminated shaking water bath. Shaking is most important in uptake experiments. Smith (1960c) discovered that the algal zones and medullae of dissected disks (see next Section) of *Peltigera polydactyla* absorbed more when shaken than when floated on solutions of glucose. Typically, a sample of 10, 7 mm disks or 100 mg fresh weight of lichen is placed in 3 ml of distilled water containing 10 μCi $NaH^{14}CO_3$ with a temperature of 18°C and a continuous daylight fluorescent illumination of 500 ft candles.

I. Dissection experiments

Harley and Smith (1956) developed a technique for separating the algal layer of *Peltigera polydactyla* from the underlying medulla by dissecting 7 mm disks of lichen under a low power binocular microscope using a fine scalpel (Swann Morton no. 15). It was found easiest to cut the disks into half before dissection but even so it took 10 to 15 min to dissect 5 disks. The samples, therefore, have to be small as slow metabolic changes can occur in disks waiting to be dissected. These can be minimized if the disks are taken from the experimental conditions, washed with, and placed on, distilled water at 0°C. Using this technique Smith and Drew (1965) have shown that radioactive carbon fixed during photosynthesis in the algal layer can move to the medulla within 10 min. In *Lobaria scrobiculata* the algal layer is very brittle and uneven but can be removed by chipping small pieces off with a fine scalpel. This gives a more exact separation of the medulla from the algal layer than is possible with *Peltigera polydactyla* but one does not have an intact algal layer and so medullae of dissected disks have to be compared with whole disks.

This technique with minor modifications has been extended to lichens containing green algae; for example *Lobaria pulmonaria* (L.) Hoffm. (*Myrmecia sp.*), *Dermatocarpon minitaum* (*Hyalococcus sp.*) and *Roccella fuciformis* D.C. (*Trentepohlia sp.*) (Richardson *et al.*, 1967; 1968). It has been found that ^{14}C fixed by the algae of these lichens during photosynthesis passes much more slowly to the medulla than in the lichens containing blue-green algae.

J. Inhibition of carbohydrate transfer

This method was first developed for studying the lichen symbiosis by Drew and Smith (1967a) and has become termed the "inhibition technique". It involves placing lichen samples on solutions of $NaH^{14}CO_3$ in the light and including in the medium a high concentration (usually 1 or 2% w/v) of the non-radioactive form of the carbohydrate which is thought to move between the symbionts. The radioactive carbohydrate released by the photosynthesizing alga is unable to compete for entry to the fungus with the much higher concentration of non-radioactive substance and it therefore diffuses into the medium. The amount of ^{14}C appearing in the medium is taken as a measure of the amount that would have moved between the symbionts. ^{14}C only appears in the medium if the non-radioactive carbohydrate used is the compound which moves between the symbionts (or one very similar to it, e.g., glucose and 2-deoxyglucose; ribitol and arabitol). In such cases there is also a great reduction in the amount of ^{14}C appearing in specifically fungal metabolites as compared with samples of lichens photosynthesizing on solutions of $NaH^{14}CO_3$ without added carbohydrates.

By feeding a pulse of ^{14}C to a lichen it is possible to build up a picture of the rate of carbohydrate transfer between the symbionts of lichens which cannot be dissected. For example samples of *Xanthoria aureola* (200 mg fresh weight) were allowed to photosynthesize for two hours on ^{14}C sodium bicarbonate solution (18°C, 500 ft candles). They were then washed with distilled water and allowed to continue photosynthesis in distilled water without ^{14}C sodium bicarbonate. After increasing time intervals, samples were removed from the distilled water and placed on 2% ribitol (this is the compound which moves between the symbionts in this lichen) in the light when it was assumed that the residual ^{14}C available for transfer would diffuse into the medium. At the end of 24 h all samples were killed in 80% ethanol and the amount of radioactivity in both media and tissues determined. The total amount of ^{14}C available for transfer from alga to fungus was assumed to approximate to the amount released by the tissues of samples transferred immediately from ^{14}C sodium bicarbonate to ^{12}C ribitol media. The amount of ^{14}C moving from alga to fungus during a particular time on distilled water will be equivalent to the total amount available for transfer minus the amount released subsequently in ^{12}C ribitol. Thus the rate at which an initial pulse of ^{14}C moves from alga to fungus can be calculated. It is most important to note that this calculated rate refers only to ^{14}C and not to total carbohydrates. This is because there is evidence that the size of the pools of mobile carbohydrates vary in different lichens. Hence it will only be possible to compare the efficiencies of the various symbionts in supplying their fungal components with carbohydrates when suitable methods are available for measuring the absolute pool size of mobile carbohydrates in the alga and the specific activity of the carbohydrates moving between the symbionts. Gas-liquid chromatography is now being used to try and solve this problem (Jackson-Hill, personal communication).

K. Electron microscopy

1. Sectioned material

The fine structure of lichens, lichen algae and lichen fungi has received little attention.

Bednar and Juniper (1965) made some preliminary studies on *Xanthoria parietina* whilst Drew and Smith (1967b) examined *Peltigera polydactyla*. These studies did not reveal fungal haustoria penetrating the algal cells in these lichens. The technique used by the above workers was to fix small pieces of lichen in $\frac{1}{2}\%$ gluteraldehyde in 0·1 M sodium cacodylate buffer (pH 7·3) for 48 h at 50°C in the dark. After washing in buffer, post-fixation was carried out in 2% osmium tetroxide in cacodylate buffer for 2 h at 0°C. The tissue was dehydrated in graded ethanol series followed by two half hour washes in propylene oxide and embedded in an epon mixture (Juniper, 1962). Sections were cut on a Cambridge Huxley microtome, mounted on an unsupported 400 mesh grid, and immersed for 10 min in lead citrate which is prepared by placing 1·33 g $Pb(NO_3)_2$, 1·76 g Na_3 $(C_6H_5O_7)$. $2H_2O$ and 30 ml distilled water in a 50 ml volumetric flask. The resultant suspension is shaken vigorously for 1 min and allowed to stand with intermittent shaking in order to insure complete conversion of lead nitrate to lead citrate. After 30 min 8·0 ml 1 N NaOH is added, the suspension diluted to 50 ml with distilled water and mixed by inversion. Lead citrate dissolves and the staining solution is ready for use. The pH of the staining solution was routinely found to be $12·0 \pm 0·1$. Faint turbidity, if present, is usually readily removed by centrifugation (Reynolds, 1963). The sections were then washed in distilled water, dried and examined in an AEI EM6 microscope. The most difficult part of this procedure was sectioning the lichen material which proved very hard.

Chervin and Baker (1968) examined sections of *Usnea rockii* and *U. pruinosa*. The fungal hyphae of these lichens were polymorphic and displayed structural differences in their walls which were apparently unrecorded in fungal hyphae. Four different fungal cell types were found: (a) the hyphae of the algal zone; (b) large medullary cells; (c) thick-walled medullary cells; and (d) the cells of the chondroid axis. Close contact between fungus and alga was seen and intra-membranous haustroial penetration of the algal cells was demonstrated in *Usnea rockii*. Their best electron micrographs were obtained from material (2 mm lichen segments) fixed with 2·5% phosphate buffered gluteraldehyde (pH 7·4) for two hours followed by 2% unbuffered potassium permanganate for two hours. Dehydration was carried out in graded series of tertiary butyl alcohol (with 2 hours between changes) followed by propylene oxide. The tissue was embedded in maraglas (Pease 1964) and then sectioned with glass or diamond knives. The sections were mounted on collodion and carbon coated grids, then post-stained in lead citrate (Reynolds 1963) and finally observed with a Phillips EM 75 or Hitachi HU-11A electron-microscope.

Moore and McAlear (1960) and Ahmadjian (1967a) using an electron microscope also found haustoria making direct contact with the algal protoplasts in a number of lichens, e.g., *Cladonia cristatella* and *Lecidea sp*. A number of investigators using the light microscope have found haustoria in nearly all lichens examined (e.g., Geitler, 1963; Danilov, 1910) but others have only found them in very few lichens (Mameli, 1920). There is evidence that the occurrence and abundance of haustoria in some lichens may be related to treatment of material before examination. Specimens kept moist and dark or dried slowly (or left in laboratory) may show abnormal algal fungal relationships. It is hoped that further electron microscope studies will determine the exact morphological relationships between the symbionts in many different lichen genera.

2. *Surface characters*

The Stereoscan electron microscope has been found highly satisfactory for the close examination of surfaces of thalli (fig. 3) but the examination of spores has not been successful to date as they are extruded with mucilage which is difficult to remove.

The dried lichen material is mounted on a stub using an adhesive such as Durofix or stuck to double sided Sellotape. A quantity of gold is evaporated onto the specimen under vacuum (10^{-4} torr) to give a coat about 100 Å thick. The layer of gold must be continuous with the surface of the stub in order to ensure that the whole surface of the specimen is at earth potential. The material is then ready for examination by a Cambridge Instrument Stereoscan mark II. (P. W. James, personal communication.) Reznik *et al* (1968) have published elegant stereoscan pictures of rhizinae in *Parmelia caperata*, *P. perlata* and *Lobaria pulmonaria*. The ends of the rhizinae in *Parmelia* were flattened and enlarged. By contrast in *Lobaria* the rhizinae were fixed to the substratum by the tips of single hyphae.

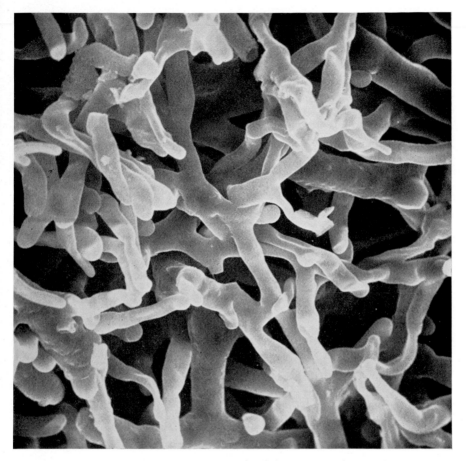

Fig. 3. A stereoscan electron micrograph of the upper surface of the tomentose of *Peltigera polydactyla*.

L. Lichen chemistry

A very large literature has accumulated describing the various "lichen acids". The technique for their extraction, isolation and study (Asahina and Shibata, 1945) fall more in the realms of organic chemistry than in microbiology. However, chemists should bear in mind the remarks in the foregoing section on sampling lichen material. Recent reviews have been published including information on lichen acids and related substances by Haynes (1966), Ahmadjian (1967a), Hale (1967) and Huneck (1968).

They may now be characterized by crystal tests, fluorescence analysis, UV absorption spectra and thin layer chromatography (Hale, 1967; Santesson, 1967). In addition the biosynthetic pathways of some lichen acids have been examined using radioactive isotopes (Fox and Mosbach, 1967; Mosbach, 1967; Maass and Neish, 1967).

M. Transplantation of lichen thalli

It has proved very difficult to maintain lichens in a healthy condition or get significant growth in thalli transferred to the laboratory (Scott, 1956, 1960). However, under natural conditions the successful transplantation between different habitats has been achieved. Brodo (1961) excised cores of tree bark on which a specimen of lichen was growing and took this to a different area but similar habitat. The core was fixed in a prepared hole in the new host tree by means of grafting wax. Brodo found that the wax had little effect on the continued growth of the lichen but bleeding of the host tree was often a problem with crustaceous species. Le Blanc and Rao (1966) have used a similar technique to transplant lichens to industrial areas to study the effect of atmospheric pollution on them.

Hale (1954) removed clumps of soil (200 cm^2) with intact lichen thalli from Baffin Island to North Western Connecticut. After two years, one lichen *Cetraria islandica* had made new growth, two other species of this genus survived but the other lichens had died.

Richardson (1967b) removed thalli of *Xanthoria parietina var ectanea* from a sea shore rock using the back of a scalpel and transported them in a polythene bag 150 miles to an inland habitat, a farm roof, where *Xanthoria aureola* grew. The transplanted lichens survived and grew. Over an eighteen month period they showed changes in morphology which indicated that *Xanthoria parietina var ectanea* was indeed only a form of *Xanthoria parietina* sensu stricto.

Such techniques should be especially valuable in determining the exact taxonomic status of critical species and chemical strains of lichens within such genera as *Ramalina*, *Xanthoria*, *Physica* and *Parmelia*.

VI. DISCUSSION

It is now possible, using isotopes, to examine in the laboratory the nitrogen, phosphorus and carbon metabolism of a lichen thallus collected from natural habitats (Bond and Scott, 1956; Feige, 1967; Millbank and Kershaw, 1969; Richardson *et al.*, 1968).

In addition advances in cultural techniques enable such studies to be carried out on the fungus and alga separately. This is of great value when assessing the interaction between two micro-organisms and will no doubt be further exploited in future. The study of the symbionts immediately after removal from the symbiosis and the nature of the changes they undergo during subsequent pure culture are especially important. Thus many techniques have been developed for studying the lichen symbiosis but a large number of outstanding problems remain. For example it is remarkable that the mobile carbohydrate moving from alga to fungus in lichens containing five genera of green algae, which were not all closely related, has been found to be a sugar alcohol. Unfortunately so little is known about the carbohydrate metabolism of terrestrial green algae, that it is not clear whether the potentiality for polyol production is a common and general character of certain groups of algae or is restricted to those genera which enter into the lichen symbiosis. Recent advances in techniques enable polyols to be easily studied and identified (Lewis and Smith, 1967) hence it is to be hoped that the photosynthetic pattern of a wide range of green algae will be examined to see if they produce these compounds.

At present time it is not known why lichen algae such as *Trebouxia* are so slow growing in culture on inorganic media and yet within the lichen must supply the complete thallus with all the carbohydrate necessary for growth. It may be that the lichen fungus produces growth promoting substances such as indole acetic acid which have been shown to stimulate the growth of algae (di Benedetto and Furnari, 1962; Ahmad and Winter, 1968). Certainly it is known that some free living fungi can excrete IAA-like substances (Went and Thimann, 1937). In return lichen fungi are known to require biotin and thiamine; Bednar and Holm-Hansen (1964) showed that *Coccomyxa sp.* from *Peltigera aphthosa* excreted much greater amounts of biotin in culture than a free living species of *Chlorella*. Such relationships have yet to be confirmed experimentally in the complete thallus. In addition it is not clear what induces lichen algae to release carbohydrate. The fungal partner might excrete a substance such as a lichen acid which would increase the permeability of the algal cell. The fact that only a few of the many intracellular substances are released from the cells may indicate that more than a simple change in permeability of the algal membrane is involved. Since *Nostoc* has a mucilagenous sheath in culture and *Trebouxia* a gelatinous coat (Drew and Smith 1967b; Ahmadjian, 1959) which is not present within the thallus, it is possible that the carbohydrate moving to the fungus results from some modification of the cell wall metabolism of these algae. However, in the case of *Trebouxia*, this seems unlikely since the mobile carbohydrate, ribitol, has never been identified as a cell wall component of organisms other than bacteria.

With regard to the taxonomy of the unicellular green algae of lichens, much work is required to determine both the delimitation of genera and what constitutes a species. This applies particularly to the genus *Trebouxia* and furthermore the relationship of this genus to free living genera such as *Chlorococcum* is obscure. Such algae do not have many taxonomically useful morphological characters and the development of biochemical characters might enable great progress.

It is obvious that a variety of morphological and behavioural modifications are manifest when an alga and a fungus come into symbiotic association. For example the fungi, when lichenized, produce vegetative reproductive structures, isidia and soredia, with specialized dispersal mechanism (Bailey, 1966) and perfect fruiting states. These are not formed by pure cultures of the fungus though a few isolates have produced pycnidia: e.g., *Calicium sp.* (Moller 1887), *Cladonia cristatella* (Ahmadjian, 1966); and others developed conidia; e.g., *Cladonia squamosa* (Werner, 1927), *Baeomyces rufus* (Tilden-Smith, 1957), *Lecidea sylvicola* and *Phaeographina fulgurata* (Ahmadjian 1963), *Anaptychia ciliaris* (Warēn, 1919–1920).

In this connection it is important to remember that the term lichen encompasses at its extremes two different types of association between alga and fungus with a continuous gradation in between. On the one hand there are the lichens in which there is not only a physiological interplay between the symbionts but a resulting new morphological entity quite unlike either separated component. At the other extreme there are a number of lichen genera, e.g. *Leptorhaphis*, in which the thalli of some species always contain algae, others never do, while a few species are sometimes found containing algae. The morphology of many such lichens closely resembles that of free living saprophytic ascomycetes. Although it is not certain whether such species are lichens that are losing their ability to obtain all carbohydrate from the alga or are free living fungi evolving the symbiotic state, it is possible that the isolated fungi from these might be induced to form conidia more easily than the morphologically more complex lichens. Such crustaceous lichens may be found in the genera which Hale (1967) places in the Pleosporales, Hysteriales and Graphidiaceae and their relationships to free living forms might be examined by comparing both the conidial states in culture and perfect states on the collected lichen. Conidia have often been observed and described on isolates from free living Hysteriaceous fungi (Lohman, 1932, 1933, 1934, 1937; Bisby, 1941). In studies on lichen symbionts it would greatly help if all isolates were lodged in central collections and it is worth noting that the Cambridge Collection of Algae and Protozoa hold a number of cultures of lichen algae whilst the Commonwealth Mycological Institute, Kew, has some cultures of lichen fungi. It has been the experience of the latter institute that lichen fungi do not survive well under oil at low temperature and it seems best to store them at about 10°C without oil but subculturing every six months. Recently it has been found that the vegetative mycelium can be lyophilized and 50 representative cultures of lichen fungi have been deposited at the American Type Culture Collection (Ahmadjian, personal communication). It seems that cultures of lichen algae can also be preserved by freeze drying (Richardson, unpublished). Lichens thus pose many intriguing problems of physiology, morphogenesis and taxonomy. Every recent study indicates the need for further work and it is to be hoped that the expanding interest in these plants will be continued with the application of new techniques.

ACKNOWLEDGEMENTS

I would like to thank Dr. D. C. Smith, Mr. P. W. James and Professor V. Ahmadjian for their many helpful suggestions and comments on this article and for making available hitherto unpublished data.

I am also grateful to Mr. P. W. James and the Electron Microscope Unit of the British Museum (Natural History) for permission to use their stereoscan electron micrograph of *Peltigera polydactyla*.

REFERENCES

Ahmad, R. M., and Winter, A. (1968). *Planta*, 78, 277–286.
Ahmadjian, V. (1959). *Svensk bot. Tidskr.*, 53, 71–80.
Ahmadjian, V. (1961). *Bryologist*, 64, 168–179.
Ahmadjian, V. (1962a). *Am. J. Bot.*, 49, 277–283.
Ahmadjian, V. (1962b). *In* "Physiology and biochemistry of algae" (Ed. R. A. Lewin), pp. 817–822. Academic Press, New York.
Ahmadjian, V. (1963). *Scient. Am.*, 208, 122–132.
Ahmadjian, V. (1966a). *In* "Symbiosis" (Ed. S. Mark Henry), pp. 35–98. Academic Press, London.
Ahmadjian, V. (1966b). *Science, N.Y.*, 151, 199–201.
Ahmadjian, V. (1967a). "The lichen symbiosis". Blaisdell, London.
Ahmadjian, V. (1967b). *Phycologia*, 6, 128–160.
Ahmadjian, V., and Reynolds, J. T. (1961). *Science N.Y.*, 133, 700–701.
Anderson, K. A., and Ahmadjian, V. (1962). *Svensk bot. Tidskr.*, 56, 501–506.
Asahina, Y., and Shibata, S. (1954). "Chemistry of lichen substances". Japan Soc. Promotion Sci., University of Tokyo, Tokyo.
Bailey, R. H. (1966). *J. Linn. Soc.*, 59, 479–490.
Bailey, R. H., and Garrett, R. M. (1968). *Lichenologist*, 4, 57–65.
Bednar, T. W. (1963). Ph.D. Thesis, University of Wisconsin.
Bednar, T. W., and Holm-Hansen, O. (1964). *Pl. Cell Physiol., Tokyo*, 5, 297–303.
Bednar, T. W., and Juniper, B. E. (1965). *Expl. Cell Res.*, 36, 680–683.
Beijerinck, M. W. (1890). *Bot. Zbl.*, 43, 725–781.
Bertsch, A., and Butin (1967). *Planta*, 72, 29–42.
Beschel, R. E. (1961). *In* "Geology of the Arctic" (Ed. G. O. Raasch). University of Toronto Press, Toronto.
Bianchi, D. E. (1964). *J. gen. Microbiol.*, 35, 437–444.
Bisby, G. R. (1941). *Trans. Br. mycol. Soc.*, 25, 127.
Bond, G., and Scott, G. D. (1956). *Ann. Bot.*, 19, 67–77.
Bonnier, G. (1889). *Annls. Sci. Nat.*, 9, 1–34.
Brightman, F. H. (1960). *Biology hum. Affaires*, 26, 1–5.
Brodo, I. M. (1961). *Ecology*, 42, 838–841.
Chervin, R. E., and Baker, G. E. (1968). *Can. J. Bot.*, 46, 241–245.

Danilov, A. N. (1910). *Izv. imp. S.-Peterb. bot. Sada*, **10**, 33–70.

Degelius, G. (1964). *Acta Horti gothoburg.*, **27**, 11–55.

di Benedetto, G., and Furnari, F. (1962). *Bollettino dell'Istituto di Botanica dell'Università di Catania*, **3**, 34–38.

Drew, E. A. (1966). D.Phil. Thesis, University of Oxford.

Drew, E. A., and Smith, D. C. (1966). *Lichenologist*, **3**, 197–201.

Drew, E. A., and Smith, D. C. (1967a). *New Phytol.*, **66**, 379–388.

Drew, E. A., and Smith, D. C. (1967b). *New Phytol.*, **66**, 389–400.

Feige, B. (1967). Ph.D. Thesis, University of Warzburg.

Follmann, G. (1965). *Umschau*, **12**, 374–377.

Fox, C. H. (1967). *Physiologia Pl.*, **20**, 251–262.

Fox, C. H., and Mosbach, K. (1967). *Acta chem. scand.*, **21**, 2327–2330.

Fritsch, F. E. (1961). "The structure and reproduction of the algae", Vol. 1. C.U.P., Cambridge.

Geitler, L. (1963). *Öst. bot. Z.*, **110**, 270–280.

Gilbert, O. L. (1965). *In* "Ecology and the industrial society" (Ed. G. T. Goodman, R. W. Edwards and T. L. Lambert), pp. 35–49. Oxford.

Hale, M. E. (1954). *Bryologist*, **58**, 244–247.

Hale, M. E. (1957). *Mycologia*, **49**, 417–419.

Hale, M. E. (1967). "The biology of lichens". Edward Arnold, London.

Hanson, W. and Palmer, H. (1965). *Hlth Physiol.*, **11**, 401–406.

Harley, J. L., and Smith, D. C. (1956). *Ann. Bot.*, **20**, 513–543.

Haynes, F. N. (1966). *In* "Viewpoints in biology" III (Ed. J. D. Carthy and C. L. Duddington), pp. 64–115. Butterworths, London.

Herisset, A. (1946). *C.r. hebd. Séanc. Acad. Sci., Paris*, **22**, 100–102

Huneck, S. (1968). "Progress in phytochemistry" (Ed. L. Reinhold and Y. Liwschitz), Vol. 1, pp. 224–345.

Juniper, B. E. (1962). *Nature, Lond.*, **194**, 1296.

Korzybski, T., Gindifer, Z. K., and Kurylowicz, W. (1967). "Antibiotics: origin, nature and properties", Vol. 2, pp. 1419–1437. Pergamon Press, London.

Kratz, W. A., and Myers, J. (1955). *Am. J. Bot.*, **42**, 282–287.

Le Blanc, F., and Rao, D. N. (1966). *Bryologist*, **69**, 338–345.

Lestang Laisné, G. de (1966). *Revue bryol. lichen.*, **34**, 346–369.

Lewis, D. H., and Smith, D. C. (1967). *New Phytol.*, **66**, 185–204.

Lilly, V. G., and Barnett, H. L. (1951). "Physiology of the fungi", p. 464. McGraw-Hill, New York.

Lohman, M. L. (1932). *Pap. Mich. Acad. Sci*, **17**, 229–288.

Lohman, M. L. (1933). *Mycologia*, **25**, 34–42.

Lohman, M. L. (1934). *Am. J. Bot.*, **21**, 314.

Lohman, M. L. (1937). *Pap. Mich. Acad. Sci.* **23**, 155–162.

Maass, W. S. G., and Neish, A. C. (1967). *Can. J. Bot.*, **45**, 59–72.

Mameli, E. (1920). *Atti Ist. bot. Univ. Lab. crittogam. Pavia*, **17**, 147–154.

Millbank, J. W., and Kershaw, K. A. (1969), *New Phytol.*, **68**, (3), in press.

Möller, A. (1887). *Untersuchungen aus dem Botanischen Institut Koniglichen Akademie zu Münster, i.W.*, 1–52.

Moore, R. T., and McAlear, J. H. (1960). *Mycologia*, **52**, 805–807.

Mosbach, K. (1967). *Acta chem. scand.*, **21**, 2331–2334.

Pearson, L. C., and Skye, E. (1965). *Science, N.Y.*, **148**, 1600–1602.

Pease, D. C. (1964). "Historical techniques for electron microscopy". Academic Press, New York and London."

Peat, A. (1968). *Archiv. Microbiol.*, **61**, 212-222.
Quispel, A. (1943). *Recl Trav. bot. néerl.*, **40**, 413–541.
Reess, M. (1871). *Mber. dt. Akad. Wiss. Berl.* 523–533.
Reid, A. (1960). *Biol. Zbl.*, **79**, 129–151.
Reid, A. (1960). *Biol. Zbl.*, **79**, 657–678.
Reynolds, E. S. (1963). *J. biophy. biochem. Cytol.*, **17**, 208.
Reznik, H., Peveling, E., and Vahl, J. (1968). *Planta*, **78**, 287–292.
Richardson, D. H. S. (1967a). D.Phil. Thesis, University of Oxford.
Richardson, D. H. S. (1967b). *Lichenologist*, **3**, 386–391.
Richardson, D. H. S., and Smith, D. C. (1966). *Lichenologist*, **3**, 197–201.
Richardson, D. H. S., and Smith, D. C. (1968a). *New Phytol.*, **67**, 61–68.
Richardson, D. H. S., and Smith, D. C. (1968b). *New Phytol.*, **67**, 69–77.
Richardson, D. H. S., Smith, D. C., and Lewis, D. H. (1967). *Nature, Lond.*, **214**, 878–882.
Richardson, D. H. S., Jackson-Hill, D., and Smith, D. C. (1968). *New Phytol.*, **67**, 469–486.
Santesson, J. (1967). *Acta chem. Scand*, **21**, 1162–1172.
Scott, G. D. (1956). *New Phytol.*, **55**, 111–116.
Scott, G. D. (1957). Ph.D. Thesis, University of Glasgow.
Scott, G. D. (1960). *New Phytol.*, **59**, 374–381.
Skye, E. (1968). *Acta Phytog. Suecica*, **52**, 123.
Smith, D. C. (1954). D.Phil. Thesis, University of Oxford.
Smith, D. C. (1960a). *Ann. Bot.* **24**, 52–62.
Smith, D. C. (1960b). *Ann. Bot.*, **24**, 172–185.
Smith, D. C. (1960c). *Ann. Bot.*, **24**, 186–199.
Smith, D. C., and Drew, E. A. (1965). *New Phytol.*, **64**, 195–200.
Starr, R. C. (1964). *Am. J. Bot.*, **51**, 1013–1044.
Stahl, E. (1877). "Beiträge zur Entwicklungsgeschichte der Flechten". Leipzig.
Stålfelt, M. G. (1939a). *Planta*, **29**, 11–31.
Stålfelt, M. G. (1939b). *Bot. Notiser*, **1939**, 176–192.
Svenson, G., and Kurt, L. (1965). *Hlth Physiol.*, **11**, 1393–1400.
Tilden-Smith, G. (1957). Diploma of Imperial Coll. Dissert. University of London.
Thomas, E. A. (1939). *Beitr. KryptogFlora Schweiz*, **9**, 1–206.
Tomaselli, R., Luciani, F., and Furhari, F. (1963). *Bollettino dell'Istituto di Botanica dell'Università di Catania*, **4**, 111–116.
Trelease, S. F., and Trelease, H. M. (1935). *Am. J. Bot.*, **22**, 520–542.
Warēn, H. (1920). *Öfvers. finska VetenskSoc. Förh.*, **6**, 1–79.
Went, F. W., and Thimann, K. V. (1937). "Phytohormones", p. 113. Macmillan, New York.
Werner, R. G. (1927). Academic Dissert. Mulhouse, University of Paris.
Zacharias, E. (1900). *Abh. Geb. Naturw., Hamburg.* **16**, 3–50.
Zehnder, A., and Gorham, P. R. (1960). *Can. J. Microbiol.*, **6**, 645–660.

CHAPTER XI

Actinomycetes

S. T. Williams

Hartley Botanical Laboratories, University of Liverpool, England

AND T. Cross

*Postgraduate School of Studies in Biological Sciences, University of Bradford,
England*

I. ISOLATION OF ACTINOMYCETES

A. General problems and principles

Difficulties encountered when isolating actinomycetes have probably
contributed to the comparative neglect of these micro-organisms in certain
fields of research. In comparison with their main competitors, bacteria and

fungi, they have certain deficiencies. Their rate of radial growth on culture media is lower than that of fungi, and their rate of cell production is generally lower than that of bacteria. Therefore methods for their isolation must be designed to compensate, at least partially, for their generally poor competitive ability under laboratory conditions. Many of the methods for isolating actinomycetes may also be employed for enumerating them in their various natural habits. These two processes are often, but not always, carried out by a single technique. Sometimes conditions conducive to the development of maximum numbers of actinomycetes also permit growth of high numbers of bacteria and fungi, thus making it difficult to isolate pure cultures. On the other hand, an efficient isolation method may allow lower numbers of actinomycetes to develop, but in a higher proportion relative to other micro-organisms.

Although actinomycetes are a relatively small group of micro-organisms, they occur in many diverse habitats, and many techniques have been described for isolating them. Here, some of the methods commonly used to isolate them from various habitats will be described.

B. Isolation from soil

1. *Isolation of general soil populations*

Many workers isolate "actinomycetes" as a group from soil, but unless specialized techniques are used, these isolates consist of a very high percentage of *Streptomyces* strains, which are the most numerous group in soil. Therefore in this Section we are mainly concerned with methods for isolating streptomycetes.

(a) *The soil-dilution-plate technique.* This well known method for isolating and counting soil micro-organisms is frequently used for soil actinomycetes. The basic procedure has been described many times and a detailed coverage was given by Johnson *et al.* (1959). Here we will confine ourselves to discussion of the various modifications that have been suggested for improving the efficiency of this method for isolation of actinomycetes. Generally the diluent used is sterile water, and with most soils suitable plates can be obtained by using dilutions of soil in water between 1 in 10^3 and 1 in 10^6.

Soil may be pre-treated in various ways to increase the numbers and/or proportion of actinomycetes in it before dilutions are prepared. Simply air-drying the sample will reduce the numbers of vegetative bacterial cells while allowing many actinomycete spores to survive. Tsao *et al.* (1960) air-dried soil, mixed it with $CaCO_3$ and incubated it for several days at 28°C. Agate and Bhat (1963) attempted to suppress bacteria and fungi by incubation of soil at 110°C for 10 min, but this will also kill heat-sensitive actinomycete propagules. Enrichment of soil with substrates, such as powdered chitin and pollen membranes, which are readily utilized by most streptomycetes, followed by incubation for 7 days at 25°C can increase the numbers of these organisms by a factor of 100 or more. Such methods, although not suitable for the ecologist interested in the *status quo* in natural soil at the time of sampling, are useful for workers screening large numbers of actinomycetes for some particular biochemical property. Soil suspensions once prepared may be subjected to differential centrifugation to try to separate actinomycete spores from other propagules (Řeháček, 1956). Soil suspensions were centrifuged for 20 min at a relative centrifugal force of 904 g at the surface and 1609 g at the base of the cuvette. The supernatant was said to yield pure cultures of actinomycetes after 10 days' incubation in a suitable agar medium. In a comparative study of some of these pre-treatments, El Nakeeb and Lechevalier (1963) found that the $CaCO_3$ treatment gave highest counts of actinomycetes, whereas centrifugation gave numbers lower than those from untreated soil suspensions.

The procedure used to incorporate the diluted soil suspensions into agar medium can also influence the efficiency of the dilution-plate technique. If samples of the soil suspensions are pipetted into a Petri dish and the molten medium then poured in and mixed, spread of bacteria can occur between the bottom of the dish and the medium during the incubation period. Similar problems can arise if the suspension is spread over the surface of solidified medium. If surface colonies are particularly required, plates can be inoculated with the soil suspension by use of a fine spray. Spread of bacteria in moisture films is most effectively and simply discouraged by incorporating samples of the soil suspension in the molten medium (at 45°– 48°C) before pouring into the dishes. An alternative to this is the use of 2 or 3 layers of agar. A basal layer of agar (often water agar) is allowed to set in the dish and the medium inoculated with the soil suspension is poured on to this layer. After the layer of medium has set, another unseeded layer may be applied, thus giving a "sandwich plate".

Finally the efficiency of the dilution-plate technique is markedly influenced by composition of the nutrient medium. Many recipes for media, designed to encourage the growth of soil actinomycetes rather than other soil micro-organisms, have been suggested. Generally the best carbon sources are glycerol, starch and chitin, with casein, asparagine and arginine as organic nitrogen sources. Many media also contain an inorganic source of nitrogen, usually nitrate, and phosphate; the reaction of media should be near to neutrality. Details of some suitable media tested by Williams and Davies (1965) are given below. The various media listed in this Chapter may be sterilized by autoclaving for 15 min at 121°C unless stated otherwise.

Glycerol–arginine medium (*Porter* et al., *1960*)

Glycerol	20 g
L-Arginine	2·5 g
NaCl	1·0 g
CaCO₃	0·1 g
FeSO₄.7H₂O	0·1 g
MgSO₄.7H₂O	0·1 g
Agar	20·0 g
Distilled water	1 litre

N.B. This medium relies on soil K^+ for that ion.

Starch–casein medium (*Küster and Williams, 1964*)

†Soluble starch	10·0 g
Casein (vitamin-free, Difco)	0·3 g
KNO₃	2·0 g
NaCl	2·0 g
K₂HPO₄	2·0 g
MgSO₄.7H₂O	0·05 g
CaCO₃	0·02 g
FeSO₄.7H₂O	0·01 g
Agar	20·0 g
Distilled water	1 litre

† Glycerol may be substituted.

Colloidal chitin medium (*Lingappa and Lockwood, 1962*)

Colloidal chitin	1·0–2·5 g
Agar	20·0 g
Distilled water	1 litre

Colloidal chitin

Wash crude unbleached chitin alternately for 24 h at a time with N NaOH and N HCl (usually 5–6 times) then with 95% ethanol 3–4 times), until foreign matter is removed. This process removes about 40% of the original material and gives a white product; moisten 15 g of the cleaned chitin with acetone and dissolve it in 100 ml of cold concentrated HCl by stirring for 20 min in an ice bath. Filter the thick syrupy solution through a thin glass-wool pad in a Büchner funnel into 2 litres of stirred ice-cold distilled water, so precipitating the material as a fine colloidal suspension. The residue may be redissolved and refiltered, usually 3 or 4 times, until no more chitin is precipitated. Wash the precipitated chitin in 5 litres of distilled water 4–5 times, the remaining acid being neutralized with dilute NaOH. The suspension may then be stored in a refrigerator, after the dry weight has been determined, and suitably diluted for preparing agar media.

Isolation plates should be incubated at a temperature of 25°–28°C for mesophilic actinomycetes from soil and water. Growth of *Micromonospora* strains is encouraged by incubation at 28°–30°C. These and all other media used for isolating soil actinomycetes cannot, of course, prevent the development of bacteria and fungi on the plates. Unfortunately many soil fungi (e.g., *Trichoderma viride*, *Mucor* spp.) have a high growth rate and the presence of a single colony on a plate can prevent the isolation of actinomycetes in a pure state. Therefore, further steps must be taken to prevent or reduce fungal growth and this can be achieved with almost complete success by the use of antifungal antibiotics. Cycloheximide (Actidione) has been used by Dulaney *et al.* (1955), Corke and Chase (1956, 1964), Corbaz *et al.* (1963) and Porter *et al.* (1960), who also used nystatin and pimaricin. Williams and Davies (1965) found that a combination of nystatin and cycloheximide (each at 50 μg/ml of medium) inhibited most soil fungi, while having no deleterious effect on actinomycetes; use of cycloheximide alone was not so effective. Use of antifungal antibiotics is therefore an essential precaution in the isolation of soil actinomycetes. Suppression of bacteria on isolation plates presents a greater problem, as their responses to antibiotics are similar to those of actinomycetes. Williams and Davies (1965) suggested the addition of penicillin (1 μg/ml) and polymyxin (5 μg/ml) to media to reduce bacterial development. However, this can only be achieved at the expense of those actinomycetes also sensitive to these antibiotics, so this is only a compromise solution.

To summarize, when the dilution-plate technique is used to isolate (or count) soil actinomycetes, the following points should be considered—

(i) Possible pre-treatment of the soil to increase numbers of actinomycetes.

(ii) Avoidance of water films encouraging spread of bacteria in the isolation plates.

(iii) Choice of a good selective medium.

(iv) Use of antifungal antibiotics.

(b) *Other methods*. Although the vast majority of workers have used the dilution-plate technique to isolate soil actinomycetes, there are other methods that can be applied, particularly in ecological studies. Soil may be incorporated directly in culture media in small quantities to make "soil plates" as described by Warcup (1950) for isolation of soil fungi. Nonomura and Ohara (1960) found this method to be particularly useful for isolating strains of *Microbispora*. If soil is washed and sieved through 3 sieves of pore size 1·0 mm, 0·5 mm and 0·25 mm before plating, various types of material can be recognized (e.g., root fragments, humus, mineral grains) and then plated on a suitable medium (Williams *et al.*, 1965), thus facilitating the isolation of actinomycetes from particular soil microhabitats. In both these procedures, fungal contamination can be reduced by using antibiotics and bacterial spread by drying particles thoroughly before plating. Trolldenier (1966) described a method in which a membrane filter is used. Suitable dilutions of soil in water were passed through a filter (0·3 μm pore size) which was then placed, face downwards, on the surface of solidified medium (various ones were used). Media that had been supplemented with soil were also used. Samples (5 g) of compost or soil were added to 15-ml portions of media before autoclaving. Use of such media resulted in up to five-fold increases in numbers of actinomycete colonies. After incubation for several days, filaments of actinomycetes grew up through the pores, and eventually colonies were visible on the upper surface of the membrane. Preliminary tests that we have made with this technique indicate that it can be useful, especially when used with a suitable selective medium containing antifungal antibiotics.

Soil ecologists often need to distinguish between spores and mycelium of actinomycetes in soil. None of the techniques mentioned so far give any indication of the origin of the developing colonies. Observations on the origins of colonies from soil–water suspensions (as used for dilution plates) indicate that many arise from spores (S. M. Ruddick, unpublished data; see Table I).

TABLE I

Origins of actinomycetes growing on soil dilution plates

Soil	% from spores	% from mycelium	% unidentifiable
A	61	0	39
B	60	3	37
C	58	0	48

A method for distinguishing between spores and mycelium in soil was described by Skinner (1951). Soil samples (5 g in 10 ml of saline solution) were placed in 1-oz bottles and shaken on a reciprocating machine; the vertical movement of the bottles was 9 cm, at a rate of about 265 strokes/min. The suspensions were shaken for a period of 1 h or more, samples being taken from the bottles periodically and plated. The colony counts obtained gave an indication of the relative proportions of spores to mycelium. Prolonged shaking broke up the mycelium, first into viable units then into smaller non-viable fragments; thus, after an initial rise, counts fell rapidly. In contrast, spores were not killed, and counts of these varied very little after an initial rise. A modification of this technique has been tested recently (S. M. Ruddick, unpublished data). Samples (1 g) of soil in 10 ml of sterile water were macerated in 1-oz vials for periods up to 1½ h, using a MSE homogenizer at a speed of approximately 14,000 rev/min. The suspension was sampled and plated periodically and some typical results are given in Table II. Again, evidence points to the presence of large numbers of actino-mycete spores in soil.

TABLE II

Colony counts from a natural soil and inoculated sterile soil after macera-tion for various periods (expressed as no. of colonies per plate)

	Time, min						
	0	5	10	20	40	60	90
Sterile soil + actinomycete mycelium	47·3	179·1	225·0	227·0	90·7	33·1	0
Sterile soil + actinomycete spores	27·4	56·3	59·4	60·3	58·8	62·1	59·1
Natural soil	25·3	23·6	23·0	24·0	23·5	21·2	21·6

2. *Isolation of particular genera and strains with special biochemical properties*

The techniques described so far are largely used for isolating strepto-mycetes. However, other genera of actinomycetes occur in soil, and although some of these may be detected by using the dilution-plate technique in the manner discussed, they are easily overlooked. Most of them are relatively slow growing and form only small colonies. Therefore, in some cases special approaches to the isolation of these genera have been made. Often, workers isolating actinomycetes from soil are not so much interested in obtaining a wide selection of the population as in obtaining isolates with particular biochemical attributes, e.g., with the capability to produce a particular antibiotic. Again this requires a somewhat different approach to isolation. (a) *Special modifications of the dilution-plate technique.* This method can be modified to facilitate the isolation of other actinomycete genera and those with special biochemical or physiological properties.

One method is to use relatively dilute isolation media, e.g., tap-water agar, and to incubate over several weeks during which time the plates are examined frequently and carefully with a stereoscopic microscope for small colonies with a margin of fine hyphae. The temptation to isolate the usually larger and more obvious *Streptomyces* colonies must be overcome. The efficiency of various carbon sources in media for isolating *Nocardia* species was tested by Farmer (1963); he suggested that a medium at pH 8 containing cholesterol acetate and sodium azide (0·03% (w/v)) was most selective. On media containing cholesterol as sole carbon source, Brown and Peterson (1966) isolated several *Nocardia* and *Streptomyces* species. An example of an extremely specific isolation method is the one suggested by McClung (1960) for *Nocardia asteroides*. Small amounts of soil were suspended in

5-ml samples of a medium containing inorganic salts. Into this, sterile paraffin-wax-coated glass rods were inserted and incubated at 37°C for 2 weeks, after which time *Noc. asteroides* developed on the rods in some of the tubes. The dilution-plate technique can be made highly selective for *Thermoactinomyces vulgaris* by using half-strength nutrient agar (Oxoid) containing 25 μg/ml of novobiocin + 50 μg/ml cycloheximide, isolation plates being incubated at 55°C (Cross, 1968). Antibiotic combinations, chosen specifically to facilitate the isolation of particular genera, or even species, may well be used with increasing success in the future.

Various modifications of the soil isolation-plate method can be used to isolate actinomycetes having specific biochemical properties or important in the biodegradation of structural materials. Wieringa (1966) was able to isolate a facultative autotrophic *Streptomyces* sp., which could oxidize elemental sulphur, by a modification of this technique. An inorganic basal medium of the following composition was prepared—

Inorganic basal medium

K_2HPO_4	500 mg
$(NH_4)_2.SO_4$	500 mg
$MgSO_4.7H_2O$	250 mg
$CaCl_2$	100 mg
Na_2CO_3	100 mg
Sodium silicate	0·1 ml
Fe EDTA	5 mg
Trace-salts solution (Pridham and Gottlieb, 1948)	1·0 ml
(for details see T. Booth, this Volume)	
Dialysed agar	15 g
Glass-distilled water	1 litre

Amounts (25 ml) of this medium were poured as a basal layer into Petri dishes containing 1 ml of 0·1 M HCl. After mixing and cooling, a thin layer of basal medium + polysulphide was poured onto the surface of the plates. Polysulphide solution was prepared by saturating a saturated solution of Na_2S in water with elemental sulphur. This solution was autoclaved, and 2 ml were added to 1 litre of sterile basal medium. Any H_2S formed was removed by keeping plates in a drying oven until the surfaces were completely dried. The acid from the basal layer precipitated the sulphur as a very fine suspension in the upper layer and the *Streptomyces* sp., capable of oxidizing this sulphur, was detected by the clear zone surrounding the colony. By incoporating substrates, such as starch, fat, casein and calcium carbonate, in the top agar layer, strains producing amylase, lipase, proteases and acids can be detected.

Clear hydrolysis zones around enzyme-producing colonies can be detected on media containing starch, casein and fat emulsions; acid-producing colonies are surrounded by a zone of clearing in media containing finely precipitated calcium carbonate.

Nette *et al.* (1959) isolated several actinomycetes capable of attacking rubber by covering the surface of the isolation plate with a thin layer of purified rubber dissolved in benzene; a basal medium with the following composition was used—

Basal medium

NH_4NO_3	2·5 g
Na_2HPO_4	1·0 g
KH_2PO_4	0·5 g
$MgSO_4.7H_2O$	0·5 g
$MnSO_4.5H_2O$	0·5 g
$FeSO_4.7H_2O$	Trace
$ZnSO_4.7H_2O$	Trace
Dialysed agar	20·0 g
Distilled water	1 litre

After sterilization, this medium was poured as a basal layer in dishes. It was inoculated by spreading a soil suspension over the dried surface or incorporating soil suspensions into 1·0% (w/v) tap-water agar and pouring a thin layer of this on the surface of the basal medium. The plates were then covered with a thin layer of 1–1·5% (w/v) rubber solution in benzene. The benzene was evaporated, leaving a thin film of rubber over the medium as the sole source of carbon. The rubber may be purified from additives before use by washing in 10% KOH solution and acetone. We have found that "Cow gum" (P. B. Cow, (Li-Lo) Ltd, Slough, Bucks.) in benzene gives a suitable rubber solution.

Actinomycetes are, of course, prolific antibiotic producers, and several modifications to the dilution-plate technique have been made to allow the quick and efficient isolation of producers from soil. If producers of one particular known antibiotic are required, the incorporation of that antibiotic into the medium may improve its selectivity for producing strains: this is because actinomycetes are normally resistant to their own antibiotics. Thus Waksman *et al.* (1946) showed that medium enriched with streptomycin could be used for isolating fresh producing strains of *Streptomyces griseus*. There are several ways in which actinomycetes, producing antibiotics inhibitory to selected test organisms, may be detected on dilution plates. This avoids unnecessary picking of colonies of non-inhibitory actinomycetes. Thus Kelner (1948) described a three-layered plate method consisting of a basal layer of medium suitable for the growth of the antagonist, a thin layer over this seeded with soil and finally the top layer, containing a suspension of the test organism. When colonies developed from the soil, those producing antibiotics effective against the test organism were located by the inhibition zones around them. There are many variations on this procedure, but all are based on similar principles. The drawbacks of such methods are that only the test organism can be tested on an isolation plate and that media suitable for selective isolation of actinomycetes from soil are not necessarily the most conducive to antibiotic production. As an alternative, replica plating techniques can be employed. Lechevalier and Corke (1953) described such a technique using a sterile velveteen stamp to transfer inoculum from colonies on the isolation plates to plates of different media each seeded with a test organism. Similar techniques can be applied in the search for producers of particular enzymes, substrates being incorporated in one of the medium layers.

The mass screening of millions of actinomycetes for antibiotics during recent years has resulted in the discovery of many useful compounds, but the frequency with which new antibiotic producers are being discovered is declining. It has been argued that there must be few strains now in soils that have not been isolated and screened and that the chances of finding high-yielding strains producing a useful and novel antibiotic are now remote. This may be true where conventional screening programmes are employed and where the more easily isolated streptomycetes are examined. If such screening programmes are to continue, it would seem more logical in the future to concentrate on actinomycetes from specialized habitats or to examine the slower-growing "difficult genera" that may have been ignored in the past. The alternative is to streamline the isolation stage and concentrate upon one test organism with a suitable assay method in order to screen the maximum number of isolates with minimum effort. An automated selection method has already been proposed by Falch and Heden (1965).

A few actinomycetes are capable of autotrophic nutrition. The initial isolation of these may be carried out by using a medium without organic nutrients with silica gel used instead of agar as a setting agent. A convenient and rapid method for the preparation of silica gel was described by Funk and Krulwich (1964). To confirm autotrophy, isolates must be grown in more controlled and exacting conditions. Thus Kanai et al. (1960) showed that *Streptomyces autotrophicus* could fix CO_2, using the energy released by combining O_2 and H_2, when grown in Knall gas (a mixture of O_2 and H_2).

The presence in soil of large numbers of micro-aerophilic micro-organisms with affinities to the families *Actinomycetaceae* and *Mycobacteriaceae* was demonstrated by Casida (1965). They were isolated from various soil samples by blending a 1-g portion of soil in a sterile Waring Blender with 100 ml of Heart Infusion Broth (Difco) adjusted to pH 7·8 with KOH. Tenfold dilutions were made from the blended soil dilution into Heart Infusion Broth, and 1-ml samples from the 10^{-8} and 10^{-9} dilutions were transferred to screw-cap tubes containing 1 ml of slanted 1·5% agar in water. The caps were screwed tight and the tubes incubated up to 4 weeks at 30°C. Tubes showing growth (a white opaque button of cells at the butt of the slope) were streaked on the surface of Heart Infusion Agar (pH 7·8) as slants or plates and incubated in air at 30°C. Catalase-positive cultures were discarded and the catalase negative cultures maintained in Brain Heart or Heart Infusion (pH 7·4) deep agar stabs or on Brain Heart Infusion Agar incubated at 37°C under a 95% N_2–5% CO_2 gas mixture or a pyrogallol–carbonate seal.

These interesting organisms seem to be present in soil in numbers greater than those for all other soil micro-organisms counted using conventional counting procedures. The high numbers of this micro-organism were utilized to allow its isolation from soil by diluting the soil in broth medium to a point beyond which other soil micro-organisms only rarely were present in dilutions. Subdividing the dilutions into small portions for incubation assured that in many cases each individual cell was not competing with other cells during growth. The isolated organism required media of high nutrient value for growth and screw-top tubes for isolation, but grew better in a N_2–CO_2 atmosphere when isolated. The taxonomic position of these organisms remains uncertain. They show some resemblance to the genus *Mycococcus* and an immunological relationship with *Actinomyces naeslundii*. (b) *Baiting techniques.* Isolation of certain genera that form sporangia, such as *Actinoplanes* and *Streptosporangium*, can be achieved by use of baiting suspensions of soil in water with suitable materials. Usually a small amount of soil is mixed with sterile water in a Petri dish and various baits added. After a week or more, baits may have become infected with actinomycete propagules (some of which are motile) and growth can be observed with a binocular dissecting microscope. All or part of the infected bait can then be transferred to fresh sterile water or a suitable solid medium. Couch (1954) recommended the use of water agar on which contents of individual sporangia could be dissected out to obtain pure isolates. Several materials have been used for baits, among the most successful being *Paspalum* grass and *Liquidambar* pollen. Floating materials like pollen facilitate the observation and isolation of colonies from the soil suspension. Couch (1963) estimated that using pollen as a bait, sporangia-forming actinomycetes had been detected in 66% of soil samples taken from all over the world. Kane (1966) used human hair as a bait and isolated a new genus, *Pilimelia*, from soil.

C. Isolation from water

Actinomycetes are an integral part of the microflora of fresh water, and have been isolated from lakes, rivers and water supplies where they have been implicated in the production of unpleasant flavours, odours and colours. Genera occurring in fresh water include *Streptomyces*, *Micromonospora*, *Nocardia*, *Actinoplanes*, *Streptosporangium* and *Actinomyces*. Again, use of the dilution-plate technique is common, water samples either being diluted or directly incorporated into a suitable medium. Safferman and Morris (1962) recommended the use of egg albumin and sodium caseinate agar containing cycloheximide for isolating actinomycetes from water supplies. Cross and Collins (1966) isolated several *Micromonospora* strains from samples taken at various depths in Blelham Tarn, using a starch–casein medium (Küster and Williams, 1964). The same medium was used by Willoughby (1966) to detect *Actinoplanes* sp. in this lake. Sporangia-forming types may also be isolated from water by using the baiting techniques previously described. The isolation of anaerobic *Proactinomyces* (*Actinomyces*) strains from natural waters was reported by Kalakutskii (1960). He used a rich meat peptone medium containing 0·5% glucose and 0·005% sodium hydrosulphite; plates were incubated under N_2 at 37°C.

Membrane filtration methods are useful for isolating actinomycetes from water. In the method developed by Burman (1965), the membrane is placed face downwards, after filtration, on the colloidal chitin medium of Lingappa and Lockwood, incubated for 2 h at 28–30°C and then removed. The agar plates are then re-incubated to develop the actinomycete colonies in 7–14 days. The method is particularly useful where large volumes of water are examined for the presence of actinomycetes, e.g., chlorinated river water. Where the water is heavily polluted and contains many bacteria, the same problems are encountered as when isolating from soil. In such circumstances, the use of antibacterial antibiotics incorporated in the isolation medium can help to suppress bacterial growth. One method that has proved successful with canal and pond waters is to shake the water sample with phenol (7 mg/ml) for 10 min, filter through a membrane and re-suspend the actinomycete spores by shaking the membrane with glass beads in saline before plating.

In salt water, actinomycetes do not appear to be an integral part of the microflora. Grein and Meyers (1958), using several media seeded with samples of sea water, isolated strains of *Nocardia*, *Micromonospora* and *Streptomyces*. They concluded, however, that these micro-organisms were associated with littoral sediments and were not part of the true sea water microflora.

D. Isolation of thermophilic species

Thermophilic species occur in several actinomycete genera (e.g., *Thermoactinomyces*, *Thermomonospora*, *Streptomyces*, *Pseudonocardia*, *Streptosporangium*). These are frequently found in vegetable materials where self-heating has occurred, such as hay, grain and compost, but also occur in soil, plant litter, milk, cheese, air and water. Such materials may be sampled using normal dilution procedures. Gregory and Lacey (1962, 1963) isolated many thermophilic actinomycetes from mouldy hay by shaking samples in a perforated drum in a wind of 4·2 m/sec. Liberated spores were sampled with the cascade impactor (Casella Ltd, London) and with an Andersen sampler (Andersen, 1958). The principle behind this is that passage of dry air removes the dry coated spores of actinomycetes (and some fungi) in preference to cells of bacteria. Wind is impacted on to the surface of dried agar medium containing an antibiotic to suppress fungal growth. This approach could also be useful for isolating actinomycetes from other habitats, such as soil, leaf litter and powdered milk. A simple modification is to shake the samples in a tin and to sample the air spora after allowing the larger suspended particles to settle. The small actinomycete spores remain suspended and can be sampled 30 min to 2 h after shaking.

Isolation plates must, of course, be incubated at high temperatures (45–65°C), and as a result steps must be taken to reduce desiccation of the medium during the incubation period. The atmosphere in the incubation chamber should be saturated with water. Agre (1964) recommended the incubation of plates in small, tightly closed containers in which water was present. She found that when such vessels were used, actinomycete colonies that appeared after 3–6 days incubation could be isolated, whereas when a normal incubator was used, the plates dried out before these developed. Use of silica gel in place of agar has also been suggested to reduce rate of desiccation (Uradil and Tetrault, 1959).

As thermophilic forms are represented in many genera of actinomycetes, it is difficult to find a medium suitable for isolation of all types. Some present few problems and will develop rapidly on routine media. Thus *Thermoact. vulgaris* can be easily isolated by using Oxoid nutrient agar (Küster and Locci, 1963). Using the same medium at half strength, Gregory and Lacey (1963) isolated thermophilic *Micromonospora* (*Thermoactinomyces*), *Streptomyces* and *Thermopolyspora* (*Micropolyspora*) strains from hay. Corbaz *et al.* (1963) compared the efficiency of several media for isolating thermophiles from hay. Many more colonies of *Thermoact. vulgaris* (*Micromonospora vulgaris*) occurred on nutrient agar than on yeast agar or "V8" vegetable juice agar; however, for thermophilic *Streptomyces* species the opposite was true. Other media recommended for isolation of thermophilic actinomycetes include one containing highly proteinaceous nutrients (soya bean meal, tryptic digest of casein), described by Uradi and Tetrault (1959), and a peptone-corn medium with 1% starch used by Agre (1964). A complex medium, containing dung extract, molasses and trace elements was described by Tendler and Burkholder (1961). Details of some of these media are given below. Incorporation of antibiotics to suppress thermophilic fungi is desirable.

Yeast extract agar (Pridham et al., 1957)

"Difco" yeast extract	4·0 g
Malt extract	10·0 g
Glucose	4·0 g
Agar	20·0 g
Distilled water	1 litre

Adjust to pH 7·3 with KOH

Half-strength nutrient agar (Corbaz et al., 1963)

Oxoid nutrient agar granules	14·0 g
Agar	10·0 g
Distilled water	1 litre

Medium T/1 (Craveri and Pagani, 1962)

Yeast extract	2·0 g
Soya bean meal	5·0 g
Crude maltose	20·0 g
Agar	20·0 g
Tap water	1 litre

pH 6·5

Several species of the thermophilic actinomycetes are facultative anaerobes, and their isolation from materials, such as dung and compost can be aided by using anaerobic techniques (Henssen, 1957). The growth of the aerobic *Bacillus* species commonly found in such habitats is inhibited and the method facilititates the isolation of certain species rarely encountered on isolation plates incubated aerobically.

E. Isolation from plant tissues

Unlike fungi and bacteria, actinomycetes are not important causal agents of plant disease. The only pathogens are *Streptomyces scabies* and related species, which cause scab of potatoes and other root crops. Isolation of *Strept. scabies* from the lesions it produces on the skin of tubers involves initial surface sterilization to suppress contaminating saprophytes followed by dispersion of the tissues in a suitable medium. General surface-sterilizing agents may be used, e.g., sodium hypochlorite solution or mercuric chloride solution, and Lawrence (1956) recommended treatment with a 1 : 140 phenol solution. Peelings of scabbed tubers were macerated and placed in the phenol solution for 10 min before incorporation in the medium. Alternatively, single lesions could be picked off and macerated with a mortar and pestle containing phenol solution.

Media generally used for isolating soil streptomycetes may be used, but Menzies and Dade (1959) described a selective-indicator medium for *Strept. scabies*. This was a low-nutrient medium to discourage growth and spread of bacteria, and it contained tyrosine from which *Strept. scabies* produces a black "melanin" pigment. Thus colonies of *Strept. scabies* are indicated by the presence of this pigment around them. Although the capacity to produce "melanin" is by no means confined to *Strept. scabies*, this method nevertheless is useful for early detection of the pathogen on isolation plates. Details of this medium are as follows—

Selective-indicator medium

Sodium caseinate	25·0 g
NaNO₃	10·0 g
L-Tyrosine	1·0 g
Agar	15·0 g
Tap water	1 litre

pH about 6·8

It has been suspected for some time that the endophytes in root nodules of certain non-leguminous plants (e.g., *Alnus, Causuarina, Myrica*) are actinomycetes. Observations with light and electron microscopes indicate that the endophyte has mycelium similar to that of actinomycetes, which can fragment in some cases. Many attempts have been made to isolate these endophytes. Usually nodules are surface sterilized and crushed in sterile water before plating. As a result of such methods, some actinomycetes have been isolated, including *Streptomyces* and *Nocardia* strains. However, so far no worker has succeeded in re-inoculating isolates into the plant and reproducing nodules, so there is no definite proof of the identity of the endophyte (Meyer, 1966).

Recently, Wollum *et al.* (1966) isolated 136 similar *Streptomyces* species from *Coenothus velutinus* nodules. These isolates, in the presence of sterile nodule extracts, caused the root hairs to become swollen. Unfortunately formation of nodules by these isolates could not be observed.

F. Isolation from animal tissues

Several actinomycetes are pathogenic in animal and human tissues; in addition there are some strains apparently having a harmless existence in such places as the human mouth. Many such strains are faculatively or obligately anaerobic. Details of some genera together with their oxygen requirements and habitats are given in Table III.

Although there are many variations in the methods employed of isolating these micro-organisms, certain general principles apply. Highly nutritious,

TABLE III
Oxygen requirements and habitats of some genera

Genus	Oxygen requirements	Habitat
Actinomyces	Anerobic to facultative	Many organs and tissues of man and animals
Dermatophilus	Facultative to anaerobic	Skin of animals
Nocardia	Aerobic	Various organs and tissues of man and animals
Odontomyces	Facultative to anaerobic	Oral cavity of hamsters and rats
Rothia	Aerobic to facultative	Oral cavity of man

complex media are used, incubation is at 37°C and is carried out under anaerobic or partially anaerobic conditions.

Usually, infected material is crushed and washed in sterile saline solution and samples placed on plates of a suitable medium, or stabbed into tubes of medium. Several media have been used, but among those most commonly chosen is brain–heart infusion agar. This was used for isolation of *Actinomyces* strains by Georg *et al.* (1964) and Gerencser and Slack (1967); *Odontomyces* and *Rothia* also grow well on this medium (Georg and Brown, 1967). Another complex medium often used is blood agar; both these media can be obtained ready made from media manufacturers. Howell and Pine (1965) described a synthetic medium, containing starch, for the isolation and maintenance of *Actinomyces* strains.

It is evident from the details given that most of these actinomycetes either need anaerobic conditions for growth or at least can tolerate them. Therefore, it is common practice to incubate plates under anaerobic or semi-anaerobic conditions. Some species require the presence of CO_2 for growth and an atmosphere containing 95% N_2 and 5% CO_2 (v/v) is commonly used. Thus, for example, this mixture was employed by Howell *et al.* (1959) to isolate oral strains of *Actinomyces* and by Howell (1963/4) to isolate *Odontomyces viscosus*. In addition to promoting growth of these actinomycetes, these conditions also discourage development of other contaminating micro-organisms.

Obligate aerobic forms, such as *Nocardia* strains, may be isolated by similar methods with incubation under normal conditions. Thus, for example, Stropnik (1965) isolated *Noc. asteroides* from human skin by placing scrappings on to horse blood agar and incubating at 37°C.

Finally, it is worth mentioning that actinomycetes occur in the gut flora of certain insects. Szabo *et al.* (1966) isolated *Streptomyces* species from the gut of larvae of the St. Mark's fly by grinding the gut contents in saline solution and plating dilutions on suitable media.

II. PURIFICATION OF ACTINOMYCETES

A. Purification from bacteria and fungi

Cultures of actinomycetes that are contaminated by bacteria or fungi may be purified in several ways. Purification may be attempted by using widely applicable techniques, e.g., streak plates or by methods more specifically designed for separation of actinomycetes from other micro-organisms.

1. *The streak-plate technique*

Separation is more efficiently achieved on media conducive to growth of actinomycetes, and media recommended for isolation purposes can be used. The surface of the medium should be dried to discourage spread of bacterial colonies; for separation from contaminating fungi, the medium should be supplemented with anti-fungal antibiotics. A variation of this method is the "spray plate": a suspension of the mixed culture is sprayed on to the surface of the medium and discrete colonies develop. Apparatus for spraying microbial suspensions have been described by Stansley (1947), Wilska (1947) and Stessel *et al.* (1953). Such an apparatus was recommended by Uradil and Tetrault (1959) for purification of thermophilic actinomycetes. If the contaminated actinomycete happens to be non-sporing, its mycelium should be macerated in sterile water or saline solution to increase the proportion of its viable propagules before any separation is attempted.

2. *The poured-plate technique*

The conditions outlined for the streak-plate method also apply here. It must be pointed out that when antifungal antibiotics are used for purification, some fungi are inhibited but not killed; therefore when apparently pure actinomycete colonies are picked off, their purity should always be checked by culturing them on media without antibiotics. Chances of successful separation using the poured-plate method are usually higher than with the streak plate, so the extra time and materials necessary are well worthwhile.

3. *Other techniques*

A method for separating *Streptomyces* strains from bacteria was described by Giolliti and Craveri (1957). Small pieces of medium bearing sporing streptomycetes were cut from the culture and placed at the bottom of a cotton-wool plug in a tube of solid sloped medium. The tube was then gently tapped and some spores fell on to the medium, whereas the bacteria were less easily displaced. This procedure could be repeated several times if necessary.

The membrane filter technique of Trolldenier (1966), which has been dealt with in the Section on isolation methods, could be easily adapted for purification of actinomycetes from bacteria.

Those actinomycetes with more unusual growth requirements and tolerances, e.g., anaerobes, thermophiles and autotrophs, can often be purified simply by growing the mixed cultures in the relevant specialized conditions. Contaminants are usually organisms unable to develop in these conditions.

B. Purification from other actinomycetes

An actinomycete culture may, of course, become contaminated with a propagule of another from an external source. However, there are several ways in which internal contamination may arise. Some variations that arise in a culture are non-heritable and therefore do not necessitate any purification. Such variations often result in differences, colour and lack of spore production, and may be caused by variations in culture conditions. On the other hand, some variations arising internally can become permanent and heritable. These are the result of mutations, saltations or genetic recombinations.

Contamination from without or within a culture can best be dealt with by the preparation of spore suspensions which can be diluted and incorporated into a suitable medium. The colonies developing should arise from single spores and, from them, those with the characteristics of the original culture can be selected. If no spores are produced, mycelial fragments must be used, but chances of obtaining genetically pure colonies are less, as more than one genome may be present in each piece of mycelium. Even the preparation of single-spore colonies may not be successful in the case of internal variations. When preparing such colonies, it is assumed that each spore is haploid (or at least homozygous). This may not always be so. Bradley (1959) found that, whereas spores of *Strept. griseus* appeared to contain only one genome, those of *Streptomyces coelicolor* had at least two. Similarly, Hopwood *et al.* (1963) found evidence of the development of heterogenous colonies from single spores of streptomycetes. A method for removing inactive variants from antibiotic-producing cultures was suggested by Waksman *et al.* (1946). Addition of streptomycin was used to inhibit the non-producing strains of *Strept. griseus*, whereas producing strains were resistant to their own antibiotic.

C. Purification from actinophage

Most actinomycetes, including thermophilic strains, are susceptible to attack by phage, and the infection of antibiotic or vitamin-producing strains can have catastrophic effects on yields. They can usually be purified fairly easily using established methods for obtaining phage-resistant strains of bacteria; mutations to phage resistance arise relatively frequently and spontaneously. In confluent plaques on an infected culture, resistant host colonies usually appear after continued incubation. If isolations from these are made there is a good chance of obtaining phage-free, phage-resistant strains. Repeated subcultures and platings should be made to check for purity. According to Waksman (1959), the phage-resistant strains do not normally differ in other respects from the original culture. In some cases it is possible to isolate resistant colonies by plating out the lysed broth or fermentation cultures (Hengeller *et al.*, 1965). In all cases, the phage-free culture should be checked to ensure that it has all the desired properties of the original infected culture.

The nature of the actinomycete growth medium can influence the appearance of typical plaques and in some cases can actually mask the presence of lytic phage. Kalakutskii and Babkova (1966), working with a phage-infected strain of *Micropolyspora caesia*, found that the presence of $CaCO_2$ in the medium enhanced plaque formation, whereas the incorporation of sodium citrate almost completely depressed lysis.

Actinomycetes freshly isolated from soil frequently carry lytic phage or they may be lysogenic. As actinophages are usually polyvalent, it is advisable to handle recent soil isolates or actinophage stock suspensions in laboratories away from important stock cultures.

III. CULTIVATION

A. For sporulation and transfer

1. *Media*

The various isolation media used for selecting the actinomycetes from mixed microbial populations in soil or water support the growth of these organisms, but are often formulated to depress the growth rates of bacteria and fungi. They may not therefore be the media of choice for demonstrating the typical aerial mycelium morphology of the species, or they may be unsuitable for producing high spore yields. It is therefore often necessary to subculture the strains on to specialized media that favour the development required. A universal medium that will support the abundant growth of all actinomycete strains has yet to be devised, and in most cases several media must be tried initially before suitable formulations are discovered.

Certain rich organic agar media may support abundant growth as measured visually by the amount of mycelium appearing on the surface of the agar, but the resulting aerial mycelium may be atypical or bear relatively few spores. For example, nutrient agar—

Nutrient agar

Peptone	5 g
Beef extract	5 g
NcCl	5 g
Agar	20 g
Distilled water	1 litre

will support the growth of most *Streptomyces* species, but many will not produce aerial mycelium on this medium or produce atypical white aerial hyphae. The extensive use of this medium in past years has resulted in many *Streptomyces* species being mistakenly identified as *Streptomyces albus*. Media such as yeast–glucose agar—

Yeast–glucose agar

Yeast extract	10 g
Glucose	10 g
Agar	15 g
Distilled water	1 litre

or dextrose–tryptone agar—

Dextrose–tryptone agar

Glucose	10 g
Tryptone	5 g
K_2HPO_4	0·5 g
NaCl	0·5 g
$FeSO_4.7H_2O$	0·1 g
Agar	20 g
Distilled water	1 litre

can give excellent growth of *Streptomyces* on agar slopes, but in several strains, the sporophore morphology is atypical for that particular species, and the spores rapidly lose their viability. We do not decry the use of rich organic media, and, indeed, in certain cases they are essential, e.g., for anaerobic and thermophilic species, and when formulated for a particular species they are necessary for producing high spore yields used for inoculating the primary stages of fermentation processes. They must however be used with care and the growth on the medium should be checked microscopically to demonstrate spore production.

In general, the agar media favouring the development of typical sporophores and abundant conidiospores are those with a high C/N ratio. Rather than suggest one medium, we give a short list of media that have proved useful for a wide range of species (I–IV) and a further selection that may be used with advantage in specific cases (V–XIII).

(I) *Oatmeal agar* (*Shirling and Gottlieb, 1966*)

Oatmeal	20 g
Agar	18 g
Distilled water	1 litre

Cook or steam the oatmeal in 1 litre of distilled water for 20 min. Filter through cheese cloth. Add distilled water to restore volume of filtrate to 1 litre. Add 1 ml of trace salts solution. Adjust pH to 7·2 with NaOH and add agar.

Pridham and Gottlieb trace salts solution

$CuSO_4.5H_2O$	0·64 g
$FeSO_4.7H_2O$	0·11 g
$MnCl_2.4H_2O$	0·79 g
$ZnSO_4.7H_2O$	0·15 g
Distilled water	1 litre

(II) *Bennetts agar* (*Jones, 1949*)

Yeast extract	1·0 g
Beef extract	1·0 g
NZ Amine A (Casein digest: Sheffield Farms)	2·0 g
Agar	15 g
Glucose	10 g
Distilled water	1 litre

pH 7·3

(III) *Potato–carrot agar* (*Cross* et al., *1963*)

Diced potato	150 g
Diced carrot	30 g
Tap water	1 litre

Steam the potato and carrot in 1 litre of boiling tap water for 30 min. Filter through muslin and adjust volume to 1 litre. Adjust pH to 6·5 and add 20 g of agar.

(IV) *Gause mineral salts medium I* (*Gause* et al., *1958*)

KNO_3	1·0 g
K_2HPO_4	0·5 g
$MgSO_4.7H_2O$	0·5 g
NaCl	0·5 g
$FeSO_4.7H_2O$	10 mg
Starch	20 g
Agar	30 g
Distilled water	1 litre

(V) *Hickey and Tresner's agar* (*Hickey and Tresner, 1952*)

Yeast extract	1·0 g
Beef extract	1·0 g
NZ Amine A (Casein digest: Sheffield Farms)	2·0 g
Dextrin	10·0 g
CoCl$_2$	20 mg
Agar	20·0 g
Distilled water	1 litre

pH 7·3

(VI) *Yeast extract–malt extract agar* (*Pridham* et al., *1957*)

Yeast extract	4·0 g
Malt extract	10·0 g
Glucose	4·0 g
Agar	20·0 g
Distilled water	1 litre

pH 7·3

(VII) *Half-strength Emersons potato dextrose agar*

Diced potato	100·0 g
Beef extract	2 g
Peptone	2 g
NaCl	1·25 g
Yeast extract	0·5 g
Glucose	7·5 g
Tap water	1 litre

Steam diced potato in 500 ml of water for 30 min. Filter, add other ingredients and make up to 1 litre. Adjust pH to 6·8–7·0. Add 20 g of agar.

(VIII) *Tomato paste-oatmeal agar* (*Asheshov* et al., *1952*)

"Heinz" Baby Oatmeal Food	20 g
"Contadina" Tomato Paste	20 g
Agar	15 g
Tap water	1 litre

(IX) *Corn meal salts agar* (*Cross* et al., *1963*)

Maize meal (rough ground)	50 g
Na$_2$HPO$_4$	1·15 g
KH$_2$PO$_4$	0·25 g
KCl	0·2 g
MgSO$_4$.7H$_2$O	0·20 g
Agar	20·0 g
Tapwater	1 litre

Add maize meal to 1 litre of tap water, bring to boil and steam for 30 min. Filter through muslin and make up to 1 litre. Add other ingredients and adjust pH to 6·8–7·0.

(X) *Soil extract agar* (*Gordon and Mihm, 1962*)

Peptone	5 g
Beef extract	3 g
Agar	15 g
Soil extract	250 ml
Tap water	750 ml

Prepare the soil extract by sifting air-dried garden soil through a coarse sieve and autoclaving 400 g of soil with 960 ml of tap water for 1 h at 121 °C. After the mixture cools, carefully decant it and filter it through paper. Adjust the pH to 6·8–7·0.

(XI) *Synthetic agar* (*Lindenbein, 1952*)

K_2HPO_4	1·0 g
$MgSO_4.7H_2O$	0·5 g
KCl	0·5 g
$FeSO_4.7H_2O$	0·01 g
$NaNO_3$	2·0 g
Glycerol	30·0 g
Agar	20·0 g
Distilled water	1 litre

pH 7·2

(XII) *Glycerol–asparagine agar* (*Pridham and Lyons, 1961*)

L-Asparagine (anhydrous base)	1·0 g
Glycerol	10·0 g
K_2HPO_4	1·0 g
Agar	20 g
Trace salts solution (see I)	1·0 ml
Distilled water	1 litre

pH 7·0–7·4

(XIII) *Carbon utilization medium* (*Shirling and Gottlieb, 1966*)

$(NH_4)_2SO_4$	2·64 g
KH_2PO_4	2·38 g
K_2HPO_4	5·65 g
$MgSO_4.7H_2O$	1·00 g
Pridham and Gottlieb trace salts solution (see I)	1·0 ml
Agar	15 g
Carbon source (sterilized separately by filtration)	10 g
Distilled water	1 litre

Almost all of the media given above will support the growth of strains belonging to other actinomycete genera. However, fewer species of these genera are able to grow to the same extent as *Streptomyces* on the defined or synthetic media, and such formulations should not be used for routine subculturing. The thermophilic actinomycetes prefer the fairly rich organic media, e.g., nutrient agar, yeast extract–malt extract agar, as also do the animal pathogenic species belonging to such genera as *Actinomyces*, *Nocardia* and *Dermatophilus*.

The spores of *Streptomyces* species are difficult to wet, and the preparation of a homogeneous spore suspension from a slope culture can prove difficult with certain species. The addition of a wetting agent to the suspending medium can aid the collection of an homogeneous suspension of spores. Several compounds have been used successfully, e.g., Triton X 100 (Lennig Chemicals Ltd, London, W.C.1), 0·05%; sodium lauryl sulphonate, 0·01%; Carbowax 400 (Union Carbide), 0·05–0·1%. Tendler (1959) recommended the addition of 0·1% agar to the suspending medium to prevent the settling out of spores during the pipetting of inocula.

2. *Maceration*

Where cultures have lost the ability to produce spores (e.g., degenerate *Streptomyces* strains) or in species where the spores are embedded in the vegetative colonies (e.g., *Micromonospora*), it is often difficult to obtain sufficient inoculum for use when inoculating a series of slopes or shake

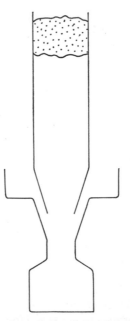

FIG. 1. Simple modified ground-glass-joint homogenizer for actinomycete colonies.

flasks. Such colonies may be homogenized in a blender to give hyphal fragments, which can be used to inoculate fresh media. The mycelium can be washed free from the growth medium by centrifugation, or, alternatively, the colonies may be separated from the agar growth medium by a film of Cellophane. Single colonies can be macerated quickly by hand in glass tubes with tight fitting glass or Teflon-coated pestles, or between the ground-glass surfaces of a modified glass joint (Fig. 1).

B. For vegetative growth

For producing vegetative hyphae in submerged culture, flasks containing the media suggested below are inoculated and incubated at a suitable temperature on a rotary shaker (200–240 rev/min) (medium-to-flask volume ratio 1 : 2·5–5·0).

Peptone–yeast extract–glucose (PYG) broth (Cross and Spooner, 1963)

Peptone	5·0 g
Yeast extract	5·0 g
Glucose	10·0 g
Casamino acids (Difco)	1·0 g
NaCl	5 g
Distilled water	1 litre

pH 7·0–7·2

Tryptone–yeast extract broth (Pridham and Gottlieb, 1948)

Tryptone	5·0 g
Yeast extract	3·0 g
Distilled water	1 litre

pH 7·0–7·2

When a defined liquid medium is required, the defined agar media given above, formulated without agar, will usually give mycelial growth in submerged culture. If the clarity of the medium is unimportant, media containing combinations of soya bean meal (0·5–2·0%), cotton seed meal (0·5–2·0%), yeast extract (0·1–0·5%) and corn steep liquor (0·1–0·5%) with a carbohydrate source, such as starch, glucose or glycerol (1·0–2·0%), can give excellent mycelial growth and also encourage the production of antibiotics, vitamins and enzymes.

C. For examination of morphology

1. *Direct methods*

The standard methods of preparing smears of bacteria on microscope slides before staining yields little information of value when applied to most actinomycetes. It can be used with advantage when examining the bacteroid genera, such as *Mycobacterium* and *Mycococcus*, but in general one should adopt techniques that disturb the colonial growth as little as possible. Smears of *Streptomyces* species show hyphal fragments and many free spores, but rarely exhibit the arrangement of the spores on the aerial hyphae which can be of major importance in taxonomy.

Direct observation of the organism growing on the surface of a suitable agar medium will usually give the information required when checking spore production or observing aerial mycelium morphology. An optical system giving magnifications between 150 and 400 × is normally required to reveal the detail, and a relatively clear agar medium should be chosen to allow maximum illumination. Condensation on the objective lens can often cause problems, especially when examining plate cultures of thermophilic strains recently removed from the incubator. Incubated plates should be cooled in the refrigerator and a long-working distance objective, e.g., 40 × long-working distance objective (Vickers Instruments Ltd) can prove particularly useful. Plate cultures will show both substrate and aerial hyphae as well as the typical conidiophores and sporangia.

Species belonging to *Actinoplanes*, *Ampullariella*, *Spirillospora*, *Amorphosporangium* and *Streptosporangium* can be examined by direct microscopy when fruiting on pollen grains floating in water or trapped between coverslip and slide. The use of phase-contrast equipment can significantly increase the contrast in such preparations.

2. *Slide and coverslip methods*

Several techniques have been proposed for examining particular actinomycete genera or which allow higher magnifications to be used. These techniques are designed to give the minimum of disturbance and can be used to follow the stages in development of the germinating spore.

The simple inclined coverslip technique (Kawato and Shinobu, 1959; Williams and Davies, 1967) can be used for a wide range of species (Fig. 2a). Glass coverslips, sterilized by autoclaving are placed at an angle of 45° into solidified medium in a Petri dish so that half the coverslip is in the medium. An inoculum from a slope culture is then spread along the line where the upper surface of the coverslip meets the agar with a fine wire needle. During incubation, the organism grows both on the medium and in a line across the upper surface of the coverslip. This line of growth remains attached to the coverslips when they are carefully withdrawn from the medium, and can be examined directly under the microscope.

The agar-cylinder method of Nishimura and Tawara (1957) can give good results with species forming aerial mycelium (Fig. 2b). Agar plugs are cut from an agar plate with a cork borer and replaced on the surface of the agar. The upper surface of the plug is lightly inoculated with spores and

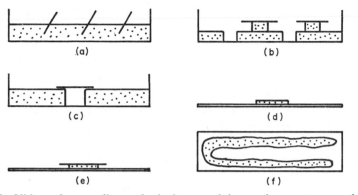

FIG. 2. Slide and coverslip methods for examining actinomycete morphology: (a), inclined coverslip; (b), agar cylinder; (c), agar trough; (d), agar block; (e), agar block and coverslip; (f), slide culture.

covered with a sterile coverslip. After incubation the coverslip can be removed with the ring of growing mycelium for direct examination. An alternative to this technique is to remove a strip of agar from a poured plate and lightly inoculate the margin of the trough with mycelium or spores. Sterile coverslips are placed over the trough and the plates incubated until growth occurs at the junction of the coverslip and agar (Fig. 2c). Plates can be examined directly on the microscope stage or the coverslips carefully removed with the adhering growth (Okami and Suzuki, 1958).

Cultivation of actinomycetes in thin films of agar on microscope slides incubated in moist chambers can often give valuable information, particularly with species bearing spores on the substrate mycelium (e.g., *Thermoactinomyces, Actinobifida, Micromonospora*) or where fragmentation of the mycelium occurs during growth (e.g., *Nocardia*). One simple method is to pipette the inoculated molten agar on to a sterile slide so that it sets in a thin film (Gordon and Mihm 1962). If several slides are prepared at the same time, single slides can be removed at intervals and examined for stages in spore germination, growth and spore formation. Sterile agar can be poured into a cool sterile Petri dish so as to set in a very thin layer. Squares of the solidified agar are cut and transferred to sterile slides and the surface lightly inoculated before incubation in a sterile moist chamber (Fig. 2d). Growth is then limited to the surface of the agar, and developmental stages, such as the transitory mycelial growth and fragmentation of *Nocardia* species, can be followed easily. The agar squares may be covered with a coverslip (Fig. 2e). Growth may then be examined directly by placing the slide on the microscope stage or the coverslip removed with adhering growth.

IV   13

The slide-culture method described by Colmer and McCoy (1950) has proved particularly useful for studying spore formation in strains of *Micromonospora*. Species of this genus produce no aerial mycelium. The spores are borne on the substrate hyphae, usually embedded within the colony, making direct observation almost impossible. In this method, the molten agar is pipetted around the edge of a sterile microscope slide (Fig. 2f). When set, the agar and slide surface in the centre of the slide are inoculated and during incubation growth occurs on the agar and in the moisture film containing diffused nutrients between the strips of agar. Such slides can be examined directly or stained after drying on a warm hot plate or over a boiling-water bath.

3. *Stains*

The hyphae and spores of most actinomycetes can be seen under the microscope without having to resort to the use of stains. They do however stain readily with Giemsa solution, crystal violet, methyl violet, haematoxylin, methylene blue and carbol fuchsin, and such stains can prove useful where the fine substrate mycelium is almost invisible in the supporting medium. For demonstrating acid fastness in species of *Nocardia*, the modified Ziehl–Neelsen method described by Gordon and Mihm (1962) is recommended—

Procedure

Smears are air-dried and immersed in carbol fuchsin, which is heated and boiled for 5 min. The slides are then washed, dipped in acid alcohol and quickly washed in water. Counter stain in methylene blue.

Carbol fuchsin

Mix 10 ml of a saturated alcoholic solution of basic fuchsin and 90 ml of a 5% aqueous solution of phenol.

Acid alcohol

Add 3 ml of conc. HCl to 97 ml of 95% ethanol.

Methylene blue

Add 30 ml of a saturated alcoholic solution of methylene blue to 100 ml of 0·01% aqueous KOH solution.

The spores of species belonging to the genera *Thermoactinomyces* and *Actinobifida* are refractile and show the resistance to simple stains exhibited by mature bacterial endospores. Where stained preparations of these organisms are required, one of the recommended bacterial spore stains should be used (see Lapage *et al.*; Volume 3a).

The cultivation and staining method described by Erikson (1947) is useful for differentiating between substrate and aerial mycelium. The culture is grown on the surface of sterile Cellophane placed on solidified agar. After incubation the Cellophane bearing growth is removed, stained for 30 min in Sudan IV, dipped in 70% ethanol and washed in water before mounting on a slide.

Sudan IV stain

Solution A

Add 0·5 g of Sudan IV to 25 ml of n-butanol, boil, cool and filter.

Solution B

Mix 4·5 volumes of n-butanol with 5·5 volumes of ethanol.

Preparation

Add 7 volumes of solution A to 9 volumes of solution B and filter before use.

The stain is retained by the aerial hyphae because of the lipid content of the outer wall, and the substrate hyphae appear almost colourless.

Corti (1954) describes an alternative method for staining sporulating cultures of streptomycetes cultivated on Cellophane. The resulting mycelium appears light yellow and the spores blue with red granules in the aerial hyphae.

Procedure

Stain for 2 min in 2 volumes of 0·1% (w/v) Bismark brown, 2 volumes of 0·1% (w/v) toluidine blue and 1 volume of saturated ammonium sulphate. Rinse carefully in distilled water, dry and examine directly or mounted in balsam.

The Cellophane cultivation technique has also been used by Giolitti and Bertani (1953) and Mikhailova (1965) when sectioning and staining actinomycete colonies.

The Rossi–Cholodny slide technique has been used for detecting the presence of actinomycetes in natural substrates such as soil or compost. Such slides after being in position for 1–3 weeks are washed gently to remove excess soil, air-dried and fixed over a low flame. The slide is then placed over a steam bath and stained for 12–15 min in phenolic rose bengal or erythrosin (Waksman *et al.*, 1939).

Stain

Rose bengal (or Erythrosin)	1·0 g
Aqueous phenol (5%)	100 ml
Calcium chloride	0·05 g

The use of certain non-toxic substances, known as fluorescent brighteners or optical brightening agents, is an interesting possibility. Several applications of such compounds were outlined by Darken (1961, 1962) and Darken and Swift (1963, 1964), who showed that spores and mycelium of streptomycetes could be labelled without any toxic effects. Preliminary work indicates that brighteners, such as the disodium salt of 4,4-bis[4-anilino-6-bis(2-hydroxyethyl)amino-s-triazin-2-ylamino]-2,2-stilbenedisulphonic acid can be used to study the morphology of actinomycetes in culture. If brighteners are incorporated in the culture medium, the organism accumulates the brightener during growth and when viewed with ultraviolet light, an intensely stained intact organism can be observed. Recent work has also suggested that fluorescent brighteners can be applied directly to fixed preparations on slides and used as an alternative to conventional stains.

D. Electron microscopy

The surface of the aerial spores of species belonging to the genus *Streptomyces* may be smooth, or may bear spines, hairs or warts. This surface configuration appears to be a remarkably constant species characteristic and has proved a reliable taxonomic aid. Direct observation of the silhouettes of whole spores viewed with transmission electron microscopes reveals the presence of these surface structures and is now regarded as an essential step in species identification. The technique is relatively simple. Formvar- or collodion-coated grids are gently pressed on to the surface of sporulating colonies and viewed under the electron microscope at magnifications between 8000 and 10,000× without further treatment (Küster, 1953; Kutzner, 1956; Preobrazhenskaya *et al.*, 1960; Tresner *et al.*, 1961; Shirling and Gottlieb, 1966). Observation of the silhouettes of whole spores of species belonging to the genera *Thermoactinomyces*, *Thermomonospora* and *Micromonospora* may also provide useful taxonomic characters (Kudrina and Maksimova, 1963; Luedemann and Brodsky, 1964; Henssen and Schnepf, 1967). The recently introduced scanning electron microscope, which has a greater depth of focus than the transmission type, provides a surface view of whole structures, such as spore chains and sporangia, as well as individual spores (Williams and Davies, 1967).

The methods of preparing thin sections of actinomycete hyphae and spores for electron microscopy are similar to those employed for other bacteria and fungi (see Greenhalgh and Evans, this Volume, Chapter IV). The application of these techniques to various actinomycete genera can be found by reference to the following papers:

Streptomyces

Moore and Chapman (1959), Stuart (1959), Glauert and Hopwood

(1960), Chen (1966), Painter and Bradley (1965), Rancourt and Lechevalier (1964).

Actinomyces

Overman and Pine (1963).

Nocardia

Kawato and Inoue (1965).

Microellobosporia

Rancourt and Lechevalier (1963).

Actinoplanes

Lechevalier and Holbert (1965).

Streptosporangium: *Spirillospora, Actinoplanes*

Lechevalier *et al.* (1966).

Dermatophilus

Gordon and Edwards (1963).

Thermomonospora

Henssen and Schnepf (1967).

E. Physiological properties

Many actinomycete species produce enzymes that diffuse into the surrounding medium giving visible digestion of certain substrates. The possession of certain enzymes can be used as an aid in taxonomy, and in certain cases may indicate an ecological rôle of the organism. In recent years, the possibility of obtaining high yields of commercially useful enzymes from actinomycetes has attracted attention and several screeing methods have been described for detecting suitable strains. Media and methods for detecting the presence of some of these enzymes are given below.

1. *Amylases*

Cultivate isolates on starch containing media (e.g., inorganic salts–starch agar or Gause mineral salts medium I) and flood with iodine after incubation to develop zones of starch hydrolysis.

2. *Chitinases*

Cultivate isolates on the colloidal chitin agar medium of Lingappa and Lockwood (1962) (see I, B) and examine plates for zones of hydrolysis after incubation.

3. Cellulases

Inoculate the surface of cellulose agar contained in tubes (e.g., $6 \times \frac{1}{2}$ in.) with the organism under test and incubate at the optimum temperature. Strains producing a cellulase give a clear zone beneath the surface growth. Suitable agar media are the carbon utilization medium of Shirling and Gottlieb (1966) (see p. 000) or the medium described by Rautela and Cowling (1966). The latter found the following method of preparation of cellulase to be most satisfactory. Whatman powdered cellulose, swollen in 85% orthophosphoric acid for 2 h at 4°C, was regenerated, washed in the cold with distilled water followed by 1% (w/w) Na_2CO_3 and then with distilled water again until neutral. Some (5 g dry weight) of the resulting suspension of cellulose particles, was added to a medium of the following composition—

Medium

$NH_4H_2PO_4$	2·0 g
KH_2PO_4	0·4 g
$MgSO_4.7H_2O$	0·89 g
Yeast extract	0·5 g
Adenine	4·0 mg
Adenosine	8·0 mg
Thiamine HCl	100·0 μg
Agar	17·0 g
Distilled water	1 litre

The medium was sterilized in test tubes and left to set in an unsloped position, to produce uniformly opaque columns. Discs of inoculum (a culture on agar media) were placed on the surface of the medium and the depth of clearing noted.

4. Proteases

Streak the culture across plates of skim-milk agar, incubate and examine the plates at intervals for clearing off the opaque casein both underneath and around the growth.

Medium

Suspend 10 g of skim-milk powder in 100 ml of distilled water. Suspend 2 g of agar in 100 ml of distilled water. Autoclave the separate suspension, mix while still molten and pour as a thin layer on the surface of 2% (w/v) water agar.

An alternative method is to streak the culture across plates of nutrient gelatin agar, incubate and develop the plates with mercuric chloride solution (15 g $HgCl_2$, 20 ml conc. HCl, 100 ml water).

Medium

Peptone	5 g
Beef extract	3 g
Gelatin	4 g
Agar	15 g
Distilled water	1 litre

pH 7·0

The special physiological tests and their interpretation for taxonomic purposes are outside the scope of this Chapter. Reference should be made to the relevant taxonomic papers of Shirling and Gottlieb (1966) for *Streptomyces*, Gordon (1968) for *Nocardia* and Gerencser and Slack (1967) for *Actinomyces*. Such tests can also be applied to the other actinomycete genera where appropriate.

F. Immunological methods

The various immunological methods described for bacteria are equally applicable to the actinomycetes. The major difficulty with this group of organisms however is the initial choice of antigenic material for producing immune serum, and this choice will be governed by the organism under study and the way in which the immune serum is to be used.

Actinomycete spores or spore homogenates have only rarely been used as the source of antigen. They do induce the formation of antibodies in the test animal, but there is little information available on their specificity, and resulting antibody titres have been reported to be relatively low. In most studies vegetative mycelium grown in submerged culture has been used for immunization. This mycelium has either been injected directly as a fine suspension or disrupted using a mortar (Krzywy, 1963), a Hughes press (Cross and Spooner, 1963) or by ultrasonic disintegration (Ludwig and Hutchinson, 1949) and re-suspended in saline. Soluble antigenic extracts can be prepared by prolonged extraction of the mycelium with phenolic saline (Coca's solution) (Pepys *et al.*, 1963). The use of such relatively crude antigens will give antisera showing many cross-reactions. Specificity may be enhanced by absorption techniques using the freeze-dried mycelium of cross-reacting strains. The alternative approach is to fractionate the mycelium initially and purify the actinomycete cell-wall antigenic components. Methods for isolating and purifying such components have been described in detail by Kwapinski (1965).

G. Influence of temperature on growth

Psychrophilic actinomycetes have not been isolated to date; most species are mesophilic, having their optimum temperatures in the range 23°– 40°C. The optimum temperature for growth can vary between species of a genus and even between strains of a single species. Most species isolated from soil and water will grow well at temperatures between 25° and 28°C, and the human and animal pathogenic species require temperatures between 30° and 37°C.

There is a large group of thermophilic actinomycetes that grow at temperatures between 45° and 70°C. The optimum temperature for the growth of these species varies widely, and many can also grow at lower temperatures, down to 37°C, and in some cases 30°C. The pigmentation, morphology and physiological properties can vary with incubation temperature, and the optimum can differ for each of these features. The incubation of slope and plate cultures at such high temperatures can result in the agar medium drying out rapidly. As already mentioned in I, D of this Chapter, this can be overcome by providing a humid atmosphere in the incubator or by sealing the cultures in a smaller closed container. Where liquid cultures are being grown with forced aeration, the air must first be humidified by passing it through water held at the same temperature.

The aluminium-block polythermostat has proved very useful for studying the effect of temperature on growth using various nutrient agar media (Cross, 1968). Similar temperature-gradient devices, varying slightly in design, temperature range and capacity have been described for studying the effect of temperature on many species of bacteria (Oppenheimer and Drost-Hansen, 1960; Sinclair and Stokes, 1963; Morita and Haight, 1964). The polythermostat consists basically of an aluminium block fitted with terminal radiators through which water is pumped at selected temperatures to give a gradient in the block. Rows of holes drilled at intervals in the block accommodate culture tubes and thermometers. Such systems, when adequately lagged and stabilized, reduce the need for a series of separate incubators to give an adequate range of incubation temperatures. They can rapidly give information on optimum, minimum and maximum growth temperatures and can also indicate the optimum temperature for enzyme, pigment or antibiotic formation. For liquid cultures, an adequate aeration system for each culture tube is essential. A suitable system has been described by Palumbo et al. (1967), consisting of a manifold for distributing the air and a temperature–humidity equilibration system, particularly necessary where high incubation temperatures are required.

IV. PRESERVATION

A. Viability of slope cultures

The great variability found in the actinomycetes, caused by mutations or the breakdown of a heterokaryotic condition on sporulation, has made it necessary to develop methods for maintaining cultures which will prevent morphological or biochemical changes. The frequent subculturing of strains on to fresh agar slopes can result in the loss of desirable biosynthetic properties and a diminution of the ability to produce aerial mycelium. This tendency to degenerate can be reduced by subculturing infrequently on to very dilute maintenance media and storing the slopes in the refrigerator. Certain strains have been stored in this way for over 3 years without showing any noticeable changes, but others have lost the ability to form spores. This method, though convenient, must be regarded as unreliable. The storage of slope cultures at very low temperatures holds out more promise. Tresner *et al.* (1960) stored slope cultures of 300 *Streptomyces* strains at −22°C for 3 years and found only 0·8% not viable over this period.

A modification of this method is to store spore suspensions in 15% glycerol at low temperatures. Such spore suspensions can be used repeatedly to provide an identical inoculum for a series of experiments, but the particular strain used should be checked for continued viability under these conditions.

The thermophilic species can usually be stored as slope cultures without taking excessive precautions. Kosmachev (1960) found that slopes plugged with cotton wool and stored in the laboratory after high-temperature incubation would retain their viability long after the agar had completely dried.

B. Freeze drying

Almost all species of actinomycetes can be freeze dried successfully and remain viable over very long periods. Freeze-dried cultures retain their ability to produce spores and pigments, and Kutznetsov *et al.* (1962) presented evidence to show that such cultures kept their antibiotic producing capacity better than did strains preserved on agar or in soil. A minimum of storage space is required, and once the culture has been freeze dried and checked for viability and sterility it needs little further attention. It is therefore the recommended method for preserving both mesophilic and thermophilic actinomycetes.

Suitable suspending media are double-strength skim milk (20% w/v) (Wiess, 1957), 10% w/v high-molecular-weight dextran (Muggleton and Ungar, 1959) or gelatin–sucrose solutions, e.g., gelatin 1% (w/v–sucrose 10% w/v. Such media can be autoclaved, and their use avoids the filtration stages necessary for media containing serum. The choice of suspending medium can influence the percentage viability after freeze drying, and where very high viabilities are required for particular species used in industrial fermentations, it may be necessary to test several alternative media. The addition of glucose or a non-reducing sugar (to 7·5% w/v) or amino-acids (e.g., sodium glutamate or lysine) can increase viability.

Very high spore concentrations in the suspending media tend to have a protective effect, and the addition of a wetting agent, as described earlier, can facilitate removal of spores from cultures.

It is essential that the spores should be mature before freeze drying, and it is advisable to choose the sporulation medium with care and to allow a generous incubation period (at least 2 weeks). Freeze-dried cultures should be re-suspended in a nutrient broth (e.g., PYG or tryptone broth) and allowed to germinate in that medium or transferred directly on to the surface of agar. A suitable recovery medium should be determined and recorded before storage of the culture.

Non-sporing strains of *Nocardia* or *Streptomyces* can be freeze dried successfully by homogenizing colonies in the suspending medium and freeze drying the mycelial fragments. Spores produced on the substrate mycelium of *Micromonospora* species and the vegetative spores of *Streptomyces* appearing after lengthy periods of submerged culture (Wilkin and Rhodes, 1955) can also be used for freeze drying.

C. Alternative methods of preservation

The other methods advocated for preserving actinomycetes involve using techniques to prevent spore germination or to reduce metabolism to an absolute minimum. Such methods require the minimum of apparatus and can be used successfully for the majority of spore-forming actinomycetes but cannot be relied upon for all strains.

Storage under mineral oil has proved successful for many *Streptomyces* and *Nocardia* strains though Frommer (1956) found that 5% of the 2300 strains maintained in this manner were no longer viable after 4–6 years' storage. Sterile and dry mineral oil should be added to the slope culture to give a layer of at least 1 cm above the agar (Pumpyanskaya, 1964). The oil can be autoclaved in 15-ml amounts for 30 min at 121°C in 1-oz screw-cap bottles containing a piece of fused $CaCl_2$. When reviving such cultures, it is advisable to pull the abstracted portion of mycelium and spores over the surface of a dried agar plate to remove any adhering oil.

An alternative method is to pipette a small volume of spore suspension on to a sterile dry substrate with a larger surface area, such as plugs of absorbent cotton wool, seived soil, sand or fine-mesh silica gel. The absorbed spores may be allowed to dry naturally by storage in the laboratory or quickly dried over a desiccant in a sealed container. Where conditions of high humidity are likely to be encountered, such cultures should be protected by covering the cap or plug with a wax film or the tube sealed in a gas flame. An earlier and quite successful method was to pipette a broth culture of the organism growing in the vegetative phase into tubes of sterilized soil. The soil can be sterilized by autoclaving the tubes on three successive days for 30 min at 121°C. The plugged tubes were incubated to allow growth and spore formation in the soil broth mixture and eventually allowed to dry.

Tresner and Backus (1957) described a simple method for preserving actinomycete cultures as herbarium specimens. Mature plate cultures were killed with formaldehyde and sealed with rubber sealers before storage at 4°C. Cultures so treated retain most of their original morphological characteristics for longer than 2 years. Dried agar cultures of streptomycetes that were 42 years old were compared with similar living strains by Pridham *et al.* (1965). None of the dried specimens was viable, but the colour of the aerial mycelium and spore surface characteristics were retained and showed a remarkable degree of similarity to their living counterparts.

D. Regeneration

Where cultures have degenerated during storage and have lost the ability to produce aerial mycelium or spores, it is sometimes possible to regenerate them or select normal-type colonies after suitable treatment. The classical method is to allow the culture to colonize a sterile fertile moist soil sample either with or without added nutrients, such as chitin or autoclaved grass cuttings. After incubation, the soil is plated out for normal-type colonies. An alternative method suggested by Narita and Tomita (1958) is to sub-culture the organism on agar containing 0·05% w/v D-glucosamine HCl.

REFERENCES

Agate, A. D., and Bhat, J. V. (1963). *Antonie van Leeuwenhoek*, **29**, 297–304.
Agre, N. A. (1964). *Mikrobiologiya*, **33**, 808–811.
Andersen, A. A. (1958). *J. Bact.*, **76**, 471–484.
Asheshov, I. N., Strelitz, F., and Hall, E. A. (1952). *Antibiotics Chemother.*, **2**, 366–374.
Bradley, S. G. (1959). *Ann. N.Y. Acad. Sci.*, **81**, 899–905.
Brown, R. L., and Peterson, G. E. (1966). *J. gen. Microbiol.*, **45**, 441–450.
Burman, N. P. (1965). *Proc. Soc. Wat. Treat. Exam.*, **14**, 125–131.
Casida, L. E. (1965). *Appl. Microbiol.*, **13**, 327–334.
Chen, P. L. (1966). *Amer. J. Bot.*, **53**, 291–295.
Colmer, A. A., and McCoy, E. (1950). *Trans. Wis. Acad. Sci. Arts. Lett.*, **40**, 49–70.
Corbaz, R., Gregory, P. H., and Lacey, M. E. (1963). *J. gen. Microbiol.*, **32**, 449–455.
Corke, C. T., and Chase, F. E. (1956). *Can. J. Microbiol.*, **2**, 12–16.
Corke, C. T., and Chase, F. E. (1964). *Proc. Soil Sci. Soc. Am.*, **28**, 68–70.
Corti, G. (1954). *J. Bact.*, **68**, 389–390.
Couch, J. N. (1954). *Trans. N.Y. Acad. Sci.*, **16**, 315–318.
Couch, J. N. (1963). *J. Elisha Mitchell scient. Soc.*, **79**, 53–70.
Craveri, R., and Pagani, H. (1962). *Annali Microbiol.*, **12**, 115–130.
Cross, T. (1968). *J. appl. Bact.*, **31**, 36–53.
Cross, T., and Collins, V. (1966). *IX Int. congr. Microbiol.* (*Moscow*), 339.
Cross, T., and Spooner, D. F. (1963). *J. gen. Microbiol.*, **33**, 275–282.

Cross, T., Lechevalier, M. P., and Lechevalier, H. (1963). *J. gen. Microbiol.*, **31**, 421–429.

Darken, M. A. (1961). *Appl. Microbiol.*, **9**, 354–360.

Darken, M. A. (1962). *Appl. Microbiol.*, **10**, 387–393.

Darken, M. A., and Swift, M. E. (1963). *Appl. Microbiol.*, **11**, 154–156.

Darken, M. A., and Swift, M. E. (1964). *Mycologia*, **56**, 158–162.

Dulaney, E. L., Larsen, A. H., and Stapley, E. O. (1955). *Mycologia*, **47**, 420–422.

El-Nakeeb, M. A., and Lechevalier, H. A. (1963). *Appl. Microbiol.*, **11**, 75–77.

Erikson, D. (1947). *J. gen. Microbiol.*, **1**, 39–44.

Falch, E. A., and Heden, C. G. (1965). *Ann. N.Y. Acad. Sci.*, **130**, 697–703.

Farmer, R. (1963). *Proc. Okla. Acad. Sci.*, **43**, 254–256.

Frommer, W. (1956). *Arch. Mikrobiol.*, **25**, 219–222.

Funk, H. B., and Krulwich, T. A. (1964). *J. Bact.*, **88**, 1200–1201.

Gause, G. F., Preobrazenskaya, T. P., Kudrina, E. S., Blinov, N. O., Rjabova, I. D., and Sveshnikova, M. A. (1958). "Zur Klassifizierung der Actinomyceten". Fischer, Jena.

Georg, L. K., and Brown, J. M. (1967). *Int. J. syst. Bact.*, **17**, 79–88.

Georg, L. K., Robertstad, G. W., and Brinkman, S. A. (1964). *J. Bact.*, **88**, 477–490.

Gerencser, M. A., and Slack, J. M. (1967). *J. Bact.*, **94**, 109–115.

Giolitti, G., and Bertani, M. A. (1953). *J. Bact.*, **65**, 281–283.

Giolitti, G., and Craveri, R. (1957). *J. gen. Microbiol.*, **17**, 649.

Glauert, A. M., and Hopwood, D. A. (1960). *J. biophys. biochem. Cytol.*, **7**, 479–488.

Gordon, M. A., and Edwards, M. R. (1963). *J. Bact.*, **86**, 1101–1115.

Gordon, R. E. (1968). *In* "The ecology of soil bacteria" (Ed. T. R. G. Gray and D. Parkinson), pp. 293–321. Liverpool University Press, Liverpool.

Gordon, R. E., and Mihm, J. M. (1962). *Ann. N.Y. Acad. Sci.*, **98**, 628–636.

Gregory, P. H., and Lacey, M. E. (1962). *Nature, Lond.*, **195**, 95.

Gregory, P. H., and Lacey, M. E. (1963). *J. gen. Microbiol.*, **30**, 75–88.

Grein, A., and Meyers, S. P. (1958). *J. Bact.*, **76**, 457–463.

Hengeller, C., Licciardello, G., Tudino, V., Marcelli, E., and Virgilio, A. (1965). *Nature, Lond.*, **205**, 418–419.

Henssen, A. (1957). *Arch. Mikrobiol.*, **27**, 63–81.

Henssen, A., and Schnepf, E. (1967). *Arch. Mikrobiol.*, **57**, 214–231.

Hickey, R. J., and Tresner, H. D. (1952). *J. Bact.*, **64**, 891–892.

Hopwood, D. A., Sermonti, G., and Spada-Sermonti, I. (1963). *J. gen. Microbiol.*, **30**, 249–260.

Howell, A. (1963/4). *Sabouraudia*, **3**, 81–92.

Howell, A., and Pine, L. (1956). *J. Bact.*, **71**, 47–53.

Howell, A., Murphy, W. C., Paul, F., and Stephan, R. N. (1959). *J. Bact.*, **78**, 82–95.

Johnson, L. F., Curl, E. A., Bond, J. H., and Fribourg, H. A. (1959). "Methods for studying soil microflora-plant disease relationships". Burgess, Minneapolis.

Jones, K. L. (1949). *J. Bact.*, **57**, 141–145.

Kane, W. D. (1966). *J. Elisha Mitchell scient. Soc.*, **82**, 220–230.

Kalakutskii, L. V. (1960). *Mikrobiologiya*, **29**, 59–63.

Kalakutskii, L. V., and Babkova, E. A. (1966). *Mikrobiologiya*, **35**, 244–246.

Kanai, R., Miyachi, S., and Takamiya, A. (1960). *Nature, Lond.*, **188**, 873–875.

Kawata, T., and Inoue, T. (1965). *Jap. J. Microbiol.*, **9**, 101–106.

Kawato, M., and Shinobu, R. (1959). *Mem. Osaka Univ. lib. Arts. Educ.*, **8**, 114–119.

Kelner, A. (1948). *J. Bact.*, **56**, 157–162.

Kosmachev, A. E. (1960). *Mikrobiologiya*, **29**, 210–211.

Krzywy, T. (1963). *Arch. Immunol. Ther. Exp.*, **11**, 521–546.
Kudrina, E. S., and Maksimova, T. S. (1963). *Mikrobiologiya*, **32**, 532–538.
Küster, E. (1953). *VIth. Int. Congr. Microbiol. (Rome)*, 114.
Küster, E., and Locci, R. (1963). *Arch. Mikrobiol.*, **45**, 188–197.
Küster, E., and Williams, S. T. (1964). *Nature, Lond.*, **202**, 928–929.
Kutzner, H. J. (1956). Diss. Landw. Hochschule, Hohenheim, 1956.
Kuznetsov, V. D., Lyagina, N. M., Sorofina, E. I., and Alyzova, L. F. (1962). *Mikrobiologiya*, **31**, 595–599.
Kwapinsky, J. B. (1965). "Methods of serological research". Wiley, New York.
Lawrence, C. H. (1956). *Can. J. Bot.*, **34**, 44–47.
Lechevalier, H. A., and Corke, C. T. (1953). *Appl. Microbiol.*, **1**, 110–112.
Lechevalier, H. A., and Holbert, P. E. (1965). *J. Bact.*, **89**, 217–222.
Lechevalier, H. A., Lechevalier, M. P., and Holbert, P. E. (1966). *J. Bact.*, **92**, 1228–1235.
Lindenbein, W. (1952). *Arch. Mikrobiol.*, **17**, 361–383.
Lingappa, Y., and Lockwood, J. L. (1962). *Phytopathology*, **52**, 317–323.
Luedemann, G. M., and Brodsky, B. C. (1964). Antimicrobial agents and chemotherapy—1963, Proc. 3rd Intersci. Conf. Washington, D.C., pp. 116–124.
Ludwig, E. H., and Hutchinson, W. G. (1949). *J. Bact.*, **58**, 89–101.
McClung, N. M. (1960). *Mycologia*, **52**, 154–156.
Menzies, J. D., and Dade, C. E. (1959). *Phytopathology*, **49**, 457–458.
Meyer, F. H. (1966). *In* "Symbiosis" (Ed. S. M. Henry), Vol. 1, pp. 171–255. Academic Press, New York.
Mikhailova, G. P. (1965). *Mikrobiologiya*, **34**, 643–647.
Moore, R. T., and Chapman, G. B. (1959). *J. Bact.*, **78**, 878–885.
Morita, R. Y., and Haight, R. D. (1964). *Limnol. Oceanogr.*, **9**, 103–106.
Muggleton, P. W., and Ungar, J. (1959). U.S. Patent 2,980, 614.
Narita, Z., and Tomita, Y. (1958). *Itsuu Kenkyusho Nempo*, **9**, 22.
Nette, I. T., Pomortseva, N. V., and Kozlova, E. I. (1959). *Mikrobiologiya*, **28**, 821–827.
Nishimura, H., and Tawara, K. (1957). *J. Antibiot. Tokyo*, **10A**, 82.
Nonomura, H., and Ohara, Y. (1960). *J. Ferment. Technol., Osaka*, **38**, 401–405.
Okami, Y., and Suzuki, M. (1958). *J. Antibiot., Tokyo*, **11A**, 250–253.
Oppenheimer, C. H., and Drost-Hansen, W. (1960). *J. Bact.*, **80**, 21–24.
Overman, J. R., and Pine, L. (1963). *J. Bact.*, **86**, 656–665.
Painter, B. G., and Bradley, S. G. (1965). *Bact. Proc.*, 5.
Palumbo, S. A., Berry, J. M., and Witter, L. D. (1967). *Appl. Microbiol.*, **15**, 114–116.
Pepys, J., Jenkins, P. A., Festenstein, G. N., Gregory, P. H., Lacey, M. E., and Skinner, F. A. (1963). *Lancet*, **ii**, 607–611.
Porter, J. N., Wilhelm, J. J., and Tresner, H. D. (1960). *Appl. Microbiol.*, **8**, 174–178.
Preobrazhenskaya, T. P., Kudrina, E. S., Maksimova, T. S., Sveshnikova, M. A., and Boyarskaya, R. V. (1960). *Mikrobiologiya*, **29**, 51–55.
Pridham, T. G., and Gottlieb, D. (1948). *J. Bact.*, **56**, 107–114.
Pridham, T. G., and Lyons, A. J. (1961). *J. Bact.*, **81**, 431–441.
Pridham, T. G., Anderson, P., Foley, C., Lindenfelser, L. A., Hesseltine, C. W., and Benedict, R. G. (1957). *Antibiotics A*. 1956/7, 947–953.
Pridham, T. G., Lyons, A. J., and Seckinger, H. L., (1965). *Int. Bull. bact. Nomencl. Taxon.*, **15**, 191–237.
Pumpyanskaya, L. V. (1964). *Mikrobiologiya*, **33**, 1065–1070.
Rancourt, M., and Lechevalier, H. A. (1963). *J. gen. Microbiol.*, **31**, 495–498.

Rancourt, M. W., and Lechevalier, H. A. (1964). *Can. J. Microbiol.*, **10**, 311–316.
Rautela, G. S., and Cowling, E. B. (1966). *Appl. Microbiol.*, **14**, 892–898.
Řeháček, Z. (1956). *Csklá Mikrobiol.*, **1**, 129–134.
Safferman, R. S., and Morris, M. E. (1962). *Robert A. Taft sanit. Engng Cent., Tech. Rep.*, **10**, 1–15.
Shirling, E. B., and Gottlieb, D. (1966). *Int. J. syst. Bact.*, **16**, 313–340.
Sinclair, N. A., and Stokes, J. L. (1963). *J. Bact.*, **85**, 164–167.
Skinner, F. A. (1951). *J. gen. Microbiol.*, **5**, 159–166.
Stansley, P. G. (1947). *J. Bact.*, **54**, 443–445.
Stessel, G. J., Leben, C., and Keitt, G. W. (1953). *Mycologia*, **45**, 325–334.
Stropnik, Z. (1965). *Sabouraudia*, **4**, 41–44.
Stuart, D. C. (1959). *J. Bact.*, **78**, 272–281.
Szabo, I., Marton, M., Buti, I., and Partai, G. (1966). *Acta microbiol., hung.*, **13**. 47–52.
Tendler, M. D. (1959). *Bull. Torrey bot. Club*, **86**, 17–30.
Tendler, M. D., and Burkholder, P. R. (1961). *Appl. Microbiol.*, **9**, 394–399.
Tresner, H. D., and Backus, E. J. (1957). *J. Bact.*, **73**, 687–688.
Tresner, H. D., Danga, F., and Porter, J. N. (1960). *Appl. Microbiol.*, **8**, 339–341.
Tresner, H. D., Davies, M. C., and Backus, E. J. (1961). *J. Bact.*, **81**, 70–81.
Trolldenier, G. (1966). *Zentbl. Bakt. Parasitkde, Abt. II*, **120**, 496–508.
Tsao, P. H., Leben, C., and Keitt, G. W. (1960). *Phytopathology*, **50**, 88–89.
Uradil, J. E., and Tetrault, P. A. (1959). *J. Bact.*, **78**, 243–246.
Waksman, S. A. (1959). "The actinomycetes", Vol. I. The Williams and Wilkins Co., Baltimore.
Waksman, S. A., Umbreit, W. W., and Cordon, T. C. (1939). *Soil Sci.*, **47**, 37–61.
Waksman, S. A., Reilly, H. C., and Johnstone, D. B. (1946). *J. Bact.*, **52**, 393–397.
Warcup, J. H. (1950). *Nature, Lond.*, **166**, 117.
Weiss, F. A. (1957). *In* "Manual of microbiological methods", pp. 99–119. McGraw-Hill, New York.
Wieringa, K. T. (1966). *Antonie van Leeuwenhoek*, **32**, 183–186.
Wilkin, G. D., and Rhodes, A. (1955). *J. gen. Microbiol.*, **12**, 259–264.
Williams, S. T., and Davies, F. L. (1965). *J. gen. Microbiol.*, **38**, 251–262.
Williams, S. T., and Davies, F. L. (1967). *J. gen. Microbiol.*, **48**, 171–177.
Williams, S. T., Parkinson, D., and Burges, N. A. (1965). *Pl. Soil.*, **22**, 167–186.
Willoughby, L. G. (1966). *J. gen. Microbiol.*, **44**, 69–72.
Wilska, A. (1947). *J. gen. Microbiol.*, **1**, 368–369.
Wollum, A. G., Youngberg, C. T., and Gilmour, C. M. (1966). *Proc. Soil Sci. Soc., Am.*, **30**, 463–467.

CHAPTER XII

Aquatic Fungi

E. B. GARETH JONES

*Department of Biological Sciences, Portsmouth College of Technology,
Portsmouth, England*

I. INTRODUCTION

Over the last twenty years considerable attention has been devoted to Ascomycetes and Fungi Imperfecti from fresh water and seawater. The pioneering work of Ingold (1942) on freshwater Hyphomycetes growing on leaves has led to the discovery of some 100 species. Most of these have been isolated and grown under laboratory conditions. Barghoorn and Linder (1944) were the pioneers in the field of marine mycology (Ascomycetes and Fungi Imperfecti). In marked contrast to the work on freshwater Hyphomycetes and Ascomycetes few marine Ascomycetes have been successfully isolated and grown under laboratory conditions. Many non-fruiting cultures have been obtained (Kohlmeyer, 1960; Jones, 1962a; 1964), and detailed physiological work of their fruiting requirements needs further study.

Emerson (1950) lists forty-one Phycomycetes known to have been cultured, while over the last eighteen years some 140 species have been added to this list. However, few of these are listed in any of the national culture collection catalogues. Mycologists have been slow to replace the standard baiting techniques (Crouch, 1939) with more up-to-date methods. Exceptions in this field have been Vischniac (1955a, b) and Fuller *et al.* (1964).

In this review, no attempt is made to include all the isolation techniques currently in use, only those considered of importance.

II. PHYCOMYCETES

A. General methods

1. *Collection of water samples and baiting in the laboratory*

Phycomycetes are found in nearly all kinds of freshwater habitats, in damp soil as well as in seawater. They occur as saprophytes or facultative parasites on dead insects, algae, twigs, seeds, dead leaves and fruits. Phycomycetes rarely develop in sufficient numbers for them to be identified in the field, consequently baiting techniques have been developed for their detection and isolation from water and soil. Water samples, soil, plant and animal debris should be collected in sterile bottles from rivers, streams, ponds and pools. Samples should be used within 24 h of collecting, before they became anaerobic.

Baiting techniques have been used successfully for many years as shown by Crouch (1939) and Sparrow (1960). Sterile Petri dishes are one half filled with autoclaved glass distilled water (as tapwater contains toxic substances) and allowed to cool. A few millilitres of the collected water or small amount of soil and debris are added to each dish. Various baits are then added and the cultures incubated and examined at daily intervals from the fourth to the fourteenth day. The following have been used as baits; autoclaved house and *Drosophila* flies, termite wings, boiled hemp seeds, boiled grass leaves, autoclaved wettable cellophane, pollen grains of various Conifers and white human hair. Fungi which grow on these baits can be isolated into pure culture by methods outlined below.

2. *Baiting in the field*

The method described above will be effective for isolating a variety of fungi, but others will escape capture. Many of these can be isolated by baiting in the field. A great variety of substrates have been used as baits, for example, hard green apples, small pears, tomatoes, grapes, plums, rose hips, green bananas, sweet corn kernels, loquets, mangoes, cashews and hawthorn berries.

The materials used as bait should be placed in small quarter-inch galvanized wire mesh baskets or plastic-covered wire baskets. These baskets are suspended in the water by nylon rope for a suitable period, determined by withdrawing samples at intervals, for example 1–2 weeks in the summer and 3–5 weeks in the winter.

Infected baits are then washed in sterile distilled water or seawater and transferred with fresh bait to freshwater or seawater. In this way further material can be obtained and the cultures are then cleaned up as indicated in Section II, A, 3.

Willoughby (1959) isolated Chytrids by baiting water with newspaper and cellophane in lakes. These baits were recovered after 2–3 weeks and placed in small beakers covered with distilled water and incubated at 25°C. The sporangia dehisce and zoospores accumulate at the surface meniscus. These zoospores were then transferred in loops to Emerson's yeast starch agar, tellurite agar or maize tellurite agar, with added antibiotics.

Perrot (1960) used a variety of baits including twigs of oak, elm, ash, birch, sycamore, alder and larch. These were immersed for four weeks and then incubated in sterile distilled water at a temperature of 3°C. These baits, alone with the low temperature proved successful in the isolation of slow growing and delicate forms, for example, *Monoblepharis* species. Fish, meat and hair were unsuccessful as baits because they became heavily colonized by bacteria. However, Dick (1961), Honeycutt (1948), Kane (1966) and Rothwell (1957) used human hair successfully as bait. Willoughby (1961) used termite wings as bait to isolate chytinophilic species from lake muds.

Baiting techniques have not been found successful for the isolation of marine Phycomycetes.

3. *Pure cultures*

The gross cultures obtained above can be cleaned up by the following methods.

Mycelium and single sporangia can be dissected out from mixed cultures, either in a Petri dish or on agar, washed in four or five changes of sterile distilled water or seawater and plated onto cornmeal agar or tellurite agar. Hypha tips can then be cut out of these cultures and transferred to fresh cornmeal or tellurite agar plates. Antibiotics can be added: 200 mg chloramphenicol per litre, or 2000 units penicillin G plus 0·5 mg streptomycin sulphate per plate.

Single zoospores can also be drawn up in capillary tubes and transferred to various nutrient media. This can be done by placing a sporangium in a cavity slide and waiting until zoospores are released. These are then drawn up in a capillary tube, transferred to a drop of sterile water (distilled or seawater) on a slide to dilute the number of zoospores. With a sterile pipette single zoospores are drawn up and plated out. It may be necessary to make several dilutions in order to pick up a single zoospore.

It has been found that media low in nutrients favour zoospore germination. If the medium is too rich then there is danger of the zoospores being plasmolysed. The following media are ideal, plain agar (3% agar), Leitner's agar (2% agar and 0·004% peptone), Foust's agar (2% agar, 0·15% maltose and 0·004% peptone) and cornmeal agar.

B. Plating out techniques

Vishniac (1956) was able to isolate marine Phycomycetes by plating seawater onto media fortified with antibiotics. The medium was as follows—

Seawater	80–100 ml
Glucose	0·1 g
Thiamine	0·2 mg
Calcium pantothenate	0·1 mg
Pyridoxamin–2 HCl	0·02 mg
Biotin	0·5 μg
Folic acid	2·5 μg
Gelatin hydrolysate	0·1 g
Liver extract	0·001 g
Nicotinic acid	0·1 mg
Pyridoxin HCl	0·04 mg
p-Aminobenzoic acid	0·01 mg
Cobalamin (B_{12})	0·05 μg
Agar	1·5 g

The plates were poured and allowed to dry. When dry, they were flooded with 2000 units of penicillin G and 0·5 mg streptomycin sulphate. The plates were then seeded with 0·2 ml seawater samples and spread over the surface with a bent sterile rod, or seeded with bits of marine algae.

Fuller et al. (1964) using modifications of Vishniac's method were able to isolate a number of marine Phycomycetes. The medium contained—

Agar	12 g
Glucose	1 g
Gelatin hydrolysate	1 g
Liver extract	0·01 g
Seawater	1000 ml

These were autoclaved together and 0·5 g streptomycin sulphate plus 0·5 g penicillin G added in a dry condition. Plates were seeded with small filaments of algae and incubated at 20°C in the dark. Plates were examined on the third day and then at intervals up to 2 weeks. The fungi isolated by this method were mainly filamentous, mycelium growing away from the algae. Hyphal tips were then transferred to new media until a pure culture was obtained.

In order to obtain spore release, small amounts of agar with mycelium were ground for 15 sec with seawater in the microcup of a Waring blender. One millilitre portions were then transferred to 125 ml flasks containing 50 ml liquid broth of the above medium, and grown on a shaker at 20°C. When growth was good, the medium was drained off and mycelium washed three times with sterile seawater. The mycelium was then placed in a crystallizing dish with 100 ml sterile seawater to discharge motile spores.

Fuller *et al.* (1966) slightly modified this procedure for isolating a parasitic *Pythium* from a red alga *Porphyra* by cutting out 2 mm square portions of the algal thallus, soaking these for 2 h in a solution containing 0·5 g penicillin G and 0·5 g streptomycin sulphate per litre seawater. These squares were then plated out in the usual way on the isolation medium.

C. Centrifugation and filtration techniques

Fuller and Poyton (1964) were able to isolate a number of monocentric Phycomycetes by continuous flow centrifugation of large water samples. By this method they were able to concentrate zoospores, normally sparsely distributed in water. They used a Servall SS-3 automatic superspeed centrifuge or a Servall RC-2 automatic superspeed centrifuge run at 5°C. A Servall KSB 3 "Szent-Gyroghi and Blum" continuous flow is also necessary. The centrifuge parts are sterilized and assembled and connected to the sterile reservoir which supplies the water sample to the centrifuge. This was run run at 27,000 g and with a flow rate of 600 ml/min in the eight-tube system or 150 ml/min in the two-tube system. The supernatant and centrifuged pellet is then mixed in the tube and poured into a sterile container. Portions of 0·2 ml are then plated onto suitable agar media containing antibiotics (for concentrations see Section II, B, p. 338)

Miller (1967) has also devised a technique for the isolation of zoospores from large quantities of water. Miller filtered water through millipore filter discs (pore size 0·8, 1·2 and 3·0 μ). The concentrate on the filter disc was resuspended in 0·5 ml water. This resuspended residue is then streaked on the surface of an agar medium containing antibiotics and low concentration of nutrients. A suitable medium contains—

Agar	30 g
Glucose	0·05 g
Peptone	0·05 g
Distilled water or seawater	1000 ml

Antibiotics are added after autoclaving, but prior to pouring the plates (for concentration see Section II, B, p. 338).

After incubation colonies of Chytrids could be picked out and transferred to a suitable Chytrid medium—

Corn meal agar	17 g
Glucose	5 g
Soluble starch	5 g
Peptone	1 g
Yeast extract	1 g
Distilled or seawater	1000 ml

Both these methods will be extremely useful in the quantitative study of the ecology of Phycomycetes, especially in determining the depth, seasonal and local distribution. Miller's technique has the limitation that large volumes of water will take a long time to filter. It is my experience that filter pores soon become clogged and filtering a litre of water can take several hours. Fragile cells are also liable to break up during membrane filtration. Ulken and Sparrow (1968), attempting to count the number of Chytrid zoospores in a lake, found centrifugation and millipore-filtered water techniques were not satisfactory. They baited known volumes of lake water with pollen grains and counted the number of grains infected with chytrids, after storage in the dark for 2 weeks.

D. Specific techniques

1. *Monoblepharales*

Perrott (1955) found that various species of *Monoblepharis* grew well on the following medium—

$NaNO_3$	2 g
KCl	0·5 g
K_2HPO_4	0·5 g
$MgSO_4$	0·5 g
$Fe_2(SO_4)_3$	0·01 g
Sucrose	15 g
Peptone	5 g
Cornmeal agar	20 g
Distilled water	500 ml
Pond water	500 ml

Sporangia were produced on 25% of the above minerals with the sucrose and peptone and zoospore release occurred on transfer to sterile water. Sexual reproduction occurred at 4 weeks on a small flamed tomato inoculated with fungus. Emerson (1950) reports 0·5% tryptone as a good medium for the growth of *Monoblepharis*.

2. *Oomycetes*

(a) *Saprolegniales*. Baiting with hemp seeds or house flies usually gives good results.

(b) *Leptomitales*. Emerson and Weston (1967) have shown that for *Aqualinderella fermentans*, a Phycomycete isolated from stagnant water, carbon dioxide is required for growth. In future isolation work, methods will have to be evolved to meet such requirements.

3. *Trichomycetes*

Lichwardt (1964) was able to isolate two species of *Smittium* by dissecting the hind guts of black flies (Simuliidae; *Simulium* and *Prosimulium*) in 0·75% sodium chloride solution. Portions of the gut are washed in sterile saline with 1000 units penicillin G (0·6 mg) and 2000 units of streptomycin sulphate (1·0 mg) per ml. Inoculum is then transferred to 10 ml agar in 6 cm Petri dish with a thin covering layer of 0·75% sodium chloride solution and added antibiotics (4000 units penicillin and 8000 units streptomycin). These fungi grew well on brain heart infusion agar diluted to 10% of the recommended strength (3·1 g Difco brain heart infusion with 15 g agar per litre water) or potato dextrose yeast extract agar (39 g potato dextrose agar and 1 g yeast extract per litre water).

Whisler (1960) isolated *Amoebidium parasiticum* by dissecting heavily infected *Cladocera* (water fleas) in sterile diluted pond water. This resulted in the release of numerous endospores which were picked up in a capillary pipette, then washed three or four times in sterile water. They were then streaked on to tryptone agar plates and clean colonies transferred to tryptone glucose broth (5 g tryptone, 3 g glucose per litre water) with added nutrients—

Machlis (1953)

Thiamine HCl	0·15 mg per litre
D, L-Methionine	0·1 g per litre
$MgCl_2$	0·001 M
KH_2PO_4	0·01 M
$(NH_4)_2HPO_4$	0·005 M
$CaCl_2$	0·0002 M
Mn as $MnCl_2$	0·5 ppm
Zn as $ZnSO_4$	0·1 ppm
B as H_3BO_3	0·5 ppm
Cu as $CuSO_4$	0·1 ppm
Fe as $FeCl_3$	0·5 ppm
Mo as $(NH_4)_6Mo_7O_{24}$	0·2 ppm
Co as $CoCl_3$	0·2 ppm

Whisler (1962) maintained cultures of *Amoebidium parasiticum* on 0·25% tryptone and 2% agar. One millilitre sterile diluted pond water was added to each test tube slope and the tube rotated daily until the agar surface was covered with plants.

4. *Predacious fungi*

Most species of predacious fungi belong to the Zoopagales or Fungi

Imperfecti. Cooke and Ludzack (1958) and Peach (1950; 1952) have reported on the presence of predacious fungi in the aquatic habitat. Peach (1950) was able to isolate such fungi by plating leaves on to maize meal agar (20 g crushed maize grains, 20 g agar and 1000 ml distilled water) and on rabbit dung agar. These are weak culture media and discourage the coarser moulds. Rough cultures may have to be kept for as long as 3 months and it is therefore necessary to conserve water.

Nematodes may have to be added to the gross cultures if the fungi are to appear. This can be done by soaking a little dried plant material, containing nematode eggs, in water. Nematodes are then separated by means of a Baermann separator. This consists of a filter funnel with a piece of rubber tubing and a clip. Material containing eel-worms is wrapped in muslin and placed in the funnel which is filled with water. The material is allowed to stand for 24 h, during which time the eelworms make their way through the muslin and sink to the bottom of the rubber tube. These nematodes are then added to the cultures. Low temperatures are best for the development of predacious fungi.

To induce trap formation in pure cultures it is best to place a small quantity of the fungus to a thriving culture of nematodes, rather than add nematodes to a pure culture of the fungus.

5. *Aquatic phycomycetes in tissues*

Perkins and Menzel (1966) were able to isolate *Dermocystidium marinum* from diseased oysters by placing diseased tissues for 48 h in thioglycolate. The tissue was then digested in 0·25% trypsin for 6–8 h at temperatures below 32°C. This was then passed through cheese cloth and centrifuged at 350 g and washed four times in sterile seawater. The pellet was resuspended in sterile artificial seawater with 0·5 mg per ml of each penicillin G and streptomycin sulphate at 30°C. Zoospores are released and used for isolating the fungus on thioglycollate medium.

6. *Phycomycetes in shells*

A number of algae and fungi grow in shells, e.g., oyster and cockle shells. These organisms are not easy to isolate. Prud'homme van Reine and Hoek (1966) and Alderman and Jones (1967) were able to isolate algae and fungi from shells with the use of the chelate disodium ethylene diamine tetraacetate (EDTA Na$_2$). Alderman and Jones placed small fragments of diseased shell to decalcify in a 5% solution of EDTA. Repeated changes of EDTA reduces the shell to a horny conchyolin wart and a gelatinous substance which is the protein matrix of the shell.

E. Zoospore liberation

This subject, while of extreme importance in culturing fungi, cannot be entered into in this work. The reader however is directed to an excellent and beautifully illustrated account by Emerson (1958; 1964).

F. Maintenance of stock cultures

Dick (1965) has described a very useful method for maintaining stock cultures of the Saprolegniaceae. The fungus is grown on nutrient agar containing the following in a litre of double glass distilled water—

Glucose	10 g
Soluble starch	5 g
Yeast extract	2 g
Na_2HPO4	0·597 g
KH_2Po_4	2·043 g
K_2TeO_3	0·1 g
Agar	12 g

and poured into a sterile Petri dish containing a Raper or van Teighem's ring. The agar in the ring is inoculated with the fungus, which 2–3 days later grows under the ring. A block of agar with hyphal tips can then be cut out.

Narrow necked 100 ml conical flasks are used for storage. They are filled with 40 ml double glass distilled water plugged with cotton wool bound in muslin. The flask is then autoclaved at 15 lb for 20 min.

Hemp seeds are sterilized by washing four or five times in hot distilled water and then tipped with the water into a small Petri dish. One hemp seed is transferred to the storage flask (while hot). When cool the flask is inoculated with a cube of agar containing hyphal tips. It may be necessary to tap the flask gently until the agar block and seed come together and the flask allowed to stand for 24 h so that the block and seed are bound together by hyphae. Members of the Saprolengniaceae kept under these conditions will remain viable for up to 15 months at 20°C while members of the Leptomitaceae kept at 5°C will remain viable for 24 months. The method is also suitable for the Pythiaceae.

Dick (1968) has found that cultures of aquatic Phycomycetes can be mailed by the following method. Two hemp seeds are placed in the centre of a filter paper 15 cm dia. and the filter paper folded in half, then into quarter and finally rolled so as to enclose the seeds. The roll is then placed in a 1 oz universal or McCartney bottle, glass distilled water added to wet and filter paper and the excess drained off. This is then autoclaved. When cool it is inoculated with a block or hemp seed culture placed on the filter paper near the seeds. The cap is screwed tight and the tube left for 24 h to make sure that the seeds are colonized before dispatch (see Fig. 1).

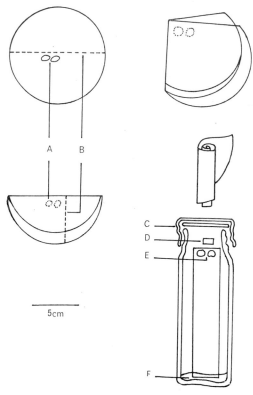

FIG. 1. A method of preparing cultures of aquatic Phycomycetes suitable for mailing. A. Two hemp seeds; B. Fold lines; C. Cap and seal; D. Inoculum; E. Filter paper enclosing hemp seeds; F. Residual water after wetting and draining (after Dick, 1968).

Webster (1965a) has designed a mechanical rocking platform for use in the maintenance of stock cultures of members of the Chytridiales and Blastocladiales (see Figs. 2 and 3). A small plug of inoculum with sporangia is transferred to a fresh tube and 2 ml sterile water added. The tubes are then placed on the platform and the machine switched on for 2–3 days. This apparatus is designed to shake the tube so as to flood the surface of the agar.

A number of aquatic Phycomycetes produce acidic products especially when grown on glucose media, e.g., *Blastocladia* (Emerson and Cantino, 1948), *Rhipidium* and *Sapromyces* (Sparrow, 1960), and *Pythiogeton* (Cantino, 1949). For this reason these fungi will need to be subcultured at frequent intervals if they are to remain viable.

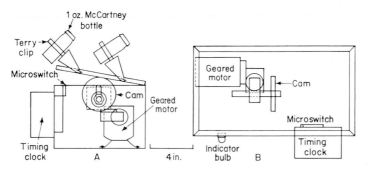

FIG. 2. Mechanical shaker for preparing cultures of aquatic Phycomycetes. A. Sectional view of apparatus; B. plan (wiring details omitted) (after Webster, 1965).

III. ASCOMYCETES

A. Hemiascomycetidae

Fell *et al.* (1960) used cores from banana stalks for the collecting of marine yeasts. Sections 2 in. long, cut from nearly ripe banana stalks with a $\frac{3}{4}$ in. dia. cork borer, were sterilized in a Petri dish and transferred aseptically to alcohol sterilized plastic vials. Four sections were placed in each vial. The vials had holes cut at either end to allow circulation around the banana cores, and submerged in the sea. These were collected after 3–10 days exposure.

Cores were then transferred to a blender with some sterile seawater. One millilitre of the resulting suspension was used to inoculate 2% glucose or 2% glucose 0·1% yeast extract plus 0·5% peptone broths, with added antibiotics. Flasks were shaken for 24 h on a rotary shaker at 25°C. Each flask was sampled at suitable time intervals and any yeasts present transferred to agar plates.

The above method can also be used to obtain yeasts from sediments, but with the necessary precaution taken to avoid contamination from the various profiles.

Fell and van Uden (1963) filtered a known volume of water through a millipore membrane, and then placed this membrane on the surface of an isolation medium in Petri dishes. The medium used contained 2% glucose, 1% peptone, 0·5% yeast extract and 2% agar per litre of seawater, the pH adjusted to 4·5 with lactic acid. Bacteria were suppressed with the addition of 10 mg chlortetracycline HCl, 2 mg chloramphenicol and 2 mg streptomycin sulphate.

FIG. 3. Mechanical shaker for maintaining and preparing aquatic fungi. A. Front view; B. Side view.

Jones and Slooff (1966) were able to isolate *Candida aquatica* from water scums by using a method described by Johnstone and La Touche (1956) for the isolation of single spores.

Phaff *et al.* (1952) isolated marine yeasts from shrimps by shaking samples of shrimps in 100 ml sterile seawater per animal. The washings were then plated out on to various media, potato dextrose agar at pH 3·5 being the best. Plates were incubated at room temperature for 5 days and colonies transferred to agar slopes for further study.

Methods used for the isolation of yeasts from polluted waters are discussed in Section VI below.

B. Euascomycetidae

1. *Marine*

Marine Ascomycetes are to be found growing on algae, leaves and culms of Angiosperms, rope and timber. Algae should be examined for ascocarps and placed in polythene bags. On return to the laboratory, the algae should be washed in sterile seawater, with added antibiotics, and incubated on sterile Kleenex tissues or filter paper. Filamentous algae rarely yield Ascomycetes, but pseudoparenchymatous types are often infected, e.g., *Pelvetia canaliculata* (*Mycosphaerella pelvetiae*, Webber, 1967; *Orcadia pelvetiana*, Sutherland, 1915a, 1915b), *Ascophyllum nodosum* (*Mycosphaerella ascophylli*, *Trailia ascophyli*, Wilson, 1951), *Fucus vesiculosus* (*Lulworthia fucicola*, Sutherland, 1916; Jones, unpublished), *Macrocystis* (*Corollospora maritima*, Meyers and Scott, 1967) *Chondrus crispus* (*Didymosphaeria danica*, Wilson and Knoyle, 1961; Kohlmeyer, 1964), *Cystoseira fimbriata* (*Thalassoascus tregoubovii*, Kohlmeyer, 1963) and *Ballia callitricha* (*Spathulospora phycophila*, Caviliere and Johnson, 1965).

Few of the Ascomycetes growing on algae have been isolated and a study of the truly parasitic species would be most interesting.

Ascomycetes are found growing on the leaves and culms of maritime Angiosperms, especially grasses. Salt marsh and intertidal plants are particularly good, e.g., *Spartina* spp. (Lloyd and Wilson, 1962; Jones, 1962b, 1963a) *Juncus* spp. (Johnson, 1956) and *Phragmites* sp. (Johnson, 1956). Plants with fungal fructifications should be collected and incubated in the same way as the algae.

The most intensively investigated group of marine Ascomycetes are the lignicolous forms. These can be collected on cordage (rope nets, mooring lines, etc.) bits of submerged or drifting wood. Care must be exercized in looking at drift material, as the fungi present need not necessarily be marine. Wooden pilings, piers and bridges should also be examined, and frequently yield fungi. Generally, badly decayed wood, especially wood soft and crumbly to the touch, yield few fungi.

The best method for the collection of lignicolous marine Ascomycetes and Fungi Imperfecti is to submerge test blocks in the sea. In this way one can be certain that the fungi colonizing the wood are marine. Dressed (or rough) test blocks of any length can be used, but for subsequent handling those measuring $6 \times 3 \times 1$ in. or $20 \times 10 \times 2$ cm have been found suitable. A variety of timbers have been used: *Pinus sylvestris* (Scots pine), *Fagus sylvatica* (beech), *Liriodendron tulipifera* (yellow poplar), *Tilia americana* (basswood), *Ochroma lagopus* (balsa) and *Acer* sp. (maple).

Test blocks may be submerged in pairs (Fig. 4a–c) or in strings of ten or twelve (Jones, 1963b). They can be suspended in various ways and the following has been found to give satisfactory results. Two holes are cut at either end of the blocks and lined with plastic tubing (Fig. 4c). The cross

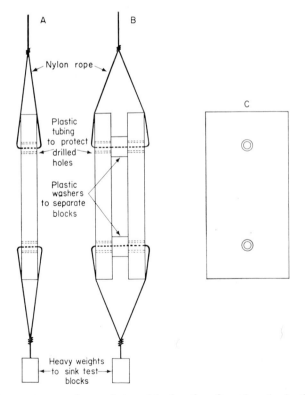

Fig. 4. Arrangement of wood test blocks for fungal colonization studies. A. Single test block in side view; B. A pair of test blocks in side view; C. Front view of test block.

cut end grains are sealed with a bituminous paint or other suitable preparation. A rope leading from the surface should be attached to the top hole and
another, carrying a weight, attached to the bottom hole (Fig 4a). If a pair
of blocks (or a string of blocks) are used then the blocks should be separated
by a $\frac{1}{4}$ in. plastic washer (Fig. 4b). After assembly the test blocks should be
sterilized by exposure to propylene oxide fumes for 18 h. This can be
done in a thick gauge polythene bag. If these are sealed, then the blocks can
be transported to the testing site in a sterile condition. The test blocks can
be suspended by a nylon rope from a fixed or a floating structure, e.g.,
rafts or buoys. They can be intertidal or completely submerged, and at
any depth required (6–10 ft below the surface is adequate). Variations of
the above method have been employed by Johnson and Sparrow (1961),
Johnson et al. (1959), Meyers and Reynolds (1958), and Woods and Oliver
(1962).

In submerging test blocks various points must be borne in mind. First
of all, if they are submerged from fixed situations, e.g., pilings, then ease
of access to facilitate removal is important. For example, if they are attached
at low water mark springs, then they are not readily accessible during neaps
or high water.

The period of submergence of the test blocks will depend on the requirement of the test and on the water temperature. In temperate climates when
water temperatures are $0°–7°C$, marine fungi will take up to 12–18 weeks to
colonize wood (Jones, 1963b) but at temperatures of $14°–20°C$ they will
appear on wood at 6–12 weeks (Jones, 1968b). In tropical conditions, exposure on a weekly basis may be required (Jones, 1968a). These times will only
give the initial colonizers, later colonizers requiring 6–12 weeks longer to
appear in temperate climates (Jones, 1963b). If the aim of the test is to investigate the colonization or succession pattern, then continuous sampling is
required, test blocks being removed every 4 or 6 weeks for up to two years
(Jones, 1963b, 1968b; Meyers and Reynolds, 1960).

In choosing a testing site a number of factors should be considered, e.g.,
tidal race, exposed or sheltered shore, local contamination from land drains
or sewage effluents, sandy or stony sea bed. All these can markedly affect the
colonization of wood blocks by marine fungi and other marine organisms.
These are details frequently ignored and make comparative work difficult
if not considered. In shallow coastal situations and sheltered bays, the test
blocks can be heavily infested with fouling organisms, such as Algae, Tuni-
cates, Polyzoa (Bryozoa), and barnacles (Jones and Eltringham, 1969). If
this occurs then test blocks should be removed and the fouling organisms
scraped off with a putty knife.

On removal, test blocks should be placed in separate sterile polythene
bags ("whirl pak") and returned to the laboratory as quickly as possible for

examination. Anastasiou (1963) found it necessary to pack his wood blocks in an ice chest to prevent drying out, and to maintain the fungi in a viable condition. Test blocks to be sent by post involving a delay of a few days before being examined, should have fouling organisms scraped off, the blocks washed in seawater and then placed in sterile bags.

Johnson and Sparrow (1961) were able to collect marine fungi on wood shavings, wood strips, cordage or pressed pulpwood sheets in tanks of running seawater or in glass cylinders through which seawater is continually pumped. Meyers (1968a) reported on the colonization of cotton cellulose filters submerged in the sea. The number of fungi obtained by these methods is low, although individual species show a high population incidence.

During the removal of test blocks, hydrographical details should be recorded, e.g., temperature, salinity, pH, oxygenation and the amount of sediment in the water. These are factors frequently ignored but very important when considering the ecology of marine fungi.

On return to the laboratory the test blocks are incubated on a layer of sterile tissues in sterile plastic boxes, e.g., sandwich boxes (see Fig. 5a). This enables fungi present as mycelium to form perithecia, and for the surface water to drain off and allow the Fungi Imperfecti to sporulate. These fungi often do not sporulate if there is a film of water on the surface of the blocks. Meyers and Reynolds (1958) and Johnson and Sparrow (1961) have

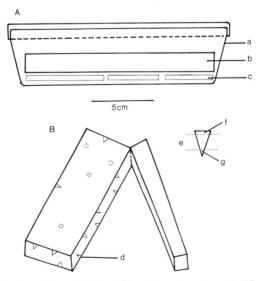

FIG. 5. A. Incubation of test block; a. Plastic luncheon box; b. Test block; c. Layer of sterile Kleenex tissues; B. Removal of plugs for isolation of fungi; d. A wedge of wood removed with a sterile scalpel; e. Outer and inner surface of plug removed for transfer to isolating medium; f. Outer surface; g. Inner surface.

described ways of drying down test blocks. Drying down of the test blocks has the effect of discharging ascospores from perithecia. Test blocks can be incubated for up to 15 weeks with the addition of small amounts of sterile seawater to prevent complete drying out.

Isolation is best done by picking up a loopful of ascospores from the spore mass often found at the tip of the perithecia and streaking out on agar media. Many of the methods described by Booth (this Volume, p. 1) can be used for the single spore culture of ascospores. Crushing perithecia on a slide and transfer of spores with pipettes or needles can be difficult as the ascospores of many marine fungi have mucilaginous appendages which tend to stick the spore to the needle.

Media suitable for the isolation of marine fungi are those low in nutrients, e.g., yeast extract glucose (0·1 g yeast extract, 1 g glucose, 18–20 g agar and 1000 ml seawater), cornmeal agar and cellulose agar (Eggins and Pugh, 1962) made up with seawater. Antibiotics should be used to supress bacteria, e.g., 0·200 g chloramphenicol per litre medium or penicillin G and strepto-mycin sulphate.

Lignicolous fungi can be isolated by removing the surface layer of decayed wood with a sterilized chisel. A sliver of wood is then removed with a second sterilized chisel or scalpel and plated on agar (Greaves and Savory, 1965).

Test blocks can be slit longitudinally and a wedge of wood cut out with a sterilized scalpel (see Fig. 5b–c). The inner and outer surface is then re-moved, flamed and plated on to agar. A cork borer may be used to remove a core as described by Meyers (1968b).

Siepmann and Johnson (1960) attempted to isolate fungi from test blocks autoclaved for one hour and exposed to propylene oxide for 24 h and then submerged in the sea, by scraping the soft surface layers of the wood into 600 ml sterile seawater. This was shaken vigorously and one or two aliquots pipetted on to agar media (with added antibiotics). The remaining suspension was diluted with three litres of sterile seawater, shaken and further plates seeded with 1 ml of inoculum. Plates were incubated at 25°C for up to 14 days. None of the fungi they isolated could be regarded as marine and un-fortunately no information was given of the fungi found fruiting on the wood prior to scraping. The fungi isolated could well have developed from conidia (terrestrial) in silt on the surface of the wood. Much greater care is needed in investigating the total number of fungi present in wood. Jones (unpublished) and Jones and Eaton (1969; cooling tower fungi) have used a similar method to determine the fungi present on submerged wood. They first of all recorded the fungi found fruiting on the wood, then scraped off the soft surface layers of wood and blended this in sterile water in an M.S.E. blender. One millilitre aliquots of this suspension were then plated out on cornmeal agar and cellulose agar. The fungi isolated were appreciably different from those recorded as fruiting on the wood. The fungi isolated were largely Fungi Imperfecti which could have been present as conidia and not necessarily active in the wood. The Ascomycetes fruiting on the wood were rarely isolated by this method. This situation is comparable to the soil fungi problem where Fungi Imperfecti are readily isolated by plating out soil, but Ascomycetes and Basidiomycetes (certainly active in soil) are rarely isolated. This indicates the need for diverse isolating techniques and some knowledge as to the role and activity of these fungi in the habitat or material being tested.

Borut and Johnson (1962) isolated fungi from estuarine sediments by collecting samples with an Ekman dredge. The dredge was surface sterilized with 5% formalin seawater or 95% alcohol. The sediment in the dredge was sliced open with an alcohol flamed spatula, and a sample removed from the central portion with a sterile scalpel. The sediment was placed immediately in sterile bottles or vials. Only one sample was taken from each dredge. The sediment was used within 24–48 h and during this time kept at 4°C.

Small amounts of the sediment were spot inoculated on to the surface of agar plates (nutrient agar, Czapek's agar, potato dextrose agar and low-nutrient Sabouraud's agar), and incubated at room temperature. Plates were examined every two days for two weeks.

Marine Ascomycetes once isolated do not fruit readily and various media have been devised to overcome this fastidiousness.

Meyers and Reynolds (1959) were able to stimulate reproduction (*Lulworthia grandispora*, *L. medusa*, *L. floridana*, *Ceriosporopsis halima*, *Corollospora maritima* and *Torpedospora radiata*) on balsa wood slips in a 0·1% yeast extract seawater broth. *Antennospora quadricornuta*, *Arenariomyces salinus* and *Lignicola laevis* did not form perithecia under these conditions, but when the fungus infested wood was transferred to aquaria of aerated seawater, fruiting occurred.

Temperature, time and substrate available markedly affect reproduction of marine Ascomycetes. Meyers and Simms (1967) showed that *Lulworthia floridana* produced perithecia at 25 days on 0·5% and 0·1% cellobiose, but require 45 days when grown on 1·0% cellobiose. They also showed that *L. floridana* needs a temperature of 25–30°C to fruit at 18 days, and that at 20°C it failed to form perithecia even at 60 days.

Kirk (1966; 1967) was successful in producing perithecia of a number of marine Ascomycetes. The medium used contained 0·3 g yeast extract, 10 μg thiamine, 0·5 μg biotin, 0·02 g succinic acid, 0·2 g KNO_3, 0·2 g K_2HPO_4, 1 g TRIS (hydroxymethylaminomethane), 18 g agar in one litre of aged natural seawater and 1 ml mineral solution (30 mg Na_2EDTA (disodium ethylenediaminetetraacetate), 0·01 mg Na_2MoO_4, 1 mg $FeCl_3$, 0·3 mg $ZnCl_2$, 0·5 mg $MnCl_2$ and 0·02 mg $CuCl_2$).

Ten millilitres of this basal medium was dispersed in 20×150 mm screw-capped tubes, each containing two white birch (*Betula papyrifera*) applicator sticks about 145 mm long (see Fig. 6a). All the ingredients were autoclaved and the tubes slanted. Tubes were incubated in the dark at 22–24°C. Perithecia are formed on and around the birch sticks. Kirk found that more ascocarps were produced at a salinity of 17–26‰ than in full seawater media.

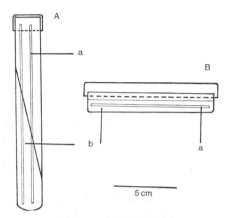

FIG. 6. Methods used to encourage marine Ascomycetes to fruit. A. Test tube method; a. Birch applicator sticks; b. Medium. B. Petri dish method.

The same method can be used for Petri dishes (see Fig. 6b). Sterilized wood strips are placed in sterile Petri dishes and autoclaved basal media added.

Johnson and Gold (1959) have described a circulating system in which seawater is pumped continually over wood blocks (or rope pieces) either inoculated directly with a pure culture of a fungus or by the introduction of a spore or conidial suspension directly into the circulating water. The apparatus is shown in Fig. 7, and sterilized prior to use.

2. *Freshwater*

Freshwater Ascomycetes are to be found growing on reed swamp plants *Scirpus lacustris*, *Eleocharis palustris*, *Juncus articulatus*, *Equisetum fluviatile*, *Phragmites communis* and *Carex riparia* (Ingold and Chapman, 1952; Ingold, 1968b), on driftwood (Ingold, 1954; 1955) and on submerged wood (Jones and Oliver, 1964) while Jones and Eaton (1968) have described Ascomycetes found growing on timber slats and wood test blocks placed in various cooling towers.

The methods used by Jones and Oliver (1964) for the collection of Asco-

IV 15

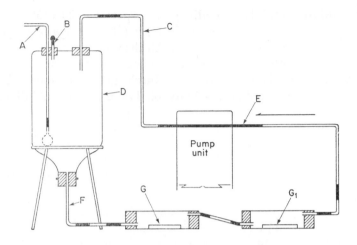

FIG. 7. A continual flow seawater system for submerged culture of marine Ascomycetes A. Aerator head; B. Air release; C. Return line of glass tubing; D. Reservoir ($3\frac{1}{2}$ gal. capacity Pyrex carboy); E. Pump unit with sequence-oscillating bars; F. Outlet line; G and G1. Test blocks (after Johnson and Gold, 1959).

mycetes on wood test blocks are similar to those outlined above (see marine Ascomycetes) and need no further discussion.

Reed swamp plants with ascocarps are incubated on damp Kleenex tissues. Mature ascocarps are dissected out of their host and crushed in a drop of sterile water. Ascospores are then transferred with a loop or pipette and plated out on agar. Many of these fungi do not fruit on agar but will do so if plant material is added, e.g., *Loramyces juncicola* and *L. macrospora* on *Eleocharis* stems (Ingold and Chapman, 1952).

Savory (1954), Duncan (1960) and Jones and Eaton (1968 and unpublished) have shown that fungi are active in the degradation of timber in cooling towers. Jones and Eaton (1968) have investigated the flora by placing test blocks in the cooling towers at various points (see Fig. 8) using methods similar to those described for the lignicolous marine Ascomycetes. Many of the Ascomycetes isolated failed to fruit. Eaton (in press) devised a sterile water circulating system (see Fig. 9) which simulates the conditions in cooling towers. The system consists of two plastic aquaria and the bottom half of a sandwich box (exit chamber). Two curtain rails are fixed to the roof of the top aquarium and from the hooks strings of test blocks are attached. The string of test blocks can be pulled back and forth with the aid of string A. Once assembled the whole system is closed and propylene oxide introduced through the water intake pipe. Sterilization takes 24 h. Sterilized water is introduced through the intake pipe and is pumped up to a distribu-

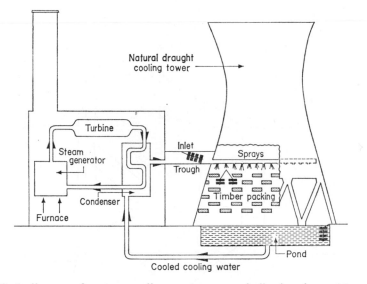

Fig. 8. A diagram of a water-cooling tower system indicating the position of the inlet trough, packing and pond (sampling areas).

ting point B, from which lead two polythene tubes punctured at regular intervals. Water sprinkles on to the surface of the test blocks and collects in the bottom aquarium before being recirculated. Test blocks can be withdrawn with the aid of string A into the exit chamber, the roller falls off the rail and into the bottom of the exit chamber. A trap is opened and the string of test blocks removed. This set up can be sterilized before use and enables test blocks to be removed aseptically. None-fruiting cultures have been induced to fruit by introducing a suspension of mycelium and water into the system and incubating for 12 or more weeks at 25°–30°C.

IV. FUNGI IMPERFECTI

Routine methods for the isolation of the Fungi Imperfecti are described by Booth (this Volume, p. 49) and many of these may be used in the isolation of aquatic hyphomycetes.

A. Freshwater

Aquatic hyphomycetes are to be found growing on the submerged leaves of various plants (Ingold, 1943; Greathead, 1961; Nilsson, 1964 and Ranzoni, 1953), submerged twigs and submerged test blocks (Jones and Oliver, 1964; Price and Talbot, 1966) and their spores are found in profusion in water scums (Ingold, 1968a; Jones, 1965b; Tubaki, 1960).

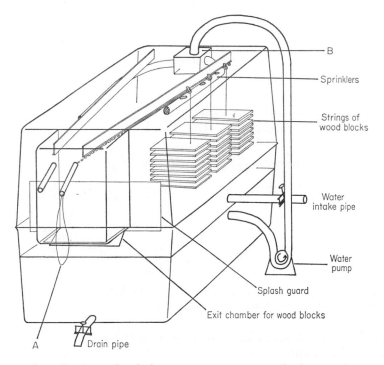

FIG. 9. A sterile water-circulating system to encourage freshwater Ascomycetes to fruit.

1. *On leaves*

Fallen and decayed leaves (*Acer, Acacia, Aesculus, Alnus, Buddleia, Cassine, Castanea, Celtis, Corylus, Crataegus, Eucalyptus, Ficus, Fraxinus, Ilex, Hedera, Podocarpus, Ulmus, Quercus* etc.) are collected from rivers, streams and ponds and placed in polythene bags. Experience has shown that brown or skeletonized leaves from well aerated rivers are well colonized by hyphomycetes. Van Beverwijk (1951; 1953) has shown that leaves covered by mud from poorly aerated water have a completely different fungal flora from those from well aerated situations.

The leaves are washed in sterile distilled water and placed in shallow dishes of water for 2 days. Rich crops of spores are soon produced. The larger spores can be picked up with a needle, small spores in fine capillary pipettes and both transferred to suitable media with added antibiotics (for concentrations see Section II, B).

Autumn and winter seem to be the best periods for collecting leaves for the isolation of these fungi, while April to June is the least productive period.

Jones and Oliver (1964) have found spores developing on wood panels submerged in the period between February and June.

Truly aquatic hyphomycetes will not produce typical spores unless submerged in water. The best results are obtained by placing strips of agar from 3–4 week old cultures in water and vigorously aerating with a stream of compressed air. Sporulation is stimulated and conidia formed in a few days.

2. On wood

The methods described by Jones and Oliver (1964) for submerging test blocks for the collection of hyphomycetes has been described in detail above (Section III, B, 1).

3. In water scums and foams

Price and Talbot (1966), Cooke (1963), Harrison (1967, unpublished), Jones and Eaton (unpublished), and Jones and Oliver (1964) have shown that hyphomycetes with inconspicuous spores do occur in river scums. Some of these grow and reproduce in the aquatic habitat. Therefore, fungi with tetraradiate or scolecosporus spores are by no means typical of the freshwater environment. Indeed, many of these tetraradiate spored fungi are undoubtedly of terrestrial origin, e.g., *Tetraploa aristata* (Ellis, 1949), *Trisulcosporium* and *Tetranacrium* (Hudson and Sutton, 1964).

Clean scums or foams can be collected by spooning the surface scum on the surface of waters in streams, rivers, ponds or lakes into sterile universal or McCartney bottles. These scums must be used within 24 h if isolations are to be made, as spores soon germinate under these conditions making single spore isolations impossible. These scums can be diluted down with sterile distilled water and 1 ml aliquots transferred and streaked onto the surface of cornmeal agar or cellulose agar, with added antibiotics. Germinating spores can then be cut out and transferred to fresh media. Webster (1959) was able to isolate *Tricellula aquatica*, which has small conidia, by drawing up conidia into a fine capillary tube and streaking out on agar plates.

Many more freshwater hyphomycetes remain to be described, as indicated by reports of unidentified spores in water scums (Ingold, 1968a; Jones 1965b; Nilsson, 1964). Systematic isolation of spores from water scums along the lines of Harrison (1967, unpublished) and the examination of different substrates, e.g., wood, bark and twine, should be attempted.

A number of aquatic hyphomycetes have been shown to have perfect stages as shown by Ranzoni (1956), Webster (1957; 1961; 1965b) and Tubaki (1966).

B. Marine

Approximately fifty marine hyphomycetes are known from wood, twine, algae and the culms of various Angiosperms, especially grasses. The methods used for their collection, incubation and isolation are similar to those outlined for terrestrial hyphomycetes (Section IV, A, 1–2), and for marine and freshwater ascomycetes (see above, Sections III, B1, and III, B, 2).

The only innovation with respect to marine hyphomycetes was that reported by Meyers and Moore (1960) for the isolation of *Nia vibrissa*. They incubated wood shavings that had been submerged for several months in the sea, in sealed Petri dishes for 2–3 weeks. The fungus produced conidia on the surface of a leathery cortex, and these could be transferred to glucose yeast extract seawater agar and cultures established.

Kohlmeyer (1967) has reported the presence of spores of marine ascomycetes and hyphomycetes in scums along shorelines. These spores can be isolated from the scum samples in the same way as described above for freshwater fungi.

C. Aero-aquatic hyphomycetes

In 1951, van Beverwijk described the first of a number of aero-aquatic fungi, and her work has been extended by Glen-Bott (1955) and Hennebert (1968). These fungi have been termed aero-aquatic as their mycelium and conidiophores are found on decaying leaves and wood submerged in water (often stagnant), while the conidia are developed above the water surface.

Decaying leaves or twigs are collected and placed in Petri dishes with sufficient water to form a thin film over them. Petri dishes are placed in plastic boxes to prevent the leaves from drying out. Conidia are soon produced on the surface of the water and these can be picked up with a needle, loop or pipette and transferred to suitable agar media (e.g., cherry agar, malt agar, yeast dextrose asparagine agar and potato dextrose agar), with antibiotics added. When clean cultures are established, water is added to the culture so that the mycelium is covered by a layer of water. Hennebert (1968) found that subcultures were easier to make if a spore suspension was made and poured on to fresh agar plates.

D. Potentially pathogenic fungi

Potentially pathogenic fungi (Ascomycetes and Fungi Imperfecti) have been isolated from the aquatic environment by Cooke (1955; *Allescheria boydii*, *Aspergillus fumigatus*, and *Geotrichum candidum* from polluted water and sewage) and Kirk (1967; *Allescheria boydii* from seawater). These fungi appear to be more widespread than anticipated and methods for their isolation are described by Waterhouse (this Volume, p. 183), Stockdale (this Volume, p. 429) and Buckley (this Volume, p. 461).

V. BASIDIOMYCETES

Two marine Basidiomycetes are known: *Melanotaenium ruppiae* (Feldmann, 1959) and *Digitatispora marina* (Douget, 1962). Ingold (1959; 1961) has described two spores with clamp connections from freshwater scums collected in Nigeria. The substrate on which they grow and the nature of the sporophore is unknown. None of these has been isolated.

VI. FUNGI IN POLLUTED WATERS, SEWAGE AND SEWAGE TREATMENT SYSTEMS

The pioneer work in the field of fungi in polluted and sewage waters has been that of Cooke (1954a–c). Cooke (1957) has indicated the importance of these fungi in the decomposition of faecal material and other wastes rich in organic content. Certain fungi, *Leptomitus lacteus*, *Geotrichum candidum* and *Fusarium aquaeductuum*, have a specialized type of metabolism which enables them to grow and thrive under these conditions of pollution. Cooke (1957) has recorded thirteen Mucorales, seventeen Ascomycetes, one Basidiomycete and 124 Fungi Imperfecti from polluted waters. Cooke (1959) listed ninety fungi found in trickle filters as well as a wide range of other organisms (Algae, Protozoa, Annelids, Insects, etc.).

1. *Collection*

Samples are best collected by grab sampling the water, bottom sediment and wet soil from the bank side of rivers. The sample should be at least 50 ml and placed in sterile containers, e.g., plastic vials or universal bottles.

Samples containing water, mud, sand and soil are spooned into the containers, closed and any dirt on the outside wiped or rinsed off. The container should not be completely air tight.

2. *Isolation*

The fungi to be isolated from the water, mud, sand and soil samples will contain a great variety of fungi from typical soil fungi, aquatic fungi, predaceous fungi to pathogenic forms and include Phycomycetes, Hemi- and Eu-Ascomycetidae and Fungi Imperfecti. The isolation of such a spectrum of organisms will involve the use of many techniques, many already described above.

Samples from polluted waters or sewage need to be diluted before isolation is attempted. For good results each plate should contain less than fifty fungal colonies. The dilution necessary to give such results will be gained by experience, but the following will serve as a guide (Cooke, 1963), liquid should be diluted 1 : 10; rich sewage may be diluted 1 : 100; sludges with 4–6% dry matter 1 : 1000 and rich, fairly dry samples with 30–60% dry matter up to 1 : 10,000.

On arrival in the laboratory a 1 : 10 dilution of the sample is made and shaken for 30 min on a rotary shaker. This will disperse the sample. On removal from the shaker, 5 ml of 1 : 10 dilution is pipetted to a second flask and a 1 : 100 dilution prepared. This is shaken by hand to obtain uniform dispersal of the sample. Further dilutions are made, if necessary, in the same way. If the dilution required is not known, then a high and low dilution should be plated out and the results compared. From this it should be possible to decide which dilution will give the best results.

Cooke (1963) found neopeptone dextrose rose bengal aureomycin agar extremely useful in the isolation of soil fungi. This medium contains 5 g neopeptone or polypeptone, 10 g dextrose, 0·035 g per litre rose bengal, 35·0 μg per ml aureomycin HCl, 20 g agar and 1000 ml distilled water. The rose bengal can be added before autoclaving but the aureomycin should be added before the agar is poured into the Petri dishes. Penicillin and streptomycin may be used if desired (see Section II, B, for concentrations), but aureomycin gives good results for sewage and sewage polluted waters.

The procedure adopted by Cooke (1963) was to melt five tubes of the neopeptone medium, add 0·05 ml aureomycin solution and 1 ml of the sample at the dilution required. The tube is then poured into a Petri dish and gently rotated to spread the medium. The plates are kept in light at room temperature for one week. The number of fungi present can be expressed as total colonies or the number of colonies of each fungal species. As it may be necessary to compare one sample with another, the number of colonies present in each sample can be correlated on the basis of dry weight. This is done by determining the dry weight of two 15 ml portions.

Aquatic Phycomycetes can be obtained by baiting with hemp seeds. The dilution sample required is prepared and poured into a Petri dish. Two sterilized hemp seeds are added and incubated at room temperature for 3–14 days. Pure cultures can be obtained from these as described in Section II, A, 3.

Cooke and Busch (1957) have reported the presence of cellulolytic fungi in polluted waters. These can be detected by plating out 1 ml of the dilution required on to cellulose agar (Eggins and Pugh, 1962). Samples of sewage, etc., can be baited with human hair, horse hair, wool or human skin to detect the presence of keratinophilic fungi. Aquatic hyphomycetes, usually found growing on decaying leaves, occur in streams, pools and ponds, and can be detected and isolated as outlined in Section IV, A,1–3. Predacious fungi are frequently present in polluted waters (Cooke and Ludzack, 1958) and these are isolated as outlined in Section II, D, 4.

Cooke (1965) and Cooke et al. (1960) have shown that yeasts are very active in polluted waters and sewage. Solutions of yeast nitrogen base (YNB, Difco) are prepared with 1% glucose added to one and another with 20% glucose. Fifty millilitres of the medium was transferred to 250 ml Erlenmeyer flasks. To a pair of flasks, one of each sugar concentrations, was added 1 or 2 ml of 1 : 10 dilution of the sample to be tested. The flasks are shaken on a rotary shaker for 60–72 h, removed and allowed to settle. This will enable yeast cells to settle, while the bacteria will remain in suspension and the filamentous fungi will either float to the surface or remain in suspension. After 4 h the flask is tilted so that a concentration of yeast cells appears at the junction of the liquid medium line and the bottom of the flask. A loopful of cells is removed and streaked on to a plate of yeast extract, malt extract glucose agar. After 3 days or less incubation at room temperature and light, colonies appear. Pure cultures are established by repeated streaking on Diamalt agar plates.

This enables various yeast species to be isolated, but gives no quantitative data. Cooke (1965) in a quantitative study of yeast populations in a sewage treatment plant used two approaches. First of all the "indicated number" (IN) technique in which single cultures of a single dilution is used, rather than replicates. The second approach is the "most probable number" (MPN) technique in which it is assumed that growth develops from a single individual, and not from a group of cells.

Samples are suspended in distilled water at a ratio of 1 : 10 and shaken on a rotary shaker for about 30 min. The media used are 1% and 20% YNB broths with 50 ml aliquots in 250 ml Erlenmeyer flasks.

For the IN method, 1 ml of the dilutions 1 : 10, 1 : 100, 1 : 10,000 and 1 : 100,000 is added to the flask or each nutrient medium. They are shaken for 64 h. Flasks are then removed and allowed to stand for 4 h so that the yeast cells settle to the bottom. Sediment from the bottom of each flask is streaked onto two plates of yeast extract glucose agar. After 2–3 days the resulting growth is restreaked on to 15% Diamalt agar plates or yeast extract, glucose plates. Individual colonies are then isolated. This enables a species list of yeasts present to be made.

For the MPN method 25 ml of each nutrient are added to 25 × 150 mm culture tubes. One millilitre of each solution is added to each tube in replicates of five. The tubes are incubated for 7 days without shaking. Yeast cells from the bottom growth are streaked on agar plates for species identification. For the quantitative work, growth is measured by turbidometry or dry weight of cells.

This method is based on enrichment in YNB with 1% and 2% glucose. A more complete estimation of the yeast population may be obtained by the use of additional media.

Cooke (1958) described a method for continuous sampling of trickle

filter populations. This involves the exposure of glass microscope slides
(3 × 1 in.) in trickle filters. These slides were removed at regular intervals
and slides placed on the surface of agar media. This was suitable for early
colonization stages, but for the latter stages the material on the slide was
scraped off. Each slide was placed in 20 ml distilled water and one side of
the slide scraped clean. (Each surface of the slide was treated separately.)
The water plus scraped material was added to 80 ml distilled water in an
Erlenmeyer flask and agitated in a Waring blender for 10–30 sec. This gave
a uniform suspension for plating onto agar. Dilutions of this suspension were
made and plated out as described above. Cooke (1959) listed ninety fungi
found in trickle filters.

The aquatic habitat has only recently received attention mycologically
and lags behind when compared with algology and bacteriology. Considerable interest lies in the oceans, estuaries and large lakes as a source of food
for a world with a human population explosion. With intensified farming of
lakes and estuaries and eventually the sea, will come the problem of diseases
of the organisms being "cultured". Epidemics of algal populations have
frequently been reported, but little work has been done in this field. Clearly
the isolation and culture of fungi from freshwater and the sea will be valuable with respect to their taxonomy (especially the Phycomycetes) and
physiology. Undoubtedly new techniques and media will enable a greater
number to be collected and isolated. The isolation and culture of marine
Ascomycetes growing on algae would be most profitable, so that the physiology of this interesting group could be investigated.

ACKNOWLEDGMENTS

I should like to thank the following for permission to reproduce their drawings
and plates: Professor J. Webster (Figs. 2 and 3), Dr. M. Dick (Fig. 1) and Mr. R. A.
Eaton (Fig. 9).

REFERENCES

Alderman, D. J., and Jones, E. B. G. (1967). *Nature, Lond.*, **216**, No. 5117, 797–798.
Anastasiou, C. J. (1963). *Nova Hedwigia*, **6**, 243–276.
Barghoorn, E. S., and Linder, D. H. (1944). *Farlowia*, **1**, 395–467.
Borut, S. Y., and Johnson, T. W. (1962). *Mycologia*, **54**, 181–193.
Cantino, E. C. (1949). *Am. J. Bot,.* **36**, 747–756.
Cavaliere, A. R., and Johnson, T. W. (1965). *Mycologia*, **57**, 927–932.
Cooke, W. B. (1954a). *Sewage Ind. Wastes*, **26**, 539–549.
Cooke, W. B. (1954b). *Sewage Ind. Wastes*, **26**, 661–674.
Cooke, W. B. (1954c). *Sewage Ind. Wastes*, **26**, 790–794.
Cooke, W. B. (1955). *Publ. Hlth Rep. Wash.*, **70**, 689–694.
Cooke, W. B. (1957). *Sydowia*, **1**, 146–175.
Cooke, W. B. (1958). *Sewage Ind. Wastes*, **30**, 21–27.
Cooke, W. B. (1959). *Ecology*, **40**, 273–291.

Cooke, W. B. (1963). A Laboratory guide to fungi in polluted waters, sewage and sewage treatment systems. U.S. Dept. Health, Education and Welfare, Cincinnati.

Cooke, W. B. (1965). *Mycologia*, 57, 696–703.

Cooke, W. B., and Busch, K. A. (1957). *Sewage Ind. Wastes*, 29, 210–217.

Cooke, W. B., and Hirsch, A. (1958). *Sewage Ind. Wastes*, 30, 138–156.

Cooke, W. B., and Ludzack, F. J. (1958). *Sewage ind. Wastes*, 30, 1490–1495.

Cooke, W. B., Phaff, H. J., Miller, M. W., Shifrine, M., and Knapp, E. B. P. (1960). *Mycologia*, 52, 210–230.

Crouch, J. N. (1939). *J. Elisha Mitchell scient. Soc.*, 55, 208–214.

Dick, M. W. (1961). *Mich. Acad. Sci., Arts Letters*, 46, 195–204.

Dick, M. W. (1965). *Mycologia*, 57, 828–831.

Dick, M. W. (1968). *Bull. Trans. Brit. mycol. Soc.*, 2, 70–72.

Doquet, G. (1962). *Bull. Soc. mycol. Fr.*, 78, 283–290.

Duncan, C. G. (1960). Report No. 2173. Forest Products Laboratory, Madison, Wisconsin.

Eggins, H. O. W., and Pugh, G. J. F. (1962). *Nature Lond.*, 193 (4810), 94–95.

Ellis, M. B. (1949). *Trans. Br. mycol. Soc.*, 32, 246–251.

Emerson, R. (1950). *A. Rev. Microbiol.*, 4, 169–200.

Emerson, R. (1958). *Mycologia*, 50, 589–621.

Emerson, R. (1964). *Am. Biol. Teach.*, 26, 90–100.

Emerson, R., and Cantino, E. C. (1948). *Am. J. Bot.*, 35, 157–171.

Emerson, R., and Weston, W. H. (1967). *Am. J. Bot.*, 54, 702–719.

Feldmann, G. (1959). *Revue. Gén Bot.*, 66, 3–6.

Fell, J. W., Ahearn, D. G., Meyers, S. P., and Roth, F. J. (1960). *Limnol. Oceanogr.*, 5, 366–371

Fell, J. W., and Uden, N. van (1963). *In* "Symposium on Marine Microbiology", (C. H. Oppenheimer, Ed.), pp. 329–340. Thomas Publ. Co., Springfield, Ill.

Fuller, M. S., Fowles, B. E., and McLaughlin, D. J. (1964). *Mycologia*, 56, 745–756.

Fuller, M. S., Lewis, B., and Cook, P. (1966). *Mycologia*, 58, 313–318.

Fuller, M. S., and Poyton, R. O. (1964). *Bioscience*, 14, 45–46.

Glen-Bott, J. I. (1955). *Trans. Br. mycol. Soc.*, 38, 17–30.

Greathead, S. K. (1961). *Jl. S. Afr. Bot.*, 27, 195–229.

Greaves, H., and Savory, J. G. (1965). *J. Inst. Wood Sci.*, No. 15, October, 45–50.

Harrison, J. L. (1967). Undergraduate project, Portsmouth College of Technology.

Hennebert, G. L. (1968). *Trans. Br. mycol. Soc.*, 51, 13–24.

Honeycutt, M. B. (1948). *J. Elisha Mitchell scient. Soc.*, 64, 277–285.

Hudson, H. J., and Sutton, B. C. (1964). *Trans. Br. mycol. Soc.*, 47, 197–203.

Ingold, C. T. (1942). *Trans. Br. mycol. Soc.*, 25, 339–417.

Ingold, C. T. (1943). *New Phytol.*, 43, 139–143.

Ingold, C. T. (1951). *Trans. Br. mycol. Soc.*, 34, 210–215.

Ingold, C. T. (1954). *Trans. Br. mycol. Soc.*, 37, 1–18.

Ingold, C. T. (1955). *Trans. Br. mycol. Soc.*, 38, 157–158.

Ingold, C. T. (1959). *J. Quekett microsc. Club.*, Ser 4, 5, 115–130.

Ingold, C. T. (1959). *Trans. Br. mycol. Soc.*, 42, 479–485.

Ingold, C. T. (1961). *Trans. Br. mycol. Soc.*, 44, 27–30.

Ingold, C. T. (1966). *Mycologia*, 58, 43–56.

Ingold, C. T. (1968a). *Trans. Br. mycol. Soc.*, 51, 137–143.

Ingold, C. T. (1968b). *Trans. Br. mycol. Soc.*, 51, 323–341.

Ingold, C. T. and Chapman, B. (1952). *Trans. Br. mycol. Soc.*, 35, 268–272.

Johnson, T. W. (1956). *Mycologia*, 48, 495–505.

Johnson, T. W., Ferchau, H. A., and Gold, H. S (1959) *Phyton, B. Aires,* **12,** 65–80
Johnson, T. W., and Gold, H. S. (1959). *Mycologia,* **51,** 89–94.
Johnson, T. W., and Sparrow, F. K. (1961). "Fungi in Oceans and Estuaries", 668 pp. J. Cramer, Weinheim.
Johnstone, K. I., and La Touche, C. J. (1956). *Trans. Br. mycol. Soc.,* **39,** 442–448.
Jones, E. B. G. (1962a). *Trans. Br. mycol. Soc.,* **45,** 93–114.
Jones, E. B. G. (1962b). *Trans. Br. mycol. Soc.,* **45,** 245–248.
Jones, E. B. G. (1963a). *Trans. Br. mycol. Soc.,* **46,** 135–144.
Jones, E. B. G. (1963b). *J. Inst. Wood Sci.,* No. 11, December, 14–23.
Jones, E. B. G. (1964). *Trans. Br. mycol. Soc.,* **47,** 97–101.
Jones, E. B. G. (1965a). *Trans. Br. mycol. Soc.,* **48,** 287–290.
Jones, E. B. G. (1965b). *Naturalist, Hull,* 57–60.
Jones, E. B. G. (1968a). *Curr. Sci.,* **37,** 378.
Jones, E. B. G. (1968b). *In* "Biodeterioration of Materials". Proceedings of the 1st International Biodeterioration Symposium, Southampton (Eds A. H. Walters, and J. J. Elphick), pp. 460–485. Elsevier Press.
Jones, E. B. G. and Eaton, R. A. (1969). *Trans. Br. mycol. Soc.,* **52,** 161–164.
Jones, E. B. G., add Eltringham, S. K. (Eds) (1969). "Marine borers, fungi and fouling organisms of wood". Published by Organization for Economic Co-operation and Development, Paris (in press).
Jones, E. B. G., and Oliver, A. C. (1964). *Trans. Br. mycol. Soc.,* **47,** 45–48.
Jones, E. B. G., and Slooff, W. Ch. (1966). *Antonie van Leeuwenhoek,* Journal of Microbiology and Serology, **32,** 223–228.
Kane, W. D. (1966). *Mycologia,* **58,** 905–911.
Kirk, P. W. (1966). "Morphogenesis and microscopic cytochemistry of marine pyrenomycetes ascospores", Nova Hedwigia, Monograph, 128 pp. J. Cramer, Weinheim.
Kirk, P. W. (1967). *Mycopath. Mycol. appl.,* **33,** 65–75.
Kohlmeyer, J. (1960). *Nova Hedwigia,* **2,** 293–343.
Kohlmeyer, J. (1963). *Nova Hedwigia,* **6,** 127–146.
Kohlmeyer, J. (1964). *Dtn Ges. Pilzkunde,* **30,** 43–51.
Kohlmeyer, J. (1967). *Trans. Br. mycol. Soc.,* **50,** 137–147.
Lichwardt, R. W. (1964). *Am. J. Bot.,* **51,** 836–842.
Lloyd, L. S., and Wilson, I. M. (1962). *Trans. Br. mycol. Soc.,* **45,** 359–372.
Machlis, L. (1953). *Am. J. Bot.,* **40,** 189–195, 450–460.
Meyers, S. P. (1968a). *Bull. Misaki Marine Biological Institute Kyoto Univ.,* No. 12, 207–225.
Meyers, S. P. (1968b). *In* "Biodeterioration of Materials" Proceedings of the 1st International Biodeterioration Symposium, Southampton (Eds A. H. Walters and J. J. Elphick), pp. 594–609., Elsevier Press.
Meyers, S. P., and Moore, R. T. (1960). *Mycologia,* **52,** 871–876.
Meyers, S. P., and Reynolds, E. S. (1958). *Bull. mar. Sci. Gulf Caribb.,* **8,** 342–347.
Meyers, S. P., and Reynolds, E. S. (1959). *Mycologia,* **51,** 138–145.
Meyers, S. P., and Reynolds, E. S. (1960). *Can. J. Bot.,* **38,** 217–226.
Meyers, S. P., and Scott, E. (1967). *Mycologia,* **59,** 446–455.
Meyers, S. P., and Simms, J. (1967). *Bull. Mar. Sci. Gulf Caribb.,* **17,** 133–148.
Miller, C. E. (1967). *Mycologia,* **59,** 524–527.
Nilsson, S. (1964). *Symb. bot. upsal.,* **18,** pp. 1–130.
Peach, M. (1950). *Trans. Br. mycol. Soc.,* **33,** 148–153.

Peach, M. (1952). *Trans. Br. mycol. Soc.*, **35**, 19–23.
Perkins, F. O., and Menzel, R. W. (1966). *Proc. natn. Shellfish Ass.*, **56**, 23–30.
Perrott, P. E. (1955). *Trans. Br. mycol. Soc.*, **38**, 247–282.
Perrott, P. E. (1960). *Trans. Br. mycol. Soc.*, **43**, 19–30.
Phaff, H. J., Mrak, E. M., and Williams, O. B. (1952). *Mycologia*, **44**, 431–451.
Price, I. P., and Talbot, P. H. B. (1966). *Aust. J. Bot.*, **14**, 19–24.
Prud'homme van Reine, W. F., and Hoek, C. van den (1966). *Blumea*, **14**, 331–332.
Ranzoni, F. V. (1953). *Farlowia*, **4**, 353–398.
Ranzoni, F. V. (1956). *Am. J. Bot.*, **43**, 13–17.
Rothwell, F. M. (1957). *Mycologia*, **49**, 68–72.
Savory, J. G. (1954). *Ann. appl. Biol.*, **41**, 336–347.
Siepmann, R., and Johnson, T. W. (1960). *J. Elisha Mitchell scient. Soc.*, **76**, 150–154.
Sparrow, F. K. (1960). "Aquatic Phycomycetes", Edn. 2, XXVI, 1187 pp. Univ. Michigan Press, Ann Arbor.
Sutherland, G. K. (1915a). *New Phytol.*, **14**, 183–193.
Sutherland, G. K. (1915b). *Trans. Br. mycol. Soc.*, **5**, 147–155.
Sutherland, G. K. (1916). *Trans. Br. mycol. Soc.*, **6**, 257–263.
Tubaki, K. (1960). *Nagaoa*, **7**, 15–29.
Tubaki, K. (1966). *Trans. Br, mycol. Soc.*, **49**, 345–349.
Ulken, A., and Sparrow, F. K. (1968). *Veroff. Inst. Meeresforsch. Bremerh.*, **11**, 83–88.
van Beverwijk, A. L. (1951). *Antonie van Leeuwenhoek*, **17**, 278–284.
van Beverwijk, A. L. (1953). *Trans. Br. mycol. Soc.*, **36**, 111–124.
Vischniac, H. S. (1955a). *Mycologia*, **47**, 633–645.
Vischniac, H. S. (1955b). *Trans. N.Y. Acad. Sci.*, **17**, 352–360.
Vischniac, H. S. (1956). *Biol. Bull.*, **111**, 410–414.
Vischniac, H. S. (1958). *Mycologia*, **50**, 66–79.
Waterhouse, G. M., and Stamps, D. J. *Soc. Appl. Bact.* (in press).
Webber, F. C. (1967). *Trans. Br. mycol. Soc.*, **50**, 583–601.
Webster, J. (1957). *Trans. Br. mycol. Soc.*, **40**, 322–327.
Webster, J. (1959). *Trans. Br. mycol. Soc.*, **42**, 416–420.
Webster, J. (1961). *Trans. Br. mycol. Soc.*, **44**, 559–564.
Webster, J. (1965a). *Trans. Br. mycol. Soc.*, **48**, 447–448.
Webster, J. (1965b). *Trans. Br. mycol. Soc.*, **48**, 449–452.
Whisler, H. C. (1960). *Nature, Lond.*, **186**, No. 4720, 732–733.
Whisler, H. C. (1962). *Am. J. Bot.*, **49**, 193–199.
Willoughby, L. G. (1959). *Trans. Br. mycol. Soc.*, **42**, 67–71.
Willoughby, L. G. (1961). *Trans. Br. mycol. Soc.*, **44**, 586–592.
Willoughby, L. G. (1962). *J. Ecol.*, **50**, 733–759.
Wilson, I. M. (1961). *Trans. Br. mycol. Soc.*, **34**, 540–543.
Wilson, I. M. (1960). *Proc. Linn. Soc. Lond.*, **171**, 53–70.
Wilson, I. M., and Knoyle, J. M. (1961). *Trans. Br. mycol. Soc.*, **44**, 55–71.
Woods, R. P., and Oliver, A. C. (1962). B.W.P.A. Annual Convention Record, pp. 145–169.

CHAPTER XIII

Air Sampling for Fungi, Pollens and Bacteria

R. R. DAVIES

The Wright-Fleming Institute of Microbiology, St. Mary's Hospital Medical School, London

I. INTRODUCTION

The atmosphere contains all the major groups of microbes ranging from the algae to the viruses. Outdoors, with the changing seasons of the year, the air contains a varying assemblage of pollen grains, a prolific fungal spora in which all the major groups are well represented, occasional spores of ferns, mosses and algal cells such as those of the terrestrial genus *Pleurococcus*. The "air spora" as it has been termed by Gregory (1952) has been studied in regions ranging from the Arctic (Pady and Kapica, 1953) to the tropical desert of Arabia (Davies, 1969). Indoors, man contaminates his environment with his activities, *Micrococci* and *Staphylococci* become airborne upon the skin scales he desquamates (Davies and Noble, 1962) whilst even talking disseminates fine droplets of saliva carrying such bacteria as the *Streptococci* from his mouth (Bourdillon *et al.*, 1948) and there is much epidemiological evidence for the aerial spread of viruses.

The methods and apparatus described in this chapter are those most commonly used to trap airborne microbes and employ either impaction, impingement or "sedimentation". Other principles are utilized such as electrostatic precipitation but since before an electrostatic attraction or electron wind can be effective, a volume of air has to be drawn into the sampler, it is for most purposes simpler to utilize the essential suction pump to produce a jet of air and direct this and the suspended particles it entrains on to a prepared surface that retains the particles. The term impactor is used to describe a device in which the sampled airstream is directed on to a solid surface such as a sticky faced microscope slide or agar medium in a Petri dish, whilst in an impinger the particles are blown into a liquid culture fluid or sterile water.

The choice of method must be governed by such features as whether sampling is to be undertaken indoors or outdoors, the concentration in which the microbes occur in the air to be sampled, whether it is essential that the microbe be grown for identification and enumeration or whether, as is the case with pollen grains and the spores of certain obligate fungal pathogens of plants, they can be recognized by their morphological features under the microscope.

II. GENERAL PRINCIPLES

Apparatus for use outdoors should have a feather edged intake nozzle and be so pivoted that the intake aperture is directed into the main air stream. If the air speed in the intake tube is greater than that of the wind, although small particles follow the stream lines converging into the orifice, large particles due to their inertia pass through the curvature of the stream lines and are underestimated in the sample (see Fig. 1). When the suction rate is too low, large particles are blown into the orifice and there are too many large particles in the sample. Ideally, particles should be sucked into the orifice at the same speed as the wind speed, a condition described by Druett (1942) as isokinetic but which is impracticable except under con-

trolled conditions of lamellar flow in a wind tunnel. A compromise solution
to this problem is that of May (1945) and later Hirst (1952), who so designed
the Cascade Impactor and Automatic Volumetric Spore Trap respectively

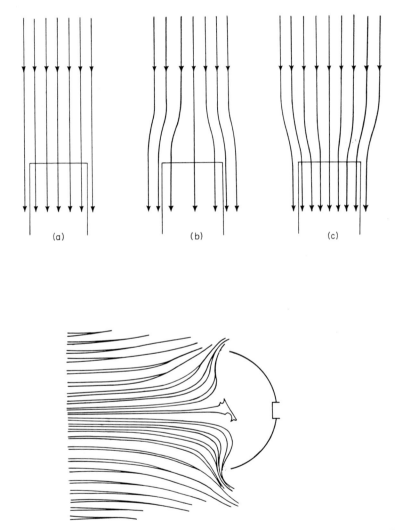

FIG. 1. (a) Isokinetic sampling (correct numbers of large and small particles in
sample), (b) sampling velocity too low (too many large particles in the sample),
(c) sampling velocity too high (too few large particles in the sample). Below: stream-
lines approaching a hemicylindrical baffle and being brought nearly to rest by the
cushion effect inside the baffle as shown by the great broadening out of the stream-
lines. (Drawn from a photograph by kind courtesy of K. R. May.)

that each is directed into wind by a vane and the air velocities in the intake tubes approximate to mean wind speed in the British Isles. With such apparatus, unless sampling is for a very brief period, anisokinetic errors tend to be averaged as the wind fluctuates above and below the mean. Anisokinetic sampling errors can be significant when studying the airborne concentrations of large particles such as pollen grains ranging from 12 to 80 μ diameter and "rafts" upon which microbes are carried, but the greater is the size of the intake orifice the smaller are these errors likely to be.

Indoors with freely falling particles in relatively still air, to avoid a deficiency of large particles in the sample, the intake orifice should be horizontally displayed and face upwards, and a radial flow occurs into it. May (1967) contrasts the high efficiency of sampling in still air with the variable errors of anisokinetic sampling in an unsteady air stream and suggests that "if moving aerosols could be brought nearly to rest about the sampling point, then high efficiency sampling over a wide range of air speed and direction might be achieved by an orifice sucking at a constant rate". In "stagnation point sampling" a baffle of suitable shape and size is fixed behind the sampling nozzle so that air approaching the baffle is brought nearly to rest by a cushion of air in the baffle. May and Druett (1953) in tests with the pre-impinger (see IV, C) found that with 30 μ dia. particles in a 6 m.p.h. wind the intake of the bulb was increased from 58 to 80% if the sample was taken through the centre of a 4 in. diameter baffle normal to the wind direction. Similarly Edwards (cited May, 1966) found the intake efficiency of the multi-stage liquid impinger (May, 1966, see IV, D) for 15 μm dia. particles in a wind of 10 m.p.h. was raised from 9·6 to 99% with a 6 in. square concave baffle. Since isokinetic sampling is usually impossible, further experimental evaluation of what May has termed "Stagnation point sampling" is highly desirable.

Apparatus for sampling airborne microbes must not only be free from collection errors but also have high retention efficiencies. It should be free from sharp bends so that the deposition of large particles on walls is minimized, the surfaces upon which particles are impacted should be highly retentive and air velocity, jet dimensions and distance from the sampling surface are particularly important variables. Since these features are easily evaluated, the retentivities of air samplers in common use are well known.

III. IMPACTORS

A. The slit sampler

In this instrument described by Bourdillon *et al.* (1941), air is sucked at high velocity through a narrow slit and the entrained particles impacted

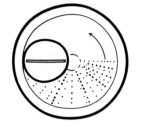

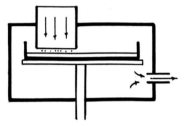

FIG. 2. The Casella slit sampler.

on to the surface of the agar medium in a Petri dish. During sampling the Petri dish is rotated on a turntable and the particles are distributed over a wide band. After suitable incubation the numbers of colonies which develop are counted and from a knowledge of the sampling rate, which is constant, and the time of exposure which may be varied, the numbers of viable microbes per unit volume of air can be calculated. Since the speed at which the dish is rotated is also known the time when particular microbes were present in the air above the sampling intake can be determined.

An improved slit sampler with accurate timing was described by Bourdillon et al. (1948b) and a large slit sampler for air containing few bacteria by Bourdillon et al (1948c). Both models with further improvements in detail and construction are manufactured by C. F. Casella and Co. Ltd., Britannia Walk, London, N.10 (see Fig. 2). In both, the aspiration rate is adjusted by reference to a built-in vacuum gauge. A time switch may be set to stop sampling after a pre-selected period, and a marker may be employed to indicate any special event during sampling. A built-in distance indicator is incorporated because the efficiency of impaction and retention is dependent on the distance between the slit and the agar surface being small and reproducible.

The collection efficiency of the original sampler was tested by Bourdillon et al. (1941) and found to be 96% for an aerosol of Staph. albus sprayed from distilled water as single cocci. From a formula given by May (1945), Bourdillon and Lidwell (1948) calculated that the impaction efficiencies for the 1 mm and 0·3 mm slit widths respectively were 90 and 98% for spheres of 1 μ dia., and unit density. To the degree of automation which makes this apparatus so simple to operate, must be added the convenience of sampling the microbes directly on to the surface of the agar medium in a Petri dish. The use of the slit sampler has contributed much to our knowledge of air hygiene and cross infection in hospitals and is illustrated in the work of Bourdillon and Colebrook (1946), Bourdillon et al. (1948), Duguid and Wallace (1948), Drummond and Hamlin (1952), Blowers et al. (1955), Williams et al. (1956) and Williams et al. (1962). Examples of its use in mycological studies are to be found in the work of Pady and Kapica (1953), Maunsell (1954), Davies (1957) and Noble and Clayton (1963).

The apparatus was designed for sampling under relatively still air conditions indoors and the intake orifice faces vertically upwards. This form, if used outdoors, cannot be operated isokinetically and sampling errors become increasingly significant as particle diameters become greater than 5 μ. The intake orifice of the slit sampler is aerodynamically similar to that of the multistage liquid impinger of May (see below) which when tested by Edwards (cited May, 1966) showed an intake efficiency of 9·6% for particles of 15 μm dia. in a 10 m.p.h. wind. In a study of the naturally occurring clouds of *Cladosporium*, Davies (1957) obtained values with a split sampler similar to those obtained with a cascade impactor (see below). Since 70% of the spores were trapped as single conidia and less than 5% as clumps of 5 or more, and Noble *et al.* (1963) reported the "median equivalent diameter"† for *Cladosporium* spores is 4·9 μ, clearly with this apparatus, aniskokinetic sampling errors are unimportant with spores as small as these.

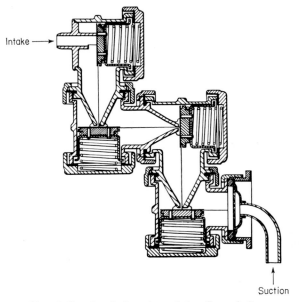

Intake →

Suction

FIG. 3. Sectional elevation of the Cascade Impactor.

K. R. May (personal communication) has designed a compact slit sampler with a wind facing nozzle and a vertical sampling dish but at present this is not available commercially. A future development of the conventional slit sampler might be an adaptor for use outdoors consisting of a wind facing baffle to produce a cushion of stagnant air over the intake orifice.

B. The cascade impactor

This apparatus described by May (1945) as an instrument for sampling coarse aerosols is essentially a system of four air-jets and sampling slides in series (see Fig. 3). When used outdoors the apparatus is freely suspended and a vane directs the feather edged intake orifice into the wind. For sampling in still air a bell mouth adaptor is fitted to the intake orifice which must

† The equivalent particle diameter is the diameter of a sphere of unit density which has a settling rate in air equal to that of the particle in question. Any diameter given without qualification in this text refers to the aerodynamic diameter.

then be faced upwards to avoid a deficiency of large particles in the sample.

The sampling rate of 17·5 litres a minute that is usually employed gives air velocities through the four jets of 2·2, 10·2, 27·5 and 77 metres per second respectively. Their performance when sampling drops of unit density is such that the maximum drop size found in the deposits behind the second, third and fourth jets have diameters of 20, 7 and 2·5 μ, each being near the minimum size impacted with 100% efficiency by the previous orifice. Impaction velocities are such that 50% of particles with diameters of 12, 4, 1·5 and 0·4 μ penetrate the respective jets. This instrument not only gives good retention for particles down to 0·5 μ dia. but traps droplets as large as 50 μ dia. and aggregates such as clumps of fungal spores without shattering them.

May's design of the cascade impactor is particularly important since it traps particles according to size in a similar manner to the human respiratory tract. Particles larger than spheres of 7 μ dia. and unit density are deposited in the upper respiratory tract, 50% of those 5 μ dia. reach the alveoli of the lung (C. N. Davies, 1963; Druett, 1967). Since the first and second stages of the cascade impactor have a 100% retention for 7·1 μ dia. spheres of unit density, it simulates the upper respiratory tract in which the naturally inhaled pollens and larger fungal spores are deposited, which in the allergic produce such symptoms as rhinitis, bronchospasm and asthma. The third and fourth stages trap such particles and spores as may be inhaled and retained in the deepest parts of the lung. Maximum retention in the lung alveoli is 50–60% for particles 1·5 to 2 μm dia. (C. N. Davies, 1952) and 50% of particles 1·5 μm dia. are trapped at the third stage of the cascade impactor.

The cascade impactor represents a milestone in air sampling techniques and it is the reference instrument by which the efficiencies of other air sampling devices are compared. The size grading slit sampler (Lidwell, 1959), portable spore trap (Gregory, 1954) and Hirst trap (see below) are based upon it. It is widely used to study dust polluted atmospheres and to establish the efficacy of sprays for spreading herbicides and insecticides. In microbiology it has been used to study the influence of particle size on infection with anthrax spores (Druett et al., 1953) and with *Pasteurella pestis* (Druett et al., 1956) and to study the spores liberated from mouldy hay associated with Farmer's Lung disease (Gregory and Lacey, 1963). Their microscopic examination of dust fractions collected with a cascade impactor when mouldy hay was shaken in a wind tunnel disclosed dense concentrations of actinomycete spores which although found to be thermophilic were never grown in anything approaching the concentrations revealed by the cascade impactor.

In the Casella version of the cascade impactor the slides of May's original version are replaced by glass discs or cover glasses, the 3rd and 4th jets are narrowed to increase impaction efficiency and a filter paper may be fitted behind the last disc. The details given above are for this model. Slides or cover glasses may be made adhesive for microbiological sampling by either smearing with petroleum jelly which is best applied hot to get an even layer, or a strip of clear plastic may be stuck on and made sticky with glycerol jelly as used by Gregory and Lacey (1963). The petroleum or glycerol jelly must not be applied too sparingly since with too thin a layer retentivity is impaired.

C. The automatic volumetric spore-trap (Hirst trap)

Hirst (1952) in experiments in the field and in a wind tunnel found most fungal spores were trapped on the slide behind the second jet of the cascade impactor and devised a continuously recording sampler based upon it. In the Hirst trap (see Fig. 4) air is drawn at 10 litres/min. through a feather edged 14×2 mm orifice which is directed into the wind by a vane. The particles in the air stream are directed on to a sticky surfaced microscope slide which is moved 2 mm an hour behind the inner edge of the orifice. On a slide exposed for 24 h the trapped particles occur in a trace measuring 14×48 mm. After suitably mounting in either glycerine jelly or lactophenol solvar (Hirst, 1953) the trace may be scanned longitudinally to obtain a daily mean or transversely, at 4 mm intervals, to trace two hourly changes in concentration.

When operated continuously this apparatus provides a complete record of the diurnal and seasonal changes in the air-spora. Under field conditions in Britain, anisokinetic sampling errors are unlikely to be large and Gregory and Hirst (1957) report trapping efficiency averages 80% but which varies with wind speed and particle size.

Identification is visual under the microscope and whilst the spores of phytopathogenic fungi such as rusts, smuts and powdery mildews which are difficult to grow in artificial culture are recorded, the conidia of such common moulds as *Aspergillus* and *Penicillium* are indistinguishable and underestimated because the impaction velocity is sufficient to trap only 50% of particles of 4 μ diameter. Noble *et al.* (1963) report that in *Aspergillus fumigatus* the spores have a medium equivalent diameter of 4 μ and that in the summer the spores have smaller equivalent diameters than in the winter. When airborne the spore is subjected to desiccation and dimensions quoted from direct measurement are usually for fully turgid spores.

The Hirst trap is robust and reliable, a Casella model with an Edwards RB4 pump* has been operated continuously by the writer for seven years.

* Edwards High Vacuum, Manor Royal, Crawley, Sussex, England.

FIG. 4. (a) Hirst spore trap.

When a Hirst trap is to be operated continuously for many months, as is necessary for seasonal surveys of pollens and spores as allergens, it is imperative that the suction source should be reliable since a pump failure and an inability to obtain an immediate replacement has ruined a number of studies. Hirst traps manufactured by Casella with Edwards RB4 pumps as the suction source have been operated successfully on my behalf under extremes of climate ranging from winter in the Alps to summer in the Arabian desert. In the Casella model air flow is controlled by an adjustable bleed and shown by a conspicuous flow meter fitted to the suction side of the trap. During continuous sampling the sole attention required is the few minutes necessary to remove the exposed slide and insert a new one every 24 h. The 8 day clockwork mechanism which moves the slide behind the intake orifice needs to be wound once each week and at that time it is convenient to switch off the suction pump for 2 or 3 minutes to insert a little oil.

A spore trap based on that of Hirst has been developed to sample continuously for periods of up to 7 days without attention. In this the particles

FIG 4. (b) Body of trap with sampling aperture removed to show slide.

are impacted on to an adhesive coated tape which is of a transparent plastic and supported on a clockwork driven drum which revolves once in 7 days at 2mm an hour. The tape is cut into 48 mm lengths and mounted for microscopic examination. Mounting is more difficult than is the case with the microscope slide made sticky with petroleum jelly or silicone grease as used in the conventional trap and large particles such as pollen grains become detached and drift in the mounting fluid when first mounted. Where the adhesive surface on the slide is petroleum jelly it can be warmed a little so that this is prevented; this is necessary if the time when pollens or spores were trapped is to be determined. Also, it is more difficult to coat an even layer of adhesive on a long length of tape than it is on a microscope slide. A more serious criticism of this seven day sampler in its present form, is that flow is metered with a "critical orifice" and after a seven day period of continuous sampling the flow may be reduced to as little as 2 litres/min.

due to an accumulation of material in the "critical orifice" (personal communications, Allit and Morrow Brown). A "critical orifice" may be recommended for use with the Anderson Sampler and is an integral part of the Porton impinger (see below) but this equipment is used for sampling over relatively short periods of time only. Attention to the conventional Hirst trap each day provides for inspection, the air flow is checked and adjusted if necessary, and in the event of a mechanical failure only one day's sample is lost whereas the failure of a 7-day version could lead to a significant loss of record.

The Hirst trap is the apparatus of choice for surveys of airborne spores and pollens in relation to inhalant allergy and because it samples continuously it overcomes the problem that the collection of air samples at a particular time or times each day can miss peak concentrations of important allergens since their diurnal periodicities are affected by variable weather factors. In this instrument the spores and pollens are impacted in a dense deposit which makes scanning less tedious and time consuming than searching an area of sticky slide upon which particles have been trapped by sedimentation and impaction by eddy diffusion. It seriously underestimates small spores such as those of *Sporobolomyces* which Evans (1965) considers an important allergen and the fitting of a narrower sampling orifice would increase impaction. However, in the polluted atmospheres of cities the environment for most of the allergic population, increasing the air speed in the orifice in addition to an increase in anisokinetic sampline errors (see II) leads to a greatly increased deposition of soot and ash on the slides which makes visual identification and counting too difficult; and the amount of soot and ash impacted can so overload the adhesive surface as to reduce its retentivity.

The use of this method of air sampling is illustrated in general accounts of the fungal air-spora by Hirst (1953), Gregory and Hirst (1957), spores as allergens Hyde (1959), Davies *et al.* (1963), Davies (1969), and in the epidemiology of apple scab by Hirst *et al.* (1955), and spore dispersal in *Ophiobolus graminis* and fungi of cereal foot rots by Gregory and Stedman (1958).

The Andersen sampler

This instrument (see Fig. 5) (Andersen, 1958) is a size grading impactor developed from the "sieve device" of du Buy and Crisp (1944). Air is drawn through a circular orifice and then through a succession of six circular plates, each perforated with 400 holes through which particles are impacted on to sterile medium in Petri dishes. The succession of plates has progressively smaller holes so that the largest particles are impacted into the first dish and the smallest into the sixth.

Air is sampled at 28·3 litres/min., retentivity is reported to be 100% for single bacterial cells, but although wall losses are said to be negligible Gregory (1961) and May (1964) report the impaction of particles on to the front of the first sieve plate. May also reports the intake cone directs a jet of air at the central zone of the first sieve and particles are impacted preferentially into the centre of the Petri dishes behind the first and second sieves. The fact that the first stage indiscriminately samples all large particles

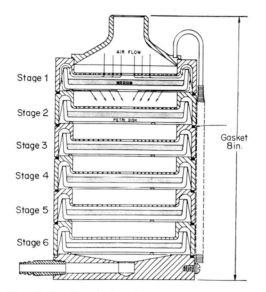

Fig. 5. Sectional elevation Andersen Sampler

Stage No., Jet size, jet velocity; Stage 1, 0·0465 in. dia., 3·54 ft/sec; stage 2, 0·0360 in. dia., 5·89 ft/sec; stage 3, 0·0280 in. dia., 9·74 ft/sec; stage 4, 0·0210 in. dia., 17·31 ft/sec; stage 5, 0·0135 in. dia., 41·92 ft/sec; stage 6, 0·0100 in. dia., 76·4 ft/sec.

down to those of 12 μ dia. diminishes its usefulness as a size grading device, since in occupied environments Noble *et al.* (1963) have shown a large proportion of pathogenic microbes are borne on particles having an aerodynamic diameter greater than 15 μ. A modification described by Lidwell and Noble (1965) comprises a pre-stage with a cut off for particles of 19 μ dia. Modifications described by May (1964) included re-designing the pattern of the hole in the top stages and the addition of another stage to extend the range of the instrument.

The Anderson sampler appears to be the only size grading device sampling on to solid culture media which is commercially available.* Plastic Petri

* Anderson Samplers and Consulting Service, 1074 Ash Avenue, Provo, Utah, U.S.A.

dishes should not be used in this apparatus since Anderson reports they consistently trapped 20% less than glass ones. The instrument is best used without the funnel shaped entry cone. Outdoors it should be operated on its side with the top sieve facing the wind. May suggests the broad multi-zone suction area might be a reasonable alternative to isokinetic sampling for particles up to 20 μm dia.

Subject to the limitations described, the use of the Andersen sampler is illustrated in studies on inhalation Anthrax by Dahlgren *et al.* (1960), the effectiveness of surgical masks by Greene and Vesley (1962), the spore content of mouldy hay by Gregory and Lacey (1963) and the liberation of organisms from textiles by Rubbo and Saunders (1963).

E. The rotorod

This instrument (Perkins, 1957) relies upon the efficiency with which small air-borne particles are deposited on narrow cylinders orientated at right angles to high velocity winds (Gregory, 1951). It consists of a pair of thin brass rods of square cross section which are whirled at a constant speed by a battery operated motor. The leading edge of each rod carries a strip of Sellotape smeared with glycerine jelly; after exposure the strip is removed, cut into four pieces and mounted in glycerine jelly beneath a cover glass on a slide. Harrington *et al.* (1959) compared a modified rotorod with membrane filters and reported its minimum collection efficiency to be 85% for particles having an equivalent diameter of 12 μ. Carter (1960), however, tested an English model against an isokinetically operated Cascade Impactor in a wind tunnel and found with *Lycopodium* spores (30 μ dia.) trapping efficiency varied between 50 and 90% according to wind speed. When compared with a slit-sampler under the relatively still air conditions of this laboratory its efficiency for *Cladosporium* was less than 50%.

Since the rotorod samples a large volume of air, for use in surveys a number would have to be run in sequence, especially in cities with a high level of pollution. The amount of labour required in preparing sampling surfaces and subsequent counting renders this method uneconomic. As a portable pollen trap however, the rotorod is particularly useful since it can be carried in an open basket through public places. Its use is illustrated in the publications of Harrington *et al.* (1959), Davies *et al.* (1963), Gregory and Lacey (1964) and Ogden *et al.* (1967).

IV. IMPINGERS

A. General features, advantages and disadvantages in comparison with impactors

The principle of sampling microbes from air by passing it through a fluid was used as early as 1854 by Thomson and Rainey (1855) who trapped microbes from the air of a cholera ward at St. Thomas's Hospital, London, by aspiration through boiled distilled water. In the modern impinger, air is sucked at high velocity through a narrow orifice which impinges the particles it entrains into the sampling fluid. It may be used to sample any airborne particulates from inert dust to viruses provided the particles are not too small for efficient impingement.

When airborne concentrations are dense, sampling with an impactor can lead to an overloading of the adhesive surface. There is a limit to the number of colonies which may be grown in a Petri dish of agar medium. When rapidly developing moulds such as *Absidia, Botrytis* and *Tricothecium* are trapped, more slowly growing fungi may be overgrown or completely inhibited. With the impinger, however, extreme ranges of airborne concentration may be sampled and examined by means of serial dilution and plating out in the manner of Miles and Misra (1938). Dilutions may be simultaneously planted on different media to achieve optimum growth for various microbes and aliquots may be stored in a refrigerator for subsequent examination. An objection to the use of impingers for air sampling has been the labour entailed in preparing the serial dilutions but nowadays, with the range of automatic devices for dispensing sterile volumes of fluid, this is of minor importance. During sampling with an impinger, due to the violent agitation of the fluid clusters of cells are broken up and the number of colonies developing in culture represents the number of viable cells in the sample. On the other hand a colony trapped by an impactor represents a dispersion unit and may have developed from either a single cell or a clump. The orifice in an impinger acts as its own constant flow metering device and once calibrated a flow meter is unnecessary. The all glass impinger may be repeatedly sterilized by either autoclaving or hot air. Its use in the study of pathogenic aerosols is illustrated in the reports of Henderson (1952) Druett *et al.* (1953, 1956) and Brackman *et al.* (1966).

Pathogenic fungi such as *Coccidioides immitis* in which the spores are easily disseminated constitute a hazard for the laboratory worker and should not be handled in a Petri dish in the open laboratory. In air sampling for fungi where *C. immitis* is endemic the impinger has the advantage that the fluid may be plated out under a ventilated hood or diluted into roller tubes (Astell Laboratory Service Co. Ltd., 172 Brownhill Rd., London, S.E.16) closed with rubber caps so that, after incubation, colony counts may be made with confidence in the open laboratory. As a result of evaporation during prolonged impingement as when dealing with microbes present in the air in very low concentration it may be necessary to replenish the fluid. After sampling, the fluid may be filtered through a membrane and the membrane then transferred to an appropriate agar medium. The presence of *Histoplasma capsulatum*, a fungal pathogen of man in the air, has been demonstrated by Ibach *et al.* (1954) who after concentrating the fluid sampled, treated the suspension with antibacterial antibiotics before injecting 1 ml into each of a series of mice. The mice were treated with antibiotics for three days to prevent the development of bacterial infection. After a month the mice were sacrificed and the spleen and lungs homogenized and cultured. Up to 2 ml of fluid from an impinger may be injected into the peritoneum of a mouse.

B. The Porton impinger

The modern instrument described by Rosebury (1947), Henderson (1952) and modified by May and Harper (1957) is illustrated in Fig. 6. It comprises a narrow glass flask with a ground neck into which fits a hollow ground glass stopper. The stopper bears a side arm to which suction is applied and through its apex an intake tube of 8 mm internal diameter runs down to terminate in a jet. The jet is a short length of capillary tubing of 1·1 mm internal bore and this acts as a critical pressure orifice. Druett (1955) reports that when the suction reaches about half an atmosphere the flow in the jet attains "sonic velocity" and increasing the suction cannot increase the jet velocity. Provided the suction is sufficient to produce a depression of mercury greater than 38·1 cm the jet gives a constant flow of 11 litres/min. and gives a very high collection efficiency for particles as small as single bacterial cells 0·5 to 1 μm dia. The flask is long, 16 cm tall, in relation to its internal diameter of 3·18 cm to reduce the likelihood of droplets being sucked out. It is normally used with 10 ml of liquid and the tip of the jet 3 cm above the bottom of the flask. With this height (recommended by May and Harper, 1957) there is little danger of delicate cells being killed from concussion due to impingement on to the bottom of the flask as occurs if the jet dips into the fluid and is only 4 mm off the bottom.

C. The pre-impinger

Because they contain fluid, impingers have to be used in a more or less vertical position and this frequently necessitates a bend in the intake tube. May and Druett (1953) report that in the standard "Porton" impinger 50% of water droplets 12 μ dia. are trapped in the bend of the curving intake tube. Since this is too high a cut off to simulate the upper respiratory tract in man they designed a pre-impinger.

The pre-impinger (see fig. 6) comprises a glass bulb with an internal diameter of 28–29 mm which is half filled with the collecting fluid. The level of the fluid is not critical. The smoothly curving neck has an internal diameter of 8 mm. The intake hole is ground out to 6·5 ± 0·25 mm in the position shown so that the axis of the air jet strikes the centre of the liquid surface at 45°. When connected to a "Porton" impinger operated at 11

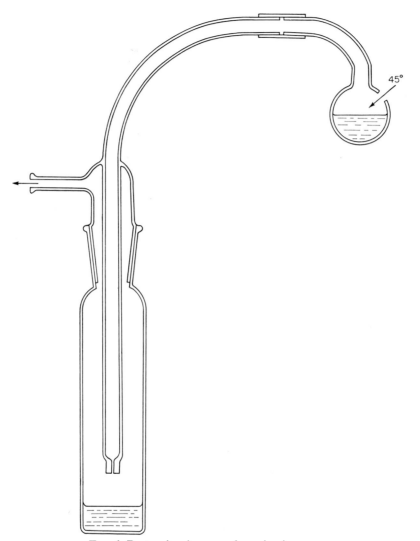

FIG. 6. Porton impinger and pre-impinger.

litres/min., 50% of water droplets 4 μm dia. penetrate the pre-impinger to be trapped in the backing impinger; the range between the smallest particle retained and the largest particle penetrating the pre-impinger at just measurable levels is reported to be 1·9 μm and 8 μm respectively.

Use of the pre-impinger not only divides the aerosol sampled into two fractions on a basis of size, but gives a more reliable result; without it

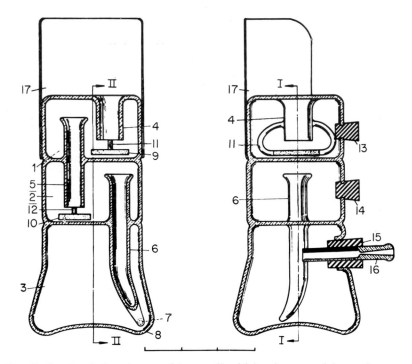

Fig. 7. Sectional elevations multi-stage liquid impinger at right angles to each other in the directions I-I and II-II. 1, 2, 3—chambers or stages; 4—air inlet tube; 5—connection for airflow from stage 1 to stage 2; 6—connection for airflow from stage 2 to stage 3; 7—nozzle; 8—annular well; 9, 10—sintered glass discs held by curved glass rods (11, 12); 13, 14, 15—rubber bungs in access holes to chambers; 16—connector for suction containing critical orifice; 17—hemicylindrical metal shield.

organisms adhering to the wall of the intake tube which are susceptible to desiccation may be lost and there is a possibility of large dry particles bouncing through the tube to give variable samples in the flask.

D. The multistage liquid impinger

The multistage liquid impinger of May (1966) has three superimposed chambers and is constructed of thick walled Pyrex glass (Fig. 7). In the first two chambers which are cylindrical and flat bottomed, air-jets impact on to sintered glass discs which are constantly wetted with sampling fluid. The third chamber is a bowl-shaped swirling impinger. The intake orifice and entry ports to the lower chambers are bell-mouthed and lead to straight sided tubes so that flow is smooth and lamellar. Suction is applied to a tube passing through the side to the centre of the bottom chamber and this incorporates a venturi-shaped critical orifice which provides a constant flow of 55 litres/min.

When air is drawn through the instrument and flow occurs over the first sintered disc, 50% of dry bacterial clusters 6 μ dia. with a specific gravity of about 1·5 are trapped on the first disc. Since the air flows out over the disc and liquid surface it is essential that the sampler is vertical and the liquid nowhere higher than the upper surface of the sinters; otherwise the outflowing film of air will encounter a standing liquid wave which will trap many particles intended for the next stage.

At the second stage where the air is accelerated to a higher velocity the same process of impaction and particle selection occurs and the 50% penetration diameter is 3·3 μ for dry bacterial particles with a specific gravity of 1·5.

The intake tube to the third chamber is smoothly tapered down into an impinging nozzle which is close to the bottom of an annular well. The nozzle lies in a plane tangential to the wall of the well and makes an angle of 45° to the vertical. The tangential component of the jet direction imparts a vigorous swirl to the liquid which ensures that impingement is always on a wetted surface, provided that the liquid volume is not less than 5 ml. The use of liquid swirl is also a feature of the Shipe impinger (Tyler and Shipe 1959). The jet gives no more than sufficient velocity for the capture of most single bacterial cells. Tests by G. J. Harper (cited May, 1966) established 80–90% of single cells of *Bacillus subtilis* and *E. coli* were retained.

The design has a number of advantages over the Porton impinger and pre-impinger. Splashing and frothing are minimized so that a high air flow can be maintained through the third chamber without liquid loss by entrainment. The flow control is downstream from the extract point and the pressure drop on the liquid is only 5 cm of mercury below ambient. The rate of liquid evaporation is therefore much lower than in the Porton impinger and the possibility of freezing when the apparatus is used in cool dry air is also reduced. It has a higher sampling rate, it gives a greater concentration of cells/unit volume of collecting fluid, it is more robust and easily transportable, and prolonged sampling does not result in so large a proportion of organism death as occurs in the Porton impinger.

Fluid suitable for the organisms of interest is pipetted into each chamber through bung holes. The two upper chambers are filled until the liquid surface is just below the upper surface of the sintered disc; usually from 7 to 10 ml are required and this varies between instruments because of the vagaries of glass blowing. The lowest stage holds 10 ml of sampling fluid. When the bungs are in place liquid cannot be spilled or transferred from one chamber to another however much the sampler is turned or inverted.

IV   17

In addition to the standard 55 litre/min. model, 20 and 10 litre/min. models are also available. The first stage of the 20 litre/min. model was designed to collect "fall-out" particles and has a 50% cut off at 10 μ; the second stage has a similar cut off to the pre-impinger (50% at 4 μ) so that its collection resembles upper respiratory retention, whereas the third stage measures lower respiratory retention. It holds 4 ml of sampling fluid/ stage. The small model which is only 8·9 cm high and is operated at 10 litres/ min. has the same particle size range per stage as the large model. It is much more compact and robust than the standard Porton impinger, yet yields more information and has the advantage of the new design. It holds 2 ml of sampling fluid/stage.

For sampling in a cross wind the apparatus is fitted with a shield or baffle. This produces a cushion of stagnant air over the intake and ensures a high capture efficiency (see Fig. 7).

Whilst an impinger with the characteristics of the Porton standard can be improvised or constructed in a laboratory with simple facilities, the multistage impinger cannot. It is manufactured by A. W. Dixon and Co., Annerley Station Road, London, S.W.20. It is a little more expensive than the Porton impinger pre-impinger combination (according to size from 22 to 30 U.S. dollars) but in view of its advantages it is highly commended.

The development of size grading devices suitable for studying airborne microbes stems in part from studies on dust concentrations in relation to pneumoconiosis and to the possible use of aerosols in chemical and biological warfare. It has been shown by Druett et al. (1953, 1956) that the infectivity of microbes such as Anthrax and *Pasteurella* for the guinea pig increases as particle diameters fall below 3·5 μ. Consequently the importance of alveolar penetration in relation to systemic infection with microbes may be somewhat over-emphasized. Noble et al. (1963) have shown a continuous distribution in sizes of airborne particles carrying bacteria and that the most common lie in the 8–16 μ range. Williams (1967) observes that apart from Tubercle, which is unable to establish itself in the upper respiratory tract, most diseases of respiratory entry result from deposition and infection in the upper respiratory tract with subsequent spread to other parts of the tract. A 12 μ particle has a free falling velocity in turbulent air of only 36·576 cm/sec (Williams, 1967) and it is the larger particles, the aggregates or microbes carried on "rafts" which are most likely to be important in the infection of wounds. In view of these features the multi-stage impinger which has a sampling rate of 20 litres/min. and which was designed to collect "fall-out" particles, with a cut-off of 50% at 10 μ, would appear to be the most useful size for most purposes in air hygiene.

V. "FALL OUT" OR SEDIMENTARY SAMPLING

When these methods are used outdoors the results are qualitative and usually expressed as the numbers deposited on an exposed surface of given area in a standard period of time. The so-called "gravity slide" and "settle plate" techniques have been amongst the most widely used methods for studying the microbial content of the atmosphere.

A. The "gravity slide"

Slides smeared with glycerine jelly, petroleum jelly or silicone grease have usually been exposed horizontally, adhesive surface upwards in a shelter such as that of Hyde and Williams (1945), so that although protected from the rain they are freely exposed to the air. The method was used by Blackley (1873) who first demonstrated that grass pollen was the cause of summer hay fever and since then has been used in many surveys on the pollen content of the air, e.g., Hyde and Williams (1945), Durham (1949) Dua and Shivpuri (1962) and Ordman (1963a). Sticky slides exposed in positions comparable to leaves were used by Ward (1882) to trap the spores of coffee rust from the air in Ceylon, and although the presence of fungal spores such as those of *Alternaria* has been noted in many surveys the method has been little used by mycologists.

Durham (1944) in a study of airborne ragweed pollen and *Alternaria* spores found inconsistencies in the results obtained when volumetric and "gravity slide" samples were taken simultaneously. Since the results he obtained with two volumetric devices, those of Hawes *et al.* (1942) and Keitt and Jones (1926) were in close agreement, he concluded the "gravity slide" method was inaccurate. He exposed slides in various positions with respect to the wind direction and discovered that a horizontal slide with its adhesive surface facing downwards caught 50% as much ragweed pollen (diameter 16 to 24 μ according to species) as a normal "gravity slide". He concluded from this that under certain conditions impaction may play a greater part than gravity in deposition.

Gregory (1950) in wind tunnel tests showed that anomalous values were obtained with horizontal glass slides. At wind speeds approaching 10 metres/ sec. as many *Lycopodium* spores were deposited on the lower surface of glass slides as on the upper. He concluded that sedimentation under gravity must play a minor part except at low wind speeds and that the efficacy of a slide trap varies with conditions, sometimes widely enough to be misleading.

Clearly, although the "gravity slide" method has provided much qualitative information in the past, quantitative sampling methods are to be preferred and it should be used only for preliminary survey work when no other method is available.

B. The settle plate

The method in which a Petri dish of agar medium is horizontally exposed to the air for a standard interval of time has been very widely used to study the fungal content of the air outdoors and the concentrations of bacteria in the air indoors. It has been criticized by du Buy *et al.* (1945) and Bourdillon and Lidwell (1948) from theoretical considerations suggested that Petri dishes used alone are of little value for particles smaller than 30 μm dia. A more recent criticism by Gregory (1952) is that surface traps are highly selective and favour the heavier spores, and Gregory and Stedman (1953) using *Lycopodium* spores in wind tunnel tests found both horizontally exposed slides and Petri dishes had low trapping efficiencies; the rim of the Petri dish affected the pattern of deposition at all wind speeds and trapping efficiency was high only at wind speeds between 1·1 and 1·7 metres/sec.

Use of the horizontally exposed Petri dish should be restricted to sampling from relatively still air when the size of the dispersion unit, i.e., the free falling velocity of the particle to be sampled is known. From the number of colonies of *Staphylococci* developing in exposed Petri dishes of serum agar, Williams (1966) calculated concentrations in the air of hospital wards from the settling rate of the particles carrying *Staphylococci* which have a mean diameter of 14 μ. Also, Gregory *et al.* (1959) and Davies (1961) have used horizontally exposed Petri dishes containing grease and oil mixtures to trap water droplets carrying fungal spores and the technique is suitable for experimental studies such as these.

Settle plates have been used to study the fungal content of the air outdoors by numerous workers including Ambler and Vernon (1951), Ainsworth (1952), Dye and Vernon (1952), Richards (1956), van der Werff (1958) and Ordman (1963b). Indoors it has been used to trap fungi by Richards (1954), and bacteria by van den Ende *et al.* (1940), Hare and Thomas (1956), Hare and Ridley (1958), Anderson and Sheppard (1959) and Williams (1966).

VI. OTHER METHODS OF AIR SAMPLING

A. Filters

Although the sampling of airborne microbes by aspiration through filters has been used since Pasteur (1861) and Petri (1888) and membrane filters which allow flow rates adequate for air sampling are available, the method is open to the objection that sampling is anisokinetic and as material is deposited the porosity of the filter changes. With particulate or fibrous filters the efficiency with which the trapped microbes may be recovered must also be considered, the membrane filter is simply placed on an appropriate medium and incubated, the alginate filter of Richards (1955) is dissolved and plated out after appropriate dilution. The butter muslin filter used by Rishbeth (1959) to detect the spores of fungal pathogens of conifers in the air of forests is simply planted on freshly cut slices of pine. In aerobiological studies the most useful application of membrane filters appears to lie in concentrating the particulates entrapped in impinger fluid and separating virus components from the other constituents of the microbiota trapped from the air.

B. The ram jet impactor for sampling from aircraft

Due to the speed of the slipstream, sampling from a fast flying aircraft by means of conventional impactors and impingers is impracticable since small deviations from isokinetic flow can cause high intake errors. The use of straight-tube ram jet impactors is recommended and their design is discussed by May (1967) who reports that a tube 1·4 cm dia. in an air stream of 200 m.p.h. would sample isokinetically at 800 litres/min. and trap particles as small as 2 μm.

C. Cyclones

The cylcone dust collectors that are used in industry have a high through put, a small pressure drop, and only retain large particles. The higher pressure drop necessary to collect small particles from air is a disadvantage for most aerobiological purposes and less demanding suction sources are adequate for the convenient techniques described. Powell (cited May, 1967) reports that a cyclone with an internal cylinder diameter of 12·5 mm and a 6 mm square entry to the volute retains all particles down to about 2 μm with suction at 75 litres/min and a pressure drop of 75 mm Hg. The use of a small cyclone to collect rust and smut spores from infected plants is described by Gregory (1961).

D. Large volume air samplers

The LVS/10K large volume air sampler (Fig. 8) designed and manufactured by Litton Systems Inc. of 2003 East Hennepin Avenue, Minneapolis, Minnesota 55413, is extremely expensive but unique in its sampling rate of between 2 and 11 cubic metres/min. The intake aperture is an upwards facing funnel shaped duct of 9 in. dia. Particles in the air stream entering the sampler pass a corona discharge where they are negatively charged and then directed on to a positively charged collector plate. A continuous flow of liquid carries the particles from the collection plate into a reservoir. Collection efficiency is over 40% for particles as small as 0·1 μ dia. with an air flow of 10 cubic metres/min. and 70% with a flow of 5·4 cubic metres/min.

The outstanding feature of this apparatus is that it concentrates particulates in the collecting fluid at a ratio of nearly a million to one and provided the corona discharge has no germicidal effects may be suitable for sampling air for microbes present in extremely low numbers. Combinations of antibiotics may be used to prevent the growth of common microbes, and in the search for viruses filters may be used to remove bacteria and fungi. Its use is illustrated in a study of virus aerosols by Gerone *et al.* (1966).

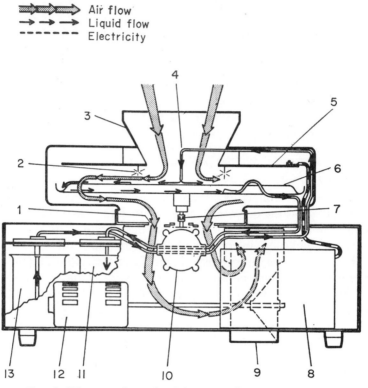

Fig. 8. Diagramatic section Litton LVS/10K Air Sampler.

1. Airflow ports
2. Corona needles
3. Inlet duct
4. Liquid input tube
5. High voltage plate
6. Collection plate
7. Multi-jaw coupling

8. High voltage power supply
9. Blower
10. Pumps
11. Return reservoir
12. Blower motor
13. Fluid reservoir

In a Multi-Slit Large-Volume Air Sampler described by Buchanan *et al.* (1968) air is drawn at 500 litres/min. through eight radial slits. The entrained particles impinge into a film of culture medium which flows over a rotating disc into an effluent container. Organisms are concentrated from large volumes of air into a small quantity of liquid at about 100,000 : 1. In comparison with an all glass impinger, the collection efficiency of this apparatus was reported to be 82% and 78% respectively for test aerosols of *B. subtilis* var *niger* and *S. marcescens*.

E. Biological air-samplers

A number of workers have made use of the animal lung and the leaves of plants as air-samplers.

The presence of dermatophyte fungi in the air of a cave was demonstrated by Lurie and Way (1957) by exposing a group of animals to the air of the cave when dust was disturbed. The lungs of the animals were subsequently removed, homogenized in saline, and portions either cultured on dextrose agar or injected intraperitoneally into mice. The mice in turn were sacrificed after 3 and 6 week periods and *Trichophyton mentagrophytes* and *Microsporum gypseum* were isolated from homogenized liver and spleen tissue. No fungus was isolated in culture from the lungs of the animals exposed in the cave.

Converse and Reed (1966) in an epidemiological study on *Coccidioidomycosis*, exposed 34 monkeys and 50 dogs in compounds open to the air in the desert near Tucson, Arizona. Although no *C. immitis* was trapped in Petri dishes of mycobiotic agar exposed to the air each day, 5 of the monkeys and 29 of the dogs became infected during the 12 month period of observation.

In a study on the infectiousness of air from a tuberculosis ward, Riley *et al.* (1962) conducted effluent air from the ward to two animal exposure chambers. The air going to the one chamber was untreated, that going to the other was intensely irradiated with U.V. The higher incidences of tuberculosis in the chamber receiving untreated air as compared with controls showed the infectiousness of the air. Air from the ward was continuously sampled by 156 guinea pigs for 2 years and 71 became infected. Since each guinea pig inhales $\frac{1}{3}$ of a cubic foot an hour they calculated the average concentration of the infective particles was 1 in 12,500 cubic feet in the first year and 1 in 4,000 in the second.

Because other microbes occur in the air in such very high concentrations in comparison with that of the tubercle bacillus and it is such a slowly developing organism in culture, it is impracticable to sample air for the tubercle with conventional impactors and impingers.

In aerobiological investigations concerned with distinct races of obligate plant pathogens such as the rusts and *Erysiphe graminis*, the only techniques at present available employ either pathogen free plants, or the use of leaf portions maintained in dilute solutions of benzimidiazole in water as described by Wolfe (1967). After the season when the conidia of *Erysiphe* occurred in the air of London had been determined by means of a Hirst trap, Davies and Wolfe (cited Wolfe, 1967) found that barley seedlings grown under pathogen free conditions and then at the appropriate season exposed to the air on a 25 metre high roof in Central London, all became infected with powdery mildew when an adjacent Hirst trap showed the conidia to be present in the air at peak concentration. It is reasonable to deduce that the conidia recorded by the Hirst trap were those of *Erysiphe graminis* and investigate the physiological race of the *Erysiphe* developing on the barley seedlings. In this experiment the nearest barley crop was over 12 kilometres away.

Powdery mildew and uredospores were trapped from the air by Hermansen *et al.* (1967) by sucking air through a tube packed with leaves detached from barley seedlings. The leaves functioned both as a filter and a substrate; after exposure the leaves were kept at 10°C overnight and then incubated at 15°C in 13 ppm benzimidiazole solution under fluorescent light.

An improvement on these techniques might be to embed leaves from barley seedlings in Petri dishes with tap water agar containing benzimidiazole. The plates could then be exposed to the air in a slit sampler, or used to assay the spores trapped in an impinger.

Crop pathogens such as Black Rust which is from time to time windborne into Britain from the endemic areas of N. Africa will be detected by the crops because although the uredospores are present in the air in extremely low concentrations, below the threshold of detection by the small number of Hirst traps in operation at a given time, the trapping potential of the total area over which the susceptible crops are grown is enormous.

The Litton sampler which appears to have the greatest suction rate of any apparatus available has a maximum capacity of 11 cubic metres a minute; but even the enormous sampling rate of this apparatus must be compared with that of the human population, a normal man at rest respiring at 8·6 litres/min. inhales the microbial content of over 12 cubic metres of air in 24 h, and the sampling rate of the Litton apparatus is but the equivalent of 1100 people engaged in light activities. Clearly, in any densely populated area the susceptible population has a greater sensitivity to airborne pathogens in low concentration than any of the sampling techniques described.

The unresolved problem of air sampling is the rapid detection of the airborne microbes which occur in extremely low concentration, the detection of the almost solitary microbe which if received by a susceptible host leads to the development of an epidemic. For studies in an epidemic, or for sampling the more common constituents of the air-spora outdoors or the microbes commonly present in the enclosed atmospheres of animal occupied places, suitable methods will be found among the techniques described.

VII. USE OF SELECTIVE MEDIA IN AIR SAMPLING

Bacteria tend to be more exacting in their nutritive requirements than fungi and consequently, media selective for genera and even species of bacteria have been developed. Apart from the use of antibiotics such as cycloheximide to inhibit saprophytic moulds in the search for airborne dermatophyte fungi there are no methods selective for groups of fungi other than the utilization of different temperatures for incubation.

In air sampling by impaction two selective media have proven value for studies in air hygiene. Barber and Kuper (1951) have described the identification of *Staphylococcus aureus* by the phosphatase reaction with a nutrient medium containing 5% horse serum and 0·1% phenylthaline phosphate. After incubation exposure to ammonia vapour turns colonies of *S. aureus* red and this medium has been widely used in studies on cross infection. Williams and Hirch (1950) who studied the *Streptococcal* content of air as an index of its pollution from respiratory sources describe a nutrient agar medium containing 5% sucrose, 0·25 mg./100 ml crystal violet and 1·0 mg/100 ml of potassium tellurite. On this medium colonies of *S. salivarius* are readily recognized by their mucoid form, other *Streptococci* may be estimated by sub-culturing a random number and the great majority of the *Micrococci* and *Staphylococci* so common in the air indoors are inhibited. Since they also report this medium underestimates the number of airborne *Streptococci* by 20–40% it is clearly inhibitory for some of the airborne *Streptococci*. The use of selective media which are partially inhibitory will fail to detect marasmic microbes which when airborne were subjected to the debilitating effects of desiccation, ultraviolet illumination, and germicidal pollutants such as that described by Druett and May (1968). The use of optimum conditions for growth with antibiotics to prevent the development of weeds is, therefore, preferable to the use of selective media.

The provision of optimum conditions for growth is particularly important in studies on the viability of microbes trapped from air and may be illustrated with the studies on *Cladosporium* by Davies (1957) and a recent observer. In the earlier study 91% of the conidia sampled were germinated within 44 hours incubation, whilst the highest percentage germination recorded by the other worker was 53% in 72 h. This discrepancy must be associated with the fact that whereas Davies utilized a slit sampler with the plate stationary and sampled the airborne conidia by impaction on to filtered Dox's agar, Harvy sampled by impaction on to silicone jelly and after lightly atomizing the slides with water, incubated them in a humid chamber. When spores are impacted on to a greased slide, a significant area of spore surface is covered with the grease (a greater area than that of the spore which appears to be in contact with the retentive surface on the slide) which greatly reduces the surface area over which moisture is absorbed. With impaction on to

Dox's agar none of the absorptive surface of the spore is occluded, the ingress of moisture is accompanied by an uptake of soluble nutrients including carbohydrates which are immediately available as respiratory substrate and development is more rapid than is the case when only water is made available. In the recent study, observations were discontinued after 72 h incubation because "mycelial development on the slide made scanning difficult", in that of Davies, when observations were discontinued after 44 h it was because the microcolonies were abundantly sporulating and the addition of conidia to those impacted from the air was imminent. Where the ability to germinate is taken as the sole criterion of viability, since even when freshly harvested some spores germinate more quickly than others, the provision of optimum conditions is necessary to encourage the laggards and stimulate rapid development so that the difference between the rate of the slowest and most quickly developing spores is reduced. Moreover, following impaction, the revival of spores seriously debilitated during flight may not be achieved with the provision of less than optimum conditions for their development.

VIII. VISUAL RECOGNITION OF FUNGAL SPORES AND POLLENS

The microscopic appearance of the commoner fungal spores and pollen grains is best learnt from prepared, labelled specimen slides. The mountant used should be that utilized for the deposits trapped from air. In the preparation of specimen slides it should be noted that the conidia produced by fungi in artificial culture may differ in appearance from those sampled from the air. Also, many fungi have more than one spore form, e.g., *Leptosphaeria* (Fig. 11a) has yellow or brown ascospores which are fusiform and septate; its imperfect state *Phoma* produces simple hyaline pycnidiospores. Palynologists frequently employ stains to facilitate the identification of pollens, but in recognizing fungal spores sampled by impaction from the spore clouds which occur naturally outdoors, the colours of the spores are particularly important.

Spore morphology, is to some extent, indicative of its mode of production. Sporangiospores produced within the sporangia of Phycomycetes, ascospores and the spores of Myxomycetes, are formed by the cleavage of protoplasm and have no point of attachment. The typical basidiospore is asymmetric, and has an apiculus which denotes the point of detachment from the basidium upon which it was borne. Similarly conidia frequently show evidence of detachment from the phialide, or other conidia such as the collar in *Scopulariopsis* and the *Aspergillaceae* (Fig. 10a). The conidium of *Botrytis* (Fig. 9c) may be asymmetric, show an apparent apiculus and be mistaken for a basidiospore. However, *Botrytis* is easily recognizable since its pale

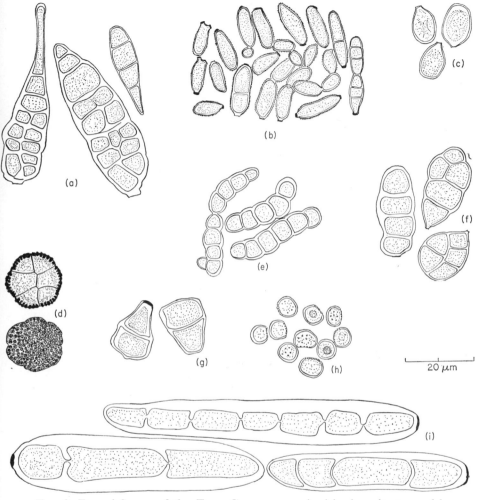

FIG. 9. Fungal Spores of the Form Genera recognized in deposits trapped by impaction.

(a) Alternaria	—	brown conidia	(Porospore)
(b) Cladosporium	—	yellow/brown conidia	(Blastospores)
(c) Botrytis	—	grey conidia	(Blastospore)
(d) Epicoccum	—	reddish-brown conidia	(Aleuriospore)
(e) Torula	—	yellow/brown thallospore	(Porospore)
(f) Stemphylium	—	brown conidia	(Porospore)
(g) Polythrincium	—	brown/grey conidia	(Blastospore)
(h) Ustilago	—	brown chlamydospore	
(i) Helminthosporium	—	yellow/brown conidia	(Porospore)

Conida are classified according to their mode of origin, e.g., Aleuriospore, Blasto-spore, Phialospore.

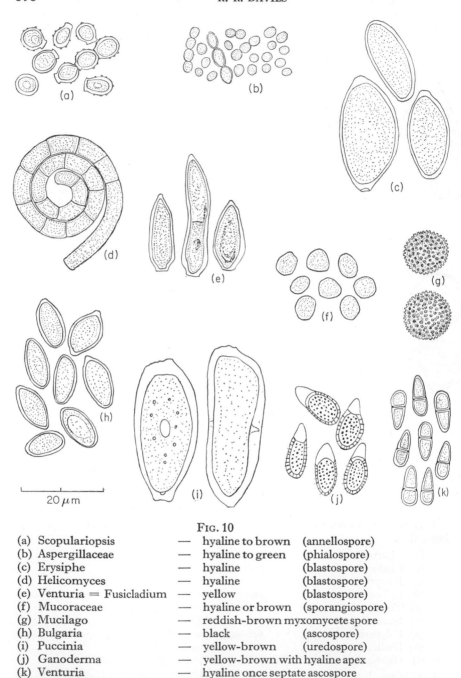

FIG. 10

(a) Scopulariopsis   — hyaline to brown   (annellospore)
(b) Aspergillaceae   — hyaline to green   (phialospore)
(c) Erysiphe   — hyaline   (blastospore)
(d) Helicomyces   — hyaline   (blastospore)
(e) Venturia = Fusicladium   — yellow   (blastospore)
(f) Mucoraceae   — hyaline or brown   (sporangiospore)
(g) Mucilago   — reddish-brown myxomycete spore
(h) Bulgaria   — black   (ascospore)
(i) Puccinia   — yellow-brown   (uredospore)
(j) Ganoderma   — yellow-brown with hyaline apex
(k) Venturia   — hyaline once septate ascospore

Series (a), (b), (c), (d) and (e) are conidiospores, (h) and (k) are ascospores, (j) is a basidiospore and (i) an uredospore.

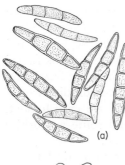

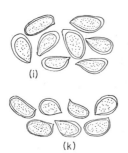
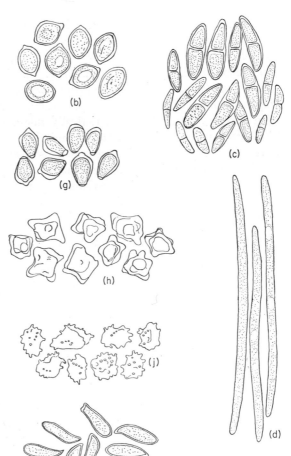

FIG. 11.

(a) Leptosphaeria	—	Fusiform septate ascospores: hyaline to yellow/brown
(b) Chaetomium	—	brown, ovoid ascospores
(c) Mycosphaerella	—	once septate hyaline ascospores
(d) Gaeumannomyces (Ophiobolus)	—	filiform hyaline ascospores
(e) Sporobolomyces	—	hyaline watery basidiospores
(f) Tilletiopsis	—	hyaline hinged conidia (blastospores)
(g) Coprinus	—	brown basidiospores
(h) Nolanea	—	hyaline basidiospores
(i) Hypholoma	—	brown basidiospores
(j) Theleophora	—	brown basidiospores
(k) Armillaria	—	hyaline basidiospores
(l) Boletus	—	yellow basidiospores

grey colour, texture and refractive index are distinct to the experienced observer. The spores of such dematiaceous fungi as *Alternaria* and *Stemphyllium* are identified by their colour, size, shape and septation and *Helminthosporium* by pseudo-septa (Fig. 9i). The most ubiquitous fungal spores in the air are those of the genus *Cladosporium*. These are frequently trapped in large clumps, and the pale yellow bean-shaped conidia, usually with rough walls and characteristically thickened ends, are easy to spot. The larger spores of *Cladosporium* may be septate and some authorities have incorrectly described these spores as *Hormodendrum*.

Ascospores vary in form from the simple hyaline symmetrical to the filiform and fusiform with variable septation. Although most basidiospores may only be grouped according to colour, some have distinct morphological features such as those of *Boletus*, *Ganoderma*, *Nolanea* and *Theleophora*. Some very small watery hyaline basidiospores belong to the genus *Sporobolomyces*, and an aid to their recognition is the fact that they are typically present in moist air when they are accompanied by the hyaline hinged conidia of *Tilletiopsis*. Basidiospores tend to be most common in the air at night, and in periods of wet weather hyaline conidia and ascospores can be particularly abundant. At present there are no simple keys for the identification of fungal spores or monograph which may be recommended, but the coloured paintings of spores in Airborne Microbes by Gregory (1961) are very useful. Drawings of some common fungal spores of the form genera recognized in deposits trapped by impaction are given in Figs. 9, 10, 11.

The identification of pollen grains has received more attention and standard books on the subject such as "Pollen Grains" by Wodehouse (1935) and "An Atlas of Airborne Pollen Grains" by Hyde and Adams (1958) are useful, and their diagnostic keys become invaluable when unknown pollens are encountered during scanning, although the descriptions in these works are those of stained defatted specimens.

The individual pollen grain is usually a single cell bounded by a two layered wall, termed the intine and exine respectively. The surface of the exine can be smooth or sculptured in a variety of ways and in most species there are areas of wall where it is thin or interrupted. These areas if elongated are called furrows, when short they are called pores. The pore or furrow membrane which covers the apertures in the exine may be thickened in places as in *Platanus* (Fig. 12) or in others such as in the grasses, the membrane is thickened over a large part of the opening to form a cap or operculum (see Fig. 13). When the area surrounding a pore is raised above the general surface of the grain it is described as an aspis (Fig. 12). The intine may be thickened and when this occurs beneath a pore or furrow it is termed an oncus (Fig. 12).

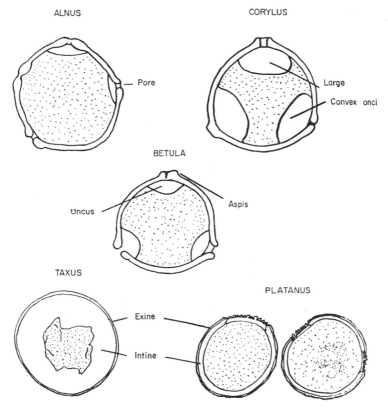

FIG. 12. Outlines of *Alnus, Corylus, Platanus* and *Taxus* pollens seen in optical section.

Such features as size, shape, the number of furrows, pores, the presence of opercula, onci and the sculpturing of the exine constitute the diagnostic criteria used in the identification of pollens. Fortunately, many common ones have distinct characters, and what may be termed the spot features of some of the more common pollens airborne in Britain are given below to illustrate the simpler principles of pollen identification.

Outlines of *Alnus, Corylus, Platanus* and *Taxus* pollens as seen in optical section are illustrated in Fig. 12.

Alnus occurs in the air from January to April and is most abundant in March. Grains are small spheroidal, 21–24 μ, aspidate, and usually have 5 pores and convex onci.

Betula occurs from March to June and is most abundant in April and May. Grains rather smaller and more spherical than those of *Alnus* and *Corylus*, 19–22 μ, aspidate, usually 3 pores and convex onci.

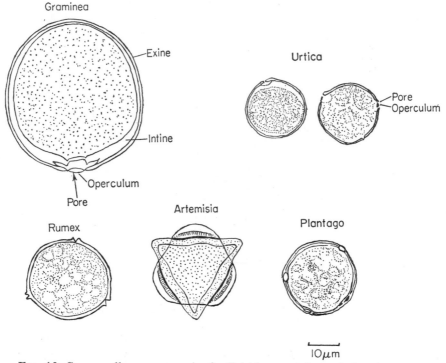

FIG. 13. Some pollens common in the British atmosphere during the summer months.

Corylus occurs from February to April and is most abundant in March. Grains subtriangular in polar view, larger than *Alnus* and *Betula*, 20–25 μ, aspidate, there are always 3 pores and very large convex onci.

Platanus occurs from April to June. Grains spheroidal to ovoid, small 17–19 μ, 3 furrows, membranes studded with small projections, intine thickened, exine pitted.

Taxus occurs from February to May and is usually most abundant in March. Grains small spheroidal, 18–29 μ, no apertures, intine very thick, protoplasm stellate.

Outlines of *Gramineae* (grasses), *Urtica*, *Rumex* and *Plantago* pollens are illustrated in Fig. 13.

Gramineae occur from April to October and are most abundant in June and July. Grains have a single pore with an operculum, spheroidal to ovoid, small to medium size, 27–35 μ, those of cultivated grasses such as wheat being larger, average 50 μ.

Urtica occurs from May to October. Grains small spheroidal, 12–15 μ, 3 or 4 pores with opercula, exine thin and raised around each pore to form an aspis, intine thin forming onci beneath the pores.

Rumex (*R. acetosella*) occurs from April to September. Grains small spheroidal, 21–23 μ, 3 or 4 pores in slit-like furrows, rounded bodies (starch grains) in cytoplasm.

Plantago occurs from May to September. Grains small spheroidal, 21–35 μ, according to species. Small scattered pores, 5–14 in number varying with species.

REFERENCES

Ainsworth, G. C. (1952). *J. gen. Microbiol.*, **7**, 358–361.
Ambler, M. P., and Vernon, T. R. (1951). *N.Z. Jl. Sci. Technol.*, **A33**, 78–80.
Andersen, A. A. (1958). *J. Bact.*, **76**, 471–84.
Anderson, K. F., and Sheppard, R. A. W. (1959). *Lancet i*, 514–515.
Barber, M., and Kuper, S. W. A. (1951). *J. Path. Bact.*, **63**, 65–68.
Blackley, C. H. (1873). "Experimental researches on the causes and nature of Catarrhus Aestivus (hay-fever or hay-asthma)". Balliere, Tindall and Cox, London.
Blowers, R., Mason, G. A., Wallace, K. R., and Walton, M. (1955). *Lancet, ii*, 786–795.
Bourdillon, R. B., and Colebrook, L. (1946) *Lancet, i*, 561–565, 601–605.
Bourdillon, R. B., and Lidwell, O. M. (1948). *Spec. Rep. Ser. med. Res. Coun.*, No. 262, 33–35. HMSO, London.
Bourdillon, R. B., Lidwell, O. M., and Thomas, John C. (1941). *J. Hyg., Camb.*, **41**, 197–224.
Bourdillon, R. B., Lidwell, O. M., and Lovelock, J. E. (1948a). *Spec. Rep. Ser. med. Res. Coun*, No. 262, 224–232. HMSO, London.
Bourdillon, R. B., Lidwell, O. M., and Schuster, E. (1948b). *Spec. Rep. Ser. med. Res. Coun.*, No. 262, 12–19. HMSO, London.
Bourdillon, R. B., Lidwell, O. M., and Thomas, J. C. (1948c). *Spec. Rep. Ser. med. Res. Coun.*, No. 262, 19–22. HMSO, London.
Brackman, P. S., Kaufmann, A. F., and Dalldorf, F. G. (1966). *Bact. Rev.*, **30**, 646–657.
Buchanan, L. M., Decker, H. M., Frisque, D. E., Phillips, C. R., and Dahlgren, C. M. (1968). *Appl. microbiol.*, **16**, 1120–1123.
Carter, M. U. (1960). *Rep. Rothamsted exp. Stn.*, 125.
Converse, J. L., and Reed, E. R. (1966). *Bact. Rev.*, **30**, 678–694.
Dahlgren, C. M., Buchanan, L. M., Decker, H. M., Freed, S. W., Phillips, C. R., and Brackman, P. S. (1960). *Am. J. Hyg.*, **72**, 24–31.
Davies, C. N. (1952). *Br. J. ind. Med.*, **9**, 120–126.
Davies, C. N. (1963). Symposium on Respiratory Physiology, *Br. Coun. med. Bull.*, **19**, 49–52.
Davies, R. R. (1957). *Trans. Br. mycol. Soc.*, **40**, 409–414.
Davies, R. R. (1961). *Nature, Lond.*, **191**, 616–617.
Davies, R. R. (1969). *J. gen. Microbiol.*, **55**, 425–432.
Davies, R. R., and Noble, W. C. (1962). *Lancet, ii*, 1295.
Davies, R. R., Denny, M. J., and Newton, L. M. (1963). *Acta allerg.*, **18**, 131–147.
Druett, H. A. (1942) "The microimpactor: an apparatus for sampling solid and liquid particulate clouds". Porton Report No. 2458, Serial No. 32, pp. 1–12.
Druett, H. A. (1955). *Br. J. ind. Med.*, **12**, 65–70.

Druett, H. A. (1967). *In* "Airborne microbes". 17th Symposium of the Society for General Microbiology (Ed. P. H. Gregory and J. L. Monteith), pp. 165–202. University Press, Cambridge.

Druett, H. A., and May, K. R. (1968). *Nature, Lond.*, **220**, 395–396.

Druett, H. A., Henderson, D. W., Packman, L., and Peacock, S. (1953). *J. Hyg., Camb.*, **51**, 359–371.

Druett, H. A., Robinson, J. M., Henderson, D. W., Packman, L., and Peacock, S. (1956). *J. Hyg., Camb.*, **54**, 37–47.

Drummond, D. G., and Hamlin, M. (1952). *Br. J. ind. Med.*, **9**, 309–311.

Dua, K. L., and Shivpuri, D. N. (1962). *J. Allergy.* **33**, 507–512.

duBuy, H. G., Hollaender, A., and Lackey, M. D. (1945). *Publ. Hlth Rep., Wash.*, Suppl. **184**, 1–39.

duBuy, H. G., and Crisp, L. R. (1944). *Publ. Hlth. Rep., Wash.*, **59**, 829–832.

Duguid, J. P., and Wallace, A. T. (1948). *Lancet, ii*, 845–849.

Durham, O. C. (1944). *J. Allergy*, **15**, 226–235.

Durham, O. C. (1949). *J. Allergy*, **20**, 255–268.

Dye, M. M., and Vernon, T. R. (1952). *N.Z. Jl. Sci. Technol.*, B**34**, 118–127.

Evans, R. G. (1965). *Acta allerg.*, **20**, 197–205.

Gerone, P. J., Couch, R. B., Keefer, G. V., Douglas, R. G., Derrenbacker, E. B., and Knight, V. (1966). *Bact. Rev.*, **30**, 576–584.

Greene, V. W., and Vesley, D. (1962). *J. Bact.*, **83**, 663–667.

Gregory, P. H. (1950). *Nature, Lond.*, **166**, 487.

Gregory, P. H. (1951). *Ann. appl. Biol.*, **38**, 357–376.

Gregory, P. H. (1952). *Trans. Br. mycol. Soc.*, **35**, 1–18.

Gregory, P. H. (1954). *Trans. Br. mycol. Soc.*, **37**, 390–404.

Gregory, P. H. (1961). "The microbiology of the atmosphere". Plant Science Monographs, p. 251. Leonard Hill (Books) London; Interscience Publishers, Inc., New York.

Gregory, P. H., and Hirst, J. M. (1957). *J. gen. Microbiol.*, **17**, 135–152.

Gregory, P. H., and Lacey, M. E. (1963). *J. gen. Microbiol.*, **30**, 75–88.

Gregory, P. H., and Lacey, M. E. (1964). *Trans. Br. mycol. Soc.*, **47**, 25–30.

Gregory, P. H., and Stedman, O. J. (1953). *Ann. appl. Biol.*, **40**, 651–674.

Gregory, P. H., and Stedman, O. J. (1958). *Trans. Br. mycol. Soc.*, **41**, 449–456.

Gregory, P. H., Guthrie, E. J., and Bunce, M. E. (1959). *J. gen. Microbiol.*, **20**, 328–354.

Hare, R., and Thomas, C. G. A. (1956). *Br. med. J.*, **2**, 840–844.

Hare, R., and Ridley, M. (1958). *Br. med. J.*, **1**, 69–731.

Harrington, J. B., Gill, G. C., and Warr, B. R. (1959). *J. Allergy*, **30**, 357–375.

Hawes, R. C., Small, W. S., and Miller, H. (1942). *J. Allergy*, **13**, 474–487.

Henderson, D. W. (1952). *J. Hyg., Camb.*, **50**, 53–68.

Hermansen, J. E., Johansen, B., and Hansen, H. W. (1967). *K. Vet. og. Landbohøisk Aarsskr.* 77–81.

Hirst, J. M. (1952). *Ann. appl. Biol.*, **39**, 257–65.

Hirst, J. M. (1953). *Trans. Br. mycol. Soc.*, **36**, 375–393.

Hirst, J. M., Storey, I. F., Ward, W. C., and Wilcox, H. J. (1955). *Pl. Path.*, **4**, 91–96.

Hyde, H. A. (1959). *J. Allergy*, **30**, 219–234.

Hyde, H. A., and Adams, K. F. (1958) "An atlas of airborne pollen grains", 112 pp. Macmillan, London.

Hyde, H. A., and Williams, D. A. (1945). *New Phytol.*, **44**, 83–94.

Ibach, M. J., Larsh, H. W., and Furcolow, M. L. (1954). *Proc. Soc. exp. Biol. Med.*, 85, 72–74.

Keitt, G. W., and Jones, L. K. (1926). *Res. Bull. agric. Exp. Stn. Univ. Wis.*, 73, 1–104.

Lidwell, O. M. (1959). *J. scient. Instrum.*, 6, 3–8.

Lidwell, O. M., and Noble, W. C. (1965). *J. appl. Bact.*, 28, 280–282.

Litton, LVS/10K. Large volume air sampler. Applied Science Division, Litton Systems Inc., 2003 East Hennepin Avenue, Minneapolis, Minnesota 55413.

Lurie, H. I., and Way, M. (1957). *Mycologia*, 49, 178–180.

Maunsell, K. (1954). *Prog. Allergy*, 4, 457–520.

May, K. R. (1945). *J. scient Instrum.*, 22, 187–95.

May, K. R. (1964). *Appl. Microbiol.*, 12, 37–43.

May, K. R. (1966). *Bact. Rev.*, 30, 559–570.

May, K. R. (1967). *In* "Airborne Microbes", pp. 60–80. The 17th Symposium of the Society for General Microbiology (Eds.. P. H. Gregory and J. L. Monteith). University Press, Cambridge.

May, K. R., and Druett, H. A. (1953). *Br. J. ind. Med.*, 10, 142–151.

May, K. R., and Harper, G. J. (1957). *Br. J. ind. Med.*, 14, 287–297.

Miles, A. A., and Misra, S. S. (1938). *J. Hyg.*, *Camb.*, 38, 732–748.

Noble, W. C., and Clayton, Y. M. (1963). *J. gen. Microbiol.*, 32, 397–402.

Noble, W. C., Lidwell, O. M., and Kingston, D. (1963). *J. Hyg.*, *Camb.*, 61, 385–391.

Ogden, E. C., Raynor, G. S., and Hayes, J. U. (1967). *N.Y. St. Mus. Sci. Service*, Prog. Rep. No. 7. pp. 1–25.

Ordman, D. (1963a). *S. Afr. med. J.*, 37, 321–325.

Ordman, D. (1963b). *S. Afr. med. J.*, 37, 325–328.

Pady, S. M., and Kapica, L. (1953). *Can. J. Bot.*, 31, 309–323.

Pasteur, L. (1861). *Annls Sci. nat.*, 4e sér., 16, 5–98.

Perkins, W. A. (1957). 2nd Semi-Annual Report, Aerosol Laboratory, Dept. Chemistry and Chemical Engng., Stanford University, 186, 1–66.

Petri, R. J. (1888) *Z. Hyg. Infekt Krankh.*, 3, 1–145.

Richards, M. (1954). *J. Allergy*, 25, 429–439.

Richards, M. (1955). *Nature, Lond.*, 176, 559.

Richards, M. (1956). *Trans. Br. mycol. Soc.*, 39, 431–441.

Riley, R. L., Mills, C. C., O'Grady, F., Sultan, L. U., Wittstadt, F., and Shivpuri, D.N. (1962). *Am. Rev. resp. Dis.*, 85, 511–525.

Rishbeth, J. (1959). *Trans. Br. mycol. Soc.*, 42, 243–60.

Rosebury, T. (1947). "Experimental air-borne infection", 222 pp. Williams and Wilkins, Baltimore.

Rubbo, S. D., and Saunders, Jane (1963). *J. Hyg.*, *Camb.*, 61, 507–513.

Thomson, D., and Rainey, G. (1855). Board of Health Reports on the cholera epidemic of 1854, pp. 119–137.

Tyler, M. E., and Shipe, E. L. (1959). *Appl. Microbiol.*, 7, 337–349.

van den Ende, M., Lush, D., and Edward, D. G. (1940). *Lancet*, *ii*, 133–134.

van der Werff, P. J. (1958). "Mould fungi and bronchial asthma I", 174 pp. Leiden, Kroese.

Ward, H. M. (1882). *J. Linn. Soc.*, 19, 299–335.

Williams, R. E. O. (1966). *Bact. Rev.*, 30, 660–670.

Williams, R. E. O. (1967). *In* "Airborne microbes", pp. 268–285. 17th Symposium of the Society for General Microbiology (Ed. P. H. Gregory and J. L. Monteith). University Press, Cambridge.

Williams, R. E. O., and Hirch, A. (1950). *J. Hyg. Camb.*, **48**, 504–524.

Williams, R. E. O., Lidwell, O. M., and Hirch, A. (1956). *J. Hyg., Camb.*, **54**, 512–523.

Williams, R. E. O., Noble, W. C., Jevons, M. P., Lidwell, O. M., Shooter, R. A., White, R. G., Thom, B. T., and Taylor, G. W. (1962). *Br. med. J.*, **2**, 275–282.

Wodehouse, R. P. (1935). "Pollen grains", 574 pp. New York and London.

Wolfe, M. S., (1967). *Trans. Br. mycol. Soc.*, **50**, 631–640.

CHAPTER XIV

Soil Fungi

GEORGE L. BARRON

Department of Botany, University of Guelph, Guelph, Ontario, Canada

INTRODUCTION

Those who have studied soil fungi either casually or critically cannot help but be impressed by the numbers and kinds of fungi which inhabit this environment. Excluding Basidiomycetes, a conservative estimate indicates that over 400 genera of fungi have been recorded from soils around the world. It is the ultimate responsibility of the soil mycologist not only to record fungi from soil but also to establish their biological role, if any. Assigning biological responsibility to a given organism has not always proved to be an easy task and, in response to the difficulties in studying this complex soil association of living and dead organisms and inorganic matter, a wealth of techniques has been devised.

Methods for analysing soil fungi may be, broadly classified, direct and indirect. With direct techniques we attempt to establish the presence of fungi and their relationship to the environment by direct observation. This can be done most simply by mounting and staining soil particles and studying them under the microscope for the presence of mycelium or hyphal fragments. The preparation of soil mounts may involve considerable disturbance of the material so that the relationship between the fungus and its environment is destroyed. Other techniques therefore involve processing and sectioning soil so that fungi may be studied *in situ* with a minimum of disturbance from the original state. Alternatively, known fungi may be added to soil and their behaviour studied by direct examination following recovery.

Indirect methods of analysis attempt to evaluate the role of a fungus by inference. Quantitative and qualitative estimates of fungi present in a given soil sample are made by processing the soil in such a way as to recover the fungi it contains. No single method recovers more than a small fraction of the spectrum of fungi present and to study a sample adequately may involve processing by a variety of techniques.

The techniques outlined below are by no means exhaustive. They do serve, however, to illustrate the diversity of the methods used in attacking the difficult and demanding problem of soil mycology. Evaluating these techniques and their results can prove an interesting and challenging exercise!

I. DIRECT EXAMINATION

A. Microscopic examination

Perhaps the simplest method of demonstrating the presence of fungi in soil is by direct microscopic examination of soil particles for the presence of hyphae. Such a method was originally described by Conn (1918) using dry mounts but without much success. Later, Conn (1922) devised an improved technique using wet mounts as follows—

A crumb of soil weighing 10 mg or less is placed on a slide and mixed with two or three drops of water. A small glass rod is dipped into a saturated aqueous solution of methylene-blue and introduced into the soil-water mixture. After a thorough mixing, large particles are removed with forceps and a coverslip placed on top. Overstaining may be compensated by adding water to one side of the slide and drawing it through by touching the other side with filter paper.

B. Impression slide

This technique was developed independently by Rossi (1928) and Cholodny (1930) and is frequently referred to as the Rossi–Cholodny slide technique. Rossi demonstrated that by pressing a clean glass slide against an exposed soil surface, an impression of the soil microflora could be obtained. Following drying, gentle heating, and staining with erythrosin, this impression could be suitably examined microscopically. Cholodny (1930) showed that if slides were left in the soil for intervals of days or even weeks then the behaviour of the microflora could be followed. Because of its simplicity and elegance the Rossi–Cholodny technique has proved valuable both as a teaching and research tool.

A slit is made in the soil with a knife or sharpened steel plate and a clean glass slide inserted into the slit. The soil is pressed gently against the back of the slide to bring it into close contact with the exposed face. The slide is left in position for selected periods and then carefully removed by drawing it away from the test surface at right angles. The outside face of the slide is wiped clean and the slide tapped gently to remove larger soil particles from the facing side. The slide is air-dried and heated gently over a low flame. Subsequent to this the slide may be suitably stained.

Conn (1932) suggested a stain made up of 1·0 g rose bengal, 0·01 g calcium chloride, and 100 ml of 5% aqueous phenol. He recommended flooding the slides with the stain while they were laid flat above a steam bath for one min.

Conn (1932) drew enthusiastic attention to the Cholodny technique and employed it to study the behaviour of micro-organisms in glass tumblers in laboratory conditions. He pointed out its value for studying behaviour in different soils and how changes in the microflora could be followed using this technique. Effects of changes in the water content or supplementing with mineral salts were brought out very distinctly using buried slides. As pointed out by Conn, a weakness in the technique is that several slides from the same soil may be quite different in appearance. Again, identification of the organisms concerned is not possible other than into broad groups, i.e., bacteria, Actinomycetes, Phycomycetes, etc.

A modified impression slide was used by Brown (1958) to estimate fungus mycelium in sand dune soils.

Nitrocellulose is thinned to a convenient consistency with amyl acetate and mixed with castor oil (5%) to reduce brittleness. The adhesive is spread over the central portion of a clean slide with a brush. The slide is then pressed lightly for 20 sec against the exposed soil surface. Excess material is removed by gently tapping the slide when the preparation is dry. The soil films are then stained for 1 h in phenolic aniline blue (see Jones and Mollison, 1948), rinsed in distilled water and dried.

Brown (1958) examined dry mounts by reflected and transmitted light using an 8 mm metallurgical objective and a compensating eyepiece. Using microquadrats ($55 \times 55 \mu$), Brown was able to express her results quantitatively by recording presence or absence of mycelium in 200 random microquadrats on each of 5 slides.

Warcup (1962) considers that there is an essential difference between soil impression slides and Rossi–Cholodny slides. He regards impression slides as indicating the situation at the time of examining the soil. Rossi–Cholodny slides on the other hand provide a surface for microbial growth after the soil has been disturbed. Warcup points out that there is strong circumstantial evidence that fungal growth on buried slides is strongly influenced by the disturbance of the soil in burying the slides.

C. Soil sectioning

1. *Resin impregnation*

Direct microscopic examination of soil as suggested by Conn (1922) has the advantage of simplicity but is of limited value because of the considerable disturbance of the soil during processing. The method tells us little of the relationships of the fungus mycelium to other soil components. To study fungus mycelium *in situ* requires processing of the soil with a minimum of physical and biological disturbance. To solve this problem, soil sectioning techniques have been developed by Alexander and Jackson (1955), Burges and Nicholas (1961), Nicholas *et al.* (1965) and others. As described by Nicholas *et al.* the technique is as follows—

Freshly collected soil samples are frozen rapidly in liquid nitrogen and subsequently stored in a deep freeze until required. Smaller samples are cut out by hacksaw from the centre of the large field samples. Single blocks ($2 \cdot 5 \times 1 \cdot 0 \times 1 \cdot 0$ cm) are cut from the field samples.

The small soil samples are thoroughly freeze-dried and then immersed in a Marco resin mixture (Alexander and Jackson, 1955). This mixture consists of Marco resin S.B. 28/c together with associated monomer C, catalyst paste H, and accelerator E. The rate of setting of the resin is controlled by altering the proportions of the various constituents in the mixture. A mixture containing 80 ml resin, 16 ml monomer, 1 g paste and 3 ml accelerator mixed together in that order formed the necessary rock-hard solid within 12 h at room temperature.

Freshly prepared resin is poured into the compartments of a polythene ice-cube tray and the freeze-dried soil samples carefully added. The ice-cube tray is then placed in a vacuum desiccator and the soil impregnated under reduced pressure (approximately 30 min at 200 mm Hg). This 30 min period is followed by a second 30 min period during which the pressure is returned to normal. The ice-cube tray is removed from the desiccator and the resin impregnated soil samples left to harden at room temperature.

Standard geological techniques of cutting, grinding, and polishing are then applied to the resin impregnated soil samples in order to obtain sections 50 μ thick which are finally mounted with Canada balsam.

2. Agar-gel impregnation

The technique described previously uses geological procedures for cutting, grinding, and polishing to prepare the finished product. Minderman (1956) has described an alternative technique for soil sectioning which allows the preparation of much thinner sections and the use of a standard microtome. The method involves dissolving out silica particles with hydrofluoric acid and impregnating with agar gel.

Soil samples are taken in the field by means of a steel cylinder 4·5 cm in length and 4·5 cm in diameter. The cylinder containing the sample is transferred to the laboratory in a plastic bag and held at $-10°C$ for at least 12 h. The temperature of the sample is increased slightly and a small cylindrical sample removed by means of a steel sampler as used by Haarløv and Weis-Fogh (1953). When the frozen soil is extruded from the cylinder it is cut into disks 4·5 mm thick which fit into holes in blocks of plastic. Each block containing a disk of frozen soil is then wrapped in a piece of copper gauze to prevent soil loss and the sample slowly immersed in a 5% solution of gelatin at 35°C. The treatment is followed by similar treatments in 10, 15, and 20% gelatin with 1–2 h at each immersion. After impregnation with gelatin the beaker containing the gelatin is cooled and the blocks cut out and placed in 10% formalin and fixed for 7 days with one change of formalin during the process. The blocks are unwrapped and the disks removed and cut in half. Both halves are placed in a perforated plastic tube and immersed in hydrofluoric acid (50%) and stored at 15°C. After one to seven days, depending on the nature of the soil, all the silica particles will be dissolved and the samples are washed in water. The washed half-blocks are again immersed in 20% gelatin, cooled, cut out, and mounted for sectioning. Before sectioning the cubes are immersed in formalin once more to fix the fresh gelatin. This is followed by immersion in 80–90% methyl alcohol for 24 h to make the gelatin hard enough to cut. Sections are cut at 7·5–10 μm. Minderman recommends Johansen's quadruple stain of Safranin, Methylviolet, Fast green, and Orange G.

D. Soil fungistasis

In the soil we have a dynamic and competitive environment in which populations of micro-organisms shift according to a number of physical and biological factors. Fluctuations in moisture or temperature, changes in the type or amount of organic material, decrease or increase of higher plant flora and the like may affect substantially the flora and fauna of the soil. Superimposed on physical and chemical changes we have the stimulations or inhibitions of micro-organisms amongst themselves.

It is generally found that the addition of spores of one or other fungus to unsterilized soil will have very little effect on the population of that fungus in the soil. If spores are recovered and examined it is frequently found that they have failed to germinate yet may still do so in distilled water or tap water. The influence of the soil in prevention of germination and growth or successful establishment of a fungus in unsterilized soil is referred to as soil fungistasis.

The most easily demonstrable facet of soil fungistasis is its effect on spore germination and most studies to date attempting to evaluate this phenomenon have used spores.

Basically the technique consists of adding spores to soil, leaving them for varying periods of time and then recovering them to study germination and viability. Techniques therefore have been designed to add and recover spores with the least amount of difficulty. It is also important to disturb the soil as little as possible before and during experimentation.

1. *Folded cellulose film*

A simple but useful technique was described by Dobbs and Hinson (1953) who demonstrated a widespread fungistasis against the spores of *Penicillium frequentans* using cellulose film.

Thin commercial cellulose film (about 20 μ thick) is cut into 2-in. squares, boiled to remove surface coatings, lightly autoclaved and dusted with test spores of the fungus while still slightly damp. In the case of slimy spores it is recommended that the spores be applied as a spray and allowed almost to dry. The square is then folded with the spores inside and partially buried in moist soil and pressed firmly to ensure intimate contact with the soil surface.

The normal test period suggested by Dobbs and Hinson was four days at room temperature but some folds were reported to show no germination after five weeks. In prolonged experiments the results may be obscured by the biological activity of the soil micro-organisms on the cellulose.

2. *Buried membrane filter*

Buried membrane filters have been used by several workers to study the activity of fungi in soil. In most cases the filter is placed between the soil and the fungus material as a buffer. Adams (1967) described a technique in which the fungus spores on the filter are placed in direct contact with the soil.

The soil under study is placed in porcelain crucibles in moist chambers. One ml spore suspension (10^5 spores/ml) is washed by filtration, resuspended, and added to a 25-mm type-HA millipore membrane filter, mounted over a filter holder and the liquid removed by vacuum filtration. A slit is made in the soil with a spatula, the filter is gently inserted, and the soil pressed against the filter. Filters marked with a grid on one side are preferred for identification of the spore bearing surface.

The filters are removed, placed on microscope slides, covered with lactophenol trypan-blue (liquid phenol, 11·2 ml; glycerine, 10 ml; lactic acid, 10 ml; distilled water, 8·8 ml; trypan-blue, 0·02 gm), and steamed 3 min over a water bath at 70°C. The filters are then placed in the lower half of a filter funnel and washed by vacuum filtration with clear lactophenol, then glycerine, and finally mounted in glycerine on slides.

Adams noted that the percentage of spores recovered in tests ranged from 55–94 with *Thielaviopsis basicola* endospores. The technique was also used to study temperature and moisture effects on spore germination in autoclaved soil and formation of chlamydospores by macroconidia and microconidia of *Fusarium solani*.

3. *Buried slide*

One of the difficulties of adding spores to soil is recovering them after convenient periods of exposure to the soil environment. A technique was described by Chinn (1953) which is simple, effective and allows both quantitative and qualitative estimates to be made of the behaviour of fungi in soil.

A spore suspension is made with 10 ml of water plus detergent using a test-tube slant culture of the selected fungus. The spores are filtered through fine copper screen to remove mycelial fragments and then mixed with 200 ml of sterile 1% water agar at 45°C. The contents are mixed and maintained at 45°C in a water bath.

Soil is sieved and held at about 50% water-holding capacity in quarter-gallon crocks half-filled with soil. Clean sterile slides are dipped into the agar spore suspension inserted vertically about one-quarter in. into the soil and allowed to set. After solidification soil is added to cover the slides. Slides are removed at selected intervals and rinsed with water. One side is scraped with a spatula the other is stained in dilute cotton blue in lactophenol for 10 min.

This technique might be improved by centrifuging and resuspending to remove the possibility of introduction of substantial quantities of nutrient material from the spore suspension.

4. *Agar disk*

The presence of a fungistatic effect in Nigerian soils was demonstrated by Jackson (1958) using a technique in which disks of agar were placed on filter paper in contact with soil.

Forty to 60 g of sieved soil are placed in a clean 85 mm Petri dish and brought to approximately 60% water-holding capacity by the addition of distilled water. Agar disks are prepared by pouring 9 ml of melted distilled-water agar into a flat-bottomed 85 mm Petri dish standing on a level surface and giving a layer of agar just under 1·5 mm in thickness. The disks are cut from the agar with a flamed 7·5 mm diameter cork-borer and removed on the tip of a flamed scalpel. Four 1 cm squares of Whatman's No. 1 filter paper are placed on the surface of the soil in each Petri dish and an agar disk placed on each square. Immediately or after a few hours incubation to allow diffusion of inhibitory substances through the agar disk, the surface of each disk is inoculated with a drop of spore suspension in distilled water on the end of a glass rod. Controls are similarly prepared but the agar disks are placed on filter paper saturated with distilled water in dishes containing no soil.

After incubation the disks are removed to microscope slides and covered with a drop of lactophenol and a coverslip and examined.

Spores were considered germinated by Jackson when they produced a germ tube of one spore diameter or more in length. Jackson preferred the agar disks to cellulose film laid on the soil as in the latter there was some-times uneven contact between the film and the soil giving unsatisfactory results. Jackson found Chinn's slide technique very useful but considered it less convenient to handle than the agar disk method and found results using buried slides more difficult to record quantitatively in routine tests. He considered Chinn's method more sensitive than the agar disk method.

5. *Smooth surface inoculation*

Lingappa and Lockwood (1963) point out that most assessments of the general fungistasis of soils have been made by placing spores on water permeable media such as agar or cellophane. These workers regard fungi-stasis as due to toxicity generated at the contact surface of the agar or cellophane with the soil as a result of enhanced metabolic activity of soil micro-organisms in this area. They suggested an alternative direct method for assaying soil fungastasis.

The method involves preparing a very smooth soil surface in a Petri dish, placing spores upon it, and subsequently staining the spores *in situ* before recovering them with a collodion film or agar.

Sifted loam soil is maintained in closed glass jars under glass wool covered with moist paper towels which maintain the soil at 25% moisture. Twenty grams of soil are placed in a small (50 × 15 mm) Petri dish, compressed, and smoothed using a small spatula with the blade bent at a 20° angle. The surface of soil must be compact and smooth. Spore suspensions are applied uniformly over the soil surface in five or six drops calculated to distribute about 400,000 spores over the dish. Spores are similarly placed on a water agar plate as a control.

Following incubation the spores and germ tubes are killed and stained by applying a few drops of aqueous phenolic rose-bengal solution (1 % rose-bengal, 5 % phenol, 0·01 % $CaCl_2$) which is allowed to diffuse into the soil.

Following killing and staining, one or more drops of collodion (1·5 % pyroxylin in 1 : 1 absolute ethanol and ethyl ether) are placed on the soil surface and allowed to spread and dry to a thin film. The film is removed with forceps and placed in a drop of mineral oil (Medicol) on a glass slide and a coverslip placed on top.

Chacko and Lockwood (1966) modified the soil surface technique for quantitative evaluation of soil fungistasis. By diluting natural soil with sterilized soil they progressively diluted the micro-organisms with a simultaneous increase in nutrients and established a dosage response curve for germination of a test fungus. Their results confirmed that fungi differ in sensitivity to soil fungistasis and that fungistasis differs in different soils and decreases with depth.

6. *Soil sandwich*

Jooste (1965) developed a method to study the germination of *Helminthosporium sativum* spores in soil which has application to other studies on soil fungistasis.

The soil is air-dried and sieved through a 1 mm mesh sieve to obtain a uniform maximum particle size. Both halves of a Petri dish are then filled with the soil and the surface of each levelled. To retain a smooth surface the surface of the soil is moistened by means of a spray gun operating at 5 lb/sq in. The amount of water applied will depend on the soil type and the nature of the test. A spore suspension is added to the lid part of the Petri dish. The number of spores must be sufficiently high to permit easy recovery after incubation. A disk of plastic or nylon mesh (approx. 2 mm) with a diameter similar to that of the Petri dish is placed on the other half and pressed down firmly. The purpose of the mesh is to separate the soil surfaces and yet provide contact and air spaces. The Petri dish is then assembled by placing the bottom half (with mesh) upside down on top of the lid seeded with spores and the two parts lightly pressed together to ensure contact between the two soil surfaces. After incubation the two parts are carefully separated and a spore print made with an agar strip. The agar strips can be incubated in a moist chamber to determine the viability of ungerminated spores.

II. INDIRECT EXAMINATION

A. Dilution plate

Of the techniques used to study soil fungi the dilution plate method is perhaps the most widely used and at the same time the most severely criticized. In its simplest form it consists of shaking up a known weight of soil in sterile water, making a dilution series with water blanks and plating out selected dilutions in agar. The method is criticized by soil ecologists as it is strongly biased in favour of the so-called high sporulators and may not represent the active spectrum of the soil fungi. On the other hand for those who are interested in teaching or who wish to demonstrate a wide range of organisms the dilution plate technique is an excellent tool. To the mycological taxonomist whose interests in the ecological significance may be of secondary importance this technique offers a more or less inexhaustible source of interesting fungi. In our studies of soil fungi in Canada we have recovered well over 200 genera of fungi using dilution plate techniques; many of these are of considerable scientific interest albeit of little ecological significance. Again fungi must not be too hastily dismissed by the ecologist just because they produce dry spores and are high sporulators.

Because of its wide acceptance and popularity there are many variations of the dilution plate technique and Johnson *et al.* (1959) gave a good description of the details of the procedure. I have outlined below an alternative technique which has proved quite successful in our own studies. We use weak water agar as the diluting agent as it keeps the dispersed particles in suspension for much longer periods of time. We also use an Osterizor to blend the samples. The Osterizor blade housing fits standard Mason-type preserving jars. The shearing forces of the blender have allowed us to recover large numbers of fungi which have a relatively low sporulating capacity or in which the spores are strongly attached to the parent hyphae.

A 25 g dry-weight equivalent of soil is added to a sterile measuring cylinder and the volume made up to 250 ml with sterile 0·15 % water agar. The soil/agar mixture is then poured into a pint Mason jar and blended for 30–60 sec. Blender blade housing and measuring cylinder are rinsed in water and sterilized by rinsing in 70 % alcohol between samples.

Five ml of this blended primary suspension are pipetted out using a wide-mouth pipette into 45 ml of 0·15% sterile, water-agar blank in a 4 oz medicine jar (narrow-mouth, screw-cap ovals). Five ml samples are transferred through a succession of agar blanks until the desired final dilution is reached. From selected dilutions (usually 1/100, 1/1000, 1/10000) 1 ml aliquots are pipetted into each of three Petri dishes for each selected dilution and 10–15 ml of cooled (45°C) agar medium added. The soil suspension is mixed with the medium by gently swirling.

To prevent excessive growth of bacteria the medium must be either acidified or supplemented with antibiotics. Very often it is advantageous to add an inhibitor to restrict the growth of the fungi to prevent colonies merging and thus facilitate identification. The medium used depends on the interests of the individual but for general purposes Martin's medium (Martin, 1950) has proved very effective.

The dilution plate may be exploited more effectively by combining it with other techniques. It may be used following soil washing and subsequent blending (Watson, 1960). It may also be modified by the use of selective agar (Dickinson and Pugh, 1965) or by use of agar medium with specific inhibitors. It may be employed subsequent to partial sterilization of the soil (Warcup, 1951) for the recovery of Ascomycetes. While we have recommended Martin's medium for general purposes Steiner and Watson (1965) found ionic surfactants to be very effective in evaluations using dilution plate counts. They found the concentration of the surfactants was not critical. Moreover more colonies developed than on Martin's medium and there was greater colony inhibition. The clear nature of the medium aided microscopic identification.

B. Syringe inoculation

As an alternative to the dilution plate technique Rodriguez-Kabana (1967) described a method using a syringe for inoculation of the plates. He considered this technique more rapid than the dilution method and also felt that it compensated for some of the deficiencies of the latter in that it eliminated much of the error due to serial dilution and allowed for a high degree of reproductibility and possible standardization.

A 25 gm equivalent of oven-dried soil is mixed with 225 ml of water in a 500 ml aspirator bottle then agitated on a magnetic stirrer. Samples are withdrawn with a syringe while the suspension is in motion. Individual drops are placed in Petri dishes and mixed with a suitable culture medium. The average weight of soil per drop is determined and colony counts related to dry weight of the soil.

C. Soil plate

The disadvantage of the dilution plate method apart from ecological considerations is that it is time consuming and somewhat tedious if large numbers of samples are being processed. It is also demanding in materials such as sterilized glassware, pipettes, etc. An alternative technique was developed by Warcup (1950) which has the advantages over dilution plates in that it is not so selective and much easier to prepare.

A soil plate is prepared by transferring a small amount of soil into a sterilized Petri dish. About 10 ml of cooled medium is added and the soil particles dispersed throughout the agar. With light soils adequate dispersal may be obtained by shaking and rotating the plate before the agar solidifies. If the soil is dry or has a high proportion of clay then it is preferable to mix the particles with a drop of sterile water on the plate before the agar is added. The amount of soil used varies with the type of soil but 0·005–0·015 g was found by Warcup to give a convenient number of colonies per plate with most soils.

Preparing soil plates as outlined by Warcup has the advantage over the direct addition of soil to previously poured plates. In this latter method a film of water soon surrounds the soil particles and this favours bacterial development even on acid media. Warcup recovered *Pythium*, *Dictyostelium*, and several Basidiomycetes using soil plates. These genera are rarely recovered using dilution plates.

D. Immersion tube

The immersion tube technique was developed by Chesters (1940) to isolate fungi from soil *in situ* on sterile agar not previously exposed to aerial contamination. This would allow the colonization of the agar under natural conditions of soil moisture and temperature.

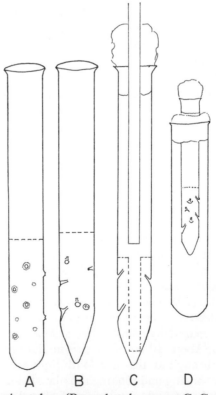

FIG. 1. Immersion tubes. (Reproduced courtesy C. G. C. Chesters.)

Immersion tubes are prepared from hard glass test tubes or from any diameter hard glass tubing. Each tube has several small holes blown in the lower half of its length. These may be flush with the wall or extend inwards as capillaries varying in length and direction as desired. Chesters described several basic types of immersion tubes. Two of these are described below.

Type 1. In this tube (Fig. 1A), nine holes are arranged in a spiral in the lower half. Each hole is about 0·5 mm in diameter and is prepared by drawing a spicule from the wall of the test tube, cutting it flush and reheating till the glass has fused round the raw edge. The diameter of the hole is adjusted with a warm waxed needle to the diameter required.

Type 2. This tube (Fig. 1B) is prepared from a pointed 6 × 3/4 in. hard-glass test-tube. It has six external openings which lead into short tapered internal capillaries made by heating a localized area of the tube wall, drawing a short capillary side tube, cutting this about 2 mm from the tube wall, carefully reheating and when the glass is molten invaginating the capillary with a warm, waxed needle point. These tubes require very careful annealing and the capillaries must be made in rapid succession.

Immersion tubes are sterilized and filled with media while enclosed within a container tube (Fig. 1D). Tubes are filled to just above the highest capillary with cool nutrient agar. If solid plant materials are used these may be packed into the immersion tube while it is enclosed in its container and the whole apparatus sterilized in an autoclave. The tube is immersed in the soil by quickly removing it from its container and pushing it to its required depth in the soil. Each tube is covered by a small specimen tube to keep the cotton plug dry. Colonies which have developed on the medium opposite the capillaries are subcultured.

A modified immersion tube technique was described by Gochenaur (1964) using a smaller size tube and soil in place of agar. Soil provides a well aerated medium whose moisture content rapidly comes into equilibrium with that of the surrounding soil. Immersion tubes containing agar are selective for rapidly growing species which can tolerate a low oxygen tension resulting from sterilization.

A rather simple and easily constructed immersion tube was described by Mueller and Durrell (1957). These workers prepared the immersion tubes from plastic centrifuge tubes which can be drilled quickly and easily with holes of any desired size.

With the aid of a template $\frac{3}{16}$ in. holes in spiral arrangement are bored through the wall of the tube and then countersunk. The tubes are wrapped spirally with Koroseal electrical tape and then filled to within $1\frac{1}{2}$ in. of the top with nutrient agar, plugged with cotton and autoclaved. In the field, holes are punctured through the tape by means of a large needle previously sterilized by means of an alcohol lamp.

E. Screened immersion plate

This technique was developed by Thornton (1952) to isolate fungi on sterile agar with as little disturbance as possible of the soil organic matter and population. The apparatus is illustrated in Fig. 2.

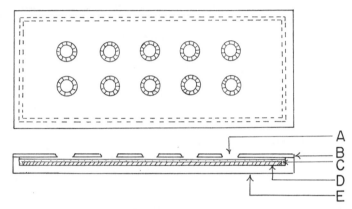

FIG. 2. Screened Immersion Plate. (Reproduced courtesy R. H. Thornton, 1952.)

The water agar is carried on a 10.5×4 cm glass slide (D) which fits closely into a shallow perspex box (E) over which is placed a screen (B) made from perspex 1 mm thick. The screen is punctured by two rows of five holes with each hole being 5 mm in diameter. The screen covers but does not touch the agar film (C). After appropriate sterilization the apparatus is assembled in a large Petri dish and 8 ml of water agar is dispensed over the slide. When the agar is cool the screen is replaced and the two lines of holes are covered by a sterile microscope slide. At the selected site a freshly exposed soil profile is trimmed with a sterile knife-edge and the cover slide removed. The immersed slide in its holder is pressed against the profile surface and held in position by two U-shaped glass clips. Soil removed during digging is replaced against the screened slides. After suitable incubation the screened slides are removed. The incubation time must not be so prolonged as to cause interference between colonies from adjacent holes in the screen.

Isolation of fungi from the slide is effected by inverting the slide over a cavity in a metal carrier under a binocular dissecting microscope and removing inocula with a fine needle.

F. Plate profile

This technique was developed by Anderson and Huber (1965) to compensate for the small area sampled using immersion tube and screened immersion plate techniques.

An autoclavable polypropylene plastic plate ($20 \times 30 \times 1.5$ cm) with 0·5 cm holes (1 cm deep) spaced at 2·5 cm intervals both vertically (7 holes) and horizontally (11 holes) is used in sampling the soil. The plates are cleaned, wrapped in aluminium foil, steam sterilized, and allowed to cool. The holes in the plate are filled aseptically with agar medium. Excess solidified agar is removed from the surface of the plate with a sterile spatula and the holes covered with autoclaved plastic electrician's tape. The plates are wrapped in aluminium foil and placed separately in large envelopes.

Soil profiles are prepared by driving a sharpened steel plate into the soil at right angles. The soil is removed from one side of the plate until the whole plate is exposed. The plate is then slid out of position. A small sterile needle is used to punch holes in the electrician's tape above the holes containing the medium and the plastic plate then placed firmly against the flat surface of the soil profile so that the top row of holes is approximately 2·5 cm below the ground line. The soil removed is then packed firmly against the back of the plastic plate and the aluminium foil placed over the plate to exclude soil and water. Plates are usually exposed five days before removal. On returning to the laboratory the tape is stripped off and each agar plug transferred to the centre of a Petri dish.

G. Hyphal isolation

In 1955 Warcup described a simple technique for recovering and plating hyphae from soil. The method depends on the fact that when a soil suspension is made many of the fungal hyphae remain with the heavier soil particles. Heavier soil particles may be separated out either by sedimentation or by using a sieve. Warcup reports that a 50 μ sieve is very useful.

The residue is examined microscopically for the presence of fungal hyphae and individual hyphae or portions of mycelium are removed by means of fine forceps. When 20–50 mycelial fragments have been transferred to a Petri plate cooled agar is added and the fragments dispersed in the usual manner. The hyphal particles are located and marked to ensure that each hyphal fragment is not associated with soil particles or in any way associated with spores. It is essential to mark hyphal fragments to examine them and ensure their origin.

Warcup recovered a number of fungi rare or absent using dilution plate techniques. He considered hyphal isolation techniques however to be selective for hyphae of wide diameter, mycelia which do not fragment readily, fungi with dark coloured hyphae.

H. Soil washing

In studies on the soil and rhizosphere it has been recognized that dilution plates and to some extent soil plates strongly favour fungi with a high sporulating capacity. Fungi such as *Penicillium*, *Aspergillus*, and *Paecilomyces* are essentially opportunists and an explosion of growth and sporulation in a micro-habitat may mask completely fungi which exist in mycelial forms or which have a low sporulating capacity. It has been adequately demonstrated by Warcup (1955a, b) that the vast majority of colonies on dilution plates are derived from spores. An important requirement in studies of soil and rhizosphere is that any method must distinguish between fungi in the active mycelial form and those that exist strictly as conidia, chlamydospores, sclerotia or some other resting or dormant state. To distinguish between resting and active states, the washing technique has been applied to both soil and plant parts. Some such techniques are simple and others quite sophisticated. Watson (1960) described the following simple technique for washing soil.

One gram equivalent of air-dried soil is placed in 200 ml of sterile water in a flask and agitated thoroughly. The soil-water mixture is allowed to settle for one min with the flask at a 45° angle and the water then decanted. Following a number of such washings the final soil residue is dispersed in water by a Waring blender and soil dilution and plate counts made to determine the fungi present.

Watson found that the percentage of counts of *Fusarium* increased from 2% to 35·8% by the 32nd washing and the number of different genera increased substantially when compared with standard dilution techniques. He found that washed soil proved far superior to unwashed soil in the soil-plate technique.

Harley and Waid (1955) used a washing technique in their studies of the rhizosphere and litter fungi of a beechwood.

Phials with screw tops of 30 ml capacity are employed for washing the samples of plant material. The apical 2 cm of beech mycorrhizas are transferred to the phials in lots of five or ten along with 10 ml of sterile water. Phials are shaken at about 300 beats per min on a shaking platform for two min. At the end of each period the water is decanted into a sterile tube and replaced with a fresh aliquot and the phial shaken again. The process is repeated up to 30 times. Aliquots of the wash water from each period are plated out to establish the fungi in the wash water. Harley and Waid noted that detachable fungal units capable of developing into mycelium are removed in the first few washes with the roots being gradually

cleaned to a low constant level by the 30th washing. They cautioned that
when washing plant material detritus may collect in the bottom of the
phial and that it is good practice to transfer the plant material to a fresh
phial after the first few washes.

Microscopic observations confirmed that colonies from repeatedly
washed roots develop primarily from fungi having a vegetative existence
on the living root surface.

A more elaborate technique for washing soil was described by Parkinson
and Williams (1961).

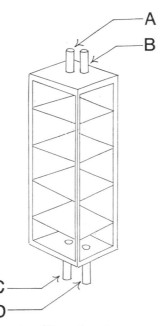

FIG. 3. Soil washing apparatus. (Reproduced courtesy D. Parkinson and S. T.
Williams.)

The apparatus used for soil washing (Fig. 3) consists of a perspex box
measuring $18 \times 6 \times 6$ cm with a removable front fact. Into the box are
fitted stainless steel sieves for the eventual separation of specific particle-
size samples of the soil. The sieves (1 mm, 0·75 mm, 0·2 mm) fit into slots
in the perspex sides of the box and are placed in descending order of mesh
size from top to bottom of the box. At the top of the box tubes, A and B are
inserted for the purpose of introducing sterile water and maintaining
atmospheric pressure respectively. An exit tube (C) at the bottom of the
box allows the emptying of the box of washing water and tube D allows
the inflow of sterile air.

Box and tubing are sterilized by rinsing in alcohol followed by sterile water. Soil is introduced into the upper compartment of the box on the mesh of largest size. The front face of the box is placed in position and clamped to prevent subsequent leakage. Tube A is connected to a reservoir of sterile water, tube B fitted with an air filter, tube C connected to a collecting container for washing water and tube D attached to an air filter and electric pump.

Washing technique

One hundred and fifty ml sterile water is run into the box. Sterile air is then pumped into the box producing vigorous agitation of the soil particles. The washing is continued for the desired interval after which the washing water is drained out. The process may be repeated as often as required.

It was pointed out by Parkinson and Williams that different soil types required different periods of washing dependent on the quantity of spores present and the physical consistency of the soil. Hering (1966) noted that organic layers of woodland soils require a long series of washings to remove most of the spores and that such washings are tedious to carry out manually. He has therefore designed and described an automatic soil-washing apparatus to perform a series of washings.

Regarding soil washing Warcup (1962) brings out an important consideration. He points out that after initial microbial attack decomposing materials become very fragile and may disintegrate into very small fragments together with free vascular tissues so that the path of decomposition is difficult to follow. Warcup properly suggests that much information may be gained by direct examination of soil fragments which may be completely lost by washing or other techniques.

I. Miscellaneous techniques

1. *Partial sterilization*

Many groups of fungi known to occur in soil are seldom or never recovered using the techniques described previously. Using soil partially sterilized by steam however, Warcup (1951) successfully demonstrated that the lower Ascomycetes are widespread in occurrence and abundant in some soils.

Samples of soil weighing 125 gm are placed in glass tumblers in a steamer at 100°C for periods of 2, 4, 6, 8, and 10 min. After the treatment the surface soil is removed to a depth of 1 cm and ten soil plates are prepared from the soil in each tumbler.

We have used a modification of this technique with some success in our own studies. In our method a primary suspension is prepared using an

Osterizor blender as described previously for the dilution plate method. Aliquots of this primary suspension are placed in 8-oz medicine bottles and immersed in a controlled temperature water bath. By varying the temperature between 50°–90°C and the time of immersion from 2–10 min a good degree of control can be achieved.

2. Flotation

This technique was devised by Ledingham and Chinn (1955) to recover the conidia of *Helminthosporium sativum* from soil.

Soil is screened to remove the larger particles and water added to bring moisture content up to 10% by weight. A 10 g sample of this soil is mixed with 5 ml of mineral oil in a watch glass, transferred to a large test tube and and 50 ml of tap water added. The mixture is shaken vigorously for about 5 min then placed in a vertical position.

Within half an hour the soil settles and an emulsion collects at the surface. A drop of the emulsion is transferred to a slide for microscopic examination. Droplets of oil and air in the emulsion are broken down and the mixture spread with a needle in a thin film.

Samples of the emulsion are obtained with pipettes which deliver drops of 0·02 ml. Pipettes can be drawn from glass tubing. Spores in at least ten drops are counted and the approximate total number of spores in the emulsion and hence soil are calculated.

3. Soil sieving

Washing soil through a graded series of sieves is standard practice in the recovery of parasitic and free living nematodes from soil. This technique has also been used effectively in special cases for the recovery of soil borne plant pathogens and other soil fungi. The sclerotia of *Phymatotrichum omnivorum* were recovered from soil by King and Hope (1932) using wet sieving techniques, as were the sclerotia of *Sclerotium rolfsii* by Leach and Davey (1938). Gerdemann and Nicolsan (1963) extracted spores of *Endogone* from soil by sieving and decanting and similar techniques were used by McCain *et al.* (1967) for the recovery of chlamydospores of *Phytophthora cinnamomi*.

Wet sieving is particularly useful for recovering large propagules such as sclerotia, zygospores, large chlamydospores, and the like.

4. Centrifugation

Ascomycetes are not normally recovered from soil in high numbers using dilution plate techniques or indeed any of the commonly used techniques with the exception of Warcup's method of partial sterilization of the soil (Warcup, 1951). To recover Ascomycetes, Paden (1967) described a modification of a technique used originally by Ohms (1957) to recover a phycomycetous mycorrhizal parasite from soil.

Ten grams of screened soil are mixed with 100 ml of water and shaken thoroughly. The suspension is allowed to stand for 30 sec and the settlings discarded. The procedure is repeated four times and results in the removal of most of the sand and part of the silt and clay. The final supernatant is centrifuged at 500 g for 5 min. The supernatant is discarded and the resultant sediment suspended in 20 ml of sterile distilled water. Ten ml of the suspension are pipetted on to density gradient columns prepared by layering sucrose solutions of specific gravity 1·38, 1·30, 1·20, and 1·15 in 1 cm layers in 25 × 100 mm, conical-bottomed, pyrex centrifuge tubes. Then a centrifugation at 350 g for 30 min results in the sedimentation of the larger mineral particles. Most spores remain in suspension and can be recovered by decanting the supernatant, diluting with three volumes of sterile distilled water and centrifuging at 1000 g for 5 min. The sediment is resuspended in 3 ml of sterile distilled water and 0·25 ml of the suspension is pipetted on to plates of 3 % water agar containing 10 ppm furfural. The plates are heated to 55°C for 30 min and then incubated in the dark at 25°C. Germinated spores are transferred to potato-carrot agar.

Paden pointed out that one of the advantages of this technique was that one could see the type of propagule from which a colony was obtained. He suggested the technique might also have value for recovering other fungi that have readily identifiable spores such as certain dermatophytes.

5. *Selective isolation*

The recovery of a specific fungus or group of fungi from the soil environment to the exclusion of all others is an interesting and challenging problem for the soil microbiologist. Some of the techniques already described have been designed for selective isolation. More commonly, however, isolation of a particular organism or group is achieved either by baiting or the use of highly selective media. It should be emphasised here that all techniques used for the recovery of soil fungi are selective to some extent, the difference is one of degree. Here, we are attempting to recover either a specific fungus or a group of fungi with ecological or physiological properties in common. A common selective technique is to bait the soil with a substrate which is particularly favourable to colonization by the fungus which subsequently may be recovered and purified. On the other hand the soil may be plated out on a medium which is fungistatic or fungicidal to all other organisms but the fungus or group under study. Selective media can be used in conjunction with most of the isolation techniques described above.

Baiting techniques have been very successful in the recovery of dermatophytes, cellulose-decomposing fungi, and predaceous Hyphomycetes amongst others but have proved particularly useful in the study of soil-borne plant pathogenic fungi.

The use of buried cellulose film as recommended by Tribe (1957, 1960) has proved very effective for the recovery of cellulolytic fungi.

Pieces of washed cellulose film (Cellophane) about 0·5 × 1·0 cm are dampened in sterile water and placed singly on coverslips to which they adhere. The coverslips are then buried vertically in the soil. On recovery the microbial material colonizing the cellulose is stained with picronigrosin in lactophenol which does not stain the cellulose. Isolation can be made directly from the cellulose immediately after its removal from soil.

In Tribe's work the soil samples were kept at laboratory temperatures and moisture content fluctuated between 40–60% water holding capacity. Soil samples were put through a 3 mm sieve before use.

For the selective isolation of cellulose decomposing fungi, Dickinson and Pugh (1965) report the use of cellulose agar. They found however that where cellulose decomposing species were abundant they could be recovered with equal facility on either cellulose agar or mud-extract agar. They did point out the advantage of the selective medium for encouraging sporulation and therefore aiding identification of members of this group.

Regarding soil baiting techniques it is clear that there is no limit to the different types of bait which may be used to "seed" the soil. Some will recover groups of fungi others only one. Yarwood (1946) described a carrot-disk technique which favoured the recovery of *Thielaviopsis basicola*.

Soils from field collections are spread over the surface of 5 mm thick carrot-root disks in Petri plates and enough water is added by atomizing to make the soil quite moist but with no free water present. After two to four days at room temperature the disks are washed to free them from soil and incubated in moist chambers. When soils containing *Thielaviopsis* are used for inoculum greyish colonies appear about six days after inoculation. These can be readily purified by streaking some of the aerial chains of spores on potato dextrose agar.

Yarwood noted that *Thielaviopsis* was isolated at seven of twelve locations tested and in none of these locations was *Thielaviopsis* observed as a pathogen of the crops growing there. Not all techniques using plant parts as bait work as beautifully as Yarwood's and the advantages of such a technique in studying the ecology of soil-borne plant pathogens are obvious.

Because of their importance as plant pathogens *Fusarium* and *Verticillium* have attracted a lot of interest using soil baiting techniques and selective media.

Nadakavukaren and Horner (1959) recommended an alcohol containing medium for the recovery of *Verticillium* from soil. Harrison and Livingston (1966) on the other hand claimed that alcohol medium was less useful in soils where *Verticillium* counts were low and used a modified air sampler.

Papavizas (1967) evaluated 18 different media which had been recommended in the literature for the isolation of *Fusarium* from soil. He considered a modified peptone/pentachloronitro-benzene medium as best both for its effectiveness and simplicity of preparation.

Singh and Nene (1965) used Czapek's agar containing malachite green (50 ppm) and Captan (100 ppm) as a selective medium for the recovery of *Fusarium* species from natural field soils. This medium apparently allows the development of *Fusarium* colonies from vegetative hyphae but is lethal to conidia and chlamydospores of *Fusarium*. The medium therefore permits quantitative estimations of the number of active hyphal propagules in the soil.

REFERENCES

Adams, P. B. (1967). *Phytopathology*, **57**, 602–603.
Alexander, F. E. S., and Jackson, R. M. (1955). *In* "Soil Zoology" (Ed. J.McKevan), pp. 433–441. Butterworths, London.
Andersen, A. L., and Huber, D. M. (1965). *Phytopathology*, **55**, 592–594.
Brown, J. C. (1958). *Trans. Br. mycol. Soc.*, **41**, 81–88.
Burges, A., and Nicholas, D. P. (1961). *Soil Sci.*, **92**, 25–29.
Chacko, C. I., and Lockwood, J. L. (1966). *Phytopathology*, **56**, 576–577.
Chesters, C. G. C. (1940). *Trans. Br. mycol. Soc.*, **24**, 352–355.
Chesters, C. G. C. (1960). *In* "The Ecology of Soil Fungi" (Ed. D. Parkinson and J. S. Waid), pp. 223–238. Liverpool University Press.
Chinn, S. H. F. (1953). *Can. J. Bot.*, **31**, 718–724.
Cholodny, N. G. (1930). *Arch. Mikrobiol.*, **1**, 620–652.
Conn, H. J. (1918). *Tech Bull. N.Y. St. agric. Exp. Stn.*, **64**.
Conn, H. J. (1922). *Soil Sci.*, **14**, 149–151.
Conn, H. J. (1932). *Zentbl. Bakt. ParositKde*, **87**, 233–239.
Dickinson, C. H., and Pugh, G. J. F. (1965). *Nature, Lond.*, **207**, 440–441.
Dobbs, C. G., and Hinson, W. H. (1953). *Nature, Lond.*, **172**, 197–199.
Gerdemann, J. W., and Nicolsan, T. H. (1963). *Trans. Br. mycol. Soc.*, **46**, 235–244.
Gochenaur, S. E. (1964). *Mycologia*, **56**, 921–923.
Haarløv, N., and Weis-Fogh, T. (1953). *Oikos*, **4**, 44–57.
Harley, J. L., and Waid, J. S. (1955). *Trans. Br. mycol. Soc.*, **38**, 104–118.
Harrison, M. D., and Livingston, C. H. (1966). *Pl. Dis. Reptr.*, **50**, 897–899.
Hering, T. F. (1966). *Pl. Soil*, **25**, 195–200.
Jackson, R. M. (1958). *J. gen. Microbiol.*, **18**, 248–258.
Johnson, L. F., Curl, E. A., Bond, J. H., and Fribourg, H. A. (1959). "Methods for Studying Soil Microflora Plant-Disease Relationships", pp. 1–178. Burgess Publishing Co., Minneapolis.
Jones, P. C. T., and Mollison, J. E. (1948). *J. gen. Microbiol.*, **2**, 54–69.
Jooste, W. J. (1965). *Nature, Lond.*, **207**, 1105.
King, C. J., and Hope, C. (1932). *J. agric. Res.*, **45**, 725–740.
Leach, L. D., and Davey, A. E. (1938). *J. agric. Res.* **56**, 619–632.

Ledingham, R. J., and Chinn, S. H. F. (1955). *Can. J. Bot.*, **33**, 298–303.
Lingappa, B. T., and Lockwood, J. L. (1963). *Phytopathology*, **53**, 529–531.
Martin, J. P. (1950). *Soil Sci.*, **69**, 215–232.
McCain, A. H., Holtzmann, O. V., and Trujillo, E. E. (1967). *Phytopathology*, **57**, 1134–1135.
Minderman, G. (1956). *Pl. Soil*, **8**, 42–48.
Mueller, K. E., and Durrell, L. W. (1957). *Phytopathology*, **47**, 243.
Nadakavukaren, M. J., and Horner, C. E. (1959). *Phytopathology*, **49**, 527–528.
Nicholas, D. P., Parkinson, D., and Burges, N. A. (1965). *J. Soil. Sci.*, **16**, 258–269.
Ohms, R. E. (1957). *Phytopathology*, **47**, 751–752.
Paden, J. W. (1967). *Mycopath. Mycol. appl.*, **33**, 382–384.
Papavizas, G. C. (1967). *Phytopathology*, **57**, 848–852.
Parkinson, D., and Williams, S. T. (1961). *Pl. Soil*, **13**, 347–355.
Rodriguez-Kabana, R. (1967). *Pl. Soil*, **26**, 393–396.
Rossi, G. M. (1928). *Italia agric* No. 4.
Singh, R. S., and Nene, Y. L. (1965). *Pl. Dis. Reptr.*, **49**, 114–118.
Steiner, G. W., and Watson, R. D. (1965). *Phytopathology*, **55**, 728–730.
Thornton, R. H. (1952). *Research*, **5**, 190–191.
Tribe, H. T. (1957). *In* "Microbial Ecology" (Eds R. E. O. Williams and C. C. Spicer), pp. 287–298. Cambridge University Press.
Tribe, H. T. (1960). *In* "The Ecology of Soil Fungi" (Ed. D. Parkinson and J. S. Waid), pp. 246–256. Liverpool University Press.
Warcup, J. H. (1950). *Nature, Lond.*, **166**, 117–118.
Warcup, J. H. (1951). *Trans. Br. mycol. Soc.*, **34**, 515–518.
Warcup, J. H. (1955a). *Nature, Lond.*, **175**, 953–954.
Warcup, J. H. (1955b). *Trans. Br. mycol. Soc.*, **38**, 298–301.
Warcup, J. H. (1960). *In* "The Ecology of Soil Fungi" (Ed. D. Parkinson and J. S. Waid), pp. 3–21. Liverpool University Press.
Warcup, J. H. (1962). *In* "Ecology of Soil-borne Plant Pathogens" (Ed. K. F. Baker and W. C. Snyder), pp. 52–67. University of California Press, Los Angeles.
Watson, R. D. (1960). *Phytopathology*, **50**, 792–794.
Williams, S. T., Parkinson, D., and Burges, N. A. (1965). *Pl. Soil*, **22**, 167–186.
Yarwood, C. E. (1946). *Mycologia*, **38**, 346–348.

CHAPTER XV

Fungi Pathogenic for Man and Animals: 1. Diseases of the Keratinized Tissues

PHYLLIS M. STOCKDALE

Commonwealth Mycological Institute, Kew, Surrey, England

Several diverse groups of fungi are capable of invading the keratinized tissues (skin, hair, nails, claws, spines, etc.) of man and animals, some growing more or less saprophytically in the dead superficial layers, others invading more deeply and usually producing a host response (Table I). Methods for the laboratory examination of infected keratinized tissues are similar for all of these fungi, and once the fungi are isolated most may be studied by techniques described elsewhere in this volume. The only important group which requires special techniques not described elsewhere is that comprising the dermatophytes, or ringworm fungi. This chapter will therefore outline some techniques for the examination of keratinized tissues for fungi, and discuss the special techniques for the study of dermatophytes. Some methods for the detection and isolation of *Dermatophilus congolensis* will also be outlined as although this is an actinomycete it is usually handled by mycologists.

TABLE I

Fungi which infect the keratinized tissues

Fungus	Tissues infected	Disease
Dermatophytes (*Trichophyton, Microsporum* and *Epidermophyton* species)	All keratinized tissues	Dermatophytosis (ringworm, tinea, favus, onychomycosis, etc.)
Candida species†	Skin, hair follicles, nails, nail folds	Candidiasis (candidosis, moniliasis)
Malassezia furfur	Superficial skin	Pityriasis versicolor (tinea versicolor)
Cladosporium werneckii, C. mansonii	Superficial skin (usually palm of the hand)	Tinea nigra
Trichosporon cutaneum	Hair (forming nodules on the hair shaft)	White piedra
Piedraia species	Hair (forming nodules on the hair shaft)	Black piedra
"Saprophytic" fungi	Nails	Onychomycosis
(*Scopulariopsis brevicaulis, Aspergillus, Cephalosporium,*	Necrotic skin (English, 1968b)	—
Fusarium species, etc.)	External auditory canal	Otomycosis

† Also infect the mucous membranes and internal tissues.
Most of the systemic pathogens may also cause skin lesions.

I. HANDLING TECHNIQUES

Most of the laboratory-acquired mycotic skin infections which have been reported have developed following contact with infected animals (Hanel and Kruse, 1967), but infection may also result from contact with infected tissues or cultures and these should be handled with caution, using strictly sterile techniques. If specimens or cultures are accidently dropped they should be covered immediately with a disinfectant such as cresol solution and left for several minutes before being wiped up. The bench should be swabbed with disinfectant when work is finished, and the workers' hands washed with soap and water. Instruments such as inoculating needles should be sterilized by flaming before they are laid down, cultures and clinical material should be autoclaved before they are discarded, and used slides, pipettes, etc. should be placed in jars of disinfectant solution. Cultures to be dried down as herbarium specimens should first be killed by exposure to formalin vapour for at least 24 h.

It is not necessary to use an inoculating cabinet, for the protection of the worker, unless a species of *Blastomyces, Coccidioides,* or *Histoplasma* is suspected, and unlike these fungi the skin pathogens may be cultured in Petri dishes, providing care is taken in handling them.

II. EXAMINATION OF CLINICAL MATERIALS

A. Collections of specimens

Swabs and samples of pus from suppurative lesions, and any specimens from skin manifestations of the systemic mycoses should be cultured and smeared on slides for microscopic examination as soon as possible after they are collected. If immediate culture is not possible they should be placed in sterile containers for transport to the laboratory. Swabs should not be sent by mail as they dry out in transit. Methods for the examination and culture of these specimens are then the same as those described by Buckley (this Volume, p. 461) in her Chapter on the examination of deep tissues.

All other specimens may be placed in sterile Petri dishes or folded into sheets of clean, smooth, preferably black paper. Paper is the most convenient when the specimen has to be sent by mail. When a Petri dish is used it is useful to place a sterile microscope slide in the dish so that greasy skin scrapings can be dislodged from the scalpel by scraping it against the edge of the slide. Placing the specimen between two glass slides sealed with adhesive tape or on lint, gauze, filter paper, or similar material on which it may become entangled is not practical. Also, specimens should not be placed in closed containers such as McCartney bottles as in these they remain moist; this encourages the growth of bacteria and saprophytic fungi which may have an antibiotic effect on the pathogen or overgrow it in primary cultures. Providing the specimen is allowed to dry out it is not essential to examine it immediately as most of the skin pathogenic fungi will survive for long periods in scales and hairs *in vitro*. The dermatophytes, for example, have been found to remain viable in skin scales for several months (Dvořák *et al.*, 1968) and often for several years.

Specimens (and cultures) being sent to other laboratories for examination should be accompanied by as many details as possible of the host (species, race or breed, country of origin, sex, age, name or number) and the disease (suspected diagnosis, length of duration, parts of the body affected, appearance of the lesions, etc.).

1. *Skin*

(a) *Scrapings.* These should be taken with a sterile scalpel over a cross-section of the lesion, not only from the active edge. It is usually recommended that lesions are swabbed with 70% alcohol before scrapings are taken, to remove surface dirt and medicaments, but this has little or no practical value, particularly if isolation media containing antibiotics are available.

Areas which have been treated with ointment should be avoided as this makes microscopic examination difficult and may inhibit the growth of the fungus.

(b) *Adhesive tape strippings.* These are useful in the diagnosis of pityriasis versicolor, but not of the more deep-seated dermatophyte infections. The sticky side of transparent adhesive tape is gently pressed on the lesion so that a thin layer of epithelial cells, usually almost continuous and only 1 cell thick, is obtained on the tape. If the process is repeated several times on the same lesion the horny layer can be removed entirely in serial layers, successive strippings being thinner and clearer. The tape may be stored until it is examined by placing sticky side down on a sheet of glass. It may then be cut into small pieces for culture or examined microscopically in several ways. Porto (1953) found that if the tape was pressed sticky side down on a clean, grease-free slide it could be examined directly, and *Malassezia furfur* could be seen in the scales with a minimum of distortion, without using any clearing, staining, or mounting agents, while Edwards and Hartman (1952) applied potassium hydroxide solution or a stain (II, B) to their strippings and examined them using the tape as a coverslip. Keddie *et al.* (1961) stressed that vinyl tape, which is more resistant to water and chemicals than cellulose tape, should be used when the strippings were to be stained. For rapid staining, they suggested that the tape should be cut into small pieces, placed adhesive side up on a slide, stained for 1 min with Hucker's crystal violet, Löffler's methylene blue, or Giemsa solution (formulae of these stains are given in most bacteriology handbooks or in Ajello *et al.*, 1963), rinsed quickly with alcohol, blotted dry, and mounted in balsam, a synthetic mounting fluid, or glycerine jelly; they found that *n*-butyl alcohol was better then ethyl or isopropyl alcohol for dehydration as it did not cause clouding of the plastic film. Keddie *et al.* also described several staining techniques for making permanent preparations, and discussed the merits and drawbacks of different types of tape and the agents which each type would tolerate during dehydration, clearing, and mounting. In more recent studies Keddie and her co-workers have prepared permanent mounts by first separating the layer of epithelial cells from the plastic film. This is essential in fluorescent-antibody studies as the thickness of the film causes optical interference (Sternberg and Keddie, 1961). Cleaned slides are dipped in 1% purified calf-skin gelatin, drained, dried, fixed in 1% formalin solution, and dried again. Pieces of adhesive tape (Scotch tape no. 681) carrying the skin strippings are pressed firmly on to the slides which are then placed in tetrahydro-furan for 10 min to dissolve the plastic vinyl film and in xylene for 30 min to remove the adhesive. The slides are then dried and stored at 4°C until stained by the periodic acid–Schiff (Part II, appendix), fluorescent anti-body, or other staining methods (Keddie and Shadomy, 1963) (II, B).

2. Nails

Pieces of nail may be clipped off with sterile scissors or scrapings taken with a solid steel scalpel; scalpels with disposable blades are too flexible. Non-dermatophytes and some dermatophytes grow in the centre or outer surface of the nail, while other dermatophytes grow in the inner layer (Walshe and English, 1966), so all levels of the infected nail should be examined. Lewis et al. (1958) considered that the use of a dental drill for collecting nail specimens (Epstein, 1945) is probably not advisable as the drill is difficult to sterilize and may also cause pulverized nail material to be disseminated in the air.

3. Hair

(a) *Clipping and plucking.* The technique for collecting hair depends on the fungus involved. In cases of piedra, hairs bearing nodules may be clipped off with scissors. In dermatophyte infections it is important to obtain the hair roots as these are attacked first by the fungus, infected hairs eventually breaking off just above the root. The lesions should therefore be scraped with a sterile scalpel and hairs, particularly the stubs of broken hairs, pulled out with tweezers. In infections by some dermatophyte species infected hairs fluoresce under a Wood's light;† animals and patients with scalp infections should therefore be routinely examined under a Wood's lamp and fluorescent hairs collected when they are present.

(b) *The hairbrush technique.* Some dermatophytes may cause only slight scaling or hair loss, or may be carried on the hair of man and animals without apparently invading the tissues. The detection of such infections and carrier states has been simplified by introduction of the "hairbrush technique". Mackenzie (1963), studying an epidemic caused by *Trichophyton sulfureum* in a residential school for girls, found that the fungus could be isolated from the girls' hairbrushes by pressing the bristles into plates of agar medium. Modifications of this method have now been devised and applied to studies of the distribution of dermatophytes in various populations. It has been found particularly useful for routine examinations during attempts to eradicate ringworm epidemics.

Ive (1966), searching for asymptomatic carriers of *Microsporum audouinii* in schools in Nigeria, used a round polythene scalp massager. The scalp

† A mercury vapour lamp with a special glass filter containing metallic compounds, principally nickel oxide, which removes short-wave ultraviolet and visible light, allowing maximum radiation at about 366·0 nm. These lamps must be shaded to protect the eyes. Commercial models are available, or bulbs and transformers can be purchased separately, and built into a box or otherwise shaded. Suitable bulbs available in Great Britain are the G.E.C. Black Bulb, Phillips H.P.W. 125 W 57236F/70, or Osram MBW/U 125 W T.M. 230–240 v.

of each child was massaged vigorously for $\frac{1}{2}$ min with a sterilized massager; this became electrostatically charged and attracted particulate matter, and was then pressed into agar medium in a Petri dish, which it fitted closely.

For examination of animals the type of brush to be used can be chosen to suit the size of the animal. For very small animals toothbrushes are most convenient (Goldberg, 1965; English, 1966), while for larger animals nail brushes, stiff nylon hairbrushes (Goldberg, 1965; Connolle, 1965) or scalp massagers can be used. Gentles *et al.* (1965) obtained higher isolation rates by forcing the bristles of plastic brushes through squares of gauze which were used to draw down debris from the base of the bristles at the time of culturing.

The sterilization of plastic and nylon brushes presents some difficulties as they cannot be autoclaved. Ive (1966) sterilized his scalp massagers by immersing them in 10% teepol solution for 24 h. Connolle (1965) found immersion in 0·1% chlorhexidine solution for 30 min satisfactory for sterilizing nylon hairbrushes; these were then washed with soap and running hot water, residual hairs were removed, and the brushes were dried and wrapped in clean brown paper ready for the next sampling.

Another modification of the technique, devised by Mariat and Tapia (1966), involves the use of small squares of wool carpet. Samples of new carpet with a short pile are cut into squares 6×6 cm. These are washed in running water for 48 h and then rinsed several times in distilled water to remove size, dressing, and insecticides, dried in an oven, placed in small paper packets, and sterilized by autoclaving for 1 h at 120°C. The area of skin or scalp being investigated is rubbed vigorously with a carpet square, which is then replaced in its paper packet until it reaches the laboratory.

The hairbrush technique is not generally suitable for examination of stabled animals as brushings from these may be so heavily contaminated with saprophytic fungi from hay and straw that the control of contamination in primary cultures is not possible (Austwick, personal communication).

It should be stressed that the hairbrush technique can demonstrate only the presence of a fungus. Invasion of the host tissues can be demonstrated only by microscopic examination, and the finding of fungal elements within the tissues.

B. Microscopic examination

1. *Direct methods*

Temporary mounts may be made by placing small pieces of the specimen in a drop of 10–20% potassium hydroxide on a slide, adding a coverslip, warming the slide on an electric hot plate or lamp or over a very low Bunsen burner flame until the tissue has cleared, and then gently pressing the coverslip to make the specimen almost transparent. Fungal elements are seen best in such preparations if examined in a fairly low light. Various modifications of this method have been advocated. Zaias and Taplin (1966) found a solution of 20% potassium hydroxide in a mixture of 60% water and 40% dimethyl sulphoxide gave better results than the traditional solution. Hyphae could be seen in scrapings after treatment with the reagent for only 1 min without heating, but the method was not satisfactory if more than 20 min elapsed between preparation on the slide and microscopic examination.

Artifacts which the novice may confuse with fungal hyphae are sometimes seen in potassium hydroxide preparations. One of the most discussed of these is the "mosaic fungus" (Weidman, 1927), an accumulation of a substance, possibly cholesterol (Davidson and Gregory, 1935 Cornbleet et al., 1943), in the intracellular spaces. Various dyes which will stain fungal elements and make them more clearly visible, but will not stain artifacts such as the "mosiac fungus", may be incorporated into the potassium hydroxide solution. One of the dyes most widely used in this way is a patented compound contained in certain brands of ink. Cohen (1954), for the demonstration of M. furfur, placed scales in a drop or two of 10% potassium hydroxide and then added a small drop of Parker 51 superchrome blue-black ink. Taschdjian (1955) used the same ink in the proportion 1 part ink: 9 parts 10% potassium hydroxide for examining scales for dermatophytes, M. furfur, and Candida albicans. Swartz and Lamkins (1964) noted that Cohen's reagent was not practical for the rapid staining of dermatophyte-infected tissues. They found a mixture of 1 part 10% potassium hydroxide, 1 part aerosol OTB 1% (an anionic wetting agent containing 0·85 g dioctyl sodium sulphosuccinate and 0·15 g sodium benzoate in 100 ml distilled water, manufactured by the American Cyanamid and Chemical Corporation), and 2 parts of Parker's "Super Quink" permanent blue-black ink, was ideal for the rapid clearing of tissues and for the staining of fungal elements. The solution was stable and could be kept for at least 2 months.

Chick and Behar (1961) used a combination of 1 ml acridine orange (1 : 1000) and 9 ml potassium hydroxide, and examined the slides under a fluorescence microscope; fungal elements fluoresced brightly.

2. Permanent preparations

Permanent mounts are useful for demonstration purposes, but their preparation is usually considered too time-consuming for diagnostic work. Most techniques involve preliminary clearing in potassium hydroxide solution. Swartz and Conant (1936) suggested clearing the specimen in 5% potassium hydroxide, washing with water in a watch glass, and then mounting it in a drop of lactophenol–cotton blue (Dring, this Volume, p. 95). If the fungus does not stain well by this method the specimen, after clearing, may be stained with 1% cotton blue in 70% alcohol and then mounted in clear lactophenol. Lactophenol will also clear tissues very slowly and narrow hairs and very thin skin scales may be mounted directly in one of the lactophenol preparations without preliminary digestion in potassium hydroxide.

Staining by the periodic acid–Schiff technique (Pt. II, appendix) after clearing in potassium hydroxide solution has been recommended for dermatophytes and other skin pathogenic fungi by several authors (Gadrat *et al.*, 1952; Sharvill, 1952; Lofgren and Batts, 1952; Taschdjian and Muskatblit, 1953). Excess potassium hydroxide is usually removed by washing the specimen in water, but Taschdjian and Muskatblit (1953) drew off excess alkali with filter paper and then immersed the specimen in 5–10% lactic acid for about 3 min, until the pH was 3–5, before mounting it in lactophenol cotton blue or fixing it on the slide by covering with 95% alcohol, allowing it to dry for several hours, and then staining by the periodic acid–Schiff technique.

Another method for preparing permanent mounts has been described by Gordon (1951a)—

Scrapings or hairs are placed on a clean slide and covered with 1 or 2 drops of methylene blue solution (prepared by diluting 1 part of a saturated 95% alcoholic solution of methylene blue hydrochloride (1·48 g/100 ml) with 9 parts of water). The dye is allowed to act for 30–60 sec, drawn off with bibulous paper, and distilled water is applied with a dropper and left on the slide until dye ceases to flow from the specimen. Alternatively the preparation may be decolorized more vigorously with 1% acetic acid followed by water. It is then dried with filter paper, 1–2 drops of 1,4-dioxane are added to dehydrate the specimen, drawn off after 5–10 sec, and when most of the remaining dioxane has evaporated a drop of resin is added and a coverslip applied. For this method, scales must be very thin.

C. Isolation

1. *General methods*

With the exception of *M. furfur*, all of the skin pathogenic fungi will grow on most routine mycological media. The medium most frequently recommended for their isolation is Sabouraud's glucose agar (III, B). This is probably the best single medium for the isolation of dermatophytes, though a few species can be identified more rapidly in primary culture on 2–4% malt extract agar (Booth, this Volume, p. 49). Some of the skin pathogenic fungi other than dermatophytes are also more easily recognizable on malt extract agar than on Sabouraud's glucose agar.

Early recognition of dermatophyte colonies on Sabouraud's glucose agar may be aided by inclusion of ink blue into the medium (III, C).

All isolation media should contain anti-bacterial antibiotics when possible; chloramphenicol (chloromycetin, Parke–Davis & Co.) at 0·05 mg/ml is widely used for this purpose as it inhibits a wide range of bacteria, and can be added to the medium before autoclaving. The isolation of dermatophytes and some other fungi is also aided by the inclusion of cycloheximide (actidione, The Upjohn Company) at 0·5 mg/ml in the medium, to suppress fungal contaminants (Georg, 1953). This antibiotic, produced by *Streptomyces griseus* (Leach *et al.*, 1947), inhibits many but not all saprophytic fungi and also a few pathogenic fungi, including some *Candida* species and *Trichosporon cutaneum*. Specimens should therefore also be cultured on a medium not containing cycloheximide. Each worker develops his own preferences regarding isolation media, but a suitable combination of media for the routine isolation of skin pathogenic fungi would be Sabouraud's glucose agar containing chloramphenicol and cycloheximide, and malt extract agar containing chloramphenicol only.

Like chloramphenicol, cycloheximide is heat stable and can be added to the medium before autoclaving. The two antibiotics are incorporated by adding to 1 litre of medium, which has been boiled to melt the agar, the following: chloramphenicol, 50 mg in 10 ml 95% alcohol; cycloheximide, 500 mg dissolved in 10 ml acetone. The medium may then be dispensed into tubes or bottles and autoclaved.

Specimens may be cultured on media in wide test tubes, bottles, or Petri dishes. Many workers prefer dishes as in these it is easier to pick off colonies before they are overgrown by contaminants and to examine them microscopically *in situ*. The specimen should be broken up into very small fragments which are then slightly embedded in the medium, about 2 cm apart. The cultures are then normally incubated at 25–28°C. For the isolation of *T. verrucosum*, particularly from animal scrapings which may be heavily contaminated, it is advisable to incubate cultures at 35–37°C (Austwick, 1954). Cultures are then examined every 3–4 days and kept for at least 2 weeks before being reported as negative. If contaminants appear the pathogen should be subcultured as soon as possible. The skin pathogenic fungi can usually be identified in primary cultures on antibiotic media, but any suspicious colonies which cannot be identified should be subcultured on to media without antibiotics. The possibility that more than one pathogenic fungus may be present in a single specimen whould also be borne in mind.

2. *Isolation and culture of* Malassezia furfur *and related species*

Pityriasis versicolor may be diagnosed by clinical and microscopical findings so that isolation of *M. furfur*, considered to be the aetiological agent of the disease, is not necessary in the routine diagnostic laboratory. Numerous reports of the successful isolation of *M. furfur* have been made, but have been either discredited or accepted with reservations (Gordon, 1951b). It has been thought that *Pityrosporum orbiculare*, isolated frequently

from pityriasis versicolor scales but also sometimes from healthy skin
(Gordon, 1951b), might be conspecific with *M. furfur*, although hyphae
like those found *in vivo* are rarely seen in cultures of *P. orbiculare*. The
morphological and fluorescent antibody studies of Sternberg and Keddie
(1961) and Keddie and Shadomy (1963) have supported this theory, which
is now becoming more widely accepted.

The medium most widely used for the isolation of *P. orbiculare* is Sabouraud's glucose agar containing cycloheximide and chloramphenicol (see
above) with a surface layer of sterile olive oil (Keddie and Shadomy, 1963).
Gordon (1951b) used 2 ml oil and recommended incubating the tubes in a
slanting position so that the agar surface is kept constantly in contact with
the oil. Growth is best at 37°C and almost as good at 30°C, visible growth
appearing in 4–5 days (Gordon, 1951b). A modification of this method
devised by Barfatani *et al.* (1964) involves culture on Sabouraud's glucose
agar containing cycloheximide, chloramphenicol, 2% olive oil, and 0·2%
Tween 80. The oil and Tween 80 are mixed together in warm water before
being added to the medium.

P. orbiculare is also reported to grow well on the tauroglycocholate medium
devised by Martin-Scott (1952) for the culture of *P. ovale*. This contains
100 g sodium tauroglycocholate, 50 g Oxoid mycological peptone, 15 g
agar, and water to 1 litre. Chloramphenicol and cycloheximide may be
incorporated for primary isolation purposes.

These three media are also suitable for the maintenance of *P. orbiculare*.
Transfers should be made every 3 weeks, and if an overlay of oil is being
used the inoculum should be streaked on the agar surface before addition
of the oil (Gordon, 1951b).

A liquid medium which does not contain oil and is stated to be suitable
for the preparation of oil-free extracts of *P. orbiculare* and for studying the
effects of antifungal agents has been described by Weary and Graham (1966).
This medium, based on Czapek-Dox broth, contains the following:

Sucrose	30 g
Sodium nitrate	3 g
Dipotassium phosphate	1 g
Magnesium sulphate	0·5 g
Potassium chloride	0·5 g
Ferrous sulphate	0·01 g
Yeast extract	2·5 g
Chloramphenicol	50 mg
Cycloheximide	500 mg
Tween 20,	75 ml
Distilled water	to 1 litre

The medium is autoclaved at 120°C for 15 min. Cultures are incubated at
30–37°C on a gyrorotary shaker.

These media are also suitable for the isolation and culture of *P. ovale*, which has less exacting nutritional requirements than *P. orbiculare*. Detailed accounts of the nutritional requirements and biology of *P. ovale* are given by Benham (1947) and Weary (1968).

P. pachydermatis does not require fatty substances for growth, and grows readily on media such as Sabouraud's glucose agar without added oil (Fraser, 1961).

D. Examination of clinical material for *Dermatophilus congolensis*

D. congolensis is an actinomycete which grows in the epidermis of sheep, cattle, and other animals, producing a disease which is known by a variety of names (lumpy wool, mycotic dermatitis, strawberry foot rot, streptotrichosis, etc.) and is characterized by the formation of thick, hard scabs. The disease has been reviewed recently by Roberts (1967) and much of the information given here comes from this review.

1. *Microscopic examination*

The diagnosis of *D. congolensis* infection is confirmed by the demonstration of the typical branching filaments, dividing transversely and longitudinally into thick bundles of coccoid forms, in the scabs. Recently formed scab material should be chopped with scissors in sterile water and smears of the suspension stained with any of the usual bacterial stains (Gram, methylene blue, etc.). The demonstration of *D. congolensis* in older scab material, in which the hyphae are sometimes completely converted into free zoospores, may be facilitated by the use of fluorescent antibody (Pier *et al.*, 1964).

2. *Isolation*

D. congolensis may be isolated from fresh scab material by inoculation of a scab suspension prepared as above on to nutrient agar, preferably containing 10% sheep blood. The plates are then incubated at 37°C for 2 days in 10% carbon dioxide and for a further day in air.

Isolation from scabs from chronic or healing lesions is more difficult because cultures may be overgrown by contaminants. Haalstra (1965) has suggested the following method for isolation from such material—

Small pieces of scabs are placed in Bijou (5 ml) bottles, moistened with 1 ml distilled water, and allowed to stand open for $3\frac{1}{2}$ h on the bench. Bottles are then transferred to a wide-mouthed jar and sealed in an atmosphere of about 20% carbon dioxide produced chemically or by burning a candle within the jar. After 15 min the bottles are carefully removed and samples taken from the surface with a loop are seeded on to 5% ox-blood agar and incubated at 37°C in approximately 20% carbon dioxide for 24–48 h. To reduce contaminants samples of three different sizes from individual scabs may be examined.

A simplification of this method was suggested by Roberts (1967) to avoid the use of carbon dioxide in the first stage. This involves immersing the scab material in physiological saline instead of distilled water, to reverse the direction of chemotaxis; the zoospores then collect at the air–liquid intersurface in response to the carbon dioxide produced by the zoospores and other bacteria in the suspension.

When secondary bacterial degradation of the scabs under wet conditions has occurred isolation and microscopic demonstration of *D. congolensis* is very difficult, but may be aided by infection of a laboratory animal. A suspension of scab material is applied to the scarified skin of a sheep or guinea pig and the animal kept in a dry environment. Scabs may be removed for microscopic examination and culture after about 1 week.

III. SPECIAL TECHNIQUES FOR THE STUDY OF DERMATOPHYTES

A. Isolation from inanimate sources

1. *Soil*

Several dermatophytes have been found to exist saprophytically in the soil, but there have been few studies of their incidence in soil using direct methods. Macroconidia of *M. gypseum* have been demonstrated in soil samples by the membrane filter technique (Gordon, 1953; Lurie and Borok, 1955), while McDonough *et al.* (1961) observed macroconidia of *M. cookei*, *T. ajelloi*, and *M. gypseum* on the surface of aqueous suspensions of soil samples previously proved by hair baiting to contain these fungi. Uzunov (1967a, b) adapted a flotation method used for detecting helminth eggs in soil. In this the soil samples are placed in a saturated solution of magnesium sulphate or sodium chloride, stirred with a rod, the mixture is passed through a metal strainer, the solution allowed to stand for about 1 h, and loopfuls of liquid from the surface examined microscopically. By this method Uzunov detected and determined the incidence of macroconidia of *T. ajelloi*, *M. gypseum*, and *M. cookei*. The method was also suitable for the demonstration of other fungi with large spores, but not small-spored species such as *Arthroderma quadrifidum* (stat. conid. *T. terrestre*), although cleistothecia of this species were found.

In most soil studies baiting techniques have been used. Mandels *et al.* (1948) and Cooke (1952) isolated *M. gypseum* by burying pieces of wollen cloth in the soil, but the method which has been used most frequently is the hair-bait technique, a modification of the method described by Karling (1946) for the isolation of chytrids and first used for dermatophytes by Vanbreuseghem (1952).

For baiting, the soil samples are placed in Petri dishes, approximately half-filling the dishes, moistened with a little sterile water if necessary, and small pieces of the baiting material, previously sterilized by autoclaving, are scattered over the soil surface. The dishes are then incubated at room temperature or 25°C and examined at frequent intervals. Each sample should be baited with as many different keratin sources as possible, as some species appear to have a preference for specific keratins. Horse hair, for example, is frequently overgrown with *T. ajelloi* with the exclusion of other species, and members of the *M. gypseum* complex appear to be isolated more readily using human hair (Stockdale, unpublished observations).

Keratin baits rapidly become overgrown with many keratinolytic and non-keratinolytic fungi; the colonization of hair by soil fungi has been described by Griffin (1960). Many of the non-keratinolytic fungi are not inhibited by cycloheximide, so that in order to obtain pure cultures of the dermatophytes it is advisable to obtain isolates derived from single spores. Failing this, very small inocula should be taken from several parts of the growth on the bait, and spread out on media containing cycloheximide and chloramphenicol.

Somerville and Marples (1967) studied the effect of soil enrichment on the isolation of keratinophilic fungi by hair baiting; 100 g samples of soil were mixed with 1 g of autoclaved cowhorn shavings and the mixture stored in plastic bags at 20°C for 14 days before being baited. This enrichment increased the yield of *M. gypseum* from 15% to 78%, but had little or no effect on *T. ajelloi* and *T. terrestre*. The authors suggested that cowhorn had a specific effect on *Microsporum* species, and that enrichment with other forms of keratin might increase the recovery of other keratinophilic species.

2. *Air*

Dermatophytes have been isolated from the air of dermatological clinics (Friedman *et al.*, 1960; Clayton and Noble, 1963), laboratories (Evolceanu and Donciu, 1963), and residential schools (Mackenzie, 1961). They may be isolated by the usual air-sampling techniques (see Davies, this Volume, p. 367) using media containing cycloheximide and chloramphenicol.

3. *Surfaces*

Scales and hairs from infected hosts may contaminate surfaces such as floors, particularly those of communal bathing places, bedding, and furniture thus providing an important reservoir of dermatophyte infection. Efficient sampling of such surfaces is therefore important in epidemiological studies. The most simple qualitative method of sampling, to determine the presence of a dermatophyte at a site, is to rub the area with a small piece of sterile cloth which is then pressed on to the surface of plates of agar medium containing cycloheximide and chloramphenicol. Mackenzie (1961) used sterile

gauze pads 6·5 cm² to rub an area of approximately 1 ft², and bacteriological swabs to collect samples from cracks in floors and walls where the
gauze pad method was impractical. The pads were pressed firmly on to the
agar surface and the plates incubated for 48 h before the pads were removed; after removal of the pads the plates were incubated for a further
12 days. Drouhet *et al.* (1967) adapted Mariat and Tapia's "carpet-square
technique" (II, A), using a rectangle of moquette 7+5 cm to rub an area of
about 60 cm².

A more quantitative method of sampling floors was described by Gentles
(1956). This method involves the use of a tool consisting of a wooden disc
approximately ½ in. thick and 3 in. dia. fixed by a wooden rod 4 in. long to
the centre of the lid of a metal container 7 in. high and 4½ in. dia. The disc
is padded with non-absorbent cotton wool and covered with velvet. The can
containing the disc is sterilized by autoclaving before use. To collect a
sample the pad is pressed firmly on the floor with a slight rotatory movement
and then replaced in the container for transport to the laboratory. Gentles
used each pad to inoculate three plates of isolation medium by pressing it
firmly on to the surface of the medium of each plate in turn. The number of
colonies of dermatophytes which develop on the plates indicates the amount
of infected material present on the area sampled by the disc.

T. mentagrophytes has also been isolated from dust from the floor of a
schoolhouse by culture of the livers and spleens of mice inoculated intraperitoneally with a suspension of vacuum-cleaner sweepings (Bocobo and
Curtis, 1958), and directly, by suspending the dust in normal saline, shaking
allowing to settle for 3 h, and plating the top layers of the suspension on to
Sabouraud's glucose agar containing chloramphenicol and cycloheximide
(Bocobo *et al.*, 1964).

B. Culture, maintenance, and preservation

1. *Routine culture*

The dermatophytes will grow well on many of the routine mycological
media, their main requirement for luxuriant growth being an organic
nitrogen source. Most species utilize sucrose poorly or not at all, and glucose
is the carbohydrate usually used in media for these fungi. The medium most
frequently used for routine culture is a modified form of Sabouraud's
glucose agar, one of the three media devised by Sabouraud (1910)—

Sabouraud's glucose (or maltose) "proof" medium

Glucose (or maltose)	40 g
Peptone	10 g
Agar	18 g†
Distilled water	to 1 litre

Sabouraud's "conservation" medium

Peptone	30 g
Agar	18 g†
Distilled water	to 1 litre

These media must be sterilized very carefully, autoclaving at 120°C for 10 min or 110°C for 15 min being recommended.

Sabouraud noted that the dermatophytes were extremely variable in their macroscopic appearance on different media, or on the same medium when different brands of peptone or sugar were used. He therefore specified that his media should contain only certain brands—glucose "massée de Chanut", maltose "brute de Chanut", and peptone "granulée de Chassaing" —and he proposed that these constituents should be adopted internationally. These brands are no longer available but as Sabouraud attached considerable importance to the macroscopic appearance of dermatophytes and gave very concise descriptions of them on his three media it is desirable to reproduce these media as closely as possible. The sugars can be replaced satisfactorily with purified products, but care must be taken in selecting a peptone as many of the brands suitable for bacteriological media are completely unsuitable for dermatophytes. A peptone in general use in Great Britain is Oxoid Mycological Peptone (Carlier, 1948), while satisfactory brands of Sabouraud's glucose agar in dehydrated form are also available.

Emmons (in Emmons *et al.*, 1963) considered that it is not necessary or desirable to use 4% glucose and suggested a modification of Sabouraud's glucose agar containing only 20 g glucose/litre. He also suggested that the pH should be adjusted to 6·8–7·0 rather than lowered to 5·6 as is often recommended, as a pH near neutral is better for some fungi, and the acid media once used for suppression of bacterial contaminants can be replaced by media containing antibiotics. A preliminary comparison of the traditional Sabouraud's glucose agar and Emmons' modification of it has revealed only small differences in the texture and growth rate of several dermatophyte species on the two media.

Enrichment of Sabouraud's glucose agar with 0·3–0·5% Difco yeast extract may be found beneficial, particularly for the isolation and culture of species with vitamin requirements.

Sabouraud's maltose and conservation media are now rarely used except in taxonomic studies. The latter, devised for maintaining dermatophytes and preventing pleomorphism, is particularly useful for obtaining deeply pigmented colonies of *M. persicolor*.

Malt extract agar (p. 49), containing 2 to 4% malt extract, is also useful

† Sabouraud recommended 18 g agar/litre, but a concentration of between 15 and 20 g agar/litre may be found optimal, depending on the quality of the agar.

for the culture of dermatophytes. On it, growth of most species is slower and less luxuriant than on Sabouraud's glucose agar, but sporulation is often increased. The reverse pigmentation of fungi such as *T. rubrum* and *T. mentagrophytes* var. *erinacei* is also more pronounced on this medium. It is useful for the maintenance of glabrous *Trichophyton* species and for many species which become rapidly pleomorphic on Sabouraud's glucose agar, but it is less suitable than the latter medium for a few species, including *Epidermophyton floccosum, Nannizzia incurvata* (stat. conid. *M. gypseum sensu lato*) and *N. gypsea* (stat. conid. *M. gypseum sensu lato*), which tend to degenerate rapidly on malt extract agar. Potato dextrose agar (Booth, this Volume, p. 49) is also useful for stimulating sporulation in some species, such as *T. soudanense*, and bacteriological nutrient agar containing 1% glucose is useful for obtaining crateriform, sporing colonies of *T. mentagrophytes* var. *quinckeanum*.

Most dermatophytes grow best at 25–28°C and poorly or not at all at 37°C, an exception being *T. verrucosum* which grows most rapidly at 35°–37°C. Most species will grow over a wide pH range, the optimum being 6·8–7·0.

2. *Chemically defined media*

A variety of basal media have been used in nutritional studies. Glucose is the usual carbohydrate source and has been used in concentrations varying from 20 to 50 g/litre. One of the potassium phosphate salts and magnesium sulphate are also included in all media, and sometimes traces of other salts. A complex mixture of amino acids such as that provided by peptone or casein hydrolysate is essential for good growth, though most species will grow to some extent on media containing only ammonium salts or asparagine as the nitrogen source. Recent studies of the utilization of single amino acids by dermatophytes include that of Chattaway *et al.* (1962). Most species are autotrophic for vitamins but some require thiamine, inositol, or nicotinic acid (Table II). Trace metal requirements of some species have been discussed by English and Barnard (1955), who outlined techniques for studying this aspect. The most recent general reviews of the nutritional requirements of dermatophytes are given by Stockdale (1953) and Drouhet and Mariat (1953).

The medium devised by Georg and Camp (1957) for the differentiation of certain *Trichophyton* species by their vitamin and amino acid requirements (III, C) provides a simple basal medium for nutritional studies. These authors found that the addition of traces of other minerals to the medium had no visible effect on the growth of dermatophytes. Another synthetic medium, stated to compare favourably with Sabouraud's glucose medium for the culture of several dermatophytes, was described by Sartory *et al.* (1950), and consists of—

Monopotassium phosphate	1 g
Calcium chloride	0·1 g
Magnesium sulphate	0·5 g
Potassium silicate	0·3 g
Glucose	40 g
Glutamic acid	6·5 g
Folic acid	0·003 g
Nicotinamide	0·012 g
Distilled water	to 1 litre
Agar	20 g

These chemically defined media are not suitable for obtaining typical cultures for morphological studies.

3. *Production of trichophytin*

Media and methods for the production of trichophytin are described by Barker and Trotter (1960), Codner *et al.* (1961), and Barker *et al.* (1962).

4. *Culture on keratin*

(a) *Morphological studies.* The form of keratin used most frequently in these studies is hair which, because of its structure, is attacked in four well-defined stages, viz. cuticle lifting by flat fronds of eroding mycelium, erosion of the cortex, penetration of the cortex by perforating organs which usually originate from the fronds, and colonization of the medulla (English, 1963; 1968a). Different dermatophytes vary in the extent to which they complete these stages, and the property was used to differentiate between atypical isolates of *T. rubrum* and *T. mentagrophytes* by Ajello and Georg (1957), who considered that *T. rubrum* never produced perforating organs. Later, English (1963) found vestigial perforating organs in isolates of *T. rubrum* and occasional true, fully developed perforating organs in some isolates.

For morphological studies blonde hair should be used. This is important even if the hair is to be dissolved in potassium hydroxide, as the pigment granules do not dissolve and will obstruct the picture. The hair should be washed in water if it is dirty, but it is not necessary to degrease it. It should then be dried, cut into small pieces about 1 cm long, and sterilized.

The method of sterilization used is of some importance as it may affect the structure of the keratin. Autoclaving can oxidize cystine and rupture salt bridges (Mathison, 1964), although Chesters and Mathison (1963) found that autoclaving wool for 10 min at 115°C did not increase its susceptibility to digestion with trypsin. Exposure for 18 h to $\frac{1}{2}$ atm of ethylene oxide vapour (Noval and Nickerson, 1959) is the most satisfactory method available, though this also alters the structure to a limited extent (Weary *et al.*, 1965). It seems unlikely that the method of sterilization is an important factor in morphological studies, although for critical work it is presumably best to use keratin which has been denatured as little as possible.

A plate of agar medium may then be inoculated with the fungus and pieces of hair scattered over the inoculum, or bundles of the sterilized hair may be placed in a shallow layer of sterile distilled water in a suitable container, such as a Petri dish or conical flask, inoculated with a heavy suspension of spores or clumps of mycelium of the fungus being tested, and incubated at 25°–28°C. Pieces of hair should then be removed at regular intervals and mounted in a lactophenol preparation for microscopic examination. Cotton blue eventually stains such preparations very heavily and a lighter staining dye such as picric acid (incorporated in lactophenol by using a saturated solution of picric acid in place of water when compounding the medium) or lactophenol without any dye may be found most suitable, particularly in the preparation of permanent mounts.

A method for examining the developmental morphology of perforating organs has been described by English (1968a). Bundles of hair removed from the cultures are placed in Khan tubes, 0·5 ml of 20% potassium hydroxide is added, and the tubes are warmed over a low Bunsen flame for a few minutes and then allowed to stand for 1 h. This treatment dissolves the hair completely. A drop of the remaining mycelial suspension is then pipetted out onto a slide, a coverslip added and gently pressed down, and the preparation examined with a phase contrast microscope. This method is unsuitable for hair reaching the final stages of breakdown, when the mass of mycelium is so thick that it causes distortion of the phase contrast picture. As the hair is completely eliminated by the treatment, whole hair at the same stage of attack must also be examined in lactophenol, in order to determine the position of the fungal structures in relation to the hair.

(b) *Biochemical studies*. Keratinized tissues contain, in addition to several types of keratin molecules, a complex mixture of other organic compounds. A fungus may therefore grow on such tissues and disrupt them without necessarily producing keratinase. Treatment such as extraction, autoclaving, and ball milling may also denature the keratin molecule. In biochemical investigations of the breakdown of keratin the choice and treatment of the keratin source is therefore very important. This aspect has been reviewed in detail by Mathison (1964) and Chesters and Mathison (1963).

Doubt has been expressed that the dermatophytes are able to digest keratin, but keratinolytic activity has now been demonstrated for a few species (Chesters and Mathison, 1963; Weary *et al.*, 1965) using chromatographic and other techniques to analyse the breakdown products which accumulate after growth of the fungus in a medium containing keratin. Recently Yu *et al.* (1968) have described a method by which they isolated and purified an extracellular keratinase from *T. mentagrophytes*.

5. *Maintenance and preservation*

Most dermatophytes are maintained with difficulty because of their tendency to become "pleomorphic", i.e., to mutate with the formation of a whitish, fluffy mycelium bearing atypical microconidia and few or no macroconidia. This mutation occurs most rapidly when cultures are stored at relatively high temperatures and on media with a high sugar content (Sabouraud, 1910; Klein, 1964a, b). Cultures which have become overgrown with pleomorphic mycelium can sometimes be recovered by pulling aside the fluffy surface growth and subculturing from the typical growth beneath, preferably by single spore isolation. Some species, such as *E. floccosum*, are particularly difficult to keep in good condition, and it is advisable to keep important strains of these in several different ways.

(a) *Storage at room temperature.* Cultures kept at room temperature must be subcultured frequently, the fast-growing and granular species at least once a month and those with slow-growing glabrous colonies every 2–3 months. Preferably they should be kept in a cool, dark place and on a selection of media. In addition to the usual agar media the soil agar (20 g soil, 2 g agar, 100 ml water) recommended by Vanbreuseghem and Van Brussel (1952) may be useful.

(b) *Storage in the refrigerator.* Most species may be stored in a refrigerator at 5°–10°C and subcultured every 4–6 months. Exceptions are *M. audouinii*, *E. floccosum*, *T. violaceum*, and *T. schoenleinii*, which do not survive refrigeration for this length of time.

(c) *Deep-freeze storage.* This appears to be the most generally successful method for keeping dermatophytes. Ajello *et al.* (1963) recommended culturing the isolate on agar medium in a screw-capped bottle, allowing it to develop for 8–10 days, then screwing down the cap tightly and storing the culture at −20°C or slightly lower. Meyer (1955) found that most of the isolates she studied survived when small fragments of culture were placed in human blood plasma or litmus milk and stored at −22° or −52°C for 2 years.

There are no published data available on the survival of dermatophytes under liquid nitrogen, but it seems likely that this method would be at least as successful as deep-freeze storage.

(d) *Lyophilization.* This is usually most satisfactory for species with granular colonies which produce large numbers of conidia. *E. floccosum* can be lyophilized only if very young cultures (not more than 5 days old) are used, and even then the survival rate is usually poor. Species with poorly sporing or glabrous colonies do not usually survive lyophilization, but Salazar Leite and Antunes (1955) found that no loss of viability occurred after 2 years when several species including *T. violaceum* and *T. schoenleinii* were suspended in 4% peptone solution and lyophilized in ampoules containing 1 ml of the suspension, the ampoules being stored at 4°C.

(e) *Storage under mineral oil*. Dermatophytes remain viable for long periods under mineral oil, but although pleomorphism is delayed it is not prevented (Ajello 1951; Little and Gordon, 1967), and cultures should be renewed about once a year. The method is most suitable for species with glabrous colonies. *T. violaceum* loses its violet pigment under oil, but reverts to normal when subcultured.

(f) *Maintenance in water*. This method, which has been advocated by Castellani in a series of papers (Castellani, 1961, 1963 *et passim*), involves placing small fragments of a culture on agar medium into tubes or bottles containing about 6–10 ml sterile distilled water, taking care to transfer as little of the agar medium as possible. If plugged tubes are used they may be sealed to prevent evaporation. The cultures are then stored at room temperature. Castellani recommended subculturing and renewing water cultures once a year, but claimed that the fungi will keep almost indefinitely by this method providing the water level is maintained.

(g) *Maintenance of cleistothecial isolates*. It has been found that during routine subculture of fertile isolates of heterothallic dermatophytes the isolates frequently degenerate to sterile, single mating type strains. Kwon-Chung (1967a) found that segregation of + and − components occurs during colony development of such isolates, resulting in + and − sectors, so that only one mating type is sometimes selected during subculture. She therefore recommended that to ensure maintenance of both mating types the isolates should be transferred as masses of spores and mycelium scraped from the whole surface of the culture or, alternatively, that the fertile isolates should be maintained on soil extract-hair agar (III, D) by transferring cleistothecia. Cleistothecial cultures can also be preserved by deep-freeze storage or lyophilization.

C. Special identification techniques

Some dermatophytes are difficult to identify rapidly by their appearance in culture on routine media, and a number of biochemical and other tests have been devised to aid in the identification of these in the diagnostic laboratory. A selection of these techniques is given below, while a recent review of the subject is given by Murray (1968).

1. *Clearing of ink blue*

This technique, devised by Baxter (1965), is based on the fact that dermatophytes and related keratinophilic fungi increase the pH of the medium while many saprophytic fungi make the medium more acid. Ink blue, which changes from blue to colourless at between pH 6·5 and 7·2, is incorporated in Sabouraud's glucose agar at 0·05%. A zone of decoloration occurs around colonies of dermatophytes and related species after growth for about 7 days at 25°C on the medium. This method can be used to help decide whether an isolate is a dermatophyte or not, and has been adapted to aid recognition of dermatophytes in primary cultures by incorporation of ink blue into the isolation medium (Quaife, 1968).

2. *Vitamin Requirements*

The media used for these tests are the basal medium (*a*) without added vitamins, (*b*) with added thiamine, (*c*) with added inositol, (*d*) with added thiamine and inositol, and (*e*) with added nicotinic acid. The basal medium is that described by Georg and Camp (1957), and consists of—

10% Casein hydrolysate, vitamin free	25 ml
(or Casamino acid, vitamin free	2·5 g)
Glucose	40 g
$MgSO_4$	0·1 g
KH_2PO_4	1·8 g
Agar (Difco-Bacto)	20 g
Distilled water	1 litre

This is heated to dissolve the agar, adjusted to pH 6·8, and 2 ml of the appropriate stock vitamin solutions are then added to 100 ml quantities of the basal medium. The resultant media are then tubed, autoclaved at 120°C for 10 min, slanted, and stored at 5°C.

The stock vitamin solutions are—

(i) thiamine: 10 mg thiamine hydrochloride in 1000 ml distilled water, pH adjusted to 4–5.

(ii) inositol: 250 mg isoinositol in 100 ml distilled water.

(iii) nicotinic acid: 10 mg in 100 ml distilled water.

These may be sterilized by autoclaving at 120°C for 10 min and stored at 5°C.

For the tests, acid-clean glassware must be used and only a very small fragment of culture of the test fungus may be used as inoculum, to avoid carry-over of vitamins.

Some vitamin requirement patterns which have been found useful for the routine identification of dermatophytes are shown in Table II.

3. *Amino Acid Requirements*

This test is based on the fact that *T. megninii* has a complete requirement for the amino acid histidine; it can thus be distinguished from *T. gallinae*, with which it is sometimes confused (Georg and Camp, 1957). The basal medium is similar to that used for testing vitamin requirements except that 1·5 g NH_4NO_3 is substituted for the casein hydrolysate.

The isolates being tested are cultured on the basal medium with and without added histidine. Histidine is incorporated by adding to 100 ml of basal medium, which has been heated to melt the agar, 2 ml of a stock histidine solution (containing 150 mg L-histidine in 100 ml distilled water, autoclaved at 120°C for 10 min and stored at 5°C).

4. Enzymatic activity

(a) *Hydrolysis of urea*. Philpot (1967) devised the following modification of Christensen's medium (Christensen, 1946) for testing dermatophytes for urease activity—

Peptone (Oxoid mycological)	1 g
NaCl	5 g
KH_2PO_4	2 g
Glucose	5 g
Agar	20 g
Distilled water	to 1 litre

These are dissolved by heating, and 6 ml phenol red solution (0·2% in 50% alcohol) are added. The medium is then autoclaved at 115°C for 15 min, cooled to 50°C, 100 ml urea are added (20% aqueous solution, sterilized by filtration), and the medium is distributed into sterile bottles and sloped.

TABLE II

Vitamin requirements of some Trichophyton species
(adapted from Georg and Camp, 1957)

Species	Growth on basal medium alone	Growth on basal medium†			
		+ inositol	+ thiamine	+ thiamine and inositol	+ nicotinic acid
T. verrucosum					
84%	0	(±)	0	4+	—
16%	0	0	4+	4+	—
T. schoenleinii	4+	4+	4+	4+	—
T. concentricum					
50%	4+	4+	4+	4+	—
50%	2+	2+	4+	4+	—
T. tonsurans	± to 1+	—	4+	—	—
T. mentagrophytes	4+	—	4+	—	4+
T. rubrum	4+	—	4+	—	—
T. equinum var. equinum	0	—	—	—	4+
T. equinum var. autotrophicum	4+	—	—	—	4+

† ± = trace of submerged growth about the inoculum; 4+ = maximum growth for the series of tubes, growth on other tubes being judged by comparison.

After growth of urease positive strains on this medium at 26°C for 7 days a deep red colour develops throughout the medium.

The test has been suggested as an aid in the differentiation of atypical isolates of *T. mentagrophytes*, which splits urea rapidly, and *T. rubrum*, which splits it slowly or not at all. It may also be useful for distinguishing between *T. schoenleinii* (positive) and *T. concentricum* (Rosenthal and Sokolosy, 1965) and *T. gallinae* (positive) and *T. megninii* (Murray, 1968).

(b) *Hydrolysis of other substances.* Rosenthal and Sokolsky (1965) found that *T. verrucosum* does not hydrolyse tyrosine, and thus differs from *T. schoenleinii* and *T. concentricum*. Rippon (1967) tested a large number of dermatophyte isolates for their ability to clear particulate elastin (1%) suspended in nutrient agar; one interesting finding with possible future application was that all + mating strains of *N. fulva* (stat. conid. *M. fulvum*) are elastase positive and all − strains are elastase negative.

5. *Pigment Production*

The pink to red pigment produced by *T. rubrum* is a characteristic of importance in the identification of this species. Some isolates fail to produce the pigment readily and various media have been advocated for stimulation of pigment production. Bocobo and Benham (1949) recommended cornmeal agar enriched with 1% dextrose and this has been widely used, but Ajello and Georg (1957) have noted that production of a pink to red pigment on this medium cannot be used as an absolute criterion in the identification of *T. rubrum* as some isolates produce only a yellowish pigment on the medium, while some isolates of *T. mentagrophytes* also produce a yellow, pink, or red pigment on cornmeal dextrose agar. More recently Baxter (1963) has recommended a medium containing 0·25% lab-lemco, 0·5% dextrose, and 2% agar. He found that this was more satisfactory than cornmeal dextrose agar for stimulating pigment production by *T. rubrum*. Also, on his medium none of the isolates of *T. mentagrophytes* which produced a red pigment on Sabouraud's glucose agar produced a red pigment which could be confused with that of *T. rubrum*.

Baxter noted that cottonwool-plugged tubes are the most suitable culture vessels for studying pigment production as they allow easy diffusion of oxygen which is required by the pigment forming system. He found that pigment production was most rapid when the inoculum is placed near the top of the tube, onto the thin part of the agar, possibly because of the greater oxygen tension in this part of the slope, and that it is maximal at pH 6·5–7·5 and inhibited by light. Some nutritional factors which affect pigment production are also discussed by Silva (1953).

Baxter also found that it is not possible to differentiate dermatophytes which produce a reddish pigment by extracting the pigment and examining it chromatographically using a chloroform–petroleum ether system.

6. *Culture on Rice Grain Medium*

This was suggested by Conant (1936) as a possible method for differentiating *M. audouinii*, which discolours the grains apricot buff but fails to produce any aerial mycelium, from *M. canis*, which produces a dense, heavily sporing aerial mycelium. The medium is made by placing 8g polished rice grains and 25 ml distilled water in cottonwool-plugged conical flasks, and autoclaving at 120°C for 15 min.

7. *Analysis of Cell Walls and Contents*

Murray (1968) noted that although a good deal of work has been done on the chemical constitution of dermatophytes the results were unlikely to be of much help to the taxonomist. He briefly reviewed some differences which have been reported. More recently Shechter *et al.* (1968a, b) have studied the culture filtrate proteins of several dermatophytes by disc electrophoresis, and analysed their results by numerical taxonomic methods. They found that their results were consistent with current conventional taxonomy of the dermatophytes (Landau *et al.*, 1968).

8. *Serological Studies*

A comparative immunoelectrophoretic analysis of antigens of seventeen dermatophyte species was made by Andrieu *et al.* (1968). Their findings supported certain taxonomic opinions, such as the placing of *M. ferrugineum* in *Microsporum* rather than in *Trichophyton*, but did not allow resolution of certain other problems, such as the precise status of *M. langeronii* in relation to *M. audouinii*.

D. Methods for obtaining the sexual states of dermatophytes

This is a relatively new field of research and sexual states have been described for only a few species. All of these have been placed in two genera of the Gymnoascaceae, *Arthroderma* and *Nannizzia*. One of the most up-to-date reviews of the subject is given by Kunert and Otčenášek (1968), while Ajello (1968) lists all known species.

All but one (*A. ciferrii*) of the sexual states described so far are heterothallic, and the technique most widely used for obtaining ascocarps (cleistothecia) in preliminary investigations has involved the culture of compatible mating strains together on a mixture of soil and keratin. This was the medium used by Nannizzi (1927) when he discovered the first known sexual state of a dermatophyte, and was used again by Dawson and Gentles (1959), Stockdale (1961), and others when interest in the subject was revived. In more recent studies various agar media have been found suitable for inducing cleistothecium formation.

The few dermatophytes with known sexual states vary considerably in their requirements for cleistothecium formation and when a new species is being investigated it should be grown on as many media and in as many varying conditions as possible.

1. *Culture on Soil and Keratin.*

Dawson and Gentles (1959) and Dawson *et al.* (1964) found unsterilized, natural soil more suitable than soil which had been autoclaved for 1 h at 126°C on three successive days. Reinoculation of autoclaved soil with a suspension of soil fungi and bacteria a few days before inoculating with the dermatophytes gave results almost as good as when untreated soil was used (Dawson *et al.*, 1964). Stockdale (1963) found sterilized soil unsuitable for cleistothecium formation by *N. fulva*, but in most other investigations she and other authors have used autoclaved soil. Success with this may have been due in part to incomplete sterilization of the soil or to recontamination during inoculation or examination of the cultures. Weitzman and Silva–Hutner (1967) found some soils unsuitable for obtaining cleistothecia, but no critical studies of this aspect have been published.

Most keratin sources will support cleistothecium formation, but the most widely used and generally successful are horse tail or mane hair, child's hair, or the outer feathers of chickens. Adult human hair is not generally satisfactory (Dawson *et al.*, 1964), but can be used for *N. fulva* (Stockdale, 1963). Dawson *et al.* (1964) found that preliminary extensive extraction with water or ether, autoclaving, and pigmentation did not affect the efficiency of the hair as a substrate for cleistothecium formation.

Cultures on soil and keratin may be set up as follows: soil is broken down into fine particles and any stones, vegetable matter, etc. removed. The soil is then dispensed into McCartney (1 oz) bottles, loosely filling them to the neck; this provides the correct quantity of soil for an 80 mm dia. Petri dish. The bottles are then autoclaved, autoclaving for at least 1 h at 126°C on three successive days being necessary for complete sterilization of the soil. The keratin is washed with water, cut into small pieces not more than 2–3 cm long, and autoclaved in wide-necked bottles for 15 min at 120°C. When required the soil is emptied into a Petri dish and moistened with a little sterile water if necessary, several pieces of keratin are scattered over the soil, and the isolates being tested are inoculated on to the soil surface. When compatible mating strains are being cultured together both should be inoculated in several places on the plate, in close proximity to each other, to ensure that both become established and that they have the maximum opportunity of coming into contact.

2. *Culture on Agar Media*

Most dermatophytes form cleistothecia sporadically or not at all on routine mycological media. An exception is *A. simii* (stat. conid. *T. simii*) which forms cleistothecia abundantly on Sabourand's glucose agar (Stockdale *et al.*, 1965).

Agar media incorporating keratin, and sometimes also soil, have been used successfully by several workers. De Vroey (1964) found grain media of the following composition satisfactory for *M. gypseum*—

Grains (ground and sieved)	3 g
Agar	2 g
Water	100 ml

This was autoclaved for 20 min at 120°C, 10 ml quantities were poured into Petri dishes, and a few sterile horse hairs were added. Oat, barley, millet, maize, and niger (*Guizotia abyssinica*) grains gave best results. Cleistothecia were also formed on niger grain medium without horse hairs and on a medium containing 20 g soil in place of the grains. Addition of neopeptone inhibited the production of cleistothecia, as did increasing the quantity of medium in the plate to 20–30 ml.

Weitzman (1964) obtained cleistothecia of several species on Difco cornmeal agar with dextrose containing powdered hair. Children's hair was ground to a fine powder in a ball-mill for 10–12 h, approximately 250 mg of the powder was suspended in 15 ml distilled water and autoclaved at 126°C for 20 min, and a few drops of the suspension were pipetted across the centre of the agar plate. Compatible mating strains were then inoculated on opposite sides of the hair. Fertile cleistothecia were formed consistently, but more slowly, and a greater percentage of empty cleistothecium-like structures were formed than on soil and hair.

Kwon-Chung (1967b), in her studies of the genetics of *Nannizzia* species, obtained abundant cleistothecia on a medium of the following composition—

Clear soil extract	1 litre
Glucose	2 g
Yeast extract	1 g
Agar	15 g

About 2 g of finely chopped, autoclaved human hair were mixed with 300 ml portions of this medium just prior to pouring and the pH was adjusted to 6·8.

As media containing pieces of hair were not suitable for slide cultures Widra (1966) and Rhodes *et al.* (1967), for studies on ascosporogenesis in *N. grubyia* (stat. conid. *M. vanbreuseghemii*), devised a particle-free medium composed of equal quantities of Difco double-strength neutral wort agar and a soluble fraction of human hair. Either a "total protein" fraction, obtained by dissolving the hair by peracetic acid–ammonium treatment, or an "α-keratose" fraction, obtained from the protein fraction by precipitation with 1N hydrochloric acid and redissolving in M/10 ammonium hydroxide, were suitable.

In attempts to eliminate soil and hair, and to obtain a more easily reproducible and handled medium, Weitzman and Silva-Hutner (1967) investigated various modifications of the alphacel medium described by Sloan *et al.* (1960) and found the following medium satisfactory—

$MgSO_4.7H_2O$	1 g
KH_2PO_4	1 g
$NaNO_3$	1 g
Hunt's tomato paste	10 g
Beech-nut baby oatmeal	10 g
Agar	18 g
Distilled water	1 litre

The pH was adjusted to 5·6 with sodium hydroxide and the medium autoclaved at 121°C for 20 min. Tomato paste appeared to have a slightly stimulating effect and the salts were essential for some species, possibly acting as a buffer.

When compatible mating strains are being cultured together on agar media they should be inoculated not more than about 2 cm apart. Stockdale *et al.* (1965) found that when mating strains of *A. simii* were inoculated more than 40 mm apart the formation of cleistothecia was greatly reduced.

3. *Physicochemical Requirements*

The only critical study of these aspects is that of Dawson *et al.* (1964). They found that the temperature range for cleistothecium formation was more restricted than that required for growth, being 10°–24°C for *A. quadrifidum*, 15°–28°C for *A. uncinatum* (stat. conid. *T. ajelloi*), and 22°–30°C for *N. incurvata*. These species were not critical in their pH and moisture requirements, but cleistothecium formation was markedly reduced by constant bright light. In yellow, orange, red, or green light cleistothecia were formed as profusely as in darkness, but blue light was inhibitory. In most other studies cultures have been incubated in the dark at about 25°C. These conditions are probably suitable for most species, except some, such as *A. quadrifidum*, from soil which may have a slightly lower optimum temperature.

4. *Typing Compatible Mating Strains*

Stockdale (1968) found that when compatible mating strains of *A. simii* were cultured on Sabouraud's glucose agar together with isolates of other dermatophytes, the 2 inocula being placed about 20–25 mm apart, a zone of denser mycelium occurred at the junction of the colonies in some crosses, any one isolate stimulating either the + or the − strain of *A. simii*, never both. The reaction varied from negative to strongly positive depending on the species being tested. This reaction was used to type mating strains of *N. persicolor* (stat. conid. *M. persicolor*) before cleistothecia were obtained for this species (Stockdale, 1967). A similar reaction was noted by Kwon-Chung (1967b) to occur between *N. grubyia* and other *Nannizzia* species, cultured together on soil extract–hair agar. It is possible that by using compatible mating strains of several different species as the tester strains, the reaction could be used to type possible mating strains of many more species without a known perfect state.

5. *Taxonomic Considerations in Relation to the Species Concept*

Before a new perfect or imperfect dermatophyte species is described, isolates should be checked by crossing with mating types of similar, perfect species when they are known. In this way the identity of unusual variants of *N. incurvata* and *N. gypsea* have been identified (Hejtmánková–Uhrová and Hejtmánek, 1965; Gordon and Lusick, 1965; Varsavsky *et al.*, 1966). It is also desirable to use more than one tester strain of each mating type, as Varsavsky *et al.* (1966) found that not all $+ \times -$ crosses within the same species are compatible. Even though these steps are taken, lack of ability to mate should not be taken as absolute proof that an isolate represents a new taxon, as mutants are often sexually sterile (Stockdale, 1964; Weitzman 1964).

In such mating studies it is also desirable to use cultures of imperfect species derived from single microconidia, as heterokaryosis has been shown to occur in dermatophytes (Kwon-Chung, 1967b) and microconidia are usually, though not always, uninucleate (Emmons, 1931). Kwon-Chung (1967b) also found that heterokaryotic, binucleate ascospores may occur in *Nannizzia* species, and recommended that single ascospore isolates should be obtained by dissection of asci containing eight ascospores, rather than by random selection of free ascospores.

E. Pathogenicity studies

1. *Animal Inoculation*

The traditional method for establishing an experimental ringworm infection in a laboratory animal is to clip very short the hair on the flank, scarify the exposed skin by scraping it with a scalpel, and rub into the prepared site either culture material or a heavy spore suspension in saline of the dermatophyte being used. Weitzman *et al.* (1967) have described a multi-puncture technique which they found gave a more standardized inoculum, was easier to use for inoculating a large number of animals, and more closely simulated a means by which infection may be acquired naturally than did the traditional method.

Kaplan and Georg (1957) devised a rubber collar to prevent the animal from licking or biting the inoculation site. This is made from sheet rubber such as the inner tube of a car tyre. The outer edge of the collar must extend beyond the tip of the animal's mouth, a disc of about 4½ in. dia. being satisfactory for an average-sized guinea pig. This collar may be stretched and slipped over the animal's head.

It is difficult or impossible to produce infections in experimental animals with dermatophytes with a narrow host range or low pathogenicity.

2. Inoculation of Egg Membranes

A technique for studying the *in vivo* growth of dermatophytes on the chorioallantoic membrane of the developing chick has been described by Partridge (1959), who used the method for studying the effect of serial animal passage on the morphology and pathogenicity of *T. rubrum*. The method is of advantage in such studies as growth and production of lesions by dermatophytes on the membrane are more rapid than those on other laboratory animals.

REFERENCES

Ajello, L. (1968). *Sabouraudia*, **6**, 147–159.
Ajello, L., and Georg, L. K. (1957). *Mycopath. Mycol. appl.*, **8**, 3–17.
Ajello, L., Grant, V. Q., and Gutzke, M. A. (1951). *A.M.A. Archs Derm. Syph.*, **63**, 747–749.
Ajello, L., Georg, L., Kaplan, W., and Kaufman, L. (1963). "Laboratory Manual for Medical Mycology." Public Health Service Publication No. 994, U.S. Government Printing Office, Washington, D.C.
Andrieu, S., Biguet, J., and Laloux, B. (1968). *Mycopath. Mycol. appl.*, **34**, 161–185.
Austwick, P. K. C. (1954). *Vet. Rec.*, **66**, 423–425.
Barfatani, M., Munn, R. J., and Schjeide, O. A. (1964). *J. invest. Derm.*, **43**, 231–233.
Barker, S. A., and Trotter, M. D. (1960). *Nature, Lond.*, **188**, 232–233.
Barker, S. A., Cruickshank, C. N. D., Morris, J. H., and Wood, S. R. (1962). *Immunology*, **5**, 627–632.
Baxter, M. (1963). *Sabouraudia*, **3**, 72–80.
Baxter, M. (1965). *J. invest. Derm.*, **44**, 23–25.
Benham, R. W. (1947). *In* "Biology of Pathogenic Fungi" (W. J. Nickerson, Ed.), pp. 63–70. Chronica Botanica Company, Waltham, Mass.
Bocobo, F. C., and Benham, R. W. (1949). *Mycologia*, **41**, 291–302.
Bocobo, F. C., and Curtis, A. C. (1958). *Mycologia*, **50**, 164–168.
Bocobo, F. C., Miedler, L. J., and Eadie, G. A. (1964). *Sabouraudia*, **3**, 178–179.
Carlier, G. I. M. (1948). *Br. J. Derm.*, **60**, 61–63.
Castellani, A. (1961). *J. trop. Med. Hyg.*, **64**, 60–63.
Castellani, A. (1963). *Mycopath. Mycol. appl.*, **20**, 1–6.
Chattaway, F. W., Toothill, C., and Barlow, A. J. E. (1962). *J. gen. Microbiol.* **28**, 721–732.
Chesters, C. G. C., and Mathison, G. E. (1963). *Sabouraudia*, **2**, 225–237.
Chick, E. W., and Behar, V. S. (1961). *J. invest. Derm.*, **37**, 103–104.

458 PHYLLIS M. STOCKDALE

Clayton, Y. M., and Noble, W. C. (1963). *Trans. a. Rep. St John's Hosp. derm. Soc.*, *Lond.*, 49, 36–38.
Christensen, W. B. (1946). *J. Bact.*, 52, 461–466.
Codner, R. C., Cruickshank, C. N. D., Trotter, M. D., and Wood, S. R. (1961). *Sabouraudia*, 1, 116–122.
Cohen, M. N. (1954). *J. invest. Derm.*, 22, 9–10.
Conant, N. F. (1936). *Archs Derm. Syph.*, 33, 665–683.
Connolle, M. D. (1965). *Sabouraudia*, 4, 45–48.
Cooke, W. B. (1952). *Mycologia*, 44, 245.
Cornbleet, T., Schorr, H. C., and Popper, H. (1943). *Archs Derm. Syph.*, 48, 282–287.
Davidson, A. M., and Gregory, P. H. (1935). *J. Am. med. Ass.*, 105, 1262–1264.
Dawson, C. O., and Gentles, J. C. (1959). *Nature, Lond.*, 183, 1345–1346.
Dawson, C. O., Gentles, J. C., and Brown, E. M. (1964). *Sabouraudia*, 3, 245–250.
De Vroey, C. (1964). *Annls Soc. belge Méd. trop.*, 44, 831–840.
Drouhet, E., and Mariat, F. (1953). *In* "Symposium. Nutrizione e Fattori di Crescita, 7–11 Settembre 1953", 113–150. Instituto Superiore di Sanita', Roma.
Drouhet, E., Marcel, M., and Labonde, J. (1967). *Bull. Soc. fr. Derm. Syph.*, 74, 719–724.
Dvořák, J., Hubálek, Z., and Otčenášek, M. (1968). *Archs Derm.*, 98, 540–542.
Edwards, R. W., and Hartman, E. (1952). *Lloydia*, 15, 39.
Emmons, C. W. (1931). *Mycologia*, 23, 87–95.
Emmons, C. W., Binford, C. H., and Utz, J. P. (1963). "Medical Mycology". Henry Kimpton, London.
English, M. P. (1963). *Sabouraudia*, 2, 115–130.
English, M. P. (1966). *Sabouraudia*, 4, 219–222.
English, M. P. (1968a). *Sabouraudia*, 6, 218–227.
English, M. P. (1968b). *Br. J. Derm.*, 80, 282–286.
English, M. P., and Barnard, N. H. (1955). *Trans. Br. mycol. Soc.*, 38, 78–82.
Epstein, S. (1945). *Archs Derm. Syph.*, 3, 209.
Evolceanu, R., and Donciu, G. (1963). *Dermato-Vener.*, 8, 49–54.
Fraser, G. (1961). *Trans. Br. mycol. Soc.*, 44, 441–448.
Friedman, L., Derbes, V. L., Hodges, E. P., and Sinski, J. T. (1960). *J. invest. Derm.*, 35, 3–5.
Gadrat, J., Bazex, A., and Dupré, A. (1952). *Bull. Soc. fr. Derm. Syph.*, 59, 375–377.
Gentles, J. C. (1956). *J. clin. Path.*, 9, 374–377.
Gentles, J. C., Dawson, C. O., and Connolle, M. D. (1965). *Sabouraudia*, 4, 171–175.
Georg, L. K. (1953). *A.M.A. Archs Derm. Syph.*, 67, 355–361.
Georg, L. K., and Camp, L. B. (1957). *J. Bact.*, 74, 113–121.
Goldberg, H. C. (1965). *Archs Derm.*, 92, 103.
Gordon, M. A. (1951a). *A.M.A. Archs Derm. Syph.*, 63, 343–346.
Gordon, M. A. (1951b). *Mycologia*, 43, 524–535.
Gordon, M. A. (1953). *J. invest. Derm.*, 20, 201–206.
Gordon, M. A., and Lusick, C. A. (1965). *Archs Derm.*, 91, 558–562.
Griffin, D. M. (1960). *Trans. Br. mycol. Soc.*, 43, 583–596.
Haalstra, R. T. (1965). *Vet. Rec.*, 77, 824–825.
Hanel, E., and Kruse, R. H. (1967). "Laboratory-acquired Mycoses". Miscellaneous Publication 28, Department of the Army, Fort Detrick, Frederick, Maryland.
Hejtmánková–Uhrová, N., and Hejtmánek, M. (1965). *Mycopath. Mycol. appl.*, 25, 183–194.

Ive, F. A. (1966). *Br. J. Derm.*, **78**, 219–221.

Kaplan, W., and Georg, L. K. (1957). *Mycologia*, **49**, 604–605.

Karling, J. S. (1946). *Am. J. Bot.*, **33**, 751–757.

Keddie, F., Orr, A., and Liebes, D. (1961). *Sabouraudia*, **1**, 108–111.

Keddie, F., and Shadomy, S. (1963). *Sabouraudia*, **3**, 21–25.

Klein, D. T. (1964a). *J. gen. Microbiol.*, **34**, 125–130.

Klein, D. T. (1964b). *Mycologia*, **56**, 656–661.

Kunert, J., and Otčenášek, M. (1968). *Česká Mykol.*, **22**, 56–67.

Kwon-Chung, K. J. (1967a). *Sabouraudia*, **6**, 37–41.

Kwon-Chung, K. J. (1967b). *Sabouraudia*, **6**, 5–13.

Landau, J. W., Shechter, Y., and Newcomer, V. D. (1968). *J. invest. Derm.*, **51**, 170–176.

Leach, B. E., Ford, J. H., and Whiffen, A. J. (1947). *J. Am. chem. Soc.*, **69**, 474.

Lewis, G. M., Hopper, M. E., Wilson, J. W., and Plunkett, O. A. (1958). "An Introduction to Medical Mycology", 4th Edition. The Year Book Publishers, Chicago.

Little, G. N., and Gordon, M. A. (1967). *Mycologia*, **59**, 733–736.

Lofgren, R. C., and Batts, E. E. (1952). *Stanford med. Bull.*, **10**, 96–98.

Lurie, H. I., and Borok, R. (1955). *Mycologia*, **47**, 506–510.

Mackenzie, D. W. R. (1961). *Sabouraudia*, **1**, 58–64.

Mackenzie, D. W. R. (1963). *Br. med. J.*, **1**, 363–365.

Mandels, G. R., Stahl, W. H., and Levinson, H. S. (1948). *Text. Res.*, **18**, 224–231.

Mariat, F., and Tapia, G. (1966). *Annls Parasit. hum. comp.*, **41**, 627–634.

Martin-Scott, I. (1952). *Br. J. Derm.*, **64**, 257–273.

Mathison, G. E. (1964). *Annls Soc. belge Méd. trop.*, **44**, 767–791.

McDonough, E. S., Ajello, L., Ausherman, R. J., Balows, A., McClellan, J. T., and Brinkman, S. (1961). *Am. J. Hyg.*, **73**, 75–83.

Meyer, E. (1955). *Mycologia*, **47**, 664–668.

Murray, I. G. (1968). *J. gen. Microbiol.*, **52**, 213–221.

Nannizzi, A. (1927). *Atti Acad. Fisiocr. Siena*, **10**, 89–97.

Noval, J. J., and Nickerson, W. J. (1959). *J. Bact.*, **77**, 251–263.

Partridge, B. M. (1959). *J. invest. Derm.*, **32**, 605–619.

Philpot, C. (1967). *Sabouraudia*, **5**, 189–193.

Pier, A. C., Richard, J. L., and Farrell, E. F. (1964). *Am. J. vet. Res.*, **25**, 1014–1020.

Porto, J. A. (1953). *J. invest. Derm.*, **21**, 229–231.

Quaife, R. A. (1968). *J. med. Lab. technol.*, **25**, 227–232.

Rebell, G., Taplin, D., and Blank, H. (1964). "Dermatophytes. Their recognition and identification". Dermatology Foundation of Miami, Miami.

Rhodes, H. J., Potter, B., and Widra, A. (1967). *Mycopath. Mycol. appl.*, **33**, 345–348.

Rippon, J. W. (1967). *Science, N.Y.*, **157**, 947.

Roberts, D. S. (1967). *Vet. Bull.*, **37**, 513–521.

Rosenthal, S. A., and Sokolsky, H. (1965). *Derm. int.*, **4**, 72–79.

Sabouraud, R. (1910). "Maladies du Cuir Chevelu. III.-Les Maladies Cryptogamiques. Les Teignes". Masson et Cie, Paris.

Salazar Leite, A., and Antunes, M. M. (1955). *C.r. Séanc. Soc. Biol.*, **149**, 1813–1815.

Sartory, A., Sartory, R., Meyer, J., and Schurmann, W. (1950). *C.r. Hebd. Séanc. Acad. Sci., Paris*, **230**, 1608–1609.

Sharvill, D. (1952). *Br. J. Derm.*, **64**, 329–333.

Shechter, Y., Landau, J. W., Dabrowa, N., and Newcomer, V. D. (1968a). *Sabouraudia*, **6**, 133–137.

Shechter, Y., Landau, J. W., Dabrowa, N., and Newcomer, V. D. (1968b). *J. invest. Derm.*, **51**, 165–169.

Silva, M. (1953). *Trans. N.Y. Acad. Sci.*, *Ser. II*, **15**, 106–110.

Sloan, B. J., Routien, J. B., and Miller, V. P. (1960). *Mycologia*, **52**, 47–63.

Somerville, D. A., and Marples, M. J. (1967). *Sabouraudia*, **6**, 70–76·

Sternberg, T. H., and Keddie, F. M. (1961). *Archs Derm.*, **84**, 999–1003.

Stockdale, P. M. (1953). *Biol. Rev.*, **28**, 84–104.

Stockdale, P. M. (1961). *Sabouraudia*, **1**, 41–48.

Stockdale, P. M. (1963). *Sabouraudia*, **3**, 114–126.

Stockdale, P. M. (1964). *Annls Soc. belge Méd. trop.*, **44**, 821–830.

Stockdale, P. M. (1967). *Sabouraudia*, **5**, 355–359.

Stockdale, P. M. (1968). *Sabouraudia*, **6**, 176–181.

Stockdale, P. M., Mackenzie, D. W. R., and Austwick, P. K. C. (1965). *Sabouraudia*, **4**, 112–123.

Swartz, J. H., and Conant, N. F. (1936). *Archs Derm. Syph.*, **33**, 291–305.

Swartz, J. H., and Lamkins, B. E. (1964). *Archs Derm.*, **89**, 89–94.

Taschdjian, C. L. (1955). *J. invest. Derm.*, **24**, 77–80.

Taschdjian, C. L., and Muskatblit, E. (1953). *A.M.A. Archs Derm. Syph.*, **68**, 579–582.

Uzunov, P. Ya. (1967a). *Vest. Derm. Vener.*, **41**, (9), 59–61.

Uzunov, P. (1967b). *In* "Recent Advances of Human and Animal Mycology" (L. Chmel, E. Hegyi and J. C. Gentles, Eds.), 51–53. Publishing House of the Slovak Academy of Sciences, Bratislava.

Vanbreuseghem, R. (1952). *Annls Soc. belge Méd. trop.*, **32**, 173–178.

Vanbreuseghem, R., and van Brussel, M. (1952). *Annls Soc. belge Méd. trop.*, **32**, 169–172.

Varsavsky, E., Weitzman, I., and Reca, M. E. (1966). *Sabouraudia*, **4**, 242–243.

Walshe, M. M., and English, M. P. (1966). *Br. J. Derm.*, **78**, 198–207.

Weary, P. E. (1968). *Archs. Derm.*, **98**, 408–422.

Weary, P. E., Canby, C. M., and Cawley, E. P. (1965). *J. invest. Derm.*, **44**, 300–310.

Weary, P. E., and Graham, G. F. (1966). *J. invest. Derm.*, **47**, 55–57.

Weidman, F. D. (1927). *Archs Derm. Syph.*, **15**, 415–450.

Weitzman, I. (1964). *Mycologia*, **56**, 425–435.

Weitzman, I., and Silva-Hutner, M. (1967). *Sabouraudia*, **5**, 335–340.

Weitzman, I. Silva-Hutner, M., and Kozma, I. (1967). *Sabouraudia*, **5**, 360–365.

Widra, A. (1966). *Mycopath. Mycol. appl.*, **30**, 141–144.

Yu, R. J., Harmon, S. R., and Blank, F. (1968). *J. Bact.*, **96**, 1435–1436.

Zaias, N., and Taplin, D. (1966). *Archs Derm.*, **93**, 608–609.

CHAPTER XVI

Fungi Pathogenic for Man and Animals: 2. The Subcutaneous and Deep-seated Mycoses

HELEN R. BUCKLEY

Division of Laboratories and Research, New York State Department of Health, Albany, New York, U.S.A.

Diseases caused by fungi may be superficial, subcutaneous or deep-seated in nature. This section is concerned with the organisms responsible for the subcutaneous and deep-seated mycoses, techniques for their isolation and problems encountered in their identification. Table I lists the diseases and their causative agent or agents.

I. GENERAL TECHNIQUES

A. Handling of pathogenic fungi

Utmost care should be taken in the handling of *Histoplasma capsulatum*, *H. duboisii*, *Blastomyces dermatitidis*, and *Coccidioides immitis*. Inhalation of spores from these fungi can be lethal to the laboratory worker. When dealing with these fungi an inoculating cabinet should always be used. A specimen that is suspected of containing any of these organisms should

TABLE I

Subcutaneous and deep-seated mycoses

Disease	Causative agent or agents†
Actinomycosis	*Actinomyces bovis, A. israelii*
Aspergillosis	*Aspergillus fumigatus, A. terreus, A. flavus, A. niger. A. nidulans, A. sydowii*
Blastomycosis	*Blastomyces dermatitidis*
Candidiasis	*Candida albicans* and other *Candida* spp.
Chromoblastomycosis	*Phialophora verrucosa, P. pedrosoi, P. compacta, P. jeanselmei, Cladosporium carrionii*
Cladosporiosis	*Cladosporium trichoides*
Coccidioidomycosis	*Coccidioides immitis*
Cryptococcosis	*Cryptococcus neoformans*
Histoplasmosis	*Histoplasma capsulatum, H. duboisii, H. farciminosum*
Mycetoma	
(a) Actinomycetoma	*A. bovis, A. israelii, Nocardia asteroides, N. madurae, N. pelletieri, Streptomyces somaliensis*
(b) Eumycetoma	*Madurella mycetomi, M. grisea, Allescheria boydii, Phialophora jeanselmei, Pyrenochaeta romeroi, Cephalosporium acremonium, Leptosphaeria senegalensis, Fusarium* sp.
Nocardiosis	*Nocardia asteroides, N. brasiliensis, N. caviae*
Paracoccidioidomycosis	*Paracoccidioides brasiliensis*
Phycomycosis	
(a) Subcutaneous	*Basidiobolus ranarum*
(b) Systemic	*Absidia ramosa, A. corymbifera, Rhizopus microsporus, R. oryzae, R. arrhizus*
Rhinosporidosis	*Rhinosporidium seeberi*
Sporotrichosis	*Sporothrix schenckii*
Torulopsosis	*Torulopsis glabrata*
Mycotic abortion	*Aspergillus fumigatus, A. terreus, A. flavus, A. nidulans, Allescheria boydii, Absidia ramosa, A. corymbifera, Mortierella* spp., *Rhizopus* spp., *Candida* spp.

† In this era of immunosuppressive and corticosteroid therapy any fungus, capable of growth at 37°C, may be involved in a disease process.

always be inoculated into cotton-plugged tubes and never on to Petri dishes. Slide cultures should be avoided since spores from these organisms are readily airborne and hazardous. A strict sterile technique should be maintained when dealing with any of the pathogenic fungi.

B. Collection and preservation of clinical material

1. *Pus, material from draining sinuses and unopened sinus tracts, and material aspirated from subcutaneous abscesses*

These should be taken aseptically, placed into sterile containers and taken immediately to the laboratory for investigation.

2. *Sputum specimens*

Sputum specimens should always include prior lavage of the mouth with an antiseptic. These specimens should always be fresh; a 24 h sputum specimen is not desirable for mycological study since (*a*) the bacterial overgrowth may inhibit the growth of the fungus present, (*b*) some organisms, such as *Candida*, multiply readily and their presence cannot be assessed, (*c*) the overgrowth of saprophytic flora impedes growth of the pathogens.

3. *Urine*

Urine should be collected by catheterization or as mid-stream sample to avoid contamination from the vaginal tract which normally harbours yeasts such as *Torulopsis glabrata* and *Candida albicans*.

4. *Cerebrospinal fluid*

At least 3 ml of CSF is required for mycological studies. The more fluid collected the better the chance of isolating the organism.

5. *Blood, bone marrow, biopsy material*

Blood, bone marrow, biopsy material and scrapings from ulcers are collected aseptically and brought to the laboratory for investigation as quickly as possible.

6. *Facilities for cultivation*

If these are not available, the material should be posted to the laboratory in the quickest manner possible. Sputum should be mixed with a penicillin–streptomycin solution or chloramphenicol (see Appendix), except in the case of suspected actinomycete infection where the causal organisms are susceptible to antibacterial antibiotics. A small amount of sterile saline may be added to tubes containing swabs to avoid desiccation. All material should be refrigerated until posted.

C. Examination of clinical material

1. *Direct mounts*

Direct mounts from the pathological material often reveal the causative agent. This is done by mounting a portion of the specimen in 10–20% NaOH or KOH teasing out and placing a cover slip on top of the preparation and gently heating over a bunsen burner.

The following method has been found useful in distinguishing between mucoraceous fungi and other types (Balogh, 1964). The material is first stained by the following method: dissolve 20 g of potassium hydroxide in 50 ml distilled water. Cool and add 50 ml of *Parker Super Permanent Blue-*

Black Quink (note: other colours and brands of ink are ineffective). Mount small pieces of material in stain, tease out, cover with a cover slip and apply gentle pressure, with warming if necessary. The mucoraceous hyphae in tissues such as bovine placentae, foetal stomach contents and alimentary ulcers, etc., generally take up the stain rapidly in their walls, which turn bright greenish blue. Certain other fungi will take up the stain more slowly, needing 8–24 h to reach maximum intensity (e.g. *Candida albicans*), whilst a few stain only faintly if at all (e.g. *Aspergillus fumigatus*). Cerebrospinal fluid should be spun down and the sediment examined in Nigrosen or India Ink. The supernatant is kept for serological investigation (see Section II, G). Urine should also be centrifuged and examined directly for the presence of fungal cells.

2. *Staining techniques*

Tissues for mycological examination are examined macroscopically for the presence of granules or grains. The specimen should be divided, one-half for cultural studies and the other fixed for histology. Stains used routinely in medical mycology include haematoxylin and eosin and the periodic acid–Schiff. The Gomori-Grocott silver stain is also excellent for the detection of fungi in tissues, especially when few organisms are present. Kinyoun acid-fast stain is preferred for the actinomycetes. These stains appear in the Appendix. The use of a particular stain is often dependent upon the personal choice of the investigator. It must be emphasized that the crucial confirmatory evidence of a mycotic infection is the demonstration of the organism within the affected tissue.

3. *Preparation of tissues for cultivation*

Biopsy specimens should be ground aseptically. If grains are obvious they are picked out and washed several times in sterile saline. If the specimen is contaminated it should be soaked for a few hours in a penicillin–streptomycin solution before cultivation except in cases of actinomycete-mediated infections. Sputum should be examined for fungal plugs and flecks. These are usually easy to see on macroscopic examination if the sputum is put into a sterile Petri dish. Pus or material from draining sinuses should also be investigated for granules or grains.

D. Cultivation

Table II lists the diseases which may be encountered, along with the media of choice for cultivating the causative agent. Medium with and without cycloheximide should always be used for isolation since some pathogenic fungi are sensitive to this antibiotic.

TABLE II

Media used for the isolation of pathogenic fungi

Disease	Media	Incubation temperature (°C)
Actinomycosis	Blood agar	37†
	Brain-heart infusion (BHI) glucose agar	37†
	Brain-heart infusion glucose broth	37†
	Brewer's thioglycollate broth	37
Aspergillosis	Glucose–peptone agar with and without antibiotics	25 and 37
Blastomycosis	Glucose–peptone agar	25
	Blood agar	37
	Cystine heart hemoglobin (CHH)	37
	BHI	37
Candidiasis	Glucose–peptone agar with and without antibiotics	25 and 37
Chromoblastomycosis	Glucose–peptone agar	25
Coccidioidomycosis	Glucose–peptone agar with and without antibiotics	25
Cryptococcosis	Glucose–peptone agar	25 and 37
Histoplasmosis	Glucose–peptone agar	25
	Blood agar	37
	BHI	37
	CHH	37
Eumycetoma	Glucose–peptone agar with and without antibiotics	25 and 37
Actinomycetoma	Glucose–peptone agar	25 and 37
Nocardiosis	Glucose–peptone agar	25 and 37
Paracoccidioidomycosis	Glucose–peptone agar	25
	Blood agar	37
	BHI	37
	CHH	37
Phycomycosis	Glucose–peptone agar with and without antibiotics	25 and 37
Rhinosporidiomycosis	The organism has never been cultured	
Sporotrichosis	Glucose–peptone agar	25
	Blood agar	37
	BHI	37
	CHH	37

† Incubated anaerobically with or without CO_2.

1. *Urine*

Urine, catheterized or mid-stream samples; dilution plating is advised when yeast infections are suspected. Ahearn *et al.* (1966) suggest 10^5 yeasts/ml of urine to be indicative of infection.

2. *Biopsy specimens*

Biopsy specimens should be placed on to the appropriate media and the cultures examined for up to 3 weeks.

3. *Cerebrospinal fluid*

Cerebrospinal fluid is spun down and the sediment cultured on to glucose peptone agar.

4. *Inoculation of mice*

Inoculation of mice with suspensions of material submitted for study is often advantageous in the case of suspected histoplasmosis. Six mice are inoculated intraperitoneally with approximately 0·2 ml of the specimen. The mice are sacrificed at weekly intervals after 3 weeks, cultivating spleen, liver and lungs.

II. SPECIAL TECHNIQUES

A. Identification

1. *Yeasts*

Yeasts are defined as organisms in which the unicellular condition is predominant and vegetative reproduction is generally by budding. Because of the ubiquity of some of these organisms, it is extremely important to demonstrate their presence in direct examination of the material. Mackenzie (1965) states that an assessment of the relative proportions of Y (yeast) and M (mycelium) forms should be made since this may provide information on the degree of involvement of the fungus in the pathological condition. This is confirmed by Hurley and Stanley (1969). Using murine kidney epithelial cells, they indicate the greater importance of the M phase in the progression of the candida lesion because of the more widespread involvement of cells by organisms with rapid growth rates, *but* they do not demonstrate a qualitative difference in the cytopathic effects of M and Y. They suggest that *in vivo* the Y form may initiate infection, as it does experimentally, and that the M phase is associated with extension of the lesion. The filamentous phase may be induced by exposure of the fungi to cellular exudates on transudates.

Table III lists characteristics of some yeasts frequently encountered in medical mycological laboratories.

TABLE III

Some criteria for the identification of yeasts

Organism	Corn meal†	Fermentation				Assimilation					Other aids
		Gl	Ma	Su	La	Gl	Ma	Su	La	KNO₃	
Torulopsis glabrata	No mycelium	+	−	−	−	+	−	−	−	−	Cells are small, round to oval
Torulopsis famata	Usually lacking mycelium. Chains of blastospores may be present	−	−	−	−	+	+	+	−	−	Cells are small, round to oval
Candida albicans	Mycelium + chlamydospores	+	+	−	−	+	+	+	−	−	Germ tube production in serum
Candida stellatoidea	Mycelium + chlamydospores	+	+	−	−	+	+	−	−	−	Germ tube production in serum
Candida parapsilosis	Mycelium	+ or ±‡	−	±‡	−	+	+	+	−	−	"Giant" mycelium often present
Candida tropicalis	Mycelium	+	+	+	−	+	+	+	−	−	Ring and layer of bubbles
Candida krusei	Mycelium	+	−	−	−	+	−	−	−	−	Pellicle
Candida guilliermondii	Mycelium sometimes well developed, sometimes rudimentary	±	−	+	−	+	+	+	−	−	
Candida pseudotropicalis	Mycelium	+	−	+	+	+	−	+	+	−	
Cryptococcus neoformans	No mycelium or chains of spores	−	−	−	−	+	+	+	+	−	Urease +, capsule, inositol +
Trichosporon cutaneum	Blastospores + arthrospores	−	−	−	−	+	+	+	−	−	Urease +
Geotrichum	Arthrospores	−	−	−	−	+	−	−	−	−	Not a yeast since reproduction is not by budding, solely by mycelial fragmentation into arthrospores

† For full description, see Lodder and Kreger van Rij (1952).
‡ A bubble may be formed.

To identify a yeast the following criteria are basic—

(*a*) Production of pseudomycelium and/or mycelium.
(*b*) Fermentation: glucose, maltose, sucrose and lactose.
(*c*) Assimilation: glucose, maltose, sucrose, lactose and potassium nitrate.

(a) *Production of pseudomycelium and/or mycelium.* The genus *Candida* is noted for its production of pseudomycelium and/or mycelium. *C. albicans* is recognized by its ability to produce large thick-walled chlamydospores. There are many media proposed for the induction of chlamydospores. Common to all of them is their low nutritive value and high carbon/nitrogen ratios. Surface reacting agents are often added. Cornmeal agar, an infusion medium prepared from ground maize meal, is the one of choice for the production of mycelium in yeasts. The addition of 1% Tween 80 to either rice or cornmeal agar enhances the production of chlamydospores of *C. albicans* (Gordon and Little, 1963).

The use of tetrazolium, bismuth sulphate and molybdenum, in which coloured colonies are supposed to be indicative of *C. albicans*, can provide only presumptive identification. The author can find no advantage to advocate routine use of these media.

The genus *Trichosporon* is noted for its production of both mycelium and arthrospores.

The genus *Torulopsis* is characterized by smaller cells (this is a generalization, since many species of *Torulopsis* have cells that reach the same dimensions as those of species of *Candida*). Most species do not produce pseudomycelium.

The genus *Cryptococcus* is generally represented by encapsulated species. Members of this genus are non-fermentative.

(b) *Fermentation.* Fermentation in the field of yeast taxonomy indicates the production of gas. There is no need to use an indicator. The amount of sugar solution utilized in the fermentations is 3%, with the exception of raffinose which is 6%. Fermentation by yeasts results in the formation of alcohol and CO_2. These tests are standardized by the use of Durham tubes.

(c) *Assimilation.* The observation of growth, i.e., an increase in the number of yeast cells in a complete medium containing one compound as a sole source of carbon or nitrogen.

2. *Aerobic actinomycetes*

Based on the work of Becker *et al.* (1965) and Gordon (1966) the following criteria can be used for the identification of aerobic actinomycetes.

Cell walls of all aerobic actinomycetes contain glucosamine, muramic acid, glutamic acid and alanine.

Major amounts of the following are found in the respective groups—

  I. 11-diaminopimelic acid (DAP) and glycine
> *Streptomyces somaliensis*
> *S. paraguayensis*
> *Streptomyces* spp. (nonpathogenic)

 III. meso DAP
> *Nocardia madurae*
> *N. pelletieri*

 IV. meso DAP, arabinose and galactose
> *N. asteroides*
> *N. brasiliensis*
> *N. caviae*
> " *N. farcinica* "

Table IV gives some of the criteria for the identification of the aerobic actinomycetes.

TABLE IV

Identification of the aerobic actinomycetes based on cell-wall type and decomposition properties

Organism	Cell-wall type	Casein	Tryosine	Xanthine	Urea	Decomposition of†
N. asteroides	IV	−	−	−	+	
N. brasiliensis	IV	+	+	−	+	
N. caviae	IV	−	−	+	+	
"*N. farcinica*"	IV	−	−	−	−	
N. madurae	III	+	+	−	−	*N. madurae*
N. pelletieri	III	+	+	−	−	+ acid pro-
S. somaliensis	I	+	+	−	−	duction in
S. paraguayensis	I	+	+	+	+	arabinose and adonitol

† Two weeks at 27°C.

3. *Technique for mycelial phase–yeast phase conversion*

The dimorphic fungi, *Histoplasma capsulatum*, *Histoplasma duboisii*, *Blastomyces dermatitidis*, *Paracoccidioides brasiliensis* and *Sporothrix schenckii*, are characterized by their ability to form the tissue phase (yeast phase) *in vitro*.

Histoplasma: probably the easiest method of obtaining the yeast phase is via animal inoculation, rather than directly from the mycelial culture. This is accomplished through inoculation of mice intraperitoneally with ground mycelial phase and culturing the spleen, liver and lungs on to a cystine heart haemoglobin (CHH) agar, or brain heart infusion agar, and incubated at 37°C. The mice are sacrificed at intervals of 2–5 weeks after inoculation.

B. dermatitidis: mice inoculated as above will show needle line nodules and nodules scattered throughout the omentum, mesentery and peritoneal surface.

P. brasiliensis and *S. schenckii* are inoculated into guinea pigs intra-testicularly; orchitis develops and the pus is inoculated onto CHH or BHI agar and incubated at 37°C.

The yeast phase of any of these organisms is easily converted to the mycelial by altering incubation temperature to 25°C.

4. *Agents of chromoblastomycosis*

To identify some of these dematiaceous pathogens it is necessary to observe three types of sporulation. *Phialophora pedrosoi* and *P. compacta* exhibit (*a*) *Cladosporium,* (*b*) pseudo-acrotheca and (*c*) phialophora types of sporulation. To observe these types it is necessary to make slide cultures on cornmeal and Czapek agar. Carrión (1940), later confirmed by Silva (1958), observed that cornmeal agar stimulates the production of the pseudo-acrotheca type and Czapek agar favours the cladosporium type of sporulation.

Table V lists the agents of chromoblastomycosis and the types of sporulation attributed to each. The disease form is identical for any of the causative agents. Carrión (1950) gives an excellent description of this disease and its clinical evolution.

TABLE V
Agents of chromoblastomycosis

Organism	Types of sporulation
P. pedrosoi	Cladosporium, phialophora, pseudo-acrotheca
P. compacta	Cladosporium, phialophora, pseudo-acrotheca
P. verrucosa	Phialophora only (flaring mouth)
Cladosporium carrionii	Cladosporium only

B. Animal mycoses

Aspergillosis and candidiasis are common amongst warm-blooded animals (Ainsworth and Austwick, 1959; Austwick, 1962, 1966; and Austwick *et al.,* 1960). Barron (1955) reported an epidemic of bovine mastitis due to *Cryptococcus neoformans.* Outbreaks of cerebral mycetoma in trout due to a dematiaceous mould have been reported by Carmichael (1966).

The methods employed in the isolation of fungi from mycotic infections in animals do not differ greatly from those used in medical mycology. There is a greater emphasis on aspergilli as pathogens and certain diseases, e.g. adiaspiromycosis (*Emmonsia* spp.), are not yet recorded in man. There is also a greater problem of contamination than with human material and strict control of saprophytic fungi and bacteria by means of antibiotics in the media is necessary.

C. Mycotic abortion

Aspergillus fumigatus and the other fungi listed in Table I under mycotic abortion are unquestionable abortifacients. According to Austwick (1964) "the placenta normally shows retention of the maternal caruncles with thickened cotyledon margins and intercotyledonary areas. Microscopical examination shows extensive invasion of maternal and foetal tissue by hyphae and field cases show destruction of the epithelia, the foetal villi and the cryptal connective tissue. The foetus may show skin lesions and in one in four cases of placental infection hyphae are present in the stomach content."

Most of the abortifacient fungi have been isolated from hay, straw or air (Austwick, 1963). Because of this it is necessary to examine the placenta for gross lesions and to see the hyphae in direct examination of the tissues.

D. Isolation of pathogenic fungi from nature

1. *Use of niger seed–creatinine media.*

Use of niger seed-creatinine media for the isolation of *Cryptococcus neoformans* (Staib and Seeliger, 1968).

Niger seed (*Guizotia abyssinica*) pulverized	50 g
H_2O	1000 ml

Boil 20 min, filter through gauze and reconstitute to 1000 ml. Then add—

Dextrose	20 g
KH_2PO_4	1 g
$MgSO_4$	0·5 g
Creatinine	1 g
Agar	15 g

Autoclave 30 min at 110°C. For a selective medium add—

Diphenyl	1 g
Ethyl alcohol absolute	20 ml

and the following antibacterials per ml of solution—

Penicillin	20 units
Streptomycin	40 units
Chloramphenicol	1 g

Streak material on to plates. Results usually noted after 4 days at 25–27°C. Colonies of *C. neoformans* are dark brown.

2. *Intravenous mouse inoculation* (Denton and DiSalvo, 1964, 1968)

Suspend a heaped tablespoonful of material in 50 ml of saline containing 5,000 units of penicillin, 5 mg of streptomycin per ml, and 0·2% Tween 80, allowing to settle for 1 h; remove a 7–9 ml sample from the bottom of the supernate at the surface of the sediment. Each of five adult Swiss mice are inoculated intravenously in the tail vein with 0·75 ml of the supernate. The mice are autopsied after 3 weeks. For *C. neoformans* the liver, lungs, spleen and brain are cultured onto brain heart infusion agar with penicillin and streptomycin at 37°C. For isolation of *Blastomyces* and *Histoplasma* the tissues are cultured onto Sabouraud's media containing 100 units of penicillin, 0·1 mg of streptomycin, and 1 mg of cycloheximide per ml at 25°C; also at 37°C on brain heart infusion agar (do not attempt to isolate *Histoplasma* or *Blastomyces* at 37°C with cycloheximide since the yeast phases of these organisms are sensitive to this antibiotic).

3. *Direct isolation method*

A direct isolation method for cultivating *C. neoformans* from soil has been described by Swatek *et al.* (1967).

4. *Air filtrates*

Ajello *et al.* (1965) recovered *C. immitis* from the air by inoculation of mice with air filtrates.

E. Tissue cultures

Although this technique is not used routinely in medical mycology, it is used as a tool to study species of fungi in which the *in vivo* and *in vitro* form differ. Stanley and Hurley (1968) and Hurley and Stanley (1969) have reviewed the literature and have utilized the growth of *Candida* spp. in murine renal epithelial cells as a tool to study the pathogenicity of members of this genus. Grose *et al.* (1968) used mouse fibroblasts for the identification of *C. neoformans* isolated from Columbian bats.

F. Preservation of fungal cultures

For a review of this subject, see Martin (1964). Lyophilization appears to be the favourite means of preservation of both moulds and yeasts. Hwang (1966) reports a much higher percentage of recovery using liquid nitrogen refrigeration. Scheda and Yarrow (1966) note that if subculturing is the means of maintenance of yeasts, a glucose medium with a light inoculum is better than a malt agar with heavy inoculum since this medium favours the selection of mutants able to utilize more carbon sources than the original strain. For recent discussions on the preservation of fungal cultures, see Onions, this Volume, p. 113; Tuite (1968); Hwang (1966); Hwang and Howells (1968).

TABLE VI

Serological tests

Disease	Test	Antigens	Interpretation
Aspergillosis	1. Agar gel	Acetone or alcohol mycelial extracts	Allergic aspergillosis
	2. Complement fixation (Longbottom & Pepys, 1964; Pulmonary mycetoma, 1968)	Broth precipitates Broth filtrates	Aspergilloma
Blastomycosis Coccidioidomycosis Histoplasmosis	1. Skin test	Broth filtrates Whole yeast phase Mycelial phase	Past or present infection Titre rises gradually during infection and falls on improvement
	2. Complement fixation		
	3. Gel precipitin	Mycelial extracts Broth concentrates	Precipitins appear early, disappear rapidly, positive in chronic infection
Paracoccidioidomycosis	1. Skin test	Mycelial extract	More surveys required
	2. Complement fixation	Broth filtrate	Titre rises gradually during infection and falls on improvement
Candidiasis	1. Agglutination	Whole yeast cells	Not informative on their own
(Taschdjian et al. 1967; Pepys et al. 1967)	2. Precipitation	Cytoplasmic extract ("somatic antigen")	Precipitins to Candida spp indicative of systemic infection
		Cell wall antigen glucomannan	Precipitins indicative of allergic candidiasis

TABLE IV (*continued*)

Disease	Test	Antigens	Interpretation
Cryptococcosis (Gordon & Vedder, 1966)	Agglutination	(a) latex sensitized with positive *C. neoformans* rabbit globulin (b) Whole cell agglutination	(a) Antigen titre indicates active infection. Titre correlates with severity of infection
(Bloomfield, *et al.* 1963)			(b) Antibody titre in body fluids appears as the antigen disappears, i.e., during treatment. This is a method of detecting the efficacy of treatment. Antibody is present in localized chronic infection
Chromoblastomycosis (Buckley & Murray, 1966; Buckley, 1968)	Gel precipitin	Mycelial extracts	Precipitins present in all cases of chromoblastomycosis studied and appear to disappear after successful therapy
	Latex agglutination	Mycelial extracts	High titres in patients with chromoblastomycosis (at present not enough information available to evaluate the efficacy of this test)
Mycetoma (Mahgoub, 1964; Murray & Mahgoub, 1968) Actinomycetoma	Gel precipitin	Mycelial alcohol or acetone extracts	Serum from patients with mycetoma precipitate with extracts from the causal organisms
Eumycetoma	Gel precipitin	Broth concentrates or precipitates Mycelial extracts	Precipitins disappear upon excision of the lesions and diminish during course of successful therapy

G. Serological tests

Table VI lists the mycoses and the serological tests available as aids for their diagnosis. There is not enough information available to evaluate serological tests in sporotrichosis, cladosporiosis and phycomycosis.

REFERENCES

Ahearn, D. G., Jannach, J. R., and Roth, F. J., Jr. (1966). *Sabouraudia*, 5, 110–119.
Ainsworth, G. C., and Austwick, P. K. C. (1959). "Fungal Diseases of Animals". Commonwealth Bureau of Animal Health, Review Series 6, Commonwealth Agricultural Bureaux, Farnham Royal, Bucks.
Ajello, L., Maddy, K., Crecelius, G., Hugenholtz, P. G., and Hall, L. B. (1965). *Sabouraudia*, 4, 92–95.
Austwick, P. K. C. (1962). *Lab. Invest.*, 11, 1065–1072.
Austwick, P. K. C. (1963). *Recent Prog. Microbiol.* (Symposia held at the 8th International Congress on Microbiology, 1962), 8, 644–651.
Austwick, P. K. C. (1964). *Proc. R. Soc. Med.* 57, 412.
Austwick, P. K. C. (1966). *In* "The Fungus Spore", Colston Papers No. 18. Proceedings of the 18th Symposium of the Colston Research Society, pp. 321–338. (M. F. Madelin, Ed.). Butterworths Scientific Publications, London.
Austwick, P. K. C., Gitter, M., and Watkins, C. V. (1960). *Vet. Rec.*, 72, 19–20.
Balogh, N. (1964). *Dt. tierärztl. Wschr.*, 71, 327–330.
Barron, C. N. (1955). *J. Am. vet. med. Ass.*, 127, 125–132.
Becker, B., Lechevalier, M. P., and Lechevalier, H. A. (1965). *Appl. Microbiol.*, 13, 236–243.
Bloomfield, N., Gordon, M. A., and Elmendorf, D. F., Jr. (1963). *Proc. Soc. exp. Biol. Med.*, 114, 64–67.
Buckley, H. R. (1968). Ph.D. Thesis, University of London.
Buckley, H. R., and Murray, I. G. (1966). *Sabouraudia*, 5, 78–80.
Carmichael, J. W. (1966). *Sabouraudia*, 5, 120–123.
Carrión, A. L. (1940). *Puerto Rico J. publ. Hlth. trop. Med.*, 15, 340–361.
Carrión, A. L. (1950). *Ann. N.Y. Acad. Sci.*, 50, 1255–1282.
Denton, J. F., and DiSalvo, A. F. (1964). *Am. J. trop. Med. Hyg.*, 13, 716–722.
Denton, J. F., and DiSalvo, A. F. (1968). *Sabouraudia*, 6, 213–217.
Gordon, M. A., and Little, G. N. (1963). *Sabouraudia*, 2, 171–175.
Gordon, M. A., and Vedder, D. K. (1966). *J. Am. med. Ass.*, 197, 961–967.
Gordon, R. E. (1966). *J. gen. Microbiol.*, 45, 355–364.
Grose, E., Marinkelle, C. J., and Striegel, C. (1968). *Sabouraudia*, 6, 127–132.
Hurley, R., and Stanley, V. E. (1969). *J. Med. Microbiol.*, in press.
Hwang, S. (1966). *Appl. Microbiol.*, 14, 784–788.
Hwang, S. (1968). *Mycologia*, 60, 613–621.
Hwang, S., and Howells, A. (1968). *Mycologia*, 60, 622–626.
Lodder, J., and Kreger van Rij, N. J. W. (1952). "The Yeasts". North Holland Publishing Co., Amsterdam.
Longbottom, J. L., and Pepys, J. (1964). *J. Path. Bact.*, 88, 141–151.
Mackenzie, D. W. R. (1965). *In* "Symposium on Candida Infections" (H. I. Winner and R. Hurley, Eds.), pp. 26–43. E. S. Livingstone Ltd., Edinburgh and London.
Mahgoub, E. S. (1964). *Trans. R. Soc. trop. Med. Hyg.*, 58, 560–563.
Martin, S. M. (1964). *A. Rev. Microbiol.*, 18, 1–16.

Murray, I. G., and Mahgoub, E. S. (1968). *Sabouraudia*, 6, 106–110.
Pepys, J., Faux, J. A., Longbottom, J. L., McCarthy, D. S., and Hargreave, F. E. (1967). Symposium Allergologicum, Prague, May 8, 1967, pp. 191–194.
Pulmonary Mycetoma. Leading Article. (1968). *Lancet*, 2, 439–441.
Scheda, R., and Yarrow, D. (1966). *Arch. Mikrobiol.*, 55, 209–225.
Silva, M. S. (1958). *Trans. N. Y. Acad. Sci.*, 21, 46–57.
Staib, F., and Seeliger, H. P. R. (1968). *Mykosen*, 11, 267–272.
Stanley, V. E., and Hurley, R. (1967). *J. Path. Bact.*, 94, 301–315.
Swatek, F. E., Wilson, J. W., and Omieczynski, D. T. (1967). *Mycopath. Mycol. appl.*, 32, 129–140.
Taschdjian, C. L., Kozinn, P. J., Okas, A., Caroline, L., and Halle, M. A. (1967). *J. infect. Dis.*, 117, 180–187.
Tuite, J. (1968). *Mycologia*, 60, 591–594.

APPENDIX

Kinyoun acid-fast stain for *Nocardia*

Kinyoun carbol fuchsin

Basic fuchsin	4·0 g
Phenol crystals	8·0 g
Alcohol (95%)	20 ml
Distilled water	100 ml

Procedure

  (i) Kinyoun carbol fuchsin 3 min (at room temperature with no heating).
 (ii) 3% HCl in 95% alcohol 45 sec.
(iii) Methylene blue 30–60 sec.
 (iv) Wash quickly.
  (v) Blot dry.

Periodic-acid Schiff stain

1. *Periodic acid or chromic acid*
   1% aqueous solution.

2. *Sublimate alcohol fixative*
   8 g mercuric chloride ($HgCl_2$) in 60 ml of 95% ethyl alcohol. Add 50 ml distilled water.

3. *Schiff's leucofuchsin reagent* (*Lillie modification*)

Basic fuchsin	1 g
Potassium metabisulphate ($K_2S_2O_5$)	1 g
Hydrochloric acid (HCl)	1·6 ml
Distilled water up to	100 ml

Mix and shake for approximately 2 h. Decolorize by adding 0·25 g of fresh activated charcoal. Shake 2–3 min. Filter. If filtrate is not clear, repeat decolorization. Refrigerate in a dark bottle. This solution may be used as long as it remains colourless.

Method

  (i) Fix, dehydrate, embed, section, deparaffinize, hydrate to water.
 (ii) 1% aqueous periodic acid for 10–15 min at room temperature (if *Histoplasma* is suspected 10 to 15 min at 50°C in a water bath).
(iii) Wash gently in running tap water for 2–5 min.
 (iv) Immerse in Schiff's leucofuchsin reagent for 10 min at room temperature. Avoid exposure to direct light during this period.
  (v) Immerse in sulphite rinse, change rinse 3 times at 5 min intervals.
 (vi) Wash in running water for 2 min.
(vii) Dehydrate in alcohol.
(viii) 0·5% fast green in 95% alcohol for 5–15 sec.
 (ix) 95% alcohol 2–3 min (2 times).
  (x) Carboxylene 5 min.
 (xi) Xylene 2–3 min.
(xii) Xylene 5 min.

Gomori–Grocott stain

  (i) Fix, dehydrate, embed, section, deparaffinize, hydrate to water, and wash in distilled H_2O.
 (ii) Oxidize in 5% aqueous chromic acid for 1 h.
(iii) Wash in tap H_2O for a few seconds.
 (iv) Rinse in 1% aqueous sodium bisulphite to remove residual chromic acid.
  (v) Wash in tap H_2O 5–10 min.
 (vi) Wash well in 3 or 4 changes distilled H_2O.
(vii) Put in methenamine silver (at 58°C) for 30–60 min until slide turns yellowish brown. Rinse in distilled H_2O and check for adequate silver impregnation.
(viii) Rinse in 6 changes of distilled H_2O.
 (ix) Tone in 0·1% aqueous gold chloride for 2–5 min.
  (x) Rinse in distilled H_2O.
 (xi) Put in 2% aqueous sodium thiosulphate for 2–5 min.
(xii) Wash thoroughly in tap H_2O.
(xiii) Counterstain in light green solution for 30–45 sec.
(xiv) Dehydrate in alcohol, clear in xylol and mount.

Methenamine–silver nitrate solution

Borax 5% solution	2 ml
Distilled water	25 ml

Mix, then add—

Methenamine–silver nitrate stock sol	25 ml
Silver nitrate, 5% aqueous	5 ml
Hexamethylenetetramine $(CH_2)6N$ 3% solution	100 ml

A white precipitate forms which dissolves on shaking. Store in refrigerator.

Stock light green solution

Light green S.F. (yellow)	0.2 g
Distilled H_2O	100 ml
Glacial acetic acid	0·2 ml

Working light green solution

Light green, stock solution	10 ml
Distilled H_2O	50 ml

Antibiotic supplements

Penicillin and streptomycin. 20 units of penicillin and 40 units of streptomycin are generally sufficient to suppress bacterial growth. The stock solutions are prepared as follows:

Penicillin. Add 20 ml sterile physiological saline to a vial containing 200,000 units crystalline potassium penicillin, giving a concentration of 10,000 units per ml. Take 10 ml of this and add to 40 ml of saline to give a stock solution containing 2,000 units per ml. Add 1 ml of this to 100 ml cooled medium to obtain 20 *units per ml* final concentration.

Streptomycin. Dissolve 1 g (1 million units) in 20 ml sterile saline to give a solution of 50,000 units per ml. Take 4 ml of this and add to 46 ml sterile saline. This forms the stock solution containing 4,000 units per ml. Add 1 ml stock solution to 100 ml cooled medium to obtain final concentration of 40 *units per ml*.

Chloramphenicol (Chloromycetin, Parke–Davis). 0·05 *mg per ml* of medium suppresses bacterial growth in agar and the antibiotic may be added before autoclaving.

CHAPTER XVII

Methods Used for Genetical Studies in Mycology

J. L. JINKS AND J. CROFT

Department of Genetics, University of Birmingham

I. INTRODUCTION

Fungi are one of the most versatile of the groups of organisms which are widely used for genetical research. They owe their popularity to a unique combination of technical advantages which are second only to those of bacteria and bacterial viruses, and to a chromosomal organization which is comparable to that of higher organisms. But while genetical studies of fungi have led to rapid and substantial contributions to genetical knowledge (see Fincham and Day, 1965; Esser and Keunen, 1967), particularly to our understanding of gene structure and function and of recombination, little use has been made of genetical methodology to gain a better understanding of the fungi themselves. Nevertheless, studies of this kind which have taken place have made valuable contributions to our knowledge of such diverse properties of fungi as their breeding systems, systems of variation, population structure and dynamics and host–parasite relationships. It is with the application of genetical methodology to these problems rather than with the study of mutants specifically induced for genetical investigations that we shall be concerned in the present chapter.

II. THE NATURE OF GENETICAL VARIATION

The variation that is encountered among independent wild isolates of the same species or among spontaneous variants of a single isolate, that is what might be called naturally occurring variation, falls into two broad categories. These are *discontinuous*, where the variants fall into a few distinct types which can be readily and unambiguously classified, and *continuous*, where the variants do not fall into discrete classes. In any one varying sample one or both types of variation may be present (Croft and Simchen, 1965). Investigations of discontinuous variation present no special methodological problems since they may be treated like the major differences that are widely used in genetical studies. The continuously varying differences present special problems at the level of experimental design, analysis and interpretation. The approach that is necessary for a proper study of continuously varying differences has the advantage that it can also accommodate any discontinuous differences, whether known or suspected, that may be present in the material.

Examination of the natural variation encountered in fungi shows that most affect metrical traits such as rate of growth, degree of pigmentation, density of sporulation and of fruit body formation, age of colony when sporulation and fruiting commences and when the spores reach maturation, and so on. These differences are not only subject to considerable modification by the conditions under which they are observed but they also require careful measurement. To compare colonies for such characters all the conditions under which they are grown must be as uniform as possible and the experimental material must be randomized over all the uncontrollable sources of variation in the experimental conditions. To obtain an estimate of the magnitude of this error variation and at the same time to increase the precision of the experiment it is essential that all of the colonies are replicated and the replicates independently randomized (Mather, 1943).

Where the variation is continuous it is necessary to decide which measurements will most accurately reflect this variation. The decision is usually empirical, being based on considerations such as the relative ease and reliability of measurement and the extent to which these measurements discriminate between different colonies. An overriding consideration, however, might be the likely importance of the characteristic being measured to the fungus when growing under natural conditions.

Before embarking on a genetical investigation of natural variation it is necessary to ensure that the wild isolates and strains derived from them have been appropriately obtained and maintained. Take, for example, two extreme situations. If we wish to investigate naturally occurring heterokaryons it is essential that the inoculum isolated from nature and the inoculum used for propagation in the laboratory would maintain the heterokaryotic condition if originally present. On the other hand, before carrying out a cross between two homokaryotic strains it is essential that they should be stable and true breeding as in general it is no use investigating the difference between two strains if one or both are segregating or sectoring for the character under investigation. Knowledge of the number and origin of the nuclei in the alternative inocula available (hyphae, asexually and sexually produced spores, etc.), will ensure that these requirements are satisfied. Also, where a comparative experiment is being performed, it is important that the inocula used are standardized for type, size and physiological state.

Irrespective of whether variation is continuous or discontinuous and has its origin as naturally occurring differences in nature or in the laboratory its heritable basis must be one of three kinds: (i) nuclear gene differences; (ii) numerical or structural chromosome differences; (iii) extrachromosomal or cytoplasmic differences. Although some insight into the underlying cause of variation may be gained by merely observing its origin and subsequent behaviour during propagation the only unambiguous evidence as to its cause must come from a genetical investigation. In fungi such an investigation may be attempted at three levels: (a) somatic; (b) parasexual; (c) sexual.

The reliability of the information which can be gained from a somatic analysis depends on the degree of detail with which the behaviour of the nuclei in the growing mycelium has been established by cytological investigation. From a knowledge of the numbers and origins of the nuclei in the different kinds of cells and spores produced by a colony, predictions can be made about the kinds of propagants that would result from the various inocula obtainable from that colony. These predictions will vary in a characteristic way according to the nature of the colony and of the variation. Thus they will differ according to whether it is a homokaryon or a heterokaryon, a nuclear gene variant or a cytoplasmic variant, or whether it has a normal or abnormal chromosome complement and according to the level of ploidy of the chromosome complement.

A parasexual analysis can be used to investigate differences between pairs of homokaryotic isolates. It will establish whether or not they are controlled by nuclear genes and locate the genes on particular chromosomes. It is the only method available for locating genes in imperfect fungi but its successful application requires quite sophisticated genetical techniques.

The analysis of the sexual progeny of a cross between a pair of isolates is prospectively capable of establishing the basis of any difference between them no matter what its cause. The utility of sexual reproduction in distinguishing between different causes and the detail of the information which may be obtained will depend on whether or not reciprocal crosses can be made and whether or not the products of individual meioses, the tetrads, are amenable to analysis. Sexual analysis of course permits the location of nuclear genes on particular chromosomes.

In addition to providing evidence of the nature of the differences between isolates, genetic analyses can throw considerable light on the nature of the action and interaction of the factors responsible for these differences. Thus it is possible to deduce the dominance and epistatic relationships of the genes controlling a character by comparing the relative properties of hetero-karyons and their homokaryotic components, heterokaryotic dikaryons and their monokaryotic components, heterozygous diploids and their homo-karyotic haploid components and the relative performances of homokary-otic parents and their homokaryotic sexual progenies.

Though the mycologist may usually be interested in the study of naturally occurring variation it is often necessary to provide mutations in the strains under study for use as genetic markers. In the heterothallic species the mating-type alleles often provide useful markers, but in the homothallic species the selection of spontaneous or induced mutants is essential. The most suitable mutants for this purpose are readily scored and should have little or no pleiotropic effects on the gross phenotype. One of the best examples of such a mutant is white conidial colour in *Aspergillus nidulans* (Yuill, 1939). In other species or for more sophisticated analyses various biochemical mutants, such as auxotrophic or drug resistant mutants may be used (Beadle and Tatum, 1945; Pontecorvo *et al.*, 1953). Suitable markers can be readily induced in most species by the use of various mutagens but care must be taken to ensure that the mutant strains produced in this way do not contain further unwanted mutations and, particularly, chromosomal rearrangements (Kafer, 1965).

III. HETEROKARYOSIS

Heterokaryosis, the association in cytoplasmic continuity of nuclei of unlike genetical constitution, is important for two reasons. Firstly it is a component of the life cycle and hence part of the genetic system of wild fungi, and secondly it can be used as a genetical tool for studying, for example, gene structure and function in complementation tests. Although the free living dikaryotic stage of the heterothallic basidiomycetes is strictly heterokaryotic (Jinks and Simchen, 1966), its special role in the mating behaviour of these fungi makes it more appropriate that it should be discussed under sexual systems of variation (Section V).

A. Isolation and identification

It follows from the definition of heterokaryosis that an isolate may be judged to be heterokaryotic if more than one distinct kind of nucleus can be shown to have been associated in cytoplasmic continuity within its mycelium. In practice this requires that two or more distinct homokaryons can be extracted from a colony of the isolate initiated from a single cytoplasmic unit, such as a single hyphal tip or spore. This test excludes the possibility that the isolate is merely an association of two homokaryotic mycelia.

Even in a heterokaryotic colony not all hyphae and spores will be heterokaryotic; indeed if the spores are uninucleate they will always be homokaryotic. In general, therefore, the first step must be to propagate the isolate by a number of single hyphae. If all of the resulting colonies are identical with each other and with the original isolate then any one may be used for further testing. If, however, they differ in appearance or in growth then further testing must be restricted to the colonies that most nearly resemble the original isolate.

The most satisfactory test for heterokaryosis consists of growing a sample of the asexual spores produced by one of the colonies. The heterokaryotic state is most readily diagnosed where the spores are uninucleate. In this case if the resulting colonies fall into two classes on any characteristic the isolate is prospectively a heterokaryon (Jinks, 1952a and b). For final proof it should be demonstrated that these two classes breed true in further asexual propagations, and it should be possible to resynthesize a heterokaryon with similar properties to the original isolate from pairs of individuals from these two classes (Hansen and Smith, 1932; Jinks, 1952b). Should the colonies belonging to one or both classes segregate again or show any signs of sectoring then an alternative explanation, for example cytoplasmic inheritance or chromosome aberration, should be suspected rather than heterokaryosis.

The demonstration of the heterokaryotic nature of an isolate can become more complicated in species possessing multinucleate or multicellular asexually produced spores. In these species an unambiguous diagnosis of heterokaryosis must depend on an understanding of the origin of the nuclei contained in a single spore (Buxton, 1954; Ishitani and Sakaguchi, 1955; Subak Sharpe, 1956; Huebschman, 1952). Where the multinucleate or multicellular condition is secondary, i.e. all of the nuclei result from the mitotic divisions of a single nucleus which enters the spore initial, then the diagnostic tests are the same as in species with uninucleate spores (see for example *Fusarium oxysporum*, Garber, *et al.*, 1961). Where, however, the nuclei of a multinucleate or multicellular spore arise from two or more primary nuclei some of the spores could transmit the heterokaryotic condition. In such cases we must expect three classes of colony in the asexual progeny of an heterokaryotic isolate, the two pure breeding homokaryotic components and the heterokaryon itself (Hansen and Smith, 1932; Ishitani and Sakaguchi, 1956; Jones, 1965).

The failure to recover two or more distinct classes among the asexual progeny of an isolate is not necessarily proof of its homokaryotic nature. A number of situations could produce such a result. (i) One of the homokaryotic components may be lethal or semi-lethal. Thus, in the heterokaryotic state mutants that would be defective in the homokaryotic state could arise and be preserved because of the presence of the wild-type dominant alleles in the nucleus of the other component. (ii) One of the homokaryotic components may be present at too low a frequency to be recovered in the size of sample used. Thus if the rarer component constituted 1 in 10 of all the spores produced then a sample size of 50 to 100 would be needed to give a reasonably high probability of recovering both homokaryons in the sample. If the relative proportions are even more extreme the sample size would need to be correspondingly larger to achieve the same degree of reliability in distinguishing between a homokaryon and a heterokaryon. (iii) One or both homokaryons may not be recovered in samples of asexual spores from a heterokaryon of a species where the number of primary nuclei which enter the spore initial is large. If neither homokaryon is recovered the heterokaryon will appear to be pure-breeding. If one homokaryon only is recovered along with the heterokaryon the wrong conclusion may be drawn, particularly where the homokaryon is recovered with a very low frequency. (iv) Since we are concerned with naturally occurring variation in the homokaryotic or heterokaryotic state the differences in phenotype upon which any classification depends will be largely of a quantitative or continuously varying kind. The recognition of different classes in the asexual progeny of an isolate may therefore require that the colonies be grown in a properly randomized and replicated experiment under controlled environmental conditions. Even so clearly defined differences in characters such as colonial morphology, degree of pigmentation or sporulation or in rate of growth may not always emerge although a statistical analysis may show that significant differences between colonies for one or more of these characteristics are present. In such cases it may be necessary to confirm any tentative classification by further experimentation.

Finally, an important consequence of heterokaryosis in homothallic ascomycetes is that a proportion of the sexual fruiting bodies produced by the heterokaryotic mycelium contain ascospores which segregate for the genes which distinguish the component homokaryons (Pontecorvo et al., 1953). The finding of hybrid fruiting bodies might thus be regarded as further evidence of the heterokaryotic nature of an isolate.

B. Artificial synthesis and properties

Some insight into the occurrence and role of heterokaryosis in nature can be gained by studying heterokaryon formation and the properties of heterokaryons in the laboratory. However, little insight can be gained by confining the studies to the mutant derivatives of a single isolate, and indeed information obtained in this way may be misleading when extrapolated to natural populations (Caten and Jinks, 1966).

An example of a simple method for detecting heterokaryon formation between different wild isolates consists of pairing two homokaryons that differ for one mutation affecting spore colour (Grindle, 1963a and b). In general the presence of mutations affecting spore colour in one or both isolates, unlike the comparable situation in which auxotrophic mutations are used, does not influence the outcome of the test for heterokaryon formation (Caten and Jinks, 1966).

The test consists of inoculating two isolates, differing in respect of spore colour, side by side, and, after incubation, looking for evidence of heterokaryon formation along the junction between the two resultant mycelia. Recognition of heterokaryon formation depends ultimately on the spore colour difference and it is essential that mutants, media, temperature and duration of incubation are all chosen to give good sporulation and a clear colour difference. Whenever this test has been applied it has been shown that a wild isolate invariably forms a heterokaryon with its own spore colour mutants (Grindle, 1963a and b; Jones, 1965; Caten and Jinks, 1966). This property can thus be used as a control to check that the experimental conditions are appropriate for detecting heterokaryon formation.

The evidence that heterokaryon formation has occurred will vary with the species. In those that have uninucleate asexual spores and autonomous action of the genes controlling spore pigmentation, heterokaryon formation leads to the production of mixed heads of spores; that is, chains of spores of different colours arising from the same hypha. These occur, for example, in *Aspergillus nidulans* and can be detected along the line of contact between two colonies which form heterokaryons with each other. In species with primarily multinucleate spores or non-autonomous action of the genes controlling spore pigmentation heterokaryon formation will be detected only if two isolates, each differing from wild-type by complementary spore colour differences, are paired in which case it leads to the production of spores with wild-type pigmentation (Jones, 1965). However, since such spores can arise by syntropism when complementary homokaryons are grown in mixed culture it is essential that samples of spores with the wild-type pigmentation are grown to confirm their heterokaryotic origin. The extent to which heterokaryons formed in this way establish themselves will vary also. For example, within the genus *Aspergillus*, in *A. nidulans* heterokaryon formation is limited to a narrow band along the junction between the two isolates being tested and is indicated by the presence of only a low proportion of heterokaryotic conidial heads along this junction (Grindle,

1963a and b). In *A. amstelodami* on the other hand large heterokaryotic sectors regularly develop as a result of the adjacent inoculation of the two isolates and in some cases the whole of the resultant growth is a stable heterokaryotic colony (Caten, personal communication).

Tests for heterokaryon formation of this general kind have been made between many pairs of wild isolates in a number of species belonging to the genera *Aspergillus*, *Fusarium*, *Penicillium* and *Sordaria*. These have shown that most combinations of wild isolates will not form heterokaryons with one another under conditions where mutant derivatives of the same wild isolate invariably form heterokaryons with each other and with the original isolate. On the basis of this test the wild isolates of each species so far examined may be divided into groups (see for example, Grindle, 1963a and b). All isolates belonging to the same group will form heterokaryons with each other, but they will not form heterokaryons with members of other groups. These groups have been referred to as heterokaryon compatibility groups. A similar situation exists in laboratory strains of *Neurospora crassa* where heterokaryons may be formed only when the pair of strains used carry identical alleles at each of a number of heterokaryon incompatibility loci (see Davis, 1966 for review).

In *A. nidulans* the 30 or so heterokaryon incompatibility groups so far identified within England and Wales have been examined in some detail. A number of experiments have shown that as a general rule, with rare exceptions, members of the same group have similar genotypes to each other, but different to members of other compatibility groups. This has been shown by experimental situations where isolates are simply compared for continuously variable characters such as rate of growth or cleistothecial density under a single environmental condition (Butcher, 1968), under many environmental conditions where comparisons of the interactions of these characters with environmental variation give a very sensitive test (Perkins and Jinks, 1968a and b; Butcher *et al.*, personal communication), or by a comparison of the genetic variance displayed by progenies of crosses between members of the same group and members of different groups (Jinks *et al.*, 1966). Though it is not yet certain whether genetic dissimilarity per se results in heterokaryon incompatibility or whether specific heterokaryon incompatibility loci are involved as in *Neurospora crassa*, it appears that heterokaryons are formed only between pairs of wild isolates having similar genotypes in *A. nidulans* and also in the related species *A. amstelodami*, *A. terreus*, *A. versicolor* (Caten and Jinks, 1966) and in *A. heterothallicus* (Kwon and Raper, 1967). This finding clearly has considerable bearing, not only on the frequency with which heterokaryons are likely to form in the wild, but also on their effectiveness and utility once they are formed.

The general properties of heterokaryons arise directly from the presence of nuclei of unlike genetical constitution in cytoplasmic continuity. These properties are: (i) interaction between the gene differences in the two (or more) kinds of nuclei which may be due to dominant or epistatic gene action; (ii) the ability of the different kinds of nuclei to occur in varying relative frequencies; (iii) the ability of pairs of the different kinds of nuclei to fuse either as a regular step in sexual reproduction or irregularly in the vegetative stage, thus initiating the parasexual cycle (see Section IV).

The existence of interactions and their nature can be investigated by comparing the properties of the heterokaryon with those of its component homokaryons. Such comparisons, if conducted under controlled conditions using replicate observations and a randomized experimental layout, will reveal whether or not the heterokaryon is intermediate, more like one component than the other or exceeds both components for any particular property. In interpreting the results, however, it is necessary to take other observations into consideration. For example, a greater similarity between the heterokaryon and one homokaryon for a particular property may arise in two ways which may or may not be related. On the one hand it may be due to the one component containing most or all of the dominant alleles, but on the other hand it may be due to the greater relative frequency of nuclei of that component in the heterokaryon. The two may be related in that, because of its selective advantage, the component containing most of the dominant genes is present in a higher frequency in the heterokaryon. Further insight may be gained by estimating the relative frequencies of the two kinds of nuclei in the heterokaryon.

Two methods of estimating nuclear proportions have been used. The first is based on the proportions of the two kinds of homokaryons recovered from the asexual spores produced by the heterokaryon. The assumption underlying this method is that the two kinds of spores will be produced in proportion to the relative frequencies of the two kinds of nuclei in the vegetative mycelium. Where the spores are primarily multinucleate it is also usually necessary to assume that the two kinds of nuclei are distributed at random among the spores. Neither assumption necessarily holds (Barron, 1962; Prout *et al.*, 1953). But while a serious failure of these assumptions would make the spore frequencies unreliable estimates of the nuclear proportions they could still be useful for comparative purposes. Thus differences in the proportions of the two kinds of nuclei in different heterokaryotic mycelia would be expected to reflect differences in the proportions of the two kinds of homokaryotic spores they produce (Jinks, 1952b). The second method depends on being able to recognize the two kinds of nuclei cytologically so that direct counts of the nuclei in the vegetative mycelium can be made. This has only been attempted once using a forced heterokaryon between a haploid and a heterozygous diploid strain of *Aspergillus nidulans* (Clutterbuck and Roper, 1966). While this approach avoids the weakness of the first method it presents its own technical difficulties and it is clearly inapplicable to the study of naturally occurring or unforced heterokaryons.

Heterokaryosis is potentially a system of vegetative variation and adaptation in that a heterokaryon can assume different phenotypes ranging from that of one homokaryotic component to that of the other by undergoing corresponding changes in the relative proportions of its nuclear components. These properties may be investigated in a naturally occurring or unforced heterokaryon by growing it in a range of environments and in each environment comparing its properties with those of the homokaryotic components. At the same time independent replicate estimates of the nuclear proportions should be obtained from samples of the asexual spores produced by the heterokaryon in each environment (Jinks, 1952a; Buxton, 1954). In choosing the environmental treatments it should be borne in mind that those which have a differential effect on the growth of the two homokaryons are most likely to lead to changes in the nuclear proportions. Before concluding that the proportions differ between environments it must be shown that they are more or less constant in any one environment and hence characteristic of heterokaryons growing under those conditions.

Finally, when two homokaryotic strains fuse to form a heterokaryon the cytoplasms of the two strains mix but the two strains retain their nuclear identity within the heterokaryon because, except for the rare occurrence of the parasexual cycle (see Section IV), there is no opportunity for reassortment or recombination of chromosomal genes. Thus, especially in species which produce uninucleate conidia, it is possible to extract the two homokaryons unchanged as far as their chromosomal genome is concerned, but associated with samples of the mixed cytoplasm from the heterokaryon. This property allows for the somatic reassortment of cytoplasmically inherited characters among the homokaryotic asexual progeny of a heterokaryon and is the basis of a test, the heterokaryon transfer test, which has been used to detect the cytoplasmic inheritance of a trait in a number of species of fungi (e.g. Jinks, 1954, 1964; Wright and Lederberg, 1957).

IV. SOMATIC RECOMBINATION

Until about twenty years ago sexual reproduction was thought to be the only mechanism for carrying out a genetic analysis of fungi. Previously (Stern, 1936) it had been demonstrated that in the fruit fly, *Drosophila melanogaster*, recombination could take place in the somatic cells, apparently as a result of crossing over at mitosis. In recent years a considerable number of examples of somatic recombination, that is genetic recombination and segregation in the vegetative cells, have been described in fungi (see Fincham and Day, 1965; Esser and Keunen, 1967 for general texts).

The occurrence of somatic recombination is quite distinct from that of the sexual system of recombination and it may be considered as an alternative to sexual reproduction. It is thus of particular importance to studies of the many imperfect species where it provides the only possible method of carrying out a genetic analysis. Such studies have been undertaken in a number of imperfect species.

In the majority of cases which have been studied in detail somatic recombination has been found to take place during a sequence of events which has been termed the parasexual cycle (Pontecorvo, 1954). This cycle of events was discovered, and is most fully understood, in *Aspergillus nidulans*. The essential steps of the cycle are: (i) heterokaryosis; (ii) the fusion of pairs of unlike nuclei to form the heterozygous diploid (like nuclei presumably also fuse but the resultant homozygous diploids are difficult to detect); (iii) mitotic crossing over; and (iv) mitotic haploidization (see Roper, 1966 for review). Once the heterokaryotic state has been established the other steps in the parasexual cycle are all relatively rare events. In *A. nidulans* for example, about one per 10^6 conidia formed on a forced heterokaryotic colony is found to be the heterozygous diploid. In the isolated diploid colony mitotic crossing over and haploidization occur at frequencies of about 10^{-2} and 10^{-3} respectively per nuclear division. Because of these very low frequencies it is necessary to use selective or inductive techniques at the various stages of analysis.

In genera such as *Aspergillus, Penicillium, Fusarium, Verticillium* and a number of others, a complementing pair of auxotrophic mutants may be used to maintain a forced heterokaryotic state, and particularly where a species produces uninucleate conidia, to select the rare prototrophic diploid conidia from among the auxotrophic haploid spores. Initially camphor was used to induce an increased frequency of diploid formation in the heterokaryon (Roper, 1952), but in general the selective technique is adequate. However, ultraviolet light has been used with considerable effect to increase the frequency of recovery of diploids in *Aspergillus oryzae* and *A. sojae* (Ishitani *et al.*, 1956). In yeasts a diploid stage forms a regular part of the life cycle of most species and no selective technique is required, it being sufficient that haploid strains of compatible mating-types are grown together. In the Basidiomycete, *Ustilago maydis*, the heterokaryotic (dikaryotic) phase is an obligate parasite, but a facultatively saprophytic diploid sporidial phase can be isolated from infected host plants (Holliday, 1961). In *U. violacea* the whole process of dikaryon formation and diploid selection can occur on artificial media (Day and Jones, 1968). In other

Basidiomycetes prototrophic selection of diploids from various hetero-karyotic mycelia has been successful, e.g. from common A heterokaryons of *Coprinus lagopus* (Casselton, 1965).

The diploid state can be confirmed in a number of ways. In those cases where it has been examined the diploid nucleus contains twice the quantity of DNA as the haploid nucleus. In many cases, e.g. in *A. nidulans*, the diploid conidium is significantly larger than the haploid, though in the multinucleate conidia of *A. sojae* the diploid spores are of a similar size to the haploid spores but contain on average only half the number of nuclei. In *U. maydis* sporidial size cannot be used as a ploidy test, but "solopathogenicity", i.e. the ability of a single sporidial culture to infect the host plant as opposed to the two compatible haploid sporidial cultures normally required, is a reliable indicator of diploidy. In all cases, however, the presence of somatic segregation leading to the recovery of all recessive markers present in the original pair of haploid strains provides the best confirmation of the diploid state.

Somatic segregation results from the independent processes of mitotic crossing over and of haploidization. In some cases, e.g. *Verticillium albo-atrum* (Hastie, 1964, 1968) the spontaneous rates of occurrence of these events is high and special techniques for their detection are not required. Where possible very efficient total isolation procedures (e.g. replica plating in *Ustilago* spp. or in yeasts) can be used to screen very large numbers of individuals, again making special techniques unnecessary. However, in most species both mitotic crossing over and haploidization are sufficiently rare events to make selective or inductive methods essential to any extensive analysis.

Mitotic crossing over has been studied in considerable detail in *Ustilago maydis* and in *Aspergillus nidulans*. In the former species various agents (ultraviolet light, mitomycin C and 5-fluorodeoxyuridine) have been used to increase the rate of mitotic crossing-over (Holliday, 1961, 1964; Eposito and Holliday, 1964) and this approach has been used to obtain information about the processes of DNA replication and crossing-over. Similar effects are also found in *A. nidulans* (Mopurgo, 1963; Beccari *et al.*, 1967) and also in *Saccharomyces cerevisiae* (Roman and Jacob, 1958; Eposito, 1968) but, if possible, for purposes of genetic analysis such procedures are best avoided because many of the agents used are known mutagens and addi-tional mutagenic events could upset the analysis (see Kafer, 1963).

As with crossing-over at meiosis, mitotic crossing-over can be used to establish the linear order of genetic loci and of the centromere within a chromosome. In *A. nidulans*, where mitotic mapping was first demon-strated (see Pontecorvo and Kafer, 1958), the spontaneous rate of mitotic crossing-over is too low for total isolation methods to be used and various nutritional, drug resistant or visual markers have been employed in order to select the rare recombinants. As a simple example to illustrate the mechanism of mitotic crossing-over we may consider a diploid where one of the linkage groups is marked as follows:

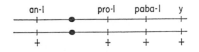

Crossing-over occurs at the four strand stage and is followed by random mitotic segregation of the centromeres. Thus, in the above case a cross-over between the *pro* and *paba* genes produces one of the following two pairs of daughter nuclei—

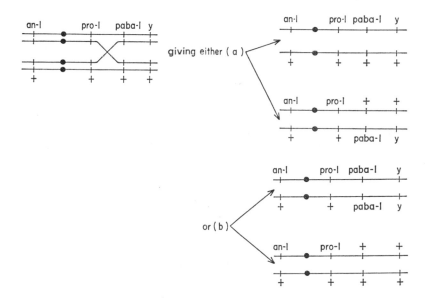

Of these four possible division products three will remain undetected, but the fourth which has become homozygous for the *y* allele now has yellow spores and produces a small patch or sector which is clearly visible against the green spored background of the diploid colony. In such selected recombinants markers distal to the point of cross-over become homozygous, thus enabling one to order the loci relative to the centromere and also to estimate the frequency of recombination between the pairs of alleles. But while this method permits the mapping of chromosome arms in relation to the centromere, it does not permit the mapping of whole chromosomes as linkage across the centromere remains undetected.

In *A. nidulans* it has been possible to make some comparison of the maps obtained by mitotic analysis with those obtained by meiotic means. In all cases the linear order of the loci has been similar, though the two methods give different relative recombination frequencies between the pairs of loci. It should be noted that, as selective methods are used, only relative map distances can be obtained from mitotic analysis and in order to compare to meiotic maps the absolute map distances of the latter should be converted to relative distances. A comparison of relative map distances for the markers used in the previous example is given below:

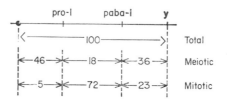

The process of mitotic haploidization provides what is, for purposes of genetic analysis, perhaps the most useful feature of the parasexual cycle, namely a very quick assignment of loci to linkage groups (Forbes, 1959). This procedure is based on the principle that, at haploidization, there is independent assortment of markers on non-homologous chromosomes but complete or near complete linkage of markers on homologous chromosomes among a sample of haploid mitotic segregants. The completeness of this linkage will depend on the frequencies of mitotic crossing-over and of haploidization. For example, in *A. nidulans* both events occur rarely and their occurrence in one nuclear lineage is very rare. Thus, upon haploidization, there is virtually no recombination of linked markers. On the other hand, in *A. niger*, the frequency of occurrence of both events in one nuclear lineage is higher and a considerable proportion of the products of mitotic haploidization show recombination between linked markers (Lhoas, 1967). In this case it is thus possible to determine the linear order of linked markers and to obtain estimates of mitotic map distances as well as to assign markers to linkage groups.

It is again necessary to employ selective methods in order to detect the haploid segregants. Thus, referring again to the *A. nidulans* diploid discussed earlier not all of the spontaneous yellow sectors produced on the diploid colony will be diploid recombinants. A proportion will be haploids containing the homologue of that particular linkage group marked by the *y* allele. However, only one such selector marker is required, whereas one distal marker on each chromosome was required for a full analysis of diploid segregants resulting from mitotic crossing-over. The process of recovering haploid segregants from diploid strains has been greatly facilitated by the discovery that a number of agents, of which the most widely used is *p*-fluorophenylalanine, induces haploidization at a greatly increased rate without the induction of any other side effects which would interfere with the analysis (Lhoas, 1961, 1967). This agent has been used very successfully in a number of species, for example *Aspergillus* spp., *Penicillium* spp. and *Ustilago violacea*.

There is good evidence that the parasexual cycle exists in higher Basidiomycetes, for example in *Coprinus lagopus* (Casselton, 1965). Somatic recombinants have been recovered from haploid-haploid dikaryons, in which case it is always assumed that the recombination process occurs in a transient, though undetected, diploid phase. The diploid monokaryon is very stable, but in haploid-diploid or diploid-diploid dikaryons haploidization occurs at a much higher frequency (Casselton and Lewis, 1966). In *Schizophyllum commune* it has been suggested that somatic recombination may result from processes other than the parasexual cycle. The very high rate of recombination resulting from crossing over found among haploid segregants has led to the conclusion that a meiosis-like process may occur within the vegetative cells (see in Raper, 1966a for review).

To the present time the major use of somatic recombination has been for formal genetic analysis. Its importance for this purpose is beyond doubt, but it must be emphasized that there is, as yet, not enough information available to allow an assessment of its importance in natural populations. General laboratory use of the parasexual cycle and somatic recombination for the study of naturally occurring differences in, for example, *A. nidulans* and related species, has been restricted by the presence of the heterokaryon incompatibility system (see Section III). This system prevents the formation of heterokaryons between pairs of unrelated strains and therefore prevents the selection of the heterozygous diploid in these cases. The parasexual cycle has been used to demonstrate that differences in pathogenicity in *Fusarium* are under genetic control (Buxton, 1956, 1962) and in *Puccinia* the somatic recombination of pathogenicity and uredospore colour factors within the host plant has been demonstrated (Ellingboe, 1961).

V. SEXUAL SYSTEMS OF RECOMBINATION

Sexual reproduction in fungi consists essentially of the bringing together of haploid nuclei in pairs, the fusion of each of these pairs of nuclei to form a diploid nucleus and, usually immediately after fusion, the meiotic division of the diploid nucleus leading to the re-establishment of the haploid state. The process of meiosis also allows the recombination and reassortment of chromosomally located genes to take place and the haploid progeny of a cross between two haploid strains therefore are likely to include strains displaying new combinations of those genetic factors for which the parental strains differ. Thus, in most fungi, the process of genetic analysis by way of the sexual system consists of the crossing of pairs of haploid strains and the analysis of the variation found in the resultant haploid progenies.

Sexual methods of analysis are complicated, however, by the very wide variety of life cycles and breeding systems found in fungi. From this very wide variety of systems the geneticist is able to choose his species according to the type of study he wishes to carry out. It is by the exploitation of particular aspects of the sexual system of certain species that the fungi have made such a large contribution to genetics, for example, the increase in our understanding of the process of recombination that has come from the study of those species in which tetrad analysis is possible. The mycologist on the other hand may find the extent of his genetic study limited by the breeding system or cultural characteristics of the species or group of species in which he is interested.

The sexually breeding fungi may be divided into two broad groups, namely homothallic and heterothallic (Whitehouse, 1949). In the homothallic fungi the complete sexual cycle occurs upon a single homokaryotic mycelium, for example, on a colony of *Aspergillus nidulans* grown from a single uninucleate haploid conidium. The heterothallic fungi on the other hand require the interaction of two morphologically or physiologically differentiated mycelia for the completion of the sexual cycle.

In homothallic species the haploid progeny which normally result from the sexual spores do not display any segregation. They are the result of a true selfing situation as the haploid nuclei which fuse to form the premeiotic diploid nucleus are genetically identical. However, it is possible to make crosses between two haploid strains in homothallic species. One method which has been widely used is to force the nuclei of the two different strains to exist together within the same hyphae by the use of complementary auxotrophic mutants, that is, to form a forced heterokaryon. The presence of the two types of nuclei within the same hypha ensures that at least a proportion of the sexual fructifications produced by that mycelium have originated from initials containing a pair of differing haploid nuclei and the resultant haploid progeny will thus segregate for the genes at which the two parental strains differ (Pontecorvo *et al.*, 1953). This method is the most efficient when crossing two strains which are independent mutant derivatives of a single strain, or are members of the same heterokaryon compatibility group. It cannot be used when the two parental strains are members of different heterokaryon compatibility groups because of the inability to form even a forced heterokaryon in these cases. However, it is not necessary to set up a balanced heterokaryotic state in order to carry out a cross in a homothallic species. It has been found in *Aspergillus nidulans* that by growing a dense mixture of both parental strains, even if they are heterokaryon incompatible, together on a complete medium that a low proportion of the fruit bodies which form are of hybrid origin (Jinks *et al.*, 1966). The genetic control of the proportion of hybrid fruit bodies which form by this method has also been demonstrated (Butcher, 1968). In homothallic species it is necessary to include a genetic marker, preferably a visual mutation such as a change in spore colour, in one of the strains in order to be able to distinguish the progenies of hybrid origin from the two possible classes of selfs.

In the heterothallic fungi it is not necessary to include genetic markers in the material, at least not while carrying out preliminary genetic investigations, as the mating-type factors themselves fulfil this function. Also, the mating system controlled by these factors ensures the crossing of compatible pairs of haploid strains. Thus, whereas in the homothallic species it is necessary to analyse each fruit body individually in order to separate those of hybrid origin from those of selfed origin, in heterothallic species completely random samples of spores can be taken from any cross as they will all have a hybrid origin.

These sexual compatibility systems themselves impose some degree of restriction upon the crossing programmes which can be carried out in heterothallic species. In the simple two allele bipolar system (that found in *Neurospora crassa* and *Saccharomyces cerevisiae* for instance) the progeny of a cross fall into two groups and it is possible to cross only one half of the progeny with the other half. In tetrapolar systems found in many basidiomycetes the progeny fall into four compatibility groups and inbreeding is further restricted to one quarter. An example of a study in which the experimental design is restricted in this way has been described by Simchen and Jinks (1964).

The majority of species of fungi which have been used most widely in genetical research have been chosen because of the relatively straightforward nature of their breeding systems. Some of the kinds of difficulties which can arise are illustrated by the following examples. In *Aspergillus nidulans* the majority of fruit bodies, or cleistothecia, are each initiated by one single pair of haploid nuclei. As a result all of the spores in a hybrid cleistothecium have been derived by meiotic divisions of heterozygous diploid nuclei (see Pontecorvo *et al.*, 1953 for details). But in other species, for example *Aspergillus glaucus*, a number of initial pairs of haploid nuclei may be involved in the developing fruit body so that the spores in a single cleistothecium may be in part of hybrid origin and in part of selfed origin (Caten, 1968, 1969). In heterothallic species it is of no consequence if the fructifications arise from multiple initials as they will all be hybrid in origin.

As a second example, in *Neurospora crassa* each meiotic division leads to the production of eight haploid nuclei, each of which becomes the nucleus of a haploid uninucleate spore. In *Neurospora tetrasperma* however eight haploid nuclei form as in *N. crassa* but they become included in pairs, usually of compatible mating-types, in each of four binucleate haploid spores (Sansome, 1946). Thus it is essential that the origins of the nuclei found in each spore is well understood before a genetic analysis can be attempted. It is not possible here to discuss all of the various and diverse breeding systems found in fungi, but the reader is referred to the general texts and reviews of Raper (1966a and b), Burnett (1968) and Esser and Keunen (1967).

The analysis of the haploid progeny of a cross may be approached by two general methods, analysis of random spores, and analysis of a sample of the products of individual meioses, the tetrads. Not all species are amenable to tetrad analysis and in all cases the degree of technical difficulty involved in analysing tetrads (microdissection for example), except in special cases, will tend to make the sample of tetrads examined small. Though analysis of only two tetrads will probably reveal whether a particular trait is being inherited in a simple Mendelian fashion, the analysis of tetrads will generally be less efficient than random spore analysis as regards information obtained per spore grown when the inheritance of continuous variation is being examined (see Pateman and Lee, 1960). The methods used, both in random spore and tetrad analysis, to establish linkage relationships between different genes are well documented and cannot be discussed here. Again the reader is referred to the excellent general texts and reviews which describe these methods (for example Pontecorvo *et al.*, 1953; Emerson, 1966; Fincham and Day, 1965; Esser and Keunen, 1967).

The analysis of continuous variation found in the haploid progenies of crosses between pairs of wild strains can reveal considerable information about the amount and nature of variation found in natural populations. If the measurements are made in properly replicated and randomized experiments estimates of the genetic variance can be made and these will reflect the degree of genetic heterogeneity between the parental strains (see for example Pateman, 1955; Lee and Pateman, 1961; Jinks *et al.*, 1966; Nelson and Sherwood, 1968). A comparison of the distribution of variation in the progeny with the parental values will show if the increasing and decreasing alleles controlling the character in question are associated or dispersed in the parents (Croft and Simchen, 1965) and will reveal the presence of non-allelic genetic interaction (Butcher, 1969). Selection experiments will provide further information about the nature of the variation (Pateman, 1959; Lee, 1962; Simchen, 1966; Papa *et al.*, 1966, 1967).

In species where it is possible to form stable diploid associations (for example, *Saccharomyces cerevisiae*) or stable dikaryotic associations (as in many basidiomycetes) it is possible to study gene interaction at a more complex level and dominance and other heterozygous interactions can be accounted for. Dikaryons of heterothallic basidiomycetes may be isolated from nature by regeneration from the non-sporogenous material of the fruiting bodies. Equally they may be produced in the laboratory by the pairing of compatible monokaryons. Where the dikaryons are isolated from nature their monokaryotic components may be extracted from uninucleate oidia produced by the dikaryon, for example, *Collybia velutipes* (Brodie, 1936) or by surgical or other treatments, for example from the dikaryotic chlamydospores of *Coprinus lagopus* (Lewis, 1961).

Gene action and interaction in dikaryons can be investigated where dikaryons and their monokaryotic components are available. At the simplest level the investigation requires only that the vegetative growth characteristics of dikaryons and their monokaryotic components are compared in a randomized and replicated experiment. A correlation analysis, for each character scored, between the dikaryons and their monokaryotic components will then provide information on the underlying gene action controlling the character (Simchen and Jinks, 1964). If the correlation is high then the genes in the monokaryotic components are acting additively in the dikaryotic associations. If the correlation is low then there are three possible explanations, (i) the environment is playing a large part in determining the character, (ii) the genes in the monokaryotic components are acting in a non-additive way in the dikaryotic associations, and (iii) the genes which control the character in the monokaryons are different from those which control it in the dikaryotic state. Using the simple design only the first explanation can be investigated further. Thus by examining the differences between replicates we can compute the contribution that environmental factors are making to the total variation among dikaryons and among monokaryons. The first explanation can only apply if this proves to be large, otherwise the alternative explanations must be responsible.

More useful information about gene action and interaction can be obtained by pairing monokaryons in a systematic way. For example, since dikaryons are formed only between compatible monokaryons the latter, in the tetrapolar species, must be divided into two groups within which compatible matings can be made. All possible dikaryons can then be made between the members of each of the two groups and the characteristics of both monokaryons and dikaryons scored in a randomized, replicated experiment. Standard biometrical genetical analyses then allow the variation among the dikaryons to be partitioned into genetical and environmental portions and the former into that due to the additive and non-additive action of the genes. Where the two groups of compatible monokaryons are random samples of the compatible monokaryons obtained from the sexual progeny of a single fruiting body the analyses further permit the non-additive variation to be partitioned into components attributable to the dominance action of the genes and their epistatic effects (for example, Simchen and Jinks, 1964). Thus we can recognize all possible kinds of gene action and interaction among the genes contributed by the two monokaryotic components to the characteristics of the dikaryon.

By using this experimental and analytical approach to investigate gene action and interaction in dikaryons produced by pairing compatible monokaryons isolated from the sexual progeny of the same fruiting body, from different fruiting bodies from the same, and from different localities the nature of the selective forces operating within and between populations and genetic isolating mechanisms can be investigated (Simchen, 1967). Further insight into the direction and intensity of selection can be gained by comparing the monokaryons and dikaryons isolated from wild populations with the monokaryotic progenies produced by the dikaryons made by randomly pairing compatible monokaryons from the progenies. Providing that the viability of the basidiospores is high the latter will be a sample of the monokaryotic and dikaryotic population that would exist in the wild if in the next generation all monokaryons had an equal probability of survival and an equal probability of combining with all other compatible monokaryons. The monokaryons and dikaryons isolated from the population on the other hand are a sample of those monokaryons which successfully survived and formed a dikaryotic association capable of reaching sexual maturity. Any difference in the means and variances of these samples can, therefore, be attributed to the effects of selection at the monokaryotic and dikaryotic stages of the life cycle.

Finally cytoplasmically inherited traits may be studied by way of the sexual cycle. A cytoplasmically inherited character may not appear at all in the sexual progeny or may appear in non-Mendelian proportions, for example, the neutral and suppressive "petite" mutants of *Saccharomyces cerevisiae* (see Ephrussi and Hottinguer, 1951, Ephrussi *et al.*, 1955). Where reciprocal crosses can be made either by way of the fertilization process as in *Neurospora crassa* (Mitchell and Mitchell, 1952) or by way of nuclear migration as in *Coprinus lagopus* (Day, 1959) then uniparental transmission of the character is good evidence for the cytoplasmic location of the factors determining that character.

REFERENCES

Barron, G. L. (1962). *Can. J. Bot.*, **40**, 1603–1613.
Beadle, G. W., and Tatum, E. L. (1945). *Amer. J. Bot.*, **32**, 678–685.
Beccari, E., Modigliani, P. and Morpurgo, G. (1967). *Genetics*, **56**, 7–12.
Brodie, H. J. (1936). *Amer. J. Bot.*, **23**, 309–327.
Burnett, J. H. (1968). "Fundamentals of Mycology". Arnold, London.
Butcher, A. C. (1968). *Heredity*, **23**, 443–451.
Butcher, A. C. (1969). *Heredity*, **24**, 621–631.
Buxton, E. W. (1954). *J. gen. Microbiol.*, **10**, 71–84.
Buxton, E. W. (1956). *J. gen. Microbiol.*, **15**, 133–139.
Buxton, E. W. (1962). *Trans. Br. mycol. Soc.*, **45**, 274–279.
Casselton, L. A. (1965). *Genet. Res.*, **6**, 190–208.
Casselton, L. A., and Lewis, D. (1966). *Genet. Res.*, **8**, 61–72.
Caten, C. E. (1968). *Aspergillus News Letter*, **9**, 12–13.
Caten, C. E. (1969). *Aspergillus News Letter*, **10**, 23.
Caten, C. E., and Jinks, J. L. (1966). *Trans. Br. mycol. Soc.*, **49**, 81–93.
Clutterbuck, A. J., and Roper, J. A. (1966). *Genet. Res.*, **7**, 185–194.
Croft, J. H., and Simchen, G. (1965). *Amer. Nat.*, **99**, 451–462.
Davis, R. H. (1966). In "The Fungi" (Ed. G. C. Ainsworth and A. S. Sussman), Vol. 2, pp. 567–588. Academic Press, New York.
Day, A. W., and Jones, J. K. (1968). *Genet. Res.*, **11**, 63–81.
Day, P. R. (1959). *Heredity*, **13**, 81–87.
Ellingboe, A. H. (1961). *Phytopathology*, **51**, 13–15.
Emerson, S. (1966). In "The Fungi" (Ed. G. C. Ainsworth and A. S. Sussman), Vol. 2, pp. 513–566. Academic Press, New York.
Ephrussi, B., and Hottinguer, H. (1951). *Cold Spring Harb. Symp. quant. Biol.*, **16**, 75–85.
Ephrussi, B., Hottinguer, H., and Roman, H. (1955). *Proc. nat. Acad. Sci. U.S.A.*, **41**, 1065–1071.
Eposito, R. (1968). *Genetics*, **59**, 191–210.
Eposito, R. E., and Holliday, R. (1964). *Genetics*, **50**, 1009–1017.
Esser, K., and Keunen, R. (1967). "Genetics of Fungi". Springer-Verlag. Berlin.
Fincham, J. R. S., and Day, P. R. (1965). "Fungal Genetics", 2nd Ed. Blackwell, Oxford.
Forbes, E. (1959). *Heredity*, **13**, 67–80.
Garber, E. D., Wyttenbach, E. G., and Dhillon, T. S. (1961). *Amer. J. Bot.*, **48**, 325–329.
Grindle, M. (1963a). *Heredity*, **18**, 191–204.
Grindle, M. (1963b). *Heredity*, **18**, 397–405.
Hansen, H. N., and Smith, R. E. (1932). *Phytopathology*, **22**, 953–964.
Hastie, A. C. (1964). *Genet. Res.*, **5**, 305–315.
Hastie, A. C. (1968). *Molec. Gen. Genetics*, **102**, 232–240.
Holliday, R. (1961). *Genet. Res.*, **2**, 231–248.
Holliday, R. (1964). *Genetics*, **50**, 323–335.
Huebschman, C. (1952). *Mycologia*, **44**, 599–604.
Ishitani, C., and Sakaguchi, K. (1955). *J. gen. appl. Microbiol.*, **1**, 283–297.
Ishitani, C., and Sakaguchi, K. (1956). *J. gen. appl. Microbiol.*, **2**, 345–400.
Ishitani, C., Ikeda, Y., and Sakaguchi, K. (1956). *J. gen. appl. Microbiol.*, **2**, 401–430.

Jinks, J. L. (1952a). *Heredity*, **6**, 77–87.
Jinks, J. L. (1952b). *Proc. R. Soc. B*, **140**, 105–145.
Jinks, J. L. (1954). *Nature, Lond.*, **174**, 409–410.
Jinks, J. L. (1964). "Extrachromosomal Inheritance". Prentice Hall, New Jersey.
Jinks, J. L., Caten, C. E., Simchen, G., and Croft, J. H. (1966). *Heredity*, **21**, 227–239.
Jinks, J. L., and Simchen, G. (1966). *Nature, Lond.*, **210**, 778–780.
Jones, D. A. (1965). *Heredity*, **20**, 49–56.
Kafer, E. (1963). *Genetics*, **48**, 27–45.
Kafer, E. (1965). *Genetics*, **52**, 217–232.
Kwon, K., and Raper, K. B. (1967). *Amer. J. Bot.*, **54**, 49–60.
Lee, B. T. O. (1962). *Austr. J. Biol. Sci.*, **15**, 160–165.
Lee, B. T. O., and Pateman, J. A. (1961). *Austr. J. Biol. Sci.*, **14**, 223–230.
Lewis, D. (1961). *Genet. Res.*, **2**, 141–155.
Lhoas, P. (1961). *Nature, Lond.*, **190**, 744.
Lhoas, P. (1967). *Genet. Res.*, **10**, 45–61.
Mather, K. (1943). "Statistical Analysis in Biology". Methuen, London.
Mitchell, M. B., and Mitchell, H. K. (1952). *Proc. nat. Acad. Sci. U.S.A.*, **38**, 442–449.
Mopurgo, G. (1963). *Genetics*, **48**, 1259–1263.
Nelson, R. R., and Sherwood, R. T. (1968). *Phytopathology*, **58**, 1277–1280.
Papa, K. E., Srb, A. M., and Federer, W. T. (1966). *Heredity*, **21**, 595–613.
Papa, K. E., Srb, A. M., and Federer, W. T. (1967). *Heredity*, **22**, 285–296.
Pateman, J. A. (1955). *Nature, Lond.*, **176**, 1274–1275.
Pateman, J. A. (1959). *Heredity*, **13**, 1–21.
Pateman, J. A., and Lee, B. T. O. (1960). *Heredity*, **15**, 351–361.
Perkins, J. M., and Jinks, J. L. (1968a). *Heredity*, **23**, 339–356.
Perkins, J. M., and Jinks, J. L. (1968b). *Heredity*, **23**, 525–535.
Pontecorvo, G. (1954). *Caryologia* (*Suppl.*) **6**, 192–200.
Pontecorvo, G., Roper, J. A., Hemmons, L. M., MacDonald, K. D., and Bufton, A. W. J. (1953). *Advan. Genet.*, **5**, 141–238.
Pontecorvo, G., and Kafer, E. (1958). *Advan. Genet.*, **9**, 71–104.
Prout, T., Heubschman, C., Levene, H., and Ryan, F. J. (1953). *Genetics*, **38**, 518–529.
Raper, J. R. (1966a). "Genetics of Sexuality in Higher Fungi". Ronald, New York.
Raper, J. R. (1966b). In "The Fungi" (Ed. G. C. Ainsworth and A. S. Sussman), Vol. 2. pp. 473–511. Academic Press, New York.
Roman, H., and Jacob, F. (1958). *Cold Spring Harb. Symp. quant. Biol.*, **23**, 155–160.
Roper, J. A. (1952). *Experientia*, **8**, 14–15.
Roper, J. A. (1966). In "The Fungi" (Ed. G. C. Ainsworth and A. S. Sussman), Vol. 2. pp. 589–617. Academic Press, New York.
Sansome, E. R. (1946). *Nature, Lond.*, **157**, 484–485.
Simchen, G. (1966). *Heredity*, **21**, 261–263.
Simchen, G. (1967). *Evolution*, **21**, 310–315.
Simchen, G., and Jinks, J. L. (1964). *Heredity*, **19**, 629–649.
Stern, C. (1936). *Genetics*, **21**, 625–730.
Subak Sharpe, H. (1956). Ph.D. Thesis. University of Birmingham.
Whitehouse, H. L. K. (1949). *Biol. Rev. Cambridge Phil. Soc.*, **24**, 411–447.
Wright, R. E., and Lederberg, J. (1957). *Proc. nat. Acad. Sci. U.S.A.*, **43**, 919–923.
Yuill, E. (1939). *J. Bot.*, **77**, 174–175.

CHAPTER XVIII

Autoradiographic Techniques in Mycology

R. L. Lucas

Department of Agricultural Science, University of Oxford

I. INTRODUCTION

Autoradiography has been used in mycology for investigations into the uptake and metabolism of nutrients by fungi, for examining the distribution of elements within mycelial systems, for the detection of isotopically labelled mycelia, and for studying translocation. It has been used in plant pathology and in investigations into the behaviour of mycorrhizas and lichens where the movement of nutrients from host to fungus and from fungus to host has been demonstrated autoradiographically. The technique has also proved valuable in determining the influence of infecting mycelium on the distribution of nutrients within infected cells, tissues and organs. Detection of isotopically labelled mycelium within the tissues of infected host plants has enabled recognition of the extent and distribution of invading, labelled mycelium.

II. METHODS

Uptake of nutrients by fungi has been widely studied using radio-isotopes, but relatively few workers have used autoradiography as a tool. However, King and Isaac (1964) have used the technique in studies on the uptake of glucose and of glycine by *Rhizoctonia solani* Kuhn. Shaw (1968) reports the use of film-stripping by the method of Pelc (1956) to determine the uptake and distribution of tritiated leucine, cytidine and uridine by mycelium of rust infecting flax.

The ability of fungi to absorb radio-isotopes has been exploited to a much greater extent in the detection of labelled mycelial systems and in the study of translocation within hyphae. Grossbard (1958) and Grossbard and Stranks (1959) grew colonies of *Armillaria mellea, Fusarium culmorum, Helminthosporium sativum, Pellicularia filamentosa, Phycomyces blake-*

sleeanus, Phytophthora cactorum, Rhizoctonia solani, Rhizopus stolonifer and *Verticillium albo-atrum* in broth culture or on agar medium incorporating either 60-cobalt or 137-caesium. An autoradiograph of mycelium of *Rhizoctonia solani* submerged in agar containing 60-cobalt shows a high grain density in that part of the image corresponding with the hyphae and the region immediately surrounding them compared with the much lower grain density corresponding to the medium two or three microns and more from the hyphae. This implies a ready motility of cobalt within the medium and an efficient sorption system associated with the mycelium. When sporangia of *Rhizopus stolonifer* were harvested from above a medium containing 60-cobalt and autoradiographed, it was shown that the nuclide was present in them, a result which Grossbard interprets as proving translocation. While such data certainly demonstrate the uptake of cobalt from the medium and its presence in the sporangia, they do not of themselves prove translocation in the strict sense of the term (Lucas, 1960). Progressive accumulation of the isotope in the sporangia after differentiation is necessary to demonstrate a movement of 60-cobalt through the mycelium and sporangiophores and into the sporangia as distinct from the acquisition of the nuclide from the medium during extension of the hyphae and initiation of the sporangia. Such a quantitative change is difficult to establish by autoradiography unless the proportional change is large. Indeed, to establish a net movement of an isotopically labelled element demands the determination of specific activities and pool sizes at both source and sink, and this is completely outside the scope of autoradiography. What can be established by such a relatively simple technique is the movement of a specific isotope and the consequent inference of a pathway whereby translocation *sensu stricto* may occur.

The ability of fungal hyphae to absorb both caesium and cobalt has been used by Grossbard and Stranks to detect the presence of mycelia growing in soil. They took samples of soil into which mycelia of *Pellicularia filamentosa* or *Rhizopus stolonifer*, previously labelled with 60-cobalt, had grown, ground them up, and autoradiographed smears prepared from the finely ground soil.

Inference of translocation resulting from the appearance of radioactive tracers at a site remote from their application has been drawn by Grossbard and Stranks (*loc. cit.*) for *Pellicularia filamentosa*, using 60-cobalt, and for *Phycomyces blakesleeanus*, using both 60-cobalt and 137-caesium, as well as for *Rhizopus stolonifer* in the instance referred to earlier. For *P. filamentosa*, the technique employed was an adaptation of the method used by Schütte (1956). A small dish of nutrient agar incorporating 60-cobalt was placed inside a large one filled with similar agar lacking 60-cobalt. The inner dish was inoculated with the fungus and the mycelium grew over the edge on to "cellophane" strips placed on the outer, inactive, medium. These strips, together with the mycelium, were removed and autoradiographed, and after exposure for 3 weeks, an image of the mycelium was obtained. However, hyphae removed from the cellophane failed to produce an image, which seems to imply that any cobalt transported by the mycelium is liable to leach out of the hyphae. *Phycomyces blakesleeanus* was grown on media containing either 60-cobalt or 137-caesium, and individual sporangiophores or tufts were removed and fixed to glass slides with collodion before preparing autoradiographs by Pelc's film stripping method (for 137-caesium) or by using X-ray film (for 60-cobalt). The sporangiophores and sporangia gave intense images in both instances.

These investigations used autoradiographic techniques to establish translocation of both 60-cobalt and 137-caesium in the sense of Schütte (1956). However, although they do clearly show that material acquired at one point may subsequently appear at another, they do not distinguish between the uptake of a tracer by growing mycelium, followed by the movement of that tracer as a consequence of the extension growth of the hyphae, and the uptake of a tracer at one point in an already established mycelial system and its translocation through the hyphae, independent of growth, to another part of the mycelium. This is a distinction made by Lucas (1960) and examined in detail by Robinson (1963). He showed that the fungi he investigated fell into two well-defined categories, those which were capable of translocating phosphorus through an already established mycelium and those which did not achieve the movement of phosphorus through pre-formed hyphae. These latter organisms were, however, able to absorb phosphorus in the form of orthophosphate, and having absorbed it, they carried it forward during the normal process of growth, so that some of the phosphorus taken up by young hyphae was transported onwards during their subsequent elongation. *Ophiobolus graminis* and *Pythium ultimum* were found to behave in this manner (Robinson, 1963) and furthermore were found not to release previously absorbed 32-phosphorus into the medium when these organisms were grown on an unlabelled substrate.

This pattern of behaviour presents a means of labelling mycelium in a manner which enables the subsequent recognition of hyphae deriving from it, even after a period of growth in the presence of hyphae derived from other sources. Consequently the technique of isotopic labelling of inocula of pathogens followed by detection using autoradiography has been used by Robinson (1963) and by Robinson and Lucas (1963) in studies on the rate of spread of *Ophiobolus graminis* infecting wheat, *Agropyron repens* and *Agrostis stolonifera*. To this end, inoculum discs of *O. graminis* labelled with 32-phosphorus were prepared as follows. Into each of a number of Petri dishes was put one millilitre of sterile 0·1% KH_2PO_4 solution incorporating 75 μCi of 32-phosphorus, and to this was added 20 g of potato dextrose agar. When the medium had cooled, a sterile disc of cellophane 4·5 cm in diameter was laid on the surface in each Petri dish. The centre of each cellophane disc was inoculated with *Ophiobolus graminis* and incubated at 23°C. When the mycelium had reached the edge of the disc, the cellophane was peeled off the medium under aseptic conditions, floated on sterile distilled water to remove any contaminating 32-phosphorus, and transferred to sterile PDA. The mycelium on the cellophane disc, which had acquired 32-phosphorus from the labelled PDA, continued to grow and to extend over the surface of the medium in the second Petri dish to which it had been transferred. This mycelium contained the radio-active isotope within the hyphae, but did not release 32-phosphorus into the medium. Discs, one centimetre in diameter, cut from this mycelium consisted of radio-actively labelled mycelium on unlabelled medium, and were used to inoculate segments of rhizomes and stolons of *Agropyron repens* and *Agrostis stolonifera* respectively growing in unsterile soil. After three, four and five weeks the plants growing from these segments were removed from the soil, washed, and autoradiographed at −10°C.

Material was prepared for examination by mounting it on boards covered with Kleenex tissue. The mounted plants were then covered with 50-gauge terylene film, on top of which was secured a sheet of Kodirex X-ray film. Exposure for 72 hours was sufficient for the plants harvested three weeks after inoculation, but for later harvests this period had to be extended to compensate for the combined effects of decay of 32-phosphorus and dilution of the isotope within the mycelium as a consequence of extension growth of the hyphae.

Images on the autoradiographs indicated the extent of mycelial spread from the point of inoculation, and allowed positive identification of the hyphae deriving from the inoculum even in the presence of lesions on the root systems resulting from infection by fungi of either the same or different species occurring naturally in the soil. Results indicated a rate of spread of *Ophiobolus graminis* along the root systems of these grasses of up to 10 mm per week. A variant of this technique has been used by Robinson (1963) to examine the rate of spread of *Pythium ultimum* through soil using a source of the fungus similarly labelled with 32-phosphorus and using lettuce seedlings sown at various distances from the inoculum to test for the presence of the pathogen. Young plants were removed, washed and autoradiographed not only to detect the presence of the pathogen but also to identify its origin positively with the inoculum.

Isotopic labelling of mycelium to achieve positive identification of hyphae deriving from an inoculum depends for its success on several factors. The isotope must be absorbed by the mycelium and must be carried forward in extending hyphae; it must remain within the hyphae and must not be released into the extra-mycelial environment; it must have a sufficiently long half-life to allow its recognition throughout the course of the experiment. In these respects 32-phosphorus has proved satisfactory for many, but not all fungi. Some organisms, while satisfying the first criterion, do not fulfil the second requirement, but allow the ready loss of previously absorbed phosphate. Thus pathogens such as *Corticium solani* may readily be grown on media containing 32-phosphorus and produce a mycelium labelled with it. But if such a mycelium is used to inoculate, for example, seedlings of sugar-beet, it is seen that the isotope is released from the mycelium and passes into the host plant, all parts of which may be shown by autoradiography (Robinson *loc. cit.*) to contain 32-phosphorus. This constitutes a means of demonstrating that not all movement of nutrients between host and pathogen is from green plant to fungus, but that some nutrients may pass, at least some of the time, from fungus to host.

One of the limitations in the value of autoradiography using 32-phosphorus is imposed by the relatively high energy of β-emission which leads to low resolution in the resultant autoradiographs. A recent development, however, (Gruverman and Davidson 1968) is the commercial production of 33-phosphorus, and this opens up new possibilities. This isotope of phosphorus emits β-rays with a maximum energy of 0·25 Mev. and a half-life of 25 days, compared with 14 days for 32-phosphorus. Consequently high resolution autoradiography is now a practical possibility using 33-phosphorus, and suggests a means of overcoming one of the difficulties encountered in field experiments on the spread of pathogenic fungi by allowing the isotopic labelling and autoradiographic detection of individual hyphae in experiments over longer periods than is possible with the relatively short half-life of 32-phosphorus.

Changes in distribution of absorbed nutrient with time have been followed using autoradiography. This technique has been of particular value in investigating the movement of nutrients between host and parasite, and Baldacci and Betto (1962) examined phosphorus uptake by *Uromyces appendiculatus* on *Phaseolus vulgaris* by this means. Whole leaves were infected, and 48 hours before inoculation the plants were labelled with 32-phosphorus. Autoradiographs of leaves taken over successive 24-hour periods at first show accumulation of 32-phosphorus in veins and veinlets, followed by accumulation in regions where haustoria are beginning to form, and ultimately the greatest accumulation is found in the spores of the pathogen. A significant feature is that the accumulation in the zone of haustorial development precedes macroscopic appearance of lesions by 24 hours.

While it may be tempting to interpret such results as indicating a progressive accumulation of phosphorus by diseased tissue and then by the pathogen itself, the evidence does not, in fact, justify such a conclusion. What is shown is an accumulation of that phosphorus which was available *at the time of labelling*. Only if free movement of phosphorus can take place in both directions between host and parasite would the observed changes in accumulation of 32-phosphorus be evidence of such progressive accumulation by diseased tissue and subsequently by the fungus itself. The findings of Robinson (1963) and of Robinson and Lucas (1963) show that this is by no means universal in host-parasite associations.

Investigations into the movement of carbon compounds between host and parasite have also used the techniques of autoradiography. Thrower (1965) followed the time course of the process of carbon accumulation in leaflets of *Trifolium subterraneum* infected with *Uromyces trifolii*. Autoradiographs of leaflets taken immediately after subjecting them to $^{14}CO_2$ revealed a high level of 14-carbon within the uredia and a zone of depressed activity surrounding them. The processes of accumulation are, however, complex and will be summarized. In leaflets harvested after 6 and 12 h exposure to $^{14}CO_2$, there was little accumulation in uredia compared with the surrounding tissues in which a distinct accumulation zone was evident. After 48, 72 and 96 h, there was progressively greater labelling in the uredia. Thrower interpreted these results as indicating in the first instance dark fixation in the uredia of some $^{14}CO_2$ and subsequently an accumulation of labelled material in tissues surrounding the uredia during the first 24 h. This material is then thought to be mobilized and translocated through the mycelium into the uredia and into the spores.

Yuen (1969) has studied carbon fixation and translocation in a similar situation in leaf discs of *Tussilago farfara* infected with *Puccinia poarum*. In this material, the parasite forms well-defined aecidial clusters which are suitable for manipulation and autoradiography.

Leaf discs were exposed to $^{14}CO_2$ using a double-dish technique which entailed placing the discs on moist filter-paper in a small open Petri dish fixed with sellotape in the centre of the lower half of a larger Petri dish. A measured amount of NaH $^{14}CO_3$ was then put in the angle between the side wall and the floor at one side of the larger Petri dish and 10% lactic acid was similarly placed at the other side. A ring of Vaseline was placed in the angle between the side-wall and the roof of the Petri dish lid which was then put onto the base so that the Vaseline formed a continuous seal. The whole assembly was then gently tilted in order to mix the lactic acid with the NaH $^{14}CO_3$ and thus to subject the leaf discs in the inner dish to $^{14}CO_2$ (Fig. 1).

Exposure of discs to $^{14}CO_2$ for a period of 2 h in the light followed by a further period of 20 h during which the discs were held either in the light

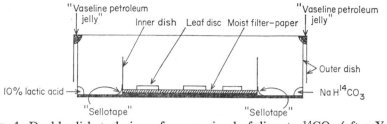

FIG. 1. Double-dish technique for exposing leaf-discs to $^{14}CO_2$ (after Yuen, 1969).

or in the dark, but in the absence of $^{14}CO_2$, before autoradiography resulted in quite different pictures. Subsequent exposure to light showed almost uniform distribution of 14-carbon whereas subsequent dark experience led to the highest levels of 14-carbon being found in the pustules of the aecidia. A further series of experiments investigated discs from healthy leaves, from the healthy parts of rusted leaves, and from parts of leaves bearing pustules. These discs were exposed to 10 μCi of $^{14}CO_2$ in the light for 2 h, using the double dish technique. They were then divided into two groups, one of which was autoradiographed intact after drying, and the other was extracted and the carbon compounds were analysed chromatographically. Results showed heavy labelling in the healthy tissues of both healthy and infected leaves, but a preponderance of 14-carbon accumulated in the pustules of the rusted discs. This latter situation could have arisen either as a result of dark fixation of $^{14}CO_2$ in the pustules or of translocation from the host or as a result of a combination of both processes, and one of the limitations of autoradiography alone is that it cannot be used to distinguish between such processes. However in combination with the valuable technique of infiltration developed by Drew and Smith (1967) for investigation into the metabolism of lichen symbionts it was possible to show in this particular instance that the accumulation of 14-carbon in the pustules was the result of translocation of photosynthate from the surrounding host tissue. Other autoradiographic studies on the movement of 14-carbon from host to parasite are summarized in an important review by Smith et al. (1969) and essentially support the conclusions drawn by Thrower and by Yuen.

The use of tritiated compounds allows high-resolution microautoradiography of structures acquiring the tritium since this nuclide emits β-radiation of very low energy. Bhattacharya and Shaw (1967) describe a technique for feeding tritiated leucine, cytidine, uridine or thymidine to leaves of wheat infected with *Puccinia graminis* var. *tritici*.

After feeding with the radioactive solutions for 16–18 h tissues were treated for 1 h with a non-radioactive solution of the appropriate compound at a concentration 100 times higher than the labelled solution and then fixed and sectioned. Slides were coated with melted Kodak NTB-3 liquid emulsion and after exposure for 5 days they were developed. Subsequently they were stained and prepared as permanent microautoradiographs.

Results indicate a much higher level of incorporation of tritium from leucine, cytidine and uridine in mesophyll cells of infected zones than in cells of adjacent uninfected tissue. Heavy accumulation by fungal hyphae and developing uredospores was also found. There was no detectable incorporation of tritium into either host or fungus from tritiated thymidine. Confirmation of the incorporation of tritium deriving from cytidine and uridine into RNA was provided by digestion with cold perchloric acid, which removes RNA, before microautoradiography. Subsequent exposure showed little labelling. Tissues fed with tritiated leucine were similarly treated with hot trichloracetic acid to remove all the nucleic acids. Subsequent microautoradiography of such sections showed heavy labelling indicating that the tritium from this source had been incorporated into protein.

Such techniques involving a combination of low-energy β-emitters as tracers detected by autoradiography constitute a powerful tool in the search for information regarding the movement of materials between host and parasite and the fate of nutrients after absorption, but cannot, of themselves, reveal the nature of the compound into which the isotope may have been transformed.

REFERENCES

Baldacci, E., and Betto, E. (1962). *Agrochimica*, **6**, 231–245.
Bhattacharya, P. K., and Shaw, M. (1967). *Can. J. Bot.*, **45**, 555–563.
Drew, E. A., and Smith, D. C. (1967). *New Phytol.*, **66**, 389–400.
Grossbard, E., (1958). *Nature, Lond.*, **182**, 854–856.
Grossbard, E., and Stranks, D. R. (1959). *Nature, Lond.*, **184**, 310–314.
Gruverman, I. J., and Davidson, J. (1968). *Atomlight*, No. 66, 7–9.
King, M. K., and Isaac, P. K. (1964). *Can. J. Bot.*, **42**, 812–821.
Lucas, R. L. (1960). *Nature, Lond.*, **188**, 763–764.
Pelc, S. R. (1956). *Int. J. appl. Radiat. Isotopes*, **1**, 172.
Robinson, R. K. (1963). A study of some aspects of the biology of certain root-infecting fungi. Oxford University: D.Phil. thesis.
Robinson, R. K., and Lucas, R. L. (1963). *New Phytol.*, **62**, 50–52.
Schütte, K. H. (1956). *New Phytol.*, **55**, 164–182.
Shaw, M. (1968) unpublished communication to First International Congress of Plant Pathology, London, July 1968.
Smith, D., Muscatine, L., and Lewis, D. (1969). *Biol. Rev.*, **44**, 17–90.
Thrower, L. B. (1965). *Phytopath. Z.*, **52**, 269–294.
Yuen, C. K. L. (1969). The movement of carbohydrates from green plants to fungal symbionts. Sheffield University: Ph.D. thesis.

CHAPTER XIX

Fluorescent Techniques in Mycology

T. F. PREECE

Agricultural Botany Division, School of Agricultural Sciences,
The University, Leeds, England

I. INTRODUCTION—AUTOFLUORESCENCE

Fluorescent materials emit visible long-wave radiation when stimulated by shorter wavelength visible light or by ultraviolet radiation. Thus if ultraviolet light arrives at a fluorescent object it "shines in the dark". In addition, the wavelength of UV light is less than visible light, and since the resolving power of an objective depends on the N.A. of the lens system and the wavelength of light being used, smaller objects can be resolved using a UV microscope than when a normal illuminant is used (Cruickshank, 1960). Many materials and objects are fluorescent, some brilliantly and others less so. The light emitted may be such that the objects appear to be of various colours. Lignin is intensely autofluorescent, usually appearing yellow white. Polysaccharide material, such as the callose plugs of pollen tubes, is similarly fluorescent and is a valuable tracer at low magnification (Myra Chu Chou, 1969, personal communication). Chlorophyll—and thus chloroplasts—has a lovely red fluorescence. Other compounds produced as the result of mechanical damage (bruising of potatoes) or the invasion by a bacterium or fungus, fluoresce strongly. Thus *Pseudomonas phaseolicola* infected bean seeds may be detected by UV light (Wharton, 1967). Both organism and host may produce fluorescent materials such as coumarins. *Phytophthora infestans* produces scopoletin in culture and as a result of host-parasite interaction (Austin and Clarke, 1966). It has been useful to the medical mycologist to know that *Microsporum*-infected hairs show a brilliant greenish fluorescence on the scalps of ringworm cases; *Trichophyton* infected hairs do not so fluoresce. A good practical description is given by Keddie (1947). Conidia of the powdery mildews fluoresce *in situ* on leaves. An example is *Podosphaera leucotricha*, the causative fungus of apple mildew, while G. Barnes (1969, personal communication) recorded intense autofluorescence of conidia and conidiophores of *Erysiphe communis* on the leaves of red clover. Some basidiomycetes, e.g. *Phlegmacium*, are auto-fluorescent and may be detected on herbarium sheets by scanning with UV light (Gams, 1962). Runner hyphae of *Ophiobolus graminis* exhibit a distinctive reddish fluorescence (Preece, 1968, unpublished).

Thus several plant tissues are themselves autofluorescent. Autofluorescence has been recorded in each of the major taxonomic groups of fungi (Phycomycetes, Ascomycetes, Basidiomycetes, and the Fungi Imperfecti) either in isolation or on host tissue. Therefore careful checking for auto-fluorescence is essential in any microscopical work involving induced fluorescence (e.g. the use of the fluorescent antibody technique). In particular comparative photomicrographs of the treated and untreated fungus should be taken as a routine procedure.

II. FLUORESCENT DYES AND STAINS

The use of fluorescent materials on the stage, in advertisements and in road safety devices has made everyone familiar with their macroscopic possibilities as differentiating agents. Because it is still visible to the naked eye at very low dilution, fluorescein is useful for tracing the source of contamination of water supplies. But mycologists have not yet fully explored the possibilities of fluorescent materials as microscopic tracers. The course of very minute tubules may be shown up in the retina of the eye by fluorescein (Dollery, 1963). Tetracycline is fluorescent and does not harm spermatozoa, so it is an excellent *in vivo* label for their study (Ericsson and Baker, 1967). Other possibilities include the use of compounds which become fluorescent after enzymatic modification inside tissue (Rotman and Papermaster, 1966). Auramine and several other materials stain the tubercle bacillus, *Mycobacterium tuberculosis*, and allow the use of low magnifications in the search for the organism in suspect material. Since it is so much smaller than fungal mycelium, an auramine stained slide of this organism in sputum is very useful as a check in microscopic work with fungi, especially if darkground illumination is being used with the UV light.

Acridine orange has been used in attempts to distinguish dead and living materials of *Plasmodiophora brassicae* (Budzier, 1965; Keyworth, 1962) and other fungi. A combination of the use of this material and potassium hydroxide has been reported valuable in the microscopic diagnosis of fungal infections of skin and hair using UV microscopy (Chick and Behar, 1961). The comparison of amounts of DNA stained with acridine orange and examined by UV microscopy may be useful in studies of the progress of human infections (Bruce and Kumar, 1967).

III. FLUORESCENCE BRIGHTENERS

If the fluorescent materials are themselves colourless, they are called optical brighteners and we have become familiar with their use in washing powders, in the whitening of paper and the "bleaching" of near white foodstuffs. But they are also good microscopic reagents, being useful tracer materials for bacteria and fungi. Some are themselves coumarins. Some of these brighteners are derived from diaminostilbene. No generalizations are possible at this stage in the development of these materials, and trial and error experiments are the only course. For example, trials with "Calcofluor White M2R, New" a water-soluble diaminostilbene disulphonic acid (Cyanamid) showed that although it would usefully brighten hyphae of *Botrytis cinerea* grown on agar incorporating low concentrations of the material, even the lowest concentration which had any effect markedly inhibited mycelial growth (Preece, 1968, unpublished). But clearly Paton and Andrews (1964) were able to grow the bacterium *Salmonella anatum* easily in concentrations of 0·05% of this particular compound, and 21 days later bacteria which had been grown on the brightener agar and suspended in saline, were still fluorescent. Wilson (1966) reports work with detached conidia of *Botrytis cinerea* and the observation of their distribution in the xylem of artificially inoculated tomato stems. Spores of *Penicillium*, *Verticillium* and *Aspergillus* also stained successfully in a solution of brightener. Wilson based his work on that of King (1951), Darken (1961; 1962) and Darken and Swift (1963; 1964). His technique (H. M. Wilson, 1969, personal communication) is as follows. All the compounds are made up as 3% solutions (weight to volume) in 20% (by volume) aqueous glycerol. Ten millilitres of brightener solution are added to 90 ml of base medium and organisms are directly subcultured onto this. Stock cultures are not maintained on media containing brightener. Hyphal macerates are prepared by washing mycelial mats from either liquid or solid brightener medium in buffered saline until the discarded liquid shows no fluorescence (3–5 changes). Mats are then homogenized in the volume of buffered saline required. Spore inocula are best prepared from cultures grown on solid media without brighteners. Spores are removed by flooding plates, filtering the resulting suspension through muslin, centrifuging at 2000 rpm for ca. 3 min and then resuspending in brightener solution (1 volume brightener to 2 volumes saline). This is left for 24 h in a cold room and thereafter centrifuged and washed in buffered saline until the supernatant is free from fluorescence when examined in UV light. No germination of spores occurs under these conditions. Suitable buffered saline for this work is made up as follows:

NaCl,	8·5g
Na2HPO4 (anhydrous)	1·07g
NaH2PO4.2HO,	0·39
Distilled water	1 litre

Adjust pH to 7·1 by 0·01 M phosphate.

The following optical brighteners are recommended by Wilson for fungal mycelium and spores: Tinopal 4BMT (Geigy), Tinopal BOPT (Geigy), Calcoflour White RW (Cyanamid), Calcofluor white PMS Conc. (Cyanamid), Photine LV (Hickson and Welch), Uvitex S2R High Conc. (Ciba), Leucophor C (liquid form) (Sandoz), Fluolite RP (ICI Ltd.)

A microscope (such as is available from Vickers Instruments Ltd., York) capable of being used with a mercury vapour lamp (e.g. Osram HBO 200) fitted with primary and secondary filters is desirable. 20× or 25× apochromatic objectives in conjunction with a condenser of high N.A. will give maximum resolution. Suitable filters are an Exciter filter UG 1/1·5 mm which passes UV used in conjunction with either UV absorption filter GG3/1 mm or GG13/1 + 3 mm + Wratten foil 2B. These filters are designed for darkground work but when used with a brightfield condenser, pass enough red light to provide a useful background.

IV. FLUORESCENT ANTIBODIES

Direct microscopic observation of the sites of antibody antigen reactions (see Preece, this Volume, Chapter XXII) is possible using this technique. It combines the sensitivity of the immunological reaction with the brilliant picture obtainable with fluorescence using UV light. It is a means of identifying fungal mycelium and also a serotaxonomic tool which has not yet been exploited by mycologists (Preece, 1966, 1968; Preece and Cooper, 1969). The diagrams (Fig. 1) show cryptically how the procedure follows on after antiserum production with the conjugation of the antibody with a fluorochrome, usually fluorescein–isothiocyanate (F.I.T.C.) In the direct test, the antibody takes the fluorochrome with it to the site of the antibody antigen reaction—the surface of the fungal hypha or spores. In the indirect or sandwich test the reaction of unlabelled antibody with the fungus is first allowed to occur, and F.I.T.C. labelled anti-rabbit globulin, usually from a commercial source, is applied later. The anti-rabbit globulin reacts with any rabbit globulin which may be attached to the surface of the fungus, and since the anti-rabbit globulin is labelled with F.I.T.C., in

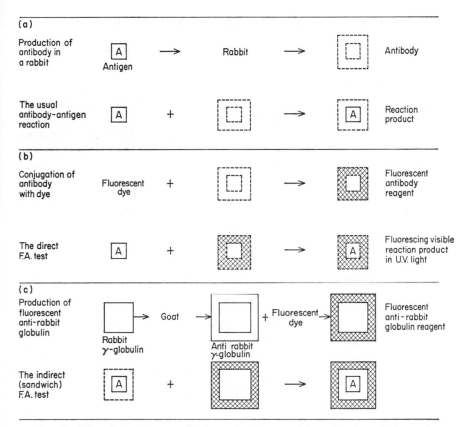

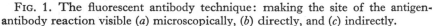

FIG. 1. The fluorescent antibody technique: making the site of the antigen-antibody reaction visible (*a*) microscopically, (*b*) directly, and (*c*) indirectly.

positive reactions a fluorescing visible reaction product will be seen as in the direct test. When viewed under UV light the indirect reaction will make a spore, for example, appear larger since two coats of antibody have been applied (Fig. 1). The use of fluorescent antibody techniques for the identification of fungi have now been reported for the following genera—

Aspergillus, Polyporus, Fomes, Phytophthora, Botrytis, Candida, Histoplasma, Cryptococcus, Blastomyces, Sporotrichum, Saccharomyces, Streptomyces, Cladosporium, Puccinia, Ceratocystis, Nocardia, Torulopsis, Coccidioides, Paracoccidioides, Dictyostelium (Refs. in Preece and Cooper, 1969; Nairn 1969), *Absidia* (Kawamura, 1969), *Trichophyton* (Stuka and Burrell, 1967), *Pythium* (Morton and Dukes, 1967), *Arthrobtrys* (Eren and Pramer, 1966). In addition (Preece, unpublished), reagents have been prepared at

Leeds to *Fusarium*, *Mycosphaerella*, *Ophiobolus*, *Pyrenochaeta*, and *Venturia*.

A flow diagram (from Preece, 1968) shows the principal steps in the production and use of F.I.T.C. labelled antibody for use in the direct reaction (Fig. 2). Details of the procedures used for *Botrytis* are given in Preece and Cooper (1969), but Nairn (1969) has reviewed the literature over a wide field and his book is quite indispensible to any mycologist using the technique. He gives complete citations for 1,500 references to the use of fluorescent antibodies; only 39 concern fungi; of these 33 are about medically important species.

After the preparation of antigen and production of antibodies, the entire antiserum or separated gamma globulin is conjugated with F.I.T.C. (Baltimore Biological Laboratories). The dye is dissolved in acetone and slowly added to the serum on an icebath at 5°C. The mixture is stirred for 24 h in the cold. Excess dye may be removed using activated charcoal (150 mg/ml) or a Sephadix column (Pharmacia Fine Chemicals Inc.) G 25 column. The fluorescent antibody passes through the column much more quickly than does the unreacted F.I.T.C. If not done previously, the globulin fraction of the conjugated serum is equally satisfactorily separated out now using ammonium sulphate (Campbell *et al.*, 1963) followed by dialysis of the labelled globulin, using borate buffered saline at pH 7·8. Dialysis is usually complete in 24 h as shown by testing for sulphate with a 2% barium chloride solution. Small quantities are stored in screw cap containers at −20°C. Staining the fungus on glass slides is aided by growing the fungus on the slides (Booth, this Volume, p. 20). Best results are obtained from young (six-day-old) slide cultures (Preece and Cooper, 1969). Several fungi can be inoculated round the edge of a small coverglass and the test reagent used more economically. The slide is carefully dried in an incubator at 37°C after the agar has been discarded, dipped in acetone for 20 min to fix the mycelium, and the F.I.T.C. labelled reagent left on the slide in a moist chamber for 30 min or longer. The reagent is washed off by two 10 min changes of buffered saline, pH 7·1 (Nairn, 1969), followed by a rinse in distilled water and rapid drying at 37°C. In the indirect or sandwich method (Fig. 1) unconjugated reagent is allowed to react with the mycelium for 30 min followed by the application for 30 min of a commercially available anti-rabbit-globulin from goat serum conjugated with with F.I.T.C. (e.g., Difco), and then proceeding as in the direct method. Careful comparative photomicrographs (e.g., Ektachrome with an exposure time of 10 min) are the best way of recording the results. It is still essential that standard identification methods are used to check F.A.T. results, especially in the early stages of the use of the method with a particular fungus. Fungal morphogenesis may also be studied by this technique. Goos and Summers (1964) and Day (1969, unpublished) record that growth can be induced after the application of labelled antisera in several fungal genera. Thus the source of new materials in germ tubes may be shown to be other than the spore coat itself, and the possible use of homologous antisera as blocking layers to the outward passage of materials from hyphae may be explored microscopically. But the principal use of the technique will concern problems of distinguishing similar fungi. As emphasized later in the Chapter on immunological techniques in mycology (Preece, this Volume, Chapter XXII) the real need is for basic sero-taxonomic research on the fungi as a group of organisms. The rewards, for example in the Fungi Imperfecti, are likely to be very great.

TABLE I

Flow diagram: production and use of F.I.T.C. labelled antibody for the detection of plant pathogenic fungi

Preparation of antigens	Isolate fungus in pure culture
	Bulk up in liquid medium
	Filter off mycelium; grind.
Production of antibodies	Inoculate rabbit with formalized suspension of mycelium
	Detect extent of reaction with antigen by agar gel diffusion.
Conjugation with	Add dilute serum to F.I.T.C. in the cold
Fluorescein-iso-	Remove unreacted F.I.T.C. with charcoal
Thiocyanate	Cross absorption may be necessary.
Specific staining	Fix slide culture of fungus with acetone
	Stain some preparations with F.I.T.C. conjugated antiserum
	Wash in buffered saline; dry.
Microscopy	Mount slide in glycerol-saline
	Oil to condenser (UV inert immersion oil)
	Examine using UV light and dark-ground.
Photomicrography	35 mm film colour or black and white.

REFERENCES

Austin, D. J., and Clarke, D. D. (1966). *Nature, Lond.*, **210**, 1165–66.

Bruce, R. A., and Kumar, V. (1967). *Lab. Pract.*, **16**, 316–7.

Budzier, H. H. (1956). *NachrBl. dt. PflSch. ut.dienst., Berl.*, N.F., **10**, 33–35.

Campbell, D. H., Garvey, J. S., Cremer, N. E., Sussdorf, D. H. (1963). "Methods in Immunology". W. A. Benjamin Inc., New York.

Chick, E. W., and Behar, V. S. (1961). *J. invest. Derm.*, **37**, 103–106.

Cruickshank, R. (1960). "Mackie and McCartney's Handbook of Bacteriology", 10th Ed., pp. 101–102. Livingstone, Edinburgh.

Darken, Marjorie A. (1961). *Appl. Microbiol.*, **9**, 354–360.
Darken, Marjorie A. (1962). *Appl. Microbiol.*, 10, 387–393.
Darken, Marjorie A., and Swift, M. E. (1963). *Appl. Microbiol.*, **11**, 154–156.
Darken, Marjorie A., and Swift, M. E. (1964). *Mycologia*, **54**, 158–162.
Dollery, C. T. (1963). *Photogr. J.*, **103**, 319–323.
Eren, J., and Pramer, D. (1966). *Soil. Sci.*, **101**, 39–45.
Ericsson, R. J., and Baker, V. F. (1967). *Nature, Lond.*, **214**, 403–404.
Gams, W. (1962). *In* "Soil Organisms" (Ed. J. Doeksen, and J. Van der Drift). p. 204. North Holland, Amsterdam,
Goos, R. D., and Summers, D. F. (1964). *Mycologia*, **56**, 701–703.
Kawamura, A. (1969). "Fluorescent Antibody Techniques and their Applications". University Park Press, Manchester.
Keddie, J. A. G. (1947). *Hlth Bull. Dep. Hlth Scotl.*, **5**, 66–68.
Keyworth, W. G. (1963). *Ann. Rep. Veg. Res. Sta. for 1962*, p. 56. Wellesbourne, Warwicks., U.K.
King, J. (1951). *Jl. R. microsc. Soc.*, **71**, 338–341.
Morton, D. J., and Dukes, P. D. (1967). *Nature, Lond.*, **213**, 923.
Nairn, R. C. (1969). "Fluorescent Protein Tracing", 3rd Edn. Livingstone, Edinburgh and London.
Paton, A. M., and Ayres, J. C. (1964). *Nature, Lond.*, **204**, 803–804.
Preece, T. F. (1966). *J. gen. Microbiol.*, **42**, v (Proceedings).
Preece, T. F. (1968). *In* "Chemotaxonomy and Serotaxonomy" (Ed. J. G. Hawkes), pp. 111–114. Academic Press, London.
Preece, T. F., and Cooper, Dorothy J. (1969). *Trans. Br. mycol. Soc.*, **52**, 99–104.
Rotman, B., and Papermaster, B. W. (1966). *Proc. natn. Acad. Sci. U.S.A.*, **55**, 134–141.
Stuka, A. J., and Burrell, R. (1967). *J. Bact.*, **94**, 914–918.
Wharton, A. L. (1967). *Ann. appl. Biol.*, **60**, 305–312.
Wilson, H. M. (1966). *Science*, **151**, 212.

CHAPTER XX

Electron Microscopy

G. N. GREENHALGH

Hartley Botanical Laboratories, University of Liverpool, England

AND L. V. EVANS

Department of Botany, University of Leeds, England

I. INTRODUCTION

The techniques involved in preparing fungal material for examination by the transmission electron microscope are basically those which have been developed for other biological materials although some specimens, such as thick-walled spores, may require special treatment. A good introductory handbook for the instruction of beginners in basic EM techniques is that by Mercer and Birbeck (1966), whilst a general description of the microscope itself and of the principles involved in its use may be obtained from Grimstone (1968). More comprehensive and detailed information can be found in, for example, Pease (1964), Kay (1965) and Sjöstrand (1967). There is also a useful review by Wischnitzer (1967).

It must be stated, however, that anyone wishing to embark for the first time on a research project involving electron microscopy would be well advised to spend a period of time (at least several weeks) at a laboratory where such work is routinely carried out.

Although the source of illumination in the electron microscope is not light but an electron beam, the optical principles of the electron microscope parallel those of the light microscope. The beam of electrons is passed between a series of electromagnetic lenses by means of which it is focused. Such a lens or lenses, placed between the illumination source (in this case a filament) and the specimen, focuses an image of the source onto the specimen in a manner exactly corresponding to that of the condenser of a light microscope. Denser parts of the specimen stop some of the electrons and remove them from the beam. A further set of electromagnetic lenses on the opposite side of the specimen produces a greatly magnified image in a manner corresponding to that of the objective and eyepiece of the light microscope. The final image is made visible to the eye by being projected onto a specially prepared fluorescent screen. This can be replaced by a photographic plate or film when an observation is to be recorded. If the image is of interest then it must be photographed immediately as, with time, the passage of the electron beam through the specimen may change it considerably or even destroy it. It is necessary to present only very thin objects for examination or the thickness of the specimen will prevent any electrons passing through. It is also essential that the interior of the microscope be in a state of high vacuum as collision of the electrons with gas molecules will cause their deflection and dispersal, thereby preventing formation of an image.

These requirements rule out observations on living material in general, although dry dormant spores can sometimes be examined directly for observation of surface details. Alternatively a cast or replica of the surface of an otherwise opaque specimen may be used for more detailed study of surface features. Small protoplasmic objects such as zoospores can be killed on a specially prepared carrier and examined directly after drying. In such cases the techniques of shadow casting can usefully enhance the clarity of external features and give a three-dimensional appearance. For most other types of biological material, especially those in which the objects of study are intracellular, the usual mode of approach involves preparation of thin sections of fixed material.

In principle the preparation of sections for electron microscopy is not different from that for light microscopy, although the requirements are so much more stringent (section thickness being less than a hundredth of that usual in histology) that every part of the process has had to be modified. The purpose of fixation—to ensure rapid killing and preservation of the specimen with minimal distortion and maximal capacity to withstand the subsequent procedures—is exactly the same as for other forms of microtomy. Several new fixative reagents have been discovered which are suitable for stabilizing and preserving the very small protoplasmic structures which can now be seen. These include potassium permanganate and certain aldehydes, notably glutaraldehyde and acrolein, in addition to older cytological fixatives used in simplified form, such as osmium tetroxide. A number of freezing techniques, including freeze-drying and freeze-substitution, have also been devised in an attempt to "fix" tissues without the use of chemical reagents (see Pease, 1964; Sjöstrand, 1967). After chemical fixation the material is washed and dehydrated using procedures closely resembling those used for paraffin-wax embedding, except that the length of time required for the various stages, notably the periods needed in each of the graded alcohols during dehydration, is commonly much shorter. The embedding process itself, however, is substantially different since paraffin wax cannot be used being in general too soft to withstand cutting at less than a thickness of several microns. For electron microscopy, where section thickness is less than 0.1μm, it has been found by experiment that certain high-polymer plastics and resins provide an embedding medium of the type required. One or other of these is infiltrated into the material in the unpolymerized form (i.e., as a fluid monomer) and is then polymerized, usually by means of a chemical activator (added before infiltration) and heat to give a hard block. The block is trimmed to a suitable size and sectioned with an ultramicrotome using a glass or diamond knife. It is not possible to handle directly the ultrathin sections obtained. These adhere to one another to form a ribbon which floats on the surface of a liquid (e.g., distilled water) which is held in a container (trough) fixed to the back of the knife. They are removed by contact with a prepared carrier, a small circular copper mesh (grid) normally with holes at a density of 100 or 200 to the square inch. This supports the sections and hold them in the microscope. Even this degree of support is generally insufficient and it is usual to coat the grid with a thin electron-transparent layer of collodion, Formvar or carbon.

As with sections to be examined with the light microscope it is customary to increase the visibility of the embedded specimen by treatment of the section with a stain or stains. Biological material consists mainly of elements of low atomic number which have a relatively poor power of electron absorption or deflection so that the variation in density across the section does not normally provide sufficient contrast in the final image. The section is therefore treated with specific salts of heavy metals (e.g., lead) in aqueous solution which combine with certain substances present in the specimen (proteins, lipids, etc.). In this way atoms of high atomic number (i.e., high electron-deflecting potential) are added to parts of the section resulting in the required increase in image contrast.

With a modern electron microscope, resolution of the order of 1 nm may be obtained routinely with well-prepared biological specimens, although with some inorganic materials even higher resolutions can be successfully exploited. By contrast, the limit of resolution with the light microscope is about 200 nm.

Most EM investigations of fungi carried out to date have utilized examination of thin sections. This account is thus mainly concerned with the preparation of material for sectioning and the treatment of sections before examination. The techniques of direct preparation (including shadowing), carbon-replication and freeze-etching will be dealt with briefly, together with a short account of the preparatory techniques involved in the use of the scanning electron microscope. All the methods described are now standard practice for fine structural studies in both the plant and animal kingdoms and have been successfully used in mycology.

II. FIXATION

Fixation for electron microscopy differs in several ways from that current in ordinary cytology even when some of the same reagents are used. The complex mixtures normally employed by light microscopists are replaced by single chemicals to which a limited number of additives are introduced for special purposes. The most important of these is a buffer to stabilize the pH. The osmolarity of the fixative is also important in that swelling or shrinking of cells must be prevented (see Maser *et al.*, 1967). For marine organisms it is usual to add sucrose to control osmotic properties, while for freshwater organisms traces of calcium or other salts are sometimes needed to stabilize membranes or other specific components of the cell (e.g., Lessie and Lovett, 1968). The process of fixation is frequently carried out at 0°–4°C using, for example, glass centrifuge tubes packed about with ice or by keeping the fixing vessel in a refrigerator. This is held to minimize undesired post-mortem changes and may reduce the rate of removal of some substances: fixatives do not necessarily render substances so insoluble that they will not slowly leach out. Since many excellent preservatives of fine structure have only a poor penetrating capacity the size of piece which can be effectively fixed is usually much smaller (less than 1 mm³) than with light microscopical fixation. Larger pieces will be incompletely fixed at the centre. This limitation can sometimes be overcome by cutting up the specimen before embedding and discarding the central portions. Care must always be taken to bring the desired parts of the specimen as quickly as possible into contact with the fixative, otherwise artefacts may form before death of the cells, and where possible the material should be cut up in the fixative. Needless to say the specimen itself should be fresh and in the best possible condition since moribund material will give misleading results. The length of time needed for fixation varies with the reagent and can sometimes be very short, i.e., 10 min with potassium permanganate or 20 min with osmium tetroxide. Usually the optimum length of time as well as the optimum pH and, indeed, choice of buffer and fixative must be determined by experiment.

Dehydration is carried out by means of a series of alcohols of increasing strength, 30%, 50%, 70%, 80%, 95% and absolute analar ethanol being commonly used for this purpose. It can be carried out more rapidly than was previously the case in light microscopy, in part owing to the smaller size of pieces of fixed material. It is rarely necessary for the specimen to spend more than 30 min in any one grade of alcohol and in certain cases 10 min is preferable (this is especially the case with plain osmic fixation). There is some latitude in the process, but in general harder materials need at least 20 min in each liquid. Once started the process of dehydration should not be interrupted until the higher concentrations (95% or absolute alcohol) have been reached as many cell components dissolve and disappear with prolonged treatment in dilute ethanol. Absolute alcohol is, however, a reasonably safe reagent in which fixed material can usually be left overnight without damage. Alternatives to ethanol include acetone/water mixtures. Tertiary butanol (after Johansen, 1940) has been used with *Pythium* (Marchant, 1968), but there seems to be no special advantage in this. In general it is impractical to transfer the material from one solvent bath to another; the liquid should be poured or pipetted out of the tube and replaced with the next. Before dehydration the fixative must be washed out with one or two changes (a few minutes each) of water or buffer. Washing needs to be thorough but not prolonged. This is especially important with osmium tetroxide, which will react with the ethanol. Acetone does not react with residual osmium.

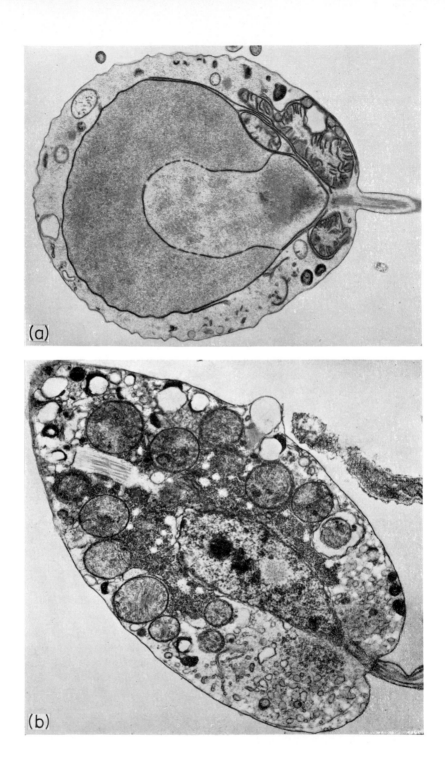

The fixatives in current use on fungal material are osmium tetroxide, potassium permanganate, glutaraldehyde and, occasionally, acrolein. Great care should be taken when handling fixatives, especially osmium tetroxide and acrolein (a potent tear-gas), to keep fumes away from the face and eyes and to prevent the liquids making contact with the skin. It is generally advisable to use a fume cupboard; in the case of acrolein this is essential. Pipettes should be used with rubber bulbs. Disposable polythene gloves are useful but affect manual dexterity and are perhaps more important when handling the embedding resins.

A. Osmium tetroxide

Osmium tetroxide is probably the most widely used fixative and gives good "general" results (fig. 1b). It is still the best choice of reagent for many organisms. Penetration of tissue is slow but reaction is rapid. It reacts with protein solutions (stabilizing them as gels), unsaturated lipids and phospholipids. It is usually used at 1 or 2% in buffer, at 0°C, for a period of time varying from 20 min to 1 h or more. Because contact with rubber bungs and other organic materials will cause contamination it should be made up in glass-stoppered bottles. The crystals are slow to dissolve and the solution should therefore be prepared and left to stand at room temperature for at least one day before use. The glass ampoule containing the crystals is thoroughly cleaned and then broken in the water or buffer solution within the bottle to be used for storage. The solution is stored in the refrigerator in the dark.

B. Potassium permanganate

First used by Luft (1956), potassium permanganate has rather special properties. It is a vigorous oxidizing agent; almost all cytoplasmic inclusions except the phospholipid membranes are oxidized and removed. It is thus not a general-purpose fixative but is useful when studying membrane systems, which appear in stained sections as a series of black lines against a grey homogeneous background (figs. 1a, 2, 12b, etc.). It may cause fragmentation of membranes, so that in general material should be examined in other fixatives as well. Penetration is relatively slow. Since the oxidative reaction is a progressive one, time is important and experiments with

---

FIG. 1. (a) *Blastocladiella emersonii*, longitudinal section of zoospore; KM_nO_4; × 14,500 (Fuller, 1966, reprinted by permission). (b) *Rhizidiomyces apophysatus*, longitudinal section of zoospore; O_sO_4; × 16,000 (Fuller and Reichle, 1965, reprinted by permission). *Both preparations stained in 1% uranyl nitrate during dehydration in acetone.*

variable fixation times on specific specimen materials may be necessary. It has been widely used for a variety of fungi, often as a 2% buffered solution at 0°C for 2 h, but satisfactory results have been obtained using a 2% buffered or unbuffered solution for much shorter times at room temperature. The solution deteriorates and it is therefore advisable to make up fresh solutions each time. Pease (1964), referring to animal tissue, concludes that the best results are obtained if dehydration is carried through rapidly.

C. Glutaraldehyde

In a study of the properties of several aldehydes as fixatives for electron microscopy and cytochemistry, Sabatini et al. (1963) concluded that glutaraldehyde and acrolein (acrylic aldehyde) were the best morphological fixatives. Glutaraldehyde is on the whole easier to use and for many purposes gives excellent results. It penetrates tissues rapidly and therefore allows slightly larger pieces of material to be fixed than is possible with osmium tetroxide. Used for a short time without post-osmication enzyme activity can be retained but lipids will leach out during dehydration resulting in negative images of cytoplasmic membranes after sectioning. Followed by osmium tetroxide made up as under IIA it is the best general fixative (fig. 3) and is especially useful for cytoplasmic tubules in the spindle and elsewhere.

Glutaraldehyde is supplied commercially as a 25% solution and for use as a fixative is diluted to 2–6% with buffer. In order to remove all traces of the fixative after glutaraldehyde treatment, the material has to be washed thoroughly with the buffer in which the fixative was prepared. Three changes of buffer over one and a half hours, decreasing the osmolarity if necessary, are usually satisfactory. Glutaraldehyde is easily oxidized to glutaric acid and if the pH is lower than about 5·5 it may be necessary to treat a small sample of the stock 25% solution with barium carbonate (or calcium carbonate) to remove the glutaric acid. Alternatively the stock solution may be stored over barium carbonate. It is now possible to obtain commercially EM-grade glutaraldehyde, purified and stabilized, for which no pretreatment should be necessary.

D. Acrolein (acrylic aldehyde)

Acrolein (acrylic aldehyde) gives more rapid structural penetration and fixation than glutaraldehyde but is otherwise rather similar in its results (fig. 4). Again, it is necessary to post-fix with osmium tetroxide. The acrolein is used as a 3–6% buffered solution. Alternatively it may be used as a 2–5% solution in 2–5% glutaraldehyde in, for example, 0·025 M phosphate buffer. We have successfully used this combined fixative (post-fixing with osmium tetroxide) on Ascomycete material (fig. 10). As mentioned above, acrolein must be used with an efficient fume cupboard and, as the solution is unstable, it has to be re-distilled at intervals. The difficulties of handling are considerable, and unless problems of fixative penetration arise, it seems preferable to use glutaraldehyde. However, as a fixative prior to embedding in glycol methacrylate (see p. 535) for high-resolution light microscopy, acrolein cannot be bettered.

E. Buffers

It is usual, although not invariable, to make up the fixative in a buffer to stabilize the pH. On the whole plant material is more tolerant than animal material of an acid fixative, although it is usual to standardize the pH near to neutrality within a range not exceeding 7·5–6·5. The most commonly used buffers are: acetate veronal, phosphate, collidine and cacodylate. Instructions for making these various buffers will be found in Mercer and Birbeck (1966). Acetate veronal was used by Palade (1952) for osmium tetroxide, but it has recently been applied to *Pythium ultimum* (Marchant, 1968) at pH 6·5 for both 2% potassium permanganate and 6% glutaraldehyde. Phosphate buffer used at or near pH 7 is often the buffer selected for terrestrial or freshwater fungi (e.g., Greenhalgh and Evans, 1968, for asci of *Hypoxylon*; Ho *et al.*, 1968, for zoospores of *Phytophthora*); it is however precipitated by seawater. For this reason cacodylate buffer often replaces it for marine organisms (see also Peat and Banbury, 1967, for sporangiophores of *Phycomyces* fixed in acrolein). Cacodylate buffer is stable and easy to use but contains arsenic which can be absorbed through the skin; care in its use is therefore necessary. There is evidence that it should not be used if enzyme work is to be carried out on the sections. Phosphate buffer is not very stable and is sometimes replaced by s-collidine (Bennet and Luft, 1959), which is more stable but difficult to purify. If a sufficiently purified commercial product is available this can usefully be tested if fixation is otherwise unsatisfactory. It has been used satisfactorily by Brenner and Carroll (1968) with 2·5% glutaraldehyde on hyphae of *Ascodesmis sphaerospora*.

Fixation without a buffer is sometimes successful, although it is not recommended without test. Potassium permanganate is, however, frequently used without buffer (e.g., ascospores of *Neurospora*, Lowry and Sussman, 1968, fig. 9). It seems impossible at the moment to draw up general rules about the use of buffers except to restate that care should be taken to select a buffer made up to a concentration which will give the desired total osmolarity when added to the fixative (Maser *et al.*, 1967). Experimentation with different buffers is usually worthwhile.

F. Suggested fixation schedules

1. *Osmium tetroxide*

Prepare a 2% solution in 0·1 M phosphate buffer at pH 7, store in a re-frigerator in darkness. Use a 10 ml round-bottomed centrifuge tube con-taining just enough fixative to cover the material easily: a depth of about 1 cm at the bottom of the tube is usually enough. Cool in ice before use. Insert the material rapidly into the fixative, plug the tube and leave on ice in a refrigerator for 20 min to 1 h. Pour away the fixative (centrifuging first if necessary) taking care to keep the fumes away from the face. Rinse (pour on, pour off) in buffer and fill up the tube with cold 30% ethanol; the tube need no longer be plugged. Change ethanol mixtures (for details see p. 521) every 10 min until 95%, when the tube should be removed from the refrig-erator and allowed to warm to room temperature. A temporary cover of aluminium foil can conveniently be used to prevent evaporation or con-tamination with atmospheric moisture. Change to absolute ethanol after 15–40 min or longer. The material is now ready for infiltration and embedding. Material fixed in osmium tetroxide alone is shown in fig. 1b.

2. *Potassium permanganate*

Prepare 2% potassium permanganate solution in distilled water. The solution must be fresh. Fix as above for 10 min to 2 h, according to require-ment, either cooled on ice or at room temperature as found best by experi-ment. Wash several times in cold distilled water. Dehydrate slowly (30 min in each ethanol mixture) or quickly as determined by test. Material fixed in potassium permanganate alone is shown in figs. 1a, 2.

3. *Glutaraldehyde–osmium tetroxide*

Prepare a 4% solution of glutaraldehyde in the selected buffer (probably phosphate) at pH 7 and a 2% solution of osmium tetroxide in the same buffer. Store both in a refrigerator, though only the osmium tetroxide need be in darkness. Fill a round-bottomed 10 ml centrifuge tube with the glutaraldehyde fixative and cool on ice. Insert the material and leave for at least 2 h (may be left up to 15 h). Pour away the glutaraldehyde and wash the material in three half-hourly changes of buffer. Add just enough of the osmium tetroxide solution to cover the material (1 cm depth at the bottom of the tube is usually sufficient). Leave on ice for 5 h or overnight. Pour off the osmic solution and fill the tube to the top with buffer. Leave for a few minutes. Dehydrate slowly through the ethanol mixtures, used cold,

---

Fig. 2. *Pyronema domestica*, section through ascus and ascospore; KM_nO_4; ×13,000 (Reeves, 1967, reprinted by permission).

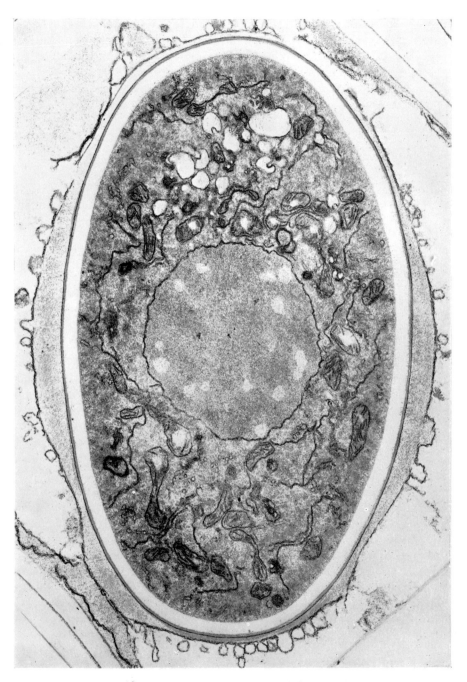

leaving 20–30 min in each. Remove from the refrigerator when 95%
ethanol has been reached and allow to warm to room temperature. At this
point the tube should be covered with a cap of aluminium foil. Change to
absolute ethanol. The material is now ready for infiltration and embedding.
Alternatively it may be left overnight in 95% or absolute ethanol. Material
fixed in this way is shown in fig. 3.

4. *Acrolein*

CAUTION: acrolein is extremely dangerous and an efficient fume cupboard
is essential.

Prepare 0·1 M phosphate buffer (pH 6·5–7·0), and 5% acrolein and 2%
osmium tetroxide in this buffer and cool in the refrigerator. If the acrolein
solution is not clear it must be redistilled. Fill a round-bottomed 10 ml
centrifuge tube with the acrolein fixative and insert suitable small pieces
of the material. Cover and leave for 12–24 h in the refrigerator. Pour away
the acrolein and wash the material in three half-hourly changes of the cold
buffer. Add the osmium tetroxide fixative and proceed as in Part 3. Material
fixed in this way is shown in fig. 4.

III. EMBEDDING

The earliest plastic adopted for electron microscopy (methyl or butyl
methacrylate) did considerable damage to the material by swelling during the
act of polymerization. It also tended to vaporize in the electron beam leaving
parts of the specimen imperfectly supported. It has been in consequence
largely abandoned (see, however, glycol methacrylate, p. 535) except for
special purposes. The most frequently used alternatives are Araldite,
Vestopal and Epon. It is possible that Maraglas resin may also become
popular, although it has been little used for fungi as yet.

Epon, Araldite and Maraglas are all epoxide resins and should be handled
with care as they may have a carcinogenic effect and are known to cause
skin reactions in some people. They should not be allowed to contact the
skin (disposable polythene gloves should be used) and are best made up and
used in disposable containers of some kind such as polythene beakers, measuring cylinders, etc. Polythene syringes (without needles) may also be used.
If the resin does get on to the skin it should be washed off immediately with
hot water and soap (solvents tend to spread the resin and should only be
used if absolutely necessary and then in small quantity). Discarded Epon
should never be put down the sink and should only be thrown away after

FIG. 3. *Hypoxylon fragiforme*, section through ascus and developing ascospore
wall; glutaraldehyde and OsO_4; × 60,000 (Greenhalgh and Evans, 1968).

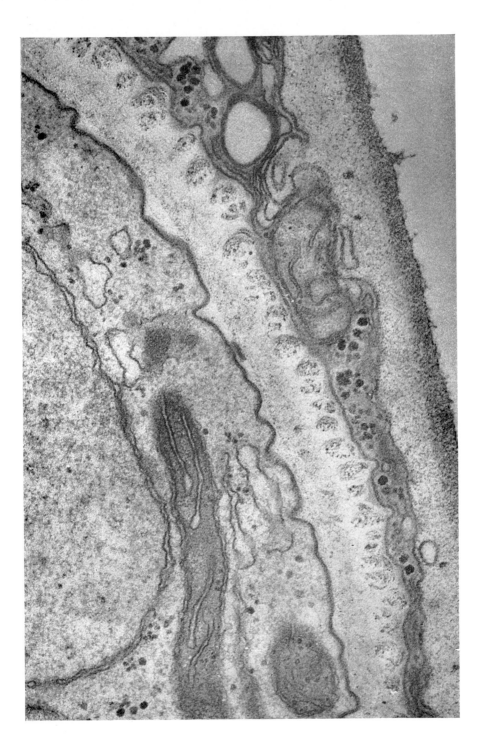

polymerization in the oven. Many of the substances used in the resin mixtures are very viscous and measuring out quantities may be difficult. Such liquids should be poured down the centre of the measuring cylinder, avoiding contact with the sides, the other substances being added to the same container and the whole thoroughly mixed.

A. Epon embedding

Epon mixtures based on the resin commercially available as Epikote 812 (Epon 812) are amongst the most widely used embedding media. The resin itself is mixed with suitable hardeners and an accelerator is added to speed up the process of polymerization. The actual hardener and the quantity used will affect the hardness of the final block. The Epon mixture may be conveniently made up as two stock solutions (A and B) using the proportions given originally by Luft (1961) or those quoted below—

Solution A

Epikote 812	25 ml
Dodecenyl succinic anhydride (DDSA)	50 ml

Solution B

Epikote 812	50 ml
Methyl nadic anhydride (MNA)	50 ml

The two solutions should be stored in a dry place at room temperature. The solutions are mixed, when required, in varying proportions, according to the degree of hardness required in the final block as follows—

Soft	80% A, 20% B
Standard	60% A, 40% B
Medium	50% A, 50% B
Hard	30% A, 70% B

During the mixing of these two solutions, 2% of an accelerator, 2,4,6-trimethylaminomethyl phenol (DMP 30) *or* benzyl dimethylamine (BDMA) is added. A bulk quantity, say approximately 400 ml, can be mixed using a motor fitted with a glass-rod impeller, stirring thoroughly for at least 5 min, the accelerator being added gradually during the process. Convenient quantities for preparing a bulk amount, based on the soft mixture given above, are as follows—

Epikote 812	150 ml
DDSA	240 ml
MNA	30 ml
BDMA (DMP 30)	8·2 ml

Fig. 4. *Hypoxylon fragiforme,* longitudinal section through ascus apex; 5% acrolein and OsO_4; × 30,000 (Greenhalgh and Evans, unpublished).

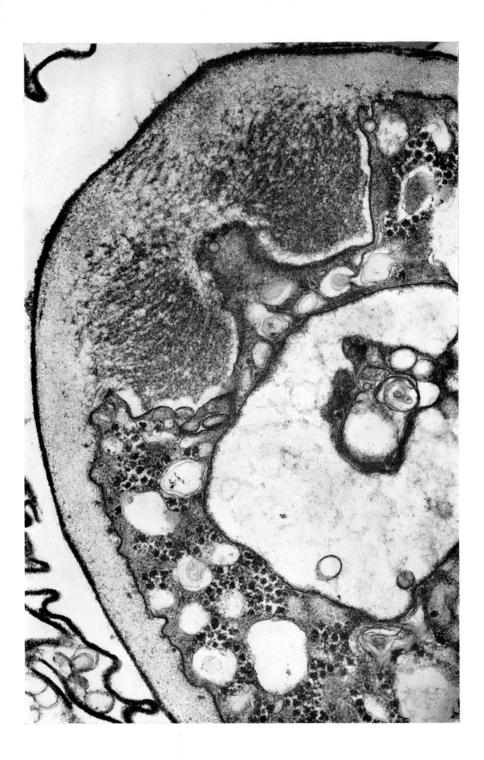

We have found this mixture (originally developed for other organisms) very satisfactory for fungal material. It can be cut readily with a glass knife.

If the Epon is mixed in bulk it should be stored in a deep freeze in small 10 ml glass phials with polythene tops. At very low temperatures (−35°C) polymerization is, to all intents and purposes, prevented and the mixture will keep indefinitely (Pease, 1964). It will, however, keep satisfactorily for several weeks at −10°C although polymerization will proceed slowly. When needed, one of the phials is removed from the deep freeze and allowed to warm up to room temperature (allow one hour for this) before it is opened, thus preventing condensation on the Epon, which must on no account be contaminated with water or water vapour. Unused Epon mixture should not be returned to the deep freeze but should either be polymerized in an oven at 60°C before being thrown away or retained for use in preparing blocks for cutting (see p. 540).

It is necessary after dehydration to remove the ethanol (or equivalent) with a transitional solvent which can be mixed in all proportions with the plastic mixture. Propylene oxide (1:2-epoxypropane) is the usual solvent used for transfer from ethanol to Epon, Araldite or Maraglas. It is a rather toxic and volatile liquid which should not be inhaled. After dehydration, material is soaked in two changes of propylene oxide (15–30 min in each) and then placed in a mixture of equal parts of propylene oxide and the final Epon mixture. It is left in this mixture in an open tube overnight (5–15 h) during which time much of the propylene oxide evaporates leaving the material in almost pure Epon. It is then ready for embedding. This can conveniently be carried out using either small flat-bottomed polythene dishes† (1½ in. dia.) or small plastic embedding capsules. These capsules, available commercially, are moulded to give a short cylindrical block with a pointed tip. The former method is to be preferred if orientation of the specimen before sectioning is important. Clean fresh Epon is poured into the dish or capsule and the specimen, broken up if required, inserted with forceps or a pipette, taking as little as possible of the propylene oxide/resin mixture with it. If a pipette is used it should have a fairly wide point and be wetted inside with Epon before use or the material will adhere to the inner surface of the glass. The Epon mixture is highly viscous and the material may take some time to sink, although this normally happens during the period in the first oven. It is put into an oven at 35°C for 12–24 h after which it is passed through a second oven at 45°C for a few hours and finally left to polymerize at 60°C for one or two days. It should then be cooled slowly. The resulting hard disc or block of Epon is then removed from its

† The small polythene "dishes" obtainable from "do-it-yourself" shops for use as recessed finger-grips in sliding cupboard doors are very suitable.

container. Its subsequent treatment is described on p. 539. Glass tubes contaminated with the Epon mixture should be washed first with propylene oxide (do not pour down the sink), then with acetone, followed by 95% alcohol (in which they may be left) and finally with hot soapy water.

B. Other embedding media

Araldite embedding mixtures, which have similar properties to Epon mixtures, may be made up with the same accelerator (DMP 30) and stored in sealed tubes in the deep freeze for several weeks in the way already described. The mixture given below contains a plasticizer, dibutyl phthalate, which controls the hardness of the final block. For harder blocks the quantity of plasticizer is reduced—

Araldite M	20 ml
DDSA (964B)	20 ml
DMP 30 (DY. 064)	0·6 ml
Dibutyl phthalate	2 ml

These are the proportions recommended by Mercer and Birbeck (1966). They advise adding the DMP 30 with a clean pipette and washing this through with the dibutyl phthalate to ensure complete incorporation. The mixture is warmed to 50°C and thoroughly stirred or shaken, then stored in tubes. Polymerization is carried out as for Epon.

Vestopal W was introduced in connection with a new method of fixation for bacterial nuclear material (Ryter and Kellenberger, 1958) and is a polyester resin similar in its embedding characteristics to Araldite. It gives a hard block with fine grain and sections which stain easily but is difficult to cut. The resin is not soluble in ethanol and dehydration is in acetone. Material is then passed through a series of Vestopal W/acetone mixtures of increasing Vestopal content before going into the final Vestopal mixture (containing initiator and activator). The final mixture has the following composition (Pease, 1964)—

Tertiary butyl perbenzoate (initiator)	1%
Cobalt naphthenate (activator)	1%
Vestopal W	98%

If the initator and activator are mixed together directly they form an explosive mixture and they should be added to the Vestopal W independently. It is possible to dehydrate in ethanol using styrene as the transitional solvent (Kurtz, 1961; for details see Glauert, 1965a).

Maraglas epoxy resin was introduced by Freeman and Spurlock (1962) as an embedding medium which sections more easily than Epon, shows less background grain and has low viscosity giving good impregnation. As far as we are aware there has been little utilization of this resin in mycological work but preliminary studies indicate its usefulness. The original technique has been modified by Spurlock *et al.* (1963) and by Erlandson (1964). The mixture recommended by Erlandson is as follows—

Maraglas 655	72 %
Diepoxide flexibilizer (DER 732)	16%
Dibutyl phthalate	10%
BDMA	2%

These must be thoroughly mixed. The hardness may be altered by varying the proportions of DER 732, adjusting the amount of Maraglas to bring up the total to 100 parts. A hard block is given by 10 parts of DER 732. Polymerization is by heating at 60°C for 48 h, and maximum hardness is reached after a further few hours at room temperature.

Fig. 5. *Hypoxylon fragiforme*, longitudinal section through perithecial ostiole; embedded in glycol methacrylate and stained in toluidine blue; light micrograph, × 700 (Greenhalgh and Evans, unpublished).

Finally, mention must be made of the use of glycol methacrylate as an ideal embedding medium for material intended for histochemical or enzymic work in conjunction with light microscopy. It is permeable to aqueous stains. Details of the technique have been summarized by Feder and O'Brien (1968) and successful application of the method to algal, pteridophyte and angiosperm material is recorded. The first example known to us of its successful use with a fungus is shown in fig. 5.

Sections (0·5–3 μm) may be cut with an ultramicrotome and glass knife and examined individually with the light microscope after treatment with various stains (e.g., toluidine blue 0, Schiff reagent, acid fuchsin, etc.). The quality of fixation and the thinness of the sections results in much improved resolution compared with that of conventional paraffin-embedded material. Acrolein fixation is recommended for material to be embedded in glycol methacrylate, although glutaraldehyde may be used with less successful results. Formaldehyde fixation has proved very satisfactory with certain algae. It is usually necessary to purify commercial glycol methacrylate; details of this process and the complete sequence of section preparation and examination, etc., are given by Feder and O'Brien. Glauert (1965a) describes a dehydration schedule involving a series of solutions of glycol methacrylate in water, thus avoiding treatment with alcohols and possible extraction of substances. Sections prepared in glycol methacrylate may be examined with the EM but will be of inferior quality to Epon-embedded material. If it is intended to examine sections from one block with both the EM and the light microscope then it is probably best to prepare thick (up to 2 μm) Epon sections. These may be examined with phase or interference microscopy or stained with, for example, toluidine blue, although the staining will be much less successful than with glycol methacrylate. Details of Aquon (a water-soluble component extracted from Epon 812) and Durcupan (based on a water-soluble epoxy resin) are given by Glauert (1965a). Both of these may be used for such work but they have certain disadvantages and have not, as far as we know, been applied to fungi.

IV. GRID PREPARATION

The electrolytically prepared copper grids designed to hold sections should be cleaned before use and coated with a thin collodion, Formvar or carbon supporting film. Hard blocks, not readily cut with a glass knife, may give ribbons which can be mounted on uncoated grids. Alternatively, there is at least one embedding resin now available commercially for which it is claimed that no supporting film is necessary. Collodion, although transparent to electrons at thicknesses easily obtainable, is no longer widely used as a supporting film because of its capacity to shrink and stretch with changes of temperature experienced in the electron beam. Formvar is better in this respect and may be used alone. The Formvar film should be toughened in the coating unit as described on p. 538. Carbon support films are stronger and Formvar is frequently employed as a foundation film onto which carbon is deposited, the Formvar then being dissolved away (see p. 538). For whole mounts it is often useful to retain the Formvar as well as the carbon.

The grids are cleaned in a small Petri dish by flooding them successively with the following liquids in order—

   (i) Two changes of analar chloroform.
  (ii) Two changes of absolute alcohol.
 (iii) Two changes of distilled water.
 (iv) Two changes of absolute alcohol.
  (v) Two changes of chloroform.

They are then dried on clean filter paper in a 3 in. Petri dish. This process is perhaps a counsel of perfection, but using small pipettes to add and remove the liquids it is not time-consuming and guarantees clean, grease-free grids. Individual grids should be handled with fine watchmaker's forceps.

The following method of preparing carbon films has been found to give very good results and is strongly recommended (for other methods of preparing support films see Bradley, 1965a).

Four main stages are involved: the production of a Formvar film on glass; the transfer of this film to the grids; the deposit of a thin layer of carbon over the Formvar; and the removal of the Formvar leaving the carbon film intact on the grid. The fourth step is omitted if both films are required.

Stage 1. A 0·5 % solution of Formvar in fresh analar chloroform is prepared (the Formvar powder should be stored over phosphorous pentoxide or calcium chloride in a desiccator). Several new microscope slides are cleaned thoroughly with non-chlorinated scouring powder and tap-water and rinsed several times with distilled water. They are then stored between sheets of filter paper. When ready, a clean slide is blotted dry and carefully wiped with Velin tissue to remove all dust particles, etc. The Formvar solution is poured into a cylindrical separating funnel held in a clamp over a beaker. To prevent evaporation of chloroform the funnel is fitted with a cap to the underside of which is fixed a spring clip. The clean slide is fixed into the clip at one end, lowered into the Formvar solution and allowed to stand for a few seconds. The solution is drained into the beaker and the slide left in the covered funnel until dry. It is then placed on a sheet of filter paper and the film scored around the edge with a mounted needle.

Alternatively the slide may be held at one edge with blunt forceps, lowered into a beaker of Formvar solution and then smoothly withdrawn.

Stage 2. The film can then be floated off by slowly introducing the slide (at a shallow angle) into a bowl of clean water and at the same time breathing on it (*not* blowing) as it becomes immersed. During the operation the slide may be supported on the edge of the bowl, which should be brim full. The resultant floating film must be free of interference colours, lines or marks. Clean grids are then placed individually on the floating film, dull side downwards, to give about three rows of ten grids each, close together.

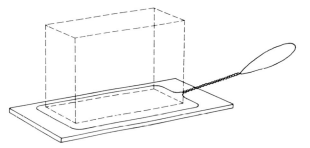

Fig. 6. Diagram to show method of transferring Formvar film from wire loop to microscope slide, using a rectangular cardboard tube placed inside the loop.

This film, together with the **grids**, has to be removed from the surface of the water. The simplest method is probably to push the film down into the water with the flat surface of a clean microscope slide or with the curved surface of a clean boiling tube, or McCartney phial, held horizontally. When under water the slide is smoothly and quickly turned over and lifted out. It will be found that the film adheres to its surface. If the tube is used, it can be immersed and withdrawn in one steady movement. Care should be taken to avoid **trapping air bubbles** between the glass and the film. The Formvar film can also be picked up by simply lowering on to it a piece of clean Parafilm (paraffin wax) held horizontally, which is then removed without pushing the film down into the water. Alternatively, the following method, which requires practice, but gives excellent results, may be used.

A rectangular wire loop (with handle), slightly smaller in area than the film, is constructed from copper wire. The wire of the loop is given slightly roughened edges and the whole is stored under analar chloroform. The loop is dipped into the water beneath the floating film and slowly removed at a shallow angle to the surface so that the film is caught on it with the grids held in the space contained by the loop. The loop, with its adherent film, is then dried off for 15 min, at 40–60°C, during which time more film can be prepared in the same way.

Stage 3. If the film has been picked up on a glass or wax surface, the individual grids can simply be picked off, after drying, with fine forceps. If, however, the film is held on a loop as described above, then it must be transferred to a clean slide. A small rectangular cardboard tube is prepared, of such cross-section that it will just pass smoothly through the wire loop. The loop is placed on clean (detergent washed) microscope slide with the grids (dull side upwards) in contact with the glass. The cardboard tube is placed over the film and within the loop (fig. 6), which is then lifted around the tube leaving the film and grids on the slide. Alternatively, it may be possible to separate the film from the loop simply by scoring around the inside edge with a mounted needle so that the film remains on the slide. The slide, with grids, may then be stored in a desiccator. The Formvar film must be free from wrinkles, folds, tears, etc., and should have a smooth surface when examined with reflected light under the microscope. Such grids may be used for examining sections directly but before use the Formvar should be toughened by exposing the film to an ionizing discharge in the coating unit (5 kV for about 5 sec at low gas pressure). Alternatively, they may be coated with carbon for additional support, after toughening the Formvar in the way described. A thin carbon layer (about 10 nm) is deposited under vacuum over the Formvar film. The thickness of the film may be roughly estimated by including within the coating chamber of the apparatus a piece of white glazed porcelain with a drop of vacuum oil on its surface (see Bradley, 1965a). The carbon deposit will be visible on the poreclain surface and not on the oil-drop, and when apparent as a light smokey-grey deposit the thickness will be sufficient for general use.

Stage 4. If the Formvar is to be removed (as will normally be the case when the grids are to be used for sections), the grids are taken carefully from the slide and placed gently (coated, or dull, side up) on a platform of fine wire mesh submerged in analar chloroform in a Petri dish. This dissolves away the Formvar, leaving the carbon film intact. Very carefully, with as little disturbance as possible, the chloroform is removed with a pipette so that the grids remain on the platform. They are removed individually, dried over a lamp for a few seconds, placed on filter paper in a Petri dish (film upwards) and stored in a desiccator.

V. SECTION CUTTING

Detailed instructions on section cutting are usually provided by the manufacturer of the ultramicrotome selected and the techniques are fully discussed by Pease (1964), Glauert and Phillips (1965), and Sjöstrand (1967). It is sufficient here to deal with a few general points. The beginner should

become expert in the preparation and use of glass knives. These are prepared from strips of plate glass using glass-breaking pliers or, preferably,
a knife-making machine (e.g., the LKB knife-maker). Although fairly
expensive, a knife-maker does facilitate the production of quantities of good-
quality glass knives and may be considered essential. The knives should
be prepared immediately before use to minimize contamination and deterioration of the cutting edge during storage, and each knife edge should be
examined carefully with a binocular microscope before use and the best
selected. The cutting edge of a glass knife deteriorates rapidly during
sectioning and any one part of the edge will only give a few sections. If
long ribbons of high quality serial sections are required, then the investigator will need to use a diamond knife. These are expensive and easily
damaged but used with care will give first class results over a long period.

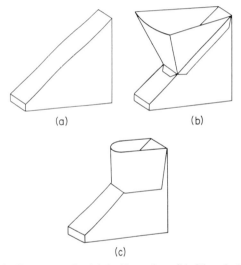

(a)   (b)

(c)

FIG. 7. (a) Glass knife prepared with knife-maker. (b) Glass knife fitted with metal
trough. (c) Glass knife fitted with trough prepared from adhesive tape.

Diamond knives are purchased in a mount with a water trough attached,
but with glass knives a trough must be fixed to each one and sealed with
dental wax to prevent leaking. Metal troughs may be purchased for this
purpose (fig. 7b) and these can be used repeatedly but it is easier to make a
trough by wrapping a short strip of Scotch insulating tape (no. 33) around
the back of the knife and completing the seal with wax (fig. 7c).

Before sectioning can begin it is necessary to prepare the block. If a
conical capsule has been used for embedding, the tip should be trimmed with
a razor blade to provide the smallest cutting face possible. Trimming should

be done with the minimum number of cuts necessary and should give a rectangular or trapezoidal cutting face such that the largest side is not longer than about 1 mm (preferably less). Sectioning is done with the long side parallel to the knife-edge. If the material has been embedded in a flat-bottomed dish the resin disc may be examined with a dissecting microscope and suitable fields ringed with Indian ink. These are then cut out singly (using a fine saw), stuck on to short lengths of plastic (acrylic) rod with epoxy glue or residual Epon mixture and hardened in an oven at 60°C. The cutting face is then trimmed as above. It should be noted that the surface of a block is always slightly harder than the interior. It is, therefore, often helpful to remove the top layer before serious sectioning begins. Blocks which are too soft may be incompletely polymerized and further heating to 60°C will be necessary. If this has no effect then the block can be temporarily hardened by cooling in the refrigerator immediately before sectioning.

The sections float off the knife-edge onto the liquid held in the trough and should remain together to form a ribbon as stated in the introduction. The liquid is usually freshly distilled water (preferably from a glass rather than a polythene bottle) but other liquids may be used. Epon sections are collected on distilled water. The liquid must be level with the knife-edge so that the sections float freely off the edge. If the meniscus is too high the block itself will pick up water from the trough. It is, of course, essential that the knife and trough be scrupulously clean.

Since sections become compressed by the act of cutting, causing objects circular in section to appear oval, it is usual to stretch them, before mounting, by exposure for a few seconds to the vapour of a suitable solvent. A glass rod moistened with trichlorethylene held a few millimeteres above the sections (care being taken not to touch the knife-edge) until they expand is recommended.

Section thickness is recognizable in a general way by the interference colours generated in the ribbon by light from above. The thinnest sections (less than 60 nm) appear grey, but for general purposes one should try to obtain either pale gold or a light silver (about 80 nm). All other colours (green, blue, purple) indicate sections which are too thick and these should be discarded.

Sections should be mounted as nearly as possible aligned with the grid bars since serial sections can rarely be observed if they cross the field diagonally. A coated grid (held with fine forceps) is either brought down from above on to the floating ribbon which will then adhere to the dry surface, or, alternatively, the grid may be positioned in the water below the ribbon and lifted through the water surface bringing the ribbon with it. If the grid is uncoated this method is satisfactory, but with coated grids particles in or on the water can be trapped between the grid and the ribbon and the former method is therefore preferable. The grids are then left to dry on filter paper in a Petri dish, a process which takes only a few minutes.

VI. STAINING

The purpose of staining is to increase selectively the electron opacity of the specimen and thus improve image contrast by suitable treatment with the salts of heavy metals. Both osmium tetroxide and potassium permanganate have some staining action, mainly affecting the phospholipids (e.g., membranes) so that material fixed in these is to some extent stained during fixation. It is, however, usual to increase the contrast still further by staining again. This may be done before dehydration (e.g., Carroll, 1967; and fig. 8), during dehydration (e.g., Fuller and Reichle, 1965; fig. 1b) or after sectioning. The latter is usually preferred.

There are several stain preparations currently in use, the best known being phosphotungstic acid, lead citrate and uranyl acetate. Phosphotungstic acid is important in negative staining of unembedded material (useful for examining very small isolated structures where very high resolution is needed; see Horne, 1965), but is not as effective with Epon sections as with methacrylate embedding media. Details will be found in Mercer and Birbeck (1966). Stains containing lead, e.g. lead tartarate (Millonig, 1961), are commonly used and are perhaps the most satisfactory. Lead citrate (Reynolds, 1963) is easy to prepare and is a good general stain (figs. 3 and 4) which keeps well. Staining of cytoplasmic membranes, nucleic acids and glycogen granules is good. It is made up according to the following schedule: Add 1·33 g lead nitrate ($Pb(NO_3)_2$) and 1·76 g sodium citrate ($Na_3(C_6H_5O_7)$ $2H_2O$) to 30 ml of glass distilled water in a 50 ml volumetric flask. Shake this vigorously for a minute then allow to stand, with occasional shaking, for 30 min to ensure complete conversion of all lead nitrate to lead citrate. Add 8·0 ml 1N NaOH and dilute the suspension to 50 ml with glass distilled water. Mix by inversion. The lead citrate will dissolve and the stain may be used. It can be stored in glass or polythene bottles for up to six months, but old solutions may become turbid and should be centrifuged.

Many workers prefer a double stain often involving treatment with 2% aqueous uranyl acetate followed by a lead salt (e.g., 20 min/10 min respectively). The pH of the uranyl acetate solution may be increased to 4–5 by adding drops of 1N NaOH (Mercer and Birbeck, 1966) after which it is filtered and allowed to stand. Gimenez-Martin et al. (1967) have recently described a staining schedule in which the lead salt and uranyl acetate are combined and used in 70% acetone during dehydration. It is claimed that this method (used on *Phalaris* root tips) greatly reduces contamination, and the sections may of course be examined as soon as they are obtained.

Staining of sections can be conveniently carried out by floating the grids, section side downwards, singly on fresh drops of the reagent placed on the clean surface of a sheet of dental wax (glass surfaces always seem to produce contaminants). It is important that each grid should float centrally on its drop otherwise the sections will invariably become contaminated. Alternatively, the grids may be immersed in the drop by insertion from the side with the aid of a pair of fine forceps. Such grids should lie section side upwards; dirt in suspension will fall upon them but surface contamination of the drop is reduced. When staining is complete the grids must be washed in several changes of distilled water or, in soft-water areas, may be held edgeways under a running tap regulated to give an even flow of cold water, for 10–15 sec. The grid is then placed section side upwards on filter paper in a Petri dish to dry. Care must be taken to avoid letting drops of liquid from between the tips of the forceps falling on to the section surface as it is laid down or else contaminants will be deposited. This may be prevented by removing residual water from between the tips of the forceps with filter paper. The specimen dries rapidly and in a few minutes can be inserted into the microscope. For a general discussion of staining methods, including autoradiography, see Glauert (1965b).

VII. SPECIAL APPLICATIONS AND TECHNIQUES

A. Zoospores and small specimens

Objects which are too small to be handled individually (such as zoospores; see also Section VIIIA) may be squirted violently in suspension with a fine pipette into fixative in a 10 ml centrifuge tube. As little as possible of the medium containing the spores should be transferred so as not to dilute the fixative. If the zoospore suspension is thin it is gently centrifuged before fixation, and the medium can then be poured or pipetted away and fixative poured onto the spores. Towards the end of fixation the material is centrifuged (or recentrifuged) to form a pellet which undergoes the remainder of the processing in the centrifuge tube. With a reasonable speed of centrifugation (approx. 6000–8000 r.p.m. for 2–3 min) the pellet should remain intact during the subsequent washing and dehydration, but if not the material can be re-centrifuged into a pellet at any time. The pellet is finally either cut up during the last stages of dehydration or during the transfer into Epon, as convenient. If the material is harmed by

---

FIG. 8. *Hypoxylon fragiforme*, section through developing ascospore; KM_nO_4, stained with uranyl acetate before dehydration; × 25,000 (Greenhalgh and Evans, unpublished).

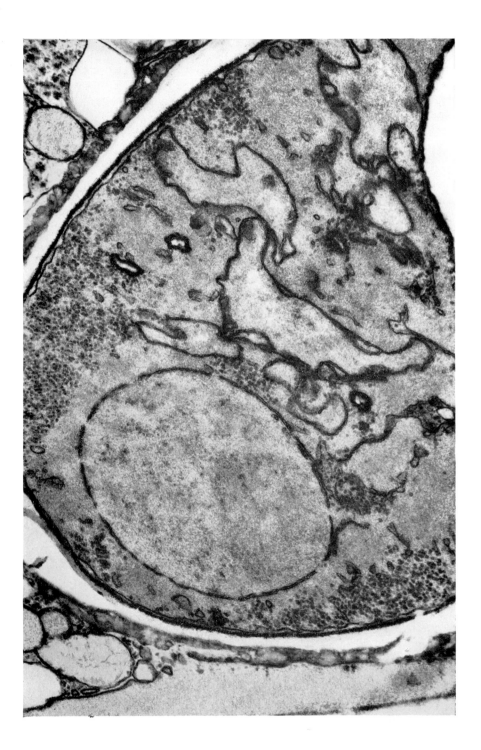

centrifuging at a speed of the order given above, then it may be centrifuged at a speed just sufficient to bring down the material without compact-
ing it (1000–2000 r.p.m.). It can then be embedded in agar, eliminating
the need for re-centrifugation at each stage, although this is likely to make

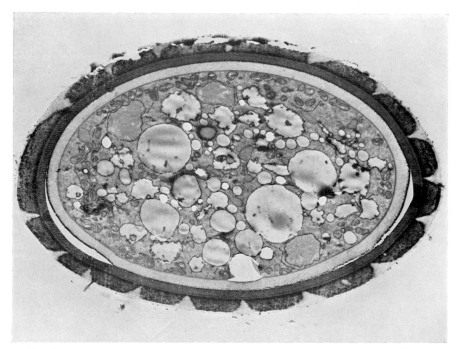

FIG. 9. *Neurospora tetrasperma*, section through germinating ascospore; KM_nO_4;
× 7000 (Lowry and Sussman, 1968, reprinted by permission).

section cutting more difficult. The process is as follows: after centrifuging,
excess liquid is poured off and the centrifuge tube with the fixed material
is placed in a water bath at 45°C which already holds a tube with 2% liquid
agar. A small drop (less than 0·3 ml) of this liquid agar is transferred to the
centrifuge tube with a warm pipette and the tube shaken quickly. The
agar containing the material is immediately run out on to a cold microscope slide. It is allowed to set and cut into small pieces which are taken
through the remaining stages of dehydration. The technique of handling
described above may be applied to all small objects which cannot be handled
individually (for further details see Glauert, 1965). A different technique
was used by Ho *et al.* (1968) to handle zoospores (of *Phytophthora*). During
fixation the zoospores were filtered on to a millipore disc which was
momentarily dipped into buffered liquid agar. The millipore disc dissolved

during dehydration in acetone, freeing the thin agar film and the contained zoospores.

"Glutaraldehyde followed by osmium tetroxide seems to be the fixative of choice for zoospores and is being used wherever possible in our current studies" (Fuller and Reichle, 1968). This is the present situation; the method gives results which are of high quality. In earlier studies, however, other fixatives were employed and at least one of these (potassium permanganate) is still used and gives valuable information (e.g., Fuller and Reichle, 1965).

B. Thick-walled spores

When dealing with thick-walled spores, penetration of the fixative or embedding medium through the wall is often a difficulty. Fixatives such as

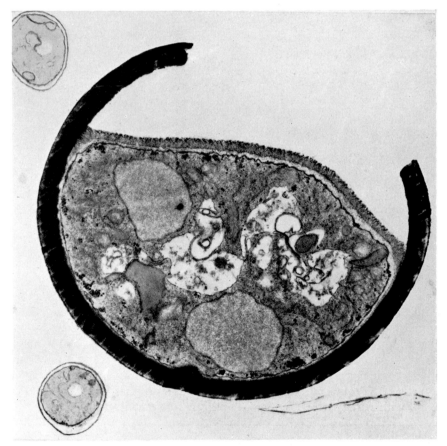

FIG. 10. *Daldinia concentrica*, section through germinating ascospore; 5% acrolein in 5% glutaraldehyde and OsO_4; × 7000 (Greenhalgh and Evans, unpublished.)

acrolein and glutaraldehyde penetrate well and will successfully enter spores where osmium tetroxide cannot do so, but this does not solve the problem of the embedding medium. The best procedure is to fracture or crack the spore wall whilst it is in the fixative. For example, Ekundayo (1966) fixed sporangiospores of *Rhizopus arrhizus* by breaking them in fixative in a disintegrator. Lowry and Sussman (1968) have described a

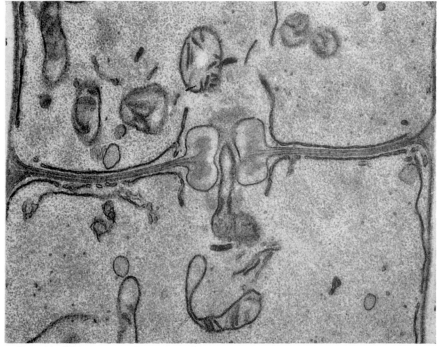

FIG. 11. *Rhizoctonia solani*, longitudinal section through hypha showing septum; KM_nO_4; × 26,500 (Bracker and Butler, 1963, reprinted by permission).

method using 2 in. square pieces of plate glass which are pressed together with parallel-jaw glass pliers. The spores (ascospores of *Neurospora tetrasperma*) were suspended, after centrifugation, in a small volume of 2% potassium permanganate and a drop placed between the pieces of glass. Pressure was applied until most of the spores were cracked (shown by microscopic examination). They were allowed to remain in the fixative for 2 h (fig. 9). This technique can be carried out with glass microscope slides (e.g., with rust spores; Henderson, personal communication).

FIG. 12. *Pythium ultimum*, section through hypha showing: (a) Golgi apparatus; glutaraldehyde and O_sO_4; × 64,000 (Grove *et al.*, 1967, reprinted by permission. (b) Lomasome; KM_nO_4; × 58,000 (Bracker. 1967 reprinted by permission).

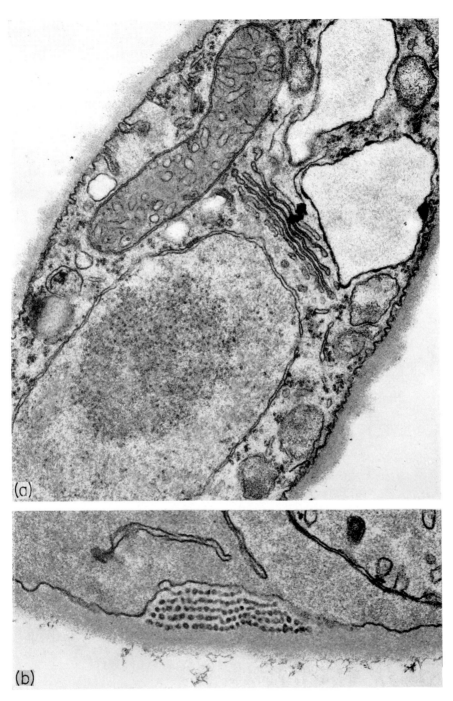

(a)

(b)

Thick-walled spores which have germinated are more easily fixed and embedded (fig. 10) though difficulty may be encountered, as with dormant spores, in cutting the thick spore wall without damage.

C. Vegetative hyphae

Small blocks of agar containing hyphal tips or older hyphae may be cut as radial segments from colonies in Petri dishes and fixed, dehydrated and embedded immediately (e.g., Brenner and Carroll, 1968, on hyphal structure in *Ascodesmis sphaerospora*). To avoid the presence of agar, which

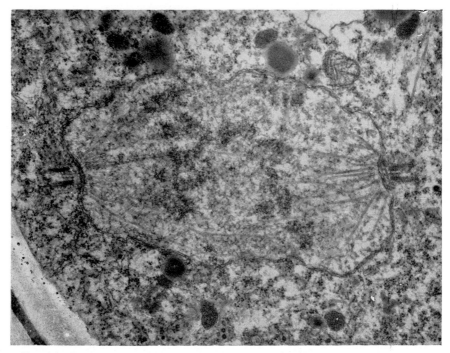

Fig. 13. *Catenaria anguillulae*, early anaphase nucleus; glutaraldehyde and O_8O_4, stained in 1% uranyl acetate during dehydration; × 26,000 (Ichida and Fuller, 1968, reprinted by permission).

causes difficulties during embedding and sectioning, Bracker and Butler (1963) grew cultures of *Rhizoctonia solani* on the surface of autoclaved cellophane placed over potato dextrose agar. Squares of cellophane were then cut out and the mycelium floated off and fixed (fig. 11). Alternatively the mycelium (over cellophane) may be flooded with fixative and then portions of it transferred to fresh fixative in phials (e.g., Grove *et al.*, 1967, on hyphae of *Pythium ultimum*; fig. 12a).

A modification of this method, permitting accurate longitudinal orientation of hyphae, has been developed by Grove and Bracker (personal communication) and applied to several fungi including one from each class with excellent results. Portions of mycelium from the colony margin, together with pieces of cellophane are fixed at room temperature and embedded in Araldite or an Epon/Araldite mixture. The final block can easily be made to split in the plane of the cellophane which is then removed with

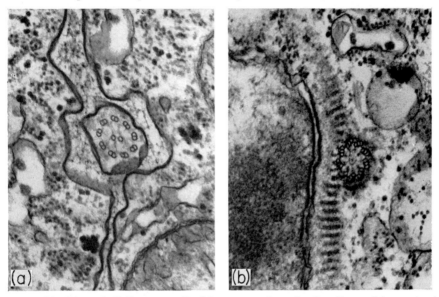

FIG. 14. *Blastocladiella emersonii*: (a) cross section of a flagellum; (b) proximal part of basal body. Both glutaraldehyde and OsO_4; × 54,300 (Lessie and Lovett, 1968, reprinted by permission).

a razor blade to expose the hyphae near the block surface. Several thick sections are removed to give a sectioning face after which suitably orientated hyphal tips can be selected by examination with a binocular microscope. The block is then precisely orientated to ensure longitudinal sections of the hyphal tips. In this standardized technique, Grove and Bracker used cacodylate buffer at pH 7·3 for the fixative (2·5 % glutaraldehyde followed by 1 % osmium tetroxide in the same buffer). The final embedding mixture was polymerized in a nitrogen atmosphere.

D. Microtubules and nuclear division

The presence of microtubules in the spindle and cytoplasm of plant and animal cells has been detected with the use of glutaraldehyde followed by osmium tetroxide (e.g. Fuller and Calhoun, 1968, on microtubules in the motile cells of Blastocladiales; and Ichida and Fuller, 1968, on mitotic spindle tubules in *Catenaria anguilulae*; fig. 13). For the study of internal flagella structure the same fixatives should be used, e.g. Lessie and Lovett, 1968, who investigated the differentiation of zoospores within the sporangia of *Blastocladiella emersonii*; fig. 14. In this study traces of calcium chloride and magnesium chloride were added to the osmium tetroxide and ethanol respectively; this helps to preserve microtubules and may stabilize membrane structure.

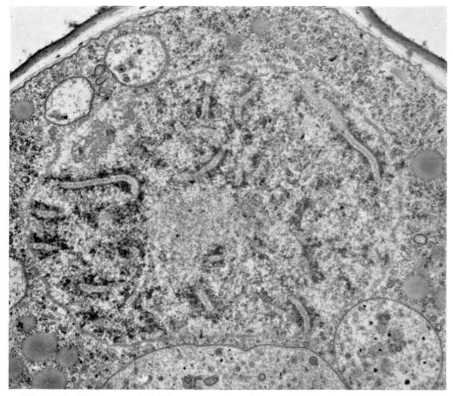

FIG. 15. *Physarum globuliferum*, section through spore showing synaptinemal complexes; glutaraldehyde and OsO_4; × 19,400 (micrograph by H. C. Aldrich, Dept. of Botany, University of Florida, Gainesville. See Aldrich, 1967).

The presence of the synaptinemal complex (fig. 15) in reproductive cells has been used to diagnoze the occurrence of meiosis in studies on slime moulds (Carroll and Dykstra, 1966; Aldrich, 1967) and fungi (Westergaard and Wettstein, 1966). These structures are known from other organisms and represent the paired chromosomes of pachytene. They are not found at other stages. Fixation should be carried out in glutaraldehyde followed by osmium tetroxide.

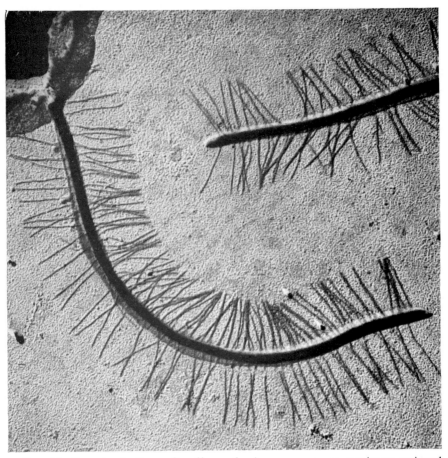

FIG. 16. *Saprolegnia ferax*, flagellum of primary zoospore; specimen stripped from glass after staining for light microscopy; shadowed direct preparation; × 10,000 (Manton *et al.*, 1952, reprinted by permission).

VIII. OTHER TECHNIQUES

A. Direct preparation

This involves the drying down of the fixed specimen or part of it on to the surface of a coated grid (preferably "ionized" Formvar, see p. 538, with carbon) and subsequent shadowing. Small whole spores may be dried down in the unfixed condition and examined directly as silhouettes in order to detect surface irregularities, spines, etc. Zoospores in suspension are best fixed by the exposure of a drop of the suspension, on a grid, to the vapour from 2% osmium tetroxide for about 30 sec. When the cells have settled, which may take a further 30 sec, the liquid drop is carefully drawn off the grid with a piece of filter paper (or with an absorbent dental point) leaving

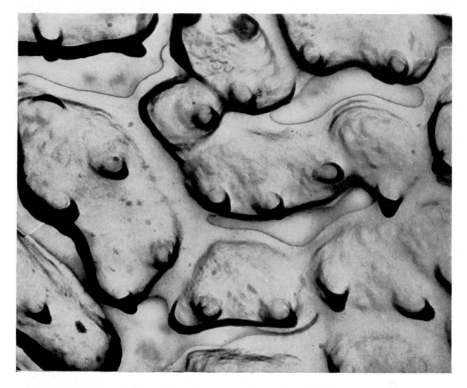

FIG. 17. *Phragmidium tuberculatum*, surface of aecidiospore; surface replica; × 24,000 (micrograph by D. M. Henderson, The Royal Botanic Garden, Edinburgh).

the zoospores behind. If gas fixation causes some of the cells to burst then liquid fixation should be tried, although this may also result in bursting or shrinkage and often causes over-fixation. One drop of 2% osmium tetroxide is added to 0·5 ml of zoospore suspension on a slide and rapidly mixed with it. A drop of the mixture is immediately transferred to a grid with a fine pipette and the cells allowed to settle. After removal of the liquid and drying, it may be found that salts crystallize out from the culture solution. These can

FIG. 18. *Saccharomyces cerevisiae*, budding vegetative cell; freeze-etched preparation; × 16,500 (Moor, 1964, reprinted by permission).

easily be dissolved by immersing the grid in distilled water without disturbing the organic materials which stick firmly to the grid surface. This must be done before the preparation is shadowed. Shadowing involves the evaporation of a metal (e.g., platinum, platinum/carbon, gold/palladium) under vacuum from a point above and to one side of the specimen. In this way metal atoms are not deposited on the side of the specimen away from the source (the "shadow"), thus increasing contrast in the final image (for details see Bradley, 1965b). If the shadowing angle is known, then information on the height of the specimen, etc., may be obtained from the length of the shadow.

These methods are frequently employed in the examination of zoospore flagella. For example, Manton *et al.* (1952) used them to examine zoospores of *Saprolegnia* and developed a means by which a particular specimen could be examined with the light microscope and then transferred (using a cellulose nitrate stripping film) to a coated grid for further examination with the electron microscope (fig. 16).

When preparing dried specimens in this way the exact amount of organic material which can be tolerated in suspension, without obscuring the field of view or splitting the supporting film after insertion into the microscope, has to be learned by experience. In general it is safer to have too little than too much.

B. Surface replicas

These are used when it is wished to examine the fine detail of the surface of an opaque object. A thin film (20–50 nm) of some suitable material of low molecular weight (usually carbon) is deposited under vacuum on the specimen surface. This film, which replicates the surface, is removed and mounted on a copper grid. Before examining, it is usually shadowed obliquely with metal to accentuate the surface detail. The process described above produces a single-stage replica. Where possible the single stage is preferred, but it often involves the destruction of the specimen. Alternatively, a more laborious two-stage technique may be employed. An impression of the surface is made in some suitable material and a replica of this surface impression separated off and examined. Replica techniques have recently been applied to basidiospores of various agarics (Bigelow and Rowley, 1968). Spores were treated with 50% acetone to expand and unfold any collapsed cells, and drops of this suspension placed on the surface of a freshly cleaved

---

Fig. 19. (a) *Aspergillus fumigatus*, surface of conidium; × 33,600 (micrograph reproduced by permission of W. M. Hess, Dept. of Botany, Brigham Young University, Provo, Utah). (b) *Penicillium camemberti*, surface of conidium; × 45,000 (Hess *et al.*, 1968, reprinted by permission). Both preparations freeze-etched.

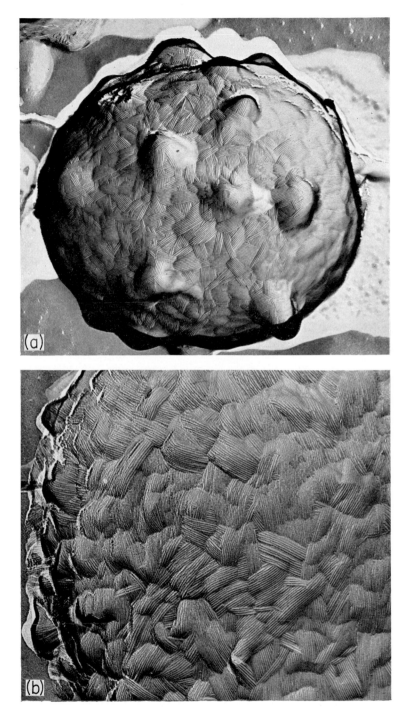

mica strip, air-dried and coated with carbon. The carbon film was scored into small squares and floated off onto a water surface (see Anderson, 1956). The basidiospores were dissolved and the film cleaned by placing the pieces in a saturated solution of chromic acid (Chromerge) for 10–90 min, then washing in hot water followed by glass distilled water. They could then be mounted and examined. Several variations of the method are briefly discussed by Bigelow and Rowley (1968). A similar technique has been successfully applied to spores of the rust fungi (D. M. Henderson, personal comunication; Fig. 18). For a detailed description of other replica techniques see Bradley (1965b).

C. Freeze etching

This complex technique, essentially a freeze-drying process, allows the EM study of frozen cells which have not been treated with fixatives or other substances which might produce artefacts. The original description of the method was by Steere (1957). Moor and Muhlethaler (1963) have developed the technique and were able to produce micrographs of frozen-etched yeast cells (*Saccharomyces cerevisiae*). These cells were not killed by the freeze-drying, implying that the final photographs represent a truthful picture of the structure within the living cell (fig. 18). The process basically involves five preparatory steps (Moor, 1964, 1965)—

(i) The specimen, soaked in 20% glycerine, is rapidly frozen to about $-100°C$ causing physical stabilization and inactivation.

(ii) The specimen, frozen into a block, is cut with a cooled knife, thus exposing a surface through it. The material actually splinters away from the surface exposing fracture planes, so that most of the exposed surface never comes in contact with the knife. The surface follows some membranes (giving surface views of nuclei, etc.) and cuts across others.

(iii) The exposed surface is etched by subjecting it to a brief period of freeze-drying, causing sublimation of the ice around the specimen and thus bringing the structure into greater relief.

(iv) The etched surface is shadowed with metal, then coated with carbon to stabilize the replica film.

(v) The frozen specimen is thawed and the replica released to float on water. It may be cleaned with acids or alkalis to remove specimen fragments. The clean replica is mounted, dried and examined.

Fig. 20. (a) *Scopulariopsis brevicaulis*, annellophores; × 6500. (b) *S. brevicaulis*, conidia; × 6500. (c) *Penicillium* sp., conidia, ×1350. (d) *Penicillium* sp., conidia ×2700. All preparations photographed on the Cambridge "Stereoscan" microscope (Greenhalgh, unpublished).

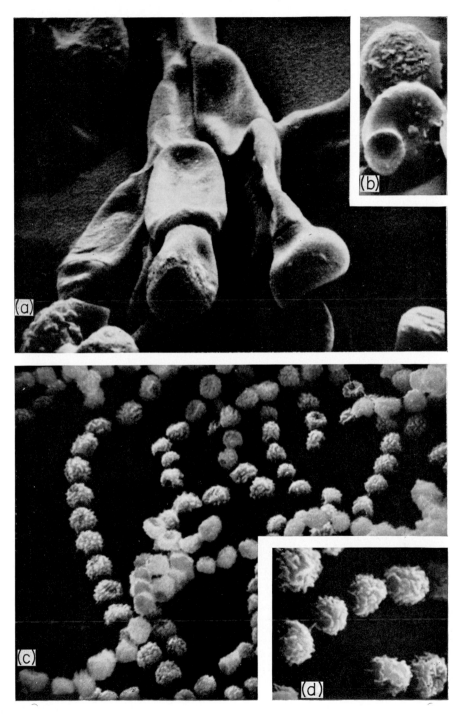

This technique has recently been used to examine both the internal structure of *Penicillium* conidia (Sassen *et al.* 1967) and their surface characteristics (Hess, *et al.*, 1968). Similar surface features were also demonstrated, using the same methods, for conidia of *Aspergillus* species (fig. 19).

IX. SCANNING ELECTRON MICROSCOPY

The scanning electron microscope (SEM) differs from the transmission electron microscope in providing surface views of whole structures. The specimens examined may be quite large (up to 1 cm across) and the images obtained are especially valuable in that the depth of focus available gives three-dimensional information (figs. 20, 21). Thus hyphae, spores, etc., can be viewed as solid objects, as if by surface lighting. Used in the scanning mode, ultrathin sections of perfectly embedded material would simply appear as plane surfaces. The SEM can be used in the transmissive mode with the aid of a special specimen stage, but the resolution obtainable (10–15 nm) is not critical enough for detailed examination of internal cell structure, or of replicas from freeze-etched specimens when used in the scanning mode. The technique of ion-beam etching, recently introduced, permits the investigation of subsurface structures by alternately eroding away the surface and re-examining. This can be done without removing the specimen from the microscope. However, the technique is still at an early stage of development and its use with biological material is severely limited by the difficulty of precisely controlling the amount of etching and of interpreting the images obtained. The greatest potential of the SEM in mycology seems at the moment to be in investigations of surface characters of hyphae and spores, etc., in developmental studies on the formation and detachment of exogenous spores and in the detection of hyphae and spores on the surface of various substrates (wood, leaves, soil particles, roots, etc.) in ecological studies.

The image is obtained by scanning the specimen with a beam of electrons which on striking the specimen cause the emission of secondary electrons. Many of these secondary electrons diffuse into the specimen but a proportion escape from the surface and it is these that are normally used to produce the image. They are attracted towards and pass through a grill which is positively charged. The final image is eventually produced on

---

FIG. 21. (a) *Hypoxylon serpens*, stroma surface; × 1400. (b) *H. rubiginosum*, ascospores; × 3700. (c) *H. rubiginosum*, perithecial ostiole exposed by splitting away of stroma; × 630. All preparations photographed on the Cambridge "Stereoscan" microscope ((a) and (c), Greenhalgh, unpublished; (b) Greenhalgh and Evans, 1968).

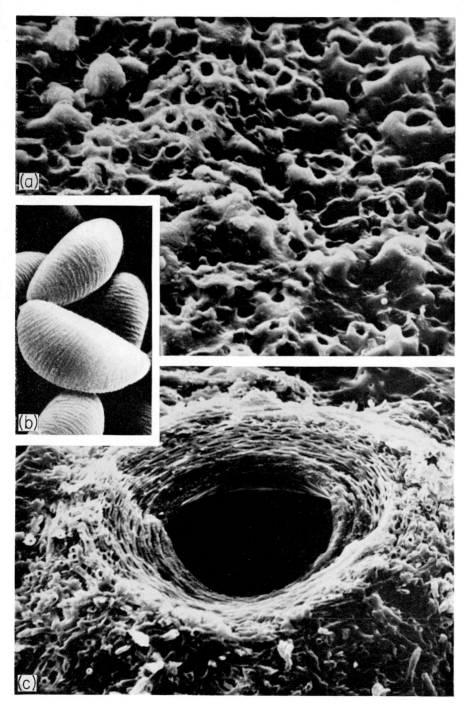

the screen of a cathode-ray tube and may be photographed directly with an external camera from a second screen. In addition to the secondary electrons emitted by the specimen some primary electrons are reflected from the surface. Images may be obtained from primary electrons but these have too high a contrast level for normal use.

Specimens are stuck to the plane surface ($\frac{1}{2}$ in. dia.) of metal specimen stubs with a suitable adhesive such as "Durofix" (which dries quickly and has negligible vapour pressure under the vacuums used). The specimen, if non-metallic, is then coated under vacuum with a thin film (about 50 nm thick) of metal, normally gold–palladium alloy. This is essential since the scanning beam will generate a charge on uncoated biological material, which in turn will cause distortion of the image. The metal film simply allows any charge to leak away or escape via the specimen stub. To eliminate specimen charging the specimen must be coated from all sides, giving a complete coating layer, and a special jig is needed to rotate the stub during coating. Specimens with a complex surface topography are difficult to coat satisfactorily, and it may be possible to improve metallic contact with the stub by careful application of silver paint. Experience so far suggests that specimens are best examined immediately after coating.

Thus, given a coating unit with suitable specimen jig, the preparation of specimens is a fairly rapid and simple process. Small pieces of, for example, wood or other substrate material bearing the specimen can be stuck directly on to the stub, coated and photographed within about 30 min, provided the material is such that a vacuum can be obtained quickly. Several stubs may be coated at once.

If detached conidia, discharged ascospores, etc., are to be examined, they may be collected conveniently on a small square of filter paper which is then stuck to the specimen stub and coated. The spores will be held in place between the cellulose fibres. Soil particles may be sprinkled directly on to the stub, any loose material being shaken off when the adhesive is dry. Alternatively the sticky surface of the stub may be applied directly to the soil surface and then removed together with the adherent particles (Gray, 1967). Hyphae within empty xylem vessels, etc., might be detected by splitting open the wood and examining the split surface. Williams and Davies (1967) in studies on the intact sporing structures of actinomycetes, have developed a method for transferring material from Petri-dish cultures to the specimen stubs with a minimum of disturbance. Sterile circular glass coverslips ($\frac{1}{2}$ in. dia.) were inserted halfway into solidified agar medium at an angle of 45° and the organism inoculated along the line of insertion. After incubation for a suitable period, the coverslips were carefully removed with their adherent mycelium, stuck to the specimen stubs, coated and examined. This technique is likely to be only suitable for fungal cultures which produce a very low aerial mycelium. It is also possible to remove thin vertical slices of agar and fungus from the culture, placing these laterally on the stub and cutting away most of the agar. Whether these should be allowed to dry out slowly before coating or placed directly in the coating chamber will depend probably on the material and could be checked beforehand.

The fact that specimens have to be coated and examined under vacuum means that some distortion of delicate structures is almost inevitable and this should be borne in mind when interpreting results. Hyphae may collapse and become wrinkled, whilst soft surface material may shrink around more dense internal substances or structures. It should be possible to prevent this collapse by using a freeze-drying process prior to coating. Dehydration or fixative treatment before coating has been found to reduce collapse of actino-mycete hyphae and spores (S. T. Williams, personal communication) and of fungal structures. This problem is obviously not so acute with specimens such as thick-walled spores, but it is in general a useful plan to carry out a parallel series of observations with the light microscope, if possible, as a check.

In all probability data obtained with the SEM on problems such as those of spore development, will be of greatest value when considered in relation to parallel studies on the same structures using thin-section techniques.

X. SUPPLIERS

Suppliers of most of the chemicals and special apparatus mentioned in the text are listed below. Most chemicals may be obtained from Taab Laboratories (52 Kidmore End Road, Emmer Green, Reading, Berks) and apparatus and accessories from LKB Instruments Ltd. (LKB House, 232 Addington Road, S. Croydon, Surrey), but some other suppliers also provide a range of reagents and accessories.

Araldite	Ciba (ARL) Ltd., Duxford, Cambridge, England. Taab Laboratories.
Carbon rods, etc.	Morganite Crucible Co., Battersea Works, London, S.W.11. Johnson Matthey Ltd., 78–83 Hatton Garden, London, E.C.1.
Dental wax	LKB Instruments Ltd.
Embedding (BEEM) capsules	LKB Instruments Ltd.
Epikote 812 (Epon 812)	Taab Laboratories. G. T. Gurr Ltd., Carlisle Road, The Hyde, London, N.W.9.
Forceps	Southern Watch and Clock Suppliers Ltd., Precista House, 48–56 High Street, Orpington, Kent.

Formvar	Shawinigan Ltd., Marlow House, Lloyds Ave., London, E.C.3. Taab Laboratories.
Freeze-etching apparatus	Balzers High Vacuum Ltd., Northbridge Road, Berkhamsted, Herts.
Glutaraldehyde	Taab Laboratories
Grids	Smethurst Highlight, Sidcot, Heaton, Bolton, Lancs. Polaron Instruments Ltd., Delviljem House, 4 Shakespeare Road, Finchley, London, N.3. LKB Instruments Ltd. Taab Laboratories.
Grid boxes	LKB Instruments Ltd. Polaron Instruments Ltd.
LKB Knife-maker	LKB Instruments Ltd.
Maraglas	LKB Instruments Ltd.
Osmium tetroxide	Taab Laboratories G. T. Gurr Ltd.
Plate glass strips	LKB Instruments Ltd.
Platinum/carbon rods	Le Carbone, A.D. Battery Works, South Street, Portslade, Sussex. (Societé le Carbone—Lorraine, Paris, France.)
Phosphotungstic acid	The British Drug Houses Ltd., B.D.H. Laboratory Chemicals Division, Poole, England. Taab Laboratories.
Shadowing materials (platinum, gold/palladium wire, etc.)	Johnson-Matthey Ltd.
Ultramicrotomes	LKB Instruments Ltd. (LKB Ultratome). Shandon Scientific Co. Ltd., 65 Pound Lane, Willesden, London, N.W.10 (Reichert Microtome). V. A. Howe Co. Ltd., 88 Peterborough Road, London, S.W.6 (Porter Blum Microtome). Cambridge Instrument Co. Ltd., 13 Grosvenor Place, London, S.W.1 (Huxley Microtome).
Velin-tissue	General Paper and Box Manufacturing Co. Ltd., Treforest Trading Estate, Pontypridd, Glam.

ACKNOWLEDGMENTS

The authors are grateful to Professor I. Manton, F.R.S., for reading the manuscript and for many valuable suggestions, and to Mr. D. Stratton and Mr. K. Oates for helpful advice. They also wish to express their gratitude to those mycologists who have generously contributed micrographs and the publishers who have given permission to reprint many of these.

REFERENCES

Aldrich, H. C. (1967). *Mycologia*, **59**, 127–148.
Anderson, T. F. (1956). *In* "Physical Techniques in Biological Research" (Eds. G. Oster and A. W. Pollister), Vol. 3, pp. 177–240. Academic Press, New York.
Bennet, H. S., and Luft, J. H. (1959). *J. biophys. biochem. Cytol.*, **6**, 113–114.
Bigelow, H. E., and Rowley, J. R. (1968). *Mycologia*, **60**, 869–887.
Bracker, C. E. (1967). *A. Rev. Phytopath.*, **5**, 343–374.
Bracker, C. E., and Butler, E. E. (1963). *Mycologia*, **55**, 35–58.
Bradley, D. E. (1965a). *In* "Techniques for Electron Microscopy" (Ed. D. Kay), pp. 58–74. Blackwell, Oxford.
Bradley, D. E. (1965b). *In* "Techniques for Electron Microscopy" (Ed. D. Kay), pp. 96–152. Blackwell, Oxford.
Brenner, D. M., and Carroll, G. G. (1968). *J. Bact.*, **95**, 658–671.
Carroll, G. C. (1967). *J. Cell Biol.*, **33**, 218–224.
Carroll, G. C., and Dykstra, R. (1966). *Mycologia*. **58**, 166–169.
Ekundayo, J. A. (1966) *J. gen. Microbiol.*, **42**, 283–291.
Erlandson, R. A. (1964). *J. Cell. Biol.*, **22**, 704–709.
Feder, N., and O'Brien, T. P. (1968). *Am. J. Bot.*, **55**, 123–142.
Freeman, J. A., and Spurlock, B. O. (1962). *J. Cell Biol.*, **13**, 437–443.
Fuller, M. A. (1966). *In* "The Fungus Spore" (Ed. M. F. Madelin), pp. 67–84. Butterworth, London.
Fuller, M. S., and Calhoun, S. A. (1968). *Z. Zellforsch. mikrosk. Anat.*, **87**, 526–533.
Fuller, M. S., and Reichle, R. (1965). *Mycologia*, **57**, 946–961.
Fuller, M. S., and Reichle, R. (1968). *Can. J. Bot.*, **46**, 279–283.
Gimenez-Martin, G., Risueño, M. C., and López-Sáez, J. F. (1967). *Experimenta*, **23**, 316–319.
Glauert, A. M. (1965a). *In* "Techniques for Electron Microscopy" (Ed. D. Kay), pp. 166–212. Blackwell, Oxford.
Glauert, A. M. (1965b). *In* "Techniques for Electron Microscopy" (Ed. D. Kay), pp. 254–310. Blackwell, Oxford.
Glauert, A. M., and Phillips, R. (1965). *In* "Techniques for Electron Microscopy" (Ed. D. Kay), pp. 213–253. Blackwell, Oxford.
Gray, T. R. G. (1967). *Science, N.Y.*, **155**, No. 3770, 1668–1670.
Greenhalgh, G. N., and Evans, L. V. (1968). *Jl R. microsc. Soc.*, **88**, 545–556.
Grimstone, A. V. (1968). "The Electron Microscope in Biology." Arnold, London.
Grove, S. N., Morré, D. J., and Bracker, C. E. (1967). *Proc. Indiana Acad. Sci.*, **76**, 210–214.
Hess, W. M., Sassen, M. M. A., and Remsen, C. C. (1968). *Mycologia*, **60**, 290–303.
Ho, H. H., Zachariah, K., and Hickman, C. J. (1968). *Can. J. Bot.*, **46**, 37–41.
Horne, R. W. (1965). *In* "Techniques for Electron Microscopy" (Ed. D. Kay), pp. 328–355. Blackwell, Oxford.
Ichida, A. A., and Fuller, M. S. (1968). *Mycologia.*, **60**, 141–155.

Johansen, D. A. (1940). "Plant Microtechnique." McGraw-Hill, New York.
Kay, D. (Ed.) (1965). "Techniques for Electron Microscopy", 2nd Ed. Blackwell, Oxford.
Kurtz, S. M. (1961). *J. Ultrastruct. Res.*, 5, 468–469.
Lessie, P. E., and Lovett, J. S. (1968). *Am. J. Bot.*, 55, 220–236.
Lowry, R. J., and Sussman, A. S. (1968). *J. gen. Microbiol.*, 51, 403–409.
Luft, J. H. (1956). *J. biophys. biochem. Cytol.*, 2, 799–801.
Luft, J. H. (1961). *J. biophys. biochem. Cytol.*, 9, 409–414.
Manton, I., Clarke, B. and Greenwood, A. D. (1952). *J. exp. Bot.*, 3, 206–208.
Marchant, R. (1968). *New Phytol.*, 67, 167–171.
Maser, M. D., Powell, T. E., and Philpotts, E. W. (1967). *Stain Technol.*, 42, 175–182.
Mercer, E. H., and Birbeck, M. S. C. (1966). "Electron Microscopy", 2nd Ed. Blackwell, Oxford.
Millonig, G. (1961). *J. biophys. biochem. Cytol.*, 11, 736–739.
Moor, H. (1964). *Z. Zellforsch. mikrosk. Anat.*, 62, 546–580.
Moor, H. (1965). "Balzers, High Vacuum Report", No. 2. Balzers, Liechtenstein.
Moor, H., and Muhlethaler, K. (1963). *J. Cell Biol.*, 17, 609–628.
Palade, G. E. (1952). *J. exp. Med.*, 95, 285–298.
Pease, D. C. (1964). "Histological Techniques for Electron Microscopy", 2nd Ed. Academic Press, New York.
Peat, A., and Banbury, G. H. (1967). *New Phytol.*, 66, 475–484.
Reeves, F. (1967). *Mycologia*, 59, 1018–1033.
Reynolds, E. S. (1963). *J. Cell. Biol.*, 17, 208–212.
Ryter, A., and Kellenberger, E. (1958). *J. Ultrastruct. Res.*, 2, 200–214.
Sabatini, D. D., Bensch, K., and Barrnett, R. J. (1963). *J. Cell. Biol.*, 17, 19–58.
Sassen, M. M. A., Remsen, C. C., and Hess, W. M. (1967). *Protoplasma*, 64, 75–88.
Sjöstrand, F. S. (1967). "Electron Microscopy of Cells and Tissues", Vol. 1. Academic Press, New York.
Spurlock, B., Kattine, V., and Freeman, J. (1963). *J. Cell. Biol.*, 17, 203–204.
Steere, R. L. (1957). *J. biophys. biochem. Cytol.*, 3, 45–59.
Westergaard, M., and Wettstein, D. (1966). *C. r. Trav. Lab. Carlsberg*, 35, 261–286.
Williams, S. T., and Davies, F. L. (1967). *J. gen. Microbiol.*, 48, 171–177.
Wischnitzer, S. (1967). *Int. Rev. Cytol.*, 22, 1–61.

SUPPLEMENTARY REFERENCES ADDED IN PROOF

The following list represents what is hoped will be a useful selection from the body of literature published since this account was submitted.

Bracker, C. E. (1968). *Mycologia*, 60, 1016–1067. (A detailed account of the fine structure of sporangiospore development in *Gilbertella persicaria*.)
Buckley, P. M., Wyllie, T. D., and DeVay, J. E. (1969). *Mycologia*, 61, 240–250. (The fine structure of the conidia and their development in two species of *Verticillium*.)
Cantino, E. C., and Truesdell, L. C. (1970). *Mycologia*, 62, 548–567. (Aspects of the fine structure of zoospores of *Blastocladiella emersonii*.)
Desjardins, P. R., Zeutmeyer, G. A., and Reynolds, D. A. (1969). *Can. J. Bot.*, 47, 1077–1079. (Direct preparations of zoospores of *Phytophthora palmivora* to show flagellar hairs.)
Ehrlich, M. A., and Ehrlich, H. G. (1969). *Can. J. Bot.*, 47, 2061–2064. (The development of uredospores in *Puccinia graminis*.)

Grove, S. N., Bracker, C. E., and Morré, D. J. (1970). *Am. J. Bot.*, **57**, 245–266. (An analysis of the fine structure of hyphal tip growth in vegetative hyphae of *Pythium ultimum*.)

Hawker, L. E., and Goodey, M. A. (1968). *J. Gen. Microbiol.*, **54**, 13–20. (A study, utilizing both the scanning and the transmission microscope, of zygospore wall development in *Rhizopus sexualis*.)

Hawker, L. E., Thomas, B. and Beckett, A. (1970). *J. gen. Microbiol.*, **60**, 181–190. (Mature and germinating spores of *Cunninghamella elegans*, investigated by thin-section, freeze-etching and scanning techniques.)

Juniper, B. E., Cox, G. C., Gilchrist, A. J., and Williams, P. K. (1969). "Techniques for Plant Electron Microscopy." Blackwell, Oxford. (A very useful handbook.)

Meek, G. A. (1970). "Practical Electron Microscopy for Biologists." Wiley Interscience. New York and London.

Motta, J. J. (1969). *Mycologia*, **61**, 873–886. (A study of somatic nuclear division in *Armillaria mellea*.)

Robards, A. W. (1970). "Electron Microscopy and Plant Ultrastructure." McGraw-Hill, London. (A book designed for students, with an introductory section dealing with the electron microscope and a general account of preparation techniques).

Spurr, A. R. (1969). *J. Ultrastruct. Res.*, **26**, 31–43. (An account of a low-viscosity epoxy resin embedding medium. Infiltration occurs more readily than with other resins because of the low viscosity. It has been used for hard and soft tissue with equal success and has given very good results with fungal spores that have proved difficult to embed using other resins. Sections are tough under the electron beam and can be used on 200-mesh grids without a supporting film.)

Stocks, D. L., and Hess, W. M. (1970). *Mycologia*, **62**, 176–191. (Dormant and germinating basidiospores of a *Psilocybe* species, examined by thin-section and freeze-etching techniques.)

Sussman, A. S., Lowry, R. J., Durkee, T. L., and Maheshwari, R. (1969). *Can. J. Bot.*, **47**, 2073–2078. (The fine structure of cold dormant and germinating uredospores of *Puccinia graminis* var *tritici*.)

Wells, K. (1970). *Mycologia*, **62**, 761–790. (A detailed account of nuclear division in the ascus of *Ascobolus stercorarius*.)

Young, T. W. K. (1969). *J. gen. Microbiol.*, **55**, 243–249. (A study of the spores of two species of *Mycotypha*, involving a single stage carbon replica technique.)

Young, T. W. K. (1970). *Trans. Br. mycol. Soc.*, **54**, 15–25. (The fine structure of the spore wall of *Linderina*, examined by thin section and replica techniques and by direct preparations of wall material after physical and chemical disintegration.)

CHAPTER XXI

Chemical Tests in Agaricology

Roy Watling

Royal Botanic Garden, Edinburgh, Scotland

I. HISTORICAL: INTRODUCTION AND SCOPE OF STUDY

Chemical tests have been frequently used in the taxonomy of the basidio-mycetous fungi; indeed with careful use and rigorous interpretation they have been utilized to suggest and substantiate possible lines of phylogeny. The use of chemical tests has not only met with opposition, it still even raises eyebrows. Such opposition is unfortunate, for the identification of the higher fungi is difficult enough by conventional means and every additional character is welcome. In no way do the tests replace the age old characters of veil, gill-attachment etc., but complement them; after all, it was not long ago that the use of the microscope for the identification of the higher fungi was scorned.

On the publication of Singer's "Agaricales in Modern Taxonomy" (1951) the theory and practice basic to the study of the agarics, indeed the larger fungi as a whole, were brought together in one volume and in this simple way revolutionized agaricology. The emphasis placed by Singer on chemical tests is reflected in the fact that in the very first couplet of his generic key the result of a chemical reaction is used as a deciding factor; no less than four major divisions are decided in this way. Keys to genera of higher fungi by Smith and Shaffer (1964) similarly use, but to a lesser extent, chemical reactions. However, it must be conceded that the importance of results from chemical reactions are somewhat reduced for we know little of how the constituents vary as the fruit-body develops and senescence approaches, or how they are affected by environmental conditions. In the latter case there are several qualitative field observations on differences in the auto-oxidation reactions typical of certain boletes, but only for the commercial mushroom do we have any idea as to the variation in content of a single chemical, i.e. tryptophane, with production of fruit-body (Hughes, 1958); indole compounds such as tryptophane are widespread in the fungi and may be important taxonomically (Worthen *et al.*, 1962).

Just as with other characters of the fruit-body, such as basidiospores, cystidial shape and tramal colour and organization, the chemical test varies in importance from group to group and depends on both the chemical reagent and the chemical reaction. Thus agaricologists on the one hand do not find it imprudent to accept species with their spores blueing in iodine solutions, i.e. amyloid, and those with spores not blueing, within a single generic circumscription, as is customary in *Mycena* and *Dermoloma*, and more recently in *Squamanita* (Bas, 1965), although on the other hand a vast amount of supporting evidence would be required to include in the Russulaceae an agaric which lacked amyloid ornamentation on the spores.

Some embarrassment has been experienced in preparing this paper, particularly when it accompanies articles describing sophisticated techniques so alien to the methods used in the testing of agarics with chemical reagents; indeed there is little in this account to compare with the admirable work by Asahina (1936 onwards) and Shibata (1964) in the sister science of lichenology. However, there is an advantage for investigators can be reminded of the potentiality of the agarics as tools for research.

Agaricologists must always remember that it was due to the foresight of lichenologists that some of their most valuable chemical tests were devised. Lindt treated various lichens with sulphovanilin as early as 1885 although it was much later that Arnould and Goris (1907) utilized the same reaction in agaricology. Now this same sulpho-aldehyde test has been extended to include several reagents (Boidin, 1951). The sulpho-aldehyde test was introduced for locating albuminous substances (Reich and Mikosch, 1890) but generally one rarely knows what substrates are concerned in many of the colour changes produced in the tests nor the pathways through which the reactions pass, let alone the relationships between the chemical reaction and their correlation with morphological characters (see Josserand, 1930). Little advance has been made even in the last decade.

Birkinshaw (1965) has discussed some of the chemicals isolated from a wide range of fungi including some constituents of agarics but the information is not easily applicable to understanding chemical tests. It appears that many of the substrates could be shunt metabolites, off the main biochemical pathways, or constituents of pathways as yet undiscovered. It is fairly certain that with some reagents we test for the presence of laccase, tyrosinase and phenol oxidase systems (Bourquelot and Bertrand, 1896); other tests may simply identify chemicals used for food storage, respiration and defence, and photo-reception (?).

This absence of knowledge concerning the chemical reactions does not, however, reflect the long and extensive work which has been carried out on several series of chemical compounds to be found in agaric fruit-bodies. Indeed certain poisonous principles and pigments, particularly those of bright colour and those which change on contact with air, have stimulated great interest and resulted in a vast array of publications.

Although the reactivity of certain chemicals with fungus tissue had been observed as early as 1872, when alkali was shown by Müller to produce an intense purple colour with *Polyporus (Hapalopilus) nidulans*, little was done to determine the nature of the chemicals involved until Kögl and his co-workers commenced work in the 1920s (1926, 1930–1944). This lack of experimental work was probably due to difficulty in obtaining sufficient quantity of the starting material and its perishable nature. Pastac, however, (1942) has drawn the major amount of this early information together and Heim (1942) discussed the possible taxonomic implications.

Amanita-haemolysin and Amanita toxin (= phalline pro parte) were the first of a whole series of poisons to be isolated from agaric fruit-bodies (Kobert, 1891; Ford, 1906; etc.) a search which took on wider dimensions with the work of the Wielands and their collaborators (1941); a full review and history of this subject has been plotted by Tyler (1963).

Chromatographic techniques have been applied to the pigments of *Cortinarius* species by Gabriel (1962) and a preliminary account of the pigments of *Russula* species, with possible taxonomic implications, has been offered by Watson (1966). An investigation of the bright pigments of Hygrophoraceae has now been carried out (A. H. Smith, personal comm.) and tends to draw the species closer together, whereas some separation of *Amanita muscaria* might be necessary after examining results from pigment analysis (Talbot and Vining, 1963); this confirms field observations. Similarly Heinemann and Casimir (1961) and Catalfomo and Tyler (1961) have analysed amino-acids of *Agaricus* spp. and *Amanita* spp. respectively and this has resulted in extremely interesting conclusions. Fries (1958) has carried out whole fruit-body extract analysis whilst Heim and Romagnesi (1934) have plotted the occurrence of allontoic acid using data of Fosse and Brunel, Frerejacque

(1939) the occurrence of mannitol, and Brown, Malone, Stuntz and Tyler (1962) the occurrence of muscarine in several agarics.

Recently more rigorous techniques have been introduced to the study so that intergradations can now be detected between so-called species of *Amanita* by analysing isazole derivatives (Benedict, Tyler and Brady, 1966) and *Amanita* section *Phalloideae* could possibly be more naturally grouped by reference to the indole derivatives (Tyler, 1961; cf. Worthen *et al.*, 1962). Brunel in Romagnesi (1948) has described the distribution of urea in many agarics and how this might reflect taxonomic relationships; Tyler, Benedict and Stuntz (1965) have extended these observations. Another genus which has caught the imagination of Tyler and his school is *Inocybe* and their results have been expressed in a chemotaxonomic key which has been briefly tested (Robbers, Brady and Tyler, 1964); this is the only time as far as the author is aware that such an ambitious treatment has been attempted. Robbers *et al.* (1964) suggest that from their observations primary metabolites have a limited value as taxonomic markers whereas secondary metabolites may turn out to be of very great use. Work by Worthen, Snell and Dick (1965) supports this view, for they have shown that alkaloids may be of use in the study of the Boletaceae. Hegnauer (1962) has attempted a compilation of the major chemical constituents of the higher fungi.

Although many of these studies are proving to be useful taxonomically the majority are as yet unrelated to the colour changes and reactions so universally relied upon by the field agaricologist. Thus only recently have the chemical compounds isolated been correlated with classification and possible characters observable in the fresh material, e.g. for Boletaceae, Paxillaceae and *Polyporus* (*Inonotus*) by Edwards and co-workers (1961; 1964; 1967) and for *Psilocybe* by Singer and Smith (1958) and Benedict, Tyler and Watling (1967). The distribution of psilocybin will be mentioned again (II, A), but it is important here to note that it or precursors may be tested for in the field with the use of metol.

Many of the chemical processes used by the chemist, particularly mass extraction, indeed the reactions noticed during isolation, can be directly related to field notes but as yet not all the information has been extracted from the vast range of chemical literature (e.g. Edwards and Elsworthy, 1967). The value of the results of the earliest chemists, and of many of those working today, is marred by doubts as to the identity of the fungus used. It cannot be over-emphasized that future work should be backed by an authorative identification and a herbarium collection. This is being done by some chemists but alas too few.

Loesecke (1871) detected the release of HCN from fruit-bodies and Josserand (1943) and Locquin (1944) (cf. Langeron, 1952) have since assessed the importance of this fact in taxonomy, but unlike many of the compounds mentioned above the release of HCN produces a very distinct field character —the cyanic odour, which when present is extremely useful. Later it will be discussed how other smells might be related to chemical tests. Brian (1951) has tabulated the antibiotics found in various fungi including agarics, and Valadon and Mummery (1968) the taxonomic significance of carotenoids; these may be future sources of information.

As long ago as 1919 Wager described the fluorescence of chemicals found in the agaric fruit-body; further work has been carried out by May (1961). Examination under long wave, ultraviolet light is a means of spotting groups of chemical compounds with similar configuration. It is a much more useful technique than ultraviolet absorption spectrum analysis for here isolates are required in a high degree of purity. Fluorescence may indicate a possible approach to the locating of the chemical substrates which are being analysed by histochemical techniques. Thus Disbrey and Watling (1967) have described and extended techniques developed by the late Dr. Max Barrett by which the chemical substrates are not isolated but located in situ by sophisticated reactions already available to the human pathologist.

Microchemical tests as sophisticated as those used by the lichenologist (i.e. Wachmeister's techniques, see Hale, 1967) have not been developed as yet in agaricology although as will be described later the ability of the flesh to produce crystals (II, A) with an ammoniacal solution has been used in the study of the Bolbitiaceae. Microscopic dyes in part express chemical constitution; the metachroism exhibited when using cresyl blue (Kühner, 1934), cotton blue and iodine-complexes (see II, B) give intriguing results which can be correlated with other chemical tests and morphology.

For too long now agarics have been thought of by the majority solely at species level, never at family level, and this perhaps has hindered their study in the University syllabus where the higher fungi are often skipped over with a possible fungus foray and later in the year a look at a couple of pickled fruit-bodies. However, the use and understanding of chemical tests will give new stimulation to their study.

The fungi covered by this account are the agarics and boleti as outlined by the New Check List of British Agarics and Boleti (1960—Dennis, Orton and Hora) although the author is aware that some of the fungi are not closely related to the true agarics and would be more naturally included in other major fungal groups, e.g. *Cantharellus* and *Russula*. However, everyone readily recognizes the gilled and poroid agarics as a group and this seems the most useful treatment for the present paper. Certain true relatives of the agarics which are cyphellaceous and secotiaceous have been excluded but there is no doubt the chemical tests to be discussed below are equally applicable. Because of the diverse nature of the fungi under consideration the

TABLE I

Chemical reagents and reaction results

In order to eliminate duplication and the difficulty of relating the fungi to the chemical reagents, the information is arranged in roughly complementary chemical groups.

Chemical group	Restrictions on material	Reaction result if positive	Family on which reagent is commonly used
1. *Simple acids/alkalis*			
(a) Sulphuric acid; H_2SO_4	Fresh material (and basidiospores)	Purplish, red to orange discoloration	*Amamitaceae, Agaricaceae, Cortinariaceae, Gomphidiaceae, Coprinaceae, Lepiotaceae*
(b) Hydrochloric acid; HCl	Fresh material (and microscopic structures)	(a) shades of red (b) green	*Boletaceae Coprinaceae Cortinariaceae*
(c) Nitric acid; HNO_3	Fresh material	Yellow to rust	*Boletaceae (Agaricaceae* —coupled with Aniline, see below)
(d) Caustic Potash/Caustic Soda KOH/NaOH	Fresh material (and basidiospores)	Variable—see text	Universal
(e) Ammonia and Ammoniacal solutions $NH_3 : NH_4^+$	Fresh and herbarium material	Variable—see text	Universal
2. *Simple heavy metal compounds*			
(a) Iron salts: Ferrous sulphate; $FeSO_4$. Ferrous ammonium sulphate; $Fe\ SO_4(NH_4)_2SO_4$ Ferric Chloride $FeCl_3$ Ferric alum; $(NH_4)_2 SO_4 . Fe_2 (SO_4)_3$	Fresh material (and carefully dried material)	(a) green (b) reddish (c) purple to lilac or violet (d) blue to grey (to blackish) (e) variable	*Russulaceae Tricholomataceae Boletaceae*

Chemical group	Restrictions on material	Reaction result if positive	Family on which reagent is commonly used
(b) Mercury complexes: Mercuric chloride HgCl₂ with HNO₃ Millon's reagent	Fresh material Fresh material	(a) grey (b) rose (c) blue-green (d) rose-red (b) orange to yellow	*Cortinariaceae* *Russulaceae*
(c) Silver salts: Silver nitrate; AgNO₃	Fresh material	Black	*Cortinariaceae*
(d) Copper salts: Copper sulphate; CuSO₄	Fresh material	Green/blue	*Boletaceae*
3. *Rare earth and uranium complexes*			
(a) Thallium Tl-4 see Appendix	Fresh material	(a) orange-yellow (b) rose-lilac rose (c) green (d) blue-violet	*Cortinariaceae* *Tricholomataceae*
(b) Uranium: K₂U₂O₇. 6H₂O	Fresh material	orange to rust	*Boletaceae* *Cortinariaceae*
(c) Cercium: (NH₄)₂ Ce(N₃)₆ Ceric ammonium nitrate	Fresh material	yellow to dark orange	*Russulaceae*
4. *Simple complexes containing halogens*			
(a) Iodine solution i.e. Iodine in potassium iodide	Fresh material	(a) yellow (b) olivaceous (c) green (d) brown; see text	*Cortinariaceae*
(b) Melzer's reagent	Fresh and herbarium material (and microscopic structures): see text	(a) blue, black, amethyst or grey (b) purple brown (c) yellow or golden: see text	*Boletaceae* (Universal as micro-chemical reagent)
(c) Calcium hypochlorite; Ca(OCl)₂	Fresh material	(a) yellow (b) decolourization (c) black (d) blue	*Boletaceae* *Cortinariaceae* *Tricholomataceae*

TABLE 1 (continued)

Chemical group	Restrictions on material	Reaction result if positive	Family on which reagent is commonly used
(d) Eau de Javel; KOCl	Fresh material	(a) yellow→orange	Cortinariaceae Tricholomataceae
(e) Perchloric acid; HClO₄	Fresh material	(b) white	Boletaceae
(f) Bromine water; Br	Fresh material	(a) blue, rose to violet	Cortinariaceae
		(b) yellow-olive	Amanitaceae
5. Organo-metallic compounds			
(a) Lead acetate (CH₃.COO)₂ Pb. 3H₂O	Fresh material	(a) dark green	Boletaceae
		(b) violaceous	
(b) Methyl chloro-antimonate (CH₃)₂Cl.Sb	Fresh material	Grey	Russulaceae
(c) Ethyl chlorostannate (C₂H₅)₃ Cl.Sn	Fresh material	Yellow brown	Amanitaceae
6. Aliphatic and aromatic aldehyde complexes			
(a) Sulphoformol KHCHO: H₂SO₄	Fresh material (microscopic structures in herbarium material)	Brown	Tricholomataceae Russulaceae
(b) Sulphobenzaldehyde C₆H₅CHO: H₂SO₄	Fresh material (microscopic structures in herbarium material)	Black	Russulaceae
(c) Sulphovanilin C₆H₃(OCH₃)(OH).CHO:H₂SO₄	Fresh material	Bright red (negative if rose, carmine or brown)	Russulaceae Boletaceae, Paxillaceae
(d) Chlorovanilin C₆H₃(OCH₃)(OH).CHO :HCl	Microscopic structures in herbarium material	(a) blue	Russulaceae
	Fresh material (microscopic structures in herbarium material)	(b) ± rose Bright red	Russulaceae
(e) Sulphonaphthaldehyde C₁₀H₇.CHO:H₂SO₄	Fresh material	(a) blue (b) rose See Boidin, 1951	Russulaceae

Chemical group	Restrictions on material	Reaction result if positive	Family on which reagent is commonly used
7. *Aliphatic compounds*			
(a) Lactic acid; $CH_3CH(OH)COOH$	Fresh material	Purple red	*Boletaceae*
(b) Formalin; $HaCHO$	Fresh material	(a) bluish	*Russulaceae*
		(b) green to grey	*Boletaceae*
		(c) red to orange-red	*Paxillaceae*
			Gomphidiaceae, etc.
(c) Ethanol; C_2H_5OH	Fresh or carefully dried material	Red-carmine	*Gomphidiaceae*
(d) Acetic acid; CH_2COOH	Fresh material	Green	*Boletaceae*
8. *Phenols and aromatic acids*			
(a) Phenol; C_6H_5OH	Fresh material	Chocolate (within twenty minutes of application)	*Russulaceae*
(b) Resorcinol; $C_6H_4(OH)_2$ 1,3-dihydroxybenzene	Fresh material	Pink	*Boletaceae, Amanitaceae Russulaceae*
(c) Catechol; $C_6H_4(OH)_2$ 1,2-dihydroxybenzene	Fresh material	Pink to rust	*Boletaceae*
(d) Quinol; $C_6H_4(OH)_2$ 1,4-dihydroxybenzene	Fresh material	Greyish pink	*Russulaceae*
(e) Guaiacol; $C_6H_4(OCH_3)OH$ 1,hydroxy-2,methoxybenzene	Fresh material	Salmon to orange, rose or flushed vinaceous	*Russulaceae Amanitaceae Tricholomataceae*
(f) Pyrogallol; $C_6H_3(OH)_3$ 3,4,5-trihydroxybenzene	Fresh material	Yellow to brownish-yellow	*Russulaceae*
(g) Phloroglucinol; $C_6H_3(OH)_3$ 1,3,5-trihydroxybenzene	Fresh material	Brown	*Russulaceae*
(h) Orcinol; $C_6H_3(CH_3)(OH)_2$ 3,5-dihydroxytoluene	Fresh material	Orange-red to yellow-brown, to lilaceous	*Russulaceae*
(i) Thymol; $CH_3C_6H_3(OH)CH(CH_3)_2$ 2,hydroxy-4,methylisopropyl-benzene	Fresh material	Light brown	*Russulaceae Tricholomataceae*

TABLE I (continued)

Chemical group	Restrictions on material	Reaction result if positive	Family on which reagent is commonly used
(j) Gallic acid; $C_6H_2(OH)_3COOH$ 3,4,5-trihydroxybenzoic acid	Fresh material	Yellow to medium brown	*Russulaceae* *Tricholomataceae*
(k) Indantrione hydrate; $C_6H_4COCOCO.H_2O$	Fresh material	Rose-orange	*Russulaceae*
(l) α-naphthol; $C_{10}H_7OH$	Fresh material	Deep indigo or purple	*Russulaceae*
9. Nitrogen-containing aromatics			
(a) Aniline; $C_6H_5NH_2$	Fresh flesh	Red to copper red	*Russulaceae* *Boletaceae*
	Fresh gills	Discolouring with grey or faintly coloured halo	*Russulaceae*
(complexed with mineral acid)			*Agaricaceae* (coupled with HNO_3 in Schaeffer's reaction, see below)
(b) Phenyl hydrazine $C_6H_5NHNH_2$	Fresh material	Yellow to orange with or without rosy brim	*Russulaceae, Boletaceae* *Tricholomataceae*
(c) o-aminophenol $C_6H_4(OH).NH_2$	Fresh material	Deep reddish brown	*Russulaceae*
(d) p-phenylene diamine $C_6H_4(NH_2)_2$	Fresh material	(a) blue (b) green to yellow green (c) lilac to black	*Russulaceae*
(e) p-methyl aminophenol sulphate $(CH_3NH.C_6H_4OH)_2H_2SO_4$	Fresh material	(a) lilaceous to violet (b) yellow	*Russulaceae* *Tricholomataceae*
(f) Tyrosine $C_6H_4(OH)CH_2CH(NH_2)COOH$	Fresh material	Red then black	*Amanitaceae*
(g) Benzidine $NH_2C_6H_4.C_6H_4NH_2$	Fresh material	(a) rose then carmine (with H_2O_2 intense black) (b) orange (c) blue to blue black (d) green to blue green	*Boletaceae* *Paxillaceae* *Russulaceae*
(h) Sulpho-benzene diazonium chloride $[C_6H_4SO_3HN_2]Cl$	Fresh material	Red, orange to yellow	*Russulaceae, Boletaceae* *Amanitaceae*

Chemical groups	Restrictions on material	Reaction result if positive	Family on which reagent is commonly used
(i) 2-4, diaminophenolhydrochloride $(NH_2)_2C_6H_3OH \cdot 2HCl$	Fresh material	Brown	*Russulaceae*
(j) n-dimethylamino-antipyrine	Fresh material	Lilac	*Russulaceae* *Tricholomataceae*
10 *Mixtures containing aniline and/or phenol*			
(a) Phenol: Aniline	Herbarium or fresh material	Brown to black	*Russulaceae*
(b) Aniline: Nitric Acid = Schaeffer's Reaction	Fresh material and carefully dried material	Flame red at the intersection of lines drawn with each liquid on cap cuticle	*Agaricaceae*
(c) Lactophenol (Lactic acid: Phenol)	Fresh material	Dark red	*Boletaceae*
(d) Schiff's reagent	Fresh material	Red violet to blue	*Tricholomataceae* etc.
11. *Crude extracts*			
(a) Gum Guaiac	Fresh material	Blue green, blue to purple	*Russulaceae*
(b) Tannin	Fresh vegetative mycelium	Yellow to medium brown	*Pleurotaceae* *Strophariaceae* (also *Polyporaceae*)
12. *Compounds used with mycelium in culture*			
	See appendix.		

chemical tests are equally diverse yet, and this is rarely appreciated, the same reagent may give similar reactions in different fungi despite the presence of unrelated substrates.

In order to give as broad a picture as possible it has been reluctantly decided to limit the account to only three families of chemical tests, for simply to list the whole series of reagents available now to the agaricologist would serve no useful purpose. The reagents to be discussed are aqueous solutions of alkali, aqueous solutions of simple inorganic iron salts and iodine complexes. These have been chosen from the vast collection of reagents which have been used to date, reagents such as α-naphthol, metol, tyrosin, silver nitrate, and sulpho-aldehyde, because they illustrate some of the problems which are experienced in this field of study and also because unlike many reagents they can be used with fresh and dried material alike and both micro- and macro-scopically. However, for completion Table I lists all the reagents used extensively in mycology with an indication of the colour reactions expected when positive results are obtained.

II. SPECIFIC PROBLEMS AND PRINCIPLES

A. Reactions of alkalis

It seems logical for the account to commence with the use of alkali, for not only is it a useful laboratory reagent but it was just this chemical which indicated that colour changes can be produced in agaric tissue by chemicals; this was the work of Harlay (1896) using *Lactarius turpis*.

The range of reactions that alkalis promote equals that of Melzer's reagent although this is not widely appreciated. However, unlike Melzer's reagent alkalis are of very great use in the pigmented spored groups particularly when dealing with reactions of the spore wall, e.g. darkening of the spores of members of the Bolbitiaceae, accentuating or producing an olive-purple hue in spores of *Stropharia*. Similarly, with microscopic structures alkalis have been utilized in the classification of the types of pleurocystidia (= facial cystidia) and have helped to redefine genera; cystidia may be encrusted with secretions or crystals and frequently these deposits darken or become yellow in alkali. Pleurocystidia which possess contents yellowing in ammonia are called chrysocystidia and the reaction in the main parallels the high affinity for dye which these same cells have when immersed in cotton blue. It is not possible at the moment to say whether all chrysocystidia are developmentally the same; in many instances they appear to be directly connected to vascular hyphae (laticifers). Chrysocystidia are to be found in *Stropharia* and *Hypholoma* but not in the closely related genus *Psilocybe* as now understood. Other genera, e.g. *Panaeolus*, possess some members exhibiting chrysocystidia and others which do not; frequently chrysocystidia are more noticeable when dried material is revived.

Alkalis have a much wider spectrum of reaction than when simply employed in studying spores and cystidia. Thus in certain *Cystoderma* spp. the outer layer of the pileus is composed of globose to elliptic cells from a veil and this darkens considerably in the presence of alkalis; similarly in certain *Crinipellis* spp. the hairs of the pileus turn grey and in certain *Xeromphalina* spp. the hymenophoral trama turns a rich red-brown.

Potassium (or sodium) hydroxide in a 2% aqueous solution is used in routine agaricology when examining dried material for it allows the hyphae and spores to swell to something like their original size and shape; a 3% aqueous ammoniacal solution has similar properties. It is therefore doubly important that we ascertain the reactions brought about by such routine reagents. Gelatinized areas or walls take on a silvery appearance in potash and their distribution can be more easily seen. However, instead of enhancing observations certain fungal pigments decompose in the presence of alkali, e.g. those of *Hygrophorus marginatus*; it has been suggested that because of this fact and possibly other correlated characters this fungus and its close relatives should be separated from the true hygrophori. When a drop of alkali is placed on the surface of the pileus a bleached area is produced.

In contrast other fungal processes which would take place naturally in the field appear to be speeded up in the presence of alkali, e.g. the yellowing of *Hygrophorus chrysaspis*, and this very fact helps to separate *H. chrysaspis* from *H. eburneus*, a close relative. Ammonia placed on the gills of *Russula drimeia* produces a pinkish carmine spot similar to the pinks and purples produced in members of the *Paxillaceae* and *Boletaceae* (Bataille, 1909; Sartory and Bertrand, 1914); golden yellow and olives, however, develop in members of the *Bolbitiaceae*.

Very vivid reactions often take place when alkali is placed on the cortical tissue of the fruit-body, e.g. rich blues in xerocomoid boletes and certain *Thelephora* spp. (Aphyllophorales). The turquoise blues produced naturally in the stipe base of various *Leccinum* spp. when exposed to the air immediately change to chrome yellow or greenish yellow on the application of alkali; a similar greenish yellow may appear spontaneously at the apex of the same stipe. Using a microprobe the pH of the tissue at the base of a stipe can be found to be vastly different from that at the apex. More observations therefore are required along these lines for it may be later shown that the greenish yellow of the stipe apex where the pH is neutral or high is simply another manifestation of the turquoise blue colour-change at the base. A similar phenonemon is experienced with other reagents in other groups, such as a faster reaction being recorded for the stipe base than the stipe apex.

Sodium carbonate solutions have sometimes been used instead of sodium hydroxide (Moser, 1960). Both sodium carbonate and hydroxide have been used in lichenology; rich purples etc. are produced due to the presence of certain anthraquinones in the tissues.

In *Cystoderma subvinaceum* only the veil turns greenish in solutions of alkalis while a similar coloration is observed in the tissue of the unrelated *Lepiota badhamii* and *Collybia alkalivirens*, hence the latter's epithet. No one knows whether these are due to the same, similar or quite different chemical substrates.

It appears that a good family character of the *Bolbitiaceae* is the fact that the basidiospores of all its members darken in alkali solutions; normally ammoniacal solutions are used. Thus spores of *Agrocybe* spp. turn from a rich yellow brown to "dresden brown" and those of *Bolbitius* and *Conocybe* spp. turn from a rich yellow to "ferruginous" in the presence of ammonia. However, in this same family ammoniacal solutions have been used to assess the potential of contents of the flesh of *Conocybe* spp. to produce long, thin, microscopic crystals in the supernatant. Singer (1962) has recorded a similar reaction in *Phaeomarasmius* (*Cortinariaceae*) and the present author, similar crystals in a mount of a *Galerina* spp. close to *G. helvoliceps*. However, this reaction has been found to be rather variable even in a single taxon, as has been hinted at by the original discoverer (Kühner, in several footnotes, 1935). Thus if formed in the mountant the crystals are useful additional evidence for an identification but a negative reaction is inconclusive. The characters and reactions of these crystals which may be simply oxalates will be dealt with in a separate paper.

I would like to suggest a tie-up between some observations made when using alkali and some of those made when using iodine solutions. Melzer's reagent (see II, B) in certain tougher basidiomycetous fungi, e.g. stipitate hydnums, produces a blue-green coloration, a colour also produced in the presence of potassium hydroxide; both are associated with a darkening of the tissue followed by an apparent discoloration to dirty olive. The blue-green colour was located in granules attached to the hyphal wall as in the agaric *Anthracophyllum*. Thus if only Melzer's reagent is used confusion results, for are the granules then really amyloid (see II, B).

The close relationship between the genera *Psilocybe* and *Stropharia* has already been mentioned and the results from the use of ammoniacal solutions has helped to redefine the two genera. Some of the members of these genera, however, also stain blue at the base of the stipe when handled or bruised; a parallel reaction is found in some members of the *Bolbitiaceae*, e.g., *C. cyanopus*. Some of these same species have been linked with the rituals of central American Indians and more recently they have been employed by students at parties because of their psychedelic properties (Benedict, Tyler and Watling 1967). In both *Conocybe* and *Psilocybe* the blue reaction appears to be connected with the presence of psilocybin (and psilocin) in the tissue (Benedict, Brady, Smith and Tyler, 1962; Benedict, Tyler and Watling, 1967; Hoffman *et al.*, 1959). A fungus described originally from Scotland by Orton and placed by him (1964) in *Stropharia* frequently exhibits a blueing stipe base, a character accentuated when grown in culture. It too contains psilocybin; it also lacks chrysocystidia and correlatory characters place this species closer to *Psilocybe* than *Stropharia*. The same treatment has been suggested for *Stropharia cubensis*, another blueing member of this same complex (Singer and Smith, 1958).

It is reported that metol produces a violet colour in certain of these species with the blueing foot but I have as yet been unable to substantiate this in all cases. Singer (1962) also reports that gum guaiac intensifies the blue reaction but evidence is available which suggests this is a secondary blueing imposed on the first, for it can be obtained in non-blueing species also. Thus gradually if observations are expanded many of the reactions correlated with other characters can be related to some substrates and, just as important, divorced from other substrates.

B. Reactions of iodine complexes

The reactions which have caused most impact on the classification of the *Agaricales* are those brought about by the action of Melzer's reagent (a mixture of chloral hydrate and an iodine complex) on the walls of fungus cells. It has separated some fungi which for very long periods have stood close together, even placed within a single genus, taxa such as *Lentinellus* and *Lentinus*, *Panus* and *Panellus*; indeed some genera now find themselves in quite different families!

Iodine has been used in various forms for generations; Moens (1963) has admirably covered the history of the use of iodine and Henry (1948–50, 1954–61) the techniques involved. Moser in his work on *Cortinarius* (1960) uses iodine in an aqueous solution of potassium iodine, and this gives a purple brown coloration when applied to the tissue of certain species, e.g. *Cortinarius sphagniphilus*; Gilbert has used iodine vapour (1929). However, much more commonly used is a similar solution to that of Moser's but to which has been added chloral hydrate. First used by Melzer and Zvara (1927) it has universally replaced the alcoholic solution of iodine or the zinc chloride–iodine complex (see Henry, 1954). The chloral hydrate appears simply to act as a clearing agent but more work is required on this for the very statement clearing agent is rather vague and we know little of the action of chloral itself on fungus tissue.

The darkening by iodine of the spore ornamentation of members of the Russulaceae helps to put this family apart from other agarics, for although superficially similar ornamentation is found in *Melanoleuca* and *Leucopaxillus*, both in the Tricholomataceae, correlatory flesh characters are absent. This same reaction coupled with other chemical reactions and similarities in tissue structure leads to the Russulaceae being directly connected with certain hypogeous Gasteromycetes. The splitting of the genus *Amanita* (Imler, 1948) and *Mycena* (Kühner, 1938) into more manageable units based on the so-called amyloidity of the basidiospore-wall is common practice and indeed a valuable method; "amyloid" spores in *Mycena* were noted in *Mycena tenerrima* as early as 1887 by Rolland (see Kühner and Maire, 1934).

Three terms are currently used to describe the reaction or colour changes which occur when using Melzer's reagent: (i) amyloid; (ii) dextrinoid (= pseudoamyloid); (iii) non-amyloid. One sees immediately a direct parallel here with the action of iodine on starch and its degradation products although little work has been carried out on the constituents of agaric basidiospores. Simple tests carried out suggest in fact that the ornamentation of many agaric basidiospores does not contain starch and so the parallel is in terminology alone; because of this Singer (1962) discussed the introduction of the word amylaceous. Locquin (1943) and Josserand (1941) both have attempted to isolate preferentially certain components of the spore wall but as yet the work has been limited and only now is keener interest being taken in these fungal "polysaccharides"; it is up to the mycologist to utilize and manipulate the chemists' results.

The three terms mentioned above do not really reflect the complexity of the colour changes which are recordable when using Melzer's reagent on different agaric tissues; thus an amyloid hypha or spore depends on the fungus tested for it may vary in colour from deep blue-black or violaceous indigo, as in *Chroogomphus*, to pale grey as in certain *Dermoloma* spp. and can be developed one is told in certain problematical *Mycena* spp. by hydrolysing in aqueous solutions of dilute acid. Nevertheless I find a similar although probably unrelated reaction to this and that of the spore wall in the tissue of certain boleti and this has assisted in the separation of critical species. The intensity of colour may be simply one of quantity of chemical substrate, molecular alignment within the structure or on the other hand due to the presence of quite unrelated compounds.

The dextrinoid reaction first noted by Metrod (1932) in *Lepiota* spp. is more consistent in the colour induced for it is a deep red brown with a hint of purple; unlike the amyloid reaction the colour produced depends on how long the material has been kept dry. Some *Mycena* spp. produced a purplish red colour in their trama with iodine and this is sometimes described as pseudoamyloid (dextrinoid) and on other occasions as amyloid (Kühner, 1938). The non-amyloid reaction is also quite variable ranging from lemon to clear golden yellow and this may again be due to quantity of substrate present; in some cases it is simply due to the presence of barriers such as gelatinized tissue (see Disbrey and Watling, 1967).

An exciting phenonemon has come to light in the last few years; this is that spores of several agarics have a high affinity for cotton blue and that amyloid tissue is mutually exclusive to the cyanophilic reaction; the reaction has been investigated by Kotlaba and Pouzar (1964). In some cases the definition of agaric families is pretty shaky because of the presence of intermediate or isolated species but this test, the results of which parallel the dextrinoid reaction, has assisted in detecting relationships. Disbrey and Watling (1967) have found a parallel reaction using other stains, e.g. eosin, and these observations support those of Kühner (1934) using an aqueous solution of cresyl blue. In discussion it has been suggested that lipids may be responsible for the low affinity to dyes by amyloid spores and high affinity by dextrinoid spores.

It is well known that basidiospores undergo very distinct structural changes as they develop (Corner, 1948; Perreau 1967; etc.) particularly when the spore wall is laid down and pigmentation commences. Some spores are dextrinoid when the basidiospores are immature but do not react when mature and ready for dispersal; this reaction is again supported by the results from using cotton blue. Other spores have been recorded as being partially dextrinoid, perhaps the end near the germ-pore, or irregularly dextrinoid. As in other agarics, members of the Bolbitiaceae possess a very thin, although very resistant membrane at the base of the germ-pore; it is possible that the structure of this membrane controls the entry of certain dyes into the spore, inhibits them or allows them to penetrate only slowly. It is interesting to speculate that this membrane also inhibits in nature the entry of toxic material from the substrate when the spore is dormant.

Amyloidity occurs in coloured spores also, although because of the difficulty in observing small shifts in colour it has not been used to any great extent in classification. The brown-spored agarics with amyloid spores turn slightly olivaceous tawny due to masking whereas certain boletes take on a chestnut coloration because of their dextrinoid character. Donk (1965) considers that in the Agaricales the amyloid/dextrinoid character has been overrated at generic level; this may be true when one considers the work of Mme. Perreau (1967) for the uniformity of basidiospore structure is impressive. Because an area or zone in a given spore possesses affinity for iodine one must not be tempted to overweigh the character. Certainly this becomes very obvious when familiar genera are recognized when abroad but on closer examination the members are found to differ only in amyloidity of the spores or hyphal walls. There are comparatively few species with amyloid spores in the Agaricales as a whole and surely the character should be treated with the same common sense as spore-wall colour. Thus Tricholomataceae contains pigmented and non-pigmented spored agarics; the Volvariaceae on the other hand contains only pink-spored fungi.

The results from studying the action of Melzer's reagent on spores has assisted in tracing some relationships in the *Russulaceae*; the same reaction, but this time in the *Gomphidiaceae*, has not only helped to trace possible phylogenetic relationships but groups the constituent species into more natural units and also allows the vegetative hyphae to be traced through the soil and onto the very area of ectotrophic mycorrhiza with which the fruit-body is connected (Miller, personal communication).

Sometimes the amyloidity may be fleeting, that is to say the spore is amyloid when taken from a fresh fruit-body but its colour is lost within an hour or so, a phenonemon which leads directly to the reaction which has been termed the "Imler reaction", in honour of Louis Imler who introduced this test for the study of bolete flesh (1950). When Melzer's reagent is placed on the cut surface of a bolete two distinct groups of species can be noted, those species with what can be called "amyloid" tissue in the base of the stipe and those with "non-amyloid" tissue; in the boletes only once has a true dextrinoid reaction been seen in many hundred tests. Although quite simple as a test it is a very powerful tool and assists in the separation of critical species. The amyloid reaction does not always have to be due to amyloid hyphal walls as an overall blue cast to the mount has often been observed with no localization of amyloid material. This coloration may be permanent but more usually it turns blue-green and finally fades through a series of olives. However, in one species *Boletus calopus* the septa stain intense blue-black and here the colour is indeed localized (Miller and Watling, 1968); a unique reaction as far as the author is aware. A. H. Smith (personal communication) has observed a non-localized colour change with chloral hydrate alone so care must be taken in interpreting the vast array of results using Melzer's reagent. Henry has made a very valuable contribution by tabulating all the known reactions between iodine and higher fungi (1954–1961) as well as those using other halogen complexes (1950).

It must always be borne in mind that many boletes undergo natural auto-oxidation processes when internal areas are exposed to the air (Bourquelot and Bertrand, 1896) or violently react on the addition of mixtures of organic substrates (see Table I); the fleeting reaction with Melzer's reagent which contains chloral hydrate may sometimes be a reflection of this reaction. Some of the auto-oxidation reactions are related to those brought about by placing an alcoholic solution of gum guaiac on flesh—that is to say assaying presence or absence of phenol–oxidase systems (cf. reaction between potato discs and gum guaiac). Many such tests are available to the agaricologist, too many to discuss here (see Appendix).

C. Reactions of simple iron salts

The amyloid and alkali reactions are applicable to both dried and fresh material but many reagents for reproducible results are only successful with the use of fresh tissue, e.g., those reactions dependent on enzymatic systems. However, although iron salts are best used with fresh material they have been successfully utilized with carefully preserved dried material. There are few field agaricologists who do not have a crystal or two of ferrous sulphate or iron alum tucked away in their foray jacket. However, students frequently think that magical powers are bestowed upon the agaricologist when he produces such crystals, when he rubs them on the tissue, observes and then pronounces.

The use of iron compounds has recently come to the fore with the delimitation of the Gomphaceae by Donk (1965), who bases the family in part on the dark green colour produced by the fruit-body tissue when treated with ferrous sulphate. The reagent is usually applied to the fresh flesh of the fruit-body and Singer (1962) has listed five different colour reactions: (i) negative as with *Russula cyanoxantha* (ii) a variable reaction as with *R. ferrotincta*, (iii) an olive green to blackish green as in *Russula xerampelina* and its allies, *Lactarius volemus*, *Clitocybe* spp. and *Paragyrodon sphaerosporus*, (iv) a pink or salmon reaction as found in the majority of *Russula* spp., and (v) a blue green or slate colour as found in certain boletes, particularly members of the genus *Leccinum*. To these the present author must add the deep violet coloration imparted to the flesh of *Boletus pallidus* (and certain *Clavaria* spp.), violaceous buff in certain lactarii and violaceous grey (Bas, personal communication) to black (Pearson, personal communication) in some *Leccinum* spp. Slight differences in shade of violet may be observed in different parts of the tissue, e.g. *Lactarius rufus* (see also II, A).

What the nature of the reaction is has not been assessed; whether the violet in *Boletus pallidus* is different from the violaceous grey in *Leccinum* or the same as the violet produced in *Clavaria* spp. remains to be seen. It should be borne in mind that depsides, 2- or 3-orcinol carboxylic acids in an ester linkage and characteristic of lichens, turn purple with ferric chloride solutions (Hale, 1967).

In the case of the *Russula xerampelina* group the reaction of iron salts parallels the reaction of aniline and of phenol and may be directly connected to a volatile constituent which was recognized and used by early taxonomists in delimiting this same group, i.e. the fishy odour; it is very significant here that *Lactarius volemus* also gives a green reaction and when mature smells of fish.

Ferrous sulphate is generally used in a 10% aqueous solution but even in this concentration it is a powerful reducing agent and during reaction becomes itself oxidized to a ferric salt; the latter can then react with any phenolic compounds present in the fungus and many such compounds have been identified in closely related fungi (cf. Birkinshaw, 1965). In the laboratory catechol gives a green colour with ferric chloride, resorcinol a violet-blue colour, phenol a clear purple, etc. certainly very close to field observations. These reactions may well follow certain auto-oxidative pathways which take place spontaneously in many agarics to produce the familiar vivid blues, purples, greens and reds but these themselves have been little studied and we do not possess correlatable evidence, e.g. all *Russula* spp. which turn black with age do not differ from those non-blackening species, for they all produce with ferrous sulphate the familiar normal rust or salmon colour.

Ferrous sulphate crystals often become encrusted with anhydrate which becomes intermixed with ferric compounds or the solution becomes cloudy and brown due to the presence of ferric hydroxide. In order to stabilize the crystals so that reproducible results are obtained it has been found useful to keep the crystals on pads of cotton wool moistened in dilute aqueous ammonium sulphate; this is now routine practice in Edinburgh. Keeping the crystals in this way ensures the presence of the ferrous ion. With careful observations it has been detected that with certain tissues sulphuric acid, with which ferrous sulphate has often been stabilized, itself gives a distinctive reaction. A similar phenomenon is experienced when using sulpho-aldehyde, the reaction in many bloetes being that of the sulphuric acid and not of the sulpho-aldehyde complex; this may also explain certain anomalous reactions obtained with certain tests, solvents and not reagents reacting with the tissue, e.g. butanol the solvent for benzidine.

The 10% sulphate solution has, however, been replaced by a ferric chloride solution in some laboratories. These findings must be very carefully interpreted because ferric chloride can give quite different results to ferrous sulphate (Josserand, 1948); substances such as catechol react very differently when ferric ions are in the presence of ferrous iron and again react differently if in alcoholic solution. This latter observation is very important for Smith (1966), when using ferrous reagents records the colour change, then applies ethanol and again records the resultant colour. Catechol although having a complicated oxidation pathway, is a well-defined chemical; what the reactions of ferrous sulphate with the as yet undescribed phenolic compounds present in agarics have yet to be discovered.

III. RESTRICTIONS AND PRECAUTIONS: AVAILABILITY AND STABILITY OF REAGENTS

The last two points mentioned above can only lead one to emphasize the care necessary when applying reagents to the flesh of agarics, indeed any higher fungi. Ammonia is frequently used on the flesh or cuticle of the fruit-body, particularly in the study of boletes, but carelessly slopping the reagent close to a ferrous sulphate rubbing will have remarkable results, in fact it will convert all *Russula* spp. for instance into one of the *Russula xerampelina* group! Fumes from reagents are often noxious and can easily mislead; unexpected reactions may take place between reagents one or both of which are volatile, which may be easily demonstrated by the copious white fumes produced when hydrochloric acid fumes and ammonia fumes intermix.

Fungus tissue is composed of filamentous hyphae and because of this surface tension and capillary action tend to draw together liquids placed on the surface of a cut section; this encourages reaction to take place. It is therefore important to keep tested areas well away from one another; thus there would be high activity of phenol when close to tissue tested with aniline.

The range of chemicals available to the agaricologist has recently been extended considerably, many are potentially useful taxonomically but on a very serious note some are potentially carcinogenic, e.g. benzidine, or explosive, e.g. perchloric acid. I am unable to believe that the indiscriminate use of these latter chemicals will advance knowledge enough to warrant their widespread use. Similarly with the employment of concentrated sulphuric acid in the identification of *Amanita phalloides*; is the use of such a corrosive liquid necessary? Fries (1821) and his contemporaries, after all, had little difficulty in recognizing this taxon and separating it from closely allied agarics on purely morphological grounds.

It is considered essential to raise two further points, firstly that certain organic reagents proposed for use in identification by some authorities undergo rapid auto-oxidation and therefore care must be taken in the interpretation of any results; although results obtained with their use may be constant and diagnostic such reagents have disadvantages when used in the field. Linked in part with this is the fact that certain reactions are only thought of as positive if they take place within a certain time and if this is a long period auto-oxidation is more likely to occur. Secondly some reagents such as formalin undergo polymerization or parallel chemical change and require frequent replacement. Methanol can be used for stabilization of formalin solutions but more information is really required about the actions of methanol with fungus tissue before such a stabilizer is used universally.

Some reactions are considered positive if they take place in a short time, whilst others only develop their distinctive reaction after a lengthy period, e.g. formalin on bolete tissue. Thus *A. crocea* becomes purple within minutes with phenol, whilst *A. fulva* takes anything up to half an hour; only the former is considered a positive result.

Further caution should be exercised in relation to the age of the fruit-body or whether or not it is sterile; it is well known that changes take place in the chemical constituents of a fruit-body during growth and that the production of some chemicals are only initiated by spore-production. There are some field observations available that suggest both these states can alter the expected colour change of a given chemical reaction (cf. Wells and Kempton, 1968). Many reactions can also affect the naturally spontaneous auto-oxidation processes found in certain agarics, either by speeding them up as the action of alkali on *Hygrophorus chrysaspis*, or hindering, or masking them as with bolete tissue and alkali, or certain *Agaricus* spp. and hydrochloric acid. Water-soaked tissues frequently give entirely different reactions to that of mature, sound flesh, probably due to bacterial activity breaking down certain compounds and supplying others; the importance of pH of the flesh has been already briefly mentioned.

A very useful test for distinguishing certain members of the Agaricaceae is the so-called Schaeffer reaction; indeed the reaction helps to connect this family to certain gasteromycetous forms. The reaction is said to be positive if a flame colour is produced on the cap at the intersection of a line drawn with a glass-rod dipped in aniline and a line of concentrated nitric acid. How many mycologists one wonders work with a clear, straw-coloured liquid for their stock aniline and how many work with a dark brown syrup? If the latter, then one is tempted to suggest that a little of the reagent is treated with nitric acid in the absence of agaric tissue! Purity of reagents should always be a pre-requisite of this study.

Just as methyl chloro-antimonate simply tries to translate the acrid taste of certain *Russula* spp., so the Schaeffer reaction is correlated with the almond smell of certain *Agaricus* spp.—in fact *Agaricus sylvicola* emphasizes this rather nicely, for populations are often found with a rather inconclusive reaction and these same populations possess only a slight indication of almonds.

Odours of members of the Bolbitiaceae are extremely constant and reproducible in culture through several generations. They are less variable than spore size and cystidial dimensions and there is every belief that a chemical reaction based on such odours would be very constant; after all, odours are simply volatile chemical constituents of the fungus tissue and, with tastes, ought to be considered in conjunction with the chemical characteristics of the fungus.

Removal of organic compounds from contact with chemical reagents is of primary importance, for not only do some decompose more rapidly in their presence but some do so so quickly they are a source of danger; some reagents, e.g. $AgNO_3$, are best kept in dark bottles. Coupled with the purity of the reagent is its strength because as has been pointed out earlier, time of a colour change and the intensity of change are important and the majority of chemical reactions are intimately connected with concentration of reagents; standard formulae should be adhered to, for some reagents give quite different results when in dilute aqueous solution, e.g. H_2SO_4. An Appendix supplied to this chapter endeavours to list the majority of those reagents used in agaricology with their recommended concentrations and formulae. Where differences between authors exist the authority has been indicated, as have the originators of the reagent or reagent mixture. Such a comprehensive list has not been attempted before and it is hoped that the formulae suggested are used in order that the work of different authors can be directly compared; the colour terminology used in describing reactions is important (cf. Henry, 1948–50).

The actual treatment of the fresh material is an equally important facet of this study. It has been the custom in the past simply to rub or drop the chemical on to the flesh, gills or surface of the fruit-body. However, on drying such treated tissue collapses and cannot be revived, thus making the specimen useless for the herbarium; by capillary action the chemicals cover a larger area than intended and draw together different chemicals as already outlined. It is therefore suggested that with the fleshy fungi the tissue is cut into small cubes and added to about half its volume of the chosen liquid held in a test-tube, or cut into discs which can be tested in a welled-slide.

Unless the chemicals are helpful in separating critical taxa the present author doubts whether many of the reagents tabulated in this account will be of widespread usage; many tests separate species which can be isolated on other more conventional lines but some have the added disadvantages that they are difficult to obtain.

IV. APPLICATION IN OTHER GROUPS OF THE BASIDIOMYCETES

For about the same period of time covered above chemical reagents have been used in other groups of the higher fungi not only for assisting in the recognition of structural units e.g. setae, but also in the delimitation of species in the field and laboratory. The history and isolation of secondary metabolites from the fruit-bodies has followed a similar pattern as that already outlined; alkali, Melzer's reagent and iron salts having played their part in the classification of the non-agaricoid fungi, e.g. Hymenochaetaceae (alkali), Bondarzewiaceae (Melzer's reagent) and Gomphaceae (iron salts). *Clavariadelphus* has been considered closely related, if not classifiable in the *Gomphaceae*: in their treatment of the genus Wells and Kempton (1968) moistened their dried material with isopropyl alcohol before treatment with iron salts. Observations in Edinburgh using agaricoid fungi indicate that phenolic compounds may react quite differently in the presence of this compound, than if treated alone; further work is desirable for chemical differences do exist between the two groups of fungi.

Perhaps of more importance in this group of fungi has been the emphasis on cultural techniques (see article on culturing hemobasidiomycetes) and many of the tests mentioned particularly in the introduction have been modified to suit this new demand, e.g. gum guaiac, tannin test, cotton blue (see Nobles, 1965, etc., and Appendix); sulpho-aldehyde (Boidin, 1951).

V. CONCLUSIONS

It is impossible to agree fully with Singer (1962) when he states that "this [lack of information] does not render them [chemical reactions] any less valuable from the taxonomic point of view", for until we know more about the substrates we test for, how they vary and their distribution in other organisms we cannot place full reliance on them. Nevertheless the results of chemical reactions are very promising and should not be spurned; it is up to the agaricologists to make their conclusions water-tight by carrying out careful and widespread observations and enlisting, whenever possible, the help of both chemists and biochemists. Once correlatable with more conventional characters such as anatomy and morphology, as well as being reliable, they can then be added to the vast array of ammunition now available to the agaricologist for exploring and better understanding relationships.

Unfortunately many chemical tests simply separate taxa which can be isolated quite easily by conventional characters such as veil and ring structure, viscidness etc. Thus the pinkish hue produced when sulphuric acid is run onto the gills of *Amanita phalloides* separates it from its close relatives but so can its cap colour and veil form; ammonia used in the same way however will allow the olivaceous and yellow forms of the normally purple-capped *Russula drimeia* to be picked out from amongst the other greenish *Russula* species. These examples show that in certain cases assistance is offered by carefully selecting the tests to confirm a suspected species' autonomy; they can therefore be useful spot or field tests.

In future chemical tests must be critically assessed as to their worthiness, for like other characters a reagent useful in one group of organisms might be of very reduced importance in another, possibly quite closely related, group; we must not allow chemical tests to run away with the taxonomy but they must be integrated into it. They must be used as an additional tool in identification for fungi, are simply constructed and possess but few characters useful in classification when compared with the number utilized in a group of flowering plants of equal size. Lichenologists have used chemical reagents for over 100 years now and it has been suggested that species separation in the lichens is at present far too narrow; perhaps the mycologist can gain from the lichenologist's experience. At the moment the majority of agaricologists need to treat the subject of chemical reagents less flippantly and assist in the accummulation of a less chaotic set of data than at present available.

ACKNOWLEDGMENTS

My thanks to Dr. Ray Edwards, University of Bradford, for comment on chemical data is here noted.

REFERENCES

Henry (1948–1961) has given an extensive bibliography for the period 1811–May 1953; although not entirely complete this covers all the major articles required in any further reading.

Ainsworth, G. C., and Sussman, A. S. (1965). "The Fungi, an Advanced Treatise", Vol. 1. Academic Press, London.
Arnould, L., and Goris, A. (1907). *Bull. Soc. Mycol. Fr.*, **23**, 174–178.
Asahina, Y. (1936). *J. Jap. Bot.*, **12**, 516. (A series of articles on identification of lichen substances appeared regularly in this journal until 1940.)
Bas, C. (1965). *Persoonia*, **3**, 331–359.
Bataille, F. (1909), *Bull. Soc. mycol. Fr.*, **25** (1), 81.
Benedict, R. G., Brady, L. R., Smith, A. H., and Tyler, V. E. (1962). *Lloydia*, **25**, 156–159.
Benedict, R. G., Tyler, V. E., and Brady, L. R. (1966). *Lloydia*, **29**, 333–342.
Benedict, R. G., Tyler, V. E., and Watling, R. (1967). *Lloydia*, **30**, 150–157.
Birkinshaw, J. H. (1965). *In* "The Fungi" (Ainsworth, G. C. and Sussman, A. S., Eds.), Vol. I, 179–228. Academic Press, London.
Boidin, J. (1951). *Bull. Soc. Nat. Oyonnax*, **5**, 72–79.
Bourquelot, E., and Betrand, G., (1896). *Bull. Soc. mycol. Fr.*, **12**, 1, 18, 27.
Bousset, M. (1939). *Bull. Soc. linn. Lyon*, **8** (6), 154–158.
Brian, P. W. (1951). *Bot. Rev.*, **17**, 357–430.
Brown, J. K., Malone, M. H., Stuntz, D. E., and Tyler, V. E. (1962). *J. Pharm. Sci.*, **51**, 853–856.
Brunel, A. (1936), see Singer, R. (1962).
Catalfomo, P., and Tyler, V. E. (1961). *J. Pharm. Sci.*, **50**, 689–692.
Corner, E. J. H. (1948). *New Phytol.*, **47**, 22–51.
Disbrey, B., and Watling, R. (1967). *Mycopath. Mycol. appl.*, **32**, 81–114.
Donk, M. A. (1965). *Persoonia*, **3**, 199–324.
Edwards, R. L., Lewis, D. G., and Wilson, D. V. (1961). *J. Chem. Soc. (C)*, 4995–5002.
Edwards, R. L., and Kale, N. (1964). *J. Chem. Soc. (C)*, 4084.
Edwards, R. L., and Elsworthy, G. C. (1967). *J. Chem. Soc. (C)*, 410–411.
Ford, W. W. (1906). *J. exp. Med.*, **8**, 437–450.
Ford, W. W., and Clark, E. D. (1914). *Mycologia*, **6**, 167–191.
Frerejacque, M. (1939). *Rev. Mycol.*, **4**, 89–100.
Fries, E. (1821). "Systema Mycologicum". E. Marituis, Uppsala.
Fries, N. (1958). *K. Vetensk samh. Upps. Handl*, **2**, 5–16.
Gabriel, M. (1962). *Bull. Soc. mycol. Fr.*, **78**, 359–366.
Gilbert, E. J. (1929). *Bull. Soc. mycol. Fr.*, **45**, 141–144.

Hale, M. E. (1967). "The Biology of Lichens". Arnold, London.
Harlay, V. (1896). *Bull. Soc. mycol. Fr.*, **12**, 156–159.
Hegnauer, R., (1962). "Chemotaxonomie der Pflanzen". Birkhaüser, Basel.
Heim, R. (1942). *Bull. Soc. Chim. biol.*, **29**, 48–79.
Heim, R. (1957). "Les Champignons d'Europe". N. Boubée, Paris.
Heim, R. (1960). *Revue Mycol.*, **25**, 224–235.
Heim, R., and Romagnesi, H. (1934). *Revue Mycol.*, **26** (1), 164–174.
Heinneman, P., and Casimir, (1961). *Bull. Soc. mycol. Fr.*, **26**, 24–33.
Henry, R. (1939). *Bull. Soc. mycol. Fr.*, **55**, 166–195.
Henry, R. (1943). *Revue Mycol.*, **8** (N.S.), 22–25.
Henry, R. (1948–50). *Revue Mycol.*, suppl. to vol. 13–15.
Henry, R. (1954, 1958–61). *Revue Mycol.*, suppl. to vol. 19–26.
Hoffman, A., Heim, R., Brack, A., Kobel, H., Frey, A. O. H., Petrzilka, T., and
    Troxler, F. (1959). *Helv. Chem. Acta*, **42**, 1557–1572.
Hughes, D. H. (1958). *MGA Bull.*, **99**, 86–88.
Imler, L. (1948). *Bull. Soc. mycol. Fr.*, **64**, 50–52.
Imler, L. (1950). *Bull. Soc. mycol. Fr.*, **66**, 177–203.
Josserand, M. (1930). *Bull. Soc. linn. Lyon*, **9**, (7), 43–44.
Josserand, M. (1941). *Bull. Soc. Mycol. Fr.*, **56**, 1–38.
Josserand, M. (1943). *Bull. Soc. linn. Lyon*, **12**, 156–158.
Josserand, M. (1948). *Bull. Soc. mycol. Fr.*, **64**, 1–32.
Kobert, R. (1891) see Ford and Clark (1914).
Kögl, F. (1926). *Justus Liebigs Annln. Chem.*, **447**, 78. (This is the first of a series
    of articles that appeared irregularly in this journal.)
Kögl, F., and Erxleben (1930). *Annalen*, **11**, 479. (This is the first of a series of articles
    that appeared irregularly in this journal.)
Kögl, F., and Quackenbush, F. W. (1944). *Recl. Trav. Chim.*, **63**, 251.
Kotlaba, F., and Pouzar, Z. (1964). *Feddes Rep.*, **69**, 131–142.
Kühner, R. (1931). *Bull. Soc. linn. Lyon*, no. **16**, 122–130.
Kühner, R. (1934). *C.r. hebd. Séanc. Acad Sci., Paris*, **198**: (1)–(3).
Kühner, R. (1935). "Le Genre Galera". P. Lechevalier, Paris.
Kühner, R. (1938). "Le Genre Mycena". P. Lechevalier, Paris.
Kühner, R., and Maire, R. (1934). *Bull. Soc. mycol. Fr.*, **50**, 9–24.
Kühner, R., and Romagnesi, H. (1953). "Flore Analytique des Champignons Sup-
    érieurs". Massen & Cie, Paris.
Langeron, M. (1952). "Précis de Mycologie". Masson & Cie, Paris.
Loesecke, A. (1871). *Arch. Pharm, Berl.*, **36**.
Lindt, W. (1885). *Z. wiss. Mikrosk.*, **11**, 495.
Locquin, M. (1942). *Bull. Soc. linn Lyon*, **11**, 1942.
Locquin. M. (1943). *Bull. Soc linn. Lyon*, **12**, 110–112, 122–128.
Locquin, M. (1944). *Bull. Soc. linn. Lyon*, **13**, 151–157.
May, C. (1961). *Plt. Dis. Reptr.*, **45**, 777.
Melzer, V., and Zvara, J. (1927). *Arch. prírodov. Výžk. Čech.*, **17**, 1–126.
Metrod, G. (1932). *Bull. Soc. mycol. Fr.*, **48**, 324.
Micka, K. (1954). *Česká Mykol.*, **8**, 165–167.
Miller, O. K., and Watling, R. (1968). *Notes R. Bot. Gdn. Edinb.*, **28** (3), 317–325.
Moeller, F. H. (1950). *Friesia*, **4**, 1–60.
Moeller, F. H. (1952). *Friesia*, **4**, 135–220.
Moens, J. (1963). *Sterbeeckia*, **3**, 1–10.
Moser, M. (1955). "Kleine Kryptogamenflora Mitteleuropa". G. Fischer, Jena.

Moser, M. (1960). "Die Gattung Phlegmacium". J. Klinkhardt, Bad Heilbrunn.
Müller, H. (1872). see Singer, R. (1951).
Nobles, M. K. (1965). *Can. J. Bot.*, **43**, 1097–1139.
Orton, P. D. (1964). *Notes R. Bot. Gdn. Edinb.*, **26**, 43–66.
Pantidou, M. E. (1962). *Can. J. Bot.*, **40**, 1313–1319.
Parrot, A. G. (1966). *Bull. Soc. mycol. Fr.*, **82**, 333–351.
Pastac, I. A. (1942). *Revue. Mycol.*, *Mém. hors*, Series No. 2, 1–111.
Perreau, J. (1967). *Annls. Sci. nat. Bot. Biol. Veg.*, **8**, 12th Series, 641–745.
Purvis, M. J., Collier, D. C., and Wall, D. (1964). "Laboratory techniques in Botany". Butterworths, London.
Reichl, C., and Mikosch, C. (1890). *Jabresber. Oberrealschule II Bez., Wien*, 34–35.
Robbers, J. E., Brady, L. R., and Tyler, V. E. (1964). *Lloydia*, **27**, 192–202.
Rolland, L. (1887). *Bull. Soc. mycol. Fr.*, **3**, 134.
Romagnesi, H. (1948). *Bull Soc. mycol. Fr.*, **64**, 53–100.
Sandor, R. (1956). *Z. Pilzk.*, **22**, 97–103.
Sandor, R. (1959). *Z. Pilzk.*, **25**, 103–112.
Sartory, A., and Betrand. G. (1914). *C.r. hebd. Séanc. Acad. Sci. Paris*, **76**, 363–364.
Schaeffer, J. (1952). "Russula–Monographie". J. Klinkhardt, Bad Heilbrunn.
Shibata, S. (1964). *In* "Beitrage zur Biochemie und Physiologie von Naturstuffen". Jena.
Singer, R. (1951). Agaricales in Modern Taxonomy, Lilloa, **22**, (1949), 1–832.
Singer, R. (1962). 2nd edition, J. Cramer, Weinheim.
Singer, R. (1965). "Die Röhrlinge". J. Klinkhardt, Bad Heilbrunn.
Singer, R., and Smith, A. H. (1958). *Mycologia*, **50**, 262–303.
Smith, A. H. (1966). *Mem. N. Y. bot. Gdn.*, **14**, 1–178.
Smith, A. H., and Shaffer, R. (1964). "Keys to Genera of Higher Fungi". University of Michigan, Ann Arbor.
Talbot, G., and Vining, L. C. (1963). *Can. J. Bot.*, **41**, 639–647.
Tyler, V. E. (1964). see Robbers *et al.*
Tyler, V. E. (1961). *Lloydia*, **24**, 71–74.
Tyler, V. E. (1963). *Prog. Chem. Toxicology*, **1**, 339–384.
Tyler, V. E., Benedict, R. G., and Stuntz, D. E. (1965). *Lloydia*, **28**, 342–353.
Valadon, L. G., and Mummery, R. S., (1968). *Nature*, **217**, 1000.
Wager, H. (1919). *Trans. Brit. mycol. Soc.*, **6**, 158–164.
Watson, P. (1966). *Trans. Brit. mycol. Soc.*, **49**, 11–17.
Wells, V., and Kempton, P. E. (1968). *Mich. Bot.*, **7**, 35–37.
Wieland, H., and Coutelle, G. (1941). *Justus Liebigs Annln Chem.*, **48**, 270.
Worthen, L. R., Stessel, G. R. J., and Youngki, H. W. (1962). *Econ. Bot.*, **16**, 315–318.
Worthen, L. R., Snell, W. H., and Dick, E. (1965). *Lloydia*, **28**, 44–47.

APPENDIX

Recommended concentrations of chemical reagents. The important characters of the reagents are included in parentheses.

1. *Simple acids and alkalis*
(a) H_2SO_4 : Sp. gr. 1·84 (a *clear, colourless, oily* liquid) tends to absorb ammonia etc. from atmosphere; replace at commencement of each collecting season.
(b) HC1. : Sp. gr. 1·18 (a *clear, colourless* liquid), see note and restrictions above.

iv   24

(c) HNO_3 : Sp. gr. 1·42 (a *clear, colourless* liquid), see note and restrictions above.

(d) KOH (or NaOH) : 10% aqueous solution (a *clear, colourless, oily* liquid). Tends to absorb carbon dioxide from atmosphere; prepare a fresh solution regularly. 5% and 2% aqueous solutions are required in microscopy.

(e) NH_4OH : Sp. gr. 0·88 (a *clear, colourless* liquid giving off *pungent fumes*) 10% aqueous solution; 2% aqueous solution is required in microscopy. Must be replaced regularly.

2. *Simple heavy-metal compounds*

(a) Ferrous sulphate, $FeSO_4.7H_2O$: crystal, or 10% aqueous solution to be replaced regularly (a *pale green, clear* liquid, which is easily oxidized to ferric sulphate which decomposes further, becoming turbid due to deposition of ferric hydroxide complexes). A 1 : 2 solution with ammonium sulphate will make a stable solution or the crystals can be kept in a tube on a pad of cotton wool dampened with an aqueous solution of ammonium sulphate. (See Melzer and Zvara, 1927.) Ferrous ammonium sulphate $(NH_4)_2SO_4.FeSO_4$; crystal or 10% solution is a stable solution frequently used to replace ferrous sulphate. Ferric-alum, $(NH_4)_2SO_4.Fe(SO_4)_3$, ammonium ferric sulphate; crystal or 10% aqueous solution (a *clear, lilaceous* solution), more stable and used generally for the preparation of solutions of ferric iron. Ferric chloride solution (a *clear, orange brown* solution) containing 60% w/v $FeCl_3$: easily hydrolysed to hydroxide and must be replaced regularly.

NOTE: often referred to as perchlorure de fer.

(b) Mercury complexes: 1 g of powdered mercuric chloride $HgCl_2$, in 10 ml nitric acid (sp. gr. 1·42); only use when freshly prepared (Moser, 1960).

Millon's reagent (*see Micka*, 1954; *Sandor*, 1959; *Purvis, Collier and Walls*, 1964)

Hg	56 ml
HNO_3 (Sp. gr. 1·42)	94 ml

Heat until mercury is dissolved then add to 100 ml of distilled water; replace regularly.

(c) Silver salts; Silver nitrate, $AgNO_3$; aqueous solution (a *clear, colourless* liquid). Must be kept in a dark vessel and away from organic material with which it rapidly reacts.

(d) Copper salts: Cupric sulphate, $CuSO_4.5H_2O$; crystal or 10% aqueous solution (a *clear, pale blue* solution) (Parrot, 1966).

3. *Rare earth and uranium complexes*

(a) Thallium: a colourless solution—

Tl-4 (*Henry*, 1943)

Thallium oxide	1·5 g
HCl (sp. gr. 1·18)	80 drops
HNO_3 (sp. gr. 1·42)	80 drops

Agitate slowly whilst adding 1 g Na_2CO_3 in 10 ml water; allow to stand for 3–4 days.

(b) Cerium: $(NH_4)_2Ce(NO_3)_6$, 20 g in 30 ml of 20% nitric acid (a *dark yellow* liquid) (Micka, 1954).

(c) Uranium: $K_2U_2O_7.6H_2O$, 10% aqueous solution (a *yellow* liquid).

4. *Simple complexes containing halogens*

(a) Iodine solution (a *deep orange yellow* solution)—

Iodine	1 part
KI	2 parts

Mix with 150 parts water and retain in dark bottle (Moser, 1955; see also Henry, 1954).

(b) Melzer's reagent or solution (a *rich red brown* solution) (after Melzer and Zvara, 1927)—

Iodine	1·5 g
KI	5 g
CCl₃CHO	100 g
Water	100 ml

(c) Bleaching powder, $Ca(OCl)_2$: saturated aqueous solution (a *cloudy white* liquid): commercial bleach is of greater stability and gives parallel results, otherwise replace very regularly and keep away from sunlight.

(d) KOCL : L'eau de Javel (a *slightly turbid* liquid)—

$Ca(OCl)_2$	1 g
K_2CO_3	2 g

Add to 20 ml water and shake thoroughly before allowing to stand; replace regularly.

(e) $HClO_4$: constant boiling point mixture with water; obtainable from reliable chemical distributors (a *clear faintly coloured* liquid) (Henry, 1956).

NOTE: *very unstable and may decompose spontaneously with explosion.*

(f) Bromine solution prepared by adding liquid bromine to water (a *reddish* liquid); in presence of light decomposes to give hydrobromic acid and hypobromous acid (Henry, 1950).

5. *Organo-metallic compounds*

(a) $CH_3COO_2Pb.3H_2O$: saturated aqueous solution (a *clear, colourless* liquid).

(b) $(CH_3)_2Cl.Sb$: solution in methanol (see Singer, 1951).

(c) $(C_2H_5)_3Cl.Sn$: 5 g in 5 ml 95% ethanol to which has been added 25–30 drops of conc. HCl (see Singer, 1951).

6. *Aliphatic and aromatic aldehyde complexes*

(a) Sulpho-formol: 1/1 ratio of 40% aqueous solution of formaldehyde and 70% sulphuric acid. Prepare fresh each day of use (Arnould and Goris, 1907); see restrictions on formal in 7b.

(b) Sulpho-benzaldehyde: 1/1 ratio of benzaldehyde and 70% sulphuric acid (Boidin, 1951); see restrictions above.

(c) Sulpho-vanilin: vanilin (0·5 g) in 4 ml H_2SO_4 and 2 ml distilled water (Arnould and Goris, 1907); see restrictions above.

(d) Chloro-vanilin: vanilin (0·5 g) in 4 ml HCl and 2 ml distilled water (Schaeffer, 1952); see restrictions above.

(e) Sulpho-α-naphthaldehyde (Sandor, 1959).

7. *Aliphatic compounds*

(a) $CH_3CH(OH)COOH$ (a *syrupy, almost colourless* liquid) (Singer, 1965).

(b) HCHO: 40% aqueous solution of BP formaldehyde (a *clear, colourless* liquid); replace regularly by fresh solution. (A solution of formaldehyde in water is stabilized by the presence of methanol—more field observations are required.)

(c) C_2H_5OH (a *clear, colourless* liquid).

(d) CH_3COOH: 33% aqueous solution (a *clear, colourless* liquid) (Heim, 1960).

8. *Phenols and aromatic acids*

(a) C_6H_5OH : 2% aqueous solution (a *clear, almost colourless,* liquid).

(b–d) Dihydric phenols : 2% aqueous solutions (*clear, almost colourless* liquids). NOTE: quinol = hydroquinone.

(e) $C_6H_4(OCH_3)OH$: a saturated aqueous solution (a *clear, almost colourless* liquid) (Kühner and Romagnesi, 1953).

(f–g) Trihydric phenols: 5% aqueous solutions (*clear, colourless to very pale buff coloured* liquids).

(h) $C_6H_3(CH_3)(OH)_2$: 50% solution in ethanol (Micka, 1954; Sandor, 1959).

(i) $CH_3C_6H_3(OH) CH(CH_3)_2$: 50% solution in ethanol (a *clear, colourless* liquid). Cf. 12. below.

(j) $C_6H_2(OH)_3COOH$ (a *clear, colourless* liquid).

(k) $C_6H_4CO.CO.CO.H_2O$: 1% solution in n-butanol (a *clear, colourless* liquid). Unstable, replace regularly. May be prepared in ethanol.

9. *Nitrogen-containing aromatics*

(a) $C_6H_5.NH_2$: either (i) an aqueous solution or (ii) an oil—

   (i) Agitate a few drops in water and filter (a *clear, almost colourless* liquid) (Kühner and Romagnesi, 1953).

   (ii) Aniline oil (a *clear, oily* liquid, almost *colourless* when freshly distilled) (Singer, 1962).

Insure the aniline oil *remains clear* and *straw yellow* for it soon darkens to a reddish brown colour on storage.

(b) $C_6H_5NH.NH_2$: 1% solution in 25% acetic acid (*colourless, syrupy* liquid which *slowly darkens* on contact with the air) (Micka, 1954). NOTE: *Caustic properties.*

(c) $C_6H_4(OH).NH_2$: 5% w/v solution in ethanol; make up fresh immediately before use (*greyish brown powder* which *slowly darkens* on contact with air).

(d) $C_6H_4(NH_2)_2$: 5% w/v solution in ethanol; make up fresh immediately before use. Solution (a *purple brown* powder which *rapidly decomposes* on exposure to air) decomposes within the day. A more stable solution can be made by adding 10 g of Na_2SO_3 to 1 g phenylene diamine in a mixture of 100 ml of distilled water and 2 ml of liquid detergent. James (personal communication) is experimenting with the use of orthodianisidiane in place of phenylene diamine; its use in mycology should be tested.

(e) $(CH_3NH.CH_4OH)_2H_2SO_4$ (=metol): 1 g in 20 ml of water (*almost colourless,* water soluble *crystalline* solid which will *darken* on exposure to air) (Bousset, 1939).

(f) $C_6H_4(OH)CH_2CH(NH_2)COOH$: dilute solution in ethanol. A stronger colour change is obtained if the reaction is carried out in an ammoniacal solution (Sandor, 1959).

(g) $NH_2C_6H_4C_6H_4NH_2$: 1% solution in 10% acetic acid (*clear, slightly brown* tinted liquid); readily decomposes and requires replacing at least once a week. Prepare by mixing first in glacial acetic acid and then diluting proportionally (Micka, 1954). NOTE: This is a proven carinogen and although the reagent gives a wide series of reactions its use is not advised.

(h) $[C_6H_4SO_3H.N_2]$ Cl: solution, (a) 1·4 g sodium nitrite (Na NO_2) in 100 ml of water (a *clear colourless* liquid) solution (b) 0·5 g (sulphanilic acid) $NH_4C_6H_4SO_3H$ in 90 ml of water and 10 ml of an aqueous solution of sodium hydroxide. Mix solutions on substrate (Micka, 1954).

(i) $(NH_2)_2C_6H_3OH.2HCl$ (= amidol): 1 g in 20 ml of water (*almost colourless, crystalline* solid which freely dissolves in water but will *darken* on exposure to air).

(j) *n*-dimethylamino-antipyrine (= pyramidon): a saturated aqueous solution (Heim, 1957).

10. *Mixtures containing aniline and/or phenol*
(a) Phenol/aniline: 3 drops of aniline and 5 drops of 1·84 sulphuric acid in 10 ml of 2% phenol (Schaeffer, 1952; Henry, 1939).
(b) Aniline: nitric acid (=Schaeffer's Cross-Test or Schaeffer reaction). 'When positivea flame red colour is developed at the intersection of a line drawn with aniline on the fruit-body's cap and a line drawn with 1·42 nitric acid (Moeller, 1950; 1952).
(c) Lactophenol (a *clear, almost colourless, syrupy* liquid) (Pantidou, 1962)—

Phenol	100 g
Lactic acid	100 ml
Glycerine	100 ml
Water	100 ml

(d) Schiff's reagent: aqueous solution of rosaniline which has been reduced to a pale straw-coloured liquid by sulphur dioxide (Micka, 1954)—

Basic fuschin	0·025 g
Water	100 ml
HNO_3 (sp. gr. 1·42)	20 drops

Add 1% $NaHSO_3$ solution until straw-coloured.

11. *Crude extracts*
(a) Gum guaiac: freshly prepared 20% solution of resin in 70% ethanol (a *dark red brown resin* giving a red brown solution) (Bourquelot and Betrand, 1896). Sandor (1959) reports that the solution can be stabilized by replacing the ethanol by glyceryl diacetate. For vegetative mycelium 0·5 g in 30 ml of 95% ethanol; filter off residue (Nobles, 1965).
(b) Tannin (a *cream powder*): see 12 below.

12. *Compounds used with mycelium in culture*
See Kaarik, Studia Forestalia Suecica No. 31, 1965. All solutions at a concentration of 0·1 M in 95% ethanol; ferulic acid; vanilline; eugenol; *p*-quinone; *o*-anisidine; 2,5-dimethylaniline; *o*-toluidene; naphthylamine; 8-oxyquinolene; induline; *p*-cresol; and those appearing in Table I 8(a)–(j) inclusive; 9(a), (c)–(g) inclusive; 10(c); 11(a), (b).

CHAPTER XXII

Immunological Techniques in Mycology

T. F. Preece

Agricultural Botany Division, School of Agricultural Sciences,
The University, Leeds, England

I. INTRODUCTION

Many mycologists regard immunological techniques together with other procedures of protein analysis as being parts of other more difficult disciplines. Whereas medical mycologists have utilized these techniques (Ainsworth, 1952) in the detection and identification of those specialized Fungi Imperfecti which affect man together with Actinomycetes, there is still a sizeable problem, both of communication and experience, for the botanically trained mycologist or plant pathologist. There are, however, now hopeful signs that the barriers between available techniques and botany are breaking down. The Systematics Association recently held a successful symposium on serotaxonomy (Hawkes, 1968) and the proceedings, which contain much technical detail, might well serve as background reading for a mycologist, reminding him of the scope of protein analysis tools, including immunological ones, now available to him. Theoretical concepts have been succinctly reviewed by Quinn (1968) in an attractive modern book. In particular, Quinn gives a glossary of terminology which mycologists with no training or experience in this field will find useful to have within arm's reach. A practical text, giving details of handling animals and the formulae of solutions is already available (Campbell, Garvey, Cremer and Sussdorf, 1963), and a much more detailed research handbook (1245 pages) of experimental immunology with extensive bibliographies has recently been produced, (Weir, 1967). The best account of techniques used with bacteria in medical and veterinary laboratories is the latest edition of Mackie and McCartney's Handbook of Bacteriology (Cruickshank, 1960).

II. THE ANTIBODY–ANTIGEN REACTION IN MYCOLOGY

The basis of immunology is the antibody–antigen reaction (Fig. 1) and for a century this reaction has been used in the identification of micro-organisms. It is highly specific and of high sensitivity and reproducability. It occurs between certain specific components (typically protein) of an organism, called its antigens, with corresponding portions of the gamma-globulin fraction of the serum of animals known as antibodies. These antibodies become detectable in the serum of infected or artificially inoculated animals such as goat, rabbit or human being (Table I). Immunologists have laboured long to discover the site of antibody production in these animals, and the actual nature of antibody producing cells. We now

TABLE I

The antigen–antibody reaction: three examples of the same reaction occurring between a fungus and a rabbit

1. *NATURAL INFECTION* with a fungus pathogenic to the rabbit. The fungus is a mosaic of
   ANTIGENS⟶ANTIBODIES—detectable by *in vitro* tests on the rabbits blood serum.
   (typically proteins)
   Causes synthesis of other specific proteins (typically gamma globulins) which are

2. *ARTIFICIAL INOCULATION* of the rabbit with a living or dead suspension of the fungus which is pathogenic to the rabbit.
   Gamma globulin produced in response may be detected in the rabbits serum. As with (1) this is now an antiserum to the fungal antigenic components.
   (i.e. ANTIGENS)⟶(i.e. ANTIBODIES)

3. *ARTIFICIAL INOCULATION WITH FOREIGN ANTIGENIC MATERIAL* such as plant cells; the mycelium or components of a saprophytic or plant parasitic fungus.
   The foreign antigens likewise produce the same type of response, and the antisera produced are valuable *in vitro* specific reagents for these antigens.
   ANTIGENS⟶ANTIBODIES

N.B.: No such system involving gamma globulin synthesis has yet been found in plants infected by fungi (nor in fungi infected with viruses). The immunity or resistance of plants to infection is, it seems, based on other biophysical and bio-chemical systems, but the subject needs modern investigation.

know that a type of white blood cell, resembling plasma cells and called macrophages are the basis of antibody production. The close link between the advance of knowledge and the utilization of new techniques is illustrated by the fact that the use of radioactively tagged antigens (Haurowitz and Crampton, 1952) and the use of fluorescent-labelled antibodies (Coons and Caplan, 1950) were needed before the cellular location of the antibody–antigen reaction could be revealed.

No system involving gamma-globulin synthesis has so far been found in plants or fungi. The "immunity" or resistance of plants to infection is, it seems, based on other biochemical systems, but the subject urgently needs study. The antigenic components of plants or fungi can be artificially injected into animals and antibodies produced in the animals' serum (antiserum production). Useful reports of the use of the use of plant material as antigens are given by Hawkes (1968) and Vaughan (1968). Antibody–antigen reactions can be examined qualitatively and quantitatively *in vitro* by visible agglutination, precipitation and complement fixation (Cruickshank, 1960) reactions. These reactions occur in quantity only when the antibodies in the antiserum are reacting with the homologous (particular and specific) antigens. The homologous antigens are identical to those used to prepare the antiserum, which is thus itself a reagent for identifying particular antigenic components of organisms. In the case of fungal infections of man (Kligman and Delameter, 1950) or of a rabbit (Table I) fungal antibodies may be demonstrated in the serum or other material from infected subjects by these procedures.

A fungus has a pattern (a particular and specific pattern) of many antigens—both of a polysaccharide and of a proteinaceous nature. The bulk of the data we have about antigen patterns or mosaics concerns bacteria; by comparison we are ignorant of the antigen mosaics of fungal genera, species and strains. Systematic serological research on fungi has not yet begun. A useful summary table of fungal species on which we have immunological data has been provided by Seeliger (1968). Information on *Aspergillus*, *Pullularia* and *Cladosporium* is given there but the picture emerges of work on isolated species or genera of especial importance in industry, human pathology or plant pathology.

III. PREPARATION OF ANTIGENS

As Pepys and Longbottom (1967) point out, various methods of growing the fungus and extracting the antigen can be adopted. In a stimulating review of recent work on medically important fungi they record their own outstanding success with thermophilic Actinomycetes. With *Micropolyspora faeni*, for example, they extracted the liquid containing the soluble antigens from solid culture media by freezing and thawing. More commonly, however, the fungus, isolated in pure culture, is bulked up in a liquid medium. *Aspergillus* spp. and other saprophytes grow well in the medium devised by Schmidt (1960) and Schmidt and Bankole (1965). This avoids the use of protein or polysaccharide material and is prepared as follows: a basal salts medium contains—

K_2HPO_4	1·0 g
$MgSO_4.7H_2O$	0·5 g
$FeSO_4.7H_2O$	0·02 g
$MnSO_4.4H_2O$	0·01 g
Distilled water	800 ml

The carbon source is glucose (2·0 g in 100 ml) and the nitrogen source dicalcium glutamate (4·0 g in 100 ml), making a total volume of one litre. The three solutions are best sterilized separately, the pH varying little from 7·3. For particular fungi, modifications in the constituents of the culture medium, the pH or the incubation temperature may be needed. The most useful general medium may be that devised by Burrell, Clayton, Gallegly and Lilly (1966) for *Phytophthora* species. It yields a good growth of an array of parasites (e.g. *Mycosphaerella*, *Ophiobolus*) and saprophytes (*Penicillium*). It is made up as follows—

D-glucose	10·0 g
$(NH_4)_2SO_4$	2·0 g
Fumaric acid	2·0 g
KH_2PO_4	1·0 g
$MgSO_4.7H_2O$	0·5 g
$Fe^{+++}(FeNH_4(SO_4)_2.12H_2O)$	0·2 mg
$Zn^{++}(ZnSO_4.7H_2O)$	0·2 mg
$Mn^{++}(MnSO_4.H_2O)$	0·1 mg
Thiamine hydrochloride	100 μg
Distilled water	1 litre

The pH of this medium is adjusted to 6·0 and it is then autoclaved.

Venturia inaequalis grows well in 10% malt extract or a semi-synthetic medium devised by Kirkham (1957), but the addition of histidine (Cook, 1969, personal communication) seems beneficial, Gooding (1966) reports success with a glucose–glycine–inorganic salts medium for *Fomes annosus* antigen preparations. It seems reasonable to suggest that any medium or procedure which will give rapid, copious growth is satisfactory. Although the use of media containing more complex substances which may be antigenic seems to be asking for non-specificity problems, the use of "Panmede" (Paines and Byrne Ltd, Greenford, U.K.), a digest of ox liver, produces a more copious growth more rapidly than other media; for example, *Botrytis cinerea* (Preece and Cooper, 1969) which subsequently gave highly specific antisera when inoculated into rabbits. Old or slowly grown mycelium is quite unsatisfactory as antigenic material. The mycelium is filtered off from the culture medium with a Buchner funnel, washed quickly in saline and, if not used, immediately frozen at − 20°C. The dry weight of an aliquot is determined and the material prepared for injection as follows. Either whole mycelium or a saline extract (soluble antigens) of the mycelium may be used. The saline extract is to be preferred, being clear and much easier to use in precipitin tests and agar gel diffusion plates. An eventual concentration of about 1 mg dry weight of mycelium per 3 ml has given satisfactory results. Grinding with a pestle and mortar using aseptic technique, is to be preferred. Many other methods including ultrasonic disintegration may result in an intractable gelatinous mass. Drying the piece of mycelial mat on filter paper before grinding is helpful. If soluble antigens are to be used, the ground up material is centrifuged and the deposit discarded. Before inoculations begin the suspension or extract is heated to 80°C for 10 min and the inoculum tested for contamination by plating on a suitable agar medium.

Since we do not know either the nature or location of fungal antigenic components, there are many possible variations of this procedure and there is a real need for basic large scale studies of a wide range of fungal antigens.

IV. PRODUCTION OF ANTIBODIES

As with antigen production, the described routine for the inoculation of rabbits for antiserum production varies from author to author. New Zealand White rabbits in good condition are usually satisfactory. Due to variation between animals, three rabbits should be used for each fungus. Intravenous injection on four occasions at weekly intervals of 1·0 to 1·5 ml of antigens has been found satisfactory with some fourteen different genera. The rabbit box described by Campbell, Garvey, Cremer and Sussdorf (1963) is useful to hold the rabbits during inoculation. Sterile disposable 5 ml syringes with 23 g needles are used for injecting the material into the external marginal ear vein dilated by massage. By the fifth week, i.e., two weeks after the inoculations commenced, a suitable level of antibodies is usually obtained. Final bleedings may be made by cardiac puncture using a suitable immobilization board (Campbell et al., 1963, loc. cit.) or by bleeding from an ear vein under reduced pressure (Sandiford, 1965). Blood samples are refrigerated overnight and the serum separated in a centrifuge. Contamination with water should be avoided at all stages as it lyses red blood corpuscles and is often contaminated with bacteria. The serum is stored at − 20°C. Various preservatives are available but with care are unnecessary. Adjuvants are much used by bacteriologists, such as Freunds adjuvant, to boost the titre of the serum. Their use sometimes leads to unpleasant local reactions in relatively unskilled hands and comparison of records leads to the conclusion that they are not usually helpful. Perhaps they should be reserved for use if inadequate antibody production is obtained with antigenic extract alone.

Small test bleeds from the ear vein are useful to test the serum for antibody content so that the large volume of serum can be collected at maximum antibody production. A tube agglutination test can be set up as follows. Using $3 \times \frac{1}{2}$ in. test tubes, 1·8 ml of saline (0·85%) is placed in the first tube and 1 ml in subsequent tubes; 0·2 ml of the serum is added to the first tube, mixed and 1 ml transferred to tube two. This serial dilution procedure is continued, thus giving dilutions of 1/20, 1/40, etc. 1 ml of the antigen preparation is added to each tube and the tubes incubated in a water bath at 37°C for 18 h; by careful comparison with control tubes the last dilution in which agglutination occurs is noted as the titre of the antiserum. Preece and Cooper (1969) record a titre of 1/640 with *Botrytis cinerea*. Schmidt and Bankole (1965) record titres of either 1/640 or 1/1280 with an array of fifteen different strains of *Aspergillus flavus* to which they produced antisera. The globulin fraction of the serum may be separated by the use of ammonium sulphate, which seems more foolproof than the ethanol precipitation procedure. The globulin fraction is freed of salt by dialysis, using borate buffered saline at pH 7·8. Dispensing the antiserum or separated globulin into small bijou bottles is desirable since then only a small quantity need be thawed out for use on each occasion. Repeated freezing and thawing causes antisera to lose potency.

In Britain, any mycologist wishing to use rabbits for antiserum production should consider whether he should hold a licence in connection with the Cruelty to Animals Act of 1876. Advice should be sought from the Under Secretary of State, Whitehall, London S.W.1. It would seem highly desirable that anyone raising antisera in animals should apply for such a licence beforehand.

V. THE USE OF ANTISERA TO FUNGI

Agglutination tests using mycelial suspensions of filamentous fungi as antigen are difficult if not impossible to read meaningfully because of the problem of producing standard suspensions of material. Yeasts are, however, amenable to this procedure. Classical *precipitin tests*, titrating the soluble antigen, have been the method of choice for mycologists. Two types of precipitin reaction between antibodies and antigens have been used. Either a ring test where the antiserum is overlayed with antigen extract and a precipitate forms at the interface or more usually mixing the two reactants together in a tube. Classical work with precipitin tests was reported during the period 1930–1940. Good examples are the work of Matsumoto (1929) on *Aspergillus*, or that of Beck (1934) on the genus *Ustilago*. Beck reported that the antisera to monosporidial strains of *Ustilago* were specific in ring tests. The antiserum to *Ustilago zeae* showed no cross reactions at all with other species. Cross reactions did occur between other species, indicating that closely related species and strains and physiological forms of fungi would require more sensitive procedures for their differentiation.

Since 1940, methods for the characterization and classification of protein materials have developed dramatically following the work of Tiselius on moving boundary electrophoresis (Everall and Wright, 1962, personal communication). In particular immunochemical analysis of protein in gels has provided a powerful tool for the microbiologist. These agar *gel diffusion tests* are a logical and ingenious extension of the tube precipitin test. Performing the precipitin reaction in a gel combines the effect of precipitating antibodies with that of the varying rates of diffusion of the antigens in the gel. The interpretation of the results of the test has been elaborated by Ouchterlony (1961) and is also the subject of a useful book by Peetoom (1963), concerned mainly with the medical applications of the technique. A further refinement of the precipitin test is *immunoelectrophoresis*. As distinct from simple double diffusion this procedure identifies a protein in two ways simultaneously; by its electrophoretic (and free) mobility and by its antigenic groupings. Again the technique is the subject of a book giving the minutiae of the procedures (Crowle, 1961). Yet another refinement of the precipitin test is *inert particle agglutination*. Inert carriers such as bentonite or latex are used to adsorb soluble antibody, and provide a very sensitive means of detecting antibody–antigen reactions. A useful account of the technical work using this procedure with an Actinomycete, *Thermopolyspora polyspora* is given in Proctor (1967).

Antisera or purified globulins can also be used in direct microscopic studies (see Preece, this Volume, p. 509) as in the fluorescent antibody technique. Ferritin-labelling and radioisotope-labelling has been used to reveal the site of specific antibody reactions in electron micrographs. Such studies on *Candida albicans* were described by Metzger and Smith (1962).

VI. AGAR GEL DIFFUSION TESTS WITH FUNGAL ANTIGENS

Good illustrations of typical patterns of the precipitates obtained in agar plates with extracts from plants are given by Vaughan (1968) and for

606 T. F. PREECE

Fomes annosus by Gooding (1966). The optimum proportions of antibody are important for the success of the test, but once these have been determined by trial, antigens from many different fungal sources can be easily compared and the patterns of precipitin line interaction utilized in determining relationships among the antigens of fungi. Of the many variants of this test compared, the best results are obtained using the procedure of Gooding (1966). The medium, used in scratch free Petri dishes consists of—

Ionagar (Oxoid)	8·0 g
Sodium Azide (NaN₃)	10·0 g
Distilled water	1 litre

15 ml of agar are used per plate. One centre well and six outer wells are cut out with a 8 mm dia. cork borer. Diluted antigen may be used in the outer wells and undiluted serum in the inner well, and vice versa. Suitable first dilutions to be tested could be 1/2, 1/4, 1/8, etc. 0·1 ml of the material is used in each well. Incubation at room temperature for 6 days followed by placing in the refrigerator at 4°C overnight has given satisfactory precipitin lines with a range of fungi (Preece, 1966) being tested prior to conjugation with fluorescent dyes. Typical of the usefulness of the test are the results of Morton and Dukes (1967) in which they demonstrated the clear differentiation of *Pythium aphanidermatum* from *Phytophthora parasitica* var. *nicotiana* and *Ph. parasitica*. Staining with mercuric bromphenol blue reveals the precipitin lines in critical detail against minimal background staining (LaVelle and Van Alten, 1965).

REFERENCES

Ainsworth, G. C. (1952). "Medical Mycology". Pitman, London.
Beck, E. C. (1934). *Can. J. Res.*, **10**, 234–238.
Burrell, R. G., Clayton, C. W., Gallegly, M. E. and Lilly, V. G. (1966). *Phytopathology*, **56**, 422–426.
Campbell, D. H., Garvey, J. S., Cremer, N. E., and Sussdorf, D. H. (1963). "Methods in Immunology". W. A. Benjamin Inc., New York.
Coons, A. H., and Caplan, M. H. (1950). *J. exp. Med.*, **91**, 1–13.
Crowle, A. J. (1961). "Immunodiffusion". Academic Press, London.
Cruickshank, R. (1960). "Mackie and McCartney's Handbook of Bacteriology", 10th ed. Livingstone, Edinburgh and London.
Gooding, G. V. (1966). *Phytopathology*, **56**, 1310–1311.
Haurowitz, F., and Crampton, C. F. (1952). *J. Immun.*, **68**, 73–85.
Hawkes, J. G., Ed. (1968). "Chemotaxonomy and Serotaxonomy". Academic Press, London and New York.
Kirkham, D. S. (1957). *J. gen. Microbiol.*, **16**, 360–373.
Kligman, A. M., and Delameter, E. D. (1950). *A. Rev. Microbiol*, **4**, 283–312.
La Velle, A., and Van Alten, P. J. (1965). *Stain. Technol.*, **40**, 347–349.
Matsumoto, T. (1929). *Trans. Br. mycol. Soc.*, **14**, 69–88.
Metzger, J. F., and Smith, C. W. (1962). *Lab. Invest.*, **11**, 902.

Morton, D. J., and Dukes, P. D. (1967). *Nature, Lond.*, **213**, 923.

Ouchterlony, O. (1961). *In* "Immunochemical Approaches to Problems in Mircobiology" (Eds. M. Heidelberger, O. J. Plescia and R. A. Day.), pp. 5–9. Institute of Microbiology, State University, Rutgers, New Brunswick, New Jersey, U.S.A.

Peetoom, F. (1963). "The Agar Precipitation Technique and its Application as a Diagnostic and Analytical Method". Oliver and Boyd, Edinburgh.

Pepys, J., and Longbottom, Joan L. (1967). *In* "Handbook of Experimental Immunology" (Ed. D. M. Weir), pp. 813–843. Blackwell Scientific Publications, Oxford.

Preece, T. F. (1966). *J. gen. Microbiol.*, **42**, v (Proceedings).

Preece, T. F., and Cooper, Dorothy J. (1969). *Trans. Br. mycol. Soc.*, **52**, 99–104.

Proctor, A. G. (1967). *In* "Progress in Microbiological Techniques" (Ed. C. H. Collins), pp. 213–226. Butterworths, London.

Quinn, L. Y. (1968). "Immunological Concepts". Iowa State University Press, Iowa, U.S.A.

Sandiford, Marjorie (1965). *J. Anim. Tech. Assoc.*, **16**, 9–14.

Seeliger, H. P. R. (1968). *In* "The Fungi, an Advanced Treatise' (Eds. G. C. Ainsworth and A. S. Sussman), Vol. III, pp. 597–624, "The Fungal Population". Academic Press, New York and London.

Schmidt, E. L. (1960). *J. Bact.*, **79**, 553–557.

Schmidt, E. L. and Bankole, R. O. (1965). *Appl. Microbiol.*, **13**, 673–679.

Vaughan, J. G. (1968). *Scient. Prog., Oxf.*, **56**, 205–222.

Weir, D. M., Ed. (1967). "Handbook of Experimental Immunology". Blackwell Scientific Publications, Oxford.

CHAPTER XXIII

A Practical Guide to the Effects of Visible and Ultraviolet Light on Fungi††‡

CHARLES M. LEACH

*Department of Botany and Plant Pathology, Oregon State University,
Corvallis, Oregon*

I. INTRODUCTION

Visible and near-visible light (200–750 nanometers) profoundly influence many aspects of the growth, development, reproduction, and behaviour of fungi. In nature, most fungi are exposed during a portion of their life cycle to a little far-ultraviolet radiation (mainly 290–300 nm), and to

† Approved for publication as Technical Paper No. 2540 of the Oregon Agricultural Experiment Station.

‡ This manuscript was read by Drs. F. H. Smith and E. J. Trione (Oregon State University) and Dr. M. Aragaki (University of Hawaii). The writer appreciates their helpful comments.

considerable near-ultraviolet (300–380 nm) and visible (380–750 nm) radiation. Historically this has not always been true for there is evidence that prior to the Silurian era the terrestial surface was also exposed to mutagenic and lethal wavelengths of ultraviolet radiation less than 290 nm (Berkner and Marshall, 1964). Indeed, much of the diversity of form and physiology of the fungi may be rooted in a reservoir of mutagenic changes originating during this early era.

Many effects of light on fungi have been reported by numerous investigators during the past century (Smith, 1936) yet with few exceptions the basic mechanisms of these photobiological phenomena are still poorly understood. The elucidation of these phenomena offers a great challenge to all sorts of mycologists whether they be taxonomists, ecologists, environmental physiologists, molecular biologists, geneticists or plant pathologists. Because of the relatively simple nature of fungi, they usually can be manipulated readily in the laboratory under exacting conditions and they can thereby serve as unique subjects in photobiological experiments. These phenomena, however, are not of interest only to photobiologists, for a better understanding of them is of considerable importance in many aspects of basic and applied biology.

This article has been written primarily as a general practical guide for those interested in the effects of light on fungi; particularly those unaware of the many ramifications of this subject. It has not been written for the sophisticated photobiologist, nor for those mainly interested in the photochemical and theoretical aspects of the subject; nor does it pretend to be a comprehensive review. The effects of ultraviolet (UV) and visible light on fungi are so numerous and the experimental approaches are so varied that it has not been possible to include in detail all the experimental procedures nor all the pitfalls that can be encountered. Fortunately, there are excellent review articles on certain aspects of this subject and these are cited. The author's own research has been on photo-induction of reproduction and on the application of the "light factor" to the detection and identification of fungi; coverage of these subjects have been somewhat expanded.

Throughout this article far-ultraviolet radiation (far-UV) refers to the wavelengths 200–300 nm, near-ultraviolet radiation (near UV) refers to wavelengths 300–380 nm, and visible light to the wavelengths 380–750 nm (Figure 4). These divisions are used by many biologists but they are rather arbitrary and generally not used by physicists.

II. GENERAL EFFECTS OF LIGHT ON FUNGI

The effects of light (UV and visible) on the fungi are quite diverse and the literature concerned with these phenomena is immense. At present there is no single review that discusses all facets of the photobiology of the fungi, yet there are some excellent general reviews that cover many of the major phenomena. The early literature is discussed by Smith (1936); and a useful guide to the extensive literature on reproduction, morphology, pigmentation, and phototropic phenomena, is provided by Marsh et al. (1959). The most recent general reviews are those by Carlile (1965) and Page (1965). The works of Pomper and Atwood (1955), Hawker (1957), Mohr (1961), Cochrane (1958), Lilly and Barnett (1951), Ingold (1962), and Beck and Roschenthaler (1960) should not be overlooked.

Each light-induced phenomenon will be described briefly in the ensuing paragraphs with emphasis on the responsible wavelengths where these are known. The reader should not be misled by the brevity of discussion for it is essential in many instances, that the cited literature be read to place these subjects in better perspective.

A. Direct effects

Pomper (1965) has classified ultraviolet effects on fungi as "genetic" and "non-genetic"; Page (1965) uses the terms "morphogenetic" and "non-morphogenetic". The categorization that I shall follow is perhaps more traditional, rather arbitrary in some instances and subject to some overlapping.

1. Mutagenic effects

The mutagenic effects of far-ultraviolet radiation on fungi have long been known (Smith, 1936). The literature prior to 1955 is reviewed by Pomper and Atwood (1955) and more recently by Pomper (1965). The action spectrum for many mutational changes resembles the absorption spectrum of deoxyribonucleic acid (DNA) with 260 nm radiation most effective. In Chaetomium globosum, however, 280 nm radiation is most effective (McAuley et al., 1949). New evidence indicates that visible light may be able to induce mutations in some micro-organisms (Webb and Malina, 1967; Leff and Krinksy, 1967).

McLaren and Shuggar (1964) discuss the theoretical and practical aspects of the rapidly expanding subject of the photochemistry of proteins and nucleic acids. Particularly significant has been the discovery that far-UV will cause the formation of thymine and other pyrimidine dimers in DNA, and that longer wavelengths of UV will break these bonds (Deering, 1962; Setlow, 1966).

Low-pressure mercury vapour lamps or "germicidal lamps" (Figure 4) have been most frequently used in mutational studies with fungi because they are both inexpensive and emit most of their energy at 253 nm which is highly efficient for inducing mutations. For monochromatic light studies and action spectrum studies, more compact and higher intensity sources of far-UV are needed such as the high pressure xenon and xenon-mercury lamps. Zetterberg (1964) discusses action spectra, dose-effect relationships, influence of other factors, as well as other important considerations in mutational studies. Fincham and Day (1963) refer to the application of far-UV to genetic studies in fungi. The article by Buttolph (1955) discusses many practical considerations useful in mutational studies and this may be supplemented by McLaren and Shuggar's "appendix" (1964) devoted to the techniques of radiation studies.

2. *Lethal effects*

The lethal effects of far-UV on fungi were first recognized during the last century (Smith, 1936). Sussman and Halvorson (1966) discuss the two principal theories proposed to explain the mechanism of killing, the "target theory" and the "indirect action theory". The target theory, they state, postulates that energy from irradiation is absorbed primarily within, or contiguous to, the key molecules of the cell; the indirect action theory assumes that the energy is absorbed by molecules inside, or near the cell, followed by production of lethal substances that affect vital elements of the cell (refer to section "Indirect effects").

The action spectra shown in Figure 1 for killing spores of *Trichophyton mentagrophytes* (Hollaender and Emmons, 1938) and cells of yeast (Pomper, 1965), are similar to the absorption spectrum for DNA with 260 nm most effective (see also the review by James and Werner, 1965, on the "Radiobiology of yeast"). The lethal effects of UV are generally attributed to wavelengths 200–300 nm (Figure 1), yet, there is some evidence that near-UV and visible light can also be lethal, at least for the bacterium *Escherichia coli* (Hollaender, 1943). Whether this might also be true for the fungi, has still to be determined.

The effect of far-UV on the survival of fungi is dependent on many factors (Sussman and Halvorson, 1966). Spores for example, are much more resistant to the lethal effects of UV than vegetative mycelium, sensitivity of spores is frequently related to colour and pigmented spores often survive longer exposures to far-UV than colourless spores; in addition thickness of spore wall may also be involved in the ability of fungi to tolerate radiation. Age can also influence sensitivity of spores and old spores of *Aspergillus melleus* are more tolerant to far-UV than young spores (Dimond and Duggar, 1941). In nature there is evidence that solar UV is an important factor in the survival of air-borne spores, particularly in long distance dispersal (Gregory, 1961).

The lethal effects of far-UV have many applications in science, industry and medicine (Rubbo and Gardner, 1965; Sykes, 1965; Koller, 1965).

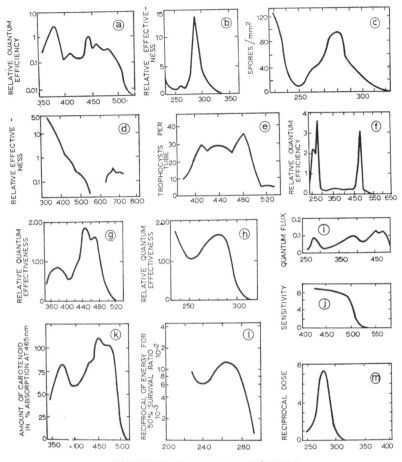

WAVELENGTH IN NANOMETERS

FIG. 1. Photo-responses of fungi: (a) sporulation in *Trichoderma viride* (Gressel and Hartmann, 1967); (b) sporogenesis in *Ascochyta pisi* (Leach, 1962b); (c) sporogenesis in *Stemphylium solani* (Herndon, 1967); (d) sporogenesis in *Physarum nudum* (Rakoczy, 1965); (e) induction of trophocyst formation in *Pilobolus kleinii* (Page, 1956); (f) inhibition of conidium formation in *Stemphylium botryosum* (Leach, 1968); (g) positive phototropic response of conidiophores of *Phycomyces blakesleeanus* (Curry and Gruen, 1959); (h) negative phototropism of conidiophores of *Phycomyces* (Curry and Gruen, 1959); (i) growth response of *Phycomyces* conidiophores (Delbruck and Shropshire, 1960); (j) spore discharge in *Sordaria fimicola* (Ingold, 1958); (k) carotenoid formation in *Fusarium aquaeductuum* (Rau, 1967); (l) killing spores (50% inactivation) of *Trichophyton mentagrophytes* (Hollaender and Emmons, 1938); and (m) mutation (50%) in *Chaetomium globosum* (McAulay et al., 1949).

The low-pressure mercury vapour lamp or "germicidal" lamp (Figure 4) is an inexpensive, efficient, and widely used source of fungicidal UV; though there are also many other sources that can be utilized (Buttolph, 1955).

3. *Photoreactivation*

Closely related to the topics of mutagenesis and the lethal effects of ultraviolet radiation is the phenomenon of photoreactivation (PR) which was discovered by Kelner in 1948 while studying the effects of far UV on *Streptomyces griseus* spores (Kelner, 1949, 1961). The phenomenon has been defined as one in which changes produced in different types of organisms by ultraviolet radiation of wavelength around 250 nm, can be counteracted if the irradiated organisms under proper conditions, are exposed to radiation of longer wavelengths in the range between 330 and 480 nm (Dulbecco, 1955). PR is not restricted to fungi and numerous photobiological phenomena in diverse organisms can be photoreactivated (Jagger, 1960). PR phenomena exhibited by fungi are discussed briefly by Pomper (1965). There are also a number of general reviews of this subject by Dulbecco (1955), Jagger (1958, 1960), Christensen and Buchman, (1961), Rupert (1964) and Setlow (1966). The theoretical and photo-chemical aspects of PR are discussed by McLaren and Shuggar (1964), Smith (1964) and Setlow (1966). There is also a more popular account by Deering (1962).

Action spectra studies indicate that most of the damaging effects of UV radiation result principally from primary absorption by nucleoproteins and, in many instances, the nucleic acid component alone. Most PR phenomena include reproduction, mutation or transformation, all of which involve the nucleus and consequently DNA protein (McLaren and Shuggar, 1964). The range of wavelengths effective in promoting PR embraces the interval 310–500 nm, but action spectra obtained are in no instance sufficiently characteristic to permit positive identification of the photo-receptor and from the accumulated data it would seem that PR receptors in different organisms are not the same. The important studies of far UV induced dimerization of thymine in DNA, as well as other pyrimidines, and the reversal of this by subsequent exposure to near-ultraviolet radiation and blue light provides new insight to the PR phenomenon. However, according to McLaren and Shuggar (1964), there are at least two authen-ticated examples of cytoplasmic photoreactivation. A photoreactivating enzyme has been obtained from *Neurospora crassa* (Terry and Setlow, 1967).

When spores of certain plant pathogenic fungi are irradiated with far-UV and then inoculated on to susceptible host plants, the number of lesions formed on leaves depends on whether the post inoculated plants are reared in light or darkness (Last and Buxton, 1955; Buxton *et al.*, 1957). In light there is photoreactivation of the damaged spores, and consequently much heavier infection occurs in light than in darkness.

In far-UV irradiation studies on fungi, the possibility of photoreactivation can be reduced by employing appropriate filters to eliminate wavelengths longer than 300 nm. McLaren and Shuggar (1964), however, suggest that PR may operate at even lower wavelengths, though this has not been demonstrated experimentally. Reversal can also be achieved by means other than light (Rupert, 1964) and this also must be considered in any far-UV irradiation experiments.

In experiments designed to induce photoreversal, radiation sources should emit strongly in the 300–500 nm range. Fluorescent "Black Light" lamps (320–420 nm) are inexpensive sources of PR effective wavelengths (Figure 4). High pressure xenon lamps are also excellent sources of PR effective wavelengths suitable for monochromatic studies (Figure 4).

4. *Growth*

The effects of light on the non-morphogenetic aspects of fungal growth are reviewed by Carlile (1965) and Page (1965). A non-morphogenetic response to light is used by Page in the sense that light influences the rate and direction of movement or growth of a structure. He classifies these responses on the basis of responding structure and whether the response is oriented or non-oriented.

(a) *Non-oriented effects of light on growth.* The actual effects of light on vegetative growth of fungi are difficult to ascertain from the confusing literature on this subject. Far too often, investigators fail to appreciate some of the pitfalls of measuring growth as discussed by Mandels (1965). Too frequently light quality, intensity, and dose response relationships are completely ignored, as are other interacting factors that affect growth. Most generally ignored is the possibility that light affects the substrate and thereby indirectly affecting growth (see under "Indirect Effects"). Dr. W. Brandt (1968) has clearly shown that the affect of near-UV on dry weight growth of *Verticillium albo-atrum* is dependent on experimental conditions, thus in shake cultures, growth is stimulated by near-UV during the log phase of growth, but, in still cultures, growth of irradiated colonies is actually less than in the dark. Brandt has also demonstrated that a single reading at the end of this type of experiment can be quite misleading. Light has no affect on the linear growth of *Pellicularia filamentosa* on an agar medium, yet in liquid cultures a significant reduction of growth (dry wt.) occurs under illumination (Durbin, 1959). Lineal growth of *Botrytis squamosa* decreases on an agar medium with increase of light intensity (Page, 1956), though Stinson *et al.* (1958) dispute these results and attribute them to temperature effects. In light studies on the growth of *Karlingea rosea* (Haskins and Weston, 1950), the carbon source significantly affects the results; thus, glucose increases growth (dry wt.) in cultures reared under illumination, while the reverse is true for cellobiose. Lingappa *et al.* (1963) report increased growth (dry wt.) when polysaccharides are used as carbon sources in illuminated cultures but not when sugars are used. There is a marked increase in growth (dry wt.) of *Blastocladiella* under visible light illumination which Cantino (1966) attributes to increased CO_2 fixation.

Under conditions of diurnal alternating light and darkness at constant temperatures, pronounced zonation frequently occurs in colonies grown on agar media. Zonation may be due to stimulation or retardation of growth caused by light, as well as to light effects on pigmentation, production of fruiting structures, and sclerotia. Jerebzoff (1965) and Carlile (1965) review this literature and discuss the action of light on endogenous and exogenous growth rhythms in fungi.

Precise qualitative and quantitative studies on the effects of light on vegetative growth have not been published, however, there is considerable circumstantial evidence indicating that wavelengths affecting growth are confined to the blue and ultraviolet regions of the spectrum. Eliminate these wavelengths and colonies appear to grow similarly to those reared in darkness.

Light also affects the non-tropic growth rate of sporangiophores of some fungi. The growth rate of the sporangiophores of *Thamnidium elegans* is negatively affected by light (Lythgoe, 1961); in contrast, the rate of conidiophore elongation of *Aspergillus giganteus* is increased by light (Gardner, 1955). Wavelengths from 300–530 nm are stimulatory for *A. giganteus*. The non-oriented growth response of sporangiophores of *Phycomyces* has been studied quite intensively (Page, 1965). The action spectrum for light stimulated growth of *Phycomyes* sporangiophores (Figure 1) shows peaks at 280, 385, 455 and 485nm; which is similar to the action spectrum for phototropism of this same fungus shown also in Figure 1 (Delbrück and Shropshire, 1960).

(b) *Oriented effects of light on growth.* Phototropism is one example of an oriented light response that occurs widely in living organisms (Withrow, 1959), including the fungi. Phototropism in the fungi, is reviewed by Carlile (1965) and Page (1965, 1968) in addition to reviews by Reinert (1959) and Banbury (1959). Both unicellular and multicellular reproductive structures frequently respond to unilateral illumination. Affected positively are coremia (Carlile *et al.*, 1961, 1962), perithecial necks (Backus, 1937;

Ingold and Hadland, 1959; Ingold, 1965), the stripes of "Agarics" (Buller, 1931, Plunkett, 1961), conidiophores (Calpouzos and Stallknecht, 1966) sporangiophores (Shropshire, 1963; Lythgoe, 1961) and asci (Buller, 1934; Ingold, 1965).

The phototropic response of sporangiophores of *Phycomyces* (Mucorales) has been particularly amenable to experimental studies and it has been studied extensively over the past half century. These studies and their theoretical implications are reviewed by Carlile (1965), Delbrück (1962) Shropshire (1963), Reichardt (1961), and Thimann and Gruen (1960). More recent literature on phototropism in *Phycomyces* is that by Forssberg *et al.* (1964), Shropshire and Gettens (1966), and Castle (1966a, 1966b). The sporangiophore of *Pilobolus* is also an excellent structure for phototropic studies and it too has received considerable attention (Carlile, 1965).

Action spectra for the positive phototropic response of *Phycomyces* conidiophores determined by Castle (1931), Bünning (1937), Curry and Gruen (1959), and Delbrück and Shropshire (1960), are in general agreement showing maximum effectiveness in the blue region of the spectrum (Figure 1). A negative response occurs when conidiophores are exposed to UV as is shown in Figure 1 (Delbrück and Shropshire, 1960). In *Phycomyces*, both the phototropic response and the light affected growth response of conidiophores appears to be mediated by the same photoreceptor.

The wavelengths of radiation most effective in initiating other types of phototropic response, such as bending of the perithecial necks of *Sordaria* (Ingold and Hadland, 1959; Ingold, 1966), the stipes of *Coprinus* (Borris, 1934) and the sporangiophores of *Pilobolus* (Page, 1962; Bünning, 1937), are fairly similar to those reported for *Phycomyces* conidiophores.

Most of the phototropic responses in fungi are positive and confined to fruiting structures, but there are a few reports of negative phototropic responses of germ tubes of *Puccinia* and *Botrytis* (Carlile, 1965, 1966; Page, 1965). Blue light is effective in inducing a negative response in *Puccinia*.

Phototaxis is another oriented response of fungi to light. Clayton (1964) in a general review defines phototaxis as a motor response elicited by light, that is caused by a temporal change of light intensity or by a non-uniform field of illumination. Though phototactic responses in the fungi appear to be rare, Kusano (1930) reports a positive light response of gametes of *Synchytrium fulgens* which is optimal at 20°C, but inoperative at 27°C. The pseudoplasmodium of the slime mould *Dictyostelium* migrates towards a 1 watt neon lamp (Bonner and Shaw, 1957). Zoospores of *Phytophthora cambivora* also respond positively to unilateral light (Petri, 1925). In none of these investigations has any attempt been made to critically analyse the light response.

It seems improbable that species of *Synchytrium*, *Dictyostelium*, and *Phytophthora* are the only fungi that exhibit phototactic responses and a critical examination of the behaviour of motile gametes and zoospores of the aquatic Phycomycetes will surely reveal this phenomenon to be more widespread.

5. *Spore germination*

Visible light triggers spore germination in some fungi, in others it can be inhibitory. In the early literature (Smith, 1936) the lethal effects caused by far-UV and the quite different categories of germination effects caused by visible radiation are not distinguished. The most recent comprehensive discussion of this subject is that by Sussman and Halvorson (1966).

The effects of light on spore germination have been studied mostly with plant pathogens, particularly certain genera of Basidiomycetes (Sussman and Halvorson, 1966). Staples and Wynn (1965) in a review of the physiology of uredospores, cite a number of examples of visible light inhibition of spore germination which occurs within the first few hours of germination, but not as the spores become older. This inhibition of uredospores is dependent on light intensity, is temperature sensitive, and in some instances is reversed by a subsequent period of darkness. *Puccinia graminis* var. *tritici* (Dillon Weston, 1932; Sharp *et al.*, 1958; Givan and Bromfield, 1964a, b) and *Phragmidium mucronatum* (Cochrane, 1945) exhibit this type of behaviour. Exposure of dry spores to light usually has no effect on their subsequent germination and they must first be hydrated. Zadoks (1967) and Zadoks *et al.* (1967) report on light inhibition of germination of uredospores of *Puccinia recondita* f. sp. *triticina*. Light and temperature are also the dominant factors in the development of infection structures of *Puccinia tritici* immediately following germination (Emge, 1958). Visible light also inhibits germination of spores of *Uromyces trifolii* both on detached leaves and on water agar (Thrower, 1964).

Light stimulates germination of spores of certain smuts and bunts, particularly species of *Tilletia* (Sussman and Halvorson, 1966). Spore germination is also stimulated in several Phycomycetes. Conidia of *Peronospora manshurica* stored under illumination in glycine show higher percentage germination than those stored in darkness. This is reversed by returning illuminated spores to darkness (Pedersen, 1964). In *Physoderma maydis*, light of very low intensity (1–12 ft candle) causes a marked increase in germination of sporangia (Hebert and Kelman, 1958). Light inhibits formation of oospores in a number of *Phytophthora* species but stimulates their germination (Romero and Gallegly, 1963; Leal and Gomez-Miranda, 1965; Berg and Gallegly, 1966).

Little detailed information is available on the precise qualitative and quantitative relations of light to germination. In most instances of stimulation, blue light has been effective (Sussman and Halvorson, 1966). In *Oidium moniliodes* (Sempio and Castori, 1950) 400–450 nm wavelengths stimulate germination, while 550–750 nm wavelengths inhibit. Germination of oospores of *Phytophthora cactorum* (Berg and Gallegly, 1966) is stimulated mainly by blue (400–480 nm) and far-red (700–1000 nm) wavelengths and least by green (500–590 nm) and near-red (600–690 nm). In *P. infestans*, blue-green wavelengths are most effective in stimulating oospore germination and least effective are red wavelengths (Romero and Gallegly, 1963). In *Physoderma maydis* germination of sporangia is stimulated most by blue light (49%) and least by yellow and red (Hebert and Kelman, 1958).

Spores, like bacteria, are amenable to monochromatic irradiation studies and there would seem to be no major obstacles to prevent the determination of accurate action spectra both for stimulation and inhibition of germination. In designing such experiments one must be aware that the photoresponse can in some instances be dependent on temperature (Gassner and Niemann, 1954), age of spores, stage of germination at which the spores are exposed, etc. These and many other facets of spore germination are well discussed by Sussman and Halvorson (1966).

6. *Spore liberation*

Spore discharge is light-dependent in a number of fungi (Ingold, 1965, 1966). The Pyrenomycete *Sordaria fimicola*, for example, shows a distinct positive light discharge reaction which is stimulated by brief periods of illumination; in contrast, spore discharge in *Hypoxylon fuscum* is inhibited by light (Ingold, 1962). Walkey and Harvey's studies (1966, 1967) on 24 species of Pyrenomycetes subjected to a diurnal cycle of light and darkness, reveal that some species discharge their spores only during periods of light while others discharge mainly in darkness.

Very little is known about the precise light requirements of the mechanism of photoinduced spore discharge. A rough action spectrum for *Sordaria fimicola* (Ingold, 1958) indicates that the wavelengths most effective in stimulating spore discharge are from 430–475 nm (Figure 1). Sporangium discharge of *Pilobolus* is also responsive to wavelengths in this same general region (Page, 1964). The apparatus described by Ingold *et al.* (1963), and Walkey and Harvey (1966) should be considered by those planning spore discharge studies.

7. *Cytological and biochemical effects*

The direct effects of light on fungi will of course always be accompanied by a multitude of cytological and biochemical changes. Undoubtedly some of the changes included in this section, are merely manifestations of the microscopic and biochemical aspects of the "larger" light-induced phenomena. As a true understanding of light-induced changes in fungi can only be achieved by the eventual exploration of these phenomena at the cytological and biochemical levels, the reader may wish to refer to the articles by Loofbourow (1948), Pollister (1955), Errera (1953) and Giese (1964) which discuss light effects at the cellular level.

(a) *Translocation, protoplasmic streaming and viscosity*. These cellular phenomena are commonly influenced by light in living organisms though reports for the fungi are rather few (Stålfelt, 1956). Three decades ago Buller (1933) discussed the effects of light on protoplasmic streaming in fungi; more recently Wilcoxson and Sudia (1968) have reviewed the few reports that exist on the effects of light on translocation in fungi. It is obvious from this review that the phenomenon is poorly understood. Dowding and Buller (1940) have observed that nuclei move from darkened regions of colonies to illuminated portions. Wilcoxon and Subbarayudu (1968) discovered that the translocation of p^{32} to sclerotia, and its accumulation in sclerotia of *Sclerotium rolfsii*, is significantly greater in light than in darkness. Plunkett (1958) has reported on the affect of light on translocation as it relates to pileus formation in *Polyporus brumalis*.

(b) *Permeability effects*. It is not uncommon for the permeability of cells of many organisms to be influenced by light (Stålfelt, 1956). Reports on permeability effects in the fungi are largely confined to far-UV studies on yeasts. Yeast cells become more permeable to p^{32} when exposed to far-UV (Hevesy and Zeradv, 1946; Swenson, 1960). The extensive studies of Loofbourow and others (Loofbourow, 1948; Webb and Malina, 1967) indicate that a variety of substances such as nucleosides, nucleotides, amino-acids, and vitamins are able to escape much more rapidly from far-UV irradiated cells than from non-irradiated cells. Swenson and Dott (1961) have summarized the literature on amino-acid leakage from UV irradiated yeast cells.

The relatively small size of most fungal hyphae, has undoubtedly been a major hurdle to the study of light effects on permeability, translocation, protoplasmic streaming, and viscosity. However, the development of new biochemical methods, particularly radioactive tracers, the advances in light and electron microscopy, and the development of microbeam techniques, offer many new "tools" with which to probe these problems.

(c) *Luminescence*. Bioluminescence occurs mainly in the Basidiomycetes (reviews by Wassink, 1948; Harvey, 1952; Chase, 1964). The early studies of Berliner (1961) on the diurnal periodicity of luminescence of three

Basidiomycetes suggested that this phenomenon is independent of daylight or darkness, however her later monochromatic light studies (Berliner and Brand, 1962) showed that both near and far-UV cause an initial depression of light emission with later recovery. Berliner postulated that a photo-labile compound necessary for light emission absorbs selectively at 288 nm, while an inhibitor of bioluminescence absorbs at 245–265 nm and 366 nm. Damage of the inhibitor leads to an eventual increase of emission over the steady state. Airth and Foerster (1960) also report that exposure of colonies to 360 nm wavelengths causes an initial decline of luminescence with eventual recovery.

(d) *Miscellaneous biochemical effects.* Ultraviolet and visible light induce a number of biochemical changes in fungi that do not readily fit into the categories of effects so far discussed. Extensive biochemical studies have been conducted by Cantino and his associates on the effects of light on the water mould *Blastocladiella emersonii* (Carlile, 1965, p. 175; Cantino, 1966, pp. 317–318). Carbon dioxide fixation is increased in colonies growing under illumination (Cantino and Horenstein, 1956). In the phenomenon of far-UV inhibition of galactozymase production by yeast, an action spectrum indicates that nucleic acids are the cellular constituent effected (Swenson, 1950). Ultraviolet radiation also influences respiration and fermentation of yeast (Loofbourow, 1948). Endogenous respiration in yeast is increased by far-UV radiation but exogenous respiration is depressed (Giese, 1942). The liberation of amino-acids nucleotides, nucleosides, vitamins, etc., from far-UV irradiated yeast, have already been mentioned under "Permeability" effects. Mertz and Henson (1967) report that light also stimulates the biosynthesis of gibberellins in *Fusarium moniliforme* (Mertz and Henson, 1967).

(e) *Pigmentation.* (Refer to Section III, "Light and Identification of Fungi".)

8. *Morphogenetic effects*

The principal morphogenetic responses of fungi to light are those of reproduction, sclerotial formation, and various gross morphological changes including spore and sporocarp morphology and zonation. These are considered separately in Section III, "Light and Identification of Fungi".

B. Indirect effects

The light effects discussed to this point have been direct effects caused by photochemical changes within the fungal cell. Not to be overlooked, however, are the indirect effects of ultraviolet and visible light that can be of considerable importance to the interpretation of results in certain irradiation studies. The changes in the extracellular milieu that have received most attention are the formation of hydrogen peroxide, ozone, and the possible decomposition of the organic components of media (Giese, 1964, pp. 203–208; Errera, 1953, pp. 91–99; and pp. 111–114; Loofbourow, 1948, pp. 110–113). The susceptibility of plants to fungal pathogens is in some instances dependent on lighting conditions. Change in susceptibility appears to be an effect of light on the plant cell rather than on the fungus.

1. *Ozone formation*

The possibility that the formation of ozone in irradiated media causes some of the biological effects of far-UV has been considered by many investigators. Ozone is formed primarily by the absorption by oxygen of UV shorter than 185 nm, though a little occurs at longer wavelengths. It is actually decomposed by wavelengths longer than 185 nm. In Giese's opinion (1964) one would not expect much ozone formation in most experiments conducted at wavelengths longer than 200 nm, but, to reduce the possibility in far-UV experiments, he recommends that a quartz dish containing an acetic acid buffered sodium chloride solution be used as a filter for the shorter wavelengths of UV.

The characteristic odour that one encounters when using lamps emitting far-UV, particularly the high intensity sources, indicates the presence of gaseous ozone. Ozone forms principally in the immediate vicinity of the lamp, or open arc, and it may diffuse into the micro-environment of the irradiated fungus where it can be toxic. Ozone at 1·0 p.p.m. completely prevents sporulation of *Penicillium italicum* and *P. digitatum* (Harding, 1968) and at 0·1 p.p.m. for 4 h, and 1·0 p.p.m. for 2 h, it stops the elongation of conidiophores of *Alternaria solani* (Rich and Tomlinson, 1968).

Adequate air circulation and the use of exhaust systems ducted to the outside atmosphere will reduce or eliminate the possibility of ozone effects (refer also to Section IV, "Health Hazards").

2. *Hydrogen peroxide and related substances*

Many workers have suspected that hydrogen peroxide (H_2O_2) is formed in media by far-UV radiation and that it may be responsible for some of the biological changes attributed to the direct effects of radiation. Giese (1964, pp. 204–205) discusses this possibility and states that H_2O_2 is formed from water only by exposure to wavelengths shorter than 185 nm, wavelengths that are seldom used on biological material. In Loofbourow's opinion (1948) one would not expect to find photo-biological effects attributable to the formation of H_2O_2 in aqueous media. Errera (1953, pp. 91–93, 111–114) agrees that water does not significantly absorb UV above 185 nm but points out that H_2O_2, and other peroxides, can occur at longer wavelengths in the presence of photosensitizing molecules. His summary of numerous reports supports this conclusion. Even under visible light these reactions can occur (e.g., ergosterol exposed to visible light can be converted to ergosterol peroxide using eosine as a photosensitizer).

Growth of *Rhizoctonia solani, Armillaria mellea,* and *Phytophthora capsici* is retarded when these fungi are grown on PDA or V-8 juice agar previously exposed to daylight fluorescent lamps (Weinhold and Hendrix, 1963). There is some evidence that this inhibition is due to peroxides. Growth of *Fusarium solani* may also be inhibited when grown on media exposed to daylight fluorescent lamps (Tousson and Weinhold, 1967). Wyss and his co-workers (Wyss *et al.*, 1948) report that the adverse effects of far-UV irradiated broth on bacteria are similar to those caused by the addition of H_2O_2 to the medium, and that these effects are negated by the addition of catalase. The mutation rate of *Staphylococcus aureus* is increased when grown on media previously exposed to far-UV (Stone *et al.*, 1947).

It is obvious from the preceding comments that under certain conditions exposure of media to visible and ultraviolet radiation can adversely influence the ability of medium to support growth of micro-organisms. The nature of these changes are generally poorly understood.

3. *Decomposition of carbohydrates and other substances*

There is some evidence that exposure of culture media to far-UV may cause the decomposition of carbohydrates and thereby result in inhibition of microbial growth. Appreciable direct photo-degradation of carbohydrates is unlikely at wavelengths longer than 200 nm though it is possible that these reactions can occur at longer wavelengths when photosensitized by other substances (Daniels, 1968). Exposure of nutrient broth to far-UV prevents the development of *Saccharomyces cerevisiae* and this inhibition can be duplicated by exposing a solution of the sugar component of the medium to far-UV, but not by exposing the inorganic components (Woodrow *et al.*, 1927). Exposure of various solutions of carbohydrates and carbohydrate derivatives to far-UV and then incorporating these into a bacterial medium, produces substances inhibitory to *Bacillus subtilis* (Blank and Arnold, 1935). Baumgartner (1936) has confirmed these latter findings and in addition reports increase of acidity in media incorporating the irradiated carbohydrates which if neutralized, allows growth of *B. subtilis*. Exposure of a special tubercle bacilli medium to diffuse daylight and to near-UV, adversely affects the growth of the tubercle bacterium (Cohn and Middlebrook, 1965). The irradiation of agar media with far-UV may also cause viscosity changes (Murty and Guha, 1963). A recent report by Barnett (1968) indicates that the inhibitory effects of light on the development of *Gonatobotryum fuscum* on its host *Graphium* sp., may be due to the destruction of vitamin B6 by light.

4. *Host susceptibility*

Susceptibility of plant cells to invasion by fungi can be altered by exposure to far-UV. When cut potato tubers are exposed to far-UV the tissues become much more susceptible to invasion by species of *Fusarium* and this is accompanied by exudation of various organic materials from the cells (Norell, 1954). The irradiated tissues are able to regain their resistance, and this recovery is temperature dependent. Far-UV irradiation of barley plants decreases their susceptibility of *Erysiphe graminis*, but similar irradiation of broad bean plants increases the number of lesions caused by *Botrytis fabae* (Buxton *et al.*, 1957). Leaf discs of chrysanthemums exposed to far-UV show an increase in susceptibility to species of *Ascochyta* (Blakeman and Dickinson, 1967). Similar exposure of sunflower leaves (*Helianthus annus*) causes them to lose their resistance to *A. chrysanthemi*. On the basis of Uehara's studies (1958) which reports the UV inactivation of phytoalexins, Blakeman and Dickinson suggest that this may be responsible for increased susceptibility of the irradiated leaf discs. Chakrabarti (1968) reports that the susceptibility of normally resistant barley plants to *Helminthosporium teres* and *H. sativum*, is lost when 5–6 day old plants are exposed to far-UV and then inoculated with these pathogens. Severity of symptoms increases with increase of UV dosage. Exposure of barley plants to the longer wavelength UV emitted from sunlamps and "Black Light" fluorescent lamps (Figure 4), has no apparent effect on susceptibility. Chakrabarti also reports that an inhibitor present in the sap of the resistant barley variety is partially inactivated by far-UV and he concludes that it is the resistance mechanism in barley leaves that is effected by UV.

Light is necessary for abundant penetration of plant tissue by the wheat stem rust *Puccinia graminis* f. sp. *tritici*. Yirgou and Caldwell (1968) report much higher penetration in plants grown under artificial light than in darkness. They conclude that there is a light mediated mechanism regulating penetration of *Puccinia* that resides in the host, and that increased penetration results from the effect of the radiation on the host rather than the fungus.

In none of these briefly summarized studies have the precise qualitative and quantitative relationships of light to susceptibility been studied. By using the leaf-disc technique described by Blakeman and Dickinson (1967) it is possible that this information can be obtained under precisely controlled conditions using monochromatic light.

III. LIGHT AND IDENTIFICATION OF FUNGI

Positive identification of fungi usually requires the presence of sexual or asexual spores, or sporocarps; lacking these, one fungus often looks much like another. Onset of reproduction is influenced by a variety of factors (Hawker, 1950; 1957; 1966; Lilly and Barnett, 1951; Cochrane, 1958) some of which are innate and not easily manipulated, while others are of an environmental nature and readily manipulated. Of the environmental factors, light is particularly effective in stimulating both sexual and asexual reproduction in many fungi (Marsh *et al*, 1959) though it is always interdependent on other limiting factors such as temperature and nutrition. Identification of fungi is also based on other light influenced characteristics such as pigmentation of mycelium and sporocarps, gross appearance of colonies in culture and the presence or absence of sclerotia. Generally these changes are morphogenetic in nature but stimulation or inhibition of mycelial growth can also be involved.

A. Induction of sporulation

1. *Light quality*

Light-induced reproduction, or photosporogenesis, is mainly caused by UV and blue wavelengths though recent studies by Lukens (1965), Rakoczy (1965), Nair and Zabka (1965), and Ingold and Nawaz (1967), indicate that in a few fungi, red light will affect sporogenesis. In spite of the immense number of reports on photosporogenesis, few investigators have precisely defined the relationship of light to sporulation. It is the more precise studies mainly conducted during the past decade, that will be emphasized in this guide.

(a) *True slime molds*. Many myxomycetes are induced to sporulate by exposing colonies to light (reviews by Gray, 1938; Alexopoulos, 1963, 1966; and v. Stosch, 1965). Certain pigmented species require light for sporulation while non-pigmented forms tend to fruit equally well in light or darkness (Gray, 1938). The wavelengths most effective in stimulating sporangium formation in *Physarum polycephalum* are restricted to blue and near-UV wavelengths (Gray, 1953). In *Physarum nudum* (Rakoczy, 1962, 1963, 1965, 1967) the most effective wavelengths of monochromatic light are from 330–540 nm, less effective are red wavelengths and ineffective are wavelengths around 560 nm (Figure 1). In *P. nudum*, the effectiveness of radiation in stimulating sporogenesis becomes greater with decrease of wavelength into the UV, however, excessive dosages of near-UV, e.g., 330 nm, will actually inhibit fruiting (Figure 2). Relatively low energy levels of blue light stimulate fruiting of *Didymium iridis*, *Physarum polycephalum*, and *P. gyrosum* while higher energy levels of red light are

iv 25

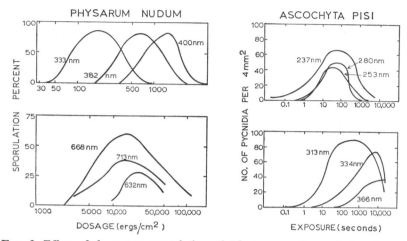

Fɪɢ. 2. Effect of dosage on sporulation of *Physarum nudum* and *Ascochyta pisi*. *P. nudum* exposed for 12 h to monochromatic light at different intensities (redrawn from Rakoczy, 1965); *A. pisi* exposed to monochromatic light at constant intensities (redrawn from Leach, 1962b).

necessary for fruiting of *D. iridis* and *P. gyrosum* (Nair and Zabka, 1965). Green light is ineffective in stimulating sporulation of the latter two myxomycetes. Sporangium formation is stimulated in *Didymium nigripes* by violet, blue, and red wavelengths of light, but not by green wavelengths (Straub, 1954; Lieth, 1956). Lack of sporulation in colonies exposed to green light is due to the actual inhibitory nature of these wavelengths rather than their ineffectiveness (Lieth, 1956; Rakoczy, 1967). Near-red light induced sporulation in *D. iridis*, is prevented by a subsequent exposure to far-red wavelengths and this phenomenon is reversible (Nair and Zabka, 1965).

It is generally accepted that pigments are intimately involved in photosporogenesis in myxomycetes. The relation of these pigments to sporogenesis are discussed in many of the preceding citations. In addition Nair and Zabka (1966) review the literature on pigmentation and sporulation.

The light relations of very few myxomycetes have been carefully studied and general recommendations for the photo-induction of fruiting cannot be presently made with any degree of certainty. Near-UV and blue wavelengths without exception, appear to be stimulatory while green wavelengths tend to be inhibitory. Near-UV is particularly effective in some myxomycetes though there is a danger of overexposing and thereby actually causing inhibition of sporulation (Figure 2). Near-UV "Black Light" fluorescent lamps with or without integral filter, would appear to be the most suited for exploratory studies (Figure 4). The relative effectiveness of a diurnal regime of light and darkness as opposed to continuous exposure, has still to be explored for the myxomycetes. The writer recommends that exploratory studies begin with a diurnal regime.

(b) *Cellular slime moulds.* The effect of light on the cellular slime moulds (*Acrasiales*) is much less well understood than for the true slime moulds. Reinhardt (1968) reports that *Acrasis rosea* needs light for sporulation, but that it is unable to fruit under continuous illumination. He does not mention which wavelengths are effective. Raper (1962) also discusses the affect of the environment on morphogenesis in the cellular slime moulds, but includes little on the precise relationships of light to sporulation.

(c) *Phycomycetes.* Within this heterogeneous group are a number of species whose reproductive processes are influenced by light. In general asexual stages are stimulated and sexual stages are inhibited by light. In making this generalization, however, one must recognize that the light relations of relatively few Phycomycetes have been carefully studied (Marsh, 1959).

Among the Oomycetes, the effects of light on the genus *Phytophthora* have received most attention (review by Lilly, 1966). Sporangial formation is stimulated by light in some species, while oogonia are generally inhibited. (Refer also to section on "Light Inhibition".) The quality of light effective in stimulating sporangial formation has not been studied with any degree of precision, but it appears to be confined to the blue region of the spectrum (Aragaki and Hine, 1963; Harnish, 1965; Merz, 1965; Lilly, 1966). In none of the studies reported to date has the effect of the UV wavelengths on sporangial production been fully explored, though it has been established that sporulation in certain species can be stimulated in the absence of UV. Trophocyst formation in the Zygomycete *Pilobolus kleinii* is stimulated by wavelengths of light less than 510 nm (Figure 1) with 400–480 nm wavelengths most effective (Page, 1956, 1962). *Choanephora cucurbitarum* exhibits a more complex light relationship. Light stimulates formation of conidia but once this process has been initiated, light then becomes inhibitory and sporulation cannot be completed unless the fungus is placed in darkness (Christenberry, 1938; Barnett and Lilly, 1950; 1952). Blue light is most effective in inhibiting sporulation of *C. curcurbitarum.* (Refer also to section on "Diurnal Sporulators".) *Syzgites megalocarpus* behaves similarly to *Choanephora* (Wenger and Lilly, 1966). Sporangium formation in a number of species of downy mildews is inhibited by light (Yarwood, 1937; Cruickshank, 1963; Uozumi and Krober, 1967; and others).

(d) *Ascomycetes.* Sporulation of many Ascomycetes can be stimulated by exposing cultures to light (Marsh *et al.*, 1959). Although the light relations of very few species have been studied precisely there is some evidence that suggests the possibility of two distinct groups. These are: (1) genera in which ascocarp formation is stimulated by only ultraviolet wavelengths, and (2) genera which respond to both blue and near-UV wavelengths.

The discomycete *Ascophanus carneus* produces apothecia in abundance when exposed to wavelengths of light 430–540 nm, but sparsely at longer wavelengths (Stoll, 1936). A similar relationship exists for perithecial production by *Leptosphaerulina briosiana* (Thomas and Halpin, 1964) as well as other species of *Leptosphaerulina* (Thomas, 1966). Many isolates of *Ophiobolus graminis* var. *avenae* produce perithecia in culture only after exposure to light that includes 400–450 nm wavelengths (personal communication from E. Weste, University of Melbourne, Australia). *Pleurage setosa* (Callaghan, 1962) forms perithecia most abundantly when exposed to blue (350–510 nm) and violet (340–420 nm) wavelengths, but not green (570–590 nm) or orange (520–720 nm). Curtis reports (1964) that visible and ultraviolet wavelengths stimulate perithecial formation in *Hypomyces solani* (*Nectria haematocca*).

The stimulatory effect of UV on reproduction was first reported by Stevens (1928, 1931). He discovered that exposure of 2–8 days-old colonies of *Glomerella cingulata* to a quartz-mercury lamp stimulated the formation of perithecia and that only UV wavelengths less than 313 nm were effective. Perithecial formation in *Ophiobolus graminis*, *Wojnowicia graminis*, and *Pleospora herbarum* is initiated by exposing actively growing colonies to near-UV fluorescent lamps (320–420 nm) and daylight fluorescent lamps (Leach, 1962a). Removal of near-UV and blue wavelengths by filters greatly reduces sporulation in these same fungi. Abundant perithecia are produced in colonies of *Diaporthe phaseolorum* after a short exposure to far-UV from an unfiltered germicidal lamp (Timnick *et al.*, 1951). Monochromatic light studies on *Pleospora herbarum* by Leach (1963, 1966) reveal that only wavelengths of UV less than 380 nm induce perithecial formation. An action spectrum for induction of photosporogenesis in this same fungus (Leach, 1966), shows two principal peaks of effectiveness at 230 and 290 nm.

To induce sporulation in light sensitive Ascomycetes, the procedures recommended for Fungi Imperfecti should be followed. (Refer to discussion on Fungi Imperfecti.)

(e) *Basidiomycetes*. Loew in 1867 (Smith, 1936) noted that species of *Coprinus* and *Sphaerobolus* require light for fruiting. A century later there are numerous reports of basidiomycetes responding to light (Marsh *et al.*, 1959; Lu, 1965, Manachére, 1967; Miller, 1967) yet none of these precisely defines the qualitative and quantitative light relationships. The studies of

Alasoadura (1963) and Ingold and Nawaz (1967) on *Sphaerobolus stellatus*, are among the more precise investigations. *Sphaerobolus* needs a fairly high intensity of light (100 lux) to stimulate optimum primordium formation (Alasoadura, 1963), but the intensity differs with the stage of development. As the basidiocarp develops, the intensity needed lessens. The quality of light effective in stimulating *Sphaerobolus* is confined to 350–500 nm wavelengths, with 440 and 480 nm wavelengths particularly effective. Blue light (400–500 nm) is necessary for overall sporophore development in *Sphaerobolus* (Ingold and Nawaz, 1967) but during the last stages of fruiting, red and far red light (640–720 nm) actually accelerate sporophore development more than blue light. Only blue wavelengths of light less than 520 nm stimulate basidiocarp formation of *Coprinus lagopus* and the light stimulus is not translocated from exposed mycelium to shaded portions of colonies (Madelin, 1956). Pileus development of *Collybia velutipes* is also stimulated by blue light of less than 470 nm (Aschan-Åberg, 1960), as is basidiocarp formation in *Cyathus stercoreus* (Garnett, cited by Lu, 1965) and *Schizophyllum commune* (Barnett and Lilly, 1952).

Stimulation of basidiocarp formation in the genus *Typhula*, is almost wholly an ultraviolet response and only 265–325 nm radiation induces sporophore formation (Remsberg, 1940). *T. incarnata* is stimulated by 250–370 nm wavelengths, but not by longer wavelengths (Lehmann, 1965).

Precise light relationships for Basidiomycetes are not easily determined experimentally primarily because of the technical problems posed by the large size of most basidiocarps. Within such genera as *Sphaerobolus*, *Cyathus*, and *Mycena*, however, there are light sensitive species that do produce small fruiting structures. Monochromatic light studies conducted on these species could well provide a much better understanding of photosporogenesis in the Basidiomycetes.

Because of the paucity of precise information on the light requirements of Basidiomycetes presently available, it is not possible at this time to make any meaningful recommendations for inducing sporulation within this diverse group of fungi. The recommendations I have made for the Fungi Imperfecti would seem to offer the best chance of success.

(f) *Fungi Imperfecti*. Many Fungi Imperfecti sporulate well when illuminated but poorly, or not at all when grown in total darkness (Marsh *et al.*, 1959). Relatively few species have been studied precisely and among these there is evidence for at least two distinct groups consisting of: (1) species that respond only to UV wavelengths; and (2) species sensitive to blue and near-UV.

Nearly forty years ago Stevens (1928, 1930) reported that UV from an unfiltered quartz-mercury arc lamp stimulated the sporulation of a number of Fungi Imperfecti. Early light quality studies on *Macrosporium tomato*, *Fusarium cepae* (Ramsey and Bailey, 1930), *Fusarium* spp. (Bailey, 1932) and *Alternaria solani* (McCallan and Chan, 1944) all clearly show that the stimulatory wavelengths of UV are confined to wavelengths less than 320 nm (approx.).

Within the past decade much more precise monochromatic light studies have been conducted on several Fungi Imperfecti. *Ascochyta pisi* (Leach, 1962b), *Helminthosporium dematioideum* (Yadav, 1962), *Stemphylium botryosum* (Leach, 1963) and *Alternaria chrysanthemi* (Leach, 1964) are all induced to sporulate by wavelengths of monochromatic UV from 230–360 nm, but not by longer wavelengths. Action spectra for photosporogenesis of *A. pisi* (Leach and Trione, 1965), *Alternaria dauci, Stemphylium botryosum* (Leach and Trione, 1966), and *S. solani* (Herndon, 1967) also show the importance of UV below 370 nm and suggest a similar mechanism for these different fungi. The action spectra for *A. pisi* and *S. solani* are shown in Figure 1. Near-UV is very effective in stimulating diverse species of Fungi Imperfecti to sporulate (Leach, 1962a; Yadav, 1962; Knox-Davis, 1965; Zimmer, 1967; Schneider, 1965; Calpouzos, 1967; Leach, 1967a; and others).

Not all Fungi Imperfecti respond solely to UV radiation. Wavelengths from 350–510 nm stimulate sporulation in species of *Trichoderma* (Figure 1), but not longer wavelengths (Bjornsson, 1959; Gressel and Hartman, 1967). These same wavelengths (approximately) induce conidium formation in *Sclerotinia fructigena*, a *Verticillium* sp. (Sagromsky, 1952b), in isolates of *Penicillium* (Sagromsky, 1952a) and *Monilia fructicola* (Jerebzoff, 1960).

To induce sporulation in Fungi Imperfecti the following recommendations are made (Leach, 1967a; Anonymous, 1968)—

1. Use "Black Light" fluorescent lamps which emit near-UV radiation in a continuous spectrum from 320–420 nm. These lamps are obtainable in 20 W, 40 W, and 80 W series from Osram, Philips (MCF/U series), Sylvania, General Electric (BLB series), and other manufacturers. The lamps are usually interchangeable with daylight fluorescent tubes and require no special circuitry. Cool white daylight fluorescent lamps also emit sufficient near-UV to induce sporulation and can be used if the other lamps are not available.

2. When space is available, place two 40 W lamps in a horizontal position 20 cm apart, and at a height of approximately 41 cm above the cultures to be irradiated. Two lamps so arranged will adequately cover an area approximately 60 by 120 cm. There is considerable latitude in the intensities that may be used.

3. Use an alternating cycle of 12 h exposure to near-UV and 12 h of

darkness. This light regime will accommodate those fungi requiring light and darkness ("Diurnal Sporulators"), as well as those capable of sporulating under continuous exposure.

4. Irradiation should start 3 or 4 days after inoculation of the plate and continue until the end of growth. It is pointless to irradiate a culture until the fungus has had time to grow and excessive exposure of the medium may have adverse effects. (Refer back to "Indirect Effects of Light on Fungi".) Old colonies are usually insensitive to light.

5. Use plastic or glass petri dishes, or other containers capable of transmitting near-UV with their lids on.

6. Irradiate colonies in a controlled temperature chamber or room. The incubation temperatures used are important but will depend on the fungus involved. Some fungi sporulate well under constant temperature while others require a diurnal fluctuation of temperature for optimum response to the near-UV stimulus. Where diurnal fluctuation of temperatures are used, these should be coordinated with the cycling of light and darkness with the high temperature during the light cycle. A constant temperature of 21°C and fluctuating temperature of 18/24°C have proved to be satisfactory for many fungi.

2. *Light intensity and dosage*

Quality of radiation effective in photosporogenesis has been emphasized in the preceding discussion, yet one cannot ignore the significant effects of dosage on sporulation. A certain minimum dosage of radiant energy (threshold dosage) is required to induce sporulation and this amount is wavelength dependent. If a fungus is exposed to stimulatory wavelengths at lower dosages no sporulation will occur. Increase the dosage above the threshold point and sporulation begins. Further increase in dosage causes more abundant sporulation up to a point at which the photochemical system presumably becomes saturated and any further increase will have no effect on sporulation.

Stimulatory wavelengths of radiation if applied at excessively high dosages may in turn inhibit sporulation (Figure 2). Inhibition of this sort is pronounced in far-UV irradiation studies (Leach, 1962, 1963, 1964) but it can also be caused by excessive dosages of near-UV and visible radiation (Rakoczy, 1963, 1965).

3. *"Diurnal sporulators"*

Many light sensitive fungi, perhaps most, sporulate abundantly when exposed continuously to illumination, others are unable to fruit unless light is followed by a certain period of darkness. Fungi exhibiting this latter behaviour have been termed "diurnal sporulators" by Leach (1967b). They need light for the induction of sporogenesis but once this process has been stimulated then light becomes inhibitory to the completion of reproduction. *Choanephora cucurbitarum* (Mucorales) produces conidia most profusely when light and darkness are alternated on a 12-hour cycle; sporulation is poor in colonies grown in darkness, or exposed continuously to high intensity visible light (Christenberry, 1938; Barnett and Lilly, 1950). Barnett and Lilly (1950) postulate that two metabolic reactions are involved in *C. cucurbitarum*, one needing light and the other inhibited by light. *Zyzygites megalocarpus* behaves very similarly to *C. cucurbitarum* (Wenger and Lilly, 1966). A number of the downy mildews (Peronosporaceae) also require periods of darkness for completion of asexual reproduction (Yarwood, 1937; Cruickshank, 1963; Uozumi and Krober, 1967).

Sporulation of the imperfect fungus *Alternaria brassicae* var. *dauci* is stimulated by light but this must be followed by darkness if conidia are to form (Witsch and Wagner, 1955). Similar relationships exist for *Helminthosporium oryzae* (Leach, 1961), *H. gramineum* (Houston and Oswald, 1946) *Stemphylium botryosum* (Leach, 1967), *Alternaria dauci* (Leach, 1967), *A. tomato* (Aragaki, 1962) and *A. solani* (Lukens, 1963).

"Diurnal sporulators" are not confined to the Phycomycetes and Fungi Imperfecti, the Basidiomycete *Coprinus congregatus* requires light for basidiocarp formation yet it is unable to produce basidiocarps if illuminated continuously (Manachére, 1967). Production of mature perithecia by the Ascomycete *Diaporthe phaseolorum* var. *batatatis* is most abundant under a regime of alternating light and darkness and few develop in darkness or under continuous illumination (Timnick *et al.*, 1951). Among the cellular slime moulds, *Acrasis rosea* is unable to develop sporangia in darkness or continuous illumination, but fruits readily when light is followed by darkness (Reinhardt, 1968).

Blue light is particularly effective in inhibiting spore formation during the "terminal phase" of the "diurnal sporulators". Conidium formation in *Choanephora cucurbitarum* is most strongly inhibited by wavelengths shorter than 571 nm with inhibition increasing toward the violet end of the spectrum (Christenberry, 1938). Wavelengths from 450–540 nm are most inhibitory to the development of conidia of *Peronospora tabacina* (Cruickshank, 1963). Photo-inhibition of the "terminal phase" of conidium formation in *Alternaria solani* (Lukens, 1963) and *A. tomato* (Aragaki, 1962) is confined mainly to 380–530 nm wavelengths, with 450 nm radiation particularly effective. In *A. solani*, blue light inhibition of conidia can actually be negated by a subsequent exposure to red light (Lukens, 1965). The action spectrum for photo-inhibition of conidia of *Neurospora* (Sargent and Briggs,

1967) is quite similar to those just mentioned for *A. tomato and A. solani*. The action spectrum for photo-inhibition of the "terminal phase" of conidium formation of *Stemphylium botryosum* (Leach, 1968) shows that far-UV is also extremely inhibitory (Figure 1). All wavelengths of radiation from 240–520 nm inhibit conidium formation of *S. botryosum* and at high enough dosages even red wavelengths are inhibitory. Most effective are 280 nm and 480 nm wavelengths.

The temperature dependency of blue-light inhibition of conidial formation has been reported for several "diurnal sporulators" belonging to the Fungi Imperfecti. Formation of conidia in colonies of *A. tomato* is inhibited by light at relatively high temperatures but not at lower temperatures (Aragaki, 1961); this is also true for *A. solani* (Lukens, 1966), *S. botryosum* (Leach 1967b, 1968) and *Helminthosporium gramineum* (Houston and Oswald, 1946). Studies on the interaction of near-ultraviolet radiation and temperature on photosporogenesis in colonies of *A. dauci, A. tomato,* and *S. botryosum* (Leach, 1967b) have revealed two distinct phases of photosporogenesis similar to those postulated for *Choanephora cucurbitarum* (Barnett and Lilly, 1950). The first phase ("inductive phase") is light induced and proceeds most efficiently at relatively high temperatures. Completion of this phase results in the formation of conidiophores. The second phase ("terminal phase") leading to the formation of conidia, is strongly inhibited by ultraviolet and blue light, generally proceeds most efficiently at lower temperatures. The length of darkness necessary for the completion of the "terminal phase" is also temperature dependent (Leach, 1968).

It should be evident to the reader from the preceding discussion that exposure of colonies to continuous illumination at a constant temperature is unsatisfactory for inducing sporulation in some fungi. For purposes of general identification the writer recommended the use of a 12 h light, 12 h dark regime, preferably with fluctuating day and night temperatures. The day temperature should be relatively high and the night temperature lower (Leach, 1967b).

4. *Interaction of light and other factors on sporulation*

Many fungi can be stimulated to fruit by exposure to light but only if other factors are not limiting. It is well known that nutrition, hydrogen ion concentration of the medium, aeration, humidity, temperature, depth of medium, age of colony, etc., can influence the onset of reproduction (Hawker, 1957). A few examples will be cited to illustrate the importance of these factors.

(a) *Temperature.* The temperature limits supporting reproduction are usually considerably narrower than those that support growth (Hawker,

1957; Cochrane, 1958). In *Ascochyta pisi*, for example, growth occurs between 5°–30°C (optimum 22°–27°C) but near-UV induced pycnidial formation is limited to 10°–25°C (Leach, 1962). An understanding of the effect of the interaction of light and temperature on sporulation is particularly important among "diurnal sporulators". The fungi *Alternaria dauci*, *A. tomato*, and *Stemphylium botryosum* (Leach, 1967a; Aragaki, 1961; Lukens, 1966) have one temperature optimum for light stimulation of sporulation ("inductive phase") and another lower optimum temperature for the completion of conidium development during darkness. In addition, the inhibitory effects of blue light on conidium formation among these same "diurnal sporulators", can in some instances be negated by lower temperatures.

(b) *Hydrogen-ion concentration.* Growth of fungi usually takes place over a fairly wide range of hydrogen-ion concentrations. Some fungi such as *Ascochyta pisi* (Leach, 1962), can be photo-induced to sporulate over a fairly wide range of hydrogen-ion concentrations, others have much more restricted ranges (Hawker, 1957).

(c) *Aeration.* The composition of the atmosphere can have a marked influence on photosporogenesis. It is not uncommon to observe poor sporulation in irradiated colonies grown in tightly capped culture tubes and abundant sporulation in the same fungi grown in Petri dishes.

(d) *Nutrition.* The nutritive composition of the substrate can markedly effect the results of photosporogenesis studies. Not infrequently a fungus grown on one medium may appear to have an obligate light requirement for sporulation, yet on another medium it will sporulate quite well in total darkness. *Pyronema omphaloides* in contrast, will not produce apothecia when grown on Czapek's (Dox) medium under illumination, yet when hyphae are allowed to grow for this medium onto water agar, apothecia form, but only under illumination.

(e) *Depth of medium.* Basidiocarp formation by a light sensitive strain of *Schizophyllum commune* studied by the writer, is markedly influenced by the depth of medium. When the medium is shallow (0·5 cm), photo-induced sporophore formation is poor and the basidiocarps are not well developed. When the medium is deeper well formed basidiocarps are formed in profusion.

(f) *Age of colony.* The age at which colonies are exposed to light in photosporogenesis studies is most important. Among the light sensitive Ascomycetes and Fungi Imperfecti, actively growing peripheral hyphae are generally the most sensitive to light and it is therefore wise to irradiate only relatively young colonies. Irradiation of old colonies that completely cover the medium usually gives negative results. In certain sensitive Basidiomycetes, it is essential that colonies be allowed to become reasonably large before irradiating to enable sufficient food reserves to accumulate within the mycelium. These reserves are necessary for the development of the relatively large basidiocarps.

(g) *Surface versus submerged cultures.* Many fungi that can be photoinduced to sporulate while growing on the surface of agar media or floating on the surface of liquid media, fail to fruit when submerged in a liquid medium. *Ascochyta pisi*, for example, produces pycnidia abundantly when exposed to near-UV on agar media but will not in liquid-shake culture under similar conditions. We have discovered that *A. pisi* will sporulate profusely in liquid culture if grown in a pyrex glass fermenter exposed to near-UV and subjected to mechanical agitation and violent bubbling of air (E. J. Trione and C. Leach, unpublished). For physiological studies colonies can be grown and irradiated on various inert or semi-inert supporting materials (cellophane, fibre glass, and nylon mesh) which are suspended at the surface of the liquid medium.

B. Inhibition of sporulation

The stimulatory effects of light on sporulation have been mainly emphasized to this point, yet, light can inhibit sporulation under certain conditions.

(a) *Inhibition by high dosages.* One form of inhibition of sporulation is caused by the exposure of colonies to excessively high dosages of light at wavelengths that normally stimulate reproduction at lower dosages (refer to "Light intensity and dosage"). This type of inhibition (Figure 2) is propably most obvious in UV experiments (Leach, 1962, 1963, 1964) but can also occur in visible light studies (Rakoczy, 1963, 1965).

(b) *Dark phase inhibition.* Among the "diurnal sporulators" light is necessary for the "inductive phase" (Leach, 1967b), but it becomes inhibitory to conidial development during "terminal phase" (refer to section on "Diurnal Sporulators").

(c) *True inhibition.* Inhibition of reproduction, other than the types just described, appears to be quite rare in fungi and largely confined to the Oomycetes. Photo-inhibition of oogonia of *Saprolegnia ferax* is caused by blue-green wavelengths; red light is only partially inhibitory (Krause, 1960). Light from daylight fluorescent lamps inhibits oogonium formation in *S. ferax*, and *S. parasitica*, but not in *Achlya americana*; most inhibitory are blue wavelengths and least inhibitory are red wavelengths (Lee and Scott, 1967). Oogonia are inhibited in colonies of *S. declina* by light intensities as low as 17–22 ft candles and this inhibition is temperature dependent (Szaniszlo, 1965). Sexual reproduction in numerous species of *Phytophthora* is inhibited by light (Merz, 1965; Harnish, 1965; Lilly, 1966). Blue wavelengths are most inhibitory and red least.

Precise action spectra for photo-inhibition of oogonium formation using monochromatic radiation, have still to be determined. To date the influence of UV wavelengths on oogonium development has been inadequately explored and future action spectrum studies should be extended to at least 290 nm.

Buller (1931) observed an inhibitory effect of daylight on basidiocarp formation of *Coprinus sterquilinus*. He reported that the development of basidiocarp rudiments less than 1 mm long were inhibited by light but as these became larger than 3 mm, light was no longer inhibitory.

C. Pigmentation

Since early times botanists have observed that fungi growing in dark tend to be less intensely coloured than those grown in light (Smith, 1936; Marsh *et al.*, 1959). Little is known about the role of pigments in fungi (Hawker, 1950; Ingold, 1962; Carlile, 1965) though it is suspected that some are involved as photoreceptors for the various photobiological phenomena described in the preceding discussions. They may also serve in a protective role by preventing damage from harmful UV radiation.

Colouration of mycelium and fruiting structure can be a most useful characteristic in the identification of fungi and the influence of light on pigmentation is therefore of considerable importance to the taxonomic mycologist.

The pigmentation of myxomycetes is influenced by light (Alexopoulos, 1963) and it is not uncommon for a plasmodium to be colourless when grown in darkness and become pigmented when exposed to light (Fergus, 1963). There has been general agreement that these pigmented substances are in some way related to sporulation as photoreceptors. Nair and Zabka (1966) report that one of the pigments present in *Physarum gyrosum* is a flavone, though there are other pigments present within this myxomycete which are not flavones. *Sordaria fimicola* produces a yellow-orange pigment within its mycelium and perithecia when exposed to light (Ingold and Marshall, 1963). Production of this pigment is stimulated by wavelengths from 320 to 540 nm (approx.); wavelengths which correspond to those effective in phototropic bending of perithecial necks in *Sordaria* species, and to those stimulatory to spore discharge. A yellow pigment which is probably a carotenoid, forms in illuminated colonies of the Basidiomycete *Sphaerobolus stellatus* (Friederischen and Engel, 1957). Both pigmentation and formation of apothecia are light dependent in *Pyronema confluens* but these appear to be distinct and unrelated phenomena

(Carlile, 1956). The orange colouration of illuminated colonies of *P. confluens* is caused by a complex of carotenoids. The water mould *Karlingea rosea* is also more intensively pigmented in light than in darkness (Haskins and Weston, 1950).

Fusarium spp. are frequently strongly coloured when exposed to light, and colourless when grown in darkness though there are a few species where the reverse is true (Zachariah *et al.*, 1956). The colour variation of colonies of *Fusarium oxysporum* is due to carotenoids and naphtha quinones (Carlile, 1956). The carotenoids are produced only in response to light but not the naphtha quinones which are mainly regulated by the C/N ratio. An action spectrum for carotenoid synthesis in *F. aquaeductum* (Figure 1), shows only wavelengths less than 520 nm to be effective, with maxima at 440/455 nm and 375/380 nm (Rau, 1967). A white spored strain of *Cochliobolus sativus* produces pink mycelium in light and colourless mycelium in darkness (Tinline and Samborski, 1959). The pink pigment, sativin, is actively produced in colonies exposed to 390–513 nm wavelengths, but not in the absence of oxygen. *Neurospora crassa* is bright orange when grown in light, but colourless in darkness. This pigmentation is caused by a complex of carotenoids (Haxo, 1949). An action spectrum for biosynthesis of carotenoids in *Neurospora* shows 430–480 nm wavelengths to be most stimulatory (Zalokar, 1955). In *Verticillium albo-atrum* melanin synthesis is inhibited by near-UV (Brandt, 1964). Melanogenesis in *Neurospora crassa* is depressed by blue-violet wavelengths of light (Schaeffer, 1953).

In summary, it is evident that wavelengths of radiation in the near-UV and blue regions of the spectrum are effective in inducing pigment formation in some fungi. Whether this is true for all fungi, has still to be determined.

D. Morphological effects

The shape, size, and structure of sporophores and spores are universally used criteria for the identification of fungi. These characteristics can often be markedly influenced by various environmental factors including light (Hawker, 1957).

1. *Spore and sporocarp morphology*

Early botanists noticed that the shape of sporophores and sporocarps of various Mucorales and Basidiomycetes could be quite different dependent on whether they were grown in light or darkness (Smith, 1936). Actually, this phenomenon is common to all classes of fungi and is probably more important than taxonomic mycologists generally realize. Hawker (1957) cites a number of reports on the effect of light on spore and sporocarp morphology; in none of these have the precise light relationships been elucidated nor have the mechanisms been explored.

Exposure of colonies of *Alternaria dianthi* to far-UV causes a decrease in spore dimensions as the dosage is increased (Joly, 1962). In colonies of *A. dauci*, as the length of exposure to near-UV is increased, conidia become longer and narrower and have more septations (Zimmer, 1967). The size and number of septations of spores of *Fusarium* species (Snyder and Hansen, 1941; Harter, 1939) and various species of Fungi Imperfecti (Johnson and Halpin, 1954) are greatly influenced by light. The diameter of the pycnidia of *Ascochyta pisi* are significantly smaller in colonies exposed continuously to near-UV than in dark reared colonies (Leach, 1962b). In addition the size and shape of conidia of *A. pisi* vary with lighting conditions. The perithecia of the Ascomycete *Melanospora destruens* are about 50% larger when exposed to diurnal illumination than those that develop in total darkness (Asthana and Hawker, 1936). *Phomopsis citri* favours the production of alpha spores over beta spores when grown in

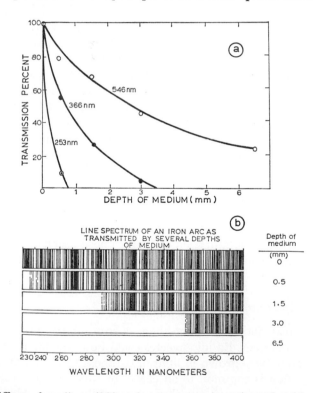

FIG. 3. Effect of medium (2% malt extract agar) on intensity (a) and quality (b) of radiation (redrawn from Leach, 1962b).

light (Koizumi, 1965). In cultures of *P. mali* both alpha and beta spores are produced in light, but only beta spores in darkness (Personal communication from Dr. T. Swinburne, Queen's University, Northern Ireland). Spores of *Mucor dispersus, Trichothecium roseum, Sordaria fimicola,* and *Coprinus heptemerus* are smaller when these fungi are grown in light (Williams, 1959). The shape of sporangia of *Phytophthora palmivora* and *P. capsici* are markedly influenced by the absence or presence of light (Hendrix, 1967).

The positioning of pycnidia and perithecia relative to surface of agar media is frequently light dependent particularly in UV studies. In darkness these fruiting structures are often borne at the surface of agar media, but under illumination they become partially or wholly submersed (Stevens, 1928; Leach, 1962b) and the higher the light intensity the deeper they lie. Ultraviolet radiation stimulates the sporulation of *Ascochyta pisi* but at high dosages at these same wavelengths, light becomes inhibitory (Figure 2). Presumably when pycnidia form submersed in the medium it is because the agar acts as a filter (Figure 3). Thus stimulation of sporulation occurs in submersed hyphae only at those depths at which the intensity is reduced below the inhibitory level. *Pyrenochaeta terrestris* behaves similarly to *A. pisi.* Colonies exposed to low intensities of near-UV form pycnidia with well developed setae on the surface of the medium, but as the intensity is increased, the pycnidia are submerged and no longer produce setae. It seems likely that this same phenomena occurs in nature among plant pathogens.

2. Gross changes

Cultures grown under illumination are often quite different from those grown in total darkness. Illuminated colonies tend to be more pigmented, have less aerial mycelium, and may be larger or smaller than comparable dark-reared colonies. The absence or presence of sclerotia may also contribute to these gross differences. Zonation is common in colonies exposed to diurnal regimes of light and darkness (Jerebzoff, 1965). This may result from a combination of light-induced changes such as differences in rate of growth and morphology of hyphae produced under day and night conditions, differences of pigmentation, zones of fruiting structures of sclerotia interspersed with vegetative mycelium, and so on. Many of these light effects have already been discussed and the following remarks will be confined to the effects of light on formation of sclerotia.

Microsclerotia of *Verticillium dahliae†* are inhibited by light (Leach,

† The common microsclerotial species of *Verticillium* is *V. dahliae* (Isaac, 1967). In the American literature the microsclerotial species is often incorrectly referred to as *V. albo-atrum* and this is true among the references I have cited.

1962a; Heale and Isaac, 1965; Brandt, 1964; Caroselli *et al.*, 1964; Kaiser, 1964). Kaiser (1964) reports that blue light (478 nm max.) is most inhibitory, with some inhibition evident in green light (540 nm max.) but none in yellow, orange or red light. Near-UV also suppresses microsclerotial formation of *V. dahliae* (Leach, 1962a; Brandt, 1964). Resting structures in other species of *Verticillium* are also influenced by light (Heale and Isaac, 1965). Sclerotia of *Botrytis squamosa* form readily in darkness (Page, 1956). Under a diurnal regime of light and darkness, zones of sclerotia are formed but if the light intensity is too high, complete inhibition of sclerotia occurs. In colonies of *Stromatinia gladioli*, fewer sclerotia occur in darkness than in light and sclerotia formed in darkness are morphologically different from those produced in light (Bjornsson, 1956). Sclerotia form only in darkness in colonies of *Aspergillus japonicus* and 1 ft candle of light is sufficient to inhibit the development of sclerotia (Heath and Eggins, 1965). Blue wavelengths (10 ft candle) effectively inhibit sclerotia but not red or orange wavelengths. The effect of light on sclerotial production by *Rhizoctonia solani* appears to vary among isolates. Some isolates require light for the development of sclerotia, others fail to produce them in light or darkness while another group produces sclerotia without regard to lighting conditions (Durbin, 1959). Sclerotial morphology of *R. solani* differs between illuminated and dark-reared cultures.

IV. PRACTICAL CONSIDERATIONS

The techniques and equipment used in photobiological research are in general not unique to fungal investigations. Fortunately there are several excellent articles on this subject and these are listed separately in the Appendix. These authors go into considerably more detail than is possible in this brief guide.

A. Light sources

The emission of most radiant energy sources used in UV and visible light studies on fungi, are heterochromatic, that is they emit a relative broad range of wavelengths. Basically the radiant energy emission of heterochromatic sources can be continuous (continuous spectrum) or discontinuous (line and band spectra) as shown in figure 4. There are also other emission patterns achieved by raising lamp pressures and use of various combinations of gases and phosphors.

1. *Solar radiation*

The sun's radiation reaching the earth's surface (Figure 4) is one readily available source of visible and near-UV radiation (290–750 nm) and it

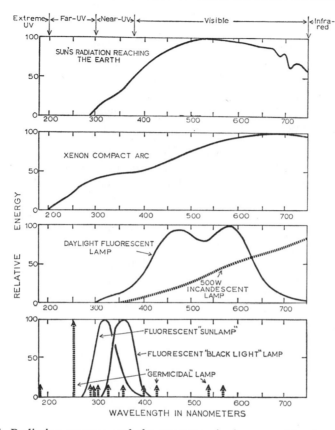

FIG. 4. Radiation sources and the spectrum (redrawn mainly from General Electric Co., Lamp Bulletin LD-1; xenon arc from Engelhard Hanovia, Inc., catalogue).

has been used in many light studies on fungi. Solar radiation, however, is an extremely variable and unreliable source and its use is not recommended for experimental studies. Under natural conditions, solar radiation is the chief stimulator of photobiological changes and an understanding of its diurnal and seasonal variation can give insight to laboratory findings as they relate to nature.

General articles on solar radiation are by Sanderson and Hulbert (1955), Robinson (1966), Koller (pp. 105–152, 1965), and Gates (1966). The article by Vezina and Boulter (1966) on the spectral composition of UV and visible radiation under forest canopies, may also be of interest to mycologists. Persons interested speculating on the possible role of mutagenic UV in the evolution of fungi, should refer to the articles by Berkner and Marshall (1964) and Barker (1968).

2. *Artificial sources*

There are many sources of UV and visible radiation ranging from simple inexpensive lamps of relatively low radiant intensity, to expensive arcs of high radiant intensity (Figure 4). These sources and their principles of operation are well covered by Koller (1965), Seliger and McElroy (1965) and Withrow and Withrow (1956). My coverage will be brief, selective and will have a personal bias.

(a) *Simple sources of far-UV* (200–300 nm). The "germicidal", or low pressure mercury vapour lamp is a very useful and widely used source of far-UV for mutation and lethal effect studies. It emits a line spectrum with approximately 85% of its total radiation emitted at 253 nm (Figure 4). Jagger (1967) recommends that a germicidal lamp with a Vycor rather than a quartz envelope be used to reduce the ozone formation caused by the emission of short wavelength UV. The lamps are usually manufactured as straight tubes in a number of sizes and are relatively inexpensive.

(b) *Simple sources of near-UV* (300–380 nm). The fluorescent "sunlamp" emits a continuous spectrum (260–400 nm) having its maximum emission at 320 nm (Figure 4). The "Black Light" fluorescent lamp also emits a continuous spectrum but at somewhat longer wavelengths (310–420 nm). Maximum emission of the "Black Light" lamp is approximately 360 nm (Figure 4). They can be obtained with or without an integral filter incorporated in the envelope. The integral filter types (e.g., General Electric Co.,† "BLB" type of "Black Light" fluorescent lamp) emits only a little visible radiation, while the other (e.g., General Electric Co., "BL" type of "Black Light" fluorescent lamp), emits considerable visible radiation in addition to near-UV. Another readily available source of long wavelength near-UV, is the "Daylight" fluorescent lamp (Figure 4). The amount of near-UV emitted from these varies considerably depending on whether they are "cool white" types, "standard", "warm white" etc. All the near-UV lamps that have been mentioned can be obtained either as tubes or bulbs and are relatively inexpensive.

(c) *Simple sources of visible light* (380–750 nm). "Daylight" fluorescent lamps are inexpensive and good sources of visible light (Figure 4). The quality of emission varies considerably among the different types of lamps as has already been mentioned. Standard tungsten filament incandescent lamps are also cheap and readily available sources of visible light (Figure 4). However, they emit considerable heat, and this can be a problem in

† Names and addresses of manufacturers of some items are provided in Section IV. Investigators should make their own inquiries as to alternatives, or refer to Jagger (1967) for a more comprehensive listing.

some studies. The low wattage incandescent lamps are poor sources of near-UV.

(d) *Complex sources of UV and visible light.* High intensity sources often operating at high pressures, are numerous and usually fairly expensive (e.g., open arcs of carbon and zirconium; enclosed arcs of hydrogen, mercury, xenon, etc.). Each of these sources has its own unique characteristics as well as advantages and disadvantages and the reader should refer to the Appendix citations for more details. High intensity compact sources of radiation that emit a continuous spectrum, are necessary for most studies utilizing monochromators. The writer has been particularly pleased with the stability and spectral qualities of a 900 watt compact xenon lamp used over the spectral region from 240–700 nm (Figure 4).

The most recent sources of high intensity and extremely pure monochromatic light, are the lasers. Their potential use in light studies on fungi appear to be rather limited.

B. Light quality

The importance of light quality in the photo-induced phenomena exhibited by fungi has been amply documented in the preceding discussions. There is evidence that several photoreceptors can be present within a single fungus, each having its own characteristic absorption spectrum and each initiating a rather specific series of events. Thus the quality of light effective in causing mutations in a species of *Stemphylium* will differ from that most effective in causing photo-reversal of these mutations and likewise this will differ from that most effective in photosporogenesis and pigmentation changes. One of the first steps in any light investigation is to determine approximately which wavelengths are effective in causing the particular phenomenon under study. This is usually achieved initially by the use of broad band filters in combination with a light source that emits over a broad region of the visible and UV spectrum. Once this preliminary information has been obtained the investigator is in a much better position to select for the most effective source of radiation and then to proceed to studies having greater precision.

1. *Isolation of spectral bands*

The usual procedure for defining the qualitative light relationships for a particular photobiological effect is to move step by step through a series of studies first using broad bands of heterochromatic light, then narrower and finally to monochromatic light studies using very narrow bands of radiation.

(a) *Broad band isolation.* This can be achieved in a rather crude manner by selecting lamps that emit over different regions of the spectrum. The investigator can progress from the far-UV through the visible spectrum by beginning with a "germicidal lamp" (Mainly 253 nm), then a "sun lamp" (260–400 nm), followed by a "Black Light" fluorescent lamp (310–420 nm) and finally a "daylight" fluorescent lamp (340–750 nm). More usually filters (glass, gelatin, liquid, celluloid, etc.) are used and these provide a simple and fairly inexpensive means of defining the biologically effective wavelengths. Where it is necessary to illuminate large areas, for example, under banks of "daylight" fluorescent lamps, large sheets of celluloid or cellophane type filters are most useful (e.g., "Cinemoid" filters, Kliegl Bros., 321 W. 50th Street, New York, U.S.A.). The transmission of "Cinemoid" filters should always be checked carefully spectrometrically, for in some, UV wavelengths are transmitted that are not

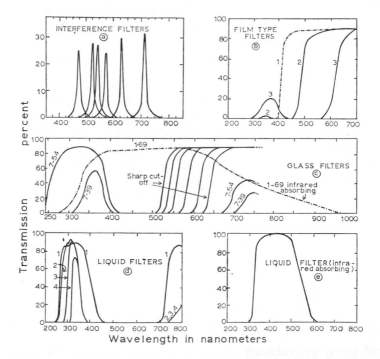

FIG. 5. Filters used in photobiology: (a) Narrow band interference filters (from catalogue of Farrand Optical Co., Inc., U.S.A.); (b) Film type filters: (1) Polyester W "Mylar" film (E.I. DuPont de Nemours and Co., Inc., U.S.A.); (2) "Cinemoid" yellow filter, (3) "Cinemoid" red filter (Kliegl Bros., New York, U.S.A.); (c) miscellaneous glass filters (from catalogue of Corning Glass Co., U.S.A.); (d) UV transmitting liquid filters: (1) 200 g/litre $CoCl_2 \cdot 6H_2O$ in 80% ethanol; (2), (3), and (4) different combinations of $CoCl_2 \cdot 6H_2O$ and $NiCl_2 \cdot 6H_2O$ (Wladmiroff, 1966); (e) infrared absorbing visible transmitting liquid filter, 100 g/litre $CuSO_4 \cdot 5H_2O$ (Koller, 1965 based on Kasha, 1948).

evident in the manufacturers' spectral data (Figure 5). There are a wide variety of glass filters (Figure 5) obtainable from a number of manufacturers (e.g., Corning Glass Works, Corning, New York, U.S.A.) and also gelatin filters (Eastman Kodak Company, Rochester, New York, U.S.A.). Sharp cut-off filters (Figure 5) are most useful for narrowing the limits of biologically effective wavelengths. Liquid filters (Figure 5) employing solutions of various chemicals, can be constructed in the laboratory and are often useful for both UV (Wladimiroff, 1966) and visible radiation studies. Large liquid filters constructed from glass and various transparent plastics, are useful for covering areas under banks of fluorescent lamps and in addition they can be used reduce radiant heat (Figure 5).

Though many filters are quite stable when used over long periods of time, in some the quality and percentage of transmitted radiation can change appreciably. Filters should be periodically checked spectrophotometrically to avoid erroneous results in light studies conducted on fungi. Solarization of glass in UV studies (Koller, 1965) may greatly reduce the percentage and quality of transmitted radiation (Figure 7).

(b) *Narrow band isolation*. Use of interference filters offers one means of obtaining narrow bands of radiation (Figure 5). Rakoczy (1963, 1965) has successfully used these filters for monochromatic light studies on the sporulation of *Physarum nudum*. Interference filters are fairly expensive and require that incident radiation be parallel.

Monochromators are the principle means of isolating narrow bands of radiation. The theory and design of these instruments is discussed by Johns and Rauth (1965a). There are two principle types of monochromators, those employing a prism, and those using diffraction gratings. A prism monochromator useful both for UV and visible light studies generally employs a quartz prism for dispersing light into its different components. A disadvantage of the prism monochromator is that its linear dispersion of light is a function of wavelength and this changes very rapidly in the UV, thus slit widths have to be changed for each wavelength in order to maintain a constant band width. Diffraction grating monochromators, show virtually no dependence of dispersion on wavelength and therefore a single slit width will provide a constant band width for all wavelengths used. One disadvantage of the grating monochromator for visible light studies is the presence of a second order spectrum which overlaps and is superimposed upon the first order spectrum. This is not a problem in the UV spectrum. Use of appropriate filters will easily eliminate the second order spectrum. Jagger (1967) highly recommends the Bausch and Lomb 500 mm grating monochromator (1200 grooves/mm) for biological studies. The writer has been most satisfied with the 250 mm model manufactured by the same company (Bausch and Lomb, Inc., Rochester 2, New York, U.S.A.).

Most commercially available monochromators are suitable only for irradiating relatively small areas of the order of several square centimeters. Large monochromators can be built using relatively inexpensive water prisms that are constructed from sheets of quartz glass filled with distilled water (Fluke and Setlow, 1954). Monk and Ehret (1956) describe a very large spectrograph suitable for biological studies.

A useful practical guide to the use of monochromators in photobiological studies is provided by Jagger (1967). Johns and Rauth (1965b) discuss the problems of "stray light" in monochromators.

The laser also provides a source of high intensity, extremely pure monochromatic light; however, the application of lasers to photobiological studies on fungi would appear to be very limited.

2. *Transmission of plastics, glassware and media*

Fungi are frequently exposed to UV or visible radiation while growing within some form of transparent or semi-transparent container. It is usually essential that the effect of containers on both the quality and intensity of radiation be known. Such information can sometimes be found in manufacturers' specifications but more usually it has to be determined spectrophotometrically within the investigator's own laboratory. Figure 6 shows the UV and visible light transmission of a few commonly used laboratory materials.

Fungi are commonly irradiated while growing on an agar or liquid

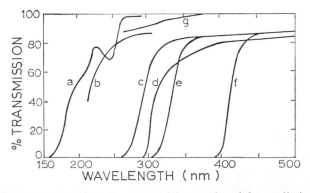

FIG. 6. Transparency of various materials to ultraviolet radiation: (a) clear fused quartz, 1 cm thick; (b) "Vycor" glass No. 791, 1 mm thick; (c) "Pyrex" petri dish lid, 2 mm thick; (d) plastic petri dish lid, 0·7 mm thick; (e) window glass, 4 mm thick; (f) Polyester W "Mylar" film; and (g) distilled water, 6 in. deep (redrawn from Koller, 1965; Leach, 1967a).

medium. Under such circumstances the medium itself may act as a filter and submerged hyphae may receive radiation qualitatively and quantitatively different from surface hyphae (Figure 3).

During studies necessitating the incubation and irradiation of colonies over long periods of time, it is usually impossible to directly expose colonies because of the resulting contamination and dehydration of the medium. To surmount these difficulties, colonies can be covered with tautly stretched "uncoated" or non-waterproof type cellophane (Held and Walker, 1957). Uncoated cellophane transmits efficiently well into the far-UV. It can be obtained from Film Department, E. I. Dupont De Nemours Co., Inc., Wilmington, Delaware, U.S.A. (No. 215 PD cellophane).

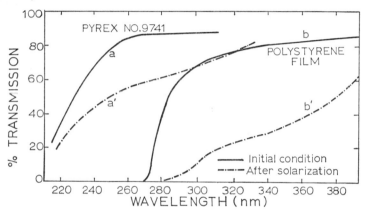

FIG. 7. Effect of exposing "Vycor" glass and polystyrene film to ultraviolet radiation (redrawn from Koller, 1965).

Some plastics and filters fluoresce when exposed to UV (French, 1965). In precise monochromatic light studies the investigator should be aware of this for it may result in the specimen being subjected to "impure" light.

C. Light measurement

No single radiation measuring device or technique is suitable for all photobiological studies on fungi. The use of a single detector may be quite satisfactory for certain types of experiments, but more frequently a combination of instruments is best employed. This subject is well covered by the articles cited in the Appendix, my discussion will be quite brief. The reader may also wish to refer to the article by Clarke (1964) in which the performances of 25 different types of radiation detectors are compared.

1. *Non-selective radiometric devices*

Thermopiles, bolometers, and radiometers are non-selective detectors which depend upon the heating effect of radiation. A thermopile, for example, responds equally well to far-UV radiation as it does to red light (Figure 8). The sensitivity and design of these instruments varies considerably. There are, for example, thermopiles designed specifically for recording solar radiation, large surface thermopiles suitable for light

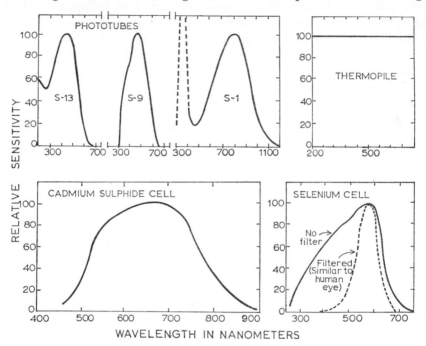

FIG. 8. Comparison of the spectral sensitivity of different types of light detectors (phototube responses from catalogue of Radio Corporation of America; cadmium sulphide cell response from catalogue of Mullard Ltd., England; selenium cell response from catalogue of Evans Electroselenium Ltd., England).

measurements under banks of lamps, vacuum, and non-vacuum types of compensated thermopiles with adjustable slits that are useful for monochromator measurements, and others. (Thermopiles are available from P. J. Kipp and Zonen, Delft, Holland; The Epply Laboratory, Inc., Newport, Rhode Island, U.S.A., Charles M. Reeder and Co., Inc., 173 Victor Avenue, Detroit, Michigan, U.S.A., Hilger and Watts Ltd., 98 St. Pancras Way, Camden Road, London, N.W.1, England.) These instruments though fairly sensitive, are both expensive and rather delicate and consequently they are sometimes used as standards from which other instruments are calibrated. Because of the sensitivity of this group of detectors to infra red radiation, it is often necessary to use heat absorbing filters when making measurements in the visible and UV spectra.

2. *Photo-electric detectors*

The spectral responses of this group of detectors are quite selective. They employ phototubes, photomultiplier tubes, photovoltaic cells, and photoinductive cells, and for effective use, their spectral response curves should be known (Fig. 8). Detectors using phototubes and photomultipliers are extremely sensitive and are essential for low intensity studies. A wide range of phototubes and photomultiplier tubes are available (refer to catalogue of Radio Corporation of America, Electron Tube Division, Harrison, N.J., U.S.A.) from which the investigator can select for maximum sensitivity to a particular band of wavelengths (Figure 8). The tubes are incorporated into various types of electronic photometers (photometers available from Photovolt Corporation, 95 Madison Avenue, New York 16, New York, U.S.A.). Tukey *et al.* (1960) have described a simple and relatively inexpensive phototube-type "illumination totalizer". This can be used to record continuously the amount of light energy from solar radiation or a bank of lamps. The instrument was designed for visible light measurements but a few simple modifications will adapt it for near-UV studies.

The photovoltaic cell type radiation detectors are relatively cheap, readily available and include the light-meters (selenium cell types) that are commonly used in photography (Figure 8). The photoconductive cell light meters (e.g., CdS cell) are also commonly used but unlike the photovoltaic cell, they require an outside source of electricity and are usually battery operated (Figure 8). The use of light meters has been extensive in fungal studies but unfortunately they have often been misused. Some of the published results based on light meter reading (usually expressed in ft candles or lux) are meaningless because their spectral selectivity (Figure 8) was ignored. Light meters are useful for measuring the uniformity of illumination under large banks of lamps, for obtaining comparable intensities in repeated experiments and for preliminary light quality studies. In light quality studies it is essential that allowances be made for the selectiveness of their spectral response. For precise studies over widely separated spectral bands, other radiation detectors are usually more satisfactory.

3. *Chemical methods*

Several radiation detectors are described that depend on solutions which undergo visible or easily measurable photochemical changes. These are

known as actinometers and are selective in their spectral response. McLaren
and Shuggar (1964) describe (in their "Appendix") procedures for several
liquid actinometers particularly useful in UV studies.

Photographic film is occasionally used for measuring light intensity.
Films are selective in their response to UV and visible radiation and the
investigator should be aware of the response of the film being used. Intensity
can be measured by measuring the degree of blackening and correlating
this with readings from another type of detector. Under carefully controlled
conditions of exposure and development, this method is capable of a high
degree of accuracy (Koller, 1965). A small piece of unexposed film placed
with dark "controls" is a useful means for detecting any accidental light
leakage.

D. Uniformity of illumination

Uniformity of illumination is important whether one is exposing a small
portion of a colony to monochromatic radiation, or growing many colonies
under fluorescent lamps. Measurements of light intensity must be made at
a number of positions to determine both the range and spatial pattern of
variation. As absolute uniformity of illumination is not always easily
achieved so replicated treatments and a statistical design are frequently
employed to account for experimental error.

Under banks of fluorescent lamps there is always considerable variation
of intensity within the illuminated area. Lowest intensities occur directly
below the ends and sides of the lamps. This can be reduced by painting
the surrounding walls of a light chamber with aluminium paint for near-UV
studies, and white paint for visible light studies. Layers of cheesecloth
(gauze) placed over the exposed colonies, can be used to balance intensities.
When space is not at a premium, colonies should be restricted to the area of
greatest uniformity directly below the central region of the lamps.

In preliminary light quality studies conducted at constant intensity,
there is always a need to balance the intensities that occur below filters.
This can be achieved either by conducting separate experiments for each
filter and varying the distance between lamps and specimen, or conducting
a single experiment in which intensities under the filters are balanced
simply by covering them with varying layers of cheese cloth (gauze).

Uniformity of intensity is particularly important in precise studies
using monochromators. This can usually be achieved by the correct
alignment and positioning of the light source and optical system of the
monochromator. In UV studies, the uniformity of the illuminated area
can be roughly ascertained by placing a piece of uranium glass at the plane
of focus. The uranium glass fluoresces brightly when exposed to UV and
this will visually reveal any major irregularities. In visible light studies a

white card can be similarly placed at the plane of focus to determine major irregularities. Uniformity of illumination can be more carefully checked by slowly moving a thermopile or photometer (equipped with a narrow slit) through the illuminated area.

In addition to the spatial variation of intensity as represented by uneven illumination, intensity may also vary with time during an experiment. Generally this is due to a change in intensity of emission of the radiation source. This can be caused by: (a) lack of a warm-up period for lamps; (b) fluctuating lamp-supply voltage; (c) aging of lamps; (d) effect of marked ambient temperature changes on a lamp output; (e) instability characteristic of a particular type of lamp; (f) a defective lamp or defective power supply; and (g) other causes.

E. Temperature control

The physiological condition of a fungus just prior to its exposure to light, and the subsequent chain of biochemical events initiated by light, are influenced by temperature. Temperature control during photobiological studies on fungi is essential.

1. *Ambient temperature*

Ambient temperature is generally controlled by the use of temperature rooms, small incubators or water baths. Temperature fluctuation should always be kept to a minimum by the use of well placed precision thermo-switches and adequate air or water circulation. In photosporogenesis studies, for example, a difference of 2°C can lead to quite different results (Leach, 1967; 1968).

Controlled "temperature platforms" upon which fungi are grown and irradiated, are most useful for light studies. In our laboratory, we use large hollow platforms (3 cm × 50 cm × 100 cm) through which controlled temperature water is circulated (Leach, unpublished). The micro-climate of fungal colonies grown within Petri dishes on these platforms, adjusts closely to the temperature of the platform and is little affected by the ambient temperature under normal conditions. "Temperature platforms" are very versatile and they can be designed for use under banks of fluorescent lamps or for small scale monochromator studies.

2. *Interaction of light and temperature*

One of the most useful means of studying the effects of the interaction of light and temperature on fungi is by employing temperature-gradient plates (Halldal and French, 1955; Larsen, 1965). Temperature gradient plates are very useful in photosporogenesis studies (Leach, 1967) and they are applicable to most photobiological studies on fungi. They may be constructed from a solid aluminium plate (e.g., $60 \times 122 \times 1\cdot6$ cm), heated at one end by circulating warm water (circulation holes drilled through aluminium plate) and by cold water at the other end.

The effect of the interaction of light and temperature on photobiological phenomena can also be studied using a series of controlled temperature incubators each equipped with lamps, or alternatively when facilities are limited by conducting a series of experiments at different temperatures using only one incubator. It is important to recognize that the output of some lamps (e.g., fluorescent lamps) vary significantly with change in ambient temperature, therefore for constant intensity it is advisable to enclose the lamps in a self contained constant temperature chamber, separated from the incubation chamber by a transparent sheet of plastic (or cellophane, quartz, etc.). Water baths are also useful in temperature–light studies on fungi (Aragaki, 1961).

3. *Problem of radiant heat*

The radiant heat emitted from lamps can significantly influence the results of light experiments by raising the temperature of the fungal microclimate over the ambient temperature, and, by causing erroneous intensity measurements when using non-selective radiometric detectors. The amount of radiant heat emitted and the problems of counteracting it will vary with the type of lamp used and other factors. The influence of radiant heat can often be reduced or eliminated by: (a) increasing the distance between lamp and specimen; (b) placing either a heat absorbing or heat reflecting glass filter, or heat absorbing water filter (Figure 5) between lamp and specimen; (c) exposing fungi on controlled "temperature platforms" as previously described (refer to "Ambient temperature" section). In addition to temperature control these platforms can be used as a "heat sink" which counteracts the radiant heat effects within the fungal micro-climate.

F. Micro-irradiation

Irradiation of fungi using a micro-beam of monochromatic light that focuses into a small spot several microns in diameter offers a potentially useful technique for photobiological studies on fungi. Micro-irradiation of living cells is employed for two main purposes (Bessis, 1965): (a) to analyse the function of cellular constituents by selectively altering them; (b) to study the effects of radiation on different parts of living cells, effects which are little understood because they are masked by more general reactions when the radiation is spread over the whole cell. I am unaware of any reports on the application of micro-beam techniques to studies on fungi. The small size of fungal hyphae and nuclei may preclude the use of these techniques in photobiological studies.

G. Age and physiological condition of mycelium

The response of a fungus to light is dependent on its physiological condition at the time of irradiation. Thus hydrated spores are more readily killed by far-UV than dry spores; young mycelium is usually more sensitive to light than old mycelium in photosporogenesis studies; inhibition of germination of uredospores by blue light occurs in the early stages of germination of uredospores by blue light occurs in the early stages of germination but not in the later stages; light is needed initially to induce sporulation in *Stemphylium botryosum* but thereafter it becomes inhibitory; *Pyronema omphaloides* can be stimulated by light to produce apothecia on a nutritionally deficient medium but not on rich medium, and so on. These are but a few randomly chosen examples to illustrate the importance of the physiological condition of the fungus in irradiation studies. The physiological condition of hyphae is influenced by age, nutrition, presence of staling products, hydrogen-ion concentration of the substrate, temperature, previous light treatment, etc., and it is important that every attempt be made to standardize and control as many of the major variables as is possible.

TABLE I

Terminology and units of radiometry and photometry

Pertains to	Radiometry		Photometry	
	Term	Units	Term	Units
Beam	Radiant energy	erg, calorie	Luminous energy	lumerg
Beam	Radiant flux	erg sec^{-1}, watt	Luminous flux	lumen
Beam	Radiant flux density	watt cm^{-2}	Luminous flux density	lumen m^{-2}
Source	Radiant intensity	watt ω^{-1}	Luminous intensity (candlepower)	lumen ω^{-1} (candle)
Source	Radiance	wat ω^{-1} cm^{-2}	Luminance (brightness)	lumen ω^{-1} cm^{-2} (lambert)
Object	Irradiance	watt cm^{-2} erg cm^{-2} sec^{-1}	Illuminance (illumination)	lumen m^{-2} (lux), lumen ft^{-2} (foot-candle)

ω = solid angle (steradians).

Reproduced from J. Jagger, 1967 (with slight modification) by permission of Prentice-Hall, Inc., Englewood Cliffs, N.J.

TABLE II

Conversion factors of radiant energy, power and intensity units

Energy	erg	joule	g-cal	whr	kg-cal
erg, d-cm	1	10^{-7}	$0\cdot239\times10^{-7}$	$0\cdot278\times10^{-10}$	$0\cdot239\times1^{-10}$
joule, w-sec	10^7	1	$0\cdot239$	$0\cdot278\times10^{-3}$	$0\cdot239\times10^{-3}$
g-cal	$4\cdot19\times10^7$	$4\cdot19$	1	$1\cdot163\times10^{-3}$	10^{-3}
whr	$3\cdot60\times10^{10}$	3600	860	1	$0\cdot860$
kg-cal	$4\cdot19\times10^{10}$	4190	1000	$1\cdot16$	1

Power	erg sec^{-1}	μw	cal min^{-1}	w	cal sec^{-1}
erg sec^{-1}	1	$0\cdot1$	$1x43\times10^{-5}$	10^{-7}	$0\cdot239\times10^{-7}$
μw	10	1	$1\cdot43\times10^{-5}$	10^{-6}	$0\cdot239\times10^{-6}$
cal min^{-1}	$6\cdot98\times10^5$	$6\cdot98\times10^4$	1	$0\cdot0698$	$0\cdot0166$
w	10^7	10^6	$14\cdot3$	1	$0\cdot239$
cal sec^{-1}	$4\cdot19\times10^7$	$4\cdot19\times10^6$	60	$4\cdot19$	1

Intensity	erg sec^{-1} cm^{-2}	μw cm^{-2}	μw mm^{-2}	w m^{-2}	cal min^{-1} cm^{-2}
erg sec^{-1} cm^{-2}	1	$0\cdot1$	$0\cdot001$	$0\cdot001$	$1\cdot43\times10^{-6}$
μw cm^{-2}	10	1	$0\cdot01$	$0\cdot01$	$1\cdot43\times10^{-5}$
μw mm^{-2}	1000	100	1	1	$1\cdot43\times10^{-3}$
w m^{-2}	1000	100	1	1	$1\cdot43\times10^{-3}$
cal min^{-1} cm^{-2}	$6\cdot98\times10^5$	$6\cdot98\times10^4$	698	698	1

Brightness	foot-lambert	lambert	c cm^{-2}	c mm^{-2}
foot-lambert	1	$1\cdot08\times10^{-3}$	$3\cdot39\times10^{-3}$	$3\cdot39\times10^{-5}$
lambert	929	1	$0\cdot318$	$0\cdot318\times10^{-3}$
c cm^{-2}, stilb	2920	$3\cdot14$	1	$0\cdot01$
c mm^{-2}	$2\cdot92\times10^5$	314	100	1

Illuminance	lux	ft-c	lumen cm^{-2}
lux, m-c	1	$0\cdot093$	10^{-4}
ft-c	$10\cdot8$	1	$1\cdot08\times10^{-3}$
lumen cm^{-2}, phot	10^4	929	1

Reproduced from R. B. Withrow and A. P. Withrow, 1956 by permission McGraw-Hill Book Co., Inc., New York.

H. Action spectra

In the "triggering" of photobiological changes in fungi, it is generally assumed that a photoreceptor compound captures radiant energy which then initiates a whole sequence of biochemical events. Action spectra are used to help identify the photoreceptor pigment (French, 1959), but they are also useful for obtaining an insight into various aspects of fungal ecology and for selecting the best light source for inducing specific responses. The problems of measuring and understanding action spectra are well discussed by Allen (1964), French (1959), Brackett and Hollaender (1959), and French and Young (1956).

I. Terminology and dosimetry

The terminology and units of dosimetry used in light studies on fungi conducted over the past century are often inconsistent and in some instances confusing. Modern workers have attempted to clarify and standardize this aspect of photobiology and the discussions by Jagger (1967), Withrow and Withrow (1956), and Seliger and McElroy (1965) are recommended. This information is summarized in Tables I and II. Craig (1964) has a worthwhile discussion on the choice of units in photobiology.

J. Health hazards in radiation studies

1. *Radiation*

Ultraviolet radiation can cause skin cancer (Blum, 1955), erythema of the skin ("sun burning") and eye damage. Most dangerous are lamps emitting far-UV but Jagger (1967) recommends that investigators should avoid exposure even to long wavelength near-UV. UV injury can be prevented by avoiding all exposure, by shielding the source of radiation, by covering the skin, wearing protective gloves and wearing face masks equipped with UV absorbing glass (Starr, 1948; Koller, 1965). The reader should refer to Buttolph (1955) and Jagger (1967) for more detailed information. The writer has found that "Mylar" (Polyester W "Mylar" film manufactured by E. I. DuPont De Nemours Co, Inc.) is an excellent shielding material for UV studies (Figure 6).

2. *Ozone*

Far-UV radiation sources such as germicidal lamps, high intensity mercury, and compact xenon lamps, produce ozone. Ozone, which can be recognized by its characteristic odour, is toxic to humans at quite low concentration (Anonymous, 1966; Svirbely and Saltzman, 1957) and it is advisable to use lamps in well ventilated rooms. High intensity radiation sources should be exhausted to the outside atmosphere by means of ducts and exhaust fans.

3. *Explosion*

Many compact high intensity sources of radiation operate at high atmospheric pressures (e.g., 30 atmospheres) and if not handled with extreme care they may explode. Eyes, hands and skin should be protected when handling these lamps and they should be operated in an explosion proof housing which deflects the force of an explosion away from the investigator.

4. Electrical shocks

Some artificial sources of UV and visible light operate at high voltages and at several amperes. These levels can be lethal and normal precautions against electrical shocks should be taken.

APPENDIX

The following sources are recommended for those readers actively engaged in photobiological research on fungi. They go far beyond the limited treatment of this "practical guide".

Jagger, J. (1967). "Introduction to research in ultraviolet photobiology". Prentice-Hall, Inc., Englwood Cliffs, New Jersey. 164 pp. (theory, equipment, conduct of experiments, mechanism of action of UV, terminology, dosimetry, etc. Very useful section listing manufacturers of light equipment such as filters, lamps, monochromators, etc.).

Koller, L. R. (1965). "Ultraviolet radiation". John Wiley and Sons, Inc., New York, London and Sydney. 312 pp. (artificial sources of radiation, solar radiation, transmission, reflection, applications and effects of UV, detectors of UV, etc.).

Seliger, H. H., and McElroy, W. D. (1965). "Light: Physical and biological action". Academic Press, New York and London. 417 pp. (Measurement and characterization of light including photometry and light sources, conversion factors, lasers, data on filters, guide to the use of thermopiles, actinometers, as well as other subjects).

Withrow, R. B., and Withrow, A. P. (1956). "Generation, control, and measurement of visible and near-visible radiant energy". *In* "Radiation Biology", Vol. III, pp. 125–258. (Ed. A. Hollaender). McGraw-Hill Co., Inc., New York and London. (Characterization of light, units of energy, sources of radiant energy, filters and optical properties of materials, measuring instruments, voltage regulators, etc.)

REFERENCES

Airth, R. L., and Foerster, G. E. (1960). *J. cell. comp. Physiol.*, **56**(2), 173–182.
Alasoadura, S. O. (1963). *Ann. Bot.*, **27**, 123–145.
Alexopoulos, C. J. (1963). *Bot. Rev.*, **29**, 1–78.
Alexopoulos, C. J. (1966). *In* "The Fungi" Vol. II. (Ed. G. C. Ainsworth and A. S. Sussman), pp. 211–233. Academic Press, New York and London.
Allen, M. B. (1964). *In* "Photophysiology", Vol. I. (Ed. A. C. Giese), pp. 83–110. Academic Press, New York and London.

Anonymous. (1966). "Preventing accidental ozone poisoning in workers". PHS No. 1526 U.S. Dept. Health, Education and Welfare.
Anonymous. (1968). "Plant Pathologists' Pocketbook". Commonwealth Mycological Institute, Kew, England.
Aragaki, M. (1961). *Phytopathology*, **51**, 803–805.
Aragaki, M. (1962). *Phytopathology*, **52**, 1227–1228.
Aragaki, M., and Hine, R. B. (1963). *Phytopathology*, **53**, 854–856.
Aschan-Åberg, K. (1960). *Physiologia Pl.*, **13**, 276–279.
Asthana, R. P., and Hawker, L. E. (1936). *Ann. Bot.*, **50**, 325–344.
Backus, Myron P. (1937). *Mycologia*. **29**, 383–386.
Bailey, A. A. (1932). *Bot. Gaz.*, **94**, 225–271.
Banbury, G. H. (1959). *Handb. PflPhysiol.*, **17(1)**, 530–578.
Barker, R. E. (1968). *Photochem. Photobiol.*, **7**, 275–295.
Barnett, H. L. (1968). *Mycologia*, **60**, 244–251.
Barnett, H. L., and Lilly, V. G. (1950). *Phytopathology*, **40**, 80–89.
Barnett, H. L., and Lilly, V. G. (1952). *Proc. W. Va Acad. Sci.*, **24**, 60–64.
Baumgartner, J. G. (1936). *J. Bact.*, **32**, 75–77.
Beck, T., and Röschenthaler, R. (1960). *Bayer. landw. Jb.*, **37**, 2–82.
Berg, L. A., and Gallegly, M. E. (1966). *Phytopathology*, **56**, 583.
Berkner, L. V., and Marshall, L. C. (1964). *Discuss. Faraday Soc.*, **37**, 122–141.
Berliner, M. D. (1961). *Science, N.Y.*, **134**, 740.
Berliner, M. D., and Brand, P. B. (1962). *Mycologia*, **54**, 415–421.
Bessis, M. (1965). *In* "Recent Progress in Photobiology" (Ed. E. J. Bowen), pp. 291–309. Blackwell, Oxford.
Bjornsson, I. P. (1956). "Effects of light on *Stemphylium, Trichoderma, Botrytis*, and certain other fungi". Thesis. Univ. of Maryland, College Park, Md.
Bjornsson, I. P. (1959). *J. Wash. Acad. Sci.*, **49**, 317–323.
Blakeman, J. P., and Dickinson, G. H. (1967). *Trans. Br. mycol. Soc.*, **50**, 385–396.
Blank, I. H., and Arnold, W. (1935). *J. Bact.*, **30**, 507–511.
Blum, H. F. (1955). *In* "Radiation Biology", Vol. II (Ed. A. Hollaender), pp. 487–528. McGraw-Hill.
Bonner, J. T., and Shaw, M. J. (1957). *J. cell. comp. Physiol.*, **50**, 145–154.
Borriss, H. (1934). *Planta*, **22**, 644–684.
Brackett, F. S., and Hollaender, A. (1959). *In* "Photoperiodism" (Ed. R. B. Withrow), pp. 41–45. Am. Assoc. Advance. Sci., Washington, D.C.
Brandt, W. H. (1964). *Can. J. Bot.*, **42**, 1017–1023.
Brandt, W. H. (1968). Private communication.
Buller, A. H. R. (1931). "Researches on fungi", Vol. 4. Longmans, Green, London and New York.
Buller, A. H. R. (1933). "Researches on fungi", Vol. 5. Longmans, Green, New York.
Buller, A. H. R. (1934). "Researches on fungi", Vol. 6. Longmans, Green, New York.
Bünning, E. (1937). *Planta*, **26**, 719–736.
Buttolph, L. J. (1955). *In* "Radiation Biology", Vol. II (Ed. A. Hollaender), pp. 41–93. McGraw-Hill, New York.
Buxton, E. W., Last, F. T., and Nour, M. A. (1957). *J. gen. Microbiol.* **16**, 764–773.
Callaghan, A. A. (1962). *Trans. Br. mycol. Soc.*, **45**, 249–254.
Calpouzos, L., and Stallknecht, G. F. (1966). *Phytopathology*, **56**, 702–704.
Calpouzos, L., and Stallknecht, G. F. (1967). *Phytopathology*, **57**, 679–681.

Cantino, E. C. (1966). *In* "The Fungi", Vol. II (Ed. A. S. Sussman and G. C. Ainsworth), pp. 283–337. Academic Press, London and New York.

Cantino, E. C., and Horenstein, E. A. (1956). *Mycologia*, **48**, 777–799.

Carlile, M. J., Lewis, B. G., Mordue, E. M., and Northover, J. (1961). *Trans. Br. mycol. Soc.*, **44**, 129–133.

Carlile, M. J., Dickens, J. S. W., Mordue, E. M., and Shipper, M. A. A. (1962). *Trans. Br. mycol. Soc.*, **45**, 457–461.

Carlile, M. J., and Friend, J. (1956). *Nature, Lond.*, **178**, 369–370.

Carlile, M. J. (1965). *Ann. Rev. Plant Physiol.*, **16**, 175–202.

Carlile, M. J. (1966). *In* "The Fungus Spore" (Ed. M. F. Madelin), pp. 175–186. Butterworth, London.

Carlile, M. J. (1956). *J. gen. Microbiol.*, **14**, 643–654.

Carlson, J. G., Gaulden, M. E., and Jagger, J. (1961). *In* "Progress in Photobiology" (Ed. B. C. Christensen and B. Buchman), pp. 251–253. Elsevier, New York.

Caroselli, N. E., Mahadevan, A., and Mozumder, B. G. (1964). *Pl. Dis. Reptr.*, **48**(6), 484–486.

Castle, E. S. (1931). *J. gen. Physiol.*, **14**, 701–711.

Castle, E. S. (1966a). *J. gen. Physiol.*, **49**(5), 925–935.

Castle, E. S. (1966b). *Science, N.Y.*, **154**, 1416–1420.

Chakrabarti, N. K. (1968). *Phytopathology*, **58**, 467–471.

Chase, A. M. (1964). *In* "Photophysiology", Vol. II (Ed. A. C. Giese), pp. 389–421. Academic Press, New York.

Christenberry, G. A. (1938). *J. Elisha Mitchell scient. Soc.*, **54**, 297–310.

Christensen, B. C., and Buchman, B. (1961). "Progress in Photobiology". Elsevier, New York.

Clarke, F. J. J. (1964). *Photochem Photobiol.*, **3**, 91–96.

Clayton, R. K. (1964). In "Photophysiology", Vol. II (Ed. A. C. Giese), pp. 51–77. Academic Press, New York.

Cochrane, V. W. (1945). *Phytopathology*, **35**, 458–462.

Cochrane, V. W. (1958). "Physiology of fungi". J. Wiley, New York.

Cohn, M. L., and Middlebrook, G. (1965). *Ann. Rev. Resp. Dis.*, **91**, 929.

Craig, R. E. (1964). *Photochem. Photobiol.*, **3**, 189–194.

Cruickshank, I. A. M. (1963). *Aust. J. biol. Sci.* **16**(1), 88–98.

Curry, G. M., and Gruen, H. E. (1959). *Proc. natn. Acad. Sci. U.S.A.*, **45**, 797–804.

Curtis, C. R. (1964). *Phytopathology*, **54**, 1141–1145.

Daniel, J. W., and Rusch, H. P. (1962). *J. Bact.*, **83**, 234–240.

Daniels, M. (1968). Private communication.

Deering, R. A. (1962). *Scient. Am.*, **207** (Dec.), 135–144.

Delbrück, M. (1962). *Ber. dt. Bot. Ges.*, **75**, 411–430.

Delbrück, M., and Shropshire, W. (1960). *Pl. Physiol.*, **35**, 194–204.

Dillon Weston, W. A. R. (1932). *Phytopath. Z.*, **4**, 229–246.

Dillon Weston, W. A. R. (1932). *Scient. Agric.*, **12**, 352–356.

Dimond, A., and Duggar, B. M. (1941). *Proc. natn. Acad. Sci. U.S.A.*, **27**, 459–468.

Dowding, E. S., and Buller, A. H. R., (1940). *Mycologia*, **32**, 471–488.

Dulbecco, R. (1955). *In* "Radiation Biology", Vol. II (Ed. A. Hollaender), pp. 455–486. McGraw-Hill, New York.

Durbin, R. D. (1959). *Phytopathology*, **49**, 59–60.

Emge, R. C. (1958). *Phytopathology*, **48**, 649–652.

Errera, M. (1953). *In* "Progress in Biophysics", Vol. III (Ed. Butler, J. A. V. and J. T. Randall), pp. 88–130. Academic Press, New York.

Fergus, C. L., and Schein, R. D. (1963). *Mycologia*, **55**, 540–548.

Fincham, J. R. S., and Day, P. R. (1963). "Fungal genetics". Blackwell, Oxford.

Fluke, D. J., and Setlow, R. B. (1954). *J. opt. Soc. Am.*, **44**, 327–330.

Forssberg, A., Stankovic, V., and Pehap, A. (1964). *Radiation Bot.*, **4**, 323–330.

French, C. S. (1959). *In* "Photoperiodism and related phenomena in plants and animals" (Ed. R. B. Withrow), pp. 15–39. Am. Assoc. Advancement Science, Washington, D.C.

French, C. S., and Young, V. M. K. (1956). *In* "Radiation Biology", Vol. III (Ed. A. Hollaender), pp. 343–391. McGraw-Hill, New York.

French, C. S. (1965). *Appl. Optics*, **4**, 514.

Friederichsen, I., and Engel, H. (1957). *Planta*, **49**, 578–587.

Gassner, G., and Niemann, E. (1954). *Phytopath. Z.*, **21**, 367–394.

Gardner, E. B. (1955). *Trans. N. Y. Acad. Sci.*, Ser. II., **17**, 476–490.

Gates, D. M. (1966). *Science N.Y.*, **151**, 523–529.

Giese, A. C. (1942). *J. cell. and comp. Physiol.* **20**, 35–46.

Giese, A. C. (1964). "Photophysiology", Vol. II. Academic Press, New York.

Giese, A. C. (1964). *In* "Photophysiology", Vol. II (Ed. A. C. Giese), pp. 203–245. Academic Press, New York and London.

Givan, C. V., and Bromfield, K. R. (1964a). *Phytopathology*, **54**, 116–117.

Givan, C. V., and Bromfield, K. R. (1964b). *Phytopathology*, **54**, 382–384.

Gray, W. D. (1938). *Am. J. Bot.*, **25**, 511–522.

Gray, W. D. (1953). *Mycologia*, **45**, 817–824.

Gregory, P. H. (1961). "The microbiology of the atmosphere". Interscience, New York.

Gressel, J. B., and Hartman, K. M. (1967). *Planta*, **79**, 271–274.

Halldal, P., and French, C. S. (1955). Carnegie Inst. of Washington Year Book No. 55, pp. 261–265.

Harding, P. R. (1968). *Pl. Dis. Reptr.*, **52**, 245–247.

Harnish, W. H. (1965). *Mycologia*, **57**, 85–90.

Harter, L. L. (1939). *Amer. J. Bot.*, **26**, 234–243.

Harvey, E. N. (1952). "Bioluminescence". Academic Press, New York.

Haskins, R. H., and Weston, W. H. (1950). *Am. J. Bot.*, **37**, 739–750.

Hawker, L. E. (1950). "Physiology of fungi". Univ. of London Press.

Hawker, L. E. (1957). "The Physiology of Reproduction in Fungi". Cambridge Univ. Press.

Hawker, L. E. (1966). *In* "The Fungi", Vol. II (Ed. G. C. Ainsworth and A. S. Sussman), pp. 435–469. Academic Press, New York.

Haxo, F. (1949). *Archs. Biochem. Biophys.*, **20**, 400–421.

Heale, J. B., and Isaac, I. (1965). *Trans. Br. mycol. Soc.*, **48**, 39–50.

Heath, L. A. F., and Eggins, H. O. W. (1965). *Experientia*, **21**, 385–386.

Hebert, T. T., and Kelman, A. (1958). *Phytopathology*, **48**, 102–106.

Held, V. M., and Walker, R. L. (1957). *Phytopathology*, **47**, 573.

Hendrix, J. W. (1967). *Mycologia*, **59**, 1107–1111.

Herndon, R. W. (1967). "The influence of ultraviolet and visible light on the sporulation of the fungus *Stemphylium solani* Weber". M.S. Thesis, Univ. of Vermont, Burlington, Vermont.

Hevesy, G., and Zeradv, K. (1946). *Acta. radiol.*, **27**, 316–327.

Hollaender, A. (1943). *J. Bact.*, **46**, 531–541.

Hollaender, A., and Emmons, C. W. (1938). *J. cell. comp. Phys.*, 13, 391–402.

Houston, B. R., and Oswald, J. W. (1946). *Phytopathology*, 36, 1049–1055.

Ingold, C. T. (1958). *Ann. Bot.*, 22, 129–135.

Ingold, C. T., and Hadland, S. A. (1959). *Ann. Bot.*, 23, 425–429.

Ingold, C. T. (1962). *Symp. Soc. exp. Biol.*, 16, 154–169.

Ingold, C. T., and Marshall, (1963). *Ann. Bot.*, 27, 481–491.

Ingold, C. T. (1965). "Spore liberation". Clarendon Press, Oxford.

Ingold, C. T. (1966). *In* "The Fungi", Vol. II (Ed. G. C. Ainsworth and A. S. Sussman), pp. 679–707. Academic Press, New York.

Ingold, C. T., and Nawaz, M. (1967). *Ann. Bot.*, 31, 469–477.

Isaac, I. (1967). *Ann. Rev. Phytopathology*, 5, 201–222.

Jagger, J. (1958). *Bact. Rev.*, 22, 99–142.

Jagger, J. (1960). *In* "Radiation Protection and Recovery" (Ed. A. Hollaender), pp. 352–377. Pergamon Press, London and New York.

Jagger, J. (1967). "Introduction to research in ultraviolet photobiology". Prentice-Hall, Inc., New Jersey.

James, A. P., and Werner, M. M. (1965). *Radiat. Bot.*, 5, 359–382.

Jerebzoff, S. (1960). *c.r. hebd. Séanc. Acad. Sci.*, *Paris*, 250, 1549–1551.

Jerebzoff, S. (1965). *In* "The Fungi", Vol. I (Ed. G. C. Ainsworth and A. S. Sussman), pp. 625–645. Academic Press, New York.

Johns, H. E., and Rauth, A. M. (1965a). *Photochem. Photobiol.*, 4, 673–692.

Johns, H. E., and Rauth, A. M. (1965b). *Photochem. Photobiol.*, 4, 693–707.

Johnson, T. W., and Halpin, J. E. (1954). *J. Elisha Mitchell scient. Soc.*, 70, 314–326.

Joly, P. (1962). *Revue. Mycol.*, 27, 1–16.

Kaiser, W. J. (1964). *Phytopathology*, 54, 765–770.

Kasha, M. (1948). *J. opt. Soc. am.*, 38, 929–934.

Kelner, A. (1949). *Proc. natn. Acad. Sci. U.S.A.*, 35, 73–79.

Kelner, A. (1961). *In* "Progress in Photobiology" (Ed. B. C. Christensen and B. Buchmann), pp. 276–278. Elsevier, New York.

Knox-Davis, P. S. (1965). *S. Afr. J. agric. Sci.*, 8, 205–218.

Koizumi, M. (1965). *Rev. appl. Mycol.*, 45, 258.

Koller, L. R. (1965). "Ultraviolet radiation". John Wiley and Sons, Inc., New York.

Krause, R. (1960). *Arch. Mikrobiol.*, 36, 373–386.

Kusano, S. (1930). *Jap. J. Bot.*, 5, 35–132.

Larsen, A. L. (1965). *Proc. int. Seed Test Ass.*, 30, 861–868.

Last, F. T., and Buxton, E. W. (1955). *Nature, Lond.*, 176, 655.

Leach, C. M. (1961). *Can. J. Bot.*, 39, 705–715.

Leach, C. M. (1962a). *Can. J. Bot.*, 40, 151–161.

Leach, C. M. (1962b). *Can. J. Bot.*, 40, 1577–1602.

Leach, C. M. (1963). *Mycologia*, 55, 151–163.

Leach, C. M. (1964). *Trans. Br. mycol. Soc.*, 47, 153–158.

Leach, C. M., and Trione, E. J. (1965). *Pl. Physiol.*, 40, 808–812.

Leach, C. M., and Trione, E. J. (1966). *Photochem. Photobiol.*, 5, 621–630.

Leach, C. M. (1967a). *Proc. int. Seed Test. Ass.*, 32, 565–589.

Leach, C. M. (1967b). *Can. J. Bot.*, 45, 1999–2016.

Leach, C. M. (1968). *Mycologia*, 60, 532–546.

Leal, J. A., and Gomez-Miranda, B. (1965). *Trans. Br. mycol. Soc.*, 48, 491–494.

Lee, P. C., and Scott, W. W. (1967). *Res. Div., Virginia Polytech. Inst., Bull.,* **2**, 35.

Leff, J., and Krinsky, N. I. (1967). *Science, N.Y.,* **158**, 1332–1334.

Lehmann, H. (1965). *Phytopath. Z.,* **53**, 255–288.

Leith, H. (1956). *Arch. Mikrobiol.,* **24**, 91–104.

Lilly, V. G., and Barnett, H. L. (1951). "Physiology of the fungi". McGraw-Hill, New York.

Lilly, V. G. (1966). *In* "The Fungus Spore" (Ed. M. F. Madelin), pp. 259–272. Butterworth, London.

Lingappa, Y., Sussman, A. S., and Bernstern, I. A. (1963). *Mycopath. Mycol. appl.,* **20**, 109–128.

Loofbourow, J. R. (1948). *Growth,* **12**, Suppl., 77–149.

Lu, B. C. (1965). *Am. Jl Bot.,* **52**, 432–437.

Lukens, R. J. (1963). *Am. J. Bot.,* **50**, 720–724.

Lukens, R. J. (1965). *Phytopathology,* **55**, 1032.

Lukens, R. J. (1966). *Phytopathology,* **56**, 1430–1431.

Lythgoe, J. N. (1961). *Trans. Br. mycol. Soc.,* **44**, 199–213.

Madelin, M. F. (1956). *Ann. Bot.,* N. S. **20**, 467–480.

Madelin, M. F. (1966). "The Fungus Spore". Butterworth, London.

Manachére, G. (1967). *Bull. Soc. mycol. Fr.,* **83**, 257–285.

Mandels, G. R. (1965). *In* "The Fungi", Vol. I (Ed. G. C. Ainsworth, and A. S. Sussman), pp. 599–612. Academic Press, New York and London.

Marsh, P. B., Taylor, E. E., and Bassler, L. M. (1959). *Pl. Dis. Reptr.,* Supple **261**, 251–312.

McAulay, A. L., Ford, J. M., and Dobie, D. L. (1949). *Heredity,* **3**, 109–120.

McCallan, S. E. A., and Chan, S. Y. (1944). *Contrib. Boyce Thompson Inst.,* **13**, 323–336.

McLaren, A. D. and Shuggar, D. (1964). "Photochemistry of proteins and nucleic acids". Macmillan Co., New York.

Mertz, D., and Henson, W. (1967). *Nature, Lond.,* **214**, 844–846.

Merz, W. G. (1965). "Effect of light on species of *Phytophthora*". M.S. Thesis, West Virginia University, Morgantown, W. Va.

Miller, O. K. (1967). *Can. J. Bot.,* **45**, 1939–1943.

Mohr, H. (1961). *Handb. PflPhysiol.,* Vol. **16**, 439–531.

Monk, G. S., and Ehret, C. F. (1956). *Radiation Res.,* **5**, 88–106.

Murty, C. N., and Guha, A. (1963). *Nature, Lond.,* **198**, 902–903.

Nair, P., and Zabka, G. G. (1965). *Mycopath. Mycol. appl.,* **26**, 123–128.

Nair, P., and Zabka, G. G. (1966). *Am. J. Bot.,* **53**(9), 887–892.

Norell, I. (1954). *Physiologia Pl.,* **7**, 797–809.

Page, O. T. (1956). *Can. J. Bot.,* **34**, 881–890.

Page, R. M. (1956). *Mycologia.,* **48**, 206–224.

Page, R. M. (1962). *Science, N.Y.,* **138**, 1238–1245.

Page, R. M. (1964). *Science, N.Y.,* **146**, 925–927.

Page, R. M. (1965). *In* "The Fungi", Vol. I (Ed. G. C. Ainsworth and A. S. Sussman), pp. 559–574. Academic Press, New York and London.

Page, R. M. (1968). *In* "Photophysiology", Vol. III (Ed. A. C. Giese), pp. 65–90. Academic Press, New York.

Pederson, V. D. (1964). *Phytopathology,* **54**(8), 903.

Petri, L. (1925). *Rev. appl. Mycol.,* **10**, 122–123.

Plunkett, B. E. (1958). *Ann. Bot.,* **22**, 237–249.

Plunkett, B. E. (1961). *Ann. Bot.*, **25**, 206–223.

Pollister, A. W. (1955). *In* "Radiation Biology", Vol. II (Ed. A. Hollaender), pp. 203–248. McGraw-Hill, New York.

Pomper, S. (1965). *In* "The Fungi", Vol. I (Ed. G. C. Ainsworth and A. S. Sussman), pp. 575–597. Academic Press, New York.

Pomper, S., and Atwood, K. C. (1955). *In* "Radiation Biology", Vol. II (Ed. A. Hollaender), pp. 431–453. McGraw-Hill, New York.

Rakoczy, L. (1962). *Acta Soc. Bot. Pol.*, **31**, 651–665.

Rakoczy, L. (1963). *Bull. Acad. pol. Sci. Chll. Ser. Sci. biol.*, **11**(11), 559–562.

Rakoczy, L. (1965). *Acta Soc. Bot. Pol.*, **34**, 97–112.

Rakoczy, L. (1966). *Acta Soc. Bot. Pol.*, **35**, 315–324.

Rakoczy, L. (1967). *Acta Soc. Bot. Pol.*, **36**, 154–159.

Ramsey, G. B., and Bailey, A. A. (1930). *Bot. Gaz.*, **89**, 113–136.

Rau, W. (1967). *Planta*, **72**, 14–28.

Raper, K. B. (1962). *Harvey Soc. Lect.*, **57**, 111–141.

Reichardt, W. (1961). *Kybernetik*, **1**, 6–21.

Reinert, J. (1959). *A. Rev. Pl. Physiol.*, **10**, 441–458.

Reinhardt, D. J. (1968). *Am. J. Bot.*, **55**, 77–86.

Remsberg, R. E. (1940). *Mycologia*, **32**, 52–96.

Rich, S., and Tomlinson, H. (1968). *Phytopathology*, **58**, 444–446.

Robinson, N. (1966). "Solar Radiation". Elsevier, New York.

Romero, S., and Gallegly, M. E. (1963). *Phytopathology*, **53**, 899–903.

Rubbo, S. D., and Gardner, J. F. (1965). "A review of sterilization and disinfection". Lloyd-Luke Ltd., London.

Rupert, C. S. (1964). *In* "Photophysiology", Vol. II (Ed. A. C. Giese), pp. 283–327. Academic Press, New York.

Sagromsky, H. (1952a). *Flora*, **139**, 300–313.

Sagromsky, H. (1952b). *Flora*, **139**, 560–564.

Sanderson, J. A., and Hulbert, E. O. (1955). *In* "Radiation Biology", Vol. II (Ed. A. Hollaender), pp. 95–118. McGraw-Hill, New York.

Sargeant, M. L., and Briggs, W. R. (1967). *Pl. Physiol.*, **42**, 1504–1510.

Schaeffer, P. (1953). *Archs biochem. Biophys.*, **47**, 359–379.

Schneider, R. (1965). *Phytopath. Z.*, **53**, 249–254.

Seifriz, W. (1943). *Bot. Rev.*, **9**, 49–123.

Seliger, H. H., and McElroy, W. D. (1965). "Light: Physical and Biological Action". Academic Press, New York.

Sempio, C., and Castori, M. (1950). *Riv. Biol.*, N.S. **42**, 287–293.

Setlow, J. K. (1966). *In* "Current Topics in Radiation", Vol. II (Ed. M. Ebert and A. Howard), pp. 195–248. North-Holland, Amsterdam.

Sharp, E. L., Schmith, C. G., Staley, J. M., and Kingsolver, C. H. (1958). *Phytopathology*, **48**, 469–474.

Shropshire, W. (1963). *Physiol. Rev.*, **43**, 38–67.

Shropshire, W., and Gettens, R. H. (1966). *Pl. Physiol.*, **41**(2), 203–207.

Smith, E. C. (1936). *In* "Biological effects of radiation", Vol. II (Ed. B. M. Duggar), pp. 889–918. McGraw-Hill Book Co., New York.

Smith, K. C. (1964). *In* "Phytophysiology", Vol. II (Ed. A. C. Giese), pp. 329–388. Academic Press, New York and London.

Snyder, W. C., and Hansen, H. N. (1941). *Mycologia*, **33**, 580–591.

Stålfelt, M. G. (1956). *In* "Radiation Biology", Vol. III (Ed. A. Hollaender), pp. 551–580. McGraw-Hill, New York.

Staples, R. C., and Wynn, W. K. (1965). *Bot. Rev.* **31**, 537–564.

Starr, R. (1948). U.S. Natn. Bureau of Standards, Circular 471.

Stevens, F. L. (1928). *Bot. Gaz.*, **86**, 210–225.

Stevens, F. L. (1930). *Zentbl. Bakt. ParasitKde*, **82**, 161–174.

Stevens, F. L. (1931). *Mycologia*, **23**, 134–139.

Stinson, R. H., Gage, R. S., and MacNaughton, E. B. (1958). *Can. J. Bot.*, **36**, 927–934.

Stoll, K. (1936). *Zentbl. Bakt. ParasitKde*, **93**, 296–298.

Stone, W. S., Wyss, O., and Haas, F. (1947). *Proc. nat. Acad. Sci. U.S.A.*, **33**. 59–66.

v. Stosch, H. A. (1965). *Handb. PflPhysiol.*, **15**(1), 641–679.

Straub, J. (1954). *Naturwiss enschaften*, **41**, 219–220.

Sussman, A. S., and Halvorson, H. O. (1966). "Spores". Harper and Row, New York.

Svirbely, J. L., and Saltzman, B. E. (1957). *A. M. A. Archs ind. Hlth*, **15**, 111–118.

Swenson, P. A. (1950). *Proc. natn. Acad. Sci. U.S.A.*, **36**, 699–703.

Swenson, P. A. (1960). *J. cell. comp. Physiol.*, **56**, 77–91.

Swenson, P. A., and Dott, D. H. (1961). *J. cell. comp. Physiol.*, **58**, 217–231.

Sykes, G. (1965). "Disinfection and sterilization". Spon, London.

Szaniszlo, P. J. (1965). *J. Elisha Mitchell scient. Soc.*, **81**(1), 10–15.

Terry, C. E., and Setlow, J. K. (1967). *Photochem. Photobiol.*, **6**, 799–803.

Thimann, K. V., and Gruen, H. E. (1960). *Beiheft zu den Zeitschr. des Schweiz. Forstv.*, **30**, 237–263.

Thomas, C. E., and Halpin, J. E. (1964). *Phytopathology* (Abstr.) **54**, 910.

Thomas, C. E. (1966). *Diss. Abstr.*, **27**(5)**B**, 1364.

Thomas, J. B. (1965). "Primary photoprocesses in biology". J. Wiley and Sons, New York.

Thrower, L. B. (1964). *Phytopath. Z.*, **51**, 280–284.

Timnick, M. B., Lilly, V. G., and Barnett, H. L. (1951). *Phytopathology*, **41**, 327–336.

Tinline, R. D., and Samborski, D. J. (1959). *Mycologia*, **51**, 77–88.

Toussoun, T. A., and Weinhold, A. R. (1967). *Can. J. Bot.*, **45**, 951–954.

Tukey, L. D., Fluck, M. F., and Marsh, C. R. (1960). *Proc. Am. Soc. hort. Sci.* **75**, 804–810.

Uehara, K. (1958). *Ann. phytopath Soc. Japan*, **23**, 230–234.

Uozumi, T., and Kröber, H. (1967). *Phytopath. Z.*, **59**, 372–384.

Vezina, P. E., and Boulter, D. W. K. (1966). *Can. J. Bot.*, **44**, 1267–1284.

Walkey, D. G. A., and Harvey, R. (1967). *Trans. Br. mycol. Soc.*, **50**(2), 241–249.

Walkey, D. G. A., and Harvey, R. (1966). *Trans. Br. mycol. Soc.*, **49**, 583–592.

Wassink, E. C. (1948). *Recl. Trav. bot. néerl.*, **41**, 150–211.

Webb, R. B., and Malina, M. M. (1967). *Science, N.Y.*, **156**, 1104–1105.

Weinhold, A. R., and Hendrix, F. F. (1963). *Phytopathology*, **53**, 1280–1284.

Wenger, C. J., and Lilly, V. G. (1966). *Mycologia*, **58**, 671–680.

Wilcoxson, R. D., and Subbarayudu, S. (1968). *Can. J. Bot.*, **46**, 85–88.

Wilcoxson, R. D., and Sudia, T. W. (1968). *Bot. Rev.*, **34**(1), 32–50.

Williams, C. N. (1959). *Trans. Br. mycol. Soc.*, **42**, 213–222.

Withrow, R. B., and Withrow, A. P. (1956). *In* "Radiation Biology", Vol. III (Ed. A. Hollaender), pp. 125–258. McGraw-Hill, New York.

Withrow, R. B. (1959). "Photoperiodism and related phenomena in plants and animals". Am. Ass. Adv. Sci., Washington, D.C.

Witsch, H., and Wagner, F. (1955). *Arch. mikrobiol.*, **22**, 307–312.
Wladimiroff, W. W. (1966). *Photochem. Photobiol.*, **5**, 243–250.
Woordrow, J. W., Bailey, A. C., and Fulmer, E. T. (1927). *Pl. Physiol.*, **2**, 171–175.
Wyss, O., Clark, J. B., Haas, F., and Stone, W. S. (1948). *J. Bact.*, **56**, 51–57.
Yadav, B. S. (1962). "Effect of ultraviolet and visible radiation on the sporulation of species of *Helminthosporium*". Thesis, Oregon State University, Corvallis, Oregon.
Yarwood, C. E. (1937). *J. agric. Res.*, **54**, 365–373.
Yirgou, D., and Caldwell, R. M. (1968). *Phytopathology*, **58**, 500–507.
Zachariah, A. T., Hansen, H. N., and Snyder, W. C. (1956). *Mycologia*, **48**(4), 459–467.
Zadoks, J. C. (1967). *Neth. J. pl. Path.*, **73**, 52–54.
Zadoks, J. C., and Groenewegen, L. J. M. (1967). *Neth. J. pl. Path.*, **73**, 83–102.
Zalokar, M. (1955). *Archs biochem. Biophys.*, **56**, 318–325.
Zetterberg, G. (1964). *In* "Photophysiology", Vol. II (Ed. A. C. Giese), pp. 247–281. Academic Press, New York.
Zimmer, R. C. (1967). "Carrot blight in southwestern Ontario and the importance of radiation and temperature in the sporulation of *Alternaria dauci*". Ph.D. Thesis, University of Western Ontario, London, Canada.

CHAPTER XXIV

Production and Use of Fungal Protoplasts

JULIO R. VILLANUEVA AND ISABEL GARCIA ACHA

Departmento de Microbiologia, Facultad de Ciencias and Instituto de Biologia Celular, CSIC. Universidad de Salamanca, Spain

I. PROTOPLASTS OF FUNGI

The living substance of which cells are composed is protoplasm. A cell membrane, which may or may not be protected by an external wall, surrounds the protoplasm. The nucleus and cytoplasm cooperate to perform the various functions characteristic of cells, and as long as these activities are carried out the cell remains alive. Fungi are frequently filamentous (moulds), at least in microscopic structure, but are sometimes single cells (e.g. many of the yeasts). Fungi are usually multicellular organisms in which the filaments are known as hyphae. Many mould hyphae are divided by crosswalls or septa into definite cells each containing one or more nuclei. Hyphae of other fungi possess few septa and appear to be continuous threads of protoplasm containing many nuclei. Structurally and chemically fungal cells possess much in common with other types of eucaryotic cells. The over-all picture is of a cytoplasm bounded by a membrane which acts as the main osmotic barrier, and surrounded by the cell wall. When living cells are plasmolysed, the cytoplasm and its membrane contract away from the cell wall. It has recently been suggested that the cell wall apart from providing protection and rigidity to the cell may have a role in regulating the ionic environment around the outside of the protoplast membrane in the intact cell (Diamond and Rose, 1970).

The term protoplast has been used since 1953 by Weibull and other workers to designate the structure remaining when the cell wall was removed from a bacterial cell. This denomination has therefore been applied to the wall-less state of a variety of micro-organisms, both eucaryotic and procaryotic, that normally have a functionally and chemically distinct cell wall. Since then the word protoplast has been used to denote equivalent forms in fungi (McQuillen, 1960). According to this a protoplast will be defined as the structure derived from a vegetative cell by removal of the entire cell wall; or alternatively as that part of the cell which lies within the cell wall and which in some organisms can be plasmolysed away from it. When residues of the cell wall remain, or when it is not known whether or not such residues remain, the terms spheroplasts (Brenner *et al.*, 1958) or "protoplasts" (in quotes) is needed.

Svihla and Schlenk (1967) when working with yeast cells used the term spheroplast in preference to protoplast. They suggest it is non-committal with respect to the completeness of removal of the cell wall; yet it suggests separation of sufficient cell wall material to yield a spherical form of the cell. Under optimal conditions the two terms would be synonymous. Osmotic sensitivity alone appears to be an insufficient criterion, as bursting of the cytoplasmic membrane occurs even after partial separation of the cell wall.

Cytological studies of the cells of fungi have revealed the presence of a cell wall surrounding the fungal protoplasm. Recent electron photomicrographs indicate that the wall is composed of an outer dense layer about $0.05\ \mu$ thick and a less dense layer of about $0.3\ \mu$. The inner part of the wall may in turn be subdivided into about three layers. Cell walls are composed of polymers of glucose and mannose, more rarely galactose, with smaller amounts of protein, lipid and chitin. A number of workers have isolated the cell walls by mechanical disintegration of the fungal cells and subsequent centrifugation and washing. Selective digestion of the cell wall by enzyme systems has been effected in several species of yeasts and filamentous fungi, resulting in the liberation of the fungal protoplasm in an intact and spherical form. Electron microscopy observations on ultrathin sections of these bodies did not show any structure corresponding to the wall of the cell (Villanueva, 1966). The cell walls isolated from these organisms by mechanical disruption could be largely dissolved by the digestive enzyme systems under the conditions used for sphere formation. These facts suggested that the spherical bodies were essentially devoid of cell wall.

The occurrence of forms of fungi resembling protoplasts under particular cultural conditions, or under abnormal conditions, is well known. From current studies it would appear that "unbalanced growth" can occur. Under many conditions suppression of wall synthesis may be effected without concomitant inhibition of protoplasmic growth. A variety of lytic organisms, mainly bacteria and actinomycetes of the soil, are able to excrete all kinds of cell wall digestive enzymes into the surrounding media (Villanueva, 1966). Hence cells entirely without walls, or lacking components of their walls, can be formed in nature, and these can survive if an appropriate medium is supplied.

A. Criteria for protoplasts

Lysis of the fungal cell wall has made it possible to obtain protoplasts and the stabilization of the protoplast suspensions has opened up a wholly new field for investigation.

The set of requirements described by Brenner et al. (1958) to define a bacterial protoplast can also be extended to fungal protoplasts. The main criterion is the complete absence of cell wall from osmotically sensitive spherical protoplasts. The spherical bodies released from yeast and mould mycelia are osmotically sensitive and appear, under the phase contrast microscope, to lack most, if not all, of the original cell wall; however, definite evidence for absence of cell wall is not available. Whether or not the protoplasts from fungus mycelia are entirely devoid of the original cell wall may be a question related to their genesis (Bartnicki-Garcia and Lippman, 1966). Thus protoplasts which arise by emergence through a pore are probably entirely deprived of cell wall. Observation of the release of protoplasts in fungi show that very often an empty wall remained after emergence of the spherical structure. The phenomenon of release in filamentous fungi and in some yeasts (not all) is similar in that a partly degraded wall is left behind. This is not always the case. It has also been observed that digestion of the cell wall may take place gradually and uniformly all around the cell wall. Then it is not possible to observe the appearance of an empty cell wall. This may be the case in those yeasts where remnants of cell walls are observed on their protoplasts.

In the case of filamentous fungi it is also possible that some undigested wall remained attached to the membrane of protoplasts formed by swelling of hyphal sections. In support of this view, studies on the digestion of isolated hyphal walls of various fungi made by Bartnicki-Garcia and in our laboratory have indicated that a small portion of the wall is not dissolved by the lytic enzyme preparations even after long incubation periods.

Cytological and chemical studies can also help to define these criteria. Sections of previously fixed protoplasts show a thin outer membrane closely adhering to the cytoplasmic ground material. No remnants of the cell wall adhered to the cell membrane. Chemical analysis of isolated cytoplasmic membrane obtained from fungal protoplasts also demonstrated the complete absence of wall components (Garcia Mendoza and Villanueva, 1965). However, inmunological studies on released protoplasts have suggested that small vestiges of cell wall may be present (Garcia Mendoza, Garcia Lopez and Villanueva, 1968). The latter results are in accordance with those of Bacon, Jones and Ottolenghi (1969) who suggested that fungal "protoplasts" in at least some cases, are not true protoplasts. Bud-scars present in isolated yeast membranes gave infrared spectra of a mixture of chitin and β (1–3) glucan, and with traces of mannan. Additional experiments carried out by Darling *et al.* (1970) suggested that, after enzyme digestion the cytoplasmic membrane of the yeasts retains a coating which is rigid during the first stages of the formation of the spherical structures but which loses rigidity when the cell is transformed into a protoplast.

The accepted criteria for defining protoplasts are loss of rigidity, resulting in a spherical form, and osmotic fragility (Villanueva, 1966). However, both criteria is valid for protoplasts as well as for spheroplasts. Thus, it might be advisable to develop specific methods for the definition of protoplasts. The freeze-etching technique allows the demonstration of remnants of the cell wall (the innermost wall layer) on the surface of the cytoplasmic membrane (Streiblova, 1968). However, the quantitative evaluation of these remnants on the protoplasts and spheroplasts seem to be rather problematic. It might also be possible that the medium used for the stabilization of protoplasts could give rise to structures resembling the innermost wall layer. As suggested by Necas, Kopecká and Brichta (1969) the danger of artifact production exists, specially at the cell surfaces where the plasma membrane is in contact with the outer aqueous medium.

Fusion of neighbouring protoplasts described by López-Belmonte *et al.* (1966) and observed in the course of microcine-matographic analysis, also argues for true protoplasts (Girbardt and Strunk, 1965).

In view of the inability to show phage receptors in fungal cell walls, it might be profitable to study protoplasts obtained by different methods for mating specificity. The retention of this capacity might point to the specificity of some wall (protein-polysaccharide) component of the residual elements surrounding the cytoplasmic membrane of the protoplast (Brock, 1961).

Summarizing, it can be said that for the time being it is difficult to define accurately what is intended by the term protoplast in fungi, since many features of protoplast structure are still obscure. For practical reasons our present working criteria for the absence of a wall arc osmotic fragility, loss of rigidity, resulting in a spherical form, and observations on the release of the protoplasts through a pore leaving behind the empty cell walls.

B. Microscopic estimation of protoplasts

Protoplasts can be examined by use of a phase-contrast microscope (1000–1500 × magnification) in films of aqueous solution. Microscopic observations on living material are usually made on wet-mount preparations or in moist chambers prepared in the following way: a layer of vaseline is applied to the edge of a cover slip and a thin layer of nutrient agar medium, when desired, is applied to the centre. A drop of the material to be examined is placed in the centre of the cover slip and the cover slip is immediately inverted on a slide in such a manner that the Vaseline forms a seal, air is trapped in the preparation, and the drop of liquid touches the slide, providing a continuous phase for the light path. Protoplast suspensions consist usually of isolated spherical bodies and exhibit the typical response to dilution with distilled water, namely, a marked swelling followed by lysis. Protoplast counts can be performed in a Thoma, Neuberger or Petroff-Hausser counting chamber. The routine procedure is to count protoplasts in squares across the ruled area of the chamber (Petroff-Hausser), starting with the uppermost left-hand square and moving two squares to the right and then down one for the next count and so on across the chamber. A second series of counts is made in the same way but starting at the lower right-hand square and proceeding to the left. A total of at least 20 squares can be counted per determination. Then the chamber can be emptied, cleaned and refilled for another count of the same sample. The total count is nearly always in excess of 100 protoplasts.

When mass estimation of protoplasts is required absorbance of the suspensions can be measured with a Beckman DU spectrophotometer, using light of 700 mμ wavelength and a 1 cm light path cuvette.

Protoplast diameters are measured with a calibrated ocular micrometer. The values obtained can be used to calculate average volumes on the assumption that protoplasts are spheres. Diameters of at least 100–150 protoplasts in the counting chamber are measured for each volume calculation, and the diameters of all the protoplasts in randomly chosen squares of the chamber are measured. Volumes calculated from direct microscopic measurements usually agree with those calculated from photographic measurements. Photographs are used only occasionally because of the difficulty in deciding which protoplasts in the micrographs are sharply in focus.

The conversion of fungal cells to protoplast bodies is usually followed by phase-contrast microscopy. By providing adequate osmotic protection during the enzymic hydrolysis of living walls of yeasts of filamentous fungi, the formation of spherical protoplast-like bodies is commonly observed. From most of the mould species, protoplasts emerge from hyphae in a bud-like manner, i.e. a variable amount of protoplasm squeezes through a small not readily visible pore in the wall, forming on the outside a growing spherical body. The final size and time of detachment of protoplasts varies greatly. Sometimes more than one protoplast is formed on a single hypha from what appears to be a continuous mass of discharging protoplasm. The liberation of large protoplasts requires the displacement of protoplasm from relatively distant hyphal areas.

Sometimes, in a more advanced digestion stage, i.e. after incubation for 24 h, when the hyphal walls are perceptibly thinner, the appearance of protoplasts by intercalary swelling is more common. This, in the case of *Phytophthora*, consists of the swelling and rounding off of short hyphal segments whose walls become progressively thinner, protoplasm from adjacent hyphal regions empties into the swellings and their resulting protoplast-like bodies can form from a single hypha. The extent of protoplast liberation, from one experiment to the next under seemingly similar conditions is widely variable. The hyphal walls of various fungi vary in their susceptibility to enzymic attack, although very often the growing tips are shown to be more susceptible to this attack and the extrusion of protoplasts often takes place preferentially at these points. Forces tending to make the protoplast assume a spherical shape apparently help in the extrusion of the body. Protoplasts of various sizes (5–50 μ in diam.) can often be seen although protoplasts are usually about 10 μ in width. Fusion among mycelial protoplasts occurs and the larger bodies often seen may originate by fusion of various smaller bodies, though the possibility of protoplast swelling cannot be ignored. In a large protoplast, streaming can easily be observed. Incubation conditions may affect the size of the bodies and protoplasts appear to be markedly affected by the nature of the substance used for stabilization. Sometimes the protoplasts swell (or grow) and eventually the spheres rupture, leaving either ghosts or clusters of vesicles and granules. Protoplasts are stable and retain their external morphology for several days when kept under stabilized conditions. There are, however, marked internal changes, such as the cytoplasm becoming coarsely granulated and the turgid protoplasts becoming contracted and crenated.

Release of protoplasts in yeast cells seemed to occur chiefly at the equatorial zone of the ovoid cells and only rarely at the polar ends. As a rule, the entire contents of one cell are extruded at one time to form one free protoplast. Sometimes during protoplast liberation the protoplasm of one cell can divide into two parts, one remaining inside the cell wall and the other giving rise to a protoplast. Sometimes when the protoplast is released from the cell wall some kind of filamentous structure remains behind. When this filament links two protoplasts it serves to originate some kind of attraction between the connected protoplasts, which later come together and fuse.

C. Electron microscopy of protoplasts

The main problem faced when trying to study protoplasts with the electron microscope is the preservation of the outer membranes and the avoidance of disintegration of the spherical bodies under the different, sometimes drastic, treatments to which protoplast must be exposed during the preparative manipulations for electron microscopy. Protoplasts are osmotically sensitive and must be stabilized by carbohydrates or salt solutions. Addition of chemicals to such suspensions during fixation may lead to the sudden disappearance of part or all the cytoplasmic contents or to complete disintegration. Protoplasts are very fragile and cannot be dried without disruption. However, the spherical bodies can be fixed under careful conditions and their form can be preserved even after fixation.

A number of fixatives have been employed with protoplasts: Osmium tetroxide (1% w/v in distilled water or in an appropriate buffer), potassium permanganate (2% w/v) and a mixture of both as well as glutaraldehyde (5% w/v) or formation (10% w/v). Bacterial protoplasts are apparently well stabilized by the addition of 0.01 M $MgCl_2$ or serum album (Thornson and Weibull, 1958). The addition of a highly charged, nonpermeating, macromolecule such as carregenin, a sulphated polysaccharide, to the fixation fluid apparently yields good results in this respect (Elbers, 1961). Very often

TABLE I

Methods used for fixation of protoplasts

Method of Fixation	Results	
In solid medium (15–20% gelatine or 1–2% agar)		
KMnO$_4$	The shape of the protoplasts is very well preserved.	Membrane systems are not contrasted.
In liquid medium (0·6 M KCl)		
KMnO$_4$ in 0·6 M KCl	Most protoplasts lyse.	Membrane systems are very well contrasted.
Formalin neutral 10, 5, 3% and KMnO$_4$ 2%	Protoplasts do not lyse.	Bad preservation of cell structure.
Formalin pH 5·4 10, 5, 3% and KMnO$_4$ 2%	Protoplasts do not lyse.	Cell structures are well contrasted.
Glutaraldehyde fixation Sentandreu and Northcote (1969)	Protoplasts do not lyse.	Bad preservation of cell structure.
Glutaraldehyde pH 5·4 and KMnO$_4$ 2%	Protoplasts do not lyse.	Bad preservation of cell structure.
Glutaraldehyde pH 5·4 and OsO$_4$ fixative	Protoplasts do not lyse.	Membrane systems are not contrasted.

the use of one fixative shows structures not detected with another, and sometimes depending on the organism the intracellular structures are badly preserved.

Necas and Havelková (1967) following the experience of other workers developed a careful study of various techniques for fixation of fungal protoplasts. The methods used for fixation of protoplasts and their results are summarized in Table I. It is interesting that the fixation with glutaraldehyde which has been found very satisfactory for normal yeast cells (Sentandreu and Northcote, 1969), although it prevents the lysis of the protoplasts, is not so good for the preservation of cytoplasmic structures in these bodies. Very similar results were found by Weiss (1963) in *Neurospora crassa* protoplasts.

Glutaraldehyde fixative method (Sentandreu and Northcote, 1969). The yeast cells or their protoplasts are fixed in a glutaraldehyde fixative (4% glutaraldehyde in 0·02 M-phosphate buffer pH 6·8, with an added salt mixture at 21°C for 1 h, and washed four times with the same buffer over a period of several hours. Post-fixation was done in 1% osmium tetroxide solution in veronal buffer (0·05 M, pH 7·0) for 1 h. The pellet was embedded in 0·75% agar blocks, dehydrated with an ascending ethanol series and embedded in araldite resin. Sections were cut with glass knives on an ultramicrotome and were collected on carbon coated copper grids and stained with uranyl acetate (saturated solution in 50% ethanol) at 60°C for 15 min, followed by lead citrate (0·09%) for 1 min. The sections are examined with an electron microscope. This method showed an unusually complex organization of its membrane system in intact cells of yeasts. Aldehyde-fixed protoplasts contained components not seen in $KMnO_4$-fixed organisms.

Potassium permanganate fixative method (Uruburu et al., 1968). Samples of cells or their protoplasts were washed with distilled water or with a solution of salt of low concentration, suspended in 5% (w/v) aqueous potassium permanganate and kept at room temperature for 2 h. The fixed protoplasts were dehydrated through 25%, 50% and 75% (w/v) acetone in water, then 100% acetone. During dehydration the material was stained overnight in 2% (w/v) uranyl acetate dissolved in 75% (w/v) acetone. Such a procedure of post-fixation was effective for the demonstration of fine details of cellular structures. The fixed and stained material was embedded in Durcupan ACM from Fluka Ab, Buchs, Switzerland, cut with an ultramicrotome and picked up on Formvar-coated grids. Sections were examined with an electron microscope.

For shadow-casting the suspensions of protoplasts were applied directly to Formvar-coated grids (after having been washed with water to free them of salts), allowed to dry and shadowed at an angle of 45° with gold-palladium in a Siemens VB 6500 evaporator. This method may be useful for studies of regeneration of protoplasts.

A similar method, used by Necas (1965), was as follows: Protoplasts were harvested from the medium by centrifugation in 0·6 M KCl and fixed with 2% KMnO₄ dissolved in 0·6 M KCl solution at 4°C for 2 h; threefold washing in distilled water and embedding the fixed material in 2% agar at 40°C; dehydration by an ascending alcohol series and embedding in a mixture of methyl and butyl-metacrylate. Ultraviolet light was used for polymerization. Ultrathin sections were prepared on a microtome and examined with the electron microscope.

D. The choice of the osmotic stabilizer

Protoplasts of fungi can be obtained by treating cells with cell wall lytic enzymes. A necessity for certain osmotic stabilizers in the medium to maintain the protoplasts has been noted. Due to the lack of external protection, protoplasts will immediately lyse during the process of release from the mother cell unless the osmotic concentration is adjusted. It has been suggested that the solute ought not be able to penetrate the osmotic barrier of the protoplast at an appreciable rate in order to protect from lysis.

A wide variety of stabilizing media has been used by different workers to obtain protoplasts from fungi. Media containing one of the solutes described in Table II have commonly been used. The compounds were usually added to a buffer solution at pH between 5·8 and 6·8. Mannitol followed by rhamnose, sorbitol and sucrose, have been preferred by most workers. Potassium chloride has produced good results with some of the fungal species tested and MgSO₄ is highly recommended as an excellent stabilizer.

The concentration of the solutes used as stabilizers varies largely; this phenomenon can be correlated to some extent with differences in internal osmotic pressure in different species. As an example the stabilization of protoplasts has been shown to require solute concentrations varying from 0·5 M rhamnose (Eddy and Williamson, 1957) for *S. carlsbergensis* and *S. cerevisiae*, to as high as 2·0 M for *S. mellis* (Weinberg and Orton, 1965).

Substances usually used for stabilization of protoplasts affect their formation. It is important to investigate the properties of the solutes not only as stabilizers but taking into consideration other important factors, such as the stimulatory effect of the solute on the lytic enzymes. Protoplast liberation by the action of the lytic enzyme of *Streptomyces* RA (Rodriguez Aguirre *et al.*, 1964) on mycelium of *Fusarium culmorum* takes place in about 30 to 45 min when fructose, sorbose, rhamnose, mannitol, NaCl or KCl (at 0·8 M) are employed as stabilizers. On the other hand when the

Substances used for the stabilization of fungal protoplasts

Stabilizer	Concentration	Organism	Author
Mannitol	0·4–0·6 M	P. cinnamomi, P. parasitica	Bartnicki-Garcia and Lippman (1966).
	0·6 M	Schiz. pombe	Necas et al. (1968).
	0·65 M	Ashbya gossypii	Tanaka and Phaff (1967).
	0·8 M	C. utilis	Garcia Mendoza and Villanueva (1962, 1964).
	0·8 M	F. culmorum, A. nidulans, T. roseum	Rodriguez Aguirre et al. (1964).
		F. culmorum	Garcia Acha et al. (1964).
	1·0 M	S. cerevisiae	Eddy and Williamson (1959).
			Longley et al. (1968).
			Svihla et al. (1961).
Sorbitol	1·0 M	C. utilis	Duell et al. (1964).
			Ottolenghi (1966).
			Lillehoj and Ottolengi (1967).
	0·63 M	S. cerevisiae	
	1·0 M	Saccharomyces strain	
Rhamnose	0·55 M	S. cerevisiae	Eddy and Williamson (1957).
Maltose	0·59 M	N. crassa	Bachmann and Bonner (1959).
	0·8 M	S. fragilis	Davies and Elvin (1964).
Sucrose	0·4–0·6 M	P. cinnamomi, P. parasitica	Bartnicki-Garcia and Lippman (1966).
	12% (w/v)	Polystictus versicolor	Strunk (1969).
Xylose, fructose	0·5 M	N. crassa	Bachmann and Bonner (1959).
Sorbose	5–10% (w/v)	N. crassa	Emerson and Emerson (1958).
	10% (w/v)	F. culmorum	Lopez-Belmonte et al. (1966).
KCl	0·3 M	Schiz. pombe	Svoboda et al. (1969).
	0·4–0·6 M	P. cinnamomi, P. parasitica	Bartnicki-Garcia and Lippman (1966).
	0·6 M	C. utilis	Svihal et al. (1961).
	0·6–0·8 M	S. cerevisiae	Holter and Ottolenghi (1960).
NaCl	0·3–1·0 M	S. cerevisiae	Tabata et al. (1965).
MgSO$_4$.7H$_2$O	0·8–1·0 M	C. utilis, B. lactis, O. suaveolens	Gascón and Villanueva (1965).
		S. cerevisiae, S. fragilis	Rost and Venner (1965).
	1·0 M	C. utilis, S. fragilis	Villanueva et al. (1969).

stabilizer used is xylose, sucrose or maltose the process of protoplast release is considerably delayed (only after 3 h incubation). Potassium and sodium chloride salts were also found to interfere with the lytic activity of the lytic enzyme produce by *Micromonospora* (Gascón *et al.*, 1965).

The concentration of the stabilizer can also affect protoplast formation. Bartnicki-Garcia and Lippman (1966) describe the liberation of protoplasts from *Phytophthora* in digestion media which contained mannitol, sucrose or KCl as osmotic regulators in phosphate buffer. Protoplasts appeared in sucrose or mannitol concentrations ranging from 0·2 M to 0·8 M with better responses at about 0·4 M to 0·6 M. Potassium chloride was similarly effective except at 0·8 M at which concentration no protoplasts were released.

II. METHODS FOR THE PREPARATION OF PROTOPLASTS

The methods used for the isolation of protoplasts from fungi can be divided into three major groups. The first of these involves the use of mechanical or autolytic methods. The second is based on the use of enzymes to dissolve the cell wall. This method is similar in principle to that developed for the isolation of bacterial protoplasts. The third makes use of the specific inhibition of cell wall synthesis without affecting the cytoplasmic components. This action results in "unbalanced growth" with the formation of abnormal walls or none at all.

For a better understanding the methods now available for the removal of the fungi cell wall during the formation of protoplasts can be classified under the following paragraphs:

1. The use of mechanical methods or autolytic enzymes.
2. The action of specific enzymes.
   (a) Snail enzymes.
   (b) Microbial enzymes.
   (c) Other enzyme preparations.
3. Metabolic disturbance with inhibition of cell wall formation.
4. Growth of cells on serum.

The main features of these methods are described to illustrate the variety of procedures used for the preparation of protoplasts.

A. The use of mechanical methods or autolytic enzymes

Necas (1956) made the interesting observation that under certain conditions, in the absence of external enzymes, suspensions of *Saccharomyces cerevisiae* became sensitive to osmotic shock and appeared to be converted to protoplasts. In autolysing cultures it seemed that the surface structures of yeast cells became digested by enzymes which are closely bound to the protoplasm and that protoplasts were released. In appropriate media these spherical forms would then break up into smaller "plasma droplets". Spherical fragile bodies from cells of *S. cerevisiae* were also liberated from cells which were grown with quartz sand or subjected to mechanical pressure by squeezing between slide and coverslip. The spherical particles containing nuclei could increase in volume and some undergo transition to normal yeast cells. The mode of regeneration differs a great deal from those known hitherto. It may be relevant that Necas obtained protoplasts from spontaneously lysing yeast whereas other workers have employed the snail digestive juice or the microbial enzymes. Since the snail digestive juice contains in high concentration a whole complex of different enzymes (Holden and Tracey, 1950) it could well degrade far more of the wall material than the autolytic method does.

Yamamoto (1962) has described the properties of the protoplast fragments of the yeast cell which may be produced by squeezing out the protoplasm through the ruptured cell wall under hypertonic conditions. These protoplast fragments may also be produced within the space surrounded by the cell wall following treatment of yeast cells with 0.1 M strontium chloride. These protoplast fragments can apparently regenerate if the cells are withdrawn from the previous medium and inoculated into normal medium. To what extent these protoplast fragments are similar to those obtained by Necas, is not clear. It is also interesting that sometimes yeasts exposed to the digestive action of cell wall lytic enzymes can squeeze out the separated contents of the cell through a hole in the cell wall yielding protoplast fragments.

B. The action of specific enzymes

The ability and efficiency of cells, using enzymes, to convert to protoplasts is dependent on the type and strain of organism, and on its physiological state. It is therefore necessary to design the method to fit the particular organism. Hence a number of methods for yeasts and filamentous fungi will be described.

1. a. *Snail enzymes*

The ability of the digestive juice of the snail *Helix pomatia* to attack cell wall of yeast has been known since 1914 by Giaja's work. A renewal of interest in this preparation was followed the reports of Eddy and Williamson in 1957, and of Myers and Northcote in the following year. The latter described the presence of several hydrolytic enzymes able to deploymerize cell-wall components. Hydrolytic activity of the snail preparations is commonly attributed to the presence of B-glucanase, emphasizing the fundamental role of the B-glucan in maintaining the structure of the cell wall. Other workers have described the action of highly purified proteolytic enzymes which completely digest cell walls. Attempts at fractionation of the lytic enzymes present in crude preparations of the snail enzyme complex have been made (Millbank and Macrae, 1964; Anderson, 1967).

b. *Source of the snail enzymes*

The enzyme preparation used usually is the "Suc-digestif d'*Helix pomatia*" obtained from L'Industrie Biologique Francaise (35 Quai du Moulin de Cage, Gennevilliers, Seine France). The juice is dispensed in ampoules and usually has as preservative an organic mercury. Therefore, it is important when trying to obtain protoplasts to add to the enzyme preparation a solution of a sulphydril-compound, i.e. 0·1 volume of 1% (w/v) cysteine HCl. Also available from the same firm is a lyophilized preparation named "Helicase". A great variation in lytic activity is usually found among different lots of these preparations. The susceptibility of different strains of fungi to cell wall digestion by snail enzyme is rather variable. It is usually found that it is much easier to obtain protoplasts from vigorously growing cultures. Moreover, the composition of the growth medium can influence the resistance of the wall to digestion by snail enzymes.

The preparation of snail digestive enzymes (Gluculase) can also be obtained from Endo Laboratories Inc., Richmond Hill, N.Y. The Gluculase ampoules have a concentration of approximately 100 mg of protein per ml.

A. *Yeast protoplasts*

Yeasts are the only group of the fungi in which quantitative conversion of cells into spherical osmotically labile bodies has been achieved by selective removal of the cell wall.

(a) *Eddy and Williamson method* (1957). Young cells of *Sacch. carlsbergensis* and *Sacch. cerevisiae* were obtained by growing them at 25°C with shaking, in a malt extract medium. The cells were harvested after about 18 h, when a population density of approximately 6×10^6 cells/ml was attained. After thorough washing in distilled water, the young yeasts were suspended at a density of about 10^8 cells/ml in 0·005 M citrate phosphate buffer (pH 5·8) containing 0·55 M rhamnose and 1·0 mg of the freeze-dried snail enzyme. After about 18 h at room temperature, or about 5 h at 25°C, more than 90% of the cells were converted into protoplasts, the remainder being completely dissolved. Despite their fragile nature, the protoplasts can be washed and centrifuged at low speeds in 0·5 M solutions of rhamnose buffered at pH 5·8 in which they appear to be stable for several hours.

(b) *Svihla* et al., *method* (1961). Protoplasts were obtained using cells of *Candida utilis*. Cells were grown in 1·5% Bacto Nutrient Broth, 0·5% Bacto Yeast Extract, and 1·5% glucose for a period of 24 h at 30°C under vigorous aeration. The cells from 20 ml of this medium were centrifuged and transferred into 100 ml of glucose, NH_4^+, and salt medium. For conversion of the yeast cells to protoplasts, 5% suspensions of washed cells were used. The quantities of snail enzyme used ranged between 0·025 and 0·2 ml of a solution containing 25 to 30 mg of dry material per ml. In early experiments, 1·0 M mannitol containing 0·05 M potassium phosphate pH 6·0 was used as an isotonic medium; four parts of this were mixed with one part of enzyme in water solution. Later 0·75 M KCl was substituted for the mannitol and results in a final concentration of 0·6 M KCl and 0·04 M potassium phosphate pH 6·0 after adding the enzyme. Digestions were carried out at 25°C with gentle agitation. The enzyme was removed on completion of the process by washing with 0·6 M KCl and centrifugation at low speed (< 1000 g).

(c) *Ottolenghi method* (1966). A diploid strain of *Saccharomyces* made from crosses of *S. carlsbergensis*, *S. chevalieri* and *S. italicus* was used. Cells were grown in a medium containing 2% glucose, 2% Bacto-peptone and 1% Bacto yeast extract. The cultures were shaken at 30°C for 15 h. The cells were harvested by centrifugation and were washed twice with 0·1 M mercaptoethanol at pH 8, twice with 0·1 M Tris buffer (pH 8), and they were then suspended in 1 M sorbitol containing 1/10 volume of the growth medium. The suspension was incubated at 30°C until plasmolysis was no longer observable (15 to 30 min), and the cells were then collected and washed three times with 1 M sorbitol. These cells were suspended in the digestion medium containing 1 M sorbitol (final concentration), citrate, phosphate or Tris buffer of ionic strength 0·05, and 25 μl of snail enzyme (containing 1/10 volume of 1% cysteine) per ml of final incubation volume. The mixture was incubated at 30°C with shaking until all the cells had been transformed into protoplasts. The time necessary for the digestion to be completed varied between 1 and 5 h.

More recently Bacon *et al.* (1969) using the same strain of *Saccharomyces* described the following method. Cells grown under the above mentioned conditions were harvested after 36 h while still growing, washed with 0·1 M mercaptoethanol, and suspended in 40 ml of 0·6 M KCl containing 0·5 ml of snail gut juice. After 20 h at 33°C the protoplasts were collected.

(d) *Longley* et al., *method* (1968). Growing cells of *S. cerevisiae* were harvested and washed three times with 67 mM KH_2PO_4, pH 4·5 and once with water. The cells (9·0 mg cells per ml) were resuspended in 5 mM citrate-phosphate buffer, pH 5·8, containing 0·8 M mannitol. One ml of the digestive juice of *Helix pomatia* was diluted with an equal volume of water, and centrifuged at $20,000 \times g$ at $0°C$. The supernatant was removed and supplemented with $0 \cdot 1$ vol of 1% (w/v) cystein HCl to inactivate the preservative (an organic mercury) present in the commercial preparation. One volume of diluted snail gut juice containing approx. 40 mg of protein/ml was added to $4 \cdot 0$ vol of yeast suspension (9 mg cells/ml) and shaken at $30°C$ for 3 h in a shaker. Examination of samples of suspension under a phase-contrast microscope showed that after 3 h at $30°C$ virtually all of the cells had been converted into protoplasts.

(e) *Villanueva* et al. *method* (1969). *Candida utilis* and *Sacch. fragilis* obtained from the Colección Española de Cultivos Tipo (Salamanca, Spain) were used for these studies. Cultures made in Hansen medium were incubated with shaking at $29°C$. Cells, harvested in the early logarithmic phase of growth, were washed twice with the stabilizer solution ($1 \cdot 0$ M $MgSO_4$ + $0 \cdot 1$ M potassium phosphate buffer, pH $5 \cdot 8$). The cells were resuspended in the stabilizer solution and lysis of cell walls was conducted using cell concentrations sufficient to give an initial absorbency of approximately $0 \cdot 5$. One ml of the enzyme preparation (snail gut juice 10 mg protein/ml or strepzyme AS, 15 mg protein/ml) was then added to 10 ml of the cell suspension and the mixture was incubated at $37°C$ with gentle shaking. The conversion into protoplasts varied with the organism used; with *S. fragilis* the process was complete in about 45 min as determined by the fall in turbidity and microscopic observations using wet-mounts; *C. utilis* required between 60–90 min for full conversion into protoplasts. Lysis of protoplasts occurred when some drops of water were added under the cover slip.

(f) Canadida tropicalis *protoplasts*. Lebeault *et al.* (1969) have obtained protoplasts from *Candida tropicalis* cultivated on *n*-tetradecane. Cells were grown in a minimal medium (NH_4Cl, $2 \cdot 5$ g; KH_2PO_4, 7 g; $NaHPO_4$, $7 \cdot 2$ g; $MgSO_4.7H_2O$, $0 \cdot 2$ g; $NaCl$, $0 \cdot 1$ g; water 1 litre) supplemented with 100 mg of yeast extract per litre. The carbon source *n*-tetradecane ($1 \cdot 5$ g/ litre). The cultures were incubated at $32°C$ with vigorous agitation and cells were harvested during the exponential growth phase. Yeast cells after being washed with $0 \cdot 9$% solution of $NaCl$ were used for protoplast studies. Cells were suspended in Tris buffer (pH $9 \cdot 3$) in a concentration of 50 to 100 mg d.wt/ml. These suspensions were used for the preparation of protoplasts by two different methods:

(a) The cells suspensions (25 ml) were incubated for 30 min at $32°C$ in the presence of $0 \cdot 5$ M sodium thioglycolate and then centrifuging. The sedimented cells were washed with a phosphate-citrate buffer (0.1 M, pH 5·8) containing sorbitol (0·7 M), mannitol (0·3 M), and EDTA (0·001 M). Cells were then suspended in 10 ml of the same solution containing 1 ml of helicase and then were incubated at 30°C with shaking.

(b) Protoplasts were exposed to the action of sodium thioglycolate as before, centrifuging at 6000 × g for 15 min, suspending the cells in 10 ml of phosphate-citrate buffer (0·05 M, pH 4·1) containing $MgSO_4$ (1 M) and adding 1 ml of helicase. Incubation was carried out as before. The formation of protoplasts was followed by the measurement of optical density at 450 mμ after dilution of samples with 500 parts of water. In marked contrast with the results described by Volfová et al. (1968) who found that protoplasts do not oxidize n-hexadecane, although intact cells do, Lebeault et al. (1969) showed that protoplasts prepared as above behave like intact cells. Those protoplasts are of great interest for studies of localization of different enzymes implicated in hydrocarbon metabolism.

(B) Protoplasts of fungi

Similar methods based on the enzymic break-down of the cell wall have been used on strains of filamentous fungi. Far less quantitative conversion into protoplasts was achieved than in the yeast studies. As in the case of yeasts, formation of protoplasts was best from young hyphae.

(a) Neurospora crassa protoplasts. Emerson and Emerson (1958) and Bachman and Bonner (1959) have obtained protoplast-like structures from mycelium of strains of Neurospora crassa by the action of commercial snail enzymes. Much better preparations are obtained using young hyphae as starting material. After about 30 to 60 min incubation at 30°C in the presence of 10% snail preparation and 20% sucrose as stabilizer, protoplasts begin to emerge through pores in the hyphal walls. Frequently, the entire contents of a hyphal compartment, between septa, is extruded as a single, large protoplast. On other occasions, the contents are extruded intermittently through one or several pores, giving rise to a number of protoplasts from one compartment. Most often, some portion of the contents of a compartment remains behind (Villanueva, 1966).

(b) Protoplasts from Polystictus versicolor. Strunk (1967, 1969) has studied the liberation of protoplasts from Polystictus versicolor by the action of snail enzyme preparations. Young hyphae were suspended in 0·15 M phosphate buffer pH 7·0 and 0·2 M Tris buffer pH 7·1 using 12% saccharose as stabilizer. The lyophylized snail enzyme preparation was added at a concentration of 20–50 mg/ml and incubation was carried out at 23°C. After a few hours of incubation protoplasts were released through pores in the cell wall of the hyphae.

2. Microbial enzymes

If we visualize the walls of fungi as heterogeneous polymers, it is obvious that any enzyme that can split it into smaller molecules will achieve lysis. But the problem is not that simple. It is well known that the wall of dead cells is susceptible to enzymes which will not attack the living cell. In life there are factors which limit the accessibility of the substrate to the enzymes.

The isolation of fungal protoplasts using microbial enzyme preparations would appear to depend on the operation of a somewhat similar group of enzymes to those present in the snail digestive juice. Enzymes of various micro-organisms have been found useful for lysis of fungal cell walls. The more active species are in the genera *Bacillus*, *Streptomyces* and *Micromonospora*. Some of the extracellular enzymes have been isolated, concentrated and purified.

An investigation of the lytic properties of Actinomycetes has revealed the production of a large number of different enzymes acting upon various cell wall constituents. Although studies on the mechanism of bacteriolysis is progressing very fast at present, the same cannot be said about fungal cell lysis. Microbial lytic enzymes active on fungal cell walls are easy to produce and should prove useful in structural studies. A survey of the lytic activities of Actinomycetes on fungal cell walls (Gascón and Villanueva, 1963), showed strains of *Streptomyces* and *Micromonospora* were most active.

The lytic action on the cell walls of fungi is due to a complex system of reactions by distinct enzymes. In the case of microbial preparations the biosynthesis of these enzymes is influenced by the composition of the growth medium and by other cultural conditions. The demonstration that the microbial preparations contain several lytic agents each with its characteristic range of activity, has made the elucidation of the mechanism of fungal lysis by micro-organisms even more complicated than had at first appeared. The characterization of the fungal wall-degrading enzymes and the products formed by the action of specific enzymes, would throw light on the problem of cell wall structure. Various workers (Furuya and Ikeda, 1960; Masschelein, 1959; Garcia Mendoza and Villanueva 1962; Nickerson, 1963; Rodriguez Aguirre *et al.*, 1964) have obtained lytic preparations from *Streptomyces spp.* which act on isolated cell wall preparations, although only under specific conditions was it possible to produce protoplasts from intact yeast cells. Cell wall preparations from *Aspergillus oryzae* were dissolved by a lytic enzyme from *Bacillus circulans* and by chitinase from *Streptomyces sp.* (Horikoshi and Ida, 1959). The cell walls of a large number of fungal species have been shown to be dissolved by lytic preparations from *Streptomyces spp.* although few attempts have been made to study the enzyme systems of the lytic organisms in detail.

(a) *Candida utilis* (*Garcia Mendoza and Villanueva*, 1962) *and other yeasts.* *Candida utilis* was grown in shake flasks on a glucose-yeast extract medium at 30°C and the cells harvested in the early exponential phase of growth (10 h). The suspension of yeast (5 mg dry wt per ml) was incubated at 37°C with shaking in the presence of an enzyme preparation called "strepzyme" obtained from *Streptomyces GM*. Mannitol 0·8 M in phosphate buffer pH 5·8 was used as stabilizing agent and the process of conversion to protoplasts required 30 to 60 min. The protoplasts were centrifuged at 3000 *g* for 15 min and washed in 0·8 M mannitol to remove the lytic enzyme preparation.

Stages in the formation of yeast protoplasts with "strepzyme" have been described (Garcia Mendoza and Villanueva, 1964). At an early stage of the digestion the protoplast may be seen to retract from both ends of the elongate cell, assuming at first a lemon shape. Later the protoplasts begin to emerge from some of the cells extruded through small holes which develop in the cell-wall. The entire contents of one cell are released at one time to form one free protoplast. The empty cell walls remain for a short time in the suspending medium and then gradually are completely dissolved, leaving a suspension of protoplasts free from debris. No correlation between the sites of cell budding and the place of the first attack of enzyme before the release of the naked protoplast was observed. It seems that the process of cell wall digestion with this lytic enzyme preparation runs parallel to that observed when using the digestive juice of the snail *H. pomatia*. With some modifications of the above method Gascón *et al.* (1965) have obtained protoplasts from *Oospora suaveolens* and *Geotrichum lactis*. Stock suspensions of whole young cells were centrifuged and the cells were resuspended in 1·0 M MgSO$_4$, instead of mannitol. The enzyme preparations derived from *Micromonospora AS* were added and the mixture incubated at 37°C with gentle shaking.

The conversion into protoplasts varied with the physiological phase of growth and with the species of organism used. The susceptibility of various species of yeasts and moulds to cell wall digestion by the "strepzyme M" was variable. The most susceptible organisms were *C. utilis*, *Torulopsis utilis*, *Geotrichum lactis* and several Gram-positive and Gram-negative bacteria.

(b) *Fusarium, Aspergillus and other fungi* (Rodriguez Aguirre *et al.*, 1964). Protoplasts can be prepared from mycelium and spores of *Fusarium culmorum*, *Aspergillus nidulans*, and *Trichothecium roseum* using an enzyme preparation from *Streptomyces* RA. Young hyphae were grown on a sucrose-salt medium for 18–20 h at 30°C in shake flasks.

For preparations of the lytic enzyme *Streptomyces* RA was grown at

28°C. To a basal medium (0.2% chitin, 0.1% K_2HPO_4, 0.05% Mg $SO_4.7H_2O$) was added 10% (w/v) washed *Aspergillus nidulans* mycelium. The final pH of the medium after autoclaving for 20 min was $6.8–7.0$. Cultures were grown with shaking in 3 l flasks containing 1 l of medium. After 8 day incubation the medium was centrifuged and the supernatants were collected. Fractionation of the enzymes present in the medium was made with ammonium sulphate. Most of the lytic activity was recovered in the fraction precipitated between $0.4–0.6$ saturation of ammonium sulphate. The enzyme was freed of salt by dialysis and then lyophilized.

For preparation of protoplasts young mycelium of *F. culmorum* (and other fungi) was suspended in 0.1 M phosphate buffer pH 5.8 with the lytic enzyme preparation (0.5 mg/ml) and 0.8 M mannitol. Digestion was continued for several hours at 28°C with gentle agitation. After $1–2$ h incubation protoplasts emerged and continued to develop for up to 5 h after which time some of the spherical bodies began to disintegrate. Frequently it was possible to observe protoplasts of various sizes ($4–15$ μ diam.) in the same microscopic field. The yield also was influenced by the nature of the stabilizing agent. Protoplasts can be obtained in a medium of $0.6–0.8$ M KCl but the digestion of the wall is rather slow, and proceeds to completion only when young cells are used. *F. culmorum* released more protoplasts than any of the other moulds tested. Further advantages of *F. culmorum* over the other species is the greater stability of the protoplasts, and the ease with which these structures can be observed under the phase-contrast microscope.

(c) *Phytophthora protoplasts*. Bartnicki-Garcia and Lippman (1966) have obtained spherical protoplast-like bodies from the mycelia of *Phytophthora cinnamomi* and *P. parasitica* on digestion of their hyphal walls with an extracellular enzyme preparation from *Streptomyces sp*. This was the first time that the artificial release of protoplasts from a fungus with cellulosic walls was described. Mycelium for the preparation of protoplasts was obtained from a $7–10$ day stationary culture in liquid V-8 juice medium (Campbell Soup Co.). A small portion of a mycelial mat was homogenized at slow speed in 15 ml of fresh V-8 solution in a Sorvall Omni-mixer for 3 min. The resulting suspension of mycelial fragments was poured into a Petri dish and incubated for $1–2$ days, at 27°C. The small actively growing pieces of mycelium were used for enzyme digestion.

A crude extracellular enzyme preparation from *Streptomyces sp*. strain QMB 814 was used. The lytic action is believed to be chiefly due to β-glucanases which cleave the major components of the wall, namely cellulose and other glucosans to soluble carbohydrate molecules of a wide range of sizes. Enzyme solutions containing 120 C $\times$ cellulase units/ml were made in 0·05 M citrate buffer (sodium citrate + citric acid; pH 5·8) or in 0·1 M phosphate buffer ($K_2HPO_4 + KH_2PO_4$; pH 6·8). Buffers were adjusted to various concentrations of sucrose, mannitol or KCl to increase the osmotic pressure. Enzyme solutions were sterilized through millipore filters. Protoplasts appeared in sucrose or mannitol concentrations ranging from 0·2 M to 0·8 M with better responses at about 0·4 M to 0·6 M. Potassium chloride was similarly effective, except at 0·8 M at which concentration no protoplasts were released. Without osmotic stabilizers no protoplasts were released; instead, amorphous masses of protoplasm were extruded. Freed mycelium-protoplasts were usually spherical, sizes ranging from 10 to 40 μ. From both *Phytophthora* species, protoplasts appeared by two distinct processes, namely a bud-like emergence or intercalary hyphal swelling. Both types of protoplast were osmotically sensitive and disintegrated upon dilution of the suspending media. Although hyphal wall digestion was always extensive, the amount of protoplast release varied widely under seemingly similar conditions.

(d) *Protoplasts from* Pythium. Sietsma *et al.* (1967) have described that enzyme preparations from two *Streptomyces spp.* released spherical protoplast-like bodies from the mycelium of *Pythium sp.* RRL 2142. The enzyme complexes were produced by growing the lytic organisms in the presence of autoclaved *Pythium* mycelium, and concentrated from the culture mixture by precipitation with acetone (2 : 1, v/v). After about 6 h incubation with the lytic preparations swellings appeared at several points on the hyphae at intercalary locations as well as terminally on branches. The swollen structures became larger and the cell wall thinner as the digestion proceeded. After about 40 h incubation little mycelial wall material remained and all of the protoplasm was transformed into spherical protoplast-like structures.

(e) *Ashbya gossypii* protoplasts were produced by β (1–3)- and β (1–6)-glucanases of a strain of *Bacillus circulans* (Tanaka and Phaff, 1967). These two enzymes were partially purified and separated from each other and tested on a number of fungal species. *Ashbya gossypii* and *Eremothecium ashbyi* were susceptible to these enzyme preparations, with rapid formation of protoplasts. Cells of these organisms were grown for 20 h on a shaker at 30°C, and the cells were harvested and washed by centrifugation. The reaction mixture contained the cells, the enzyme preparations, at pH 6·5 an 0·65 M mannitol to create an osmotically balanced environment. Protoplast formation started within 10–20 min. Young cells, primarily hyphal tips, appeared more susceptible than other cells. Several protoplasts were often formed at the tip of a single hyphae. After prolonged action of the enzyme mixture, the protoplasts started to form along the hyphal strands as well.

The cyotplasmic contents began to emerge from small pores in the walls assuming a spherical shape. The entire contents of one hyphal compartment formed usually a single protoplast. The protoplast showed great stability after several days but lysed immediately when placed in distilled water.

3. *Other enzyme preparations*

The lysis of walls of *Saccharomyces cerevisiae* by proteolytic enzymes such as trypsin, papain and by enzyme preparations from malt has been studied by Eddy (1958). These enzymes were able to digest only partially the cell walls and complete degradation never occurred. Satomura *et al.* (1960) carried out similar studies but using a combination of β-glucanase, lipase and phospholipase showing complete digestion of intact walls of yeast.

Garcia Acha *et al.* (1966*a*) succeeded in obtaining protoplasts from *Fusarium culmorum* mycelium by a combination of pancreatic lipase and strepzyme M. This last preparation alone did not form protoplasts. The *Fusarium* mycelium obtained from 2 day old cultures was previously incubated for 30 min with 200 μg/ml of pancreatic lipase enzyme (Sigma, U.S.A.) and then washed by centrifugation with a 0·8 M mannitol-phosphate buffer pH 5·8 solution. The mycelium was again resuspended in the same solution and treated with the lytic enzyme preparation (20 mg d.wt/ml) from *Micromonospora RA* (Strepzyme RA). Surprisingly it was found that the combined action of lipase and the strepzyme successfully gave rise to the rapid formation of large numbers of mycelium protoplasts.

Emerson and Emerson (1958) reported production of protoplast-like structures from a variety of strains of *Neurospora crassa* by treatment with a commercial hemicellulase preparation or crude snail hepatic juice. Hyphae were digested with 0·3–0·5% (w/v) hemicellulase microbial enzyme preparation, containing cellulase, maltase and chitinase activity (from Nutritional Biochemical Corporation, Cleveland 28, Ohio, U.S.A.) in a medium containing 2% sucrose, 5–10% rhamnose or sorbose and the standard salt mixture. Protoplasts were more easily obtained at the higher concentrations of sugars and enzyme. The spherical forms lysed in water to leave a delicate membrane.

Villanueva *et al.* (1969) reported studies carried out to assess the formation of *Candida utilis* protoplasts by a variety of enzymes which are believed to be active during the process of cell wall digestion. β (1–3)-glucanases from *Micromonospora AS* and *Rhizopus arrhizus*, β (1–4)-glucanase from *T. viride*, β (1–6)-glucanase from *Penicillium brefeldianum* and chitinase from *Serratia marcescens* were tested on young cells (Table II). *Candida utilis* and *Sacch. fragilis* cells were grown at 29°C with shaking in Hansen medium. Cells of 12 and 30 h cultures were used and washed twice with

1·0 M SO$_4$ solution used as stabilizer of protoplasts. The digestion of the cells using a variety of enzymes (β (1–3)-, (1–4)- and β (1–6)-glucanases, *trypsin, carbozypeptidase*, crude proteases, pancreatic lipase and chitinase) were conducted at 30°C according to Gascón and Villanueva (1965). The concentration of the enzymes varied between 0·03 to 0·2% (w/v), and a cell concentration of an initial absorbancy of approximately 0·5 was used. Protoplast formation was observed when using both β (1–3)-glucanases and only rarely with β (1–6)-glucanase. The addition of both preparations together gave rise to good formation of protoplasts. β (1–3)-glucanase was also very effective in inducing formation of protoplasts from young cells of *Sacch. fragilis*.

Nagasaki *et al.* (1966) described another enzyme for the formation of yeast protoplasts. The enzyme was partially purified from filtrates of *Bacillus circulans* and was named phosphomannanase (McLellan and Lampen, 1968). The preparation apparently contained β (1–3)-glucanase which is another essential component necessary to produce protoplasts from log-phase yeast. It was suggested that a third enzyme present in other lytic preparations may also be necessary to produce protoplasts from young cultures of yeast. The purified preparations of the phosphomannanase have no proteolytic activity and act on yeast cells, phosphomannans and cell walls prepared from yeast. The enzyme was found to split a mannosidic bond adjacent to a phosphodiester-linked mannose. The purification of phosphomannase is interesting for the understanding of the specific enzymatic events required to remove the cell wall of yeast. This enzyme causes separation of the cell wall from the plasma membrane (McLellan, McDaniel and Lampen, 1970). Phosphomannanase added to small amounts of snail enzyme appears to be a more effective and reliable method of preparing protoplasts from yeast than procedures previously used. This mannanase has proven useful for studying the macromolecular organization of polymers in the fungal cell walls.

Various workers have suggested that the mannan forms the outer layer of the cell wall (Eddy and Rudin, 1958; Northcote, 1963); then it is not surprising that the removal of mannan from the yeast wall is the first step in producing protoplasts. A step by step study of the enzymes required to produce protoplasts will undoubtedly lead us to a better understanding of the structure and organization of the cell wall. Previously, it was reported that β (1–3)-glucanase partially purified from *Micromonospora AS* and *Rhizopus arrhizus* would lyse walls of intact cells of yeasts with the further formation of protoplasts (Villanueva *et al.*, 1969). These glucanase preparations probably contained significant amounts of phosphomannanase in addition to other enzymes needed for the dissolution of walls. In other systems it was shown that the glucan mesh on the inner layer of the cell wall cannot be disrupted by β (1–3)-glucanase alone (McLellan and Lampen, 1968).

Cellulases and pectinases have been used by Cocking (1968) for obtaining protoplasts from plant tissues with great success. This method has as its basis the same principle as that used for the isolation of yeast protoplasts using snail enzymes. 0·5 cm root tips of tomato seedlings grown under aseptic conditions were treated at 27°C with 5% *Myrothecium verrucaria* cellulase in 0·02 M phosphate buffer pH 6·0 containing 0·6 M sucrose as an osmotic stabilizer. Microscopic examination of the root tip, mounted under a cover slip, showed the formation of numerous spherical protoplasts after 2 h of treatment with cellulase (Cocking, 1960).

Takebe *et al.*, (1968) described a method for the isolation of leaf protoplasts in which cells isolated from the leaf by prior treatment with 0·5 (w/v) pectinase (Macerozyme, Kinki Yakutt Manuf. Co., Nishinomiya, Japan) are subsequently treated with 5% (w/v) cellulase (Onozuka P 1500 from *Trichoderma viride* Kinki Yakult Manuf. Co.) to digest away the cell wall. Conditions are the same as those described by Cocking. Very large quantities of protoplasts can be isolated by using mixtures of these commercially available enzymes.

Ruesink and Thimann (1966) also described obtaining protoplasts from higher-plant tissues with a concentrated cellulase preparation (2% w/v) from *M. verrucaria* in 0·5 M mannitol or a mixture of salts. Addition of polyuronidases or hemicellulase to the cellulase preparation did not increase the yield of protoplasts. Protoplasts could be prepared from root, leaf, and callus tissue as well as from coleoptiles of *Avena*.

4. *The effect of mercapto-compounds on the formation of protoplasts*

When washed cell suspensions of *Saccharomyces fragilis* were incubated with 0·2 vol of snail enzyme in presence of 0·8 M mannitol (or maltose) at pH values in the range 5·5–7·5 no protoplasts were formed. Addition of 0·01 M β-mercaptoethanol resulted in complete conversion of the yeast cells to protoplasts within 1 h at pH 7·5 and 37°C. The rate of conversion was slightly faster at high cell densities and there was little differences of behaviour between exponential and stationary phase cells. Davies and Elvin (1964) have suggested that the site of action of the mercaptocompound is on the yeast cell and not on the snail enzyme, since cells treated with the compound at pH 7·5 and then washed free from mercaptoethanol are readily converted to protoplasts when incubated with snail enzyme in 0·8 M maltose at pH 7·5. It is possible that the increase in sensitivity to the lytic enzyme may be related to the observations of Nickerson and Falcone (1956) that reduction of -S-S-bridges in a mannan-protein complex present in cell walls of yeast may be involved in cell division. Other workers have also found that with certain yeasts which do not normally form protoplasts under the action of snail juice, protoplast formation can be induced by means of β-mercaptoethanol. The cross linkages between protein moieties via disulphide bridges may render those parts of the cell wall insoluble (Anderson, 1967).

Regardless of the eventual explanation of the mode of action Svihla and Schlenk (1967) have demonstrated the increased sensitivity of *C. utilis* and *S. cerevisiae* cells from sulphur compound supplemented medium to snail gut juice enzyme compared with cells from non-supplemented medium. It was shown that 2,3-dimercaptopropanol, L-homocysteine or 2-mercaptoethylamine as culture supplements enhanced protoplast production greatly, although the first compound produced protoplasts of poor quality and stability. Addition of the compounds to the digestion medium was ineffective.

C. Metabolic disturbance with inhibition of cell wall formation

There is evidence that some compounds interfere with biosynthetic reactions involved in cell-wall formation without affecting other reactions connected with synthesis of protoplasm. This action results in "unbalanced growth" in which cells produce an abnormal cell wall or none at all. In order to demonstrate the effect of these substances cells must be in an actively growing state and in the presence of stabilizing solutions, such as hypertonic sucrose or mannitol.

The specific inhibition of cell wall synthesis without affecting the synthesis of the cytoplasmic components has been used for preparation of protoplasts of bacteria. Similar attempts with fungi have been only partly successful. De Terra and Tatum (1961) showed that hyphae of *Neurospora* become bulgy and irregular when grown in 4% sorbose medium. Hamilton and Calvet (1964) have found that when sorbose (5–20%) was added to growth media, the osmotic mutant M 16 of *N. crassa* produced small colonies composed almost entirely of protoplasts. Cultures of the osmotic mutant were grown in 100 ml of minimal medium containing 5, 10, 15 and 20% filter-sterilized sorbose.

The composition of the minimal medium is as follows (in grams/litre):

Sucrose	20·0
Na$_2$ citrate 5·5H$_2$O	3·0
KH$_2$PO$_4$ (anhydrous)	5·0
NH$_4$NO$_3$ (anhydrous)	2·0
MgSO$_4$.7H$_2$O	$2·0 \times 10^{-1}$
CaCl$_2$.2H$_2$O	$1·0 \times 10^{-1}$
Citric acid 1H$_2$O	$5·0 \times 10^{-3}$
ZnSO$_4$.7H$_2$O	$5·0 \times 10^{-3}$

$Fe(NH_4)_2(SO_4)_2 . 6H_2O$	1.0×10^{-3}
$CuSO_4 . 5H_2O$	2.5×10^{-4}
$MnSO_4 . 1H_2O$	5.0×10^{-5}
H_3BO_3 (anhydrous)	5.0×10^{-5}
$Na_2MoO_4 . 2H_2O$	5.0×10^{-5}
Biotin	5.0×10^{-6}

Cultures were inoculated with a loop from a 24 h slant and were grown at room temperature for 2 to 3 weeks. Microscopic examination of cultures was performed by use of wet mounts. Bursting of protoplasts occurred when 1 to 2 drops of water were added under the cover slip. Similar molar concentrations of fructose, glucose, and sucrose also produced protoplasts, suggesting a relationship between inhibition of cell-wall synthesis and osmotic pressure. Although the biochemical defect in this mutant is unknown, it appears likely that it involves the cell membrane, since the membrane is thought to be the regulator of materials passing into and out of the cell. Protoplast formation may be the result of inhibition of cell-wall synthesis by loss of necessary substrates or enzymes. An alternative explanation would be the extracellular movement and activation of an autolytic enzyme. More recently Robertson (1965) has suggested that the sorbose molecule might be preventing the normal building of glucose units into the fungal wall at the apex of the cell. The spherical bodies observed in 20% fructose, glucose, or sorbose-sucrose meet the criteria for *Neurospora* protoplasts as set by Bachman and Bonner (1959); i.e., they are viable and osmotically sensitive. It was found that in lower sugar concentrations, the spherical bodies do not meet these criteria as well, since varying amounts of cell wall were present thus decreasing the osmotic sensitivity.

Cohen *et al.* (1969) recently reported the isolation of a temperature-sensitive mutant of *Aspergillus nidulans* with altered cell walls; these contain drastically reduced amounts of amino-sugars. It was observed that this isolate grew on malt extract glucose agar at 30°C but, unlike the parent strain, could no longer grow at 42°C. When grown in a liquid glucose medium the organism appeared to have lysed. Lysis could be prevented by adding 20% sucrose or 3% NaCl. With osmotic stabilizers, the conidia germinated into irregular shaped hyphae with swollen sections. These forms readily burst in distilled water. Transfer of normal germinated conidia to 42°C resulted in swellings appearing within 90 min at the tip of the hypha. It is relevant that hyphal extension takes place only at the tip. The observations suggested that the mutant forms a defective wall at the high temperature with formation of spherical bodies very similar to protoplasts. Chemical analysis of the isolated walls of the osmotic mutant in comparison with that of the walls from the parent strain showed that the mutant contains little or no chitin in the walls. These results suggest that chitin contributes to the mechanical strength of the wall, is made exclusively near the hyphal tip and is needed to form normally shaped hyphae.

It has been reported that 2-deoxyglucose interferes with glucose metabolism in yeast cells and especially with some processes connected with cell wall formation (Johnson, 1968). The mechanism of the action of the 2-deoxyglucose remains unexplained. In liquid media, growth of the protoplasts is inhibited completely when the 2-deoxyglucose is present, although the protoplasts remain alive for more than 48 h. No rigid cell wall component was found on their surfaces when examined with the electron microscope. It has been stated by Svoboda *et al.* (1969) that the 2-deoxyglucose inhibition is reversible—simple addition of glucose to the medium initiates normal metabolic processes. Of the two cell wall components, fibrillar (glucan) and amorphous (mannan-protein), the latter seems to be more sensitive to the action of 2-deoxyglucose.

Release of protoplasts in the yeast phase of *Histoplasma capsulatum* without added enzyme has recently been reported by Berliner and Reca (1970). When yeast cultures in 2 M $MgSO_4$ broth were examined after 48 h of incubation at 37°C, it was noticed that large numbers of protoplasts were being released. After the cultures were incubated for 96 h, they contained virtually pure suspensions of protoplasts. It was found that each yeast cell released only one or more protoplasts at only one point on the periphery of what appeared to be a rigid and intact cell wall. The action of 2 M $MgSO_4$ with or without 2-deoxy-D-glucose, thus appears to be limited to a very small, highly susceptible area of the cell wall, namely, the most recent formed bud scar which may be delimited by a membrane or an incomplete wall (Edwards *et al.*, 1959). This is supported by the observation that only young and actively dividing cells release protoplasts and only at one polar locus.

Attempts to prepare protoplasts of fungi by means of antibiotics and other fungicides have been made without success. A variety of antifungal antibiotics tested on actively growing cells of yeasts failed to produce osmotically fragile protoplasts (Shockman and Lampen, 1962). Griseofulvin, a compound known for its action on immature fungal walls which contain chitin (Bent and Moore, 1966), produced alterations of the permeability of the fungus, but no formation of protoplasts was detected (Kinsky, 1961).

Glycine (3% w/v) has been used to induce the formation of spherical forms of various Gram-negative bacteria. Glycine was also tried by us to convert *F. culmorum* into protoplasts. All attempts failed, although some abnormal globular, osmotically sensitive forms were observed suggesting the building of altered cell walls. Forms resembling protoplasts were not detected.

D. Growth of cells on serum

Meinecke (1960) has described obtaining "protoplasts" of fungi on cultivation of *Penicillium glaucum* on thickened serum. Cells of the fungus were inoculated in thickened guinea-pig serum between two sterilized glass plates and fastened with paraffin. After 7–14 days some abnormal globular forms connected with the hyphae were observed but it is dubious whether these sperical, ovoid bodies, can be considered as protoplasts. The same effect could be seen when *P. glaucum* was embedded into living plants (Meinecke, 1957).

III. SEPARATION OF PURE PREPARATIONS OF PROTOPLASTS

After the enzyme action, the transformation of yeast cells into protoplasts does not require precautions other than the usual gentle methods to avoid rupture of the protoplast membranes during centrifugation. Protoplasts can be purified from the contaminating media by repeated centrifugations at low speed and resuspension in an appropriate osmotic stabilizer. In order to achieve total conversion of whole cells to protoplasts young cells are usually required.

Protoplast suspensions of filamentous fungi can be partially freed of remaining hyphal filaments by filtration of the whole suspension through coarse sintered glass filters to remove most of the mycelium. *F. culmorum* protoplasts were freed from mycelial debris by passing the suspension through a 11G2 Jena fritted glass filter without pressure (Garcia Acha *et al.*, 1966a). Microscopic examination of the filtrate revealed a high concentration (10^6 to 10^8/ml) of the structures designated as mycelial protoplasts. Some of the medium sized and smaller protoplasts and most of the bigger protoplasts were retained by the filters. The few hyphal remnants which could be observed could be removed completely by low speed (500–1000 g) centrifugation. A purified preparation of protoplasts was obtained in this way but recovery was generally about 50% (Villanueva, 1966).

IV. PROTOPLASTS FROM SPORES

Studies on the lytic action of natural soil have shown that the most important survival organs of plant pathogens are resistant spores (Lockwood, 1960). Resistant forms survive in soil for long periods.

The digestion of cell walls of spores has only rarely been studied in detail. Lysis appears to be associated with the presence of lytic enzymes excreted to the surrounding medium by lytic micro-organisms. Direct attack by a species of *Verticillium* on spores of *Hemileia vastatrix* was shown by Leal and Villanueva (1962). The mould developed well when the rust spores provided the only carbon and nitrogen source, total dissolution being the last step of the lytic process. Tetrads of spores in yeasts were released after the sporulated cells were digested by snail enzymes (Johnston and Mortimer, 1959). Ascospores from mature, inoperculate asci were liberated by *Helix pomatia* enzyme treatment (Haskins and Spencer, 1962). Ascus walls of many species of yeasts were dissolved on treatment of sporulating cultures with a combination of β (1–3)- and β (1–6)-glucanases from *B. circulans* (Tanaka and Phaff, 1967). This observation is rather surprising since the ascus walls of many species were found to be much more susceptible to digestion by the microbial β-glucanases than were the walls of vegetative, non-sporulating cells. The susceptibility of ascus walls to enzymatic lysis however, varied considerably from species to species.

Osmotically fragile structures may be obtained from certain fungus spores by dissolving the wall of living cells with snail and microbial enzymes. The availability of viable conidial cells deprived of cell walls is of great interest for the better understanding of the structure and physiology of fungi and in many cases represents the most satisfactory starting point for obtaining suspensions of free "protoplasts".

Bachmann and Bonner (1959) were first to describe liberation of protoplasts from fungal spores using snail enzymes. Free protoplasts were obtained when single-celled conidia of *N. crassa* were incubated with the lytic preparation. Weiss (1963) also described obtaining conidial protoplasts of *N. crassa* and studied them by electron microscopy. Macroconidia were obtained by seeding 500 ml of appropriate medium (Vogel, 1956) in a 1 l flask with a heavy suspension of conidia washed from a slant. Incubation was at 28°C for 3 to 4 days. Production of conidia was improved by reducing the temperature for the last 24 h of incubation to about 25°C. Conidia were harvested by washing them from the medium and freeing them from the accompanying mycelium by filtration through glass wool. The enzyme preparation used was the "Suc-digestif *d'Helix pomatia*" obtained from L'Industrie Biologique Française. Free protoplasts are released when conidia are incubated in a solution containing 10% enzyme preparations, 20% sucrose and approximately 0·03 M phosphate buffer at pH 6·0. After about 1 h of incubation at 28°C the walls of conidia begin to bulge, assuming at first a lantern shape and later becoming spherical. As a rule, the entire contents of one conidium are extruded through one pore in the side wall to form one free protoplast. In many cases, the protoplasts do not emerge from the conidia, however, and remain as spherical structures within the walls which appear very weak.

We have described the formation of protoplast-like structures from spores of various moulds (Garcia Acha *et al.*, 1966a) by exposure to microbial lytic preparations which apparently attack either partially or completely the outer layers of the spores which leads to the formation of "protoplasts". The osmotically sensitive bodies do not emerge from the conidia, but remain as spherical bodies within weakened walls that virtually disappear. Later the individual cells making up multi-celled conidia are released, giving rise to isolated spherical "protoplasts".

Naked "protoplasts" of *Fusarium culmorum* conidia were obtained by the following method. Spores were obtained from cultures grown at 28°C on 100 ml of glucose-asparagine-yeast extract medium solidified with agar in a Roux bottle. Mature spores were collected after 4 days incubation by resuspending in water. Spore suspensions were mixed with two volumes of 0·1 M phosphate buffer pH 6·8 containing 0·8 M mannitol to which 1/10 volume of the "strepzyme RA" preparation was added and incubated at 28°C with gentle shaking. Under the microscope the "protoplasts" appeared to be about as dense as the contents of the conidia from which they came. The "protoplasts" were not released from the conidia but remained as spherical bodies within the weakened walls. After 15–20 h of agitation the conidial walls as well as the cross-walls which separated the various cells of the conidia became almost invisible. "Protoplasts" isolated from conidia were much more stable than those from young hyphae. In general the free "protoplasts" were not very sensitive to osmotic shock although some of them lysed after addition of distilled water. Observation under the phase-contrast microscope has shown a tenuous membrane of about the same size as the protoplast remaining after the explosive dispersion of the cell contents. These membranes seem to be less delicate than those observed after lysis of protoplasts derived from hyphae.

"Protoplasts" obtained under these conditions can germinate forming a vegetative mycelium. However, this mycelium often showed abnormalities, and at some points gave rise to the formation of protuberances. Susceptibility of the spores varies considerably with age; young spores (3–6 days) are much more susceptible than old spores (more than 15 days).

Aquatic *Phycomycetes* release during their life cycles flagellated zoospores which may be regarded in some respects as natural protoplasts (Bartnicki-Garcia and Lippman, 1966). Although zoospores are not quite spherical, and will withstand very low osmotic pressure without bursting, they seemingly lack a cell wall. Electron microscopy of thin-sectioned zoospores of related fungi indicate the absence of a cell wall (Cantino *et al.*, 1963). Questions then arise as to whether the membrane of an artificial protoplast is of the same nature as that of a zoospore.

V. USE OF PROTOPLASTS

The production of protoplasts in various micro-organisms has simplified the study of their biochemistry, morphology and genetics. The potentiality of the study of fungal protoplasts to elucidate many of the present problems associated with cell structure, growth, nutrition, biosynthesis and antibiotic action is evident. The availability of protoplasts opens up new ways in the investigation of the role of the cell wall in both physiology and genetics. In view of the osmotic fragility of most of the protoplasts obtained from yeast and moulds it seems likely that they can be used as source material for preparation of native subcellular fractions of interest for biochemical studies. The advantages are the ease of disruption of the cell by simple dilution, or removal of the stabilizer for the protoplasts, as well as the mild conditions employed.

A. For the preparation of cell membranes

The physical nature of the protoplast membrane and the protective role of the cell wall in eucaryotic and procaryotic cells are seen when protoplasts are derived from long, cylindrical bacilli or yeasts or from branching mould filaments. These protoplasts promptly assume a spherical but easily distorted shape. It is evident that the cell wall is a rigid, protecting and retaining structure, whereas the protoplast membrane is delicate and easily ruptured.

The cytoplasmic membrane of fungal cells is a distinct and separable structure from the outer envelope or cell wall. By gentle enzymatic removal of the fungal cell wall, followed by osmotic rupture of the resulting protoplast in a hypotonic solution, the contents of the cytoplasmic membrane can be removed and the resulting structure, sometimes called a "ghost", cleaned by gentle washing and centrifugations. Studies of such materials show that fungal (yeast) cell membrane typically consist of protein and lipids with small proportions of carbohydrates and ribonucleic acid. The results so far obtained do not allow the establishment of well defined structural and functional roles of the components identified, but they suggest some unique properties of the fungal plasmalemma.

A study of the activity of the fungal membranes in relation to protein synthesis would be of considerable interest, particularly the extent to which activity is correlated with the presence of ribosomes since somewhat similar crude membrane fractions from osmotically lysed bacterial protoplasts, which are known to contain ribosomes, are very active in protein synthesis (Cocking, 1965).

Protoplast membranes were usually prepared by lysis of the protoplast preparations obtained either by enzymes present in the digestive juice of the snail *Helix pomatia* or by microbial enzyme preparations (Villanueva, 1966).

Few studies have been made in the composition of protoplast membranes from fungi. Boulton (1965) was first to describe the gross composition of fractions obtained from lysed preparations of *Saccharomyces cerevisiae* protoplasts. In an attempt to isolate various cytoplasmic membranes from yeast, protoplasts were lysed under different conditions of temperature, pH, calcium and magnesium ion content. Protoplasts were obtained as described by Eddy and Williamson (1957) using as osmotic stabilizer mannitol. The protoplasts were kept at 0°C in 10% mannitol solution containing acetate buffer at pH 5. For the preparation of the membranes protoplasts (40–60 ml as packed volume) up to 48 h old were mixed with ice cold 0·025 M Tris buffer solution (350 ml) at pH 7·2 containing 1 mM Mg Cl$_2$. The mixture was allowed to stand for 30 min at 0°C and was then centrifuged for 30 min at 20,000 *g*. The pellet was suspended in a cold solution of the same buffer (50 ml) and gently homogenized by hand in a loose fitting Potter homogenizer. The product was centrifuged for 5 min at 1500 *g* and the supernatant solution after being separated from the pellet (1·5 p 5 fraction) was centrifuged for 30 min at 20,000 *g* when a further pellet (20 p 30 fraction) was obtained. The two pellets were next washed by suspending them in the cold buffered solution of Mg Cl$_2$ (12 ml for each) and centrifuging for 30 min at 20,000 *g*. The particulate fractions so obtained, each containing about 35 mg of protein, were separated and lyophilized.

Boulton has pointed out that the residues from the above treatment appear to consist principally of membranes forming vesicles of various sizes. The various membranous structures differ among themselves both in composition and enzyme content. The greater part of the membrane materials recovered from the protoplasts was presumably derived not from nuclei nor from the mitochondria, but from the outer cytoplasmic membranes, and other internal membranes including that of the primary vacuole. It seems likely that internal membranes may contribute about twice as much material as the surface membranes.

Cytoplasmic membranes of yeasts were also prepared by Garcia Mendoza and Villanueva (1967). Protoplasts of *C. utilis* were obtained by incubating a suspension of cells (5 mg dry wt./ml) at 37°C with shaking in the presence of the "strepzyme GM" preparation and 0·8 M mannitol as stabilizing agent. The protoplasts were centrifuged at 3000 *g* for 15 min and washed in the same solution of stabilizer, protoplast membranes were prepared by dilution of the protoplast suspension. An iced solution containing 10^{-4} M Tris buffer (pH 7·2) and 10^{-4} M Mg^{2+} was used and when disruption was complete a solution of the same buffer containing MgCl$_2$ was added to give a final concentration of 10^{-2} M Mg^{2+}. This cation has been shown to prevent further disintegration of the membranes. The membrane fraction was obtained by centrifugation at 15,000 g for 15 min. The dark yellow pellet was purified by washing several times with the same buffer. The cell membrane of *C. utilis* consists largely of protein and lipids each accounting for 40% of the dry weight of the membrane preparations. The total carbohydrate content was found to be 5% glucose, mannose and galactose being the only components. The total nucleic acid content of 1% corresponds to RNA.

Garcia Mendoza *et al.* (1968) carried out some further work to elucidate the structure of the cell membrane of *C. utilis*. In this report special attention was paid to the relationship between the structure and inmunology of membrane preparations which were before studied from a morphological and chemical point of view. While traces of residual cell-wall material have not been detectable in the preparations of membranes obtained by enzymatic methods, the immunological behaviour of these membranes suggests that the protoplast membrane does possess small amounts of structural cell-wall material.

Longley *et al.* (1968) made a more detailed analysis of protoplast membranes from another yeast species, *S. cerevisiae*. Growing cells were harvested by centrifugation, washed and suspended in 5 mM citrate-phosphate buffer, pH 5·8, containing 0·8 M-mannitol. One volume of snail preparation (40 mg protein/ml) was added to 4 vol of yeast suspension and shaken at 30°C for 3 h. Yeast protoplasts were separated by centrifuging the suspension for 10 min at 0°C at 600 g. The protoplasts were washed several times with phosphate buffer pH 5·8 containing 0·8 M mannitol and suspended in 100 ml of buffered mannitol (0·8 M) containing 10 mM $MgCl_2$. To prepare protoplast membranes, the protoplasts were resuspended in a small volume of buffered mannitol (0·8 M) containing 10 mM $MgCl_2$. The thick suspension was then squirted into 20 vol of ice-cold phosphate buffer, pH 7·0 containing 10 mM $MgCl_2$ and the suspension was gently stirred for 30 min. The suspension was then centrifuged at 1500 g for 30 min. The supernatant liquid was removed and the membranes were resuspended by gentle stirring in phosphate buffer, pH 7·0, at 0°C containing 10 mM $MgCl_2$ and again centrifuged at 1500 g for 30 min. The membranes were washed once more in this buffer and then twice in water. All operations were carried out at 0°C. Membranes were resuspended in 10 ml of water in a stoppered tube stored in a vacuum desiccator over silica gel at -20°C. These protoplast membranes accounted for 13–20% of the dry weight of the yeast cell. They contained on a weight basis about 39% of lipid, 49% of protein, 6% of sterol and traces of RNA and carbohydrate (glucan × mannan). The principal fatty acids in membrane lipids were $C_{16:0}$, $C_{16:1}$ and $C_{18:1}$ acids. Phospholipid fractions from membranes and whole cells had similar fatty acid compositions and contained two major and three minor sterol components.

After sudden and complete lysis of *F. culmorum* protoplasts a more or less spherical cytoplasmic membrane about the size of the protoplast, showing distinct folds, was sometimes observed (Rodriguez Aguirre *et al.*, 1964). These structures enveloped some cytoplasmic material in an advanced state of disorganization. Clean and empty membranes were never obtained. Attempts to obtain electron micrographs of these membranes failed owing to difficulties with the usual methods of fixation.

Washed protoplasts free of lytic enzymes were resuspended in cold 0·1 M phosphate buffer pH 6·0. The lysed protoplasts were sedimented at 40,000 *g* for 15 min in order to collect the membranes. However, instead of a protoplast membrane pellet a large, viscous mass was found. The addition of DNA-ase only partially eliminated the viscous material.

Some other attempts to obtain membrane preparations from filamentous fungi were described by Strunk (1969) using protoplasts of *Polystictus versicolor* prepared by using gut juice of the snail *Helix pomatia*. Lysis of suspensions of protoplasts followed by centrifugation facilitates the separation of plasmalemma.

One shortcoming of the methods of preparation of cytoplasmic membranes using protoplasts is the possible degradation of components of the plasmalemma by the action of the snail or microbial enzymes used for preparing protoplasts. Suomalainen *et al.* (1967) avoided this interference obtaining membranes after whole cell disintegration. Matile *et al.* (1967) also carried out some analysis of isolated plasmalemma based on preparations obtained after mechanical breakage of *S. cerevisiae* whole cells and subsequent fractionation of extracts using differential and density gradient centrifugation. A comparison of the overall compositions of membranes isolated with the methods described above shows considerable differences although they strongly suggest the lipoprotein nature of these membranes. The most important qualitative difference concerns the nucleic acids which are present in the preparations obtained from protoplasts and absent in membranes obtained from whole cells. The amount of carbohydrate is much lower in the protoplast membranes than in the membrane fragments from whole cells.

(a) *Enzyme activities associated with cytoplasmic membranes.* An extremely important function of the cell membrane is maintenance of a favourable intracellular environment. The cell membrane, aided by permease systems, exerts during life a truly remarkable degree of selective permeability upon which the life of the cell depends. It is well known for other systems that many of the enzymic activities of the cell, specially those concerned with the energy-yielding reactions of the cell, undoubtedly take place in the cell membrane.

Early studies on protoplast membranes of "ghosts" isolated from lysed protoplasts clearly indicated that these structures were biochemically active.

Boulton (1965) has studied the localization of a number of enzymes namely hexokinase, aldolase, pyruvate kinase, phosphoglycerate kinase, triosephosphate dehydrogenase, alcohol dehydrogenase, $NADH_2$ oxidase, $NADH_2$ diaphorase, NADH cytochrome-reductase and ATPase, connected with membrane systems. Some of these enzyme system might arise from contaminating mitochondria. Garcia Mendoza (personal communication) has also detected ATPase activity associated with the purified membrane preparation.

In purified preparations of plasmalemma isolated from whole cells, various enzymes known to be localized in the ground cytoplasm, in vacuoles or mitochondria are absent (Matile, 1969). The prominent enzyme activity associated with these preparations is that of an Mg^{++} dependent ATPase. A comparatively low specific activity of invertase has been detected occasionally in freshly prepared membranes. The author speculates that the plasma membrane is the place of reactions which result in the conversion of the internal into the external invertase (Gascón et al., 1965).

Chitin synthetase activity of yeast was demonstrated in particles derived from the cell membrane (Kinsky, 1962) and similar results were obtained by Algranati et al. (1963) who showed that the cell membrane was the site of synthesis of yeast mannan. Further investigations with these systems would not only give valuable information about the localization of enzymes in the envelopes of mould and yeast, but also contribute to our knowledge of the chemical anatomy of these structures.

B. For obtaining nuclei, vacuoles and tonoplasts

Vacuoles, nuclei, granules, inclusions or particles of various sizes are also found within the protoplast and can sometimes be seen with the phase contrast microscope. Several authors have found that even during lysis of protoplasts the vacuoles can readily be released, and centrifuged down in intact form. Eddy (1959) was first to observe that ultrasonic vibration of yeast protoplasts in certain circumstances released the main vacuoles and smaller vacuoles in a relatively intact form. These vacuoles have been shown to be surrounded by semi-permeable membranes. Svihla and Schlenk (1960) had first demonstrated the utility of *Candida utilis* spheroplasts for obtaining vacuoles after osmotic shock. Various compounds accumulate in vacuoles and an analytical exploration of isolated vacuoles was a pre-requisite for the analysis of this cellular compartment. The same group of workers have shown, that the irradiation of yeast protoplasts with about 6×10^4 ergs/mm^2 of ultraviolet light of wavelength between 250 and 270 mμ results in the breaking of the plasmalemma and the release of the main vacuole with its membrane intact (Svhila *et al.*, 1961).

More recently Indge (1968) thought that the phenomenon of metabolic lysis described by Abrams (1969) might provide the basis of an appropriate method of preparing intact vacuoles. When yeast protoplasts (2×10^7/ml) were incubated at 30°C in a solution containing mannitol (10%, w/v), citrate ions (6 mM) or EDTA (5 mM) and glucose (0·5% w/v) at pH 6·4 rapid metabolic lysis occurred. Microscope examination showed that the protoplasts, after swelling, lysed releasing spherical bodies 2–3 μ diam., each protoplast giving rise to one such body. These bodies were identified with the cell vacuole. The liberated vacuoles could be concentrated by centrifugation at 2000 *g* for 4 min, when the vacuoles tended to remain dispersed in the pellet formed in the centrifuge tube. The pellet could be resuspended in a solution containing mannitol (10% w/v) and 0·01 M imidazole HCl buffer, pH 6·4 giving a supernatant solution rich in vacuoles which could be decanted from the other cell constituents. The success of this technique in liberating apparently intact vacuoles from yeast protoplasts results from the change in the permeability properties of the protoplast membrane induced by glucose and chelating agents. The isolated vacuoles exhibited vital-staining reactions.

An alternative method for the liberation and purification of vacuoles involves flotation from suspensions of lysed protoplasts in the presence of Ficoll (Matile and Wiemken, 1967). The lysis of the protoplasts was achieved by decreasing the tonicity of the suspending medium: two volumes of a solution containing 0·025 M Tris-citrate pH 6·5, 0·025% Triton- × -100 and 1 mM EDTA were added to one volume of a suspension which contained Ca 10^9 protoplasts per ml in 0·7 M buffered mannitol. The lysis was facilitated by a gentle agitation in a Potter-Elvenhjem homogenizer. For the isolation of the liberated vacuoles, Ficoll (Pharmacia Uppsala) was added to the lysed protoplasts to a concentration of 8% (w/v), where the vacuoles assume a density lower than that of the medium. In centrifuge tubes the suspensions were overlayered with two layers, 7·8% and 7·4% of Ficoll. The isolated vacuoles form a white layer on top of the system after 25 min of centrifugation at 2000 *g*. Preparations of isolated vacuoles showed great variation in the diameters of the spheres, usually between 0·2–2 μm. It is uncertain whether these small spheres represent fragments of large vacuoles or small vacuoles which were already present in the intact cells. Hydrolytic enzymes are present in high specific activities in isolated vacuoles. Therefore it was suggested that the vacuoles represent the lysosome of the yeast cell.

Vacuoles in hyphae and in protoplasts of different species of fungi are easily visible under the phase contrast microscope. The presence of vacuoles in fungal cells and their size are affected by the composition of the growth medium, the age of the culture and possibly growth temperature and other factors. During cell wall digestion, and after the protoplasts have been released, the composition of the suspending medium affects the formation, size, and number of the vacuoles seen in the protoplasts. Some protoplasts have several vacuoles of different sizes while others show no apparent vacuole (Garcia Acha *et al.*, 1967).

Conditions for preparing vacuoles from fungal protoplasts have been established. When a 2 ml volume of *Fusarium culmorum* protoplasts in 0·8 M mannitol was diluted with 5 ml of distilled water, disruption of most of the protoplasts occurred with the release of the cytoplasmic contents. Intact vacuoles could be seen in the suspension and these could be partially recovered by centrifugation. It has also been observed that 0·2 M $MgSO_4$ in 0·1 M phosphate buffer, pH 6·5 allows bursting of protoplasts but maintains the integrity of the vacuoles. The protoplasmic contents are dispersed into the medium and can be eliminated by further washing. On further dilution (between 0·2 M and 0·15 M $MgSO_4$ in the same buffer) the vacuoles swell, ultimately bursting and collapsing to liberate the contents into the medium. A faint remnant of the tonoplast is left behind. This again adopts a spherical shape, but it is smaller in size than the original vacuole. No attempt has been made to isolate the tonoplasts. Osmotic conditions must be very precise; otherwise the vacuoles also burst. If osmotic changes are not made very gradually, breakage of protoplasts and vacuoles is simultaneous and nothing with a definite organization can be seen.

Bartnicki-Garcia and Lippman (1966) have also described how the dilution of the *Phythophthora* protoplast suspensions caused the protoplasts to swell slightly and to burst with disintegration of their membrane. Concomitant with bursting, a spherical vesicle, about the original size of the protoplast, was frequently formed. Conceivably these vesicles represent swollen cytoplasmic vacuoles and characteristically, protoplast debris remained attached to them from without.

Isolation of vacuoles from higher plant cells has also been described using gentle lysis of naked protoplasts isolated from roots and cotyledons using *M. verrucaria* cellulase in 10–20% w/v sucrose (Gregory and Cocking, 1964). Vacuoles were much less stable than protoplasts.

Nuclei might be isolated by using similar procedures to those described for vacuoles. Attempts at preparation of nuclei from shocked protoplasts have already been made. Refractile bodies associated with DNA, and probably derived from the yeast nucleus, have been isolated following brief ultrasonic treatment of protoplasts (Eddy, 1959).

C. For the preparation of mitochondria

The use of protoplasts as starting material for the fragmentation of yeast cells has at least two advantages. The mechanical force required to disperse the cytoplasm of the protoplasts is much lower than that required to rupture intact cells and the mechanical damage to the particles is correspondingly much less. Therefore osmotic bursting can be used as a mild procedure for the preparation of protoplasmic components. Intact mitochondria have been successfully prepared by osmotic shock of yeast protoplasts as described by Duell *et al.* (1964).

Sacch. cerevisiae was grown on a medium containing the following constituents per litre:

Glucose	8 g
Peptone	5 g
Yeast extract	10 g
KH_2PO_4	9 g
$CaCl_2$	0·033 g
$MgSO_4$	0·5 g
$(NH_4)_2SO_4$	6 g

Cells were grown for 10 h with continuous shaking. The cells were harvested on millipore filters (RA 47), washed with cold water, and stored at 4°C until used. Cells were suspended in 2 ml of a solution containing 0·63 M sorbitol, 0·1 M citrate-phosphate (pH 5·8), 0·03 M 2-mercapto-ethylamine HCl, 4×10^{-4} M EDTA, and 33 mg of Gluculase (preparation of snail digestive enzyme from Endo Laboratories Inc., Richmond Hill, N.Y.) per g of cells (wet weight). The suspension was thoroughly mixed in a mixer and was incubated for 30 to 60 min at 30°C with occasional stirring. The formation of protoplasts can be followed roughly by diluting a sample of the incubation mixture $10 \times$ with water and observing the number of remaining cells. After the incubation period, the cells and the protoplasts were centrifuged at 1000 *g* for 10 min at 0°C. The supernatant fluid was decanted; the precipitate was washed with 3 ml of 0·9 M sorbitol per g of original yeast, and was recentrifuged. Lysis of the protoplasts was accomplished with 0·25 M sucrose, containing 0·05 M KH_2PO_4 (pH 6·8) and 0·001 M EDTA, by mixing the protoplasts with a glass rod and then on the mixer. The cell debris and whole cells were centrifuged at 1000 *g* for 10 min at 0°C and the resulting supernatant layer was centrifuged at 5000 *g* for 10 min at 0°C. The pellet containing the mitochondria was washed with 4 ml of 20% sucrose containing phosphate buffer and EDTA per g of cells, and was recentrifuged; the supernatant layer was decanted, and the pellet was resuspended in the 20% sucrose. The normal yield was 20–25 mg of mitochondrial protein per g of cells.

The method described in the previous section was satisfactory for cells harvested during or shortly after the log phase of growth. However, the method was inadequate for cells that were well out of the log phase of growth, and for these cells a two-step method proved to be superior. A solution (2·5 ml) containing 0·14 M 2-mercaptoethylamine and 0·04 M EDTA was used per g of cells. After incubation for 30 min at 30°C the cells were centrifuged at 1000 g for 10 min at 0°C. After decantation of the supernatant layer, the pellet was resuspended in 1·7 ml of 0·72 M sorbitol, 0·02 M citrate-KH_2PO_4 (pH 5·8) and 33 mg of gluculase per g of cells. The cells were thoroughly mixed and incubated at 30°C from 15–30 min. From now, the procedure for the one-step method was followed as outlined in the previous paragraph.

Heick and Stewart (1965) have also described the preparation of a particle fraction from citrate-grown *Lipomyces lipofer*. Protoplasts were prepared by conventional methods and were centrifuged from the mannitol solution in which they had been washed after preparation, and were washed once by centrifugation with the appropriate incubation solution. After they were washed they were suspended in approximately 25 ml of incubation solution and ruptured in the French pressure cell (American Instrument Co., Silver Spring, Maryland) by decompression from pressures between 500 and 800 p.s.i. No rise in temperature was associated with this mild decompression. The ruptured protoplasts were centrifuged at 1500 g (saline media) or 2500 g (sucrose media) for 5 min. The resulting pellet was resuspended in 10 ml and both the resuspended pellet and the supernatant were recentrifuged. The two supernatants were then combined and centrifuged at 16,000–17,000 g for 10 min. The pellet was washed by centrifugation in 10 ml of incubation solution and resuspended in the incubation solution. Evidence was presented to show that the particles obtained, which morphologically resemble mitochondria seen in thin sections of yeast cells, are capable of oxidative activity with substrates of the tricarboxylic acid and of oxidative phosphorilation.

The recovery of cell fractions in good condition has sometimes been difficult owing to the instability of components arising as a result of the osmotic changes, although the activity of mitocondrial preparations from *Lipomyces* seems to be similar to that of the most active particle preparations from other yeasts. The possibility of finding an agent able to promote bursting, as has been shown in plant protoplasts with 3-indolylacetic acid (Cocking, 1961, 1965) without changing the osmotic environment, seems promising. Methods for the preparation of mitocondria from *Neurospora crassa* by mechanical means have recently been described (Weiss et al., 1970).

D. For the preparation of cell-free extracts

Lysates of protoplasts produced by diluting the suspension medium can be used for investigations of the biochemical potentialities of subcellular particles. Protein synthesis activity of protoplasmic components can be studied, isolation of highly polymerized and biologically active deoxyribonucleic acid in excellent yield may be effected. Enzymes (invertase) can also be isolated from protoplast lysates which sometimes cannot be detected in cell-free extracts prepared by other methods (Gascón et al., 1965).

E. For the study of osmotic systems

It is well known that protoplasts are osmotically sensitive and that when they are not encased within the protective, rigid cell wall, they are stable only in media containing high concentrations of impermeant solute. Despite the widespread use of wall-less protoplasts, there still remains an overall vagueness in current knowledge of the processes involved in their osmotic swelling and bursting. Probably the cells most thoroughly studied in regard to osmotic behaviour are mammalian red blood cells. It is difficult to compare the osmotic responses of microbial protoplasts with those of erythrocytes because of our limited knowledge of the former processes.

Intact cells of fungi are assumed to have an osmotic barrier permeable to small molecules, which adhere to the wall and surrounds the protoplasm. This osmotic barrier corresponds to the cytoplasmic membrane. The free protoplasts of fungi are sensitive to osmotic shock and lyse immediately when placed in distilled water. Svihla et al. (1961) have determined the optimum conditions for osmotic stability by measuring the release of radioactivity from labelled protoplasts of C. utilis placed in several concentrations of KCl. Higher stability of the spheres was found with KCl concentrations of $0 \cdot 5$–$1 \cdot 0 M$; Lillehoj and Ottolenghi (1966) have also studied the osmotic properties of Saccharomyces protoplasts. This work showed that cells grown in different osmotic pressures behave differently not only in respect to growth (increase in the osmotic pressure caused a decrease in the growth rate) but also in other physiological activities, i.e. respiratory activity of the cells is also affected by the high osmolarity of the medium.

A study of the relative effectiveness of sugars, of short peptides and of amino acids as osmotic stabilizers for bacterial protoplasts indicate that the protoplast membrane can act as a porous differential dialysis membrane and that its effective porosity increased when it is stretched during osmotic swelling. The protoplast membrane also behaves as a highly extensive structure, in contrast to membranes such as those of erythrocytes, and enormous protoplasts could be prepared by slowly dialyzing stabilizing solutes from protoplast suspensions. It appears that when protoplasts swell in hypotonic solutions, their surface membranes may become sufficiently stretched so that they admit stabilizing solutes. There is then a rapid influx of solutes and water resulting in rapid stretching of the membrane and rupture due to a process of brittle fracture. Thus, bursting generally occurs without the membrane becoming fully extended. It appears that bacterial protoplast membrane is physically different from the erythrocyte membrane despite their similar appearances in electron micrographs (Corner and Marquis, 1969). It has been observed that the osmotic fragility of protoplasts varies markedly, depending on the stage of the culture cycle at which cells were harvested. Therefore, to obtain consistent results, it is convenient to harvest cells only from cultures in the same stage of growth.

F. For studies of permeability

Permeability properties of protoplasts and spheroplasts have been rather widely studied in bacteria. Relatively little similar work has been done in fungi, although Heredia et al. (1968) have exploited the presence of the constitutive hexose transport system in protoplasts of Saccharomyces cerevisiae to study its specificity turbidimetrically.

Protoplasts of fungi (mostly yeasts) have proven to be of great value for metabolic studies, mainly those related to sugar transport. Permeability of fungal protoplasts seems to be very similar to those of intact cells. Holter and Ottolenghi (1960) found that when cells of Saccharomyces are impermeable to some sugars, such as melibiose, and sucrose, their protoplasts retain this property. This clearly suggests that the cell membrane acts as a barrier through which some sugars and other compounds are not free to pass. The efficiency of the membranes as a barrier depends presumably on the fungus. In S. cerevisiae protoplasts, penetration of some sugars may be catalyzed by a constitutive transport system (Heredia et al., 1963). Cirillo (1966) has suggested that the efficiency of such transport is much higher than penetration by simple diffusion. Yeast protoplasts have been extensively used to study the phenomenon of alteration of cellular permeability using antibiotics and other chemicals (Marini et al., 1961). Penetration of enzymes (ribonuclease) has also been studied using yeast protoplasts. Whereas ribonuclease does not affect living yeast cells, it can slowly penetrate into protoplasts digesting the volutin in the cytoplasm (Necas, 1958). In contrast Schlenk and Dainko (1966) have described penetration of ribonuclease into intact yeast cells.

Protoplasts of filamentous fungi have been much less used for permeability studies. Elorza et al. (1969) using protoplasts of Aspergillus nidulans, followed the penetration of a number of compounds and found that at least four transport systems in A. nidulans—one for sugars—remain functional after removal of the cell wall. This is consistent with the generalization (Villanueva, 1966) that the permeability properties of fungal protoplasts are similar to those of intact cells.

G. For growth studies

Protoplasts have also been used for growth studies. Emerson and Emerson (1958) and Bachman and Bonner (1959) were first to observe growth of protoplasts obtained from filamentous fungi. However, growth of hyphal protoplasts soon results in regeneration with the ultimate formation of normal mycelium. No growth was ever observed in conidial protoplasts of *F. culmorum* that always resulted in the formation of a germ tube (Villanueva, 1966). Growth of protoplasts has also been observed as a result of fusion of two independent spherical units (Strunk, 1967; Lopez-Belmonte *et al.*, 1966). In marked contrast, yeast protoplasts placed in an appropriate medium grow, increasing in DNA and RNA, but do not divide (Tabata *et al.*, 1965; Shockman and Lampen, 1962). Eddy and Williamson (1959) using media inoculated with 10^7 protoplasts per millilitre found that cell nitrogen increased four- to eight-fold in 24 h. The larger increases were obtained when a smaller inoculum was employed.

H. For tests of resistance to physical factors

Protoplasts of fungi being deprived of the protective wall, are susceptible to a variety of physical forces. Sensitivity to a number of factors such as centrifugation, shaking, sonic vibration and ultraviolet irradiation have been tested. Protoplasts can be centrifuged gently and washed, provided they are maintained in appropriate stabilizing solutions. However, the extent of cell damage is proportional to the force employed and is affected by the organism used. Damage to *Fusarium* protoplasts was very small even at 20,000 *g* for 15 min (Rodriguez Aguirre *et al.*, 1964). By contrast centrifugation at 10,000 g for 10 min produced a considerable amount of cell breakage of protoplasts of *C. utilis* (Svihla *et al.*, 1961).

The protoplasts of most fungi appear to be very sensitive to sonic vibration. An exposure of 1–2 min produced total breakage of the osmotic spheres of *Fusarium* and *Canida* species (Rodriguez Aguirre *et al.*, 1964; Svihla *et al.*, 1961).

Irradiation with 6×10^4 erg/mm^2 of ultraviolet energy at 250–270 mm causes marked damage to the cytoplasmic membrane (Svihla *et al.*, 1961).

I. For investigations of protein and enzyme synthesis

Little work has been done on the ability of fungal protoplasts to synthesize constitutive and inducible enzymes but there is no doubt that protoplasts can make proteins and various other compounds as rapidly as can intact living cells. De Kloet *et al.* (1961) followed protein synthesis by incubating protoplasts of *S. carlsbergensis*. Incorporation of amino acids was detected and inhibition of this incorporation and of the adaptive enzyme synthesis was observed when respiration was modified. Enzymes such as the ribonuclease apparently produced marked inhibition of protein synthesis when the naked protoplasts were exposed to the nuclease. Van Dam and collaborators (1964) also described a repressor of induced enzyme synthesis in naked protoplasts. The extent of incorporation of carbon from a wide variety of substrates makes it apparent that "de novo" synthesis of the cellular constituents does occur in isolated protoplasts of this species.

It has been suggested that induction of some enzymic systems in protoplasts appears to be a rare phenomenon. Volfova *et al.* (1968) using a strain of *Candida lipolytica* which oxidizes hydrocarbon adaptively, showed that protoplasts, unlike the whole cells, are not capable of oxidizing those compounds. These workers showed that their cells cultivated on glucose were able to adapt themselves to hydrocarbon oxidation and that protoplasts were not. The same kind of phenomenon was described by Duercksen (1964) in studying the induction of penicillinase with intact cells and protoplasts of *B. cereus*.

J. For studies of localization of enzymes

A wide variety of fungal protoplasts has been used from time to time by different workers to study the localization of various enzymes. The role of the cytoplasmic membrane in the secretion of enzymes by microorganisms has aroused great interest in recent years. Yeast invertase is an interesting system to study this phenomenon. Most of the enzyme is located in the wall. The secretion of invertase by fungal protoplasts has been studied by various workers. Friss and Ottolenghi (1959) showed that protoplasts in species of *Saccharomyces* from sucrose-adapted cells, released more than 74% of their invertase activity. Isolated cell walls of *S. cerevisiae* contain large amounts of invertase but the protoplasts failed to ferment sucrose while still fermenting glucose (Sutton and Lampen, 1962).

The distribution and characteristics of invertase isozymes of yeast and moulds was studied using protoplasts. The results of the distribution of the invertases in *Saccharomyces* and *Neurospora* protoplasts are markedly different. Metzenberg (1964) has shown that in *Neurospora* both isoenzymes can coexist outside the cytoplasmic membrane, although the heavy one predominates inside the protoplasts. Sentandreu *et al.* (1966) confirmed these results working with *C. utilis* protoplasts. In contrast in yeasts the heavy and light invertases correspond to the external and internal enzymes. Only the external invertase contains carbohydrate. The invertase in the protoplast inside the membrane is a smaller form (Gascón and Lampen, 1968).

Related enzymes also appear to be associated with the cell envelope. Complete lack of melibiase was observed in protoplasts of *S. cerevisiae* although the enzyme could be detected in intact yeasts (Friss and Ottolenghi, 1959).

Parallel studies to those carried out with invertase enzyme were made with acid phosphatase. Experiments with intact and disintegrated cells and application of the protoplast technique have demonstrated that acid phosphatase of baker's yeast is mainly located on the surface of the cell, whereas the alkaline phosphatase is found inside the cell. McLellan and Lampen (1963) found that yeast protoplasts are capable of synthesizing acid phosphatase, and the enzyme is released into the medium. Tonino and Steyn-Parve (1963) showed that this enzyme is located in the wall of the yeast cells although situated in two different areas of the cell wall. It is thus apparent that yeast walls may accumulate enzymes which have been liberated from the cytoplasm. It has been suggested that the acid phosphatase is of low substrate specificity. However, Rothstein and Meier, (1948) believe that a number of different acid phosphatases are located on the cell surface of the yeast. An interesting study was recently made by Nurminen et al. (1970) of the enzyme content of the isolated cell walls and of a plasma membrane preparation obtained by centrifugation after enzymic digestion of the cell walls of baker's yeast. The isolated cell walls showed no hexokinase, alkaline phosphatase, esterase of NADH oxidase activity which exist in the interior of the cell. On the other hand considerable amounts of invertase and a variety of phosphatases were found in the isolated cell walls. Enzymic digestion of the cell wall released most of these enzymes but the bulk of the Mg^{2+}-dependent adenosine triphosphatase remained in the plasma membrane preparation. This finding suggests that this phosphatase is an enzyme of the cytoplasmic membrane whereas the other enzymes are located in the cell wall outside the plasma membrane. Suomalainen's group showed that Mg^{2+}-dependent adenosine triphosphatase differs from the phosphatases with pH optima in the range pH 3–4 with regard to location, substrate specificity and different requirement of activators.

K. For studies of the mode of action of antibiotics and surface active agents

Nystatin and other polyene antibiotics, which are known to bind to the sterols of fungal membranes, produce membrane damage and cell lysis (Lampen et al., 1962; Kinsky, 1963). A number of surface active agents and related substances known to disrupt cytoplasmic membranes were tested against protoplasts of fungi. Digitonin acting on *Aspergillus* and *Fusarium* protoplasts causes immediate lysis (Rodriguez Aguirre, 1965). Sodium dodecyl-sulphate (0·05%) causes great alterations of the osmotic barrier of protoplasts followed by lysis (Rost and Venner, 1965).

L. For studies of the regeneration of the cell wall and its biosynthesis

The protoplasts of fungi appear to have all the synthetic abilities of whole cells from which they are derived, including the ability to make a new cell wall. Fungal protoplasts are very useful for the study of the biosynthesis of the cell wall. As stated by Svoboda et al. (1969), it is possible in these protoplasts to trace a gradual construction of single wall components, their arrangement in a complex structure, the regulatory mechanisms of these processes and their relation to the other cell structures.

The relation of the synthesis of the cell wall to the starting of regeneration has been studied in yeast by Uruburu et al. (1968) and by Necas and Svoboda (1966). Later the same group of workers gave some basic information about the morphology of the biosynthesis of the cell wall in protoplasts of Schizosaccharomyces pombe, a yeast that does not contain mannan (Necas, 1965). The protoplasts were prepared from actively growing cells in wort medium, using snail enzymes. Freshly isolated protoplasts were washed with a synthetic medium to remove the snail enzyme. Part of the protoplasts were then embedded in a synthetic medium, containing 30% gelatin and others were grown in wort, containing 0·3 M KCl. To obtain isolated surfaces of fresh and growing protoplasts the samples were lysed with distilled water, either directly on the grids for fresh protoplasts, or in tubes with successive treatment with trypsin to remove the remnants of cytoplasm for growing protoplasts. The surface of regenerating protoplasts was isolated mechanically and treated as above. These workers have found substantial differences in the morphology of the biosynthesis of the cell wall.

In Schizosaccharomyces pombe the new cell wall is synthesized "de novo" on the surface of the plasma membrane, as in protoplasts of S. cerevisiae. The formation of fibrils synchronously with the formation of matrix is probably the mechanism of the biosynthesis of the cell wall in normal cells. In liquid medium of protoplasts of Sch. pombe, the closely meshed envelope of fibrils ensures the necessary accumulation, making cell wall synthesis possible. The formation of fibrils ceases after a few hours, interfering with the formation of cell wall in S. cerevisiae. More recently Necas and Kopecká (1969) have studied the structural phenomena in the plasma membrane related to the formation of fibrils and the dynamics of formation of the fibrillar network by using labelled glucose. Autoradiographic studies showed that the fibrils are synthetized isotropically around the whole protoplast surface and that the fibrillar network grows by interposition of new elementary fibrils. It seems possible that the remnants of original cell walls do not serve as a necessary primer for the synthesis of fibrils.

Interesting studies were also carried out by the Necas group on the effect of some agents on the biosynthesis of new yeast cell wall. It was found that cyclohexamidine does not block, at least in the initial stage, the formation of glucan fibrils. However, it blocks the synthesis of the matrix (Necas, Svoboda and Kopecká, 1968). The difference between the action of ribonuclease and actidione was also apparent, suggesting that the plasma membrane probably takes part in the synthesis of the cell wall material and damage to it may result in irreversible inhibition of the synthesis of all wall components (Kopecká, Svoboda and Necas, 1969).

That the cell membrane was the site of synthesis of yeast mannan was demonstrated by Algranati *et al.* (1963) using a particulate fraction obtained by centrifugation of a protoplast lysate of *S. carlsbergensis*. Recently Behrens and Cabib (1968) have shown that a particulate enzyme preparation from membranes of yeast cells incorporated mannose from GDP-mannose into mannan. Later Tanner (1969) reported that a lipophilic mannosyl intermediate is the immediate precursor for mannan biosynthesis.

More recently Elorza and Sentandreu (1969) have demonstrated that the biosynthesis of the major constituents of the yeast cells is the result of two processes. One of them is independent initially of protein synthesis and is involved in the production of insoluble network of glucan (Garcia Mendoza and Novaes, 1968; Necas *et al.*, 1968). The organized particles present on the yeast membrane surface (Moor and Mühlethaler, 1963) might play an important role in this process. The amorphous mannan-peptide matrix which embeds the insoluble glucan might be synthesized in the cytoplasm in two different steps:

$$\text{Amino acids} \xrightarrow{\ \ \uparrow\ \ } \text{Protein} \xrightarrow{\quad} \text{Glycopeptide}$$
$$\qquad\qquad \text{Cycloheximide} \quad \text{Carbohydrate}$$

It has been suggested that cycloheximide that prevents regeneration of protoplasts (Soskova *et al.*, 1968) would interfere with the first step of this process. The mannan-peptides ordinarily synthesized may be carried to the site of wall synthesis in vesicles which are seen to accumulate at the region of bud formation (Sentandreu and Northcote, 1969). The contents of the vesicles could be transferred across the cytoplasmic membrane by a mechanism equivalent to reverse pinocytosis.

Little is known about the synthesis of another important constituent of the fungal cell wall, chitin. Glaser and Brown (1957) demonstrated the presence of a particulate enzyme in *N. crassa* that catalysed the synthesis of chitin from uridine diphospho-N-acetyl-glucosamine labelled with C^{14}. A similar synthetic process was shown to occur in cell-free preparations from *Venturia inaequalis* (Jaworski *et al.*, 1965). The same group of workers reported a particulate enzyme from the mycelium of *Allomyces macrogynus* able to catalyse the synthesis of chitin from the above uridine derivative (Porter and Jaworski, 1968). The enzyme was associated with both mitochondrial and microsomal fractions but it exhibited higher specific activity and greater stability in the latter. The enzyme was activated by soluble chitodextrin and N-acetyl-glucosamine in a manner similar to that observed in *Neurospora* preparations. Kinsky (1962) also suggested that chitin synthetase activity takes place in a particulate fraction that might be localized in particles obtained from the cell membrane of yeast protoplasts. In connection with these studies Bartnicki-Garcia and Lipman (1969) have shown by using an autoradiographic technique two different patterns of cell wall construction associated with two types of morphogenesis in *Mucor rouxii*. In hyphae, the cell wall was preferentially synthesized in the apical region. In spherical cells (germinating spores, yeast cells), wall formation occurred largely, if not entirely in a uniformly dispersed fashion over the entire cell periphery. The experiments were carried out by cell exposure to tritiated N-acetyl-D-glucosamine which apparently was incorporated into two main cell wall polymers-chitin and chitosan.

Fungal protoplasts formed by various methods have been utilized in regeneration studies. Practically all naked protoplasts are capable of regeneration into new normal cells, their capacity being apparently determined primarily by the physical conditions in the medium. Yeast protoplasts might be suitable material for a casual analysis of the individual phases of regeneration and of their interrelationship (Necas, 1962). Since the regeneration process in most fungal protoplasts takes place very slowly, it is possible to follow the individual stages of the biosynthesis of the new wall.

Regeneration of cell walls by naked protoplasts in nutrient solutions has been a matter of controversy (Villanueva, 1966). Yeast protoplasts have been placed in two main categories depending on their ability or inability to re-form a normal cell wall on cultivation and thereby to revert directly to cells of the parent type (Eddy and Williamson, 1959). In contrast to the protoplasts themselves, the regenerated bodies possessed aberrant walls, almost devoid of amino acid residues but with a high content in N-acetyl-glucosamine residues. Glucose and mannose are also present. Valuable contributions have been made by the Brno School in recent years in this field using autolysed yeast cells to demonstrate the ability of naked protoplats to regenerate into new normal cells.

Elorza *et al.* (1969) have recently examined a number of mutants and wild types of *Aspergillus nidulans* protoplasts for differences in rate of regeneration of mycelial form under conditions which serve to distinguish mycelial colonies of the mutants from the wild type. The regeneration of mycelial forms from protoplasts was followed microscopically under conditions with which the rate of regeneration depended upon a specific transport function. Protoplasts were prepared by use of lytic enzymes produced by *Streptomyces violaceous* MR in an aerated, chitin containing medium as modified after Elorza *et al.* (1966). Appropriate amounts of mycelium were incubated with the enzyme preparation at 30°C in the presence of 1 M NH_4Cl or 800 mM $MgSO_4$ to stabilize the newly formed protoplasts. Protoplasts were then centrifuged and resuspended in different kinds of regeneration medium. Regeneration was followed microscopically. After 20 h at 25°C, convoluted figures similar to those found in *Fusarium culmorum* (Garcia Acha *et al.*, 1966a) were seen. By 40 h, normal hyphae developed. The regeneration process has an absolute requirement for a carbon source. Nitrogen and sulphur sources did not seem to be essential, although regeneration was retarded in the absence of a nitrogen source. Omission of at least some auxotrophic requirements and addition of certain toxic substances which inhibit growth of mycelial colonies did not interfere with the process. The authors concluded that rates of regeneration should only be compared in the initial stages, because in the later stages uptake properties are partly those of intact mycelium.

Strunk (1967 and 1969) has studied regeneration of protoplasts in the fungus *Polystictus versicolor*. Once protoplasts have been transferred to a growth medium with the stabilizer they revert to typical mycelial growth. The type of growth and regeneration found vary with the osmotic strength of the medium and the cultural conditions under which they develop. Sometimes the protoplasts were seen to revert directly to normal hyphae with rapid formation of a clamp. The different possibilities of regeneration of the protoplasts which varies from direct hyphae formation to an abnormal mass that continuously grows were described (Strunk, 1969).

Washed protoplasts of *Fusarium culmorum* have also been incubated in a variety of regeneration media in order to facilitate studies of the formation of cell walls (Garcia Acha *et al.*, 1966). Preparations of protoplasts freed from mycelial debris by passing the suspension through a 11 BZ Jena fritted glass filter, were washed 3 times with the phosphate × citrate buffer (0·1 M, pH 6·5) × 0·6 M mannitol or NH_4Cl, centrifuged at 4000 g for 15 min, and the pellet resuspended and plated on media containing various supplements. The determination of reversion was made by phase-contrast microscopy in wet mount preparations and in hanging drops containing isolated protoplasts. Regeneration of protoplasts takes place in three different ways (a) by means of a chain of yeast-like forms and later originating a germ tube; (b) by direct formation of the germ tube from the protoplast; (c) through a complicated process with the formation of various globular forms and giving rise to the formation of the germ tube at the end. Maximum regeneration, about 80% of the protoplasts, was found in the presence of 2% (w/v) sucrose (or 1% (w/v) glucose × 1% (w/v) fructose) +

10% (w/v) sorbose + 0·2% agar + the mineral salts of the Czapek medium. No other sugars were able to substitute for sucrose and sorbose. Agar was the best substance for regeneration; gelatin produced an inhibition. No differences were found in the regenerative process as between protoplasts obtained by the use of snail or by microbial lytic enzymes.

M. For studies of conjugation between protoplasts

A number of fungi have been used to study the physiology of the conjugation process, or cell fusion. In *Hansenula wingei* apparently the process takes place by a softening of the cell wall followed by formation of a conjuation tube, dissolution of the cross wall, and formation of a new bud at the region of contact of the two mating types (Brock, 1961; Conti and Brock, 1965). Some yeasts are able to conjugate not only by the mating of vegetative cells, but by the fusion of a vegetative cell and an ascospore, or by the fusion of two types of ascospores (Suminame and Duckmo, 1963). It remains to be seen whether protoplasts from such cells are able to conjugate.

Attempts to demonstrate conjugation in fungal protoplasts have been made by various workers. Holter and Ottolenghi (1960) failed to obtain conjugation by mixing protoplasts from two haploid strains of opposite mating types. Protoplasts of *Polystictus versicolor* (Strunk, 1966) and of *Fusarium culmorum* (Lopez-Belmonte et al., 1966) fuse under appropriate conditions. It is not known whether this is a true conjugation process.

The finding by Brock (1959) that a protein of one strain and a wall polysaccharide of the opposite strain are needed for mating may explain the inability to obtain conjugation in yeast protoplasts. If either or both specific complementary mating components are to be found in the cell walls mating reactions should not be expected between two opposite types of wall-less protoplasts.

N. For studies of spore formation

When protoplasts of *Saccharomyces carlsbergensis* were inoculated in a medium containing glucose-yeast extract to which 0·6 M KCl was added formation of spores was observed (Holter and Ottolenghi, 1960). Formation of spores started 14 h after incubation in a starvation medium containing sorbitol and phosphate. About 1–7 spores per cell were observed. The spores burst during the early stages of formation but were resistant when fully formed. Formation of spores was observed in *Schizosaccharomyces pombe* only if the yeast cells were allowed to form zygotes before the cell walls were digested with the lytic enzymes.

ACKNOWLEDGMENTS

The authors express their thanks to Dr. Elwyn T. Reese of the U.S. Quarter-master Corps Research Laboratories, Natick, Mass., for criticism and linguistic improvements of the manuscript. Part of these investigations were supported by grants from Laboratories Lepetit and from the Ministerio de Educación y Ciencia of Spain.

REFERENCES

Abrams, A. (1959). *J. biol. Chem.*, **234**, 383–387.

Algranati, I. D., Carminatti, H., and Cabid, E. (1963). *Biochem. biophys. Res. Commun.*, **12**, 504–509.

Anderson, F. B. (1967). In "Symposium über Hefe-Protoplasten, Jena 1965" (Ed. R. Müller), p. 65. Akademie-Verlag, Berlin.

Bacon, J. S. D., Jones, D., and Ottolenghi, P. (1969). *J. Bact.*, **99**, 885–887.

Bachmann, B. J., and Bonner, D. M. (1959). *J. Bact.*, **78**, 550–556.

Bartnicki-Garcia, S., and Lippman, E. (1966). *J. gen. Microbiol.*, **42**, 411–416.

Bartnicki-Garcia, S., and Lippman, E. (1969). *Science*, **165**, 302–304.

Behrens, N. H., and Cabid, E. (1968). *J. Biol. Chem.*, **243**, 502–506.

Bent, J. J., and Moore, R. H. (1966) In "Biochemical Studies of Antimicrobial Drugs" (Eds. B. A. Newton and P. E. Reynolds), p. 102. Cambridge Univ. Press, London and New York.

Berliner, M. D., and Reca, M. E. (1970). *Science*, **167**, 1255–1257.

Boulton, A. A. (1965). *Exptl. Cell. Res.*, **37**, 343–359.

Brenner, S., Dark, F. A., Gerhardt, P., Jeynes, M. H., Kandler, O., Kellenberger, E., Klieneberger-Nobel, E., McQuillen, K., Rubio-Huertos, M., Salton, M. R. J., Strange, R. E., Tomcsik, J., and Weibull, C. (1958). *Nature, Lond.*, **181**, 1713.

Brock, T. D. (1959). *J. Bact.*, **78**, 56–68.

Brock, T. D. (1961). *J. gen. Microbiol.*, **26**, 487–497.

Cantino, E. C., Lovett, J. S., Leak, L. V., and Lythgoe, J. (1963). *J. gen. Microbiol.*, **31**, 393.

Cirillo, V. P. (1966). In "Symposium über Hefe-Protoplasten, Jena 1965" (Ed. R. Müller), p. 153. Akademie-Verlag, Berlin.

Cocking, E. C. (1960). *Nature, Lond.*, **187**, 927–929.

Cocking, E. C. (1961). *Nature, Lond.*, **191**, 780–782.

Cocking, E. C. (1965). In "Viewpoints in Biology" (Eds. J. D. Carthy and C. L. Duddington), Vol. 4, p. 170. Butterworths, London.

Cocking, E. C. (1968). *Biochem. J.*, **111**, 33P.

Cohen, J., Katz, D., and Rosenberger, R. F. (1969). *Nature, Lond.*, **224**, 713–715.

Conti, S. F., and Brock, T. D. (1965). *J. Bact.*, **90**, 524–533.

Corner, T. R., and Marquis, R. E. (1969). *Biochim. Biophys. Acta.*, **183**, 544–558.

Darling, S., Theilade, J., and Birch-Andersen, A. (1969). *J. Bact.*, **98**, 797–810.

Davies, R., and Elvin, P. A. (1964). *Biochem. J.*, **83**, 325–331.

De Kloet, S. R., van Dam, J. W., and Koningsberger, V. V. (1961). *Biochim. Biophys. Acta.*, **47**, 138–148.

De Terra, N., and Tatum, E. L. (1961). *Science*, **134**, 1066–1068.

Duell, E. A., Inowe, S., and Utter, M. F. (1964). *J. Bact.*, **88**, 1762–1773.

Duercksen, J. D. (1964). *Biochem. biophys. Acta*, **87**, 123–140.

Eddy, A. A. (1958). *Proc. R. Soc. B.*, **149**, 425–440.

Eddy, A. A. (1959). *Expt. Cell. Res.*, **17**, 447–464.

Eddy, A. A., and Rudkin, S. (1958), *Proc. R. Soc. B.*, **148**, 419.
Eddy, A. A., and Williamson, D. H. (1957). *Nature, Lond.*, **179**, 1252–1253.
Eddy, A. A., and Williamson, D. H. (1959). *Nature, Lond.*, **183**, 1101–1104.
Edwards, M. R., Hazen, E. L., and Edwards, G. A. (1959). *J. gen. Microbiol.*, **20**, 496–501.
Elbers, P. F. (1961). *Nature, Lond.*, **191**, 1022–1023.
Elorza, M. W., Arst, H. N., Cove, D. J., and Scazzochio, C. (1969). *J. Bact.*, **99**, 113–115.
Elorza, M. V., Ruiz, E. M., and Villanueva, J. R. (1966). *Nature, Lond.*, **210**, 442–443.
Elorza, M. V., and Sentandreu (1969). *Biochim. Biophys. Research Com.*, **36**, 741–747.
Emerson, S., and Emerson, M. R. (1958). *Proc. natn. Acad. Sci. U.S.A.*, **44**, 669–671.
Friss, J., and Ottolenghi, P. (1959). *C.r. Trav. Lab. Carlsberg*, **31**, 279–271.
Furuya, A., and Ikeda, Y. (1960). *J. Gen. Appl. Microbiol.*, (Tokyo). **6**, 40–47.
Garcia Acha, I., and Villanueva, J. R. (1963). *Nature, Lond.*, **200**, 1231.
Garcia Acha, I., Aguirre, M. J., and Villanueva, J. R. (1964). *Can. J. Microbiol.*, **10**, 99–101.
Garcia Acha, I., Aguirre, M. J. R., López-Belmonte, F., and Villanueva, J. R. (1966b). *Nature, Lond.*, **209**, 95.
Garcia Acha, I., López-Belmonte, F., and Villanueva, J. R. (1966a). *J. gen. Microbiol.*, **45**, 515–523.
Garcia Acha, I., López Belmonte, F., and Villanueva, J. R. (1967). *Can. J. Microbiol.*, **13**, 433–437.
Garcia Mendoza, C., and Novaes, M. (1968). *Nature, Lond.*, **220**, 1035.
Garcia Mendoza, C., and Villanueva, J. R. (1962). *Nature, Lond.*, **195**, 1326–1327.
Garcia Mendoza, C., and Villanueva, J. R. (1964). *Nature, Lond.*, **202**, 241.
Garcia Mendoza, C., and Villanueva, J. R. (1967). *Biochem. Biophys. Acta.*, **135**, 189–197.
Garcia Mendoza, C., Garcia López, M. D., Uruburu, F., and Villanueva, J. R. (1968). *J. Bact.*, **95**, 2393–2398.
Gascón, S., and Lampen, J. O. (1968). *J. biol. Chem.*, **243**, 1567–1572.
Gascón, S., and Villanueva, J. R. (1963). *Can. J. Microbiol.*, **9**, 651–652.
Gascón, S., and Villanueva, J. R. (1965). *Nature, Lond.*, **205**, 822–823.
Gascón, S., Ochoa, A. G., and Villanueva, J. R. (1965). *Can. J. Microbiol.*, **11**, 573.
Giaja, J. (1914). *C.r. Séanc. Soc. Biol.*, **77**, 2.
Girbardt, M., and Strunk, Ch. (1965). Personal communication.
Glasser, L., and Brown, D. H. (1955). *Proc. Natl. Acad. Sci. V.S.*, **41**, 253–259.
Gregory, D. W., and Cocking, E. C. (1964). *Biochem. J.*, **88**, 40P.
Hamilton, J. B., and Calvet, J. (1964). *J. Bact.*, **88**, 1084–1086.
Haskins, R. H., and Spencer, J. F. T. (1962). *Can. J. Microbiol.*, **8**, 279–281.
Heick, H. M. C., and Stewart, H. B. (1965). *Can. J. Biochem.*, **43**, 561–571.
Heredia, C. F., de la Fuente, G., and Sols, A. (1963). Atti delle VII Giornate Biochemiche Latine, S. Margherita Ligure (Genova), p. 86.
Heredia, C. F., Sols, A., and de la Fuente, G. (1968). *Eur. J. Biochem.*, **5**, 321–329.
Holden, M., and Tracey, M. V. (1950). *Biochem, J.*, **47**, 407–414.
Holter, H., and Ottolenghi, P. (1960). *C.r. Trav. Lab. Carlsberg.*, **31**, 409–422.
Horikoshi, K., and Ida, S. (1959). *Nature, Lond.*, **183**, 186–187.
Indge, K. J. (1968). *J. gen. Microbiol.*, **51**, 441–446.

Jaworski, E. G., Wang, L. C., and Carpenter, W. D. (1965). *Phytopathology*, **55**, 1309–1313.
Johnson, B. F. (1968). *J. Bact.*, **95**, 1169.
Johnston, J. R., and Mortimer, R. K. (1959). *J. Bact.*, **78**, 292.
Jones, D., and Watson, D. (1969). *Nature, Lond.*, **224**, 287–288.
Kinsky, S. C. (1961). *J. Bact.*, **80**, 889–897.
Kinsky, S. C. (1962). *J. Bact.*, **83**, 351–358.
Kinsky, S. C. (1963). *Archs Biochem. Biophys.*, **102**, 180–188.
Kopecká, M., Svoboda, A., and Necas, O. (1969). *Antonie van Leeuwenhoek*, **35**, B9.
Lampen, J. O., Arnow, P. M., Borowska, Z., and Laskin, A. I. (1962). *J. Bact.*, **84**, 1152–1160.
Leal, J. A., and Villanueva, J. R. (1962). *Science*, **136**, 715–716.
Lebeault, J. M., Roche, B., Duvnjak, Z., and Azoulay (1969). *J. Bact.*, **100**, 1218–1221.
Lillehoj, E. B., and Ottolenghi, P. (1967). *In* "Symposium über Hefe-Protoplasten, Jena, 1965" (Ed. R. Müller), p. 145. Akademie-Verlag, Berlin.
Lockwood, J. L. (1960). *Phytopathology*, **50**, 787–789.
Longley, R. P., Rose, A. H., and Knithtè, B. A. (1968). *Biochem. J.* **108**, 401–412.
López-Belmonte, F., Garcia Acha, I., and Villanueva, J. R. (1966). *J. Gen. Microbiol.*, **44**, 222–229.
Marini, F., Arnow, P. M., and Lampen, J. D. (1961). *J. Gen. Microbiol.*, **24**, 51–62.
Masschelein, C. A. (1959). *Revue Ferment. Ind. aliment.*, **14**, 59–69.
Matile, Ph. (1969). *In* "Symposium on Membranes. Structure and Function", 6th FEBS Meeting, Madrid, 1969 (Eds. J. R. Villanueva and F. Ponz). Academic Press, London (in press).
Matile, Ph., and Wiekmen, A. (1967). *Arch. Mikrobiol.*, **56**, 148–155.
Matile, Ph., Moor, H., and Mühlethaler, K. (1967). *Arch. Mikrobiol.*, **58**, 201.
McLellan, W. L., and Lampen, J. O. (1963). *Biochim. Biophys. Acta*, **67**, 324–326.
McLellan, W. L., and Lampen, J. O. (1968). *J. Bact.*, **95**, 967–974.
McLellan, W. L., McDaniel, LL. E., and Lampen, J. O. (1970). *J. Bact.*, **102**, 261–270.
McQuillen, K. (1960). *In* "The Bacteria" (Eds. I. C. Gunsalus and R. Y. Stanier), Vol. I, p. 249. Academic Press, New York.
Meinecke, G. (1957). *Zentbl. Bakt. Parasitkde*, I, **169**, 113.
Meinecke, G. (1960). *Nature, Lond.*, **188**, 426.
Metzenberg, R. L. (1963). *Archs Biochem. Biophys.*, **100**, 503–509.
Metzenberg, R. L. (1964). *Biochim. Biophys. Acta*, **77**, 455–465.
Millbank, J. W., and Macrae, R. M. (1964). *Nature, Lond.*, **201**, 1347.
Moore, H., and Mühlethaler, K. (1963). *J. Cell. Biol.*, **17**, 609.
Myers, F. L., and Northcote, D. H. (1958). *J. Exptl. Biol.*, **35**, 639–648.
Nagasaki, S., Neuman, N. P., Arnow, P., Schnable, L. D., and Lampen, J. O. (1966). *Biochem. biophys. Res. Commun.*, **25**, 158–164.
Necas, O. (1956), *Nature, Lond.*, **177**, 898.
Necas, O. (1958). *Exptl. Cell. Res.*, **14**, 216–219.
Necas, O. (1962). *Folia Biol.* (Prague). **8**, 256–262.
Necas, O. (1965). *Folia biol., Praha*, **11**, 97–102.
Necas, O., and Havelkova, M. (1967). *In* "Symposium über Hefe-Protoplasten, Jena 1965" (Ed. R. Müller), p. 55. Akademie-Verlag, Berlin.
Necas, O., and Kopecká, M. (1969). *Antonie van Leeuwenhoek*, **35**, B7.
Necas, O., Kopecká, M., and Brichta, J. (1969). *Exptl. Cell. Res.*, **58**, 411–419.

Necas, O., and Svoboda, A. (1966). *In* "Symposium über Hefe-Protoplasten, Jena 1965" (Ed. R. Müller), p. 45. Akademie-Verlag, Berlin.

Necas, O., Svoboda, A., and Havelkova, M. (1968). *Folia biol., Praha,* **14**, 80–85.

Necas, O., Svoboda, A., and Kopecká, M. (1968). *Exptl. Cell. Res.,* **53**, 291.

Nickerson, W. J. (1963). *Bact. Rev.,* **27**, 305–324.

Nickerson, W. J., and Falcone, G. (1956). *Science,* **124**, 722.

Northcote, D. H. (1963). *Pure appl. Chem.,* **7**, 669–681.

Nurminen, T., Oura, E., and Suomalainen, H. (1970). *Biochem. J.,* **116**, 61–69.

Ottolenghi, P. (1966). *C.r. Trav. Lab. Carlsberg.,* **35**, 363–368.

Porter, C. A., and Jaworsky, E. G. (1968). *Eur. J. Biochem.,* **5**, 1149–1154.

Robertson, N. F. (1965). *Trans. Br. mycol. Soc.,* **48**, 1–8.

Rodriguez Aguirre, M. J. (1965). *Ph.D. Thesis, Madrid University.*

Rodriguez Aguirre, M. J., Garcia Acha, I., and Villanueva, J. R. (1964). *Antonie van Leeuwenhoek,* **30**, 33–34.

Rost, K., and Venner, H. (1965). *Arch. Mikrobiol.,* **51**, 130–139.

Rothstein, A., and Meier, R. (1948). *J. Cell. Comp. Physiol.,* **32**, 77.

Ruesink, A. W., and Thiman, K. V. (1966). *Science,* **154**, 280–281.

Satomura, Y., Ono, M., and Fujumoto, J. (1960). *Bull. agric. chem. Soc. Japan,* **24**, 317–321.

Schlenk, F., and Dainko, J. L. (1966). *J. Bact.,* **89**, 428–436.

Sentandreu, R., and Northcote, D. H. (1969). *J. gen. Microbiol.,* **55**, 393–389.

Sentandreu, R., López-Belmonte, F., and Villanueva, J. R. (1966). Proc. 2nd Meeting FEBS, Vienna, 1965, p. 28.

Shockman, G. D., and Lampen, J. O. (1962). *J. Bact.,* **84**, 508–512.

Sietsman, J. H., Eveleigh, D. E., Haskins, R. H., and Spencer, J. F. T. (1967). *Can. J. Microbiol.,* **13**, 1701–1704.

Soskova, L., Svoboda, A., and Soska, J. (1968). *Folia Microbiol.,* **13**, 240–245.

Streiblová, E. (1968). *J. Bact.,* **95**, 700–707.

Strunk, Ch. (1966). *In* "Symposium über Hefe-Protoplasten, Jena 1965" (Ed. R. Müller), p. 213. Akademie-Verlag, Berlin.

Strunk, Ch. (1969a). *Z. allg. Mikrobiol.,* **9**, 49–60.

Strunk, Ch. (1969b). *Z. allg. Mikrobiol.,* **9**, 205–216.

Suminame, K., and Duckmo, H. (1963). *J. gen. Microbiol.,* **9**, 243–248.

Suomalainen, H., Nurminen, T., and Oura, E. (1967). *Suom. Kemistilehti B,* **40**, 323–327.

Sutton, D. D., and Lampen, J. O. (1962). *Biochim. Biophys. Acta.,* **56**, 303–312.

Svihla, G., and Schlenk, F. (1960). *J. Bact.,* **79**, 841–848.

Svihla, G., and Schlenk, F. (1967). *In* "Symposium über Hefe-Protoplasten, Jena 1965" (Ed. R. Müller), p. 15. Akademie-Verlag, Berlin.

Svihla, G., Schlenk, F., and Dainko, J. L. (1961). *J. Bact.,* **82**, 808–814.

Svoboda, A., Farkas, V., and Baner, S. (1969). *Antonie van Leeuwenhoek,* **35**, B11.

Tabata, S., Imai, T., and Termi, B. (1965). *J. Ferment. Technol.,* **43**, 221–227.

Takebe, I., Otsuki, Y., and Aoki, S. (1968). *Pl. Cell Physiol.,* **9**, 115–119.

Tanaka, H., and Phaff, H. J. (1967). *J. Bact.,* **89**, 1570–1580.

Tanner, W. (1969). *Biochim. Biophys. Res. Com.,* **35**, 144–150.

Thorsson, K. G., and Weibull, C. (1958). *J. Ultrastruct. Res.,* **1**, 412–417.

Tonino, G. J. M., and Steyn-Parvé, E. P. (1963). *Biochim. Biophys. Acta.,* **67**, 453–459.

Uruburu, F., Elorza, M. V., and Villanueva, J. R. (1968). *J. gen. Microbiol.,* **51**, 195–198.

Van Dam, G. J. W., Slavenburg, J. H., and Koningsverger, V. V. (1964). *Biochem. J.*, **92**, 48P.

Villanueva, J. R. (1966). *In* "The Fungi" (Eds G. C. Ainsworth and A. S. Sussman), Vol. II, p. 3. Academic Press, London and New York.

Villanueva, J. R., Elorza, M. V., Monreal, J., and Uruburu, F. (1969). *In* "II Symposium on Yeasts". Bratislava, 1966. p. 203 (Ed. A Kockova-Kratodhvilová). Vydavatelstvo Slovenskej Akademie-Vied.

Vogel, H. J. (1956). *Microbial Genetics Bull. No. 13*, 42.

Volfova, O., Munck, V., and Dostalek, M. (1968). *Experientia*, **23**, 1005–1006.

Weibul, C. (1953). *J. Bact.*, **66**, 688–696.

Weingberg, R., and Orton, W. L. (1965). *J. Bact.*, **90**, 82–90.

Weiss, B. (1963). *J. gen. Microbiol.*, **39**, 85–94.

Weiss, H., Jagow, G., Klingenberg, M., and Bucher, T. (1970). *Eur. J. Biochem.*, **14**, 75–82.

Yamamoto, T. B. (1962). *J. Biol. Osaka City Univ.*, **13**, 13–18.

Author Index

Numbers in *italics* refer to the pages on which references are listed at the end of each Chapter

Kawato, M., 320, *332*
Kay, D., 518, *564*
Keay, M. A., 63, *92*
Keddie, F. M., 432, 438, *459, 460*
Keddie, J. A. G., 510, *516*
Keefer, G. V., 390, *402*
Keitt, G. W., 57, 84, *92*, 296, 311, *334*, 387, *403*
Kellenberger, E., 533, *564*, 666, 667, *714*
Kelley, J., 247, *263*
Kelman, A., 618, 619, *659*
Kelner, A., 303, *332*, 614, *660*
Kempton, P. E., 588, 589, *593*
Kendrick, E. L., 206, 209, *215*
Kendrick, W. B., 23, *46*
Kent, G. C., 208, *218*
Kerken, A. E. van, 161, 168, *181*
Kern, F. D., 195, *215*
Kernkamp, M. F., 134, *149*
Kerr, N. S., 253, 254, 255, 256, 257, 258, 262, *264*
Kerr, S., 243, *264*
Kershaw, K. A., 272, 288, *292*
Keunen, R., 479, 489, 496, *499*
Keyworth, W. G., 6, 16, *46*, 510, *516*
Khair, J., 17. *46*
Keinholz, J. R., 205, 209, *215*
Kiesling, R. L., 208, *215*
King, C. J., 423, *426*
King, J., 511, *516*
King, K. P., 168, *181*
King, M. K., 501, *508*
King, T. H., 57, *92*
Kingsolver, C. H., 618, *662*
Kingston, D., 373, 375, 379, 386, *403*
Kinsky, S. C., 691, 699, 708, 711, *716*
Kingsley, R. N., 168, *179*
Kinugawa, K., 230, *235*
Kirchoff, H., 25, *46*
Kirk, P. W., 352, 358, *364*
Kirkham, D. S., 602, *606*
Kirsop, B., 173, *177*
Klarman, W. L., 53, *92*
Klebahn, H., 195, 196, *215*
Klein, D. T., 447, *459*
Kleinert, J., 168, *177*
Klemme, D., 40, *46*
Klemmer, H. W., 50, *92*, 184, 186, *192*
Kleyn, J. G., 172, *177*

Klieneberger-Nobel, E., 666, 667, *714*
Kligman, A. M., 225, *235*, 601, *606*
Klingenberg, M., 703, *718*
Klotz, L. J., 187, *192*
Knapp, E. B. P., 360, *363*
Knaysi, G., 22, *46*
Kniep, H., 225, *235*
Knight, V., 390, *402*
Knithtš, B. A., 675, 679, 697, *716*
Knox-Davis, P. S., 74, *92*, 630, *660*
Knoyle, J. M., 347, *365*
Kobel, H., 581, *592*
Kobert, R., 569, *592*
Koch, W., 230, *235*
Koenigs, J. W., 199, *215*
Kogan, S., 262, *263*
Kögl, F., 569, *592*
Kohlmeyer, J., 335, 347, 358, *364*
Kohn, S., 72, *91*
Koizumi, M., 639, *660*
Koller, L. R., 612, 641, 642, 644, 645, 646, 647, 650, 655, 656, *660*
Kondo, W. T., 212, *215*
Koningsberger, V. V., 706, 707, *714, 718*
Kopecká, M., 669, 675, 709, 710, *716, 717*
Koroleva, I. F., 166, *177*
Korzybski, T., 268, *292*
Kosikov, K. V., 173, *177*
Kosmachev, A. E., 329, *332*
Kotlaba, F., 583, *592*
Kozinn, P. J., 473, *476*
Kozlova, E. I., 303, *333*
Kozma, I., 456, *460*
Kramer, C. L., 121, *150*
Kratz, W. A., 267, *292*
Krause, R., 635, *660*
Kreger-van Rij, N, J. W., 166, 167, 172, *174, 177, 178,* 467, *475*
Kreitlow, K. W., 204, 205, *215*
Krinsky, N. I., 611, *661*
Krisova, M., 168, *175*
Kriss, A. E., 167, *177*
Kröber, H., 627, 632, *663*
Krulwick, T. A., 304, *332*
Kruse, R. H., 430, *458*
Krzywy, T., 327, *333*
Kuć, J., 60, *93*
Kudrina, E. S., 315, 324, *332, 333*
Kvehn, H. H., 70, *92*

Subject Index

A

Abscesses, fungi from, 462

Absidia spp.,
air sampling for, 381
fluorescent antibody for, 511

A. corymbifera, disease agent, 462

A. ramosa, disease agent, 462

Acaricides, 53–55, 120

Acarospora fuscata, attempted resynthesis of, 277

A. smaragdula, fungus of, 271

Acervuli, 7

Aceto-iron haematoxylin technique, 99–100

Achlya ambisexualis, medium for, 85

A. americana, photo-inhibition of sporulation of, 635

Acid phosphatase, localization of, 708

Acid-tolerant bacteria, and yeast isolation, 160

Acids, as reagents in agaricology, 572, 593–594

Acrasiales, (Cellular slime moulds),
amoebae, 238, 241, 259, 261, 262
collection of, 260
habitat, 259–260
identification of, 260
isolation of, 260
life-cycle, 259–268
maintenance of, 262
photosporogenesis in, 627, 632
shaken liquid culture of, 261
spores of, 259

"Acrasin", 259

Acrasis spp., 260

A. rosea, photosporogenesis in, 627

Acremoniella atra, 29

Acridine orange, fluorescent dye in mycology, 510, 513

Acrolein, in electron microscopy, 519, 524–525, 528

Acrylic aldehyde, *see* Acrolein

Actidione, *see also* Cycloheximide, 437
actinomycete isolation, in, 241
differential yeast media, for, 161, 272

Actinobifida spp., 173, 175

Actinomyces spp., *see also* Actinomycetes,
animal pathogenic, 262
electron microscopy of, 277
fresh water, 258
physiological tests and taxonomy of, 279

A. bovis, mycosis due to, 462

A. israeli, mycosis due to, 462

A. naeslundii, soil microbes related to, 257

Actinomycetaceae, micro-organisms with affinities to, 256

Actinomycete media, 57, 72, 80, 90, 265, 271

Actinomycetes, 247–286
actinophage and, 264–265
amylases of, 279
animal tissues, from, 261–262
cellulases of, 278
chitinases of, 277
cover slip methods for examining, 272–274
cultivation of, 265–280
electron microscopy, cultivation for, 276
enzymes for protoplast preparation from, 682
freeze drying of, 281–282
immunological methods for, 279
isolation of, 247–262
media for, 265–271
mesophilic, 279–280
mineral oil storage of, 282
morphological examination of, 272–276
pathogenic, 260–262, 430, 439–440, 439–440, 468–469
physiological properties, cultivation and, 279–281

Hymenophoral trama, 221
Hyphal isolation technique, for fungi, Homobasidiomycetes, 222
soil fungi, 419–420
Hypholoma,
chrysocystidia, 578
spore morphology, 397
Hyphomycetes,
aquatic, 335, 355–358
herbarium specimens, 137, 144
Hypochlorite, as surface sterilant, 4, 5
Hypomyces solani, 628
Hypoxylon, fixation for electron microscopy, 525
H. fragiforme, electron micrographs of, 529, 531, 534, 543
H. fuscum, spore discharge photoresponse, 619
H. rubiginosum, electron micrograph, 559
H. serpens, electron micrograph, 559
Hysteriaceous fungi, 290

I

IAA, *see* Indole-3-acetic acid.
Identification of fungi, *see also under specific names*,
light and, 625–640
Identification methods,
acrasiales, for, 260
myxomycetes, for, 244–245
yeasts, for, 170–171
Illumination, *see* Light.
"Imber reaction", 584
Immersion tube methods, for soil fungi, 416–418
Immunoelectrophoresis, in mycology, 605
Immunofluorescent (*see also* Fluorescent antibody) techniques,
yeast identification by, 161
Immunological methods,
actinomycete study by, 279
mycology, in, 599–607
Impactors, for air sampling, 368, 370–380
Impingers, for air sampling, 368, 380–386
Inclined coverslip technique, for actinomycetes, 272, 273

Incubators, for rust fungi, 197
Indexes, of preserved fungi, 147–148
Indole-3-acetic acid (IAA), 25
blue-green algal growth stimulant, 276–289
Induced dormancy, for fungal culture storage, 115
Inert particle agglutination tests, 605
Infection techniques, *see* Inoculation.
Infestations of fungal cultures, 53–55
Inhibitors, *see also specific compounds*,
in yeast isolation, 160
Ink blue clearing technique, for pathogenic fungi, 448–449, 463–464
Inoculation,
punches for, 29–32
rust fungi, techniques for, 196–200
smut fungi, techniques for, 205–219
yeast sampling, for, 164–166
Inocybe, chemotaxonomy of, 570
Inositol,
requirement for, 450
yeast isolation and, 162
Insecticides, spreading efficiency, 374
Insects, *see also* Mites,
culture collection hazards, as, 120
fungi from, 189–190
herbarium danger, as, 144
Interference filter characteristics, 644
Intine, of pollen grains, 398, 399, 400
Invertase isoenzymes, 707
Iodine complexes, as reagents in agaricology, 585
Iron salts, as reagents in agaricology, 585–586, 594
Irradiation, see Light.
Isokinetic sampling principle, 368–369, 370
Isolation of fungi, *see also under specific names*,
Acrasiales, 260
actinomycetes, 247–262
general problems and principles, 247–248
general methods for, 2–12
Homobasidiomycetidae, 220–225
myxomycetes, 243–244
phycomycetes, 183–192
yeasts, 169–171
Isazole derivatives, in agaricology, 570

S. carlsbergensis,
contaminating yeasts, detection of, 163
protoplasts of, 674, 678, 679, 706, 710, 713
S. cerevisiae,
freeze-etching for electron microscopy, 553, 556
irradiated media and growth, 623
mitotic crossing-over in, 490
myxomycete culture with, 249
protoplasts of,
cell wall synthesis by, 709
enzyme localization and, 707
membranes of, 696, 697, 698
mitochondria from, 702
permeability studies with, 705
sexual recombination in, 495
sporulation and light in, 166
streptomycin inhibition of, 161
S. chevalieri, protoplasts of, 679
S. ellipsoideus, myxomycete culture with, 253
S. fragilis,
actidione in the isolation of, 161
protoplasts of, 675, 680, 686, 687, 688
S. italicus, protoplasts of, 679
S. lactis, 170
isolation of, 163
S. mellis,
osmophilic, 164
protoplasts of, 674
S. rouxii, osmophilic, 163–164
Saccharomycoides, storage of, 174
Sach's agar, formula, 86
Salmonella anatum, fluorescence brightener for, 513
Salt tolerance, of yeast, 163
Sampling techniques,
air sampling, 367–404
aircraft, from, 389
anisokinetic errors in, 370, 375
biological, 391–392
cyclone collectors for, 389
"fall-out" method, of, 387–388
filters for, 388–389
general principles of, 368–370
impactors for, 370–380
impingers for, 380–386
isokinetic principle for, 368–369, 370

large volume, 389–390
sedimentary method of, 387–388
selective media in, 393–394
yeasts, for, 153–168
Sandeural, 513
Saprolegnia declina, photo-inhibition of sporulation of, 635
S. ferax,
zoospore electron micrograph, 551, 554
sporulation photo-inhibition, 635
S. parasitica, sporulation photo-inhibition, 635
Saprolegniaceae, *see also under specific names*
chytrids parasitic on, 188
maintaining stock cultures of, 343
medium for, 67
storage of, 125, 131
Saprolegniales, *see also under specific names,*
isolation of, 340
preservation, growth for, 118
Sapromyces, maintenance of, 344
Saprophytic associations, between yeasts and other microbes, 170
Sartory's defined medium, 86
Sativin, 637
Sawdust medium, formula, 87
Scab, of root crops, 260
Scanning electron microscopy, of fungi, 558–561
Schaeffer reaction, 588
Schiff's reagent, preparation, 514–515
Schizophyllum spp.,
fructification of, 229
genetics of, 232
preservation of, 126
S. commune, 219, 228
basidiocarp formation, 629, 634
somatic recombination in, 493
Schizosaccharomyces, storage of, 174
S. pombe,
growth pH for, 160
protoplasts of, 675, 709, 713
Schopfer's medium, formula, 87
Scleroderma aurantium, fructification of, 231
Sclerotia, of myxomycetes, 240, 243, 245